D1577302

Archives,

HIV Prevention

HIV Prevention

A comprehensive approach

Edited by

Kenneth H. Mayer and Hank F. Pizer

AMSTERDAM • BOSTON • HEIDELBERG • LONDON • NEW YORK • OXFORD
PARIS • SAN DIEGO • SAN FRANCISCO • SINGAPORE • SYDNEY • TOKYO
Academic Press is an imprint of Elsevier

Academic Press is an imprint of Elsevier
32 Jamestown Road, London NW1 7BY, UK
30 Corporate Drive, Suite 400, Burlington, MA 01803, USA
525 B Street, Suite 1900, San Diego, California 92101-4495, USA

First edition 2009

Library of Congress Cataloging-in-Publication Data
A catalog record for this book is available from the Library of Congress

British Library Cataloguing-in-Publication Data
A catalogue record for this book is available from the British Library

ISBN: 978-0-12-374235-3

For information on all Academic Press publications
visit our web site at www.elsevierdirect.com

Printed and bound in the USA

09   10   11   12   13   9   8   7   6   5   4   3   2

Working together to grow
libraries in developing countries

www.elsevier.com | www.bookaid.org | www.sabre.org

ELSEVIER   BOOK AID
             International   Sabre Foundation

Dedications

I am grateful to have had parents and relatives who instilled in me an interest in asking questions, constantly learning, working hard, and trying to make the world a better place. The deaths of many dear friends and patients because of AIDS stimulated my early work, and the desire to live in a world without AIDS informs my current efforts.

Kenneth H. Mayer

I dedicate this work to my wife, Christine, who in all ways makes it possible for me to be who I am; and to our marvelous daughter, Katie, and her daughter, Annabel, the newest joy in our lives.

Hank F. Pizer

Contents

Foreword   xi
Judith D. Auerbach

About the Editors   xvii

Notes on Contributors   xix

Acknowledgments   xxxix

Introduction   1
Kenneth H. Mayer and Hank F. Pizer

**Part I: Epidemiological and biological issues in
HIV prevention**   9

1   Current and future trends: implications for HIV
prevention   11
Vikrant V. Sahasrabuddhe and Sten H. Vermund

2   Understanding the biology of HIV-1 transmission: the
foundation for prevention   31
Deborah J. Anderson

3   HIV vaccines   53
Robert E. Geise and Ann Duerr

4   Microbicides   85
Ian McGowan

5   Using antiretrovirals to prevent HIV transmission   107
Cynthia L. Gay, Angela D. Kashuba and Myron S. Cohen

6   Male circumcision and HIV prevention   146
Ronald Gray, David Serwadda, Godfrey Kigozi and Maria J. Wawer

Part II: Behavioral issues in HIV prevention   167

7   Payoff from AIDS behavioral prevention research   169
Willo Pequegnat and Ellen Stover

8   Individual interventions   203
Matthew J. Mimiaga, Sari L. Reisner, Laura Reilly,
Nafisseh Soroudi and Steven A. Safren

9   Couples' voluntary counseling and testing   240
Kathy Hageman, Amanda Tichacek and Susan Allen

10   Updating HIV prevention with gay men: current
challenges and opportunities to advance health among
gay men   267
Ron Stall, Amy Herrick, Thomas E. Guadamuz and Mark S. Friedman

11   Reducing sexual risk behavior among men and women
with HIV infection   281
Jean L. Richardson and Tracey E. Wilson

12   Injection drug use and HIV: past and future
considerations for HIV prevention and interventions   305
Crystal M. Fuller, Chandra Ford and Abby Rudolph

13   HIV risk and prevention for non-injection substance users   340
Lydia N. Drumright and Grant N. Colfax

14   Preventing HIV among sex workers   376
Bea Vuylsteke, Anjana Das, Gina Dallabetta and Marie Laga

15   Interventions with youth in high-prevalence areas   407
Quarraisha Abdool Karim, Anna Meyer-Weitz and Abigail Harrison

16   Interventions with incarcerated persons   444
Ank Nijhawan, Nickolas Zaller, David Cohen and Josiah D. Rich

17   Preventing mother-to-child transmission of HIV   472
James A. McIntyre and Glenda E. Gray

Part III: Structural and technical issues in HIV prevention   499

18   Harm reduction, human rights and public health   501
Chris Beyrer, Susan G. Sherman and Stefan Baral

19 HIV testing and counseling   524
Julie A. Denison, Donna L. Higgins and Michael D. Sweat

20 Structural interventions in societal contexts   550
Suniti Solomon and Kartik K. Venkatesh

21 Evaluating HIV/AIDS programs in the US and
developing countries   571
Jane T. Bertrand, David R. Holtgrave and Amy Gregowski

22 Adapting successful research studies in the public health
arena: going from efficacy trials to effective public health
interventions   591
*Kevin A. Fenton, Richard J. Wolitski, Cynthia M. Lyles and
Sevgi O. Aral*

Index   619

Foreword

Judith D. Auerbach
Deputy Executive Director for Science and Public Policy,
San Francisco AIDS Foundation

As the AIDS pandemic has evolved in complex ways over the past three decades, so has the field of HIV prevention. The chapters in this very comprehensive book reflect that evolution, and reveal both the many accomplishments we have made and the persistent challenges that confront us in attempting to reduce or eliminate HIV transmission.

Perhaps most notably on the research front, basic biomedical researchers now regularly commune with clinical trials researchers and behavioral interventionists – and, once in a while, even social scientists – in interdisciplinary discussions of HIV prevention topics. Certainly, there are still separate scientific meetings and journals for various disciplines and approaches, but increasingly, there are also mixed conferences and publications, like this book, with a great deal more cross-talk than occurred in the first two decades of the response. As a result, the HIV prevention field as a whole has come to recognize that HIV is fundamentally a pathogen that is transmitted in the course of human relationships that occur and are influenced by social and cultural contexts, and that targeting only one aspect of the interacting biological, behavioral, and social features of HIV/AIDS will have limited effect. As HIV has become a more multi-disciplinary conversation, a number of interrelated issues, challenges, and opportunities have arisen that are addressed directly or indirectly in the chapters of this book.

Conducting multi-disciplinary and multi-level science

Although scientists now talk across disciplines, it remains difficult for their work to be truly interdisciplinary and multi-level. This is a function of increasingly specialized knowledge and training, but also of paradigmatic disagreements about what questions to ask, what methodologies to employ to answer them, what outcome measures to accept as "evidence", and how to interpret findings.

Most clinical trials of biomedical technologies for HIV prevention now include behavioral science components, but these more often than not are treated as "hand-maids" to clinical research. They are focused on behavioral issues of relevance to the conduct of the trial – for example, assessing acceptability of or adherence to a product (such as a microbicide candidate, female diaphragm, pre-exposure proph-ylaxis, etc.) under study, rather than on the more general behavioral and social dynamics affecting the lives of trial participants that might inform the design of the study and its likelihood of success (or failure) in the first place. And, often, when funding is tight, the behavioral components of a trial are the first to be sacrificed as they are still seen by many clinical researchers as dispensable – not necessary for testing the efficacy of a biomedical/technological strategy. But, as behavioral sci-entists point out, no strategy will be effective – nor trial of it conclusive – if people don't use it, so knowing what motivates or impedes use is essential, not discretion-ary, and must, therefore, be viewed as an integral part of efficacy trials.

Gaining consensus on appropriate methods and measures for establishing evidence of efficacy

In HIV prevention science, the randomized controlled trial (RCT) with an HIV incidence outcome measure remains the gold standard method for establishing effi-cacy among the biomedical community. But its hegemony is being challenged by social scientists who argue that experimental methods often are not appropriate for addressing social-level questions. They also note that declining HIV infection rates observed over the course of the pandemic in a number of diverse settings (e.g., San Francisco, Thailand, Uganda, and Senegal) resulted from community-driven behav-ioral and social change, not from experimental interventions, and that such com-munity-generated responses do offer observational evidence of effectiveness – even if it is not entirely clear to what specific actions the declining infection rates may be attributed. So, while RCTs remain valid and necessary for assessing the efficacy of some types of HIV prevention strategies, they will never produce the entirety of rel-evant evidence of what actually works in different modes, populations, and settings.

Rather, the HIV prevention field must come to accept a range of "ways of knowing" in which evidence is derived from different methodologies appropriate to the question and level of analysis being addressed. This includes both quanti-tative and qualitative data from such methods as RCTs, quasi-experimental inter-ventions, surveys, interviews, ethnography, content analysis, policy analysis, and program evaluation, to name a few.

Moving from efficacy studies to effectiveness studies

There are many small-scale behavioral, biomedical, and social science-based interventions that have shown efficacy in reducing risk and infection through sexual, parental, and perinatal routes in a range of population groups and settings. But we do not yet know how well their outcomes hold up over time, and what their aggregate effect is if fully implemented and scaled-up. Given that adaptation nearly always occurs in intervention replication, it is an open question how loss of fidelity to an original intervention design affects subsequent outcomes. Moreover, implementation of proven interventions at the population level involves confronting a host of individual, institutional, and societal forces that are hard to predict.

The case of male circumcision makes clear the many issues that arise in moving from efficacy to effectiveness. Three clinical trials recently substantiated a wealth of observational data showing adult male circumcision – when conducted under sterile, medical conditions – significantly reduces the risk of HIV transmission from females to males by 50 – 60 percent. Moving from efficacy trials to population-level effectiveness raises a number of issues. First, if circumcision is conducted under non-sterile conditions, there is a risk of harm to men and, potentially, no benefit for reduced HIV transmission. Second, in settings where male circumcision is not common, it will be essential to attend to cultural norms – including religious beliefs and practices surrounding circumcision. The potential for increased risk-taking among circumcised men who may believe they are fully protected from HIV infection is another concern. Any increase in risk-taking, such as decreased condom use and increased number of sex partners, could obviate the benefits of male circumcision and contribute to higher rates of HIV infection in the population. Messaging about partial efficacy and the need to continue engaging in other risk-reduction strategies is quite tricky in the face of new HIV prevention options, like adult male circumcision, that people hope – and believe – will eliminate the need for condom use. (These concerns underscore the need for multi-disciplinary research mentioned above.)

Recognizing the partial efficacy of most HIV prevention strategies, and the contextual issues that affect population-level implementation and uptake of all methods, there now is a call from many quarters to develop a "combination" approach to HIV prevention – putting together packages of efficacious interventions that are appropriate to particular settings and populations to assess their combined effectiveness.

Impact of an evolving "standard of prevention"

There is an undisputed ethical imperative to offer all trial participants (in both experimental and control arms) state-of-the-art HIV prevention information and services, which currently include behavioral risk reduction counseling and the provision of male condoms – and soon may include offering male circumcision. In trial after trial of new HIV prevention technologies, this has been shown to increase protective action (e.g., partner reduction and increased condom use) in both experimental and control arms of the study, thereby making it difficult to observe an independent effect (if there is one) of the product under study. As new strategies are shown to be efficacious and must then be included in the standard of prevention offered in trials, it will become ever-more difficult to ascertain the independent effect of investigational products; and the tension between research ethics and study design will be exacerbated.

Clinical trials resulting in null and negative findings

In recent years, a number of multi-site RCTs of promising biomedical/techno-logical interventions for HIV prevention, such as microbicides, vaccines, and the female diaphragm, have yielded either null or negative findings. That is, there was no difference between experimental and control groups with respect to HIV infection rates or, in a couple of cases, participants in the experimental arm appeared to have higher rates of infection than those in the control arm. Null findings may have been influenced by the standard of prevention mentioned above; and negative findings may be a result of the mechanisms of action of the products under study that did not appear during pre-clinical or Phase I studies.

Both kinds of outcomes raise a host of questions whose answers will affect the ability to conduct future, successful HIV prevention trials, including: how are null and negative findings communicated to and understood by trial participants and other members of their communities? How can expectations about optimal trial results be managed? How, in the face of disappointing – and even harmful – findings, can support for HIV prevention trials be maintained among communities and funders? Does a null result in a large trial of a particular product (e.g., the latex diaphragm) doom that product for any subsequent trials, even if the finding may be a result of uptake of risk reduction behaviors among all trial participants rather than product non-efficacy? If so, will we ever truly be able to know if the product itself is effective; and might we be running the risk of ruling out a potentially efficacious product?

Putting HIV prevention in the larger social context

It is easy for researchers to believe that a clinical trial is the most important thing in a participant's life, since it is the most important thing to the researcher in relation to that participant. In reality, a trial is just one feature of a participant's often complicated life; and it occurs in a social and cultural context that highly influences participants' daily lives and, ultimately, trial outcomes. For example, although most HIV prevention trials screen out women who are or say they intend to become pregnant during the course of the study, in recent trials, pregnancy rates of 10 – 80 percent have occurred. It is not simply a question of whether women are not being truthful about their intentions during recruitment;

rather, such unanticipated pregnancy rates reflect the strong pull of cultural norms about childbearing experienced by women everywhere. Perhaps some women do intend to get pregnant but do not want to miss out on a trial that they think may save their lives. Perhaps they do not intend to get pregnant, but are pressured or coerced into it by their male partners or family members. "Intention" itself may not be a normative cultural construct. In sum, pregnancy in the context of HIV prevention trials is an expression of the complex operation of gender – a major social organizing principle in all societies – in the calculus of disease prevention and fertility expectations that will continue to make preventing the sexual transmission of HIV infection a formidable challenge.

While these (and other) unresolved issues and complex challenges face us, they do not stymie us. Rather, as this book demonstrates, HIV prevention scientists, community advocates, policy-makers, and funders – individually and collectively – continue to find creative ways to fill knowledge gaps, to protect the rights and improve the health of research participants, and to rally political and financial support for an improved and enhanced response to AIDS. The impact of that response is dependent on a comprehensive, multi-disciplinary view that understands, respects, and knows how to interpret the dynamic interplay of biological, psychological, and social and cultural forces at work everywhere.

About the Editors

Kenneth H. Mayer, MD is Professor of Medicine and Community Health at Brown University, Director of the Brown University AIDS Program, and Attending Physician in the Infectious Disease Division of The Miriam Hospital in Providence, Rhode Island. He is also Medical Research Director at Boston's Fenway Community Health Center, where (since 1983) he has conducted studies of HIV's natural history and transmission. In the early 1980s, as a research fellow studying infectious diseases at Brigham and Women's Hospital, Dr Mayer was one of the first clinical researchers in New England to provide care for patients living with AIDS. In 1983, he co-authored *The AIDS Fact Book*, one of the first books about AIDS to be written for the general public. In 1984, he began one of the first studies of the natural history of HIV infection, and was subsequently funded by the federal government to study biological and behavioral factors associated with male-to-male HIV transmission. Since 1987, Dr Mayer and his colleagues have been supported by the NIH and CDC to study the dynamics of heterosexual HIV transmission and the natural history of HIV in women, and to study HIV prevention interventions, ranging from vaccines (HIVNET, HVTN) to microbicides, behavioral and other strategies (HPTN). He has been the principal investigator of four Phase I microbicide trials, including the first human trial of Tenofovir gel. He has collaborated with basic virologists and immunologists to more accurately characterize the genetics and immunopathology of HIV disease. In the late 1980s he initiated the first community-based clinical trials for people living with HIV/AIDS in New England, and helped amFAR develop its national Community-Based Clinical Trials Network (CBCTN). He was subsequently elected to the Board of Directors of amFAR and was co-chair of its Clinical Research and Education Committee; he is now a member of their Program Board. He has also served on the national boards of HIVMA and GLMA.

Hank F. Pizer, BA, PA is a medical writer, health-care consultant and physician assistant. He has written and edited 15 books and numerous articles about health and medicine that have been published in English and translated into at least 6 foreign languages. With Kenneth Mayer he co-authored the first book about AIDS for the general public, *The AIDS Fact Book* (Bantam Books, 1983), and co-edited *The Emergence of AIDS: Impact on Immunology, Microbiology,* *and Public Health* (American Public Health Association Press, 2000) and *The AIDS Pandemic: Impact on Science and Society* (Academic Press, 2005). With Chris Beyrer, he co-edited *Public Health and Human Rights: Evidence-Based Approaches* (Johns Hopkins University Press, 2007, The Director's Circle Book for 2007) and with Kenneth Mayer, *The Social Ecology of Infectious Diseases* (Academic Press, 2008), to which he also contributed. His other works cover a variety of subjects in health and medicine, including the first books for the general public on organ transplants (*Organ Transplants: A Patient's Guide*, with the Massachusetts General Organ Transplant Teams; Harvard University Press, 1991) and stroke (*The Stroke Fact Book*, with Conn Foley, Bantam Books, 1985; Courage Press and the American Heart Association) and, in women's health, on family planning (*The New Birth Control Program*, with Christine Garfink, RN, Bolder Books, New York, 1977; Bantam Books, New York, 1979) and parenting (*The Post Partum Book*, with Christine Garfink, RN, Grove Press, New York, 1979). He is currently co-founder and principal of Health Care Strategies, Incorporated, a consulting firm that provides program evaluation and management consulting services to clients in health and education. From 1984 to 1994 he was founder and President of New England Medical Claims Analysts, Incorporated, a consulting firm that provided cost containment, utilization review and coordination of benefits services to health insurers, health maintenance organizations, union health plans and self-insured companies.

Notes on Contributors

Quarraisha Abdool Karim, PhD, MS is an Associate Professor at the Department of Epidemiology, Mailman School of Public Health and in the School of Family Medicine and Public Health at the University of KwaZulu-Natal. She is an infectious diseases epidemiologist whose main current research interests are in understanding the evolving HIV epidemic in South Africa; factors influencing the acquisition of HIV infection in young women; and establishing sustainable strategies to introduce HAART in resource-constrained settings. She is Scientific Director of the Centre of the AIDS Programme of Research in South Africa (CAPRISA); Director of the Columbia University–Southern African Fogarty AIDS International Training and Research Programme (CU-SA Fogarty AITRP) and Co-Chair of the HIV Prevention Trials Network Leadership Group (HPTN). Dr Abdool Karim was responsible for establishing the South African National HIV/AIDS and STD Programme shortly after the first democratic elections in South Africa.

Susan Allen, MD MPH, DTM&H is Principal Investigator, Director of the Rwanda-Zambia HIV Research Group (RZHRG), and Professor of Global Health at the Rollins School of Public Health at Emory University, in Atlanta, GA. Dr Allen received her medical degree from Duke University, her Diploma of Tropical Medicine and Hygiene from the Liverpool School of Tropical Medicine, and her Masters in Public Health from the University of California at Berkeley. She founded and directs the Project San Francisco (PSF) in Rwanda (1986) as well as the Zambia-Emory HIV Research Project (ZEHRP) in Lusaka (1994), and Ndola and Kitwe (2004), Zambia. Promoting and expanding access to couples' VCT – which reduces HIV transmission within discordant couples by over 60 percent – is a core aim of RZHRG. Since 2002, 60,000 couples have been tested by RZHRG, and the 4 sites have enrolled and followed over 4000 HIV-discordant couples, making them the largest single-site heterosexual HIV-discordant couples cohorts in the world. In addition to the original studies on HIV disease progression and HIV prevention within discordant couples, all four sites now conduct sophisticated laboratory studies and clinical trials, including HIV vaccine clinical trials sponsored by International AIDS Vaccine Initiative.

Deborah J. Anderson, PhD is a Professor in the Departments of Microbiology and Obstetrics and Gynecology at Boston University School of Medicine, Boston, MA. Her primary research interests are mechanisms of immune defense and infection of genital tract tissues. She has published widely and received numerous awards for her research. Dr Anderson has served on a number of professional committees, and in professional organizations including the World Health Organization Human Reproduction Programme, Family Health International Technical Advisory Board, numerous National Institutes of Health (NIH) Study Sections, and the Scientific Advisory Board of the American Foundation for AIDS Research (AmFar).

Sevgi O. Aral, PhD, MS, MA is the Associate Director for Science in the Division of STD Prevention, National Center for HIV/AIDS, Viral Hepatitis, STD, and TB Prevention (NCHHSTP), CDC. In this role, Dr Aral is responsible for the oversight and direction of all scientific activities including the intramural and extramural research programs and science-program interactions. In addition to her appointment at the CDC, Dr Aral has served as a Professor of Sociology in the United States and Turkey. She has served in the role of mentor for both trainees and colleagues needing help with social science perspectives bridging the gap between clinical epidemiology and behavior. She currently serves as a Clinical Professor at the University of Washington School of Medicine. Dr Aral's work has focused on risk and preventive behaviors, gender differences, societal characteristics that influence STD and HIV rates, contextual issues, and effects of distinct types of sexual mixing on STD spread. Her research has been in both domestic and international settings, and her writings have included cross-cultural comparative analyses. Dr Aral is on the editorial boards of several scientific journals, including *Sexually Transmitted Diseases*, *AIDS Education and Prevention*, and *Sexually Transmitted Infections*. In addition, she is the Associate Editor of *Sexually Transmitted Diseases* and *Sexually Transmitted Infections*. In the past she has served multiple terms on the editorial boards of *AIDS* and the *American Journal of Public Health*. Dr Aral received her PhD and MA in Social Psychology from Emory University, and another MA in Demography from the University of Pennsylvania. She received her undergraduate degree from the Middle East Tech University in Turkey.

Judith D. Auerbach, Ph.D. Dr. Judith Auerbach is Deputy Executive Director for Science and Public Policy at the San Francisco AIDS Foundation (SFAF), where she is responsible for developing, leading, and managing SFAF's local, state, national, and international science and policy agenda. Prior to joining SFAF, Dr. Auerbach served as Vice President, for Public Policy and Program Development, at amfAR (The Foundation for AIDS Research) and as Director of the Behavioral and Social Science Program and HIV Prevention Science Coordinator in the Office of AIDS Research at the National Institutes of Health (NIH). Dr. Auerbach received her Ph.D. in sociology from the University of California, Berkeley and taught sociology at Widener University and the University of California, Los Angeles. She has published and presented in the fields of AIDS, health research and science policy, and family policy and gender, and serves on numerous professional and advisory groups, including the Council of the American Sociological Association the Global HIV Prevention Working Group, and the NIH/OAR Microbicides Research Working Group.

Stefan Baral, MD, MPH, MBA, MSc is a resident physician at the University of Toronto and a Post-Doctoral Fellow at the Center for Public Health and Human Rights in the Department of Epidemiology, at the Johns Hopkins School of Public Health. He completed his undergraduate degree specializing in immunology and microbiology at McGill University and then went on to graduate school at McMaster University researching novel molecular vaccination strategies. His medical school training was completed at Queen's University, with a focus in both public health and global health disparities. Dr Baral then pursued further graduate training at the JHSPH, specializing in epidemiology and non-profit health-care management. Since graduating, Stefan has joined the IDU Working group of the HIV Vaccine Trial Network, which is focused on designing and launching HIV Vaccine trials among injecting drug users. In addition, he has spent the last few years evaluating and reviewing the HIV epidemic among MSM in lower-income settings. Most recently, he has designed and is helping to coordinate, in equal partnership with local LGBT community groups, a multicenter cross-sectional probe of HIV prevalence, determinants of infection, and human rights contexts of MSM in four sites across Southern Africa. Dr Baral maintains a clinical practice serving the needs of at risk populations in Toronto, and has provided care in settings ranging from methadone clinics and homeless shelters in Canada to large tertiary-care institutions in Kampala.

Jane T. Bertrand, PhD, MBA is currently Director of the Center for Communication Programs (CCP) and Professor, Department of Health, Behavior & Society, at the Johns Hopkins Bloomberg School of Public Health. Prior to moving to Hopkins in 2001, Dr Bertrand was on the faculty at the Tulane University School of Public Health and Tropical Medicine, where she chaired the Department of International Health and Development from 1994 to 1999 and served as PI for the Tulane subcontract under the EVALUATION Project (1991–1996) and the MEASURE Evaluation Project (1996–2001). Dr Bertrand focused on information–education–evaluation for family planning in the 1970s, on operations research during the 1980s, and on program evaluation in the 1990s. Since joining Hopkins in 2001, she has been able to combine these interests (e.g., in publishing on the effects of communication programs on behavior change in HIV programs and teaching a highly subscribed course on fundamentals of program evaluation). She has published over 60 articles, a dozen technical manuals, and two books. Recent publications include a manual on *Evaluating HIV/AIDS*

Prevention Programs with a Focus on NGOs and *Strategic Communication in the HIV/AIDS Epidemic*. She has worked extensively in Guatemala, the Democratic Republic of the Congo, and Morocco, as well as on short-term assignments in over 30 countries, being fluent in French and Spanish.

Chris Beyrer, MD, MPH is Professor of Epidemiology, International Health and Health, Behavior and Society at the Johns Hopkins Bloomberg School of Public Health in Baltimore, Maryland. He serves as Director of Johns Hopkins Fogarty AIDS International Training and Research Program, and as Director of the Center for Public Health and Human Rights. Dr Beyrer is Associate Director for the School of Public Health of the new Johns Hopkins Center for Global Health. He currently serves as Co-Chair of the Injecting Drug Use Working Group of the HIV Vaccine Trials Network and a Senior Scientific Liaison for the HVTN. He has extensive experience in conducting international collaborative research and training programs in HIV/AIDS and other infectious disease epidemiology, in infectious disease prevention research, HIV vaccine preparedness, in health and migration, and in health and human rights. He is the author of the 1998 book *War in the Blood: Sex Politics and AIDS in Southeast Asia* and co-editor of the 2007 book *Public Health and Human Rights: Evidence-Based Approaches*. Dr Beyrer currently serves as a member of the Global Health Advisory Council of the Open Society Institute, as a trustee of the Institute for Asian Democracy, and as advisor to the International Partnership for Microbicides and the HIV Vaccine Trials Network. He has previously served as advisor to the US CDC, the Office of AIDS Research of the US NIH, the US Military HIV Research Program, the World Bank Institute, the World Bank Thailand Office, the Royal Thai Army Medical Corps and the Thai Red Cross, as well as numerous other organizations.

David Cohen, BS is a research assistant in the Department of Medicine, Division of Infectious Diseases at The Miriam Hospital and the Center for Prisoner Health and Human Rights. He graduated in 2004 with a degree in electrical engineering from Yale University, and worked for the MIT Lincoln Laboratory from 2004 to 2006 as an electrical/aerospace engineer before switching to the field of medicine. He begins the Alpert Medical School of Brown University in the fall of 2008, and is looking forward to embarking on a career in medicine.

Myron S. Cohen is J. Herbert Bate Distinguished Professor of Medicine, Microbiology and Immunology and Public Health at the University of North Carolina at Chapel Hill, Director of the UNC Division of Infectious Disease and UNC Institute for Global Health and Infectious Disease, Associate Vice Chancellor for Medical Affairs-Global Health and Associate Director of the UNC Center for AIDS Research. He serves on the Senior Leadership Group of the NIH Center for HIV Vaccine Immunology (CHAVI) and the leadership group of the NIH HIV Prevention Trials Network (HPTN). He is an Associate Editor of the journal *Sexually Transmitted Diseases* and the Editor of the comprehensive textbook, *Sexually Transmitted Diseases*. In 2005, he received an NIH MERIT Award for ongoing support of his work in HIV transmission and prevention which focuses on the role played by STD co-infections and the use of antiretroviral agents in HIV prevention. He is the author of more than 400 publications. Dr. Cohen received his Bachelor of Science *Magna Cum Laude* from the University of Illinois, medical degree from Rush Medical College and completed an Infectious Disease Fellowship at Yale University.

Grant N. Colfax, MD is Director of HIV Prevention and Research, San Francisco Department of Public Health. He received his MD degree at Harvard Medical School and completed his residency in Internal Medicine at the University of California, San Francisco. Dr Colfax is an expert on substance use and HIV risk, with most of his research focusing on non-injection substance use and sexual risk among men who have sex with men. Most recently, his research has focused on testing pharmaceutical agents to treat methamphetamine dependence. As HIV Prevention Director, he oversees a program that funds 32 community-based agencies delivering HIV prevention services to San Francisco's diverse communities. He is the author of numerous peer-reviewed papers on substance use, HIV risk, and prevention interventions.

Gina Dallabetta, MD joined the Bill & Melinda Gates Foundation in January 2005 as a Senior Program Officer on the India AIDS Initiative, Avahan. Avahan is an HIV prevention intervention working with populations most at risk in six states in India. Dr Dallabetta brings over 15 years of experience in HIV programming to Avahan, of which 13 years were spent at the Family Health International (FHI). Prior to joining the foundation, Gina was Director of the Prevention Department of the HIV/AIDS Institute of FHI based in Arlington, Virginia. The Department was responsible for sexually transmitted infections (STI), behavior-change communication, monitoring and evaluation, and related operations research in FHI activities in over 40 countries in Asia, Africa, Latin America, the Caribbean, Eastern European and the Middle East.

Anjana Das, MBBS, DCH works as Senior Technical Officer with Family Health International, India. She is a member of the STI Capacity Building team of the India AIDS Initiative (Avahan) program supported by the Bill & Melinda Gates Foundation. She is a clinician who has worked on HIV prevention and STI programs for sex workers and their clients for the past 4 years.

Julie A. Denison, PhD is a scientist with Family Health International's Behavioral and Biomedical Research Division. Dr Denison received her doctoral degree from The Johns Hopkins University, Bloomberg School of Public Health. She has conducted international research to examine the role of families in the provision of

HIV testing and counseling for young people in sub-Saharan Africa. Dr Denison has also published a meta-analysis of the effectiveness of voluntary counseling and testing as a behavior-change strategy in developing countries.

Lydia N. Drumright, PhD, MPH is a research fellow at the University of California, San Diego, Department of Family and Preventive Medicine, Division of International Health and Cross-Cultural Medicine. She received her BSc in Biochemistry and Cellular Biology from the University of California, San Diego (UCSD), Masters in Public Health in Health Education from California State University, Northridge, and PhD in Epidemiology from (UCSD). For the past 10 years Dr Drumright has worked extensively on risk behaviors associated with acquisition and transmission of sexually transmitted infections among adolescents and men who have sex with men (MSM). Most recently, her work has focused on non-injection substance use and HIV acquisition and transmission among MSM with acute and early HIV infection.

Ann Duerr, MD, MPH, PhD received her BSc from McGill University, her PhD from the Massachusetts Institute of Technology, and her MD cum laude from Harvard Medical School. She then completed a Preventive Medicine Residency at the Johns Hopkins School of Hygiene. Dr Duerr joined the US Centers for Disease Control (CDC) in 1991 as Chief of the HIV Section in the Division of Reproductive Health, National Center for Chronic Disease Prevention and Health Promotion. Under her direction, the HIV section expanded and developed a domestic and international research portfolio related to HIV and reproductive health of women. At the CDC Dr Duerr led the development of several notable multinational efforts, including: (1) the HIV Epidemiology Research Study (HERS), (2) research to increase awareness of refugee women's health, (3) the investigation of HIV transmission in Thai couples, (4) an initiative on microbicide research, and (5) the ongoing Breastfeeding Antiretrovirals Nutrition (BAN) trial. Dr Duerr has received numerous honors, including the Surgeon General's Exemplary Service Award, and the Public Health Service Special Recognition Award; she has served as a consultant to the World Health Organization (WHO) and the Joint United Nations Programme on HIV/AIDS (UNAIDS). Since 2003, Dr Duerr has been the Associate Director of the HIV Vaccine Trials Network (HVTN) in Seattle, WA. She is the author of over 100 peer-reviewed publications, and in 1995 co-edited a book entitled *HIV Infection in Women*.

Kevin A. Fenton, MD, PhD is Director of the National Center for HIV/AIDS, Viral Hepatitis, STD, and TB Prevention (NCHHSTP) at the US Centers for Disease Control and Prevention (CDC). He received his medical undergraduate degree at the University of the West Indies (Mona), his postgraduate training in Public Health Medicine at the London School of Hygiene and Tropical Medicine, Royal Free and University College Medical School, and his PhD in Epidemiology from the University of London. Prior to his work at the CDC, Dr Fenton was the Director of the HIV and Sexually Transmitted Infections Department in the United Kingdom's Health Protection Agency (HPA). He has published numerous book chapters and peer-reviewed articles on HIV and STD epidemiology, policy and sexual behavior, with a special emphasis on racial and ethnic health disparities. Dr Fenton is a Fellow of the Faculty of Public Health of the Royal Colleges of Physicians of the United Kingdom, and a Visiting Professor at University College London.

Chandra Ford is in the Department of Epidemiology at the Mailman School of Public Health at Columbia University, where she is a postdoctoral fellow in the Multidisciplinary Track of the Kellogg Health Scholars Program. Before arriving to Columbia, Dr Ford was a postdoctoral fellow in the Department of Social Medicine in the School of Medicine at the University of North Carolina, where she earned her PhD. She also holds an MPH in Health Services Administration and a Master's in Library and Information Sciences with a concentration in Health Information from the University of Pittsburgh. She has developed expertise in the study of individual (e.g., perceived racism), interpersonal (e.g., patient–provider interactions) and structural (e.g., residential segregation) factors relative to racial and ethnic disparities in HIV/AIDS.

Mark S. Friedman, PhD is an Assistant Professor in the Department of Behavioral and Community Health Sciences, Graduate School of Public Health, University of Pittsburgh. Dr Friedman received his PhD in Social Work from the University of Pittsburgh. His publications have focused on defining and measuring sexual orientation; the relationship between gender-role non-conformity, bullying and suicidality among gay youth; and antecedents of adult health problems among gay males. Dr Friedman was recently awarded a grant from the National Institute of Mental Health to develop Internet-based interventions for gay youth.

Crystal M. Fuller, PhD is an Associate Professor of Epidemiology at the Mailman School of Public Health at Columbia University and also serves as a Senior Epidemiologist at the Center for Urban Epidemiologic Studies at the New York Academy of Medicine. Dr Fuller's work has largely focused on HIV prevention and intervention research among drug users and other marginalized populations in low-income, urban communities. She has directed several federally funded, large-scale public health program and policy evaluation studies examining their impact on reducing individual and community-wide disease rates, particularly in communities where racial disparities persist. Dr Fuller also has extensive experience in the design and conduct of large cross-sectional and cohort studies, including community-based multilevel intervention trials, often utilizing a community-based participatory research approach targeting adolescent and young adult injection and non-injection drug users.

Cynthia L. Gay, MD, MPH is a Clinical Assistant Professor and Infectious Diseases Specialist at the University of North Carolina at Chapel Hill, Chapel Hill, North Carolina. She is a graduate from the University of North Carolina at Chapel Hill School of Medicine, and completed her internal medicine residency at Vanderbilt Medical Center. She obtained a MPH and completed an Infectious Diseases Fellowship at the University of North Carolina at Chapel Hill. She has provided clinical care and conducted HIV-related research in several African settings.

Robert E. Geise, MD, MPH is a Clinical Assistant Professor of Medicine at the University of Washington in the Division of Infectious Diseases and a Protocol Team Leader at the HIV Vaccine Trials Network (HVTN). Dr Geise completed his undergraduate education at Cornell University and completed an MBA in Finance at the University of Wisconsin. After 5 years in business, Dr Geise received his MD at the Medical College of Virginia. He was a Resident and Chief Resident in Medicine at the George Washington University, and completed an infectious disease fellowship at the University of Washington. He has done research in primary HIV infection, antiretroviral therapy (ART) and HIV vaccines.

Glenda E. Gray, FCP(SA) (Paeds) is an Associate Professor in Pediatrics and the co-founder and co-Executive Director for the Perinatal HIV Research Unit, based at the Chris Hani Baragwanath Hospital, and affiliated to the University of the Witwatersrand. Professor Gray has been an investigator in the field of mother-to-child transmission of HIV since 1993. She helped with the development of clinical infrastructure necessary to conduct trials across the spectrum of HIV care, prevention and treatment, including prevention of mother-to-child transmission, adult and pediatric treatment, HIV prevention, and trials of candidate HIV vaccines in Soweto, South Africa. She was awarded a Fogarty Training Fellowship at Columbia University in 1999, and completed an intensive program on clinical epidemiology at Cornell University. Evidence of the quality and significance of her work includes numerous peer-reviewed publications, invited lectures in national and international settings, and leadership roles in both the HVTN and IMPAACT. Professor Gray, together with James McIntyre, was awarded the 2002 Nelson Mandela Award for Health and Human Rights, in recognition of their research and advocacy work in the field of PMTCT. In 2004, together with McIntyre, Gray was awarded the IAPAC "Hero in Medicine" Award. She is a member of the Academy of Science of South Africa.

Ronald Gray, MBBS, MSc is the Robertson Professor of Reproductive Epidemiology at the Johns Hopkins University, Bloomberg School of Public Health. He is an epidemiologist and was the principal investigator on the trial of male circumcision for HIV prevention in men. He is co-principal investigator on the Rakai Health Sciences Program, and has conducted several studies of male circumcision for HIV prevention. Dr Gray has published over 300 papers on reproductive health and HIV.

Amy Gregowski, MHS, is a Research Associate in the Department of International Health, Social and Behavioral Interventions Program. She received her Master's degree from the Johns Hopkins Bloomberg School of Public Health. She is currently working on two HIV prevention projects funded by the US National Institutes of Health: one is in Vietnam with men who are HIV-positive and injection drug users, and the other is a multisite study in several countries in sub-Saharan Africa and Thailand.

Thomas E. Guadamuz, PhD., MHS is a postdoctoral fellow in the Department of Behavioral and Community Health Sciences and the Center for Research on Health and Sexual Orientation, Graduate School of Public Health, University of Pittsburgh. Dr Guadamuz received his PhD in Infectious Disease Epidemiology from The Johns Hopkins University, and has received NIH Fogarty and Fulbright fellowships to conduct HIV prevention research among MSM populations in Thailand. Most recently, Dr Guadamuz was a member of the Thailand MSM Study Group, where he collaborated with investigators from the US CDC, Thailand Ministry of Public Health and Rainbow Sky Association of Thailand (the first and largest Thai MSM community-based organization) to carry out the first HIV surveillance among MSM populations in Thailand.

Kathy Hageman, MPH is currently pursuing a PhD in Behavioral Science and Health Education at Emory University in Atlanta, Georgia. Ms Hageman has recently been awarded a Ruth L. Kirschstein National Research Service Awards (NRSA) Pre-Doctoral Fellowship to investigate the behavioral barriers (individual, couple, and socio-cultural) that prevent consistent and correct condom use within long-term HIV-discordant relationships, and how these barriers can be overcome. Collaborating with the two largest HIV-discordant research sites in the world, in Lusaka (Zambia) and Kigali (Rwanda), this study aids in the refinement of risk-reduction counseling messages and intervention development for this high-risk yet understudied population.

Abigail Harrison PhD is Assistant Professor (Research) at the Population Studies and Training Center, and Instructor, Department of Medicine, Warren Alpert Medical School, Brown University. She is a social demographer whose research examines HIV prevention in adolescents, gender and reproductive health, and the social and cultural processes underlying health outcomes. Her current research focuses on adolescent sexual behavior and the transition to adulthood in South Africa, as well as behavioral interventions for this population.

Amy Herrick, MA is currently a doctoral student in the Department of Behavioral and Community Health Sciences, Graduate School of Public Health at the University of Pittsburgh. She received a Master of Arts in Social Science from the University of Chicago in 2001, with a focus on sociology of gender. Using primarily a community-based approach, Amy has been working with the sexual minority youth community for the past 15 years. Amy's current research interests focus on the health disparities of young women who have sex with women, and HIV risk behaviors of transgender youth and young men who have sex with men.

Donna L. Higgins, PhD is a Technical Officer with the HIV/AIDS Department at the World Health Organization (WHO). Dr Higgins is responsible for leading WHO's global HIV testing and counseling program, and has provided program and evaluation support on the topic in multiple developing countries. Dr Higgins has expertise in the development, implementation and evaluation of individual-, group- and community-level HIV prevention strategies. Prior to working at the WHO, Dr Higgins served at the US Centers for Disease Control and Prevention.

David R. Holtgrave, PhD has since August 2005 been Professor and Chair of the Department of Health, Behavior and Society at Johns Hopkins Bloomberg School of Public Health. From 2001 to 2005, Dr Holtgrave was Professor of Behavioral Sciences and Health Education, and Professor of Health Policy and Management at Rollins School of Public Health at Emory University. He served as Director of Behavioral & Social Science Core of the Center for AIDS Research (CFAR) and Vice-Chair of the Department of Behavioral Sciences and Health Education. From 1997 to 2001, Dr Holtgrave was Director of the Division of HIV/AIDS Prevention: Intervention Research and Support in the National Center for HIV, STD and TB Prevention at the Centers for Disease Control and Prevention. The Division has major responsibilities in funding HIV prevention programs, providing technical assistance to HIV prevention service delivery organizations, conducting program evaluation studies, and performing HIV prevention intervention research. Dr Holtgrave has worked in the field of HIV prevention since 1991. From 1991 until 1995 and from 1997 to 2001, he worked at the CDC in HIV prevention; from 1995 until 1997 he was an Associate Professor and Associate Center Director at the Center for AIDS Intervention Research at the Medical College of Wisconsin. His research focuses on the effectiveness and cost-effectiveness of a variety of HIV prevention interventions, and the relation of the findings of these studies to HIV prevention policy-making. Dr Holtgrave worked on HIV prevention community planning, and on the Wisconsin HIV Prevention Community Planning group.

Angela D. Kashuba, BScPhm, PharmD, DABCP is an Associate Professor in the Division of Pharmacotherapy and Experimental Therapeutics at the University of North Carolina School of Pharmacy. She is the Director of the UNC Center of AIDS Research Clinical Pharmacology and Analytical Chemistry Core, and Director of the Analytical Chemistry Laboratory for the Verne S. Caviness General Clinical Research Center. She has worked extensively on characterizing the pharmacology of small molecules in the genital tract of men and women to inform primary and secondary HIV prevention strategies.

Godfrey Kigozi, MBChB is the senior Medical Officer for the Rakai Health Sciences Program, and was responsible for the conduct of the trials of male circumcision for HIV prevention in Rakai, Uganda. He has been first author and co-author on several papers describing the trials of male circumcision in Rakai.

Marie Laga, MD, MSc, PhD is Professor and Head of HIV/STI Epidemiology and Control Unit at the Institute of Tropical Medicine (ITM) in Antwerp Belgium. M. Laga started working on HIV/AIDS in 1984 and spent several years overseas in Burundi, Kenya, DR Congo, the UK (training at LSHTM), the US (Visiting Scientist in the AIDS program at the CDC) and in Côte d'Ivoire (as Director of "Projet Retro-CI", a large CDC-funded HIV/AIDS research and intervention program). Current HIV/AIDS activities include policy support and operational research in the areas of expansion of care and strengthening prevention strategies in developing countries, as well as development of applied training modules for HIV/AIDS control program managers. Marie Laga is author of over 140 scientific publications on different aspects of HIV AIDS in developing countries.

Cynthia M. Lyles, PhD is a mathematical statistician and Team Leader of the Research Synthesis and Translation Team within the Prevention Research Branch for the Division of HIV/AIDS Prevention, National Center for HIV/AIDS, Viral Hepatitis, STD, and TB Prevention (NCHHSTP), CDC. She received two Bachelor's degrees in Mathematics and Math-education from the University of South Florida, and her MS and PhD in Biostatistics from the University of North Carolina at Chapel Hill. While at The Johns Hopkins University, and currently at CDC, Dr Lyles has worked on domestic and international HIV epidemiology and behavioral prevention research across a range of risk populations. She has published numerous peer-reviewed articles on HIV epidemiology and behavioral prevention research. Dr Lyles has served as a consulting editor of *Health Psychology* for the evidence-based medicine and methodology section.

Ian McGowan, MD, PhD, FRCP is a Professor of Medicine at the Magee Womens Research Institute at the University of Pittsburgh, Pittsburgh, Pennsylvania. He graduated in Medicine from the University of Liverpool, and obtained his PhD from the University Of Oxford, England. He completed post-graduate training in HIV medicine and gastroenterology. After working in the pharmaceutical industry on the development of a number of antiretroviral drugs, including Viread®, Dr McGowan returned to academic research. He worked at the David Geffen School of Medicine at UCLA for 5 years, and has recently moved to the University of Pittsburgh. His research interests focus on clinical and translational aspects of microbicide development, with a specific focus on rectal microbicide development. Dr McGowan is the Co-Principle Investigator of the NIH funded Microbicide Trials Network, and is a member of the Antiviral Advisory Committee of the United States Food and Drug Administration.

James A. McIntyre, FRCOG is the co-founder and an Executive Director of the Perinatal HIV Research Unit of the University of the Witwatersrand, South Africa, based at the Chris Hani Baragwanath Hospital in Soweto, one of Africa's largest AIDS research centres working in HIV prevention, treatment and care, and HIV vaccines. Professor McIntyre leads the CIPRA-SA "Safeguard the household" collaborative South African research program, funded by the US National Institutes for Health, and the Soweto Clinical Trials Unit affiliated to the ACTG, IMPAACT and HVTN trials networks. He is an international authority on mother-to-child transmission of HIV and HIV in women, and has published widely in this field. He has served as a consultant to the WHO, UNAIDS and UNICEF, advising on pregnancy, treatment guidelines, and is a member of the Network Executive Committee of the IMPAACT. He and Professor Glenda Gray were jointly awarded the 2002 Nelson Mandela Award for Health and Human Rights, and the 2003 "Heroes in Medicine" award of the International Association of Physicians in AIDS Care (IAPAC).

Anna Meyer-Weitz, PhD is a Professor in the School of Psychology, University of KwaZulu-Natal. Her research interests are in the development, implementation and evaluation of health promotion interventions with a particular focus on adolescent health. Other areas of interest include STIs, AIDS stigma and discrimination, and mental health promotion.

Matthew J. Mimiaga, **ScD, MPH** is an Instructor in Psychiatry at Harvard Medical School/Massachusetts General Hospital, and a Research Scientist at The Fenway Institute, Fenway Community Health. He completed his Post-Doc training in Behavioral Medicine at Harvard Medical School/Massachusetts General Hospital and received his Doctorate from Harvard School of Public Health, majoring in Psychiatric Epidemiology, with minors in Infectious/Chronic Disease Epidemiology and Biostatistics, and was awarded the Harvard University Presidential Scholarship.

He received his Master of Public Health from Boston University School of Public Health, majoring in Epidemiology and Behavioral Sciences. He has co-authored more than 45 articles, chapters and other publications on HIV/AIDS and related infectious disease topics and was recently awarded a grant from NIDA to develop a behavioral treatment for crystal methamphetamine addiction in HIV-uninfected MSM. Dr. Mimiaga is also the PI (with Dr. Mayer) on a MA Department of Public Health funded study examining the social and sexual network characteristics and associated HIV risks of Black/African American MSM and is the co-PI (PI: Dr. Mayer) on a Gilead funded project to study the barriers and facilitators to implementing recent CDC guidelines on routine HIV testing in primary care settings. In addition, he is currently a member of the protocol development team for HPTN 063 (PI: Dr. Safren) – a proposal to develop international prevention trials of HIV-infected individuals in care settings.

His main research interests include HIV/AIDS, mental health and substance use disorders, psychiatric and infectious disease epidemiology, and global health.

Ank Nijhawan, MD is a Research Fellow in Infectious Diseases at The Miriam Hospital, at Brown University School of Medicine. She completed her medical degree and internal medicine training at the University of Texas Southwestern in Dallas, and completed her Infectious Disease fellowship at Massachusetts General and Brigham and Women's Hospitals in Boston. She is the recipient of an NIH T32 grant, and her primary research interests include HIV prevention and treatment in incarcerated populations and injection drug users.

Willo Pequegnat, PhD is Associate Director of International AIDS Prevention Research in the Center for Research on Mental Health at the National Institute of Mental Health (NIMH). As the Senior Prevention Scientist, Dr Pequegnat has primary responsibility for a wide range of national and international projects. Her research involves multilevel social organization and complex relationships – couples, families, communities, societal (media, policy), technological (internet, web, etc.) – in national and international settings. Dr Pequegnat has served as a Staff Collaborator (federal Principal Investigator) on four randomized clinical trials: (1) the NIMH Collaborative HIV/STD Prevention Trial, which is a community-based trial that is being conducted in five countries (China, India, Peru, Russia and Zimbabwe); (2) the NIMH Multisite HIV Prevention Trial with African-American Couples, which is a four-city preventive intervention with serodiscordant African-American couples; (3) the NIMH Healthy Living Project, which is a four-city study of prevention effort with HIV-positive men and women; and (4) the NIMH Multisite HIV/STD Prevention Trial, which was a behavioral prevention in 37 clinics in the US. She took the initiative to develop a research program on the role of families in preventing and adapting to HIV/AIDS, and chairs the only national annual international research conference on families and HIV/AIDS. She co-edited the book on this program of research, entitled *Working with Families in the Era of AIDS*. Dr Pequegnat initiated and is co-editor of *How to Write a Successful Research Grant Application: A Guide for Social and Behavioral Scientists*. She also co-edited a book on community prevention and the role of childhood sexual abuse in HIV prevention. She received her PhD in Clinical Psychology from the State University of New York.

Laura Reilly, BA is currently a research coordinator in the Behavioral Medicine Service at Massachusetts General Hospital (MGH). She received her degree in Psychology from the University of Delaware, and is currently working on a NIDA-funded grant to evaluate the efficacy of CBT for medication adherence and depression for HIV-infected individuals on methadone therapy.

Sari L. Reisner, MA is a Behavioral Science Research Associate at The Fenway Institute, Fenway Community Health. She holds degrees from Brandeis University (MA) and Georgetown University (BA). Her behavioral science research interests focus on the intersection of physical and mental health, including substance-abuse intervention development, health psychology and behavioral medicine within the context of serious illness (HIV/AIDS and cancer), and the epidemiology of mental illness and substance abuse in marginalized populations.

Josiah D. Rich, MD, MPH is Professor of Medicine and Community Health at Brown Medical School and Attending Physician at The Miriam Hospital in Providence, Rhode Island. He is a practicing internist and an infectious disease specialist. He completed medical school at the University of Massachusetts Medical School, and internship and residency at Emory University in Atlanta, Georgia. He subsequently received his MPH from the Harvard School of Public Health, and completed HIV/AIDS and Infectious Diseases fellowships at Harvard Medical School and the Brigham and Women's Hospital in Boston, Massachusetts. He provides medical care both at The Miriam Hospital Immunology Center and at the Rhode Island State Correctional Facility, where he provides infectious disease sub-specialty care. He also serves as Medical Director for the Whitmarsh House, the State of Rhode Island's only STD clinic. Dr Rich's research is on the overlap between infectious diseases and illicit substance use. He is the Principal or Co-investigator on several research grants involving the treatment and prevention of HIV infection. Dr Rich has advocated for public health policy changes to improve the health of people with addiction, including improving legal access to sterile syringes and increasing drug treatment for incarcerated populations. He is Co-Founder, along with Dr Scott Allen, of the Center for Prisoner Health and Human Rights at The Miriam Hospital Immunology Center, www.prisonerhealth.org.

Jean L. Richardson, DrPH is Professor of Preventive Medicine at the Keck School of Medicine at the University of Southern California. Her research involves the control of chronic diseases by designing and testing programs that combine psychological theory with sound public health program planning in experimental field trials. Her research has addressed reducing unsafe sexual behavior among people living with HIV disease, reducing household allergen exposure for children with asthma, increasing mammography among elderly Hispanic women, increasing cancer screening among siblings of breast-cancer cases, increasing compliance with cancer chemotherapy, documenting the effects of after-school care on the use of tobacco, alcohol and marijuana among adolescents, and examining psychological and behavioral effects among women living with HIV.

Abby Rudolph is a doctoral candidate in Infectious Disease Epidemiology at the Bloomberg School of Public Health at Johns Hopkins University. Her current research focuses on reducing HIV acquisition through interventions that reduce high-risk injecting practices, and preventing HIV transmission from HIV-positive IDUs to those in their sexual and injecting networks in Vietnam. Prior to her work in Vietnam, she conducted extensive research on evaluation of pharmacy syringe access among injection drug users in New York City.

Steven A. Safren, PhD specializes in behavioral medicine and cognitive-behavioral intervention development. He is an Associate Professor in Psychology in the Department of Psychiatry at Harvard Medical School, Director of the Behavioral Medicine Service at Massachusetts General Hospital, Director of the Cognitive Behavioral Tracks of the MGH clinical psychology internship, and a research scientist at Fenway Community Health. Dr Safren received his PhD in Clinical Psychology from the University at Albany (State University of New York) in 1998, and did his internship and postdoctoral fellowship at Massachusetts General Hospital/Harvard Medical School. Dr Safren has over 80 professional publications and has been the Principal Investigator on five federally funded NIH grants, with a major focus being on mental health and substance use aspects of HIV adherence and primary and secondary prevention.

Vikrant V. Sahasrabuddhe, **MBBS, MPH, DrPH** is Assistant Professor in the Department of Pediatrics – Division of Infectious Disease at Vanderbilt University School of Medicine, and directs the Vanderbilt-India programs at the Institute for Global Health. He received his medical degree from the University of Pune in India, and his Masters and Doctorate in International Health-Epidemiology at the University of Alabama at Birmingham. Dr Sahasrabuddhe's research interests and work span clinical epidemiology and policy research in HIV/AIDS and reproductive health in developing countries. He has spearheaded the development of NIH and CDC-funded cervical cancer prevention research and service programs for HIV-infected women in India and Zambia that have focused on the use of low-cost "screen-and-treat" strategies. Additionally, Dr Sahasrabuddhe co-directs the Vanderbilt-Meharry Framework Program in Global Health, which focuses on curricular innovation in Global Health.

David Serwadda, MBChB is Dean of the Makerere University, School of Public Health. He is the Ugandan Principal Investigator on the Rakai Health Sciences Program and on the trials of male circumcision for HIV prevention. He has contributed numerous papers on HIV prevention, including the trials of male circumcision.

Susan G. Sherman, PhD, MPH is an Associate Professor in Infectious Diseases Epidemiology at the Johns Hopkins Bloomberg School of Public Health. She uses both quantitative and qualitative methods in conducting research. Dr Sherman is a behavioral scientist and social epidemiologist whose work focuses on epidemiological studies of and socio-economic interventions with drug users. She has worked on several randomized behavioral interventions with drug users in the United States, Thailand and Pakistan. She has studied IDU dyads, social networks, gender differences in illicit drug utilization patterns and disease acquisition, and factors related to transition to injection drug use. She has also evaluated several overdose-prevention interventions in several US cities.

Suniti Solomon, MD is founder and director of YRG CARE, the largest community-based HIV tertiary-care center in South India. Dr Solomon's experience covers a wide range of aspects related to HIV infection, from biomedical to socio-economic. She has a deep interest in community education and mobilization, and leads an effort that supports a > Phase I HIV vaccine trial in Chennai, India with community education and volunteer enrollment. She is the Indian Principal Investigator of several pioneering HIV research studies supported by the US National Institute of Mental Health and the US National Institute of Allergy and Infectious Disease. She also is Director of the Southern India program of the Brown-Tufts Fogarty AIDS Training and Research Project, and a permanent board member of the Gates Foundation AIDS Initiative in India.

Nafisseh Soroudi, PhD recently completed a Clinical Fellowship in Psychiatry at Massachusetts General Hospital (MGH). Dr Soroudi received her PhD in Clinical Health Psychology from Yeshiva University, Ferkauf Graduate School of Psychology. She completed her internship at Montefiore Medical Center and her postdoctoral fellowship at Massachusetts General Hospital/Harvard Medical School. Dr Soroudi has six professional publications on behavioral medicine approaches to the study of HIV, obesity and diabetes. She served as Project Director and a protocol therapist of a NIDA-funded grant to evaluate the efficacy of CBT-AD in patients with HIV and on methadone therapy. Dr Soroudi is a clinical psychologist specializing in behavioral medicine interventions for clients and couples with chronic medical conditions, and in cognitive-behavioral therapy approaches to treatment of mood and anxiety disorders.

Ron Stall, PhD, MPH is currently Professor and Chair of the Department of Behavioral and Community Health Sciences in the Graduate School of Public Health at the University of Pittsburgh. His central research interest is the study of how social and cultural forces shape the behaviors that place individuals at higher risk for disease outcomes. Professor Stall began work in 1984 on the AIDS Behavioral Research Project, one of the first longitudinal studies of AIDS risk-taking behaviors in the world. Since that time he has published over 120 scientific papers on many different aspects of the AIDS epidemic, including methodological research, research on determinants of risk-taking behaviors and HIV seroconversion, life-course issues important to AIDS risk-taking behavior, behavioral intervention research, research on care-seeking behavior for HIV infection, and a portfolio of international research on AIDS. Professor Stall is particularly proud of his record of collaborative research conducted with AIDS community-based organizations, which include a broad range of organizations within the United States and abroad. Stall is the 1999 recipient of the Chuck Frutchey Board of Directors Award from STOP AIDS/San Francisco, is listed as one of the most highly cited behavioral science researchers in the world in the ISI Most Highly Cited website, received the 2005 CDC/ATSDR Honor Award for Public Health Epidemiology and Laboratory Research, and was inducted into Delta Omega (a public health honor society) in 2006.

Ellen Stover, PhD is Director, Division of AIDS and Health and Behavior Research at the National Institute of Mental Health. Her division supports a broad research portfolio focused on domestic and international HIV prevention along with the pathogenesis and treatment of neuropsychiatric consequences of HIV/AIDS. Its annual budget is approximately $180 million. Dr Stover received her PhD in Psychology from Catholic University, Washington, DC in 1978, and has held progressively responsible positions at NIMH over the past 36 years. She has been responsible for developing and overseeing all NIMH AIDS research programs since their inception in 1983. Her accomplishments include the convening of the NIH Consensus Development Conference that produced science-based national recommendations for preventive interventions targeting HIV risk behaviors in 1997. Among her numerous awards, in 2001 Dr Stover received the Senior Executive Service Presidential Meritorious Award for her creation of international HIV/AIDS prevention collaborations in India. Dr Stover is on the Editorial Boards of *AIDS and Behavior* and *Neuropsychopharmacology,* and of other key journals.

Michael D. Sweat, PhD is a Professor of Psychiatry and Behavioral Sciences at the Medical University of South Carolina. Dr Sweat has conducted extensive HIV behavioral prevention science research in Tanzania, India, The Dominican Republic, and a host of other developing countries. Dr Sweat's areas of expertise include HIV testing and counseling, ecologically-based community interventions, cost-effectiveness analysis, mathematical modeling, and meta-analysis and synthesis. He was earlier a researcher at the US Centers for Disease Control and Prevention, a Research Scientist at Family Health International, and an Associate Professor of International Health at The Johns Hopkins University, Bloomberg School of Public Health.

Amanda Tichacek, MPH is the Program Coordinator and Assistant Director for the Rwanda Zambia HIV Research Group (RZHRG) at Emory University in

Atlanta, GA. Ms Tichacek has been working with the RZHRG since she began her studies in epidemiology under the mentorship of Dr Susan Allen in 1996, first as a student research assistant and, after graduation, as an intern in Lusaka. After spending 3 years in the private sector, Ms Tichacek returned to work as the US-based Program Manager for the RZHRG in 2003. Using her background in microbiology and immunology, public health and business, and a 12-year history with research and investigators, Ms Tichacek coordinates activities between the largest HIV-discordant couples research sites in the world.

Kartik K. Venkatesh is currently completing his MD and PhD in Epidemiology at Brown University Medical School. As part of his graduate work sponsored by the US National Institute of Mental Health examining HIV prevention in South India, Kartik is working under the mentorship of Dr Solomon at YRG CARE, Chennai, and Dr Kenneth Mayer at Brown. Kartik is currently conducting research at YRG CARE in primary and secondary HIV prevention within clinical-care settings, and on the natural history of HIV disease in South India.

Sten H. Vermund, MD, PhD is a pediatrician and infectious disease epidemiologist. He serves as Amos Christie Chair in Global Health, and Director of the Institute for Global Health at Vanderbilt University School of Medicine. He received his undergraduate degree from Stanford University, his MD from the Albert Einstein College of Medicine, his Masters degree in Tropical Public Health from the University of London, and his PhD in Epidemiology from Columbia University, where he also trained in pediatrics. Along with over 20 years in academia, Dr Vermund worked at the National Institutes of Health from 1988–1994 as chief of the Vaccine Trials and Epidemiology Branch in the Division of AIDS (NIAID), where he was awarded the 1994 Superior Service Award – the highest civilian recognition in the US Public Health Service. In recent years he has founded two non-governmental organizations in Africa – the Centre for Infectious Disease Research in Zambia in 2000, and Friends in Global Health in Mozambique in 2006 – both of which work on HIV prevention, care and treatment.

Bea Vuylsteke, MD, PhD is a researcher at the STI/HIV Epidemiology and Control Unit of the Institute of Tropical Medicine, Antwerp, Belgium. She was trained as a physician at the University of Leuven, Belgium, where she received her MD in 1984. After her studies, she worked for Médecins sans Frontières in Mali, Chad, Mozambique and Ethiopia. In 1992 she joined the Institute of Tropical Medicine in Antwerp, where she worked as STI advisor in FHI's AIDSCAP program. Since 1999 she has been based in the Côte d'Ivoire, providing technical assistance and leadership for sex worker programs.

Maria J. Wawer, MD, MHSc is Professor in the Department of Population, Family and Reproductive Health, Johns Hopkins Bloomberg School of Public

Health, and has a joint appointment at the Columbia University Mailman School of Public Health. She is a Principal Investigator on the Rakai Health Sciences Program (RHSP) in Uganda, which she initiated with her colleagues Drs David Serwadda and Nelson Sewankambo in 1988. The RHSP conducts extensive research on epidemiological, molecular, behavioral, preventive, service delivery, clinical and treatment aspects of HIV and associated infections, and conducts one of the longest-running community-based HIV surveillance cohort studies in the world. Over 100 publications have resulted from this work. Dr Wawer has also worked on reproductive health and family planning service delivery and evaluation in Latin America, Thailand and multiple North African and sub-Saharan countries.

Tracey E. Wilson, PhD is an Associate Professor of Preventive Medicine and Community Health at the State University of New York, Downstate Medical Center. Dr Wilson is a behavioral scientist with expertise in the design, implementation and evaluation of health promotion and risk-reduction programs, with a focus in the area of HIV/STI prevention. She has directed several federally-funded trials to reduce sexual risk behaviors of men and women with HIV infection and those at risk for infection, and has published extensively in these areas.

Richard J. Wolitski, PhD is Deputy Director of Behavioral and Social Science for the Division of HIV/AIDS Prevention, National Center for HIV/AIDS, Viral Hepatitis, STD, and TB Prevention (NCHHSTP), CDC. He received his PhD in Community Psychology from Georgia State University, and his Masters degree in Psychology from California State University Long Beach. For the past 20 years he has studied HIV risk behavior and interventions to reduce this risk in a wide range of populations, including gay, bisexual and other men who have sex with men, injection drug users and their sex partners, commercial sex workers, incarcerated men, homeless persons, and people living with HIV. He has published extensively on HIV prevention and has co-edited three books. Dr Wolitski currently serves on the editorial board of *AIDS and Behavior*.

Nickolas Zaller, PhD is Assistant Professor of Medicine (Research) at The Warren Alpert Medical School and a researcher at The Miriam Hospital in Providence, Rhode Island. He completed his PhD in Public Health at The Johns Hopkins Bloomberg School of Public Health in Baltimore, Maryland. Dr Zaller completed a NIDA T32 fellowship at The Miriam Hospital. His research interest is on the overlap of infectious diseases, illicit substance use and incarceration. He is currently Project Director for a federally funded research grant linking HIV-positive and high-risk HIV-negative substance users to treatment services. Dr Zaller's work focuses on racial disparities and on developing integrated models to provide comprehensive care and services to individuals with HIV, and strategies to control and prevent the spread of bloodborne and sexually transmitted infections, such as HIV and viral hepatitis.

Acknowledgments

I am fortunate to work with very competent administrative colleagues at the Miriam Hospital and Fenway Community Health, who enable me to do research and think about how to best communicate new HIV prevention information, and would like to acknowledge the help of Lola Wright, Sue Johnson, Hilary Goldhammer and Rodney Vanderwarker in these efforts. I also am fortunate to have become part of the leadership group of the NIH-funded HIV Prevention Trials Network, and have learned a great deal from my colleagues, Sten Vermund, Quarraisha Abdool Karim, King Holmes, Tom Fleming, Deborah Donnell, Tom Coates, Mike Cohen, David Vlahov and Wafaa el-Sadr. On a daily basis, I work with many talented clinical researchers and educators in Boston and Providence who inform and inspire me, including Steve Safren, Judy Bradford, Patricia Case, Matthew Mimiaga, Conall O'Clerigh, Steve Boswell, Chris Grasso, Harvey Makadon, Charles Carpenter, Tim Flanigan, Susan Cu-Uvin, Karen Tashima and Jody Rich. These specific individuals are part of two larger amazing clinical-care and research systems, the Miriam Hospital Immunology Center of the Warren Alpert Medical School of Brown University, and The Fenway Institute of Fenway Community Health.

Kenneth H. Mayer

Introduction

Kenneth H. Mayer and Hank F. Pizer

Currently, more than 25 years have elapsed since the AIDS pandemic was first noted, and there is both good and bad news to report. On the positive side of the equation, it is possible that the global epidemic peaked some time in 2000 or 2001 and since then there have been statistically significant declines in overall HIV incidence in some of the most heavily impacted countries, like Zimbabwe, Kenya, Malawi and Cambodia. Condom use is increasing and sexual debut is being postponed in Botswana, Cameroon, Central African Republic, Chad, Côte d'Ivoire, Namibia, Rwanda, Senegal and Zambia. These are signs that public health prevention is working.

However, aside from mother-to-child transmission and occupational exposure to HIV-infected material, HIV acquisition is the consequence of two pleasurable activities: unprotected anal or vaginal sexual intercourse, and injecting recreational drugs with un-sterile equipment. It should therefore be no surprise that even though HIV is difficult to transmit and HIV prevention programs have been in place for decades, millions of new infections continue to occur annually and prevention gains are not always sustained. For example, Uganda saw a decline in new cases in the 1990s as a result of the ABC effort (abstinence, be faithful and condom distribution), but recent data show the benefits from these programs seem to have stabilized. Other countries that documented early decreases in HIV incidence because of visionary national leadership, like Thailand and Brazil, have recently seen increases in HIV incidence in some subpopulations, like men who have sex with men, suggesting that the epidemic continues to be dynamic and that current behavioral interventions and medical technologies are unlikely to fully reverse this significant global pandemic.

The final answers are not yet available as to exactly why gradual, albeit uneven, positive results from HIV public health prevention programs have been seen. Nor is it certain whether the stabilizing or slowly declining rates of new HIV infections are a temporary trend, or whether they are durable over the long run. Some

1

of the decreases in HIV incidence could either reflect an unfortunate consequence of deaths in the highest-risk populations, if people are not truly changing their behaviors. Is improvement due more to extraordinary advances in antiretroviral drugs made over the last 20 years instead of public health prevention efforts? In Chapter 5, by Cynthia L. Gay, Angela D. Kashuba and Myron S. Cohen, the case is made that today's highly effective antiretroviral medicines can do an excellent job reducing the infectiousness of HIV-positive individuals, but the ultimate impact on HIV incidence of providing more people living with HIV with access to life-saving medicines remains to be seen. Programs like the President's Emergency Program for AIDS Relief (PEPFAR) and non-profit foundations, like the Clinton Foundation, have enabled millions of HIV-infected persons in resource-constrained environments to gain access to therapy. So it is possible that behavioral changes from effective health education may be coupled with enhanced access to better drugs to create awareness that being diagnosed with HIV is not a death sentence, and that it also creates opportunities for clinicians to reinforce behavioral changes. It is possible that the modest gains that have been noted are due to some combination of the above plus other factors that have not been fully elucidated, resulting in a complex interaction of societal forces, cultural issues, individual behavioral, demographic trends, geography and economics in specific settings. Whatever the reasons – and probably only the passage of time will provide clear answers – the good news is that AIDS mortality is declining in much of the world, and the number of new cases of HIV infection also seems to be declining or at least remaining level in resource-constrained environments.

Ironically, the number of new HIV infections appears to be modestly increasing in some highly developed nations, including the United States, which may be due to the increased longevity of HIV-infected patients in care, as well as the perception that, since the epidemic is more manageable, individuals do not have to be as careful about risk-taking compared to the earliest days of the epidemic. Many research teams are at work to develop interventions that incorporate the changing realities of a mature epidemic of a disease which is increasingly treatable, but not curable.

Overall, there is reason to be guardedly optimistic, but complacency needs to be avoided. There continues to be a need to develop new prevention modalities and study ways to enhance the efficacy of promising interventions in diverse cultural settings. The goal of this text is to provide an up-to-date, comprehensive look at the state of AIDS prevention through what has been learned through evidence-based research.

Diverse HIV epidemics across the globe

With the increasing sophistication of genetic typing techniques, it appears that the HIV epidemic started in Africa with a virus that is largely asymptomatic in monkeys and then made its way to people, where it spread by direct person-to-person contact. Monkey meat is a major source of protein in several central and western African settings, and the processing of monkey meat afforded exposed individuals extensive exposure to blood and cuts in the skin that could provide a portal for host entry. Post-colonial population disruptions, urbanization in the developing world and changes in social mores helped to create the pre-epidemic conditions that enabled HIV to rapidly spread in subsequent years. In the modern world of rapid travel and population migration, it took less than a decade from AIDS being first reported in 1981 in a small number of patients in California and New York to it infecting people on every continent. The speed by which this epidemic spread is singularly striking, because HIV is not transmitted by casual touch, air, water or food, or even very efficiently when there is direct intimate contact of infected body fluids. Today there is not one HIV epidemic but numerous different ones that vary by location and population affected. There are new epidemics in the Former Soviet Union potentiated by injection drug use and a resurgence of epidemics in populations that were the first to be heavily impacted, like urban men who have sex with men (MSM) in North America, Europe and Australia. It is estimated that 33 to 45 million people worldwide are infected with HIV, including between 2–3 million children. An additional 4–7 million more people are newly infected each year. There are communities in the developing world where the epidemic is widening among women and youth. In Africa, almost half of new infections are in women and young people aged 15–24. In all, there are approximately 6000 new infections daily and about 12 million AIDS orphans. In India, new HIV infections appear to be leveling off or declining in Tamil Nadu, Maharashtra, Karnataka and Andhra Pradesh, while high rates of infection continue among urban sex workers, men who have sex with men, and intravenous drug users in the northeastern states of Manipur and Nagaland. In China, new infections appear to be on the increase in Henan, Guangdong, Guangxi and Yunnan provinces due to intravenous drug use and sex work. In Pakistan, Vietnam and Indonesia, new cases of HIV spread primarily via drug use and homosexual activity. Meanwhile, prevention efforts seem to be having a moderating effect in Thailand, Cambodia and Burma. The concentration of HIV infection differs significantly across the world. About 740,000 individuals are living with HIV in Western and Central Europe. In Eastern Europe the incidence of new cases is 210.8 per million population, but in Western Europe it is 82.5 per million. Western Europe is more affluent, has a better public health infrastructure, more open sex education and attitudes about sexuality, less stigma and fewer restrictive laws about homosexuality and sex work, and better programs for drug users, and societies have for some time accepted harm-reduction programs. In the 13 countries that make up Western Europe, the predominant mode of transmission is heterosexual contact, and new cases are often among recent arrivals from countries with generalized epidemics. Still, and very troubling, between 1999 and 2006 the number of new cases in Western Europe nearly doubled among men who have sex with men. Though the number of people living with HIV and the number of new cases is higher in Eastern Europe than in Western Europe, the good news is that in the East the epidemic peaked in 2001 at 98,526 new cases. It is nevertheless worrisome that more than one-fourth of these new cases are in young people aged 15–24, and just over 40 percent are in females. Injecting drug use is a significant underlying cause of HIV transmission in both Eastern and Western Europe, but particularly so in the East.

HIV/AIDS in the Americas has a somewhat different pattern. In the United States, the number of new cases peaked in 1992–1993 at about 80,000 per year. Until recently there have been about 40,000 new infections annually, and this is disturbing because that number has remained the same for some time despite national, state and local prevention programs; moreover, the CDC will release statistics later this year indicating a new rise in HIV infections, particularly among men who have sex with men, and among heterosexuals from racial and ethnic minority populations. Recent data seem to indicate progress among intravenous drug users. The rate of African women with HIV is disproportionately high compared to that of the general population. In Canada there are 2300–4500 new infections each year, with the largest number in men who have sex with men. The number of new cases is stable in the Caribbean region, but overall infection rates are second only to those in Africa. Central and South America have stable infection rates among men who have sex with men. While Brazil has been a regional leader in its AIDS programs, including providing access to antiretroviral medications at public expense, about two-thirds of the people living with HIV/AIDS in the region are in Brazil.

Tailoring prevention strategies

HIV transmission is a biological event of low probability, averaging less than 1 in 100 exposures, but one of very great significance. The biological factors that increase the likelihood that a single person-to-person contact will produce infection are increasingly understood. Increased concentrations of HIV in the blood (viral load) and co-infection with other sexually transmitted diseases are associated with increased infectiousness. Individual genetic susceptibility is less well understood but also plays a role, as some individuals who lack one of the receptors that HIV uses to enter cells are less susceptible to becoming infected. These insights have led to the development of certain prevention strategies based on biology, such as using antiretroviral drugs to reduce infectiousness, treating sexually transmitted diseases and circumcising men. Unfortunately, despite years of research and billions of dollars invested, an effective vaccine or microbicide has not yet been developed.

Humankind lives in a global gene pool where almost any microbe can be carried by jet travel almost anywhere in less than 24 hours. The social ecology of AIDS means that events in one part of the increasingly well-connected world community can result in HIV spread across the planet, ranging from African peacekeepers in Cambodia in the mid-1980s to sex tourists. Until an effective vaccine is developed and successfully administered globally, preventing the spread of HIV will require international cooperation. While the world community is increasingly interlocked, human behavior everywhere is modified by differing local demographics, geography, culture, economics and politics. In other words, the admonition to "Think globally, but act locally" must be heeded. Right now the epidemic is still most severe in poor countries of sub-Saharan Africa, but there is cause for concern in middle-income countries from Brazil to Iran, from Thailand to the new states formed from the old Soviet Union. Given the ability of populations to migrate and travel across borders and oceans, it is inevitable that at-risk populations will shift and new viral strains will arise. Accurate data are needed to track large migrations and changes in social norms in changing societies, which is no easy task in the best of circumstances and is much harder when tracking groups at risk for HIV, such as sex workers and their clients, the victims of human trafficking and sexual violence, refugees, migrant workers and drug users. Modern tracking methods have confirmed the spread of HIV in settings of civil war and social dislocation, such as in the wake of conflicts in the Congo and Rwanda. Failed states, civil war and social dislocation provide an ideal incubator for spreading infectious diseases, and HIV is no exception. For too long, many governments denied the HIV epidemic in their countries. They did not collect data, and failed to establish public health education campaigns to reach people at the margins or even the center of society. Time was lost, and almost certainly there were people that became infected who could have been spared. Prevention programs are needed in prisons and other institutionalized settings that are marked by high rates of mental illness, substance abuse and violence.

One message that comes through from all of the contributors to this text is that no single type of prevention program is going to work for all populations at risk and all the diverse geographic settings where HIV circulates. Prevention programs must be conceived, designed, implemented and evaluated in ways that account for local attitudes, traditions, religion, economic disparities and even political systems. While circumcision has been found to reduce the risk of HIV transmission, especially among high-risk African men, this practice is not acceptable everywhere because of cultural traditions. Thus, there are structural issues that must be confronted, most notably the subordination of women in traditional societies. Adolescents and young adults are at high-risk in areas of the developing world where HIV is common. Effective programs must account for the normal sexual drive of young people, plus the reality that young women often become sexually active with older men who have engaged in high-risk sex and/or drug use. These are just a few of the special contexts that HIV prevention programs must tackle.

Kenneth H. Mayer and Hank F. Pizer

About this book

This volume is divided into three sections. Part I, Epidemiological and biological issues in HIV prevention, covers the sciences that inform HIV public health prevention, including epidemiology, immunology and microbiology, the behavioral sciences, vaccines and microbicides, and the use of antiretroviral drugs for prevention. In Chapter 1, Vikrant V. Sahasrabuddhe and Sten H. Vermund discuss the up-to-date global epidemiological trends about HIV prevalence and incidence, with the sobering message that in some areas AIDS has reversed decades of progress in reducing mortality and increasing life expectancy. In Chapter 2 Deborah Anderson discusses the immunology and virology that impact HIV transmissibility, and in Chapter 7 Willo Pequegnat and Ellen Stover apply the principles of behavioral medicine to HIV risk-reduction. In Chapter 3 Robert Geise and Ann Duerr review the work to date on vaccine development, which so far has been disappointing because of HIV's genetic variability, ability to evade immune control, lack of a full understanding of the correlates of protection against HIV and absence of reliable animal models, and its diverse modes of transmission. In Chapter 4 Ian McGowan discusses microbicides, which are topical compounds that can be applied to vaginal or rectal mucosal surfaces to significantly reduce or prevent male-to-female transmission. Given the increasing feminization of the AIDS epidemic, an effective microbicide would offer women an important degree of control over their reproductive health. As with vaccine development the microbicide effort to date has been disappointing, although several new approaches are of great interest and new efficacy trials are underway. Chapter 5, by Cynthia Gay, Angela Kashuba and Myron Cohen, provides a comprehensive look at the three ways antiretroviral drugs can be employed to prevent HIV infection: by reducing viral load in people who know they are HIV-infected, as post-exposure prophylaxis, and as pre-exposure prophylaxis both orally and topically. Post-exposure prophylaxis is now standard in occupational settings, and it is likely that it will soon become standardized for non-occupational exposure. Meanwhile, trials to evaluate pre-exposure prophylaxis on the population level are now just coming in. It is already known that ART reduces viral load and that translates into lowering infectiousness, so there is hope that it can be adapted in ways that would be beneficial and cost-feasible for selected high-risk groups. In Chapter 6, Ronald Gray, David Serwadda, Godfrey Kigozi and Maria Wawer cover the evidence-based work that demonstrates the protective effects of circumcision and STD control, in particular for preventing HIV transmission among high-risk men.

Part II, Behavioral issues in HIV prevention, contains an array of chapters that cover what has been tried and learned from implementing and studying behavior-based prevention interventions. These have been tried in a wide array of contexts around the globe: individuals (Chapter 8, by Matthew Mimiaga, Sari Reisner, Laura Reilly, Nafisseh Soroudi and Steven Safren), dyads and groups (Chapter 9, by Susan Allen, Kathy Hageman and Amanda Tichacek), men who have sex with men (Chapter 10, by Ron Stall, Amy Herrick, Thomas Guadamuz and Mark Friedman), HIV-positive patients (Chapter 11, by Jean Richardson and Tracey E. Wilson), injecting drug users (Chapter 12, by Crystal Fuller, Chandra Ford and Abby Rudolph) and recreational drug users (Chapter 13, by Lydia Drumright and Grant Colfax), sex workers (Chapter 14, by Bea Vuylsteke, Anjana Das, Gina Dallabetta and Marie Laga), youth in high-prevalence areas (Chapter 15, by Quarraisha Abdool Karim, Anna Meyer-Weitz and Abigail Harrison), incarcerated and institutionalized persons (Chapter 16, by Ank Nijhawan, Nickolas Zaller, David Cohen and Josiah Rich), and to prevent mother-to-child transmission (Chapter 17, by James Alasdair McIntyre and Glenda Elisabeth Gray). Each of these chapters provides a broad and detailed view at the current state of prevention programs in diverse geographic and social contexts with the theme that "No one size fits all." Interventions must be tailored to and evaluated within local and regional contexts.

Part III, Structural and technical issues in HIV prevention, starts with Chris Beyrer, Susan Sherman and Stefan Baral discussing harm reduction, a human rights-based approach designed to limit the harms that result when individuals engage in high-risk behaviors (Chapter 18). Examples of these programs are clean needle exchange and providing condoms in prison. In Chapter 19, Julie Denison, Donna Higgins and Michael Sweat cover what has been learned from studying the implementation of HIV counseling and testing programs. In the early days of the epidemic, HIV testing was expensive and could only be performed in a relatively few specialized laboratories, while it took weeks to get a patient's test results. HIV testing is now fast and inexpensive. It is possible to get preliminary results at the time of testing in community settings, and thereby identify and provide immediate on-the-spot counseling, partner notification and referral. In Chapter 20, Suniti Solomon and Kartik K. Venkatesh focus on structural societal barriers to HIV prevention, especially the subordination of women in traditional societies. Here, the difficulties are enormous and the barriers to change are well-entrenched. A variety of innovative interventions are being considered and tried, like working through local wine shops or providing microcredit to women to start businesses. If there is a silver lining to the AIDS tragedy, it is that it has brought world attention to regressive and egregious social conditions, probably most of all to the mistreatment of women. Many of the contributors in this book document how effective HIV prevention programs address multiple interwoven social issues by trying to empower disadvantaged populations in addition to combating the spread of the virus. In Chapter 21, Jane Bertrand, David Holtgrave and Amy Gregowski cover program monitoring and evaluation, the essential activities that provide policy-makers, practitioners and funders with unbiased analysis as to whether programs are effectively meeting their stated goals and objectives, and at what cost. Continuous, ongoing evaluation allows for quality improvement during program implementation so that vital time and resources are not wasted. Summative evaluation provides an analysis of what worked and what didn't, based on actual program performance. This information is vital for designing and funding new programs. In Chapter 22, Kevin Fenton, Richard Wolitski, Cynthia Lyles and Sevgi Aral complement the previous chapter by discussing how government agencies and other funders work to get effective prevention concepts and programs identified, supported and put into practice. This is a synthetic process that integrates research and program support to inform, develop and fund future prevention work.

Looking ahead

The world is in the middle of the HIV epidemic – not at the beginning or the end. The underpinning of effective HIV prevention is an active, integrated and adequately funded public health community. It is a worldwide effort that includes international agencies and organizations, national and local governments, private non-profit foundations, the business community and non-governmental organizations. Positive results from continued investment and the hard work of committed professionals are anticipated, but sustained success will require time. The world's response to the HIV epidemic is unprecedented, but more needs to be done and, despite important gains, public health must not become complacent. A lesson to be learned from the re-emergence of dengue fever, malaria and other infectious diseases that until recently were well-controlled is that dropping the guard is foolish and dangerous. The challenge is to take the prevention strategies that have been shown to work and put them together synergistically with each other. There is no foolproof way to control human behavior, and no one kind of prevention approach is going to work for everyone, everywhere, until there is an effective vaccine that has been distributed globally. Public health programs will have to combine what we know and learn about the biology of HIV transmission with multidisciplinary approaches to behavioral modification that take into account local culture and traditions. The good news is progress appears to have been made, and there is a continuing high level of engagement on the part of governments, community leaders and the scientific and business communities on an unprecedented international scale. Given the reality that more than 60 million people have become infected since the start of the AIDS epidemic, with possibly more than 30 million deaths, the stakes remain high, and the price of inaction unacceptable. The editors hope that this book will engage policy-makers and scholars, as well as new generations of students, who may be able to use the information in this text as a basis for thinking of new and creative approaches to one of the most serious health problems that will continue to confront humanity well into this new century.

Epidemiological and biological issues in HIV prevention

PART

I

Current and future trends: implications for HIV prevention

1

Vikrant V. Sahasrabuddhe and Sten H. Vermund

In November 2007, the Joint United Nations Program on HIV/AIDS (UNAIDS) and the World Health Organization (WHO) revised downward the estimates of the number of persons with HIV/AIDS worldwide, based on more accurate data from revised estimates in India and Africa. Their 2007 AIDS Epidemic Update estimates that of the 33.2 million (likely range: 30.6–36.1 million) persons who were living with HIV/AIDS in 2007, 30.8 million were adolescents and adults (28.2–33.6 million) and 2.5 million children <15 years of age (2.2–2.6 million)(UNAIDS, 2007; Figure 1.1). Almost half of the persons living with HIV/AIDS (PLWHA) were women: 15.4 million (13.9–16.6 million). Sub-Saharan Africa, both historically and currently, bears 68 percent of the world's disease burden while only hosting about 12 percent of the world's population. With a huge epidemic in the general population, sub-Saharan Africa's HIV epidemic contrasts with the rest of the world that sees HIV heavily over-represented in high-risk subgroups. Sex workers and their sexual partners are over-represented as a high-risk subgroup in Asia as well as most parts of Africa, men who have sex with men (MSM) are a predominant high-risk subgroup in the HIV epidemics in the Americas, Western Europe and Australia/New Zealand, while injecting drug users (IDUs) drive the epidemic in China, southeast Asia and Eastern Europe.

The latest epidemic trend analysis suggests that the global AIDS epidemic peaked around the turn of the century (2000–2001) (Chin, 2007; UNAIDS, 2007; Figure 1.2). Slight downward trends in new cases have been observed in many countries since 2002; some observers suggest that this is due to saturation of the highest-risk pool of individuals along with death of infected persons, while others attribute declines to slow and sporadic yet widespread successes in public health prevention programs that have reduced high-risk behaviors

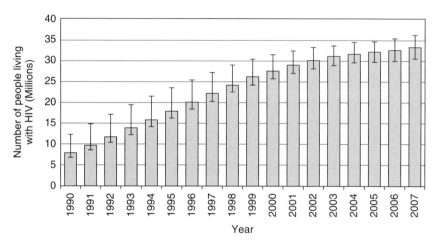

I This bar indicates the range around the estimate

Figure 1.1 Trends in the estimates of numbers of HIV-infected persons living worldwide. Source: UNAIDS (2007).

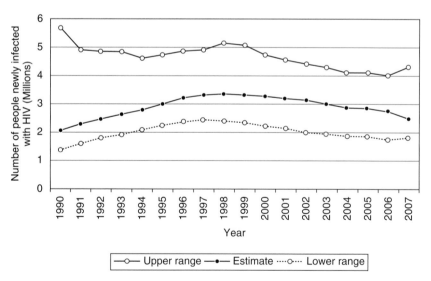

Figure 1.2 Trends in the estimates of numbers of people newly infected with HIV worldwide. Source: UNAIDS (2007).

(Meda *et al.*, 1999; Ainsworth *et al.*, 2003; Shelton *et al.*, 2006; Chin, 2007; Gregson *et al.*, 2007). The truth is likely a mix of these factors, with differential impacts of each in different venues. AIDS mortality is also decreasing in many countries, most significantly due to improved access to antiretroviral therapy for PLWHA (Ventelou *et al.*, 2008).

Estimates and projections: methodology and refinements

Case definitions and surveillance methodology have undergone significant revisions over the past 27 years since HIV was first described, partly as a result of improved scientific understanding of the biomedical aspects of the disease but also due to more robust population estimation methods (Schmidt and Mokotoff, 2003; Cleland *et al.*, 2004; Posner *et al.*, 2004; Chin, 2007). The Bangui definition by the WHO in 1986 and the United States Centers for Disease Control and Prevention (CDC)-led AIDS surveillance case definitions in 1985, 1987 and 1993 have been the backbone of AIDS case reporting based on HIV test results, clinical indications and, usually, CD4+ T cell counts (WHO, 1985; CDC, 1987, 1992). Given the reluctance of many to obtain HIV tests, and owing to the undercapacity of clinical infrastructures in developing countries, there has been significant under-reporting of AIDS cases at regional and national levels, particularly in resource-limited settings (Chin, 2007).

Population-based unlinked anonymous testing (UAT) methods have been adopted for estimating national figures. UAT has been most widely undertaken for surveillance of "sentinel" populations of pregnant women in antenatal care (ANC) clinics. This method, in use since 1989, has been applied widely, since it is operationally uncomplicated, cost-effective, sustainable, and thought to represent the general population in the most at-risk reproductive age groups. Unfortunately, this method has been shown to systematically over-represent the population HIV prevalence, since most sentinel sites have been concentrated in larger cities, under-representing rural areas. ANC clinic sampling that systematically over-samples urban women, including high-risk groups, will overestimate population HIV seroprevalence (Saphonn *et al.*, 2002; Dandona *et al.*, 2006; Chin, 2007). An improved surveillance method employed in recent years by national governments includes population-based sampling and HIV-testing through Demographic and Health Surveys (DHS) (Mills *et al.*, 2004). These estimates are significantly lower (between half to one-third) than those of the UNAIDS-promoted sentinel surveillance system. While a population-based survey is subject to bias due to under-representation (particularly for men, due to low-response rates or absenteeism), it is nonetheless more representative of the general population for estimation of national disease prevalence (Chin, 2007). "Triangulation" of multiple approaches (e.g., DHS, ANC and targeted surveys of high-risk groups) is useful to estimate most accurately the overall seroprevalence in a given venue (Bennett *et al.*, 2006).

The culmination of this process by UNAIDS and the WHO incorporated the results of population-based surveys with sentinel surveillance data in its annual report for 2007, resulting in a substantial downward revision of the world's total numbers of persons estimated to be living with and to have died due to HIV/ AIDS. In India, for example, lower than anticipated rural prevalence estimates resulted in PLWHA estimates reduced from 5.2 million to 2.5 million; similar downward revisions were made in such nations as Angola, Kenya, Mozambique, Nigeria and Zimbabwe (Dandona and Dandona, 2007; Shelton, 2007). Greater reliance on population-based surveys was accompanied by other methodological improvements in 2007, including improved HIV sentinel surveillance, expansion to more sites in relevant countries, as well as adjustments to mathematical models to accommodate a better understanding of the natural history of untreated HIV infection in low- and middle-income countries. Continuous revisions of prevalence estimates are made by UNAIDS and WHO as more data become available.

While HIV prevalence provides a snapshot of the magnitude of the pandemic, especially when accompanied by CD4+ cell-count distributions in the case of HIV surveillance, a critical indicator is the incidence of HIV infection. HIV incidence represents the number of new HIV infections in a population-at-risk per year, and can help distinguish newer from older epidemics (McDougal *et al.*, 2005; Xiao *et al.*, 2007a). Resources can be allocated where transmission rates are highest, not where they *used* to be the highest. It is also better to use incidence surveillance in prevention program evaluations than to rely on slow-to-change prevalence estimations (Bhys *et al.*, 2006; Sakarovitch *et al.*, 2007).

UNAIDS and WHO now estimate that global HIV incidence likely peaked around the turn of the century (UNAIDS, 2007). However, this broad generalization hides significantly increasing sub-epidemics in high-risk population subgroups in many countries, especially the ones with low HIV prevalence rates in their general populations. High-risk groups of MSM, sex workers and their clients, and IDUs and their sexual partners continue to be represented disproportionately in HIV incidence and prevalence, and must be the targets of intensive prevention and risk reduction efforts.

Sub-pandemics in different world regions

Sub-Saharan Africa

Sub-Saharan Africa remains the world's most affected region, with more than two-thirds of all people HIV-positive living here and the region accounting for more than three-fourths (76 percent) of all AIDS deaths in 2007 (UNAIDS, 2007). A total of 1.7 million (range: 1.4–2.4 million) became newly infected in

2007, with another 1.6 million (range: 1.5–2.0 million) dying in this region the same year (UNAIDS, 2007). Comparing revised estimates for 2007 to revised estimates for 2001, there were half a million fewer infections but 200,000 more deaths (UNAIDS, 2007). Though substantial gains in improving access to antiretroviral therapy and subsequent lengthening survival have been noted in Africa, UNAIDS estimates that at least four new infections have occurred for every person placed on antiretroviral therapy, prompting the prevention slogan, "We cannot treat our way out of the AIDS pandemic" (Shelton, 2007).

A poignant feature of the African AIDS epidemic has been the significantly higher number of women infected with HIV (approximately 61 percent of the adolescent and adult annual total, by UNAIDS estimates) and dying annually from it, unlike the rest of the world where men predominate. As the women go, so go the infants and children; over 90 percent of the children with HIV/AIDS during the pandemic have come from sub-Saharan Africa. Three regions within sub-Saharan Africa (southern, east, and west-central) have considerable differences, with southern Africa afflicted most severely. National adult HIV prevalence exceeded 15 percent in eight countries in 2005 (Botswana, Lesotho, Mozambique, Namibia, South Africa, Swaziland, Zambia and Zimbabwe). South Africa still has the greatest disease burden of any single country in the world, with >5 million PLWHA. The adult HIV prevalence has remained stable in comparatively lower-level epidemics in most of west-central Africa, while it appears to be stable or declining in many countries of east Africa. (To avoid misunderstanding, we emphasize that the rates are lower compared to southern Africa; they are still higher in east and west-central Africa than in nearly any other part of the world.) Statistically significant declines in HIV prevalence (and incidence, though data are limited) are evident in Zimbabwe, Kenya and Malawi (UNAIDS, 2007). Furthermore, positive trends in risk behaviors (increasing age of sexual debut, increased condom use, etc.) measured in behavioral surveillance are evident from population survey results in Botswana, Cameroon, the Central African Republic, Chad, the Côte d'Ivoire, Namibia, Rwanda, Senegal and Zambia. While the Ugandan HIV epidemic showed the first real decline in Africa during the 1990s, coincident with behavioral indicators suggesting the success of so-called "ABC" interventions (abstinence for youth, "be faithful" messages, and condom distribution and advocacy), current trends suggest that the incidence has stabilized in the face of stagnant rates of adoption of safer sexual behaviors (Stoneburner and Low-Beer, 2004; Bunnell *et al.*, 2006a; Hallett *et al.*, 2006; UNAIDS, 2007). Combined with the rapid population growth rate in Uganda, these trends translate to a greater number of individuals being infected with HIV each year, after over a decade of declining trends.

Asia

The epidemic in Asia is composed of multiple heterogeneous epidemics predominantly restricted to high-risk sexual and injection drug-using population subgroups. In India, the epidemic has leveled off or declined in the peninsular states of Tamil Nadu, Maharashtra, Karnataka and Andhra Pradesh (Dandona and Dandona, 2007; Fung *et al.*, 2007; Mehendale *et al.*, 2007; UNAIDS, 2007; Steinbrook, 2008). However, there, much higher prevalence rates persist in sex workers and MSM populations in urban areas across the country, as well as in IDUs in the northeast states of Manipur and Nagaland. Thus, core transmitter populations persist with high HIV rates (over 20 percent among female sex workers in Mumbai in 2007, for example) (Brahme *et al.*, 2006; Silverman *et al.*, 2006). There is a burgeoning epidemic among IDUs in Pakistan, Vietnam and Indonesia, with MSM populations also at increasing risk (Pisani *et al.*, 2003a; Bokhari *et al.*, 2007; Des Jarlais *et al.*, 2007). The epidemic seems to be declining in Thailand, Cambodia and Myanmar, as a result of years of sustained prevention efforts (Cohen, 2003a, 2003b; Saphonn *et al.*, 2005). The overlap of IDU (still the predominant mode of transmission) and sex work (especially among MSM) continues to be an important feature of the increasing epidemic in China, especially in the Henan, Guangdong, Guangxi, Xinjiang and Yunnan provinces (Ji *et al.*, 2006, 2007; Wu *et al.*, 2007; Xiao *et al.*, 2007b).

Europe, Oceania and the Americas

The HIV epidemic in Eastern Europe and Central Asia is one of the fastest growing in the world. Nearly two-thirds of the growth is due to IDU while a third is due to unprotected heterosexual intercourse (Kelly and Amirkhanian, 2003; Bobrova *et al.*, 2007; Sarang *et al.*, 2007). Ukraine in particular has seen increasing trends in its southeastern region, predominantly in IDUs but also increasingly in MSM (Booth *et al.*, 2007; Bruce *et al.*, 2007). The HIV epidemic in the Russian Federation is still growing, although not as rapidly as in the late 1990s (Moran and Jordaan, 2007; Heimer *et al.*, 2008; Tkatchenko-Schmidt *et al.*, 2008). Western Europe's epidemic is driven by MSM and IDU, but its severity may have been blunted by aggressive public health responses in the 1980s (Hamers *et al.*, 2006; Dougan *et al.*, 2007). Australia and New Zealand have been similar to Europe in their assertive response (Guy *et al.*, 2007; Rupali *et al.*, 2007).

Haiti still accounts for the largest burden among countries in the Caribbean, although there is evidence of stabilization of the epidemic, at least partially due to decreasing risky behaviors (Cohen, 2006a; Hallett *et al.*, 2006). The epidemic in the Dominican Republic and the islands and nations of the Caribbean Basin remains relatively stable, albeit at rates that exceed any other region of the world excluding sub-Saharan Africa (Figueroa, 2003; Cohen, 2006b).

The South and Central American epidemics have generally remained stable after peaking in the 1990s (Cohen, 2006c). The Brazilians have mobilized one of the most effective public health responses of any large nation worldwide, providing access to comprehensive prevention and care services, including risk-reduction efforts and antiretroviral therapy, well before PEPFAR or the Global Fund to Fight AIDS, Tuberculosis, and Malaria were even conceived. Consequently, incidence has dropped, although more than two-thirds of PLWHA in Latin America reside in Brazil (Greco and Simao, 2007). The Argentinean epidemic has shown a shift from IDU to a more heterosexually-driven epidemic; most other countries in South America have predominantly MSM populations represented in their epidemics (Aceijas *et al.*, 2006; Baral *et al.*, 2007). Behavior change is becoming apparent in some countries, like Honduras (Cohen, 2006d).

MSM populations in North America still represent about half of new infections, with African-American and other minority men estimated to represent about half of these new MSM infections (Fenton, 2007; Millett *et al.*, 2007; Pence *et al.*, 2007). White heterosexuals are under-represented in both new cases and deaths, while African-American women continue to be disproportionately over-represented (Levine *et al.*, 2007; Moreno *et al.*, 2007). Rates in IDUs have dropped, likely due to needle-exchange programs and better access to drug treatment and HIV care. The total numbers of PLWHA are increasing in North America, Australia and Western Europe due to better access to HIV treatment and care and much lower HIV death rates, without commensurate declines in HIV incidence rates (UNAIDS, 2007).

Conceptual framework for HIV prevention

HIV prevention interventions may be viewed as having three components: (1) *primary prevention*, mainly directed towards persons uninfected by HIV; (2) *secondary prevention*, which includes early detection of HIV infection to offer early prevention and therapeutic services to both acutely and chronically infected persons, with a goal of reducing their risk behaviors to decrease STI rates and HIV transmission to others (termed prevention in positives, or "positive prevention") and to treat them as indicated; and (3) *tertiary prevention*, which involves targeting persons with chronic HIV infection to reduce their death and disability levels by using antiretroviral therapy and enabling partial immune reconstitution. Most often, more than one approach for prevention is directed towards a target group, since risk behaviors for HIV are multidimensional and overlapping. A host of interventional strategies have been shown to be effective. These strategies can be applied at the micro-level (individual focused interventions such as PLWHA, sexual or needle-sharing partners, family members), meso- (community) level, and macro- (policy or structural) level (Global HIV Prevention Working Group 2007; Sahasrabuddhe and Vermund, 2007). A simple conceptual framework for prevention based on evidence of effectiveness of each strategy is presented in Table 1.1.

Table 1.1 Strategies and implementation levels for HIV prevention activities

Level	Prevention strategy and target groups		
	Primary prevention for hitherto HIV-uninfected individuals	Secondary prevention for recent HIV seroconverters/"acutely infected" individuals	Tertiary prevention for individuals with established HIV infections/AIDS
Micro level (Individual, partner(s)/family)	**Approach:** Improving access to healthcare and individual empowerment through appropriate education and socio-economic development.		
	• The "ABC" strategy • Accessible VCT services • Prevention and care services for STIs • Provision of safe, hygienic adult male circumcision services by trained providers	• IEC to reduce stigma around testing and help with coping • Improving accessibility of health care services	• Improving accessibility to ART and prophylaxis for OIs and malignancies • Emphasizing "positive prevention" • Assistance for re-entry back in workforce and society
Meso level (Community)	**Approach:** Broaden individual and family-centric interventions in a community-development framework		
	• Promoting community-based VCT services • Impacting traditions, social norms, and beliefs regarding sexuality to promote less-risky behaviors • Promotion of broader gender-equality interventions	• Improving awareness and utilization of VCT and STI treatment services • Implementation of outreach programs for hard-to-reach groups	• Access to self-help groups, gainful employment, social networks and legal recourse • IEC interventions to mitigate stigma and discrimination in communities and workplaces • Development of self-help groups and support for peer networks for coping and positive lifestyles

Macro level (Policy)	**Approach:** Developing and sustaining policy frameworks that support individual and community-based interventions		
	• Evidence-based guidelines for developing interventions for individuals • Laws and policies for poverty alleviation, socioeconomic development and gender equality	• Establish clear policy framework for addressing issues in high-risk groups [sex work, drug abusers, etc.] including options for legalization and/or licensing • Strict laws against violence and exploitation of women	• Provision of development opportunities by affirmative action • Making HIV/AIDS treatment and care a significant agenda for policy makers and legislators • Reorientation of health and social support services to improve HIV prevention and care

Adapted from Sahasrabuddhe and Vermund (2007). Notes: ABC, Abstinence, Be faithful, Correct and Consistent Condom use; VCT, Voluntary Counseling and Testing services; IEC, Information, Education and Communication; ART, Antiretroviral treatment; STI, Sexually Transmitted Infections; OI, Opportunistic Infections.

Vikrant V. Sahasrabuddhe and Sten H. Vermund

Multimodality of risk factors and impact on prevention interventions

Sexual mixing rates

HIV continues to be transmitted at unacceptably high rates globally, though some nations have seen recent declines. HIV transmission is most intense where intersections are most concentrated for high-risk human behaviors, community stigma, and societal disadvantages of the poor, women and disenfranchised subgroups. The highest-risk behaviors worldwide are unprotected heterosexual intercourse, especially in sub-Saharan Africa, South Asia and the Caribbean. Well-described risk factors include high sexual "mixing" rates via multiple concurrent sexual partnerships, transactional (commercial) sex work, sexual assaults and rape, early age of sexual intercourse, dry sex customs, genital trauma, lack of male circumcision, presence of concurrent bacterial and viral sexually transmitted infections, and non-use of condoms (Holmes *et al.*, 2004). Multiple and concurrent sexual partnerships have been widely considered responsible for the epidemic levels of HIV in general populations of sub-Saharan Africa (Epstein, 2007; Shelton, 2007).

Abstinence, be faithful, condoms (ABC)

Most prevention programs worldwide are targeted towards hitherto HIV-uninfected (negative) individuals to reduce the incidence of HIV acquisition (Salomon *et al.*, 2005). Prevention intervention activities that emphasize partner reduction have resulted in population-level declines in HIV infections; most notably the "zero grazing" policy espoused by Ugandan President Museveni that influenced Ugandans to decrease risk behaviors and prevalence markedly in Uganda in the 1990s (Stoneburner and Low-Beer, 2004; Bunnell *et al.*, 2006a; Hallett *et al.*, 2006). Counseling and interventions to reduce individuals' risk behaviors are the overarching strategies applied to all target populations and individuals. Reduction in sexual intermixing rates among the sexually active men and women from the general population may be achieved by promoting "be faithful" messages that advocate monogamy, as well as correct and consistent use of condoms (Cleland *et al.*, 2004; Stoneburner and Low-Beer, 2004). The "ABC" (**A**bstinence; **B**e faithful; **C**orrect and **C**onsistent **C**ondom use) strategy promulgated for prevention of sexual transmission of HIV is believed to have had measurable effects in reducing HIV incidence and prevalence in settings such as Uganda, Senegal, Thailand and Cambodia (Meda *et al.*, 1999; Cohen, 2003a; Hallett *et al.*, 2006).

Delaying onset of sexual debut through abstinence, coupled with skill-building and motivational counseling for esteem development, are key strategies for preventing new infections among young people (Kirby *et al.*, 2006; Maticka-Tyndale and Brouillard-Coylea, 2006). Programs targeting youth can include school, religious and recreational venue-based interventions, which are most successful if tied to local customs, traditions and social norms (Kirby *et al.*, 2006). Prevention programs for especially high-risk persons, like sex workers, IDUs, migrant laborers, clients of sex workers and MSM, are designed strategically to focus on risk and harm reduction (Pisani *et al.*, 2003b; Rekart, 2005; Wegbreit *et al.*, 2006).

Treatment of STIs, and male circumcision

One dramatic randomized clinical trial from Mwanza, Tanzania, has confirmed that treatment of STIs is an effective means for preventing HIV, but the evidence is much weaker to suggest that STI control will reduce the HIV epidemic substantially in settings with well-established HIV epidemics and where there is an especially high background prevalence of viral STIs, as with herpes simplex virus type 2 (HSV-2) (Grosskurth *et al.*, 1995, 2000; Wawer *et al.*, 1999; Kamali *et al.*, 2003; Orroth *et al.*, 2003; Sangani *et al.*, 2004; Korenromp *et al.*, 2005). The use of acyclovir suppression to downmodulate HSV-2 expression and thereby also reduce HIV-1 transmission was tried in a large, three-continent trial, HPTN 039. No effect was seen on HIV-1 transmission despite high adherence to acyclovir for 1 year by HSV-2-infected but initially HIV-1-uninfected MSM (in Peru and USA) and women (in Zambia, South Africa and Zimbabwe) (Celum, 2008). There are compelling reproductive health reasons why STI treatment services should be strengthened, especially in emerging epidemic settings where STI control can be considered a backbone of HIV prevention strategy. STI control and education is synergistic with "ABC" approaches. When a STI health-service intervention is implemented effectively, it can substantially improve the quality of sexual health care and education, contributing in turn to HIV prevention and care (Sangani *et al.*, 2004; Freeman *et al.*, 2007; White *et al.*, 2007).

Adult male circumcision has been proven to be an important once-in-a-life-time intervention for preventing HIV acquisition and transmission. There exists a wealth of observational evidence, including the three randomized clinical trials in South Africa, Uganda and Kenya, to suggest a 50 percent reduction in the risk of men acquiring HIV if they are circumcised (Weiss *et al.*, 2006; Auvert *et al.*, 2005; Bailey *et al.*, 2007; Weiss, 2007). Given its relative simplicity, infant circumcision in males should be a universal practice. Adult circumcision should be expanded rapidly to meet the growing demand from men and partners. As of this writing in early 2008, substantial efforts are being made to expand basic surgical services for adult male circumcision services as part of comprehensive HIV prevention package in high-prevalence regions of sub-Saharan Africa.

Voluntary counseling and testing: individual, dyad, network and community

The role of confidential and easily accessible voluntary counseling and testing (VCT) services is a critical primary and secondary modality for prevention (Sweat *et al.*, 2000; Vermund and Wilson, 2002; Denison *et al.*, 2007). An enabling environment for people who have seroconverted to HIV requires wider access to self-help and peer-support groups from the VCT base of referral (Latkin and Knowlton, 2005). Couples counseling may be more effective than individual counseling, since both partners can be engaged in prevention goals (Karita *et al.*, 2007). Networks can be engaged, as has been done in VCT for IDUs (Denison *et al.*, 2007). Finally, community-level advocacy of VCT and referral to care is being done in the HIV Prevention Trials Network 043 protocol in which 48 communities have been randomized into intensive and usual advocacy for VCT. Results of this trial are expected in 2011 (see also Chapter 19).

Acute infection

VCT using serology may identify persons late into illness, and there is considerable work to see whether nucleic acid amplification testing (NAAT) is practical for batch screening of very high-risk seronegative persons, as in an STD clinic (Cohen and Pilcher, 2005). It has been shown that the proportion of HIV infections that occur via acute seroconverters may constitute a high attributable fraction of the preventable HIV infections (Gray *et al.*, 2004; Wawer *et al.*, 2005). Therefore, enabling approaches for early detection of HIV through confidential testing, rapid HIV tests and effective linkages to care and social support programs can promote "acute seroconverters" to reduce risky sexual behaviors and contain the spread of HIV. The goal of "positive prevention" is the same in chronically and in acutely infected persons, but the public health risks are higher in persons infected but who have not yet seroconverted (Auerbach, 2004; Bunnell *et al.*, 2006b).

Antiretroviral drugs for prevention

In 2000 and 2001, two studies with amazingly similar results suggested that HIV viral load correlated in a continuous quantitative fashion with the risk of transmission from an infected to an uninfected heterosexual partner (Quinn *et al.*, 2000; Fideli *et al.*, 2001). Clearly, the correct and consistent use (as is the case with condoms) of antiretroviral therapy has promise for reducing HIV transmission between discordant sexual partners (the topic of the ongoing HPTN 052 clinical trial in persons with higher than the WHO-recommended CD4+ cell-count thresholds for initiating antiretroviral therapy in resource-limited nations).

Of course, if antiretroviral drugs are not used correctly, they may not protect from transmission and, further, may stimulate transmission of drug-resistant virus. Several trials are extant as of 2008 that are looking at the use of pre-exposure prophylaxis with antiretrovirals to help protect people even before they are exposed (Liu *et al.*, 2006). It will be essential to include behavior change empowerment in such work, as pre- or post-exposure prophylaxis is a stop-gap measure that is not suitable for continuation for decades without commensurate behavior change interventions (see also Chapter 5).

Gender-power and poverty dynamics

The most important community-level or structural intervention is promotion of broader gender equality in sexual relations. Laws and policies that nurture socio-economic development and poverty alleviation are additional macro-level, long-term interventions that can be expected to reduce HIV transmission (Dayton and Merson, 2000; Merson *et al.*, 2000; Hogan *et al.*, 2005; Pronyk *et al.*, 2006). If HIV is tackled as a developmental, women's empowerment and human rights issue, it is easier to imagine long-term, sustained success in HIV prevention programs (Elliott *et al.*, 2005; Elsey *et al.*, 2005; Dworkin and Ehrhardt, 2007). It remains a complex and unprecedented challenge to influence deeply entrenched traditions, social norms and beliefs around male–female issues, superstition, lack of a lifelong tradition of chemoprophylaxis and drug therapy, and community stigma. Policy-makers must create an environment that reduces risk activities as well as promoting the well-being of people living with HIV/AIDS. Gender equity and poverty alleviation are long-term goals of a truly comprehensive HIV/AIDS prevention and care agenda.

Future trends

The global stabilization of the HIV epidemic is certainly a welcome realization, but somewhat apocalyptic predictions remain valid in that four persons were newly infected for every person placed on antiretrovirals in 2007 (UNAIDS, 2007; Ventelou *et al.*, 2008). The future efforts in HIV prevention and control still need the same strong commitment through consistent and concerted responses at local, national and international levels. Investing in population-level integrated bio-behavioral surveillance systems for both prevalence and incidence is critical to help predict local, national and regional trends, to study the impact of interventions. Improved monitoring and evaluation of prevention (and treatment and care) programs are equally important, and should be cross-linked to population-level surveillance outputs. Horizontal and diagonal collaborations with other vertical disease-intervention programs, such as tuberculosis control, malaria prevention and control, family planning, cervical cancer prevention, etc., should be considered to optimize resource utilization and maximize program efficiencies (Reid *et al.*, 2004; Franceschi and Jaffe, 2007). Investing in strengthening the health infrastructure and developing health manpower resources in developing countries are critical enabling factors for successes in sustaining HIV prevention and control efforts (Vergara, 2008).

Conclusions

The days of considering prevention to be the purview of seronegative persons are over. We now see prevention as a continuum through the entire population: for low- and high-risk HIV-seronegative persons; acutely HIV-infected persons; chronically infected persons; and as a special issue for persons on antiretroviral therapy. No rigorous scientific work to test multiple interventions at different social levels (individual, dyad, network and community) and their impact on HIV transmission has been done. King Holmes has characterized this as "highly active retroviral prevention" or HARP (Holmes, 2007). HARP is an idea whose time has come in that the combination of multiple interventions, each of which is sub-optimal, may together make a huge impact on the epidemic at a community level. Large, simple trials of multiple interventions are warranted to test the validity of this hypothesis. Behavior-change advocacy and condom availability ("ABC"), STI control, male circumcision, VCT and condom advocacy, leadership in gender-power relations and stigma reduction, care and treatment (reducing viral load itself helps prevent HIV), pre-exposure prophylaxis, drug addiction and needle-exchange services, and "positive prevention" are among the measures to be considered (see Table 1.1). There are no viable options to this approach, as the world appreciates that we cannot merely "treat our way out of the epidemic."

References

Aceijas, C., Oppenheimer, E., Stimson, G. V. *et al.* (2006). Antiretroviral treatment for injecting drug users in developing and transitional countries 1 year before the end of the "Treating 3 million by 2005. Making it happen. The WHO strategy" ("3 by 5"). *Addiction*, 101, 1246–53.

Ainsworth, M., Beyrer, C. and Soucat, A. (2003). AIDS and public policy: the lessons and challenges of "success" in Thailand. *Health Policy*, 64, 13–37.

Auerbach, J. D. (2004). Principles of positive prevention. *J. Acquir. Immune Defic. Syndr.*, 37(Suppl. 2), S122–25.

Auvert, B., Taljaard, D., Lagarde, E. *et al.* (2005). Randomized, controlled intervention trial of male circumcision for reduction of HIV infection risk: the ANRS 1265 Trial. *PLoS Med.*, 2, e298.

Bailey, R. C., Moses, S., Parker, C. B. *et al.* (2007). Male circumcision for HIV prevention in young men in Kisumu, Kenya: a randomised controlled trial. *Lancet*, 369, 643–56.

Baral, S., Sifakis, F., Cleghorn, F. and Beyrer, C. (2007). Elevated risk for HIV infection among men who have sex with men in low- and middle-income countries 2000–2006: a systematic review. *PLoS Med.*, 4, e339.

Bennett, S., Boerma, J. T. and Brugha, R. (2006). Scaling up HIV/AIDS evaluation. *Lancet*, 367, 79–82.

Bobrova, N., Sarang, A., Stuikyte, R. and Lezhentsev, K. (2007). Obstacles in provision of anti-retroviral treatment to drug users in Central and Eastern Europe and Central Asia: a regional overview. *Intl J. Drug Policy*, 18, 313–18.

Bokhari, A., Nizamani, N. M., Jackson, D. J. *et al.* (2007). HIV risk in Karachi and Lahore, Pakistan: an emerging epidemic in injecting and commercial sex networks. *Intl J. STD AIDS*, 18, 486–92.

Booth, R. E., Lehman, W. E., Brewster, J. T. *et al.* (2007). Gender differences in sex risk behaviors among Ukraine injection drug users. *J. Acquir. Immune Defic. Syndr.*, 46, 112–17.

Brahme, R., Mehta, S., Sahay, S. *et al.* (2006). Correlates and trend of HIV prevalence among female sex workers attending sexually transmitted disease clinics in Pune, India (1993–2002). *J. Acquir. Immune Defic. Syndr.*, 41, 107–13.

Bruce, R. D., Dvoryak, S., Sylla, L. and Altice, F. L. (2007). HIV treatment access and scale-up for delivery of opiate substitution therapy with buprenorphine for IDUs in Ukraine – programme description and policy implications. *Intl J. Drug Policy*, 18, 326–8.

Bunnell, R., Ekwaru, J. P., Solberg, P. *et al.* (2006a). Changes in sexual behavior and risk of HIV transmission after antiretroviral therapy and prevention interventions in rural Uganda. *AIDS*, 20, 85–92.

Bunnell, R., Mermin, J. and de Cock, K. M. (2006b). HIV prevention for a threatened continent: implementing positive prevention in Africa. *J. Am. Med. Assoc.*, 296, 855–8.

CDC (1987). Revision of the CDC surveillance case definition for acquired immunodeficiency syndrome. Council of State and Territorial Epidemiologists; AIDS Program, Center for Infectious Diseases. *Morbid. Mortal. Wkly Rep.*, 36(Suppl. 1), S1–15.

CDC (1992). 1993 revised classification system for HIV infection and expanded surveillance case definition for AIDS among adolescents and adults. *Morbid. Mortal. Weekly Rep.*, 41, 1–19.

Celum, C., Wald, A., Hughes, J. *et al.* and HPTN 039 (2008). HSV-2 suppressive therapy for prevention of HIV acquisition: results of HPTN 039. *15th Conference on Retroviruses and Opportunistic Infections*. Boston, MA.

Chin, J. (2007). *The AIDS pandemic: the collision of epidemiology with political correctness*. Oxford: Radcliffe.

Cleland, J., Boerma, J. T., Carael, M. and Weir, S. S. (2004). Monitoring sexual behaviour in general populations: a synthesis of lessons of the past decade. *Sex. Transm. Infect.*, 80(Suppl. 2), ii1–7.

Cohen, J. (2003a). Asia – the next frontier for HIV/AIDS. Two hard-hit countries offer rare success stories: Thailand and Cambodia. *Science*, 301, 1658–62.

Cohen, J. (2003b). The next frontier for HIV/AIDS: Myanmar. *Science*, 301, 1650–5.

Cohen, J. (2006a). HIV/AIDS: Latin America and Caribbean. HAITI: making headway under hellacious circumstances. *Science*, 313, 470–3.

Cohen, J. (2006b). HIV/AIDS: Latin America and Caribbean. Dominican Republic: a sour taste on the sugar plantations. *Science*, 313, 473–5.

Cohen, J. (2006c). HIV/AIDS: Latin America and Caribbean. Overview: the overlooked epidemic. *Science*, 313, 468–9.

Cohen, J. (2006d). HIV/AIDS: Latin America and Caribbean. Honduras: mission possible: integrating the church with HIV/AIDS efforts. *Science*, 313, 482.

Cohen, M. S. and Pilcher, C. D. (2005). Amplified HIV transmission and new approaches to HIV prevention. *J. Infect. Dis.*, 191, 1391–3.

Dandona, L. and Dandona, R. (2007). Drop of HIV estimate for India to less than half. *Lancet*, 370, 1811–13.

Dandona, L., Lakshmi, V., Kumar, G. A. and Dandona, R. (2006). Is the HIV burden in India being overestimated? *BMC Public Health*, 6, 308.

Dayton, J. M. and Merson, M. H. (2000). Global dimensions of the AIDS epidemic: implications for prevention and care. *Infect. Dis. Clin. North Am.*, 14, 791–808.

Denison, J. A., O'Reilly, K. R., Schmid, G. P. *et al.* (2007). HIV voluntary counseling and testing and behavioral risk reduction in developing countries: a meta-analysis, 1990–2005. *AIDS Behav.*, 12, 363–73.

Des Jarlais, D. C., Kling, R., Hammett, T. M. *et al.* (2007). Reducing HIV infection among new injecting drug users in the China–Vietnam Cross Border Project. *Aids*, 21(Suppl. 8), S109–114.

Dougan, S., Evans, B. G. and Elford, J. (2007). Sexually transmitted infections in Western Europe among HIV-positive men who have sex with men. *Sex. Transm. Dis.*, 34, 783–90.

Dworkin, S. L. and Ehrhardt, A. A. (2007). Going beyond "ABC" to include "GEM": critical reflections on progress in the HIV/AIDS epidemic. *Am. J. Public Health*, 97, 13–18.

Elliott, R., Csete, J., Wood, E. and Kerr, T. (2005). Harm reduction, HIV/AIDS, and the human rights challenge to global drug control policy. *Health Hum. Rights*, 8, 104–38.

Elsey, H., Tolhurst, R. and Theobald, S. (2005). Mainstreaming HIV/AIDS in development sectors: have we learnt the lessons from gender mainstreaming? *AIDS Care*, 17, 988–98.

Epstein, H. (2007). The Invisible Cure: Africa, the West, and the Fight against AIDS. New York, NY: Farrar, Straus, and Giroux.

Fenton, K. A. (2007). Changing epidemiology of HIV/AIDS in the United States: implications for enhancing and promoting HIV testing strategies. *Clin. Infect. Dis.*, 45(Suppl. 4), S213–20.

Fideli, U. S., Allen, S. A., Musonda, R. *et al.* (2001). Virologic and immunologic determinants of heterosexual transmission of human immunodeficiency virus type 1 in Africa. *AIDS Res. Hum. Retroviruses*, 17, 901–10.

Figueroa, J. P. (2003). HIV/AIDS in the Caribbean. The need for a more effective public health response. *West Indian Med. J.*, 52, 156–8.

Franceschi, S. and Jaffe, H. (2007). Cervical cancer screening of women living with HIV infection: a must in the era of antiretroviral therapy. *Clin. Infect. Dis.*, 45, 510–13.

Freeman, E. E., Orroth, K. K., White, R. G. *et al.* (2007). Proportion of new HIV infections attributable to herpes simplex 2 increases over time: simulations of the changing role of sexually transmitted infections in sub-Saharan African HIV epidemics. *Sex. Transm. Infect.*, 83(Suppl. 1), i17–24.

Fung, I. C., Guinness, L., Vickerman, P. *et al.* (2007). Modelling the impact and cost-effectiveness of the HIV intervention programme amongst commercial sex workers in Ahmedabad, Gujarat, India. *BMC Public Health*, 7, 195.

Ghys, P. D., Kufa, E. and George, M. V. (2006). Measuring trends in prevalence and incidence of HIV infection in countries with generalised epidemics. *Sex. Transm. Infect.*, 82(Suppl. 1), i52–66.

Global HIV Prevention Working Group (2007). *Bringing HIV Prevention to Scale: An Urgent Global Priority*. Global HIV Prevention Working Group.

Gray, R. H., Li, X., Wawer, M. J. *et al.* (2004). Determinants of HIV-1 load in subjects with early and later HIV infections, in a general-population cohort of Rakai, Uganda. *J. Infect. Dis.*, 189, 1209–15.

Greco, D. B. and Simao, M. (2007). Brazilian policy of universal access to AIDS treatment: sustainability challenges and perspectives. *AIDS*, 21(Suppl. 4), S37–45.

Gregson, S., Nyamukapa, C., Lopman, B. *et al.* (2007). Critique of early models of the demographic impact of HIV/AIDS in sub-Saharan Africa based on contemporary empirical data from Zimbabwe. *Proc. Natl Acad. Sci. USA*, 104(14), 586–91.

Grosskurth, H., Mosha, F., Todd, J. *et al.* (1995). Impact of improved treatment of sexually transmitted diseases on HIV infection in rural Tanzania: randomised controlled trial. *Lancet*, 346, 530–6.

Grosskurth, H., Gray, R., Hayes, R. *et al.* (2000). Control of sexually transmitted diseases for HIV-1 prevention: understanding the implications of the Mwanza and Rakai trials. *Lancet*, 355, 1981–7.

Guy, R. J., McDonald, A. M., Bartlett, M. J. *et al.* (2007). HIV diagnoses in Australia: diverging epidemics within a low-prevalence country. *Med. J. Aust.*, 187, 437–40.

Hallett, T. B., Aberle-Grasse, J., Bello, G. *et al.* (2006). Declines in HIV prevalence can be associated with changing sexual behaviour in Uganda, urban Kenya, Zimbabwe, and urban Haiti. *Sex. Transm. Infect.*, 82(Suppl. 1), i1–8.

Hamers, F. F., Devaux, I., Alix, J. and Nardone, A. (2006). HIV/AIDS in Europe: trends and EU-wide priorities. *Euro Surveillance*, 11, E061123 1.

Heimer, R., Barbour, R., Shaboltas, A. V. *et al.* (2008). Spatial distribution of HIV prevalence and incidence among injection drugs users in St Petersburg: implications for HIV transmission. *AIDS*, 22, 123–30.

Hogan, D. R., Baltussen, R., Hayashi, C. *et al.* (2005). Cost effectiveness analysis of strategies to combat HIV/AIDS in developing countries. *Br. Med. J.*, 331, 1431–7.

Holmes, K. K. (2007). King Kennard Holmes – Chair of the Department of Global Health of The University of Washington. Interviewed by Marc Vandenbruaene. *Lancet Infect. Dis.*, 7, 516–20.

Holmes, K. K., Levine, R. and Weaver, M. (2004). Effectiveness of condoms in preventing sexually transmitted infections. *Bull. World Health Org.*, 82, 454–61.

Ji, G., Detesl, R., Wu, Z. and Yin, Y. (2006). Correlates of HIV infection among former blood/plasma donors in rural China. *AIDS*, 20, 585–91.

Ji, G., Detels, R., Wu, Z. and Yin, Y. (2007). Risk of sexual HIV transmission in a rural area of China. *Intl J. STD AIDS*, 18, 380–3.

Kamali, A., Quigley, M., Nakiyingi, J. *et al.* (2003). Syndromic management of sexually-transmitted infections and behaviour change interventions on transmission of HIV-1 in rural Uganda: a community randomised trial. *Lancet*, 361, 645–52.

Karita, E., Chomba, E., Roth, D. L. *et al.* (2007). Promotion of couples voluntary counselling and testing for HIV through influential networks in two African capital cities. *BMC Public Health*, 7, 349.

Kelly, J. A. and Amirkhanian, Y. A. (2003). The newest epidemic: a review of HIV/AIDS in Central and Eastern Europe. *Intl J. STD AIDS*, 14, 361–71.

Kirby, D., Obasi, A. and Laris, B. A. (2006). The effectiveness of sex education and HIV education interventions in schools in developing countries. *World Health Org. Tech. Rep. Ser.*, 938, 103–50, discussion 317–41.

Korenromp, E. L., White, R. G., Orroth, K. K. *et al.* (2005). Determinants of the impact of sexually transmitted infection treatment on prevention of HIV infection: a synthesis of evidence from the Mwanza, Rakai, and Masaka intervention trials. *J. Infect. Dis.*, 191(Suppl. 1), S168–78.

Latkin, C. A. and Knowlton, A. R. (2005). Micro-social structural approaches to HIV prevention: a social ecological perspective. *AIDS Care*, 17(Suppl. 1), S102–13.

Levine, R. S., Briggs, N. C., Kilbourne, B. S. *et al.* (2007). Black–white mortality from HIV in the United States before and after introduction of highly active antiretroviral therapy in 1996. *Am. J. Public Health*, 97, 1884–92.

Liu, A. Y., Grant, R. M. and Buchbinder, S. P. (2006). Preexposure prophylaxis for HIV: unproven promise and potential pitfalls. *J. Am. Med. Assoc.*, 296, 863–5.

Maticka-Tyndale, E. and Brouillard-Coylea, C. (2006). The effectiveness of community interventions targeting HIV and AIDS prevention at young people in developing countries. *World Health Org. Tech. Rep. Ser.*, 938, 243–85, discussion 317–41.

McDougal, J. S., Pilcher, C. D., Parekh, B. S. *et al.* (2005). Surveillance for HIV-1 incidence using tests for recent infection in resource-constrained countries. *AIDS*, 19(Suppl. 2), S25–30.

Meda, N., Ndoye, I., M'boup, S. *et al.* (1999). Low and stable HIV infection rates in Senegal: natural course of the epidemic or evidence for success of prevention? *AIDS*, 13, 1397–405.

Mehendale, S. M., Gupte, N., Paranjape, R. S. *et al.* (2007). Declining HIV incidence among patients attending sexually transmitted infection clinics in Pune, India. *J. Acquir. Immune Defic. Syndr.*, 45, 564–9.

Merson, M. H., Dayton, J. M. and O'Reilly, K. (2000). Effectiveness of HIV prevention interventions in developing countries. *AIDS*, 14(Suppl. 2), S68–84.

Millett, G. A., Flores, S. A., Peterson, J. L. and Bakeman, R. (2007). Explaining disparities in HIV infection among black and white men who have sex with men: a meta-analysis of HIV risk behaviors. *AIDS*, 21, 2083–91.

Mills, S., Saidel, T., Magnani, R. and Brown, T. (2004). Surveillance and modelling of HIV, STI, and risk behaviours in concentrated HIV epidemics. *Sex. Transm. Infect.*, 80(Suppl. 2), ii57–62.

Moran, D. and Jordaan, J. A. (2007). HIV/AIDS in Russia: determinants of regional prevalence. *Intl J. Health Geogr.*, 6, 22.

Moreno, C. L., El-Bassel, N. and Morrill, A. C. (2007). Heterosexual women of color and HIV risk: sexual risk factors for HIV among Latina and African-American women. *Women Health*, 45, 1–15.

Orroth, K. K., Korenromp, E. L., White, R. G. *et al.* (2003). Higher risk behaviour and rates of sexually transmitted diseases in Mwanza compared to Uganda may help explain HIV prevention trial outcomes. *AIDS*, 17, 2653–60.

Pence, B. W., Reif, S., Whetten, K. *et al.* (2007). Minorities, the poor, and survivors of abuse: HIV-infected patients in the US deep South. *South Med. J.*, 100, 1114–22.

Pisani, E., Dadun Sucahya, P. K. *et al.* (2003a). Sexual behavior among injection drug users in 3 indonesian cities carries a high potential for HIV spread to noninjectors. *J. Acquir. Immune Defic. Syndr.*, 34, 403–6.

Pisani, E., Garnett, G. P., Grassly, N. C. *et al.* (2003b). Back to basics in HIV prevention: focus on exposure. *Br. Med. J.*, 326, 138–47.

Posner, S. J., Myers, L., Hassig, S. E. *et al.* (2004). Estimating HIV incidence and detection rates from surveillance data. *Epidemiology*, 15, 164–72.

Pronyk, P. M., Hargreaves, J. R., Kim, J. C. *et al.* (2006). Effect of a structural intervention for the prevention of intimate-partner violence and HIV in rural South Africa: a cluster randomised trial. *Lancet*, 368, 1973–83.

Quinn, T. C., Wawer, M. J., Sewankambo, N. *et al.* (2000). Viral load and heterosexual transmission of human immunodeficiency virus type 1. Rakai Project Study Group. *N. Engl. J. Med.*, 342, 921–9.

Reid, S. E., Reid, C. A. and Vermunc, S. H. (2004). Antiretroviral therapy in sub-Saharan Africa: adherence lessons from tuberculosis and leprosy. *Intl J. STD AIDS*, 15, 713–16.

Rekart, M. L. (2005). Sex-work harm reduction. *Lancet*, 366, 2123–34.

Rupali, P., Condon, R., Roberts, S. *et al.* (2007). Prevention of mother to child transmission of HIV infection in Pacific countries. *Intl Med. J.*, 37, 216–23.

Sahasrabuddhe, V. V. and Vermund, S. H. (2007). The future of HIV prevention: control of sexually transmitted infections and circumcision interventions. *Infect. Dis. Clin. North Am.*, 21, 241–57, xi.

Sakarovitch, C., Rouet, F., Murphy, G. *et al.* (2007). Do tests devised to detect recent HIV-1 infection provide reliable estimates of incidence in Africa? *J. Acquir. Immune Defic. Syndr.*, 45, 115–22.

Salomon, J. A., Hogan, D. R., Stover, J. *et al.* (2005). Integrating HIV prevention and treatment: from slogans to impact. *PLoS Med.*, 2, e16.

Sangani, P., Rutherford, G. and Wilkinson, D. (2004). Population-based interventions for reducing sexually transmitted infections, including HIV infection. *Cochrane Database Syst. Rev.*, CD001220.

Saphonn, V., Hor, L. B., Ly, S. P. *et al.* (2002). How well do antenatal clinic (ANC) attendees represent the general population? A comparison of HIV prevalence from ANC sentinel surveillance sites with a population-based survey of women aged 15–49 in Cambodia. *Intl J. Epidemiol.*, 31, 449–55.

Saphonn, V., Parekh, B. S., Dobbs, T. *et al.* (2005). Trends of HIV-1 seroincidence among HIV-1 sentinel surveillance groups in Cambodia, 1999–2002. *J. Acquir. Immune Defic. Syndr.*, 39, 587–92.

Sarang, A., Stuikyte, R. and Bykov, R. (2007). Implementation of harm reduction in Central and Eastern Europe and Central Asia. *Intl J. Drug Policy*, 18, 129–35.

Schmidt, M. A. and Mokotoff, E. D. (2003). HIV/AIDS surveillance and prevention: improving the characterization of HIV transmission. *Public Health Rep.*, 118, 197–204.

Shelton, J. D. (2007). Ten myths and one truth about generalised HIV epidemics. *Lancet*, 370, 1809–11.

Shelton, J. D., Halperin, D. T. and Wilson, D. (2006). Has global HIV incidence peaked? *Lancet*, 367, 1120–22.

Silverman, J. G., Decker, M. R., Gupta, J. *et al.* (2006). HIV prevalence and predictors among rescued sex-trafficked women and girls in Mumbai, India. *J. Acquir. Immune Defic. Syndr.*, 43, 588–93.

Steinbrook, R. (2008). HIV in India – a downsized epidemic. *N. Engl. J. Med.*, 358, 107–9.

Stoneburner, R. L. and Low-Beer, D. (2004). Population-level HIV declines and behavioral risk avoidance in Uganda. *Science*, 304, 714–18.

Sweat, M., Gregorich, S., Sangiwa, G. *et al.* (2000). Cost-effectiveness of voluntary HIV-1 counselling and testing in reducing sexual transmission of HIV-1 in Kenya and Tanzania. *Lancet*, 356, 113–21.

Tkatchenko-Schmidt, E., Renton, A., Gevorgyan, R. *et al.* (2008). Prevention of HIV/AIDS among injecting drug users in Russia: opportunities and barriers to scaling-up of harm reduction programmes. *Health Policy*, 85, 162–71.

UNAIDS (2007). *2007 AIDS Epidemic Update*. UNAIDS.

Ventelou, B., Moatti, J. P., Videau, Y. and Kazatchkine, M. (2008). "Time is costly": modelling the macroeconomic impact of scaling-up antiretroviral treatment in sub-Saharan Africa. *AIDS*, 22, 107–13.

Vergara, A. and Vermund, S. (2008). Global health. In: F. D. Scutchfirld and A. W. Keck (eds), *Principles of Public Health Practice*. Clifton Park, NY: Thomson-Delmar Learning (in press).

Vermund, S. H. and Wilson, C. M. (2002). Barriers to HIV testing – where next? *Lancet*, 360, 1186–7.

Wawer, M. J., Sewankambo, N. K., Serwadda, D. *et al.* (1999). Control of sexually transmitted diseases for AIDS prevention in Uganda: a randomised community trial. Rakai Project Study Group. *Lancet*, 353, 525–35.

Wawer, M. J., Gray, R. H., Sewankambo, N. K. *et al.* (2005). Rates of HIV-1 transmission per coital act, by stage of HIV-1 infection, in Rakai, Uganda. *J. Infect. Dis.*, 191, 1403–9.

Wegbreit, J., Bertozzi, S., Demaria, L. M. and Padian, N. S. (2006). Effectiveness of HIV prevention strategies in resource-poor countries: tailoring the intervention to the context. *AIDS*, 20, 1217–35.

Weiss, H. A. (2007). Male circumcision as a preventive measure against HIV and other sexually transmitted diseases. *Curr. Opin. Infect. Dis.*, 20, 66–72.

Weiss, H. A., Thomas, S. L., Munabi, S. K. and Hayes, R. J. (2006). Male circumcision and risk of syphilis, chancroid, and genital herpes: a systematic review and meta-analysis. *Sex. Transm. Infect.*, 82, 101–9, discussion 110.

White, R. G., Orroth, K. K., Glynn, J. R. *et al.* (2007). Treating curable sexually transmitted infections to prevent HIV in Africa: still an effective control strategy? *J. Acquir. Immune Defic. Syndr.*, 47, 346–53.

WHO (1985). Workshop on AIDS in Africa: 22–25 October 1985.

Wu, Z., Sullivan, S. G., Wang, Y. *et al.* (2007). Evolution of China's response to HIV/AIDS. *Lancet*, 369, 679–90.

Xiao, Y., Jiang, Y., Feng, J. *et al.* (2007a). Seroincidence of recent human immunodeficiency virus type 1 infections in China. *Clin. Vaccine Immunol.*, 14, 1384–6.

Xiao, Y., Kristensen, S., Sun, J. *et al.* (2007b). Expansion of HIV/AIDS in China: lessons from Yunnan Province. *Soc. Sci. Med.*, 64, 665–75.

Understanding the biology of HIV-1 transmission: the foundation for prevention

2

Deborah J. Anderson

HIV sexual transmission occurs when HIV in genital secretions from an infected sex partner breech the mucosal epithelium of an uninfected partner to establish an infection. Approximately 80 percent of the 60 million people infected with HIV since the AIDS epidemic began 25 years ago were infected through sexual intercourse (UNAIDS, 2007). Paradoxically, the sexual transmission of HIV is not particularly efficient. Early studies estimated the rate of HIV sexual transmission to be one infection per 1000–2000 unprotected coital acts with an HIV-infected partner (Royce et al., 1997). Subsequent studies have provided information about individual risk factors associated with higher rates of infection. Insertive rectal intercourse runs a risk of 1/10 to 1/1600 unprotected exposures (Shattock and Moore, 2003), whereas HIV transmission via oral sex is exceedingly rare (Campo et al., 2006). The HIV transmission rate is 10-fold higher when unprotected intercourse occurs within 2 months of seroconversion of the HIV-infected partner (acute infection stage), and 4-fold higher when the infected partner is in advanced stage disease (Wawer et al., 2005). Other risk factors associated with increased transmission rates are genital infections, inflammation and pregnancy (Cohen, 2004; Gray et al., 2005; Chen et al., 2007). Factors associated with decreased infection risk include circumcision (Bailey et al., 2007; Gray et al., 2007), polymorphisms in certain genes that encode molecules that make up the HIV cellular receptor and major histocompatibility complex (Rowland-Jones et al., 2001), and highly active antiretroviral therapy

(HAART) (Cohen *et al.*, 2002). Despite improved AIDS education and condom distribution programs, increased access to STD drugs and antiretroviral therapy, 2.5 million people worldwide were newly infected with HIV in 2007 (UNAIDS, 2007). Clearly, new approaches are needed for HIV prevention.

The scientific community has worked for several years to develop HIV vaccines to be used at the population level to control the HIV epidemic, and vaginal microbicides which could provide the first major woman-controlled method to prevent HIV sexual transmission. However, progress has been painfully slow in these two areas, and the recent failure of several leading vaccine and microbicide candidates in clinical trials has sent scientists back to the drawing board to design second-generation products that better address the complexities of the HIV lifecycle, genital tract physiology and mechanisms underlying HIV transmission.

This chapter takes the viewpoint that highly effective HIV prevention strategies depend upon an in-depth appreciation of the sophisticated interactions between HIV and host mucosal immune defense. It begins with a review of the basics of the HIV lifecycle in human mucosal tissues, and conditions associated with increased levels of HIV-1 in genital secretions. Next, data from animal and *ex vivo* tissue models will be presented that provide insight into mechanisms of HIV sexual transmission, including evidence that infected cells may be important Trojan horse vectors of HIV transmission. Structural and immune defense mechanisms of the genital tract and rectal mucosae thought to be instrumental in preventing HIV transmission will be also described. Finally, novel ideas for HIV prevention that have arisen from these research findings will be presented.

The assault force: HIV in genital secretions

HIV resides in cells that populate genital and rectal tissues of HIV-infected men and women, and genital secretions can contain both cell-free HIV virions and cell-associated HIV (HIV-infected cells or cells that carry HIV virions). Epidemiological data as well as evidence from animal and *in vitro* infection studies (described below) indicate that both of these forms of virus can be infectious.

Semen is a mixture of secretions derived from several organs, including the testis, epididymis, seminal vesicles, prostate, Cowper's glands and penile urethra (Figure 2.1). HIV-infected lymphocytes and macrophages have been detected throughout the male genital tract, and are thought to be the principal source of seminal HIV (Pudney and Anderson, 1991). HIV-infected cells have been found in the testis, raising concerns about germ-line infection, but recent studies indicate that motile sperm from HIV-1 infected men do not carry HIV DNA (described in more detail below). It is likely that the lower genital tract is the principal source of seminal HIV. HIV is detected in semen from vasectomized men (Anderson *et al.*, 1991; Krieger *et al.*, 1998), in urethral and prostatic secretions (Coombs *et al.*, 2006) and in pre-ejaculate (urethral secretions released as a

Male reproductive system

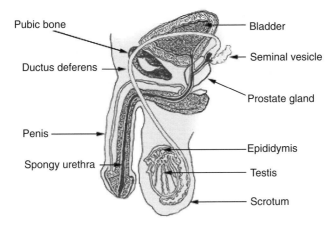

Figure 2.1 Anatomy of the human male genital tract. Source: From SEER's training website (sponsored by the NCI), http://training.seer.cancer.gov/module_anatomy/anatomy_physiology_home.html (public domain).

lubricant during sexual excitement) (Pudney *et al.*, 1992). Furthermore, seminal HIV levels are significantly elevated in HIV-infected men with urethritis (Pudney *et al.*, 1992; Cohen *et al.*, 1997, 1999; Winter *et al.*, 1999). These data suggest that the urethra is an important HIV infection site.

The male genital tract is often called an HIV reservoir or compartment. The term *reservoir* is applied to a tissue that receives a continuous inflow of virus and can produce and release virus. A *compartment* is a site in which a virus evolves independently from other tissues (Nickle *et al.*, 2003). There is evidence that the male genital tract serves as both an HIV reservoir and a compartment. A number of studies have documented higher concentrations of HIV in semen than blood in some subjects, suggesting local replication of virus in the genital tract (Xu *et al.*, 1997; Coombs *et al.*, 1998; Sadiq *et al.*, 2002), and genotypic differences in HIV quasispecies between semen and blood (Zhu *et al.*, 1996; Gupta *et al.*, 1997; Coombs *et al.*, 1998; Delwart *et al.*, 1998; Ping *et al.*, 2000; Pillai *et al.*, 2005).

The male genital tract has several unique features that may influence independent replication and evolution of HIV in this site. First, the testis is an immunologically privileged site (thought to be so to protect germ cells from immunological damage). In the healthy testis, lymphocytes and antibodies cannot penetrate the Sertoli cell barrier surrounding the seminiferous tubules (the site of germ cell maturation), and immune activity is tightly regulated by high concentrations of immunosuppressive factors such as testosterone, prostaglandins and TGF-β (reviewed in Anderson and Pudney, 2005). HIV residing in the testis would therefore be free from many selection pressures exerted by cellular and humoral immunity. Furthermore, serum antibodies do not penetrate very effectively into genital secretions (concentrations usually <10 percent than those in peripheral blood), and plasma cells residing in genital tissues produce a specialized form of immunoglobulin, called secretory IgA, which is elicited by antigenic stimulation at mucosal sites, and is abundant in mucosal secretions but scarce in blood. HIV-specific s-IgA antibodies and oligoclonal populations of CD8+ T cells have been detected in semen, indicating that the male genital tissues generate a local immune response to HIV which could influence HIV evolution at this site (Wolff *et al.*, 1992; Quayle *et al.*, 1998a; Musey *et al.*, 2003). Furthermore, antiretroviral drugs penetrate the male genital tract to varying degrees, and can be present in lower concentrations in seminal plasma than blood (Chan and Ray, 2007). Thus, it is possible that HIV is incompletely suppressed in the male genital tract under certain HAART regimens, and that replicating genital virus may develop drug-resistance mutations not seen in blood.

Similar events occur in the organs that comprise the female genital tract (Figure 2.2). HIV has been detected in macrophages and memory T lymphocytes residing in the cervix of HIV-infected women (Zhang *et al.*, 1999). The uterus is an immunologically privileged site during pregnancy, to protect the fetus from immunological rejection (Tristram, 2005), and female reproductive hormones are potent immunomodulators (Wira *et al.*, 2005a). The female genital tract is also a component of the mucosal immune system, and HIV residing in the

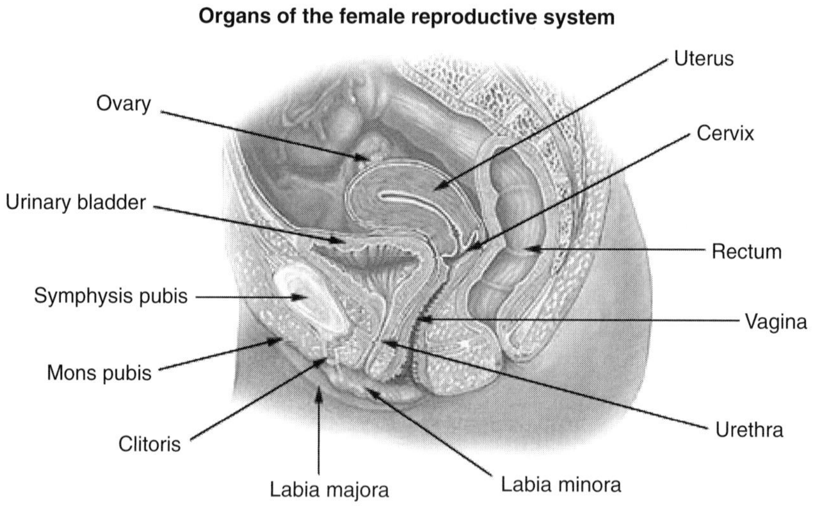

Figure 2.2 Anatomy of the human female genital tract. Source: From SEER's training website (sponsored by the NCI), http://training.seer.cancer.gov/module_anatomy/anatomy_physiology_home.html (public domain).

female genital tract may be subject to local evolutionary pressures due to the presence of mucosal s-IgA antibodies and specialized T-cell populations (Kutteh *et al.*, 2005). Recent evidence suggests that HIV can undergo divergent evolution in female genital tissues, as it does in the male genital tract (Kemal *et al.*, 2003; De Pasquale *et al.*, 2008).

Factors affecting genital HIV levels and infectiousness

HIV viral loads in blood and genital secretions are usually measured by clinical quantitative RNA assays such as Roche Amplicor or NASBA. Except for a subpopulation of individuals with disproportionately high levels of HIV in genital secretions (described in more detail below), genital HIV RNA concentrations generally correlate with those in peripheral blood (Gupta *et al.*, 1997; Vernazza *et al.*, 1997; Tachet *et al.*, 1999). Thus, conditions that affect HIV levels in blood usually have a similar effect on genital HIV concentrations. For instance, during the acute stage of HIV infection, very high concentrations of HIV are present in both blood and semen (Pilcher *et al.*, 2007). HIV in semen during the acute phase of infection may be particularly infectious as it has an R5 phenotype which is particularly suited for infection of mucosal sites, and it is not neutralized by HIV-specific antibodies which develop later. Many scientists think that seminal viremia during acute infection is a primary driving force of the HIV/AIDS epidemic (Shattock and Moore, 2003). Other conditions that affect levels of HIV RNA in genital secretions and blood include disease-stage and antiretroviral therapy. After the acute phase of infection, most HIV-infected men and women enter a period of viral latency where little HIV RNA is detectable in blood or genital secretions. During this phase, however, HIV can be intermittently detected in genital secretions, possibly as a result of episodes of genital inflammation (discussed below) (Anderson *et al.*, 1992; Krieger *et al.*, 1995; Quayle *et al.*, 1997; Coombs *et al.*, 1998, 2003). As HIV infection progresses to the acquired immunodeficiency disease stage (AIDS), HIV RNA levels are again elevated in blood and genital secretions (Pilcher *et al.*, 2007). A recent study from Uganda showed that infected partners in monogamous discordant relationships are more infectious to their partners during the acute and later stages of disease progression than during the latent phase of infection (Wawer *et al.*, 2005). Antiretroviral drug therapy, especially HAART, suppresses viral replication leading to decreased levels of HIV in blood and genital secretions. However, HIV can persist as a latent infection in genital cells, and rebound during HAART failure or withdrawal (Mayer *et al.*, 1999; Cu-Uvin *et al.*, 2006).

HIV viral load is usually lower in genital secretions than in blood, but most comparisons report a minority of patients where values are higher in semen or cervicovaginal secretions than in blood (Xu *et al.*, 1997; Coombs *et al.*, 1998; Cu-Uvin *et al.*, 2006). This may be due to independent replication of HIV in the genital tract, possibly related to activation of HIV-infected cells by inflammatory mediators produced during symptomatic or asymptomatic genital infections. HIV RNA and DNA viral loads in semen and other genital tract secretions can increase considerably (up to 20-fold) during episodes of gonococcal, non-gonococcal and chlamydial urethritis, and fall following treatment of the urethritis (Cohen, 2004). Genital lesions may be another source of sexually transmitted HIV-1, as genital ulcerative diseases have also been associated with HIV-1 viral shedding and an increased incidence of HIV-1 transmission (Celum, 2004; Wald, 2004). HIV-infected men and women with STDs and elevated levels of HIV in genital secretions may disproportionately transmit HIV-1 to their sexual partners.

The lower gastrointestinal tract/rectum is another important HIV infection site (Figure 2.3), and rectal secretions can contain high levels of HIV RNA (Zuckerman *et al.*, 2004). Thus rectal secretions may be another source of transmitted virus, although no one has yet demonstrated infectious HIV in rectal secretions (the virus could be inactivated), and the risk of HIV transmission in men that have sex with men (MSM) is higher for those that receive unprotected anal intercourse than for insertive partners (Koblin *et al.*, 2006; Goodreau and Golden, 2007).

HIV RNA copy numbers in blood and secretions are generally orders of magnitude higher than the infectious titer (a measure of the number of infectious viral particles) because much of the HIV produced is defective and/or neutralized by host factors and therefore not infectious. Factors that determine the infectiousness of HIV-1 in genital secretions include viral phenotype; opsinizing antibodies; and concentrations of factors in genital secretions that may decrease the infectiousness of HIV, including neutralizing antibodies, chemokines, proinflammatory cytokines, antimicrobial proteins and other innate immunity factors (described in more detail below).

Cervicovaginal secretions and semen contain variable numbers of CD4+ T lymphocytes and macrophages, and in HIV+ individuals these cells are often

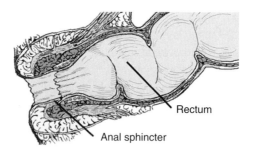

Figure 2.3 Anatomy of the human rectum. Source: From the NIDDK Image Library, www.catalog.niddk.nih.gov/imageLibrary, Image Number N00041 (public domain).

infected with HIV. Most quantitative studies have focused on cell-free HIV RNA, but a few have enumerated infected cells in genital secretions by quantifying HIV DNA copies associated with the cellular fraction (Xu *et al.*, 1997; Mayer *et al.*, 1999; Spinillo *et al.*, 1999; Tuomala *et al.*, 2003). These studies have shown that HIV-infected cells are present in cervicovaginal secretions and semen from HIV+ individuals. Although it has been speculated that sperm may transmit HIV, most recent studies have shown that viable motile sperm do not contain appreciable amounts of HIV DNA or RNA (Quayle *et al.*, 1998b). This has made it possible for Assisted Reproduction Clinics to inseminate uninfected female partners of HIV+ men with isolated, washed sperm to achieve pregnancy without seroconversion (Bujan *et al.*, 2007). In contrast, CD4+ T cells and macrophages isolated from semen from HIV+ men are highly infectious *in vitro* (Quayle *et al.*, 1997). HIV-infected cells are less vulnerable than cell-free virions to many of the mediators of immune defense in genital secretions, such as neutralizing antibodies and antimicrobial peptides, and thus may be important mediators of HIV transmission.

Establishing a beachhead: the cellular organization of genital tract and rectal tissues, and early events in HIV sexual transmission

To establish an infection in a new host following sexual intercourse, HIV must reach host target cells bearing HIV receptors, such as CD4+ T cells, macrophages, Langerhans and dendritic cells. These target cell types reside in genital mucosae (Figure 2.4). To reach them, HIV must first navigate a thick mucous layer. HIV-infected cells such as macrophages can effectively migrate through mucus (Parkhurst and Saltzman, 1994); cell-free HIV virions can diffuse through mucus, but may be trapped in it if HIV-specific antibodies are present (Olmsted *et al.*, 2001). The mucous layer is inhabited by commensal microflora, and is laden with antimicrobial molecules that can particularly inactivate cell-free virus.

Once through the mucous layer, the virus encounters the mucosal epithelium. Epithelial surfaces vary in structure depending on their location and function. The skin covering the outer genitalia (scrotum, foreskin, penis and penile meatus in men, and labia, vulva and introitus in women) is a keratinizing stratified squamous epithelium. The mucosal skin lining the inner genital surfaces (vagina in women and fossa navicularis – the opening of the penile urethra – in men) is a non-keratinizing stratified squamous epithelium. Both keratinizing and non-keratinizing stratified squamous epithelia are multilayered structures (up to 45 epithelial cells thick) and present a formidable barrier to many invading pathogens.

The thick stratified squamous epithelium transitions to a simple columnar epithelium, a single layer of polarized epithelial cells, at the cervical opening (os)

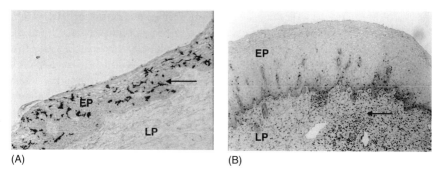

Figure 2.4 Cells in the genital mucosa that play important roles in HIV sexual transmission. A. Langerhans cells in the stratified squamous epithelial (EP) layer (shown here in the human vaginal epithelium; stained dark brown by immunohistology) can be infected with HIV-1 as well as capture and transfer HIV-1 to susceptible T cells in draining lymph nodes; B. CD4+ memory T cells in the lamina propria (LP) of genital mucosal epithelia (shown here in the human vaginal epithelium; stained dark brown by immunohistology) are the other primary target cells involved in HIV-1 sexual transmission. Source: Original photo, courtesy of Dr Anderson.

in women and at the opening of the penile urethra (fossa navicularis) in men. Thus, the epithelium lining the endocervical canal (women) and the penile urethra (men) is columnar and only one cell-layer thick. The squamocolumnar transitional areas are preferential infection sites for a number of sexually transmitted pathogens, including human papilloma virus, *Chlamydia trachomatis* and possibly HIV-1. These areas are protected by numerous mucin-secreting glands or pseudo-glands, and by a high density of immune cells.

Mucosal epithelial layers are by no means inert structures. They shed surface cells at a high rate; the rate of epithelial cell turnover is unknown for the genital tract, but is estimated to be $> 10^{11}$ cells per day from the human small intestine. Furthermore, epithelial cells can propel the mucous layer in which they are bathed through peristaltic and ciliary action. Shed cells and mucus carry with them a burden of microorganisms, and their excretion is thought to constitute a major pathogen clearance mechanism. Furthermore, as described in a later section, epithelial cells actively participate in innate and acquired immune defense functions by secreting cytokines, chemokines, defensins and other antimicrobial peptides, and performing antigen presentation functions via MHC Class II and CD1d pathways.

Physical and chemical trauma and ulcerating infections can disrupt the epithelial layer and provide access to HIV target cells in the underlying germinal layer and lamina propria. Epithelial lesions, abrasions and ulcers are major risk factors for the sexual transmission of HIV-1 and other STIs (Lederman *et al.*, 2006).

A number of studies point to the foreskin as a primary HIV infection site in men. First, the foreskin epithelium contains a high density of Langerhans cells, dendritic cells, macrophages and T lymphocytes (Hussain and Lehner, 1995; Patterson *et al.*, 2002; McCoombe and Short, 2006). Some of these cells express DC-SIGN and CCR5 and CD4 HIV co-receptors (Patterson *et al.*, 2002; Soilleux and Coleman, 2004), and are readily infectable with HIV-1 *in vitro* (Patterson *et al.*, 2002). African men with a history of sexually transmitted infections have a higher density of Langerhans cells in the foreskin, suggesting that these men may be more susceptible to HIV infection (Donoval *et al.*, 2006). The human foreskin is also a common infection site for other STD pathogens, including HPV, HSV, *Treponema pallidum* and *Haemophilus ducreyi*, which can promote HIV infection (Weiss *et al.*, 2006). Second, recent studies indicate that circumcision (removal of the foreskin) reduces HIV infection in men. For nearly 20 years observational studies have suggested that circumcision protects men from HIV infection (Weiss *et al.*, 2000) and other STDs (Weiss *et al.*, 2006), but these studies were often confounded by differences in sexual practices, hygiene, and other factors associated with religious and ethnic groups practicing circumcision. However, recent data from three randomized controlled intervention trials in Africa clearly demonstrate that male circumcision dramatically reduces HIV infection risk by 50–60 percent (Sahasrabuddhe and Vermund, 2007).

The opening of the penile urethra is another probable HIV infection site in men as numerous HIV target cells reside in this location, and the stratified squamous epithelium transitions to a thin columnar epithelium in the fossa navicularis at the opening of the urethra (Pudney and Anderson, 1995). Furthermore, this too is a common infection site for a number of sexually transmitted pathogens that increase HIV infection risk.

A variety of potential HIV host cells, including Langerhans and dendritic cells, T lymphocytes and macrophages, populate the lower genital tract of women, and are concentrated in the cervical transformation zone (Poppe *et al.*, 1998; Johansson *et al.*, 1999; Patton *et al.*, 2000; Anderson and Pudney, 2005). As mentioned above, the epithelial layer transitions from the thick stratified squamous epithelium that lines the vagina and ectocervix to a single-cell columnar epithelium at this site, indicating that the cervical transformation zone may be an important site of early HIV-1 infection. HIV-infected cells (macrophages, lymphocytes) have been detected in the endocervix of HIV+ women. Studies on HIV-1 viral loads in swabs taken from the endocervix have shown higher titers compared to those taken from the vagina (Reichelderfer *et al.*, 2000). Furthermore, cervical ectopy, a condition where the endocervical transformation zone extends into the vagina through the cervical os and is more exposed than normal, is another risk factor associated with sexual transmission of HIV (Mostad *et al.*, 1997). Other HIV infection sites, however, must also be present in the lower genital tract, because HIV has been detected in genital tract secretions of hysterectomized women (Farrar *et al.*, 1997). Genital tract infections and inflammation can disrupt the integrity of the cervicovaginal epithelial barrier and attract increased numbers of HIV-1 target cells to the cervical and vaginal mucosae. A number of sexually transmitted infections (STIs) have been associated with an increased susceptibility to HIV-1 infection in both men and women (Fleming and Wasserheit, 1999). These studies provide information on possible sites of infection in the female genital tract and potential host cells, but it is impossible, because of ethical considerations, to study directly the initial sequence of cellular interactions underlying the mucosal transmission of HIV-1 in human subjects.

Studies of HIV sexual transmission mechanisms in animal models

Higher apes can be infected with both human and simian immunodeficiency viruses (SIV), but their endangered status limits their utility for most biomedical research. Nonetheless, one study in chimpanzees demonstrated that both HIV-1 infected cells and high titers of cell-free HIV-1 were capable of transmitting infection after their atraumatic insertion into the vaginal cavity near the cervical os (Girard *et al.*, 1998). The most common model used for studies on mechanisms of HIV-1 sexual transmission has been the SIV_{mac}/rhesus macaque model. Vaginal inoculation of cell-free SIV infects female macaques via target cells in the vagina and cervix (Miller *et al.*, 1989, 1992), and male macaques can be infected by application of SIV to the opening of the penile urethra (Miller *et al.*, 1989). The vaginal transmission model has been used to study early events following SIV infection. Dendritic cells located in the cervicovaginal mucosa have been identified as initial cellular targets for the sexual transmission of cell-free SIV in macaques (Spira *et al.*, 1996). Other studies, however, have shown that the first cells to be infected are CD4+ lymphocytes located in the endocervix (Zhang *et al.*, 1999). Memory CD4+ T cells residing in mucosal tissues such as the genital tract and intestine express a high density of CD4 and CCR5 (HIV receptors), and are preferentially targeted by HIV during acute infection (Picker, 2006). Memory CD4+ T-cell populations in the intestine and genital tract are decimated during acute SIV and HIV infection (Lim *et al.*, 1993; Brenchley *et al.*, 2004; Veazey *et al.*, 2003; Quayle *et al.*, 2007). Interestingly, macaque studies have also clearly shown that SIV infects not only activated rapidly dividing CD4+ T cells, but also resting CD4+ lymphocytes. These resting infected lymphocytes were found to contain low viral copy numbers with a reduced level of viral replication, and were long-lived. It was suggested that these latently and chronically infected cells form a reservoir responsible for persistent infection. Once SIV-infected cells reach the lamina propria, they can then infect other host cells and undergo expansion to establish a localized site of viral propagation, which ultimately results in the systemic dissemination of SIV. Uptake of cell-free SIV from the macaque vagina is rapid, with dissemination of virally infected cells occurring within days following infection (Spira *et al.*, 1996; Zhang *et al.*,

1999). Most of the macaque sexual transmission studies have been conducted with superphysiological titers of cell-free SIV. However, recent studies have also shown transmission following multiple vaginal inoculations with lower doses of cell-free SIV (Ma *et al.*, 2004) or SHIV (Tasca *et al.*, 2007), or with SIV-infected cells (Kaizu *et al.*, 2006). Monkeys exposed to low-dose inoculations often show periods of transient viremia and occult infection, and localized anti-SIV cellular immune responses, before persistent infections are established. Further research is needed to delineate the mechanisms of transmission using these more physiologically relevant models.

The macaque model has been used to study the role of reproductive hormones in SIV transmission. Progesterone therapy results in an increased risk of vaginal transmission of cell-free SIV (Marx *et al.*, 1996; Smith *et al.*, 2000). A possible explanation for this effect is that the progesterone implants used for these SIV transmission studies shut down ovarian estrogen secretion, resulting in a dramatically thinned and more vulnerable vaginal epithelium. On the other hand, estrogen treatment had a protective effect against the vaginal transmission of SIV (Smith *et al.*, 2000). These data suggest that estrogen-deficient women, i.e. post-menopausal women or those on progesterone-based contraceptives, may be at higher risk for HIV infection through sexual intercourse; however, epidemiological studies designed to investigate this association have yet to show a significant effect.

Normal mice cannot be infected with HIV, but models have been developed using humanized immunodeficient mouse to study the sexual transmission of HIV-1. In one such model, the immune system of severe combined immunodeficient (SCID) mice was reconstituted by injecting human peripheral blood lymphocytes into the peritoneal cavity. Following intravaginal inoculation with HIV-infected lymphocytes but not cell-free HIV, mouse peritoneal lymphocytes showed evidence of HIV infection. Furthermore, HIV+ cells were detected in the iliac lymph nodes, demonstrating systemic dissemination of the virus (Khanna *et al.*, 2002). In another model, Rag2(−/−)gammac(−/−) mice engrafted with human hematopoietic progenitor cells sustained systemic HIV infections following vaginal or rectal challenge with R5 or X4 tropic cell-free virus (Berges *et al.*, 2008). These models require further study and validation, but could prove to be very useful for studies on early events in HIV pathogenesis *in vivo*, as well as testing novel strategies to prevent HIV-1 transmission.

In vitro studies of HIV transmission mechanisms using explant tissue models

To avoid many of the problems associated with animal models, investigators have developed *in vitro* models to study the sexual transmission of HIV. The initial studies involved HIV infection of monolayers of primary or transformed cell lines derived from the human female genital tract (Tan *et al.*, 1993; Howell *et al.*, 1997). Some of these studies showed evidence of a low level infection of epithelial cells by HIV, but the most dramatic finding was that infectious HIV-1 can be transported across a cultured layer of polarized epithelial cells by a process called transcytosis (Bomsel, 1997). This suggests that some epithelial cell layers may transport HIV into subepithelial layers where the usual HIV host cells (i.e., T cells, macrophages and dendritic cells) reside. However, this process has only been demonstrated with transformed laboratory cell lines, and may not occur *in vivo*. To more faithfully simulate *in vivo* conditions, investigators have used explants derived from fresh vaginal or cervical tissues for studies of mechanisms of HIV-1 sexual transmission (Palacio *et al.*, 1994; Collins *et al.*, 2000; Greenhead *et al.*, 2000; Maher *et al.*, 2005). The results from these studies have been conflicting, and depend on the culture conditions, which vary from lab to lab and indeed from tissue donor to tissue donor. In one report (Palacio *et al.*, 1994), HIV+ cells resembling macrophages were detected in the submucosa of tissue that had been exposed to R5 (monotropic) virus but not to R4 (T-cell tropic) strains. CD1a+ dendritic cells and CD3+ lymphocytes were not infected with HIV under these culture conditions. By contrast, Collins *et al.* (2000) detected infection in submucosal cells following addition of cell-associated HIV-1 as well as cell-free inoculums of both R4 (T-cell tropic) and R5 (monotropic) HIV-1. The T tropic strain of HIV primarily infected CD4/CXCR4-expressing lymphocytes located just beneath the mucosal epithelium, whereas R5 strains of HIV-infected CCR5+ cells presumed to be macrophages. In this explant culture system, productive infection of dendritic cells was not identified. Greenhead *et al.* (2000) reported that a primary patient isolate of HIV established infection in endocervical tissue leukocytes located in the submucosa. For multilayered ectocervical tissue, cell-free virus could bind to epithelial cells to effect penetration of the mucosa. Labeled viable cells from semen bound to the ectocervical epithelium, but failed to bind to the endocervical explants due to their entrapment in mucus secreted by these cells. Another group has used separated human vaginal epithelial sheets as a model system to study early HIV infection events (Hladik *et al.*, 2007). They showed that CD4+ T memory cells were early targets of HIV infection, and that HIV was also associated with Langerhans cells; these cells did not become infected, but migrated out of the tissue, and could presumably transfer HIV to T target cells in regional lymph nodes. These studies demonstrate the complexity of interactions that occur between HIV and cells associated with mucosal surfaces in the female genital tract.

Human foreskin has also been used in studies of HIV transmission mechanisms. The inner foreskin tissue contains numerous Langerhans cells and macrophages that can be infected with HIV (Patterson *et al.*, 2002; Donoval *et al.*, 2006). A recent study used vaginal skin to study the early events of HIV infection of the female genital tract (Sugaya *et al.*, 2004). Exposure to HIV resulted in the infection of Langerhans cells. Furthermore, infected Langerhans cells isolated from these preparations were capable of transmitting the virus to autologous proliferating CD4+ T cells. A rectal explant model has also been developed for studies on HIV infection of the rectal mucosa (Fletcher *et al.*, 2006). The rectal mucosa contains a high concentration of HIV host cells and is thin and friable, making this a highly vulnerable HIV infection site.

Blocking transmission: host immune defense

The genital microenvironment is generally inhospitable to pathogens. An important barrier to infection is provided by the acidic pH of vaginal secretions. Vaginal secretions in healthy reproductive-aged women usually have a pH of about 4.0 (Larsen, 1993), produced by the anaerobic metabolism of glycogen to acidic products by vaginal lactobacilli (Boskey *et al.*, 2001). HIV is inactivated under low pH conditions (O'Connor *et al.*, 1995); unfortunately, certain common vaginal conditions such as bacterial vaginosis are associated with an elevated vaginal pH and increased risk of HIV transmission (Spear *et al.*, 2007). Furthermore, studies performed over 40 years ago by Masters and Johnson showed that semen is alkaline and neutralizes vaginal acidity within seconds of intercourse (Masters and Johnson, 1961). The vaginal environment remains at neutral pH for over an hour after intercourse, providing a window for HIV transmission. Protection is also afforded by an array of antimicrobial molecules in genital secretions. Male and female genital secretions contain small inorganic molecules such as hydrogen peroxide and nitric oxide that rapidly inactivate HIV and other sexually transmitted pathogens. Furthermore, mucosal epithelia are protected by the innate immune system, consisting of a variety of cells that recognize microbes and their products through cellular receptors known as pattern-recognition receptors (PRRs). PRRs detect virulent microorganisms through recognition of invariant pathogen-associated molecular patterns (PAMPs). Cells that are activated by PAMPs can produce a number of antimicrobial substances, including small antimicrobial proteins (e.g., defensins and cathelicidins) and large proteins (e.g., lysozyme, azurocidin, cathepsin G, phospholipase A2, lactoferrin and type 1 interferons). By producing these antimicrobial factors, innate immunity rapidly limits the expansion of invading pathogens and provides time for more effective host adaptive immunity to be generated. (This topic was recently reviewed by the author (Anderson, 2007) and others (Wira *et al.*, 2005b; Neutra and Kozlowski, 2006). Both R5 (CCR5 tropic) and X4 (CXCR4 tropic) HIV populations are detected in genital secretions, but R5 virus is most commonly transmitted, perhaps because SDF-1 (CXCR4 ligand) is abundant in female genital secretions and semen, and blocks the CXCR4 HIV co-receptor (Agace *et al.*, 2000; Politch *et al.*, 2007).

Both the male and female genital tracts can also mount an acquired immune response, mediated by memory T and B lymphocytes and antigen-presenting cells (macrophages, dendritic cells) (Anderson, 2007). Pathogen-specific cytotoxic T cells have been isolated from the endocervix of women infected with HSV (Koelle *et al.*, 2000) and HIV-1 (Shacklett *et al.*, 2000; Musey *et al.*, 2003), and semen from HIV-infected men (Quayle *et al.*, 1998a; Sheth *et al.*, 2005). Such cells may kill pathogen-infected cells in genital tissues and limit the spread of localized infections. Antigen-specific antibody responses are also frequently detected in genital secretions following infection and immunization (Moldoveanu *et al.*, 2005; Woof and Mestecky, 2005). Genital antibodies can neutralize pathogens and may also play a role in antibody-dependent cellular cytotoxicity (ADCC) responses (Battle-Miller *et al.*, 2002; Nag *et al.*, 2004; Woof and Mestecky, 2005). It is likely that innate and acquired immunity work in concert to defend the male and female genital tracts against sexually transmitted infections.

Male and female genital tract secretions also contain cytokines and chemokines that can affect HIV-1 expression and infection mechanisms and play a role in HIV-1 transmission. Proinflammatory cytokines such as IL-1β, TNF-α, M-CSF and IL-6, are often elevated in genital secretions following infection or inflammatory stimuli. These cytokines can upregulate HIV-1 expression by activating the Nf-κB pathway, and enhance HIV-1 infection of susceptible target cells by upregulating CCR5 and CXCR4 (Kinter *et al.*, 2000; Kedzierska *et al.*, 2003). Levels and effects of proinflammatory cytokines are downregulated by other cytokines (for example, Il-10) and cytokine receptor antagonists (such as Il-1RA) that may also be found in genital secretions. In addition, type 1 interferons, SLPI and other antimicrobial protein mediators of innate immunity are potent suppressors of HIV-1 infection. Thus, HIV transmission outcome depends on the balance of factors in the genital microenvironment with competing influences on HIV replication and survival.

HIV-infected cells from genital secretions would likely be more resistant than cell-free virions to the antiviral effects of the antimicrobial proteins found in genital secretions such as SLPI, lactoferrin and lysozyme, and would not be neutralized by HIV-specific antibodies which are found in genital secretions of most infected individuals. Infected cells deposited in a host through sexual intercourse also express foreign Major Histocompatibility Complex (MHC) antigens (i.e., of the partner) and therefore could evade detection by host cytotoxic T cells primed to kill infected cells that express HIV antigens presented by self-MHC class-1 molecules. We have proposed that HIV-infected cells are Trojan horse vectors of HIV transmission (Anderson and Yunis, 1983).

Summary and future directions

- *What do we know about HIV transmission?*
  Probable mechanisms of HIV sexual transmission are: (1) infectious HIV-1 virions in genital secretions are captured by Langerhans cells or other dendritic

cells in the genital mucosae, and taken to draining lymph nodes where they are transferred to susceptible CD4+ T cells to initiate an infection cycle; (2) infectious HIV-1 virions in genital secretions directly infect macrophages or memory T cells in genital mucosa that are accessible due to inflammation or abrasion; (3) HIV-infected cells in genital secretions migrate through the epithelial layer or transfer HIV to target cells in the genital mucosa. CD4 and CCR5 are important coreceptors in HIV sexual transmission.

- *What natural factors in the genital environment regulate HIV transmission efficiency?*

Lactobaccilli in vaginal secretions of healthy cycling women maintain an acidic environment which is inhospitable to HIV and other pathogenic organisms. Both male and female genital secretions also contain antimicrobial proteins such as defensins, lactoferrin, SLPI and lysozyme, and anti-HIV antibodies which could inhibit infections. On the other hand, HIV infection is promoted by proinflammatory cytokines produced in the genital tract in response to infections and inflammatory stimuli. Reproductive hormones may affect HIV transmission through effects on epithelial thickness, mucus production and host defense mechanisms.

- *How is this information being used to develop new prevention strategies?*

Topical vaginal, penile and rectal microbicides are under development to prevent the sexual transmission of HIV-1 and other sexually transmitted pathogens. Some microbicide candidates have been inspired by natural defense mechanisms. Buffering compounds that maintain the vaginal pH at virucidal levels (pH < 4.5), genetically engineered lactobacilli, gels containing antimicrobial proteins or HIV antibodies, and estrogen creams are under development for HIV prevention. Candidate compounds that have been designed to interfere with HIV infection mechanisms include molecules that block HIV envelope proteins (e.g., gp41 ligands), cellular HIV co-receptors (e.g., CCR5 and CD4), and cell adhesion molecules (e.g., LFA-1). Evidence that the genital tract is capable of mounting local HIV-specific humoral and cellular adaptive immune responses is promoting mucosal immunization strategies to produce HIV immunity in the genital tract that will block HIV transmission.

Further research and knowledge concerning the intricate relationship between HIV and its human host will undoubtedly promote increasingly sophisticated and more effective means to prevent and treat HIV infections.

References

Agace, W. W., Amara, A., Roberts, A. I. *et al.* (2000). Constitutive expression of stromal derived factor-1 by mucosal epithelia and its role in HIV transmission and propagation. *Curr. Biol.*, 10, 325–8.

Anderson, D. J. (2007). Genitourinary immune defense. In: K. K. Holmes, P. F. Sparling, Piot, P. *et al.* (eds), *Sexually Transmitted Diseases*. New York, NY: McGraw-Hill, pp. 271–88.

Anderson, D. J. and Yunis, E. J. (1983). "Trojan Horse" leukocytes in AIDS. *N. Engl. J. Med.*, 309, 984–5.

Anderson, D. J., Politch, J. A., Martinez, A. *et al.* (1991). White blood cells and HIV-1 in semen from vasectomised seropositive men. *Lancet*, 338, 573–4.

Anderson, D. J., O'Brien, T. R., Politch, J. A. *et al.* (1992). Effects of disease stage and zidovudine therapy on the detection of human immunodeficiency virus type 1 in semen. *J. Am. Med. Assoc.*, 267, 2769–74.

Anderson, D. J. and Pudney, J. (2005). Human male genital tract immunity and experimental models. In: J. Mestecky, M. E. Lamm, W. Strober *et al.* (eds), *Mucosal Immunology*. Boston, MA: Elsevier Academic Press, pp. 1647–60.

Bailey, R. C., Moses, S., Parker, C. B. *et al.* (2007). Male circumcision for HIV prevention in young men in Kisumu, Kenya: a randomised controlled trial. *Lancet*, 369, 643–56.

Battle-Miller, K., Eby, C. A., Landay, A. L. *et al.* (2002). Antibody-dependent cell-mediated cytotoxicity in cervical lavage fluids of human immunodeficiency virus type 1-infected women. *J. Infect. Dis.*, 185, 439–47.

Berges, B. K., Akkina, S. R., Folkvord, J. M. *et al.* (2008). Mucosal transmission of R5 and X4 tropic HIV-1 via vaginal and rectal routes in humanized Rag2($-/-$) gammac($-/-$) (RAG-hu) mice. *Virology*, 373, 342–51.

Bomsel, M. (1997). Transcytosis of infectious human immunodeficiency virus across a tight human epithelial cell line barrier. *Nat. Med.*, 3, 42–7.

Boskey, E. R., Cone, R. A., Whaley, K. J. and Moench, T. R. (2001). Origins of vaginal acidity: high D/L lactate ratio is consistent with bacteria being the primary source. *Hum. Reprod.*, 16, 1809–13.

Brenchley, J. M., Schacker, T. W., Ruff, L. E. *et al.* (2004). CD4+ T cell depletion during all stages of HIV disease occurs predominantly in the gastrointestinal tract. *J. Exp. Med.*, 200, 749–59.

Bujan, L., Hollander, L., Coudert, M. *et al.* (2007). Safety and efficacy of sperm washing in HIV-1-serodiscordant couples where the male is infected: results from the European CREAThE network. *AIDS*, 21, 1909–14.

Campo, J., Perea, M. A., del Romero, J. *et al.* (2006). Oral transmission of HIV, reality or fiction? An update. *Oral Dis*, 12, 219–28.

Celum, C. L. (2004). The interaction between herpes simplex virus and human immunodeficiency virus. *Herpes*, 11(Suppl. 1), 36A–45A.

Chan, D. J. and Ray, J. E. (2007). Quantification of antiretroviral drugs for HIV-1 in the male genital tract: current data, limitations and implications for laboratory analysis. *J. Pharm. Pharmacol.*, 59, 1451–62.

Chen, L., Jha, P., Stirling, B., Sgaier, S. K., Daid, T., Kaul, R. and Nagelkerke, N. (2007). Sexual risk factors for HIV infection in early and advanced HIV epidemics in sub-Saharan Africa: systematic overview of 68 epidemiological studies. *PLoS ONE*, 2, e1001.

Cohen, M. S. (2004). HIV and sexually transmitted diseases: lethal synergy. *Top. HIV Med.*, 12, 104–7.

Cohen, M. S., Hoffman, I. F., Royce, R. A. *et al.* (1997). Reduction of concentration of HIV-1 in semen after treatment of urethritis: implications for prevention of sexual transmission of HIV-1. AIDSCAP Malawi Research Group. *Lancet*, 349, 1868–73.

Cohen, M. S., Weber, R. D., Mardh, P.-A. and Anderson, D. J. (1999). Genitourinary mucosal defenses. In: K. K. Holmes, P. F. Sparling, P.-A. Mardh *et al.* (eds), *Sexually Transmitted Diseases*. New York, NY: McGraw-Hill, pp. 173–90.

Cohen, M. S., Hosseinipour, M., Kashuba, A. and Butera, S. (2002). Use of antiretroviral drugs to prevent sexual transmission of HIV. *Curr. Clin. Top. Infect. Dis.*, 22, 214–51.

Collins, K. B., Patterson, B. K., Naus, G. J. *et al.* (2000). Development of an in vitro organ culture model to study transmission of HIV-1 in the female genital tract. *Nat. Med.*, 6, 475–9.

Coombs, R. W., Speck, C. E., Hughes, J. P. *et al.* (1998). Association between culturable human immunodeficiency virus type 1 (HIV-1) in semen and HIV-1 RNA levels in semen and blood: evidence for compartmentalization of HIV-1 between semen and blood. *J. Infect. Dis.*, 177, 320–30.

Coombs, R. W., Reichelderfer, P. S. and Landay, A. L. (2003). Recent observations on HIV type-1 infection in the genital tract of men and women. *AIDS*, 17, 455–80.

Coombs, R. W., Lockhart, D., Ross, S. O. *et al.* (2006). Lower genitourinary tract sources of seminal HIV. *J. Acquir. Immune Defic. Syndr.*, 41, 430–48.

Cu-Uvin, S., Snyder, B., Harwell, J. I. *et al.* (2006). Association between paired plasma and cervicovaginal lavage fluid HIV-1 RNA levels during 36 months. *J. Acquir. Immune Defic. Syndr.*, 42, 584–7.

Delwart, E. L., Mullins, J. I., Gupta, P. *et al.* (1998). Human immunodeficiency virus type 1 populations in blood and semen. *J. Virol.*, 72, 617–23.

De Pasquale, M., Sutton, L., Ingersoll, J. *et al.* (2008). HIV-1 replicates locally in sub-compartments of female genital tissue. In: *Proceedings of the 15th Conference on Retroviruses and Opportunistic Infections*, Boston, 3–6 February (in press).

Donoval, B. A., Landay, A. L., Moses, S. *et al.* (2006). HIV-1 target cells in foreskins of African men with varying histories of sexually transmitted infections. *Am. J. Clin. Pathol.*, 125, 386–91.

Farrar, D. J., Cu-Uvin, S., Caliendo, A. M. *et al.* (1997). Detection of HIV-1 RNA in vaginal secretions of HIV-1-seropositive women who have undergone hysterectomy. *AIDS*, 11, 1296–7.

Fleming, D. T. and Wasserheit, J. N. (1999). From epidemiological synergy to public health policy and practice: the contribution of other sexually transmitted diseases to sexual transmission of HIV infection. *Sex. Transm. Infect.*, 75, 3–17.

Fletcher, P. S., Elliott, J., Grivel, J. C. *et al.* (2006). Ex vivo culture of human colorectal tissue for the evaluation of candidate microbicides. *AIDS*, 20, 1237–45.

Girard, M., Mahoney, J., Wei, Q. *et al.* (1998). Genital infection of female chimpanzees with human immunodeficiency virus type 1. *AIDS Res. Hum. Retroviruses*, 14, 1357–67.

Goodreau, S. M. and Golden, M. R. (2007). Biological and demographic causes of high HIV and sexually transmitted disease prevalence in men who have sex with men. *Sex. Transm. Infect.*, 83, 458–62.

Gray, R. H., Li, X., Kigozi, G. *et al.* (2005). Increased risk of incident HIV during pregnancy in Rakai, Uganda: a prospective study. *Lancet*, 366, 1182–8.

Gray, R. H., Kigozi, G., Serwadda, D. *et al.* (2007). Male circumcision for HIV prevention in men in Rakai, Uganda: a randomised trial. *Lancet*, 369, 657–66.

Greenhead, P., Hayes, P., Watts, P. S. *et al.* (2000). Parameters of human immunodeficiency virus infection of human cervical tissue and inhibition by vaginal virucides. *J. Virol.*, 74, 5577–86.

Gupta, P., Mellors, J., Kingsley, L. *et al.* (1997). High viral load in semen of human immunodeficiency virus type 1-infected men at all stages of disease and its reduction by therapy with protease and nonnucleoside reverse transcriptase inhibitors. *J. Virol.*, 71, 6271–5.

Hladik, F., Sakchalathorn, P., Ballweber, L. *et al.* (2007). Initial events in establishing vaginal entry and infection by human immunodeficiency virus type-1. *Immunity*, 26, 257–70.

Howell, A. L., Edkins, R. D., Rier, S. E. *et al.* (1997). Human immunodeficiency virus type 1 infection of cells and tissues from the upper and lower human female reproductive tract. *J. Virol.*, 71, 3498–506.

Hussain, L. A. and Lehner, T. (1995). Comparative investigation of Langerhans' cells and potential receptors for HIV in oral, genitourinary and rectal epithelia. *Immunology*, 85, 475–84.

Johansson, E. L., Rudin, A., Wassen, L. and Holmgren, J. (1999). Distribution of lymphocytes and adhesion molecules in human cervix and vagina. *Immunology*, 96, 272–7.

Kaizu, M., Weiler, A. M., Weisgrau, K. L. *et al.* (2006). Repeated intravaginal inoculation with cell-associated simian immunodeficiency virus results in persistent infection of nonhuman primates. *J. Infect. Dis.*, 194, 912–16.

Kedzierska, K., Crowe, S. M., Turville, S. and Cunningham, A. L. (2003). The influence of cytokines, chemokines and their receptors on HIV-1 replication in monocytes and macrophages. *Rev. Med. Virol.*, 13, 39–56.

Kemal, K. S., Foley, B., Burger, H. *et al.* (2003). HIV-1 in genital tract and plasma of women: compartmentalization of viral sequences, coreceptor usage, and glycosylation. *Proc. Natl Acad. Sci. USA*, 100, 12972–7.

Khanna, K. V., Whaley, K. J., Zeitlin, L. *et al.* (2002). Vaginal transmission of cell-associated HIV-1 in the mouse is blocked by a topical, membrane-modifying agent. *J. Clin. Invest.*, 109, 205–11.

Kinter, A., Arthos, J., Cicala, C. and Fauci, A. S. (2000). Chemokines, cytokines and HIV: a complex network of interactions that influence HIV pathogenesis. *Immunol. Rev.*, 177, 88–98.

Koblin, B. A., Husnik, M. J., Colfax, G. *et al.* (2006). Risk factors for HIV infection among men who have sex with men. *AIDS*, 20, 731–9.

Koelle, D. M., Schomogyi, M. and Corey, L. (2000). Antigen-specific T cells localize to the uterine cervix in women with genital herpes simplex virus type 2 infection. *J. Infect. Dis.*, 182, 662–70.

Krieger, J. N., Coombs, R. W., Collier, A. C. *et al.* (1995). Intermittent shedding of human immunodeficiency virus in semen: implications for sexual transmission. *J. Urology*, 154, 1035–40.

Krieger, J. N., Nirapathpongporn, A., Chaiyaporn, M. *et al.* (1998). Vasectomy and human immunodeficiency virus type 1 in semen. *J. Urology*, 159, 820–5, discussion 825–6.

Kutteh, W. H., Mestecky, J. and Wira, C. R. (2005). Mucosal immunity in the human female reproductive tract. In: J. Mestecky, M. E. Lamm, W. Strober *et al.* (eds), *Mucosal Immunology*. Amsterdam: Elsevier Academic Press, pp. 1631–46.

Larsen, B. (1993). Vaginal flora in health and disease. *Clin. Obstet. Gynecol.*, 36, 107–21.

Lederman, M. M., Offord, R. E. and Hartley, O. (2006). Microbicides and other topical strategies to prevent vaginal transmission of HIV. *Nat. Rev. Immunol.*, 6, 371–82.

Lim, S. G., Condez, A., Lee, C. A. *et al.* (1993). Loss of mucosal CD4 lymphocytes is an early feature of HIV infection. *Clin. Exp. Immunol.*, 92, 448–54.

Ma, Z. M., Abel, K., Rourke, T. *et al.* (2004). A period of transient viremia and occult infection precedes persistent viremia and antiviral immune responses during multiple low-dose intravaginal simian immunodeficiency virus inoculations. *J. Virol.*, 78, 14048–52.

Maher, D., Wu, X., Schacker, T. *et al.* (2005). HIV binding, penetration, and primary infection in human cervicovaginal tissue. *Proc. Natl Acad. Sci. USA*, 102, 11504–9.

Marx, P. A., Spira, A. I., Gettie, A. *et al.* (1996). Progesterone implants enhance SIV vaginal transmission and early virus load. *Nat. Med.*, 2, 1084–9.

Masters, W. H. and Johnson, V. E. (1961). The physiology of the vaginal reproductive function. *West J. Surg. Obstet. Gynecol.*, 69, 105–20.

Mayer, K. H., Boswell, S., Goldstein, R. *et al.* (1999). Persistence of human immunodeficiency virus in semen after adding indinavir to combination antiretroviral therapy. *Clin. Infect. Dis.*, 28, 1252–9.

McCoombe, S. G. and Short, R. V. (2006). Potential HIV-1 target cells in the human penis. *AIDS*, 20, 1491–5.

Miller, C. J., Alexander, N. J., Sutjipto, S. *et al.* (1989). Genital mucosal transmission of simian immunodeficiency virus: animal model for heterosexual transmission of human immunodeficiency virus. *J. Virol.*, 63, 4277–84.

Miller, C. J., Vogel, P., Alexander, N. J. *et al.* (1992). Localization of SIV in the genital tract of chronically infected female rhesus macaques. *Am. J. Pathol.*, 141, 655–60.

Moldoveanu, Z., Huang, W. Q., Kulhavy, R. *et al.* (2005). Human male genital tract secretions: both mucosal and systemic immune compartments contribute to the humoral immunity. *J. Immunol.*, 175, 4127–36.

Mostad, S. B., Overbaugh, J., DeVange, D. M. *et al.* (1997). Hormonal contraception, vitamin A deficiency, and other risk factors for shedding of HIV-1 infected cells from the cervix and vagina. *Lancet*, 350, 922–7.

Musey, L., Ding, Y., Cao, J. *et al.* (2003). Ontogeny and specificities of mucosal and blood human immunodeficiency virus type 1-specific CD8(+) cytotoxic T lymphocytes. *J. Virol.*, 77, 291–300.

Nag, P., Kim, J., Sapiega, V. *et al.* (2004). Women with cervicovaginal antibody-dependent cell-mediated cytotoxicity have lower genital HIV-1 RNA loads. *J. Infect. Dis.*, 190, 1970–8.

Neutra, M. R. and Kozlowski, P. A. (2006). Mucosal vaccines: the promise and the challenge. *Nat. Rev. Immunol.*, 6, 148–58.

Nickle, D. C., Jensen, M. A., Shriner, D. *et al.* (2003). Evolutionary indicators of human immunodeficiency virus type 1 reservoirs and compartments. *J. Virol.*, 77, 5540–6.

O'Connor, T. J., Kinchington, D., Kangro, H. O. and Jeffries, D. J. (1995). The activity of candidate virucidal agents, low pH and genital secretions against HIV-1 in vitro. *Intl J. STD AIDS*, 6, 267–72.

Olmsted, S. S., Padgett, J. L., Yudin, A. I. *et al.* (2001). Diffusion of macromolecules and virus-like particles in human cervical mucus. *Biophys. J.*, 81, 1930–37.

Palacio, J., Souberbielle, B. E., Shattock, R. J. *et al.* (1994). In vitro HIV1 infection of human cervical tissue. *Res. Virol.*, 145, 155–61.

Parkhurst, M. R. and Saltzman, W. M. (1994). Leukocytes migrate through three-dimensional gels of midcycle cervical mucus. *Cell Immunol.*, 156, 77–94.

Patterson, B. K., Landay, A., Siegel, J. N. *et al.* (2002). Susceptibility to human immunodeficiency virus-1 infection of human foreskin and cervical tissue grown in explant culture. *Am. J. Pathol.*, 161, 867–73.

Patton, D. L., Thwin, S. S., Meier, A. *et al.* (2000). Epithelial cell layer thickness and immune cell populations in the normal human vagina at different stages of the menstrual cycle. *Am. J. Obstet. Gynecol.*, 183, 967–73.

Picker, L. J. (2006). Immunopathogenesis of acute AIDS virus infection. *Curr. Opin. Immunol.*, 18, 399–405.

Pilcher, C. D., Joaki, G., Hoffman, I. F. *et al.* (2007). Amplified transmission of HIV-1: comparison of HIV-1 concentrations in semen and blood during acute and chronic infection. *AIDS*, 21, 1723–30.

Pillai, S. K., Good, B., Pond, S. K. *et al.* (2005). Semen-specific genetic characteristics of human immunodeficiency virus type 1 env. *J. Virol.*, 79, 1734–42.

Ping, L. H., Cohen, M. S., Hoffman, I. *et al.* (2000). Effects of genital tract inflammation on human immunodeficiency virus type 1 V3 populations in blood and semen. *J. Virol.*, 74, 8946–52.

Politch, J. A., Tucker, L., Bowman, F. P. and Anderson, D. J. (2007). Concentrations and significance of cytokines and other immunologic factors in semen of healthy fertile men. *Hum. Reprod.*, 22, 2928–35.

Poppe, W. A., Drijkoningen, M., Ide, P. S. *et al.* (1998). Lymphocytes and dendritic cells in the normal uterine cervix. An immunohistochemical study. *Eur. J. Obstet. Gynecol. Reprod. Biol.*, 81, 277–82.

Pudney, J. and Anderson, D. (1991). Orchitis and human immunodeficiency virus type 1 infected cells in reproductive tissues from men with the acquired immune deficiency syndrome. *Am. J. Pathol.*, 139, 149–60.

Pudney, J. and Anderson, D. J. (1995). Immunology of the human male urethra. *Am. J. Path.*, 147, 155–165.

Pudney, J., Oneta, M., Mayer, K. *et al.* (1992). Pre-ejaculatory fluid as potential vector for sexual transmission of HIV-1. *Lancet*, 340, 1470.

Quayle, A. J., Xu, C., Mayer, K. H. and Anderson, D. J. (1997). T-lymphocytes and macrophages, but not motile spermatozoa, are a significant source of human immunodeficiency virus in semen. *J. Infect. Dis.*, 176, 960–8.

Quayle, A. J., Coston, W. M., Trocha, A. K. *et al.* (1998a). Detection of HIV-1-specific CTLs in the semen of HIV-infected individuals. *J. Immunol.*, 161, 4406–10.

Quayle, A. J., Xu, C., Tucker, L. and Anderson, D. J. (1998b). The case against an association between HIV-1 and sperm: molecular evidence. *J. Reprod. Immunol.*, 41, 127–36.

Quayle, A. J., Kourtis, A. P., Cu-Uvin, S. *et al.* (2007). T-lymphocyte profile and total and virus-specific immunoglobulin concentrations in the cervix of HIV-1-infected women. *J. Acquir. Immune Defic. Syndr.*, 44, 292–8.

Reichelderfer, P. S., Coombs, R. W., Wright, D. J. *et al.* (2000). Effect of menstrual cycle on HIV-1 levels in the peripheral blood and genital tract. WHS 001 Study Team. *AIDS*, 14, 2101–7.

Rowland-Jones, S., Pinheiro, S. and Kaul, R. (2001). New insights into host factors in HIV-1 pathogenesis. *Cell*, 104, 473–6.

Royce, R. A., Sena, A., Cates, W. Jr. and Cohen, M. S. (1997). Sexual transmission of HIV. *N. Engl. J. Med.*, 336, 1072–8.

Sadiq, S. T., Taylor, S., Kaye, S. *et al.* (2002). The effects of antiretroviral therapy on HIV-1 RNA loads in seminal plasma in HIV-positive patients with and without urethritis. *AIDS*, 16, 219–25.

Sahasrabuddhe, V. V. and Vermund, S. H. (2007). The future of HIV prevention: control of sexually transmitted infections and circumcision interventions. *Infect. Dis. Clin. North Am.*, 21, 241–57, xi.

Shacklett, B. L., Cu-Uvin, S., Beadle, T. J. *et al.* (2000). Quantification of HIV-1-specific T-cell responses at the mucosal cervicovaginal surface. *AIDS*, 14, 1911–15.

Shattock, R. J. and Moore, J. P. (2003). Inhibiting sexual transmission of HIV-1 infection. *Nat. Rev. Microbiol.*, 1, 25–34.

Sheth, P. M., Danesh, A., Shahabi, K. *et al.* (2005). HIV-specific CD8+ lymphocytes in semen are not associated with reduced HIV shedding. *J. Immunol.*, 175, 4789–96.

Smith, S. M., Baskin, G. B. and Marx, P. A. (2000). Estrogen protects against vaginal transmission of simian immunodeficiency virus. *J. Infect. Dis.*, 182, 708–15.

Soilleux, E. J. and Coleman, N. (2004). Expression of DC-SIGN in human foreskin may facilitate sexual transmission of HIV. *J. Clin. Pathol.*, 57, 77–8.

Spear, G. T., St John, E. and Zariffard, M. R. (2007). Bacterial vaginosis and human immunodeficiency virus infection. *AIDS Res. Ther.*, 4, 25.

Spinillo, A., Zara, F., De Santolo, A. *et al.* (1999). Quantitative assessment of cell-associated and cell-free virus in cervicovaginal samples of HIV-1-infected women. *Clin. Microbiol. Infect.*, 5, 605–11.

Spira, A. I., Marx, P. A., Patterson, B. K. *et al.* (1996). Cellular targets of infection and route of viral dissemination after an intravaginal inoculation of simian immunodeficiency virus into rhesus macaques. *J. Exp. Med.*, 183, 215–25.

Sugaya, M., Lore, K., Koup, R. A. *et al.* (2004). HIV-infected Langerhans cells preferentially transmit virus to proliferating autologous CD4+ memory T cells located within Langerhans cell-T cell clusters. *J. Immunol*, 172, 2219–24.

Tachet, A., Dulioust, E., Salmon, D. *et al.* (1999). Detection and quantification of HIV-1 in semen: identification of a subpopulation of men at high potential risk of viral sexual transmission. *AIDS*, 13, 823–31.

Tan, X., Pearce-Pratt, R. and Phillips, D. M. (1993). Productive infection of a cervical epithelial cell line with human immunodeficiency virus: implications for sexual transmission. *J. Virol.*, 67, 6447–52.

Tasca, S., Tsai, L., Trunova, N. *et al.* (2007). Induction of potent local cellular immunity with low dose X4 SHIV(SF33A) vaginal exposure. *Virology*, 367, 196–211.

Tristram, D. A. (2005). Maternal genital tract infection and the neonate. In: J. Mestecky, M. E. Lamm, W. Strober *et al.* (eds), *Mucosal Immunology*. Amsterdam: Elsevier Academic Press, pp. 1721–34.

Tuomala, R. E., O'Driscoll, P. T., Bremer, J. W. *et al.* (2003). Cell-associated genital tract virus and vertical transmission of human immunodeficiency virus type 1 in antiretroviral-experienced women. *J. Infect. Dis.*, 187, 375–84.

UNAIDS (2007). *AIDS Epidemic Update.* Geneva: UNAIDS and WHO.

Veazey, R. S., Marx, P. A. and Lackner, A. A. (2003). Vaginal CD4+ T cells express high levels of CCR5 and are rapidly depleted in simian immunodeficiency virus infection. *J. Infect. Dis.*, 187, 769–76.

Vernazza, P. L., Gilliam, B. L., Dyer, J. *et al.* (1997). Quantification of HIV in semen: correlation with antiviral treatment and immune status. *AIDS*, 11, 987–93.

Wald, A. (2004). Synergistic interactions between herpes simplex virus type-2 and human immunodeficiency virus epidemics. *Herpes*, 11, 70–6.

Wawer, M. J., Gray, R. H., Sewankambo, N. K. *et al.* (2005). Rates of HIV-1 transmission per coital act, by stage of HIV-1 infection, in Rakai, Uganda. *J. Infect. Dis.*, 191, 1403–9.

Weiss, H. A., Quigley, M. A. and Hayes, R. J. (2000). Male circumcision and risk of HIV infection in sub-Saharan Africa: a systematic review and meta-analysis. *AIDS*, 14, 2361–70.

Weiss, H. A., Thomas, S. L., Munabi, S. K. and Hayes, R. J. (2006). Male circumcision and risk of syphilis, chancroid, and genital herpes: a systematic review and meta-analysis. *Sex. Transm. Infect.*, 82, 101–9, discussion 110.

Winter, A. J., Taylor, S., Workman, J. *et al.* (1999). Asymptomatic urethritis and detection of HIV-1 RNA in seminal plasma. *Sex. Transm. Infect.*, 75, 261–3.

Wira, C. R., Crane-Godreau, M. A. and Grant, K. S. (2005a). Endocrine regulation of the mucosal immune system in the female reproductive tract. In: J. Mestecky, M. E. Lamm, W. Strober *et al.* (eds), *Mucosal Immunology*. Amsterdam: Elsevier Academic Press, pp. 1661–76.

Wira, C. R., Fahey, J. V., Sentman, C. L. *et al.* (2005b). Innate and adaptive immunity in female genital tract: cellular responses and interactions. *Immunol. Rev.*, 206, 306–35.

Wolff, H., Mayer, K., Seage, G. *et al.* (1992). A comparison of HIV-1 antibody classes, titers, and specificities in paired semen and blood samples from HIV-1 seropositive men. *J. Acquir. Immune Defic. Syndr.*, 5, 65–9.

Woof, J. M. and Mestecky, J. (2005). Mucosal immunoglobulins. *Immunol. Rev.*, 206, 64–82.

Xu, C., Politch, J. A., Tucker, L. *et al.* (1997). Factors associated with increased levels of human immunodeficiency virus type 1 DNA in semen. *J. Infect. Dis.*, 176, 941–7.

Zhang, Z., Schuler, T., Zupancic, M. *et al.* (1999). Sexual transmission and propagation of SIV and HIV in resting and activated CD4+ T cells. *Science*, 286, 1353–7.

Zhu, T., Wang, N., Carr, A. *et al.* (1996). Genetic characterization of human immunodeficiency virus type 1 in blood and genital secretions: evidence for viral compartmentalization and selection during sexual transmission. *J. Virol.*, 70, 3098–107.

Zuckerman, R. A., Whittington, W. L., Celum, C. L. *et al.* (2004). Higher concentration of HIV RNA in rectal mucosa secretions than in blood and seminal plasma, among men who have sex with men, independent of antiretroviral therapy. *J. Infect. Dis.*, 190, 156–61.

HIV vaccines

3

Robert E. Geise and Ann Duerr

As the impact of the HIV epidemic has unfolded over the last 25 years, it has become increasingly clear that the best way to control infections is through multiple prevention strategies (circumcision, microbicides, barrier contraception and education). Many have argued that a preventative vaccine will be the most effective weapon in this armamentarium. Although combination antiretroviral therapy (ART) has improved the course of disease, such therapy is not universally available, does not prevent the massive immune destruction that occurs soon after infection (Brenchley *et al.*, 2004) and does not prevent the transmission of virus – especially the disproportionately high transmission during acute infection (Brenner *et al.*, 2007). Ideally, a vaccine candidate would trigger a rapid and vigorous memory response when an individual is exposed to virus, and elicit broad protection by focusing the immune response on the conserved epitopes of the virus or by other mechanisms.

The effort to find an effective vaccine has been met with a number of hurdles: numerous modes of transmission; viral genetic diversity; and HIV's ability to evade immune control, especially attempts to induce an antibody response to cover the diverse variety of HIV antigens. Results from several large efficacy trials have been disappointing; the lead vaccine candidates did not prevent HIV infection, and results from the most recent trial suggest that the vaccine may even have enhanced acquisition of virus. Since President Clinton challenged the scientific community in 1997 to find an effective vaccine within 10 years, significant resources have been devoted to that effort. Although this goal has not yet been attained, many challenges have been overcome. In 1997, no infrastructure was in place to work toward finding an effective HIV vaccine. Since that time, such an infrastructure has been built, late-phase clinical trials are ongoing, and a vaccine that may modify disease course, if not prevent infection, could be on the horizon. Through overcoming obstacles, establishing infrastructure and instituting new measures to evaluate products, significant strides are being made in HIV vaccine science.

Transmission and immunology of HIV and associated vaccine challenges

A variety of immunologic and physiologic factors make the development of a truly preventative HIV vaccine difficult. High mutation rates, virion structure that prevents recognition by the immune system, and rapid crippling of the arm of the immune system most needed for control of virus greatly impact the ability to develop an effective antigenic stimulant that will prevent or mitigate infection.

Immunology and host defenses

The adaptive immune response is composed of two branches. The humoral branch controls infections through antibodies. Antibodies are proteins that, because of their structure, are able to attach to an antigen (such as a virus). Once antibodies have bound to an antigen via a "lock and key" interaction, the invading pathogen can be destroyed through a number of intracellular and extracellular mechanisms. By binding to pathogens before they infect their target cells, antibodies can provide "sterilizing immunity" – that is, the immune responses can prevent the establishment of any detectable infection. The cellular immune system is a more complex system that works best on clearing established infections. Antigen-presenting cells elicit responses to selected pathogen peptides. When antigen-presenting cells are detected by cytotoxic T cells, pathogen peptides containing 9 to 15 amino acids (epitopes) are displayed on the surface of infected cells in conjunction with a "self" antigen (HLA). These infected cells are then destroyed by phagocytosis, cytokine release and cell lysis.

Traditional vaccine constructs

The most common mechanism of action for vaccines is to introduce an antigen from a specific pathogen into the host, causing the vaccinated individual to develop antibodies against that antigen. When re-exposed to the same stimulus (as when exposed to the disease), the humoral immune system will recognize the infecting agent and develop antibodies that will coat the invader, and the body's immune system will then destroy or remove that pathogen. Vaccines that induce cellular immunity, to date, have not been commonly developed.

Transmission physiology and primary infection

The primary mode of transmission for HIV worldwide is sexual contact. HIV breaches the epithelial barrier at sites of inflammation or micro-abrasions in the cervico-vaginal epithelium, penile epithelium or rectal mucosa, and via contact with the cells (Langerhans and dendritic cells) that shuttle HIV from the mucosal surface to underlying target cells (Kahn and Walker, 1998). Cells infected at the mucosal surface then present HIV virus to CD4-positive lymphocytes, and virus is transported to deeper tissues. HIV can be detected in regional lymph nodes within 2 days, and then in the blood within 7 days. What follows is a burst of viremia that has a dramatic impact on the immune system, including significant destruction of CD4 memory T cells. This occurs mainly in gut-associated lymphoid tissue, where a large percentage of these cells are located, essentially crippling the prime defenses against infection (Brenchley *et al.*, 2004).

In response to natural infection, the body mobilizes the two arms of the immune system. The humoral branch develops antibodies to the various protein constituents of HIV. The complex structure of the HIV envelope protein limits antibody access to these surface proteins. Glycosolation and conformational masking effectively shield those parts of the HIV envelope involved in T-cell binding (and subsequent infection) from interaction with antibodies. In addition, escape isolates easily develop and limit antibody effect in neutralizing circulating virus (Kwong *et al.*, 1998). Rapid viral evolution changes the proteins on the envelope, making it difficult for existing antibodies to recognize these new isolates (Wei *et al.*, 2003). Antibodies develop against the new isolates, and in some individuals, broadly neutralizing antibodies develop that allow for some control (Burton *et al.*, 2005). However, there is usually a loss of this immune control. If an individual were to possess these types of broad neutralizing antibodies prior to infection, he or she might be able to prevent infection or establish viral control prior to significant immune destruction. A major challenge of HIV vaccine research is to elicit such neutralizing responses.

The cellular arm of the immune system, dominated by CD8-positive lymphocytes, can temporarily control infection through recognition of infected cells, apoptosis, and cytokine secretion. The degree of viral control can be varied based on characteristics of infecting virus as well as on host immune characteristics. Not all aspects of this variability are understood. For example, deletions in the nef gene, important for HIV viral pathogenicity through its ability to downregulate major histocompatibility complex (MHC) function, have been linked to long-term survival in HIV-infected individuals (Deacon *et al.*, 1995; Kirchhoff *et al.*, 1995; Salvi *et al.*, 1998). Also, certain HLA classes, which vary among different people, have been associated with both slow and rapid progression (Altfeld *et al.*, 2006), with these polymorphisms potentially explaining up to 15 percent of the difference in viral load set point and disease progression (Fellay *et al.*, 2007). However, in most individuals, escape viral isolates develop through mutation and viral evolution and overcome this control, leading to disease progression.

Challenges for HIV vaccine development

The factors that make it difficult for the immune system to prevent and control natural HIV infection also lead to several significant challenges in the development of an effective HIV vaccine. The greatest hurdle is developing an antigen that elicits a response that provides significant sterilizing immunity against a broad variety of viruses, especially at the initial portals of entry. In their mission to control HIV infection, these vaccines face challenges including: (1) controlling, eliminating or preventing infection of latent cellular reservoirs such as resting CD4 T cells and sanctuary sites (where virus can integrate into host cell DNA and potentially require over 60 years of effective therapy to clear (Pierson *et al.*, 2000)); and (2) controlling viral replication when faced with envelope protein variation and sophisticated impediments to immune control (such as glycosolated, protected envelope proteins). The current generation of T-cell-based vaccines may not provide sterilizing immunity, but could control viral replication and thereby mitigate the destruction of the immune system, especially memory CD4+ cells, by priming it to control virus and prevent progressive disease.

Additionally, the first large efficacy trial testing this concept, the STEP trial, showed no efficacy in preventing HIV infection or in effecting viral load set point. In fact, in one subgroup analysis there appeared to be enhanced HIV acquisition. The sponsors of the trial provided the data from the trial in a timely and transparent fashion. Public response from government, scientific and community advocacy communities was disappointment about the result, but was positive as to the way the dissemination was handled. However, future trials using similar vaccine approaches will need to overcome this question of potential enhanced acquisition.

Approaches to developing an HIV vaccine

Because of the many challenges in eliciting sterilizing immunity to HIV, the traditional threshold of vaccine efficacy of greater than 90 percent may need to be put aside (Anderson and Hanson, 2005) and a new approach to evaluating a successful vaccine may be needed. There are four key outcomes that may identify an effective HIV vaccine candidate (HIV Vaccine Trials Network (HVTN) 2007):

1. Prevention of infection
2. Transient infection with control of HIV replication
3. Amelioration of disease process
4. Decreased transmission without effect on the infected individual.

The ideal vaccine would prevent infection through complete sterilizing immunity (complete protection without detectable HIV virus at any time or any transmission to others). However, several clinical trials using antibody-based vaccines have not shown any efficacy. Although effective against laboratory viral isolates, these candidate vaccines did not have the same impact on naturally occurring isolates (Moore *et al.*, 1995).

Transient infection with control of viral replication within days to months (without detectable HIV in 6 to 12 months, regardless of serostatus) would be manifested by brief viremia with a subsequent robust cellular response. After becoming transiently viremic, cellular immunity, with the possible assistance of humoral immunity, would control viral replication and clear infection. The infected individual's immune system would remain intact; however, he or she would be HIV antibody positive and potentially infectious early in the course of the disease.

The approach utilized by most vaccines currently in clinical trials is mitigation of progression of infection with low viral loads, minimal drop of CD4+ T cells, and a long time-course or no progression to AIDS. This type of vaccine, which elicits a cell-mediated immune response capable of controlling viral replication, could have a significant impact on disease course. Control of viral replication may be reflected in a low viral load set point, an early marker of the rapidity of disease progression. If the relationship between viral load set point and disease progression were the same among vaccinees as is seen in natural infection, a drop of 0.5 log in the viral load set point could delay the time to developing AIDS by 3 years. An even larger decrease in viral load set point, a decrease of 1.25 logs, could delay the progression of AIDS by 12 years and delay the need for ART by 3 years (Gupta *et al.*, 2007).

Finally, an effective vaccine candidate might not have significant impact on the vaccine recipient but could still decrease transmission of virus, perhaps through control of virus in mucosal secretions or body fluids. This outcome could be achieved either by some form of cellular or humoral response in the mucosa that decreases or eliminates viral shedding, or by attenuating viremia to a point that it effects disease transmission without effecting disease course. Should a vaccine lead to a significant decrease in an individual's viral load, this could result in decreased transmission to others as the risk of transmission among discordant couples has been directly correlated with plasma viral load levels (Quinn *et al.*, 2000).

Measure of success and surrogate markers

Progress in the development of a successful HIV vaccine will be incremental; the current generation of candidate vaccines aims primarily to ameliorate the course of infection. Later generations will likely need to utilize a different strategy to prevent HIV infection. Primary endpoints used to evaluate the effectiveness of an HIV vaccine would be: (1) reduction of susceptibility (VE_s), (2) reduction of disease progression (VE_p) as measured by time to AIDS or death, and (3) effects of vaccine on infectiousness (VE_i). Measurement of infections in trial participants is relatively straightforward. However, measurement of VE_p is more difficult. Our ability to assess concrete clinical endpoints, such as time to AIDS or death, is very limited, because infected participants are identified early in the course of disease and ART can be initiated prior to progression to clinical endpoints. Similarly, VE_i, a measurement of an infected participant's likelihood of transmitting HIV to partners, cannot be measured in conventional clinical trials. Enrolling the partners of trial participants, or studying vaccine effects in populations, requires significantly larger investments of time and money.

Because of the difficulty in assessing VE_p and VE_i directly, a number of surrogate endpoints have been proposed (Follmann *et al.*, 2006). Potential surrogate endpoints for VE_p might be viral load set point and peak viral load after infection. Also, CD4 count and trajectory of CD4 decline, time before meeting clinical criteria to initiate ART, and duration of a suppressed viral load might act as surrogate endpoints for VE_p. Because peripheral viral load is correlated to infectivity, it or the viral load in genital tract secretions could be used as a surrogate for VE_i.

Advanced clinical trial design

Large Phase III trials to measure efficacy can be extremely costly and time-consuming. Such investment of resources can limit the number of candidate vaccines tested, and might result in delayed implementation of an effective vaccine. These time delays translate into new infections and lives lost. To overcome some of these obstacles, HIV vaccine trials have employed Phase IIB test-of-concept trials (Kim *et al.*, 2007). Phase IIB trials are designed to bridge the gap between Phase IIA trials, which evaluate immunogenicity and safety, and large Phase III trials that are designed to prove efficacy and qualify a product for licensure. Not just an over-powered Phase II trial, the Phase IIB trials are designed to test a specific concept. Key design goals of Phase IIB trials are:

- to guide future vaccine development, and not to qualify for licensure;
- to use more surrogate endpoints (e.g., viral set point as a surrogate for HIV disease progression) rather than more definitive endpoints that would be needed for licensure (Gilbert *et al.*, 2003);
- to focus on specific populations, such as higher-risk/higher-incidence populations, rather than the general population that might be the recipient of the final product;
- to use a prototype product or one that is made by a manufacturing process other than that which will be used for the licensed vaccine; and
- to remain smaller and less expensive to conduct.

Because it adds a step between Phase IIA and Phase III trials, the use of a Phase IIB trial may prolong time to licensure. However, it can also potentially prevent prolonged development of a less than optimal product. By rapidly identifying products that would not achieve the necessary efficacy, resources can be allocated towards other more viable products. This approach worked with the STEP vaccine trial. Within 3.5 years of initiating the trial, preliminary results showed that this product was not efficacious and further development of this candidate was halted. This decision allowed resources to be funneled to other promising constructs.

To even further accelerate potential product selection, the International AIDS Vaccine Initiative (IAVI) has proposed the "screening" test-of-concept trials. These Phase II trials would enroll approximately 500 individuals from very high-risk HIV acquisition groups (incidence between 4 percent and 5 percent) (International AIDS Vaccine Initiative 2006). Although they are not powered to test vaccine efficacy for prevention of HIV acquisition (VE_s), these trials are powered to detect reduction in HIV viral load among participants who become HIV infected after vaccination (VE_p). The key to the success of this concept would be evaluating a larger number of candidate vaccines at a time, allowing the results to help guide the selection of these products to be advanced more rapidly to larger trials.

Vaccine approaches and evaluations

A wide variety of both traditional and novel approaches has been used in the quest to find an HIV vaccine. Initially, research was done using attenuated virus as a vaccine candidate. These experiments, done in macaques with an attenuated SIV strain, were initially promising. Subsequently, it was demonstrated that juvenile macaques, when exposed to the same virus, rapidly progressed to AIDS and death (Daniel *et al.*, 1992; Hulskotte *et al.*, 1998). In addition, there were significant safety concerns with using either attenuated or killed HIV in humans (Whitney and Ruprecht, 2004). Concerns remain as to whether any form of SIV or HIV virus can be given to humans because: (1) it is difficult to guarantee attenuated viruses might not revert into a form that would lead to progressive disease, and (2) the possibility exists that the inactivation/killing process might not be 100 percent effective. Polio, for example, when given orally in an attenuated form, has led to viral shedding and infections in both immunocompromised vaccine recipients and close contacts of vaccinees who are not immune to polio. Because of these concerns, approaches using attenuated or killed HIV in vaccine development have not been pursued.

Vaccines to elicit neutralizing antibodies

Initial approaches to vaccine development in humans attempted to elicit antibodies that would neutralize HIV. Specifically, HIV envelope proteins were used as the initial antigen. A number of trials performed in the late 1980s and 1990s demonstrated that these vaccines were safe, well tolerated, and elicited relatively high levels of serum antibodies that were capable of neutralizing laboratory-adapted HIV strains (Belshe *et al.*, 1994; Connor *et al.*, 1998). However, when the serum of vaccinated individuals was tested against virus isolated from naturally infected individuals, the antibodies could not neutralize those viruses (Moore *et al.*, 1995). Because of these findings, the Division of AIDS (DAIDS) of the National Institutes of Health (NIH) made the decision not to pursue these vaccines in trials in the US. Private industry (VaxGen) went on to raise the capital to support efficacy testing of the gp120 subunit of the HIV envelope. These products were tested alone or in combination with other products in three separate Phase III trials. These are the only Phase III HIV vaccine clinical trials to date (see "Vaccine trials" section).

Vaccines to elicit cellular response

As the limitations of early-generation vaccines designed to elicit neutralizing antibodies became apparent, attention turned to vaccines that would stimulate cellular immunity. Again, cellular immunity, especially the actions of cytotoxic (CD8) T lymphocytes, has been shown to be important in the early control of HIV virus in natural infection. HIV, however, affects cellular immunity early in the course of infection, primarily through the early depletion of CD4+ memory lymphocytes. Error-prone replication eventually leads to the development of HIV virus that escapes these early responses, and the immune system is unable to clear HIV in infected, non-replicating, latent T cells. In natural infection, this branch of the immune system cannot eliminate HIV but can control it in some individuals.

As opposed to directly destroying virus, cellular-based vaccines work by activating a response to destroy infected cells. Antigen is initially introduced into the vaccinee, processed, and presented on cell surfaces together with the major histocompatability complex antigens. In this manner, these vaccines elicit functional CD4 and CD8 T-cell responses. Because these vaccine-induced responses would be present at the time of natural infection, it is hoped that they will provide protective responses leading to early control of HIV infection, thus preventing the rapid depletion of CD4 cells (the target cells for HIV).

In non-human primates, there is evidence that this approach would work. When depleted of CD8+ cells, viral load increases rapidly in macaques (Jin *et al.*, 1999; Schmitz *et al.*, 1999). In challenge models using SIV and a hybrid HIV–SIV (SHIV) virus, this class of vaccines did not prevent infection but did lead to control of viral replication and to prolonged survival in vaccinated animals after infection (Shiver *et al.*, 2002).

As with the vaccines that elicited an antibody response, the initial human trials involving vaccines that stimulated cellular immunity were disappointing. The Merck STEP Study, which used a trivalent vaccine (adenovirus 5 vector with a clade B gag, pol, and nef inserts), suspended further vaccinations after an interim analysis showed no efficacy in preventing infection or modifying viral load in

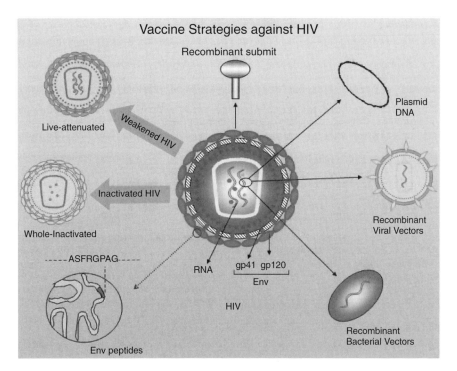

Figure 3.1 Schematic of various approaches to vaccine development that include the use of (1) attenuated or inactivated virus; (2) subunits of the virus; (3) DNA plasmids; (4) recombinant viral and bacterial vectors; and (5) Env peptides. Reproduced from Singh (2006), with permission from *Virology Journal*; ©Singh (2006).

those who became infected after vaccination (Merck, 2007). Follow-up in this trial is ongoing (see "Vaccine trials" section).

Vaccine candidates

A number of traditional and novel strategies are being explored to prevent HIV infection and/or control its impact on those who become infected after vaccination (see Figure 3.1).

Live-attenuated/killed viruses

These vaccine candidates would use an attenuated (weakened) or killed form of the virus to mimic natural infection without causing disease. Killed organisms are used in the injected polio and influenza vaccines. Intranasal influenza and oral polio use the attenuated approach.

As stated earlier, a live-attenuated or killed HIV vaccine has not been pursued due to safety concerns. Although some studies in animal models did show significant levels of protection with an attenuated virus, there have been others that have shown that animals exposed to this type of vaccine, when challenged with naturally occurring viruses, develop infection with rapid progression of disease (Daniel *et al.*, 1992; Hulskotte *et al.*, 1998). In addition, infant macaques challenged with attenuated virus had a fatal disease course (Baba *et al.*, 1995).

Peptide and subunit vaccines

This approach, which uses a protein or subunit (a small piece of the pathogen) from the HIV virus, has already been employed in the VaxGen trials. These trials used the gp120 subunit of the HIV envelope protein. The introduction of this envelope subunit induces an immune response in B cells, producing antibodies against the pathogen. Subsequently, if there is a natural infection of the host, the antibodies that have already been induced by the vaccine will assist the immune system in neutralizing the pathogen, if the right types of antibodies are induced by the vaccine. The hepatitis B vaccine works in the same manner.

Although not successful in the VaxGen trials, this method has been found to be safe. Other vaccine candidates continue to use this approach, either in a stand-alone regimen or in combination with other approaches. The RV144 trial is using such a combination approach, with the initial vaccination (the prime) using a viral vector and the second vaccination (the boost) using a subunit vaccine.

Viral vectors (Ad5, MVA, novel)

A number of live viral vectors have been used to stimulate anti-HIV responses. Live viruses as vaccine vectors work by introducing a genetic insert containing HIV genes. These genes are then processed and presented on the cell surface with the MHC I and II antigens, resulting in recognition of these antigens by CD4 and CD8 cells. Later exposure (through infection) would then stimulate (cellular) responses to recognize and eliminate infected cells.

The current lead viral-vector candidates are modified, replication-incompetent adenovirus 5 (Ad5) virus vectors. Ad5 generates strong humoral and cellular immunity to both the vector (Ad5) and transgenes (inserts) in gene therapy studies. Compared to other vectors, it can be manufactured at high titers, stimulates dendritic cells, and stimulates the innate immune system (Kim *et al.*, 2007). Although promising, Ad5 vaccines face challenges. Response to vaccines using this vector may be limited by pre-existing vector immunity, with this impact being most significant in the developing world, where a large portion of the population has been previously exposed to Ad5. Although not natural hosts for Ad5, rhesus macaques have been vaccinated with Ad5 without HIV inserts.

These animals were subsequently vaccinated with Ad5 vaccine and immuno-genicity assays were performed. The pre-existing immunity to Ad5 did attenuate the vaccine effect, but did not completely abrogate it (Shiver and Emini, 2004). Prevalence studies in South Africa show that up to 80 percent of adults tested had pre-existing immunity to Ad5 (Morgan *et al.*, 2005). To overcome the challenge of prior Ad5 immunity, several less common adenoviruses are being used as vectors to introduce immunogen. Viruses such as adenovirus serotype 26 and adenovirus serotype 35 (which occur less frequently), as well as chimeric products of several adenoviruses, are entering clinical trials in humans (Morgan *et al.*, 2005; Abbink *et al.*, 2007). In light of the results of the STEP trial, the future of Ad5-vector-based vaccines is in question; it is unknown whether these findings will apply to other Ad5 vectors. One or several of the alternative adenovirus vectors, which are generally associated with lower frequency of pre-existing immunity, are likely to be developed either for use in a heterologous prime-boost regimen or to boost a DNA product.

Currently, the leading candidate Ad5-based vaccine is under development by the NIH's Dale and Betty Bumper's Vaccine Research Center (VRC). Planned to be used in a combination regimen, the VRC adenovirus vaccine is used to boost the DNA prime (see Figure 3.2). This vector contains inserts with envelope genes

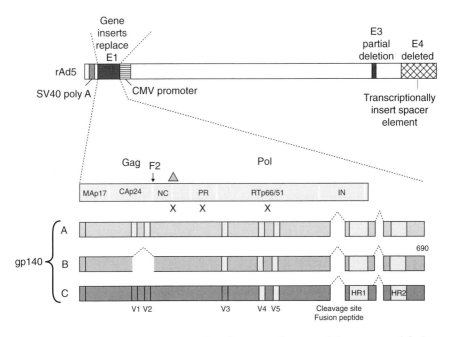

Figure 3.2 Schematic for the VRC adenovirus 5 product containing a gag–pol fusion and envelope inserts from clades A, B and C. Reproduced from Catanzaro *et al.* (2006), with permission.

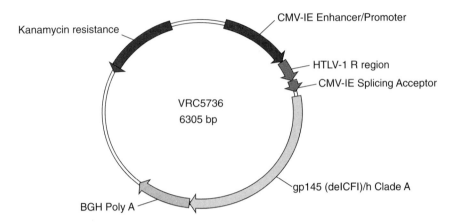

Figure 3.3 Schematic of the VRC DNA plasmid product for the clade A envelope. Similar constructs are used in a six-plasmid design for gag, pol and nef from clade B and envelope from clades A, B and C. Reproduced with permission of Gary Nabel, US National Institutes of Health (NIH) Dale and Better Bumpers Vaccine Research Center (VRC).

from clades A, B and C, as well as a fusion of clade B gag and pol genes (see Figure 3.3).

As previously discussed, a trivalent Ad5 vaccine (developed by Merck) entered clinical efficacy trials several years ago. This vaccine has been shown, in a rhesus macaque model, to be highly immunogenic. After three vaccinations, when intravenously challenged with a chimeric SIV/HIV pathogen, three of three vaccinated monkeys became infected but had a lower viral load peak, lower viral set point and higher CD4+ cell counts, and did not progress to AIDS, compared to control monkeys that did progress to AIDS (Shiver *et al.*, 2002). In later experiments, macaques that were vaccinated with two doses of an Ad5 vaccine containing an SIV gag insert, and were later challenged with SIV, became infected and failed to control viral replication. Infections occurred in animals vaccinated with a DNA-Ad5 combination regimen, and transient control of viremia was observed (Casimiro *et al.*, 2005). When advanced to human trials using inserts expressing consensus clade B constructs of the gag, pol, and nef parts of the HIV genome, Phase I and II trials showed promising results. However, a large Phase IIB test-of-concept trial, the STEP Study, was stopped early when an interim analysis showed no efficacy (see "Vaccine trials" section).

The other vectors that have undergone significant clinical testing are pox viruses (canary pox, modified vaccinia Ankara (MVA) and fowl pox). The largest ongoing current trial, RV144, uses a canary-pox-based vector containing the envelope gene of clades B and E as well as a gag/pol fusion (clade B) as the prime vaccine, followed by a gp120-based boost. This is the pox vector that has progressed furthest in trials, and is the only viral vector to have entered Phase III studies. Other pox virus vectors that have entered early-phase human trials include MVA and the New York strain of vaccinia (NYVAC).

Experience in non-human primates and in humans indicates that these vectors are safe and well tolerated, but challenges remain for the use of pox virus vectors. Vaccinia, the first vector used in this class of products, was abandoned early due to concerns that the vaccine might induce disseminated vaccinia (especially in areas with large immune-compromised populations) after disseminated vaccinia reportedly caused the death of an HIV-infected individual (Redfield *et al.*, 1987). Although immunogenicity data in non-human primates has been encouraging, these data were not replicated in human trials with second-generation products. Also, in the elderly population there is significant exposure to pox viruses, due to smallpox vaccination, that may result in attenuated immune responses. Because of possible cardiac toxicity associated with an earlier smallpox vaccination program (CDC, 2003), many pox virus products have careful cardiac monitoring during clinical trials. While first-generation pox virus candidates (ALVAC and MVAs) were poorly immunogenic, second-generation MVA and NYVAC candidates have shown good safety and immunogenicity profiles in early-phase trials.

Among other available vectors are adeno-associated virus (AAV) and vesicular stomatitis virus (VSV). AAV is not known to cause human disease (and requires a helper virus for replication). Studies using an SIV insert in macaques that were later challenged with SIV showed that the vaccinated animals had better virologic suppression and control than placebo recipients (Johnson *et al.*, 2005). Used in other treatment modalities (cystic fibrosis, hemophilia, rheumatologic disorders), AAV suffered a setback when a participant receiving AAV-based therapy for inflammatory arthritis died from disseminated histoplasmosis. Although later found not likely to have been associated with the AAV product, this event has slowed the development of this vector. A recombinant rabies–VSV product with an SIV insert has also been used in macaques that were later challenged with a pathogenic SHIV, and those that were vaccinated had attenuated disease progression (McKenna *et al.*, 2007). These potential vectors are in late-phase preclinical or early clinical testing, and will be used alone or in prime-boost regimens.

DNA vaccines

One of the newest delivery methods is the use of DNA plasmids. Pieces of HIV DNA are delivered to the cells to process without integration into the host genome. The protein products are presented on the cell surface in conjunction with MHC molecules, and can then elicit a cellular response to the HIV antigens.

Although successful in producing a focused immune response in smaller mammals and non-human primates, the response in humans has been less robust. Alternative delivery systems and improved adjuvants have been formulated in attempts to increase immunogenicity.

PAVE 100, a large clinical trial (a Phase IIB test-of-concept trial) of an HIV DNA vaccine was scheduled to open in 2008 but will not go forward as planned. Planning for further clinical testing of the regimen developed at the NIH VRC, which uses a six-plasmid DNA vaccine (expressing envelope glycoproteins of a clades A, B or C as well as expressing clade B gag, pol or nef, respectively), boosted by an Ad5 vector vaccine (see Figures 3.2 and 3.3) is ongoing.

Heterologous prime-boost

Heterologous prime-boost vaccine strategies are being used in about half of ongoing trials. This technique uses a "prime" vaccine that initially stimulates the immune system generating memory T and B cells. Later, a boost is given with a different vector, a different insert, or sometimes both. There are limited data to show that this approach allows for better immune response (International AIDS Vaccine Initiative (IAVI) 2007). Conceptually, this approach may work by circumventing immunity that might have developed to the initial vector. However, the exact mechanism is currently not known.

Delivery methods (mechanical) and adjuvants

To help improve response to these novel vaccine candidates or to assist with more efficient antigen presentation, a number of novel mechanical delivery methods have been developed. The Biojector® is one such device. It uses a single, needleless, syringe-based device that injects product under high pressure by compressed carbon-dioxide cartridges. This device is used primarily to inject DNA vaccine product into the tissues, conceptually allowing for increased cellular uptake. Immunigenicity results obtained with this technique have been inconsistent (Meseda *et al.*, 2006; Rao *et al.*, 2006; Brave *et al.*, 2007). Electroporation is a second method that is being developed to help improve the immunogenicity of DNA vaccines. Already used experimentally in gene therapy and cancer vaccines, electroporation increases the permeability of cell membranes and theoretically enhances the uptake of the immunogen into cells. In non-human primates (Luckay *et al.*, 2007), electroporation has increased immune response 10- to 40-fold at one-fifth the dose of immunogen, compared to conventional needle-and-syringe delivery.

Adding adjuvants to HIV vaccine candidates is another area of ongoing research. Many of the current vaccine strategies do not produce immune responses that are robust, long-lived and appropriately focused on production of neutral-izing antibodies or cytotoxic responses. Adjuvants could overcome this prob-lem by targeting the antigen to antigen-presenting cells, or increasing immune response by stimulating production of cytokines and costimulatory molecules, or both. Adjuvants such as polymeric microspheres (eg, polylactide-coglycolide or PLG) have been tested with HIV vaccine candidates to increase immune response through facilitating interactions with antigen-presenting cells (Singh and Srivastava, 2003). Other adjuvants being tested in conjunction with HIV vac-cine candidates attempt to increase induction of relevant cytokines and upregulate costimulatory molecules. CpG, unmethylated cytosine-guanine dinucleotides, which acts as a ligand for toll-like receptor 9 (TLR9), is one example (Kojima *et al.*, 2002). Stimulation of TLRs in turn enhances and directs the immune response. A related approach is the administration of vaccine candidates with costimulatory molecules such as Il-12, IL-15 or granulocyte-macrophage stimu-lating factor (GM-CSF) in an attempt to manipulate the immune response and increase cell-mediated immunity to the co-administered HIV vaccine antigen (Calarota and Weiner, 2004).

Consideration for insert selection

When genetic inserts are considered for vectors, the biological diversity of HIV becomes a significant issue. One the major challenges of the HIV infection and epidemic is the significant viral diversity of the target pathogen. HIV has two major types, HIV-1 and HIV-2, with HIV-1 having 10 subtypes or clades. The rate of mutation is so quick that the diversity of virus circulating within an infected individual's blood may be broader than the genetic diversity of influenza circulat-ing worldwide (Figure 3.4) (Korber *et al.*, 2001). It is also unclear that inserts based on a single clade or several limited clades will provide protection against genetically diverse isolates. Although the responses generated by natural infec-tion have been shown *in vitro* to have cross-clade reactivity, this protection has not been demonstrated in humans (Cao *et al.*, 1997; Coplan *et al.*, 2005).

Several approaches are used to select the genetic inserts that might induce the broadest immune response and hopefully provide the best protection. Three pro-posed design approaches for vector inserts are consensus sequences, ancestral sequences, and center-of-the-tree design. Consensus-sequence design uses the most common amino acid, among naturally occurring viruses, at each position in the genome for the insert being developed (Gaschen *et al.*, 2002). Ancestral sequencing is based on a phylogenetic tree developed from known HIV sequences. Using maximum likelihood analysis, the most likely genetic make-up of the "ancestor" for these circulating viruses is hypothesized, and that sequence is used to develop inserts (Gaschen *et al.*, 2002). Finally, a center-of-the-tree design looks at all isolates from a subtype, finds one that is genetically most similar (closest to

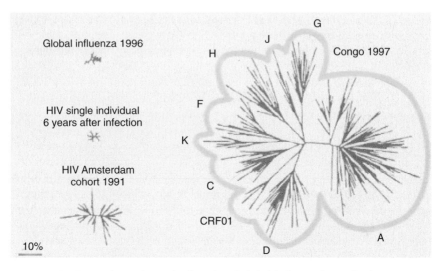

Figure 3.4 Comparison of genetic diversity of global influenza in a calendar year, HIV in an individual, a single city cohort in 1991 and a single sub-Saharan country in 1997. Reproduced under BioMed Central Open Access license agreement; originally appeared in Garber *et al.* (2004).

the center of the diversity tree) to other isolates, and uses it as the basis for insert development (Gaschen *et al.*, 2002; Mullins *et al.*, 2004).

Vaccine trials

The initiation of HIV vaccine trial research received impetus from the US government as early as 1984. At that time, Margaret Heckler, Secretary of Health and Human Services, predicted that an HIV vaccine would be in human trials within 2 years. Perhaps the greatest political push came from President Bill Clinton, who in 1997 challenged researchers to find an effective vaccine within 10 years, evoking imagery of John F. Kennedy and his challenge to the NASA science program to put a man on the moon.

There were many starts and stops in vaccine research, including attempts to develop various vaccine products. Early in vaccine research and development, attenuated or inactivated virus was considered too dangerous to use in humans. The most promising early candidate was an envelope glycoprotein gp120 candidate that induced antibody responses in vaccinated individuals. Although neutralization of virus was seen using laboratory-adapted virus, the antibodies produced after vaccination failed to neutralize virus isolated from infected individuals. Because of this, the NIH decided not to fund further research into this approach.

However, as described in this next section, these studies (VAX003 and VAX004) were performed with other funding sources.

Past trials

Over 50 candidates have been tested in Phase I trials, with only about 20 progressing on to Phase II trials. There have only been two completed Phase III trials, both of which have used recombinant envelope glycoproteins to elicit antibody response.

Initiated in the mid-1990s, the first of these trials (VAX004) studied a candidate vaccine with two clade B gp120 (AIDSVAX B/B) envelope glycoproteins as the antigenic stimulus. Conducted in North America and the Netherlands, this study enrolled individuals at high risk for sexually transmitted HIV. A parallel study (VAX003) enrolled intravenous drug users in Thailand using a candidate vaccine with one clade B and one clade E gp120 envelope glycolprotein (AIDSVAX B/E) as the antigenic stimulus. The results of both trials were disappointing. Neither showed efficacy (Flynn *et al.*, 2005; Pitisuttithum *et al.*, 2006). Immunologic studies of this product showed a robust and complex immune response that did not translate into protection (Gilbert *et al.*, 2005). In the North American and Dutch study (5095 men and 308 women at high risk due to sexual exposure), infection rates were 6.7 percent in 3598 vaccine recipients and 7.0 percent in 1805 placebo recipients. There was no difference in infection between vaccinees and those who received placebo with regard to viral loads, infecting viral strains, or the time to initiation of ART (Flynn *et al.*, 2005). In the study on intravenous drug users in Thailand, a total of 2546 volunteers were enrolled (of which 93.4 percent were male). The cumulative incidence was 8.4 percent, and there were no differences between the vaccine and placebo arms in infection rate or set-point viral load (Flynn *et al.*, 2005).

Although the trials were disappointing in their results, they were encouraging in their execution. The VAX trials showed that large-scale HIV trials could be conducted successfully in a variety of settings.

The STEP Study (Merck 023/HVTN 502) was a large Phase IIB trial using the Merck trivalent Ad5 vaccine with gag, pol and nef clade B inserts. This trial was conducted in the US, Canada, South America, Australia and the Caribbean, and enrolled 3000 individuals (1500 each vaccine/placebo) at high sexual risk for HIV acquisition. The objectives of this study were to determine if this vaccine could: (1) decrease acquisition of HIV, and (2) in those who became infected, mitigate the course of disease by decreasing the viral load set point. This study began enrollment in 2004 and completed enrollment in March 2007.

An interim analysis, done on the first 30 per-protocol HIV infections in participants with low pre-existing Ad5 immunity at time of enrollment, showed no efficacy of vaccine. Further vaccinations were halted in September 2007. In an intent-to-treat analysis of those who received at least one injection, there were 24 HIV infections in 741 volunteers who received vaccine and 21 HIV infections in 762 volunteers who received a placebo. In the subgroup of volunteers who received at least two vaccines (at week 8 of the trial) and were HIV uninfected for the first 12 weeks of the trial, there were 19 HIV infections in 672 volunteers who received vaccine and 11 HIV infections in 691 volunteers who received a placebo (Merck, 2007). Additionally, no difference was seen between the two groups in early viral load levels, a surrogate for disease progression.

After halting vaccinations, a number of *post-hoc* analyses were performed to further investigate the results of this trial. In a modified intent-to-treat analysis, there was no difference detected in those participants without prior Ad5 immunity. In the vaccine arm, there were 20 infections out of 382 participants. In the placebo arm, there were 20 infections out of 394 participants. Concerning, however, were the results that were seen in those participants with evidence of prior immunity to Ad5 at the time of the first vaccination. In that group, there were 29 infections out of 523 participants in the vaccine arm (infection rate of 5.5 percent) compared to 13 infections out of 528 participants in the placebo arm (infection rate of 2.4 percent). Details of the infections by Ad5 strata are shown in Table 3.1. Because of these findings and the desire to maintain open communications with volunteers and the community, trial participants were unblinded to treatment assignment in November 2007, and those who received the vaccine were counseled on potential enhanced risk of HIV acquisition.

A second trial using the same product in South African adults (a primarily clade C region), the Phambili study, was suspended when the interim results of the STEP study were announced. At that time, Phambili had immunized 800 participants of a planned 3000 individuals. The Phambili trial was also trying to

Table 3.1 Number of HIV infections, and incidence per 100 person years follow-up, by stratum of adenovirus 5 titer prior to first vaccination

Titer	Relative risk of HIV infections by pre-existing adenovirus type 5 titers			
	<18	18–200	201–1000	>1000
Vaccine[1]	20/382 (4.0)	8/140 (4.4)	14/220 (6.1)	7/163 (4.4)
Placebo[2]	20/394 (4.0)	4/142 (2.2)	7/229 (3.0)	2/157 (1.2)
Relative risk	1.0	2.1	2.0	3.5

Reproduced with permission of Michael Robertson, Merck, Inc.
[1]Number of infections/total number of participants at risk.
[2]Incidence = Number of infections/number at risk.

determine whether this vaccine could decrease acquisition of HIV infection and, in those who became HIV infected, determine whether it would decrease the viral load set point and thereby mitigate the course of disease. Additionally, this study was meant to investigate whether vaccine efficacy is affected by a mismatch between the clade of the vaccine insert and the clade of the circulating virus. This study began enrollment in January 2007 and was suspended in September 2007.

The data from these two trials are very complex, and analysis will continue for years. In those who became infected, HIV was not transmitted by vaccination but acquired by exposure to the virus from another individual. It is unclear if and how the vaccine caused enhanced acquisition; the disparity could have been caused by differential risk behavior in cohorts, a vaccine-induced biological event, or chance. Future vaccine constructs and trial designs will be impacted by these results and by further analyses of the data and specimens from the STEP and Phambili studies.

Ongoing and upcoming efficacy trials

There are currently several ongoing or upcoming large Phase IIB (test-of-concept) and Phase III efficacy trials. The largest trial to date is the Phase III trial RV 144 or Thai Prime-Boost Trial. This trial is sponsored by the US military, the NIH and the Thai Ministry of Health. The regimen in this trial uses a recombinant canary-pox vector (ALVAC-HIV vCP 1521) developed by Aventis Pasteur that elicits a cellular response as the initial vaccination and then boosts with the AIDSVAX B/E (a gp120 subunit vaccine) product used in VAXGEN 003. Vaccination was completed for this trial in 2006, which is now in follow-up. This study enrolled 16,000 HIV-negative individuals in Thailand (8000 each vaccine/placebo) with a planned 3-year follow-up after vaccination. The primary objective is to evaluate efficacy for the prevention of HIV acquisition. A secondary objective is to determine if those vaccinated who later acquired HIV have different disease progression as measured by HIV viral load and CD4 count over 3 years after infection (US National Library of Medicine, 2007).

(PAVE) a multinetwork collaboration involving the HVTN, IAVI, US Centers for Disease Control (CDC), US Department of Defense (DoD), and DAIDS. This group had planned a Phase IIB trial to study the VRC prime-boost regimen consisting of a six-plasmid DNA prime (containing env from clades A, B and C, as well as clade B gag, pol and nef) given in three priming doses via Biojector® and an Ad5 vaccine boost (containing env from clades A, B and C and a clade B gag/pol fusion insert). The primary objective of this study was to determine if this vaccine can decrease acquisition of HIV and, in those who become infected, decrease the viral set point and mitigate the course of disease. Due to the findings of the STEP study, this trial will not go forward as planned.

Early-phase trials

To date, over 200 trials have been performed using a variety of constructs. Of those, only five regimens have progressed to 2b or 3 trials (International AIDS Vaccine Initiative (IAVI), 2007). The early trials, which look primarily at cell-based immunity, are testing a variety of delivery vectors, HIV inserts, delivery methods, and adjuvants. Concepts being tested in these early trials include heterologous prime-boost regimens, use of novel viral vectors (e.g. rare adenovirus vectors, MVAs, other pox vectors), novel DNA plasmids, multiclade or non-clade-B inserts, addition of cytokines or other adjuvants, and use of novel delivery or mechanical adjuvant methods. A list of currently ongoing Phase II and III trials is in Table 3.2. Details about Phase I trials can be found in the IAVI (http://www.iavi.org) or AVAC (http://www.aidsvaccineclearinghouse.org) databases.

Organizations, costs, and funding

The breadth of organizations involved in the funding and operations of HIV vaccine research reflects the enormity and international scope of the need for a preventative vaccine. In an effort to coordinate the global research effort, leading HIV vaccine scientists in 2003 proposed an HIV enterprise be formed to: (1) prioritize the scientific challenges to be addressed, as well as product development; (2) devise an implementation plan for all components of HIV vaccine development; and (3) identify a plan for resources needed (Klausner *et al.*, 2003). Out of this policy paper, the Global HIV/AIDS Vaccine Enterprise was formed. The Enterprise seeks to accelerate the pace of HIV vaccine development in three ways: (1) by building a consensus of scientific priorities through the development of a continually updated scientific plan (published in 2005) that lists the roadblocks and proposed methods of addressing the issues; (2) by mobilizing donor governments, private philanthropists and other potential sources of funding to help support the priorities identified in the Enterprise's strategic plan; and (3) by improving scientific collaboration and reducing the duplication of efforts through specialized forums and group meetings (Coordinating Committee of the Global HIV/AIDS Vaccine Enterprise, 2005).

Many other agencies and networks have significant involvement in HIV vaccine development and funding. These groups are listed in Table 3.3.

Funding for HIV vaccine research is the result of the joint efforts of governmental agencies, private industry and philanthropic organizations. The biggest sponsor by far has been the US government (primarily through the NIH), which contributed approximately 70 percent of the $993 million invested in HIV vaccine research in 2006 (HIV Vaccines and Microbicides Resource Tracking Working Group, 2007). Philanthropic organizations (such as the Bill and Melinda Gates

Table 3.2 Ongoing Phase II and Phase III HIV preventive vaccine trials (as of 1 November 2007)[1,2,3]

Trial name and description	Organizer/manufacturer	Site locations	Description of vaccine
Phase III			
RV 144: A trial of live recombinant AVLAC-HIV priming with VaxGen gp120 B/E boost – enrollment complete and in follow-up	Walter Reed Army Institute of Research (WRAIR), Armed Forces Research Institute, Thai MoH, Thai AIDS Vaccine Evaluation Group. Manufacturers: Aventis Pasteur and VaxGen	Thailand	Live canary pox with a clade B env and gag-pol fusion boosted with a subunit clade B/E gp120
Phase II and IIB (test of concept)			
STEP and Phambili (HVTN 502 and 503): Multicenter test of concept trial using 3-dose regimen; enrollment and further vaccination suspended September 2007	Merck, US Division of AIDS (DAIDS), HIV Vaccine Trials Network (HVTN), South African AIDS Vaccine Initiative (SAAVI)	North and South America, Australia, and South Africa	Replicative incompetent Ad5 with clade B gag, pol and nef inserts.
Phase II			
HVTN 204: Clinical trial evaluating the safety and immunogenicity of DNA plasmid vaccine followed by adenovirus boost; enrollment complete and in follow-up.	DAIDS, HVTN. Manufacturer: VRC	North and South America, South Africa	VRC 6 plasmid DNA of ENV A, B and C plus clade B gag, pol, and nef. Boosted with Ad5 vector with a ENV A, B, and C and a gag-pol fusion insert.
IAVI A002: Placebo-control double-blind trial evaluating the safety and immunogenicity of AAV vaccine	International AIDS Vaccine Initiative (IAVI). Manufacturer: Targeted Genetics	South Africa, Uganda, and Zambia	Adeno-associated virus 2 with a gag, pol, and $\triangle$RT.
ANRS VAC 18: Randomized double-blind trial evaluating the safety and immunogenicity of LIPO-5 versus placebo	Agence Nationale de Recherches sur le SIDA (ANRS). Manufacturer: Aventis Pasteur	France	5 lipopeptides containing CTL epitopes from gag, pol, nef

[1]IAVI Database. http://www.iavireport.org/trialsdb/ (accessed 1 November 2007).
[2]AIDS Vaccine Clearing House. http://aidsvaccineclearinghouse.org/ (accessed 1 November 2007).
[3]Duerr et al. (2006).

Table 3.3 Organizations involved in HIV vaccine research

Name	Description	Location
Aaron Diamond AIDS Research Center (ADARC)	Non-profit basic and clinical science research organization affiliated with Rockefeller University	New York, NY, USA
AIDS Vaccine Advocacy Coalition (AVAC)	Non-profit community- and consumer-based organization that uses public education, policy analysis, advocacy and community mobilization to accelerate the ethical development and global delivery of AIDS vaccines and other HIV prevention option	New York, NY, USA
Armed Forces Research Institute of Medicine (AFRIMS)	Combined entity of the US and Thai militaries dedicated to infectious disease research	Bangkok, Thailand
Agence Nationale de Recherches sur le SIDA (ANRS)	French national public AIDS agency	Paris, France
Centers for Disease Control (CDC)	Part of the US Department of Health and Human Services dedicated to all aspects of public health	Atlanta, GA, USA
Division of Acquired Immunodeficiency Syndrome (DAIDS)	Division of the National Institutes of Allergy and Infectious Diseases (NIAID) dedicated to all aspects of HIV research	Bethesda, MD, USA
European Union	A political and economic community composed of 27 member states primarily located in Europe	Brussels, Belgium, and Strasbourg, France
The Global HIV Vaccine Enterprise	An alliance of independent organizations around the world dedicated to accelerating the development of a preventive HIV vaccine through a (1) shared scientific plan, (2) increased resources, and (3) greater collaboration	Seattle, USA (moving to New York, NY)

Organization	Description	Location
HIV Vaccine Trials Network (HVTN)	DAIDS-sponsored international collaboration network of clinical research sites, centralized laboratories, and statistical and data management facilities, dedicated to HIV vaccine research	Seattle, WA, USA
International AIDS Vaccine Initiative (IAVI)	Not-for-profit, public–private partnership of international clinical research sites and vaccine developers with a primary mission of developing a safe and effective HIV vaccine	New York, NY, USA
National Institute of Allergy and Infectious Diseases (NIAID)	One of 27 institutes of the NIH, NIAID's primary mission is in research to understand, treat, and prevent infectious, immunologic and allergic diseases	Bethesda, MD, USA
National Institutes of Health (NIH)	Part of the US Department of Health and Human Services, NIH is the primary federal agency for conducting and supporting medical research	Bethesda, MD, USA
South African AIDS Vaccine Initiative (SAAVI)	Was established to coordinate the research, development, and testing of AIDS vaccines in South Africa; SAAVI is part of the South African Medical Research Council and is funded by the South African government	South Africa
US Military HIV Research Program (USMHRP)	Dedicated to HIV vaccine development, prevention, disease surveillance, and care and treatment for HIV; part of WRAIR	Washington, DC, USA
Dale and Betty Bumpers Vaccine Research Center (VRC)	Part of the NIH under the NIAID, VRC was established to facilitate research in vaccine development and is dedicated to improving global human health through the rigorous pursuit of effective vaccines for human diseases (not only HIV); it was established by former President Bill Clinton as part of an initiative to develop an AIDS vaccine	Bethesda, MD, USA
Walter Reed Army Institute of Research (WRAIR)	The US Department of Defense and Army biomedical research arm, whose mission is stated as the development of the knowledge, technology and medical material that sustain the combat effectiveness of the warfighter	Washington, DC, USA
World Health Organization (WHO)	The directing and coordinating authority for health within the United Nations system, the WHO is responsible for providing leadership on global health matters, shaping the health research agenda, setting norms and standards, articulating evidence-based policy options, providing technical support to countries, and monitoring and assessing health trends	Geneva, Switzerland

Foundation), European governmental agencies and the commercial sector (primarily large pharmaceutical manufacturers) accounted for about 8–9 percent each of funding, with the remainder coming from other public sector organizations. Although steadily increasing between 1997 and 2004, US governmental funding was level from 2005 onward. Funding is spent on a variety of activities. In 2005, approximately 38.1 percent of funds were allocated to preclinical research, 25.7 percent to basic science research, 21.9 percent to clinical research, 13.1 percent to cohort and site development, and 1.3 percent to advocacy and policy development (HIV Vaccines and Microbicides Resource Tracking Working Group, 2007).

Ethical considerations

A number of ethical issues arise in the conduct of HIV vaccine research, especially issues related to trials in locations with limited resources, inconsistent medical care, and poor access to treatment and medication for HIV-infected individuals. Ethical principles of autonomy, beneficence and justice (combined with international human rights norms) impart significant responsibilities and obligations to those who conduct HIV vaccine trials (Tarantola *et al.*, 2007). As with any trial, HIV vaccine trials are obliged to use methodology in accordance with Good Clinical Practice (GPC) guidelines, follow accepted standards in the protection of human subjects, and insure, through careful monitoring, the safety of trial participants. Furthermore, many of the groups conducting these trials have taken on the added responsibilities of providing aggressive HIV acquisition and sexually transmitted infection (STI) risk-reduction counseling, treating STIs in trial participants, supplying barrier and hormonal contraception, and even arranging for or providing procedural interventions (such as circumcision), all with the goal of reducing acquisition of HIV among trial participants. An additional issue, without clear guidance for trial sponsors, involves the responsibility of medical care for participants. Most care related to trial-associated events is provided by sponsors and researchers. Beyond that, however, stakeholders differ regarding the level of care that can be provided. Sponsors and investigators are concerned about the feasibility and the financial, human and structural resources that would be needed to provide extensive medical care to a large group of participants in resource-limited settings. Additionally, the principals of equipoise and undue influence between trial participants and the rest of the hosting community might arise if participants were to receive care that was superior to that of the surrounding community.

Multiclade issues

Developing "clade-specific" vaccines would be difficult, impractical and costly. International travel is such that even if one could be vaccinated against the predominant clade in one's own region, the possibility of exposure to another clade is real. This problem evokes the ethical challenge of developing vaccines that are likely to be effective in the region where they are tested, yet have broad enough immune coverage to be effective in other regions.

Oversight and review

As in any trial, proper oversight, regulation and protection of human subjects are essential. Regulations and requirements may vary by country. It is important to have independent and competent scientific review within each country (UNAIDS, 2000).

Informed consent is crucial for prospective volunteers, and must include the following:

- reasons for the research;
- questions researchers are trying to answer;
- procedures done during the trial, and how long the trial will last;
- risks involved;
- possible benefits;
- alternative interventions;
- mention of the participant's right to leave the trial at any time (AIDS Vaccine Advocacy Coalition, 2007).

This process should be ongoing, and should involve continued counseling and updates as new information is available related to the trial product or HIV in general.

Local or central Institutional Review Boards (IRBs) or Ethics Committees constitute independent groups of scientists and lay people charged with protecting subjects' rights. By critically examining all aspects of a study prior to implementation and evaluating new information, an IRB can make recommendations or even stop trials. A Data and Safety Monitory Board (DSMB) can act as an additional group tasked with protecting participants and maintaining study integrity. This group can include scientists and statisticians, independent of the trial, as well as community members, ethicists and clinicians. The DSMB reviews study safety and efficacy data regularly during the trial. This review is usually done with knowledge of treatment assignment to ensure that the participants are not exposed to undue risk. The DSMB can recommend that the trial be modified or stopped if there are safety concerns or if certain predetermined criteria are met showing efficacy or no effect.

A number of the internationally recognized codes developed after Nazi experimentation during World War II are used in HIV vaccine trials, including the Nuremberg Code (1947), the Declaration of Helsinki (1962), The Belmont

Report (1979), the Council for International Organizations of Medical Sciences (CIOMS) Guidelines (1982), and Ethical considerations in HIV preventative vaccine research (UNIADS Guidance Document, 2000).

Confidentiality and social harms and benefits

As with any trial, confidentiality is imperative for enrollment, retention and maintaining ethical standards. Stigma associated with participation in HIV vaccine trials persists in certain communities.

Trials generally assess harm and benefit, both perceived and actual, that participants experience. Issues can include discrimination, access to care, and the impact on personal and professional relationships. Assessment of these potential social harms and benefits can help guide trial conduct and design of future trials.

Community involvement

As outlined in the UNAIDS guidance document on the conduct of HIV vaccine trials, involvement of affected communities should occur early in vaccine trial development. This relationship should not be one-way or occur in a single encounter, but should be an ongoing process of information-sharing throughout a trial. This collaboration can be accomplished through the use of focus groups, engagement of community leaders, or development of a Community Advisory Board (CAB) composed of community members who provide input, leadership and guidance at every step of the research (AIDS Vaccine Advocacy Coalition, 2007). Advantages of early community engagement include the following:

- obtaining information on health beliefs of study populations;
- getting input into protocol design that reflects community needs;
- obtaining input into appropriate informed consent documents and process;
- receiving assistance with developing risk reduction strategies;
- developing effective methods to disseminate information;
- establishing trust between the community and researchers;
- ensuring equality in choice of participants;
- guiding decisions with respect to treatment (both for vaccine related and unrelated healthcare issues) and duration of treatment; and
- equality in helping to assure resource provision and vaccine distribution. Careful, close and early community involvement is important to prevent misunderstanding and mistrust.

Counseling in high-risk populations

Careful, consistent and extensive risk-reduction counseling is imperative for ethical and successful HIV vaccine trial conduct. Trial organizers and researchers should make every effort to prevent HIV infection and high-risk behavior through: (1) in-depth counseling on risk factors associated with HIV acquisition, such as STDs; (2) counseling on the use of barrier methods, particularly condoms, as a method to prevent acquisition and transmission of HIV; (3) counseling on the sharing of drug paraphernalia and the availability of needle exchanges; and (4) explicit warnings that the vaccine being examined has not been proven to prevent, and may not prevent, HIV infection, and may make participants more prone to acquiring HIV infection. By facilitating procedures and behaviors that decrease transmission and acquisition of HIV (such as circumcision, safe-sex counseling, condom distribution, post-exposure prophylaxis, and identification and treatment of sexually transmitted infections), researchers may decrease the incidence of HIV infection within their study populations and subsequently increase the sample size necessary to reach certain endpoints and increase trial cost. However, they will meet a moral imperative to make every effort to protect participants.

Care of infected volunteers

Perhaps one of the most difficult issues facing researchers in all HIV prevention trials is the subsequent care of volunteers who become infected, especially in resource-limited settings. In 2003, leaders of the HIV vaccine research community stated their position that those performing HIV vaccine research have a moral imperative to ensure that participants who became HIV infected have access to therapy (Fitzgerald *et al.*, 2003). The logistics and ethical issues of distributing this care in resource-limited settings are complex. Issues that need to be addressed and resolved include:

- equivalence of care in local areas among vaccine trial participants and non-participants;
- funding/providing ART without further exacerbating discrepancies in resource-limited settings;
- undue influence in trial enrollment;
- adequate monitoring and care;
- assurance of uninterrupted supply of therapy; and
- the impact of insuring treatment on small trials conducted by groups with limited resources.

The future of HIV vaccine efforts

The future of HIV vaccine research will be greatly affected by the results of the current Phase IIB and III trials. The failure of the Merck Ad5 vector vaccine to prevent acquisition or reduce HIV viral load has raised questions about the utility of these vectors and T-cell vaccine approaches. During the next several years, laboratory research, and preclinical and early-phase clinical testing, will focus on these issues as well as addressing the concern over potential enhancement of HIV acquisition in those with pre-existing Ad5 immunity receiving an Ad5 vaccine vector. The information obtained about the efficacy of vaccines based on cellular immunity will guide future development. Other issues that will be addressed in trials include:

- the use of DNA plasmids as immunogens;
- appropriate viral vectors (pox virus, adenovirus) and what serotypes to use to optimize immunogenicity;
- the effect of prior immunity to certain viral vectors on immunologic and clinical response to vaccine (including efficacy);
- cross-clade efficacy and how specific vaccines protect against clades other than those from which particular inserts were developed;
- use of novel delivery devices and their effects on efficacy and immunogenicity;
- the use of adjuvants in vaccines to improve efficacy and immunogenicity; and, finally,
- the effects of vaccination on protection against infection, on mitigating infection, and on preventing transmission to others.

Conclusions

HIV vaccine development has been a challenging and daunting endeavor for scientists, public health officials and populations at risk. Many issues need to be addressed prior to the licensure of a vaccine. These issues include: (1) what impact a vaccine that mitigates disease but does not prevent infection might have, and what data would be necessary to support licensure of such a vaccine; (2) how such a vaccine would affect mucosal viral shedding and HIV transmission; (3) if efficacious, how to incorporate such a vaccine into multifaceted prevention programs with vaccine as one prong; and (4) whether a slower progression of disease would aid the implementation of treatment programs in resource-limited settings by making the number of individuals needing therapy more manageable.

The next decade will provide the opportunities for these questions to be answered, and will likely see the proposal of a number of new questions as second- and third-generation vaccines are developed. Multidisciplinary approaches with vaccination as just one strategy may ultimately lead to complete control of this pandemic.

Acknowledgment

The authors thank Katie Skibinski for editorial assistance in the preparation of this chapter.

References

Abbink, P., Lemckert, A. A., Ewald, B. A. *et al.* (2007). Comparative seroprevalence and immunogenicity of six rare serotype recombinant adenovirus vaccine vectors from subgroups B and D. *J. Virol.*, 81, 4654–63.

AIDS Vaccine Advocacy Coalition (2007). AIDS Vaccine Clearing House (http://aidsvaccineclearinghouse.org/), accessed 11-1-07.

Altfeld, M., Kalife, E. T., Qi, Y. *et al.* (2006). HLA alleles associated with delayed progression to AIDS contribute strongly to the initial CD8(+) T cell response against HIV-1. *PLoS.Med.*, 3, e403.

Anderson, R. and Hanson, M. (2005). Potential public health impact of imperfect HIV type 1 vaccines. *J. Infect. Dis.*, 191(Suppl. 1), S85–96.

Baba, T. W., Jeong, Y. S., Pennick, D. *et al.* (1995). Pathogenicity of live, attenuated SIV after mucosal infection of neonatal macaques. *Science*, 267, 1820–5.

Belshe, R. B., Graham, B. S., Keefer, M. C. *et al.* (1994). Neutralizing antibodies to HIV-1 in seronegative volunteers immunized with recombinant gp120 from the MN strain of HIV-1. NIAID AIDS Vaccine Clinical Trials Network. *J. Am. Med. Assoc.*, 272, 475–80.

Brave, A., Boberg, A., Gudmundsdotter, L. *et al.* (2007). A new multi-clade DNA prime/recombinant MVA boost vaccine induces broad and high levels of HIV-1-specific CD8(+) T-cell and humoral responses in mice. *Mol. Ther.*, 15, 1724–33.

Brenchley, J. M., Schacker, T. W., Ruff, L. E. *et al.* (2004). CD4+ T cell depletion during all stages of HIV disease occurs predominantly in the gastrointestinal tract. *J. Exp. Med.*, 200, 749–9.

Brenner, B. G., Roger, M., Routy, J. P. *et al.* (2007). High rates of forward transmission events after acute/early HIV-1 infection. *J. Infect. Dis.*, 195, 951–9.

Burton, D. R., Stanfield, R. L. and Wilson, I. A. (2005). Antibody vs HIV in a clash of evolutionary titans. *Proc. Natl Acad. Sci. USA*, 102, 14,943–8.

Calarota, S. A. and Weiner, D. B. (2004). Enhancement of human immunodeficiency virus type 1-DNA vaccine potency through incorporation of T-helper 1 molecular adjuvants. *Immunol. Rev.*, 199, 84–99.

Cao, H., Kanki, P., Sankale, J. L. *et al.* (1997). Cytotoxic T-lymphocyte cross-reactivity among different human immunodeficiency virus type 1 clades: implications for vaccine development. *J. Virol.*, 71, 8615–23.

Casimiro, D. R., Wang, F., Schleif, W. A. *et al.* (2005). Attenuation of simian immunodeficiency virus SIVmac239 infection by prophylactic immunization with dna and recombinant adenoviral vaccine vectors expressing Gag. *J. Virol.*, 79, 15,547–55.

Catanzaro, Coup, R. A., Roederer, M. *et al.* (2006). Phase I safety and immunogenicity evaluation of a multiclade HIV-1 candidate vaccine delivered by a replication-defective recombinant adenovirus vector. *J. Infect. Dis.*, 194, 1638–49.

CDC (2003). Update: cardiac-related events during the civilian smallpox vaccination program – United States, 2003. *Morb. Mortal. Wkly Rep.*, 52(21), 492–6.

Connor, R. I., Korber, B. T., Graham, B. S. *et al.* (1998). Immunological and virological analyses of persons infected by human immunodeficiency virus type 1 while participating in trials of recombinant gp120 subunit vaccines. *J. Virol.*, 72, 1552–76.

Coordinating Committee of the Global HIV/AIDS Vaccine Enterprise (2005). The Global HIV/AIDS Vaccine Enterprise: scientific strategic plan. *PLoS.Med.*, 2, e25.

Coplan, P. M., Gupta, S. B., Dubey, S. A. *et al.* (2005). Cross-reactivity of anti-HIV-1 T cell immune responses among the major HIV-1 clades in HIV-1-positive individuals from 4 continents. *J. Infect. Dis.*, 191, 1427–34.

Daniel, M. D., Kirchhoff, F., Czajak, S. C. *et al.* (1992). Protective effects of a live attenuated SIV vaccine with a deletion in the nef gene. *Science*, 258, 1938–41.

Deacon, N. J., Tsykin, A., Solomon, A. *et al.* (1995). Genomic structure of an attenuated quasi species of HIV-1 from a blood transfusion donor and recipients. *Science*, 270, 988–91.

Duerr, A., Wasserheit, J. and Corey, L. (2006). HIV vaccines: new frontiers in vaccine development. *Clin. Infect. Dis.*, 43, 500–11.

Fellay, J., Shianna, K. V., Ge, D. *et al.* (2007). A whole-genome association study of major determinants for host control of HIV-1. *Science*, 317, 944–7.

Fitzgerald, J. C., Gao, G. P., Reyes-Sandoval, A. *et al.* (2003). A simian replication-defective adenoviral recombinant vaccine to HIV-1 gag. *J. Immunol.*, 170, 1416–22.

Flynn, N. M., Forthal, D. N., Harro, C. D. *et al.* (2005). Placebo-controlled phase 3 trial of a recombinant glycoprotein 120 vaccine to prevent HIV-1 infection. *J. Infect. Dis.*, 191, 654–65.

Follmann, D., Duerr, A., Tabet, S. *et al.* (2006). Endpoints and regulatory issues in HIV vaccine clinical trials: lessons from a workshop. *J. Acquir. Immune. Defic. Syndr.*, 44, 49–60.

Garber, D. A., Silvestri, G. and Feinberg, M. G. (2008). Prospects for an AIDS vaccine: three big questions, no easy answers. In: A. Duerr *et al.* *Sexually Transmitted Diseases*, 4th edn. New York, NY: McGraw Hill, in press.

Gaschen, B., Taylor, J., Yusim, K. *et al.* (2002). Diversity considerations in HIV-1 vaccine selection. *Science*, 296, 2354–60.

Gilbert, P. B., DeGruttola, V. G., Hudgens, M. G. *et al.* (2003). What constitutes efficacy for a human immunodeficiency virus vaccine that ameliorates viremia: issues involving surrogate end points in phase 3 trials. *J. Infect. Dis.*, 188, 179–93.

Gilbert, P. B., Peterson, M. L., Follmann, D. *et al.* (2005). Correlation between immunologic responses to a recombinant glycoprotein 120 vaccine and incidence of HIV-1 infection in a phase 3 HIV-1 preventive vaccine trial. *J. Infect. Dis.*, 191, 666–77.

Gupta, S. B., Jacobson, L. P., Margolick, J. B. *et al.* (2007). Estimating the benefit of an HIV-1 vaccine that reduces viral load set point. *J. Infect. Dis.*, 195, 546–50.

HIV Vaccine Trials Network (HVTN) (2007). HIV Vaccine Trials Network (HVTN) website (http://www.hvtn.org), accessed 11-1-2007.

HIV Vaccines and Microbicides Resource Tracking Working Group (2007). HIV Vaccines and Microbicides Resource Tracking Working Group website (http://www.hivresourcetracking.org), accessed 11-1-2007.

Hulskotte, E. G., Geretti, A. M. and Osterhaus, A. D. (1998). Towards an HIV-1 vaccine: lessons from studies in macaque models. *Vaccine*, 16, 904–15.

International AIDS Vaccine Initiative (2006). AIDS Vaccine Blueprint (http://www.iavi.org/viewfile.cfm?fid=41059), accessed 11-1-2007.

International AIDS Vaccine Initiative (IAVI) (2007). IAVI Report (http://www.iavireport.org/), accessed 11-1-2007.

Jin, X., Bauer, D. E., Tuttleton, S. E. *et al.* (1999). Dramatic rise in plasma viremia after CD8(+) T cell depletion in simian immunodeficiency virus-infected macaques. *J. Exp. Med.*, 189, 991–8.

Johnson, P. R., Schnepp, B. C., Connell, M. J. *et al.* (2005). Novel adeno-associated virus vector vaccine restricts replication of simian immunodeficiency virus in macaques. *J. Virol.*, 79, 955–65.

Kahn, J. O. and Walker, B. D. (1998). Acute human immunodeficiency virus type 1 infection. *N. Engl. J. Med.*, 339, 33–9.

Kim, D., Elizaga, M. and Duerr, A. (2007). HIV vaccine efficacy trials: towards the future of HIV prevention. *Infect. Dis. Clin. North Am.*, 21, 201–17, x.

Kirchhoff, F., Greenough, T. C., Brettler, D. B. *et al.* (1995). Brief report: absence of intact nef sequences in a long-term survivor with nonprogressive HIV-1 infection. *N. Engl. J. Med.*, 332, 228–32.

Klausner, R. D., Fauci, A. S., Corey, L. *et al.* (2003). Medicine. The need for a global HIV vaccine enterprise. *Science*, 300, 2036–9.

Kojima, Y., Xin, K. Q., Ooki, T. *et al.* (2002). Adjuvant effect of multi-CpG motifs on an HIV-1 DNA vaccine. *Vaccine*, 20, 2857–65.

Korber, B., Gaschen, B., Yusim, K. *et al.* (2001). Evolutionary and immunological implications of contemporary HIV-1 variation. *Br. Med. Bull.*, 58, 19–42.

Kwong, P. D., Wyatt, R., Robinson, J. *et al.* (1998). Structure of an HIV gp120 envelope glycoprotein in complex with the CD4 receptor and a neutralizing human antibody. *Nature*, 393, 648–59.

Luckay, A., Sidhu, M. K., Kjeken, R. *et al.* (2007). Effect of plasmid DNA vaccine design and in vivo electroporation on the resulting vaccine-specific immune responses in rhesus macaques. *J. Virol.*, 81, 5257–69.

McKenna, P. M., Koser, M. L., Carlson, K. R. *et al.* (2007). Highly attenuated rabies virus-based vaccine vectors expressing simian-human immunodeficiency virus89.6P Env and simian immunodeficiency virusmac239 Gag are safe in rhesus macaques and protect from an AIDS-like disease. *J. Infect. Dis.*, 195, 980–8.

Mehandru, S., Wrin, T., Galovich, J. *et al.* (2004). Neutralization profiles of newly transmitted human immunodeficiency virus type 1 by monoclonal antibodies 2G12, 2F5, and 4E10. *J. Virol.*, 78, 14,039–42.

Merck (2007). Vaccination and Enrollment Are Discontinued in Phase II Trials of Merck's Investigational HIV Vaccine Candidate. Press release, 21-9-2007, available at http://www.merck.com/newsroom/press_releases/research_and_development/2007_0921.html . (accessed 11-1-2007).

Meseda, C. A., Stout, R. R. and Weir, J. P. (2006). Evaluation of a needle-free delivery platform for prime-boost immunization with DNA and modified vaccinia virus ankara vectors expressing herpes simplex virus 2 glycoprotein D. *Viral Immunol.*, 19, 250–9.

Moore, J. P., Cao, Y., Qing, L. *et al.* (1995). Primary isolates of human immunodeficiency virus type 1 are relatively resistant to neutralization by monoclonal antibodies to gp120, and their neutralization is not predicted by studies with monomeric gp120. *J. Virol.*, 69, 101–9.

Morgan, C., Bailer, R., Metch, B. *et al.* (2005). International seroprevalence of neutralizing antibodies against adenovirus serotypes 5 and 35. AIDS Vaccine 2005 Conference, Montreal, Canada.

Mullins, J. I., Nickle, D. C., Heath, L. *et al.* (2004). Immunogen sequence: the fourth tier of AIDS vaccine design. *Expert. Rev. Vaccines*, 3(Suppl.), S151–159.

Pierson, T., McArthur, J. and Siliciano, R. F. (2000). Reservoirs for HIV-1: mechanisms for viral persistence in the presence of antiviral immune responses and antiretroviral therapy. *Annu. Rev. Immunol.*, 18, 665–708.

Pitisuttithum, P., Gilbert, P., Gurwith, M. *et al.* (2006). Randomized, double-blind, placebo-controlled efficacy trial of a bivalent recombinant glycoprotein 120 HIV-1 vaccine among injection drug users in Bangkok, Thailand. *J. Infect. Dis.*, 194, 1661–71.

Quinn, T. C., Wawer, M. J., Sewankambo, N. *et al.* (2000). Viral load and heterosexual transmission of human immunodeficiency virus type 1. Rakai Project Study Group. *N. Engl. J. Med.*, 342, 921–9.

Rao, S. S., Gomez, P., Mascola, J. R. *et al.* (2006). Comparative evaluation of three different intramuscular delivery methods for DNA immunization in a nonhuman primate animal model. *Vaccine*, 24, 367–73.

Redfield, R. R., Wright, D. C., James, W. D. *et al.* (1987). Disseminated vaccinia in a military recruit with human immunodeficiency virus (HIV) disease. *N. Engl. J. Med.*, 316, 673–6.

Salvi, R., Garbuglia, A. R., Di Caro, A. *et al.* (1998). Grossly defective nef gene sequences in a human immunodeficiency virus type 1-seropositive long-term nonprogressor. *J. Virol.*, 72, 3646–57.

Schmitz, J. E., Kuroda, M. J., Santra, S. *et al.* (1999). Control of viremia in simian immunodeficiency virus infection by CD8+ lymphocytes. *Science*, 283, 857–60.

Shiver, J. W. and Emini, E. A. (2004). Recent advances in the development of HIV-1 vaccines using replication-incompetent adenovirus vectors. *Annu. Rev. Med.*, 55, 355–72.

Shiver, J. W., Fu, T. M., Chen, L. *et al.* (2002). Replication-incompetent adenoviral vaccine vector elicits effective anti-immunodeficiency-virus immunity. *Nature*, 415, 331–5.

Singh, M. (2006). No vaccine against HIV yet – are we not perfectly equipped? *Virol. J.*, 3, 60.

Singh, M. and Srivastava, I. (2003). Advances in vaccine adjuvants for infectious diseases. *Curr. HIV Res.*, 1, 309–20.

Tarantola, D., Macklin, R., Reed, Z. H. *et al.* (2007). Ethical considerations related to the provision of care and treatment in vaccine trials. *Vaccine*, 25, 4863–74.

UNAIDS (2000). *Ethical Considerations in HIV Preventive Vaccine Research.* UNAIDS/04.07E.

US National Library of Medicine (2007). ClinicalTrials.gov – information on clinical trials and human research studies, available at http://clinicaltrials.gov/ (accessed 11-1-2007).

Wei, X., Decker, J. M., Wang, S. *et al.* (2003). Antibody neutralization and escape by HIV-1. *Nature*, 422, 307–12.

Whitney, J. B. and Ruprecht, R. M. (2004). Live attenuated HIV vaccines: pitfalls and prospects. *Curr. Opin. Infect. Dis.*, 17, 17–26.

Microbicides

4

Ian McGowan

Increasingly, AIDS is becoming a disease of poverty and more particularly of women. It is now estimated that in sub-Saharan Africa young women are three times more likely to be HIV infected than age-matched males in the same community (Quinn and Overbaugh, 2005). The impact of the "ABC" program on HIV incidence has been controversial. It may have worked in Uganda, but it is clear that in many settings abstinence is not happening, men are not being faithful, and they refuse to use condoms. This situation is further compounded by the high incidence of coercive sexual activity, including rape, and the low socioeconomic status of women (Csete, 2004). As a consequence, it has been suggested that the ABC approach needs to be supplemented by focusing on Gender relations, Economic empowerment and Migratory behavior of male partners, or "GEM" (Dworkin and Ehrhardt, 2007). Implementation of these types of structural interventions will require decades, and there is an urgent need to develop modalities of HIV prevention that can be used by women with or without partner consent. The holy grail of HIV prevention is a safe and effective HIV vaccine. However, with the recent closure of the Merck adenovirus Phase III study due to lack of efficacy, the vaccine field remains uncertain about the future. Vaccine trials are increasingly moving towards evaluating their potential for disease modification rather than disease prevention (Duerr et al., 2006; see also Chapter 20). The unambiguous determination that male circumcision significantly reduces male acquisition of HIV infection is good news for men, but the operational roll-out and significance for women of this intervention remains unclear (Sawires et al., 2007; see also Chapter 17). In this setting, microbicides are an important prevention technology that is in a state of rapid evolution.

Microbicides are products (formulated as gels, sponges, films or rings) that can be applied to the vaginal or rectal mucosa with the goal of preventing or significantly reducing the acquisition of sexually transmitted infections (STIs), including HIV. Zena Stein first proposed the concept of a topical "virucide" that might block HIV-1 transmission in 1990 (Stein, 1990). The idea grew out of the reproductive health research community and, not surprisingly, one of the first products to be considered as a microbicide was the spermicidal agent nonoxynol-9 (N-9). Unfortunately, this agent was subsequently shown to be neither safe nor effective (Van Damme *et al.*, 2002; Hillier *et al.*, 2005) as a microbicide, and development for this indication was terminated although it is still marketed as a spermicide. Despite this disappointing start, the field continued to evaluate microbicide candidates. The first generation of microbicides included agents that were primarily detergents or surfactants, such as N-9, which damaged the viral envelope; products that enhanced vaginal defenses through mechanisms such as maintenance of vaginal acidity; or products that blocked the virus from attaching to cellular targets in the genital mucosa. Increasingly, the microbicide pipeline is linked to a growing understanding of the pathogenesis of HIV-1 transmission and the identification of viral and cell receptor targets (D'Cruz and Uckun, 2004). At this point in time there are no licensed microbicides, and the HIV prevention field is still waiting for the successful conclusion of a microbicide clinical trial to indicate that this strategy is feasible. The field is complex both from a scientific and an operational perspective. There have been a number of recent setbacks where studies have undergone premature termination because of safety concerns or lack of efficacy. These are discussed in more detail below, as they provide useful case studies regarding microbicide development.

The biological rationale for microbicides

Sexual transmission of HIV-1 is initiated when semen containing cell-free or cell-associated virus is deposited in the vagina or rectum, or when virus passes from these compartments to the insertive partner (Shattock and Moore, 2003). The exact mechanism of viral transmission remains uncertain, and may well involve multiple pathways. The vaginal epithelium is a stratified squamous epithelium that does not possess a traditional receptor for HIV-1, but the vaginal tissue underlying the epithelium contains multiple targets for the virus (Figure 4.1). These targets include mucosal Langerhans cells (dendritic cells expressing the HIV-1 CD4 receptor and the CCR5 co-receptor), T cells and macrophages. Passage of virus from the lumen to the cellular targets may be facilitated by binding of virus to dendritic cell projections that extend into the epithelial compartment, with subsequent presentation to subepithelial target cells (Shattock and Moore, 2003). A more mundane but equally likely explanation is that virus accesses the subepithelial space through epithelial breaks caused by local trauma and/or STIs. This would help explain the increased risk of HIV-1 transmission associated with the presence of concomitant STIs such as HSV-2 infection (Corey *et al.*, 2004). The morphology of the genital tract epithelium changes at the endocervical junction, where the stratified squamous epithelium transitions to a single layer of columnar

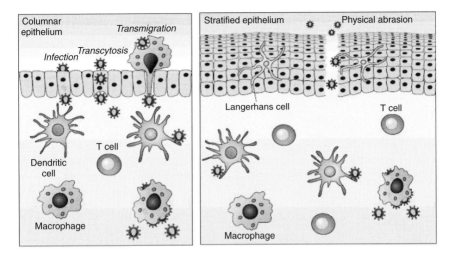

Figure 4.1 Mucosal transmission of HIV-1 probably occurs through multiple pathways. Target cells in the subepithelial area include CD4+ lymphocytes, macrophages, and dendritic cells. Mucosal inflammation and epithelial disruption secondary to STIs increase the risk of HIV-1 transmission through recruitment of additional target cells.

epithelium (Figure 4.1). This area is probably much more vulnerable to HIV-1 infection. Following initial infection, local viral replication is followed by dissemination of virus to the regional lymph nodes, at which point systemic infection is established. Animal models have suggested that initial infection can occur within 1 hour of exposure, and dissemination within 24 hours (Hu *et al.*, 2000; Veazey *et al.*, 2003). A successful microbicide will have to address these challenges. In particular, it will have to (1) outdistance the virus (protect mucosal surfaces at risk of HIV-1 transmission), (2) outlast the virus (provide an adequate therapeutic window such that virus cannot infect once local concentrations of a microbicide fall below a therapeutic level), and (3) prevent dissemination of infected cells from the local mucosa to the regional lymph nodes. Whilst these challenges may seem insurmountable, they are compensated for by the fact that HIV-1 infection is a relatively inefficient process and that the female genital tract is a relatively small anatomical area to protect.

Protecting the rectal compartment is a more challenging problem. A single layer of columnar epithelium lines the rectosigmoid. Anal intercourse is often associated with local mucosal trauma, and the subepithelial lamina propria is rich in target cells. Many of these cells have an activated phenotype, which makes them extremely susceptible to HIV-1 infection (Poles *et al.*, 2001). This helps to explain why receptive anal intercourse carries the highest risk of HIV-1 transmission. It is not clear which region of the rectosigmoid will require protection, and obviously this is a larger area to protect than the female genital tract. However, gastroenterologists routinely prescribe topical agents to patients with inflammatory bowel disease who have left-sided colitis (Haghighi and Lashner, 2004) and so there is a foundation of formulation science that may transfer to rectal microbicide development. Despite these scientific challenges, two recent monkey infection studies have shown that topical application of either cyano-virin-N or tenofovir can significantly reduce the incidence of infection following rectal exposure to SIV or RT-SHIV (monkey analogues of the human HIV virus) respectively (Tsai *et al.*, 2003; Shattock, 2006).

A detailed discussion of the molecular interactions between HIV and its cel-lular targets is beyond the scope of this chapter. However, recent insights into this process are proving critical in identifying new targets for the generation of highly specific antiviral microbicidal agents (Klasse *et al.*, 2008).

Microbicide development

A fundamental difference between the development of microbicides for HIV pre-vention and the development of drugs for treating HIV infection is that in the former case trial participants are healthy HIV-uninfected individuals who are being evaluated, whereas in the latter case participants are HIV-infected indi-viduals who may or may not have HIV-associated disease. The risk/benefit ratio for HIV prevention using microbicides is focused on developing products, essentially without any adverse safety profile, that are highly protective against HIV infection. In contrast, new HIV drugs may be licensed with known safety issues because the therapeutic value is perceived as greater than any safety prob-lem. An additional hurdle for microbicides is that any mucosal inflammation or genital irritation that occurs secondary to microbicide use might increase rather than decrease HIV transmission. In practice, this means that the ideal microbi-cide needs to be inert from a safety perspective but highly potent from a viro-logical perspective. It might be assumed that the extensive preclinical and early clinical assessment of candidate microbicide would be sufficient to exclude the suboptimal products before moving into Phase IIB/III effectiveness studies, but unfortunately this may not be the case. Subtle immunological changes in the local genital mucosa, such as recruitment of target cells and increases in cellular recep-tors for HIV, are not routinely evaluated in preclinical studies, but these potential changes may be sufficient to increase the risk of HIV acquisition. Perversely, this enhanced risk of HIV acquisition may only be recognized in Phase IIB/III studies that evaluate the effectiveness of the microbicide in preventing HIV infection. An ideal candidate will quickly show a decreased incidence of new HIV infections in the active treatment arm compared to the placebo. However, to date Phase IIB/III studies of N-9, C31G (Savvy®), cellulose sulfate and carraguard have shown no difference between active and placebo. Indeed, an increased incidence of HIV infection was observed in the active arms of the N-9, study and a trend towards increased infection was seen in the cellulose sulfate study. Despite theoretical and actual concerns about the safety profile of first-generation microbicides, three large Phase IIB/III studies have been completed (Carraguard, Population Council) or are ongoing (PRO 2000 (0.5 percent) and Buffergel, Microbicide Trials Network HPTN-035 study; and PRO 2000 (0.5 percent and 2 percent), the UK Medical Research Council (MRC) MDP-301 study). These studies undergo frequent independent review, including unblinded assessment of the clinical data if needed. In February 2008, the UK MRC announced that enrollment into the 2 percent PRO 2000 arm of the MDP-301 study would be stopped because the reviewers felt that there was insufficient evidence of benefit to justify continuing this arm. More optimistically, the second generation of microbicide candidates (primarily HIV specific reverse transcriptase inhibitors) appears to be highly potent and less likely to induce mucosal inflammation.

As discussed above, future microbicides should not cause epithelial damage, local irritation or disruption of local ecology, but should have a high selectivity index and, for certain classes, a high genetic barrier to resistance. Guidance documents for the preclinical and clinical phases of microbicide development have been published (Mauck *et al.*, 2001; Lard-Whiteford, 2004; US Food and Drug Administration, 2006). These studies need to satisfy the regulatory authorities that the sponsor is able to manufacture a product that is stable at the ambient temperatures that might be anticipated in the developing world, which has no evidence of teratogenicity (the ability to induce fetal abnormality) or carcinogenicity (induce tumors), and is safe and effective. Clinical evaluation of systemic absorption is an important consideration, especially for the reverse transcriptase (RT) inhibitor class of microbicide candidates, as this may have implications for both safety and the potential for resistance. Pregnant or lactating females and subjects with renal or hepatic impairment should be excluded from early phases of microbicide development. Clinical safety monitoring in microbicide trials usually includes physical assessment (including gynecological examination) and laboratory testing. The frequency of evaluation will be higher in Phase I/II studies, to provide an early detailed assessment of product safety profile. Phase I and II safety studies routinely involve monthly evaluations including colposcopy. Following review by a data and safety monitoring board (DSMB), the product then moves into the next phase (Phase IIB or III) to evaluate whether the product reduces HIV transmission. Here, the intensity of safety monitoring is significantly reduced (quarterly evaluations and no colposcopy). One exception is the Phase IIB/III evaluation of RT microbicides, where monthly HIV testing may be required to reduce the risk of acquired HIV resistance in participants who seroconvert while receiving study product.

Microbicide development is beginning to explore the concept of combination microbicides. In this situation, it is important to ensure that each component is contributing to activity, and that two products (or more) are not more toxic than the individual product. Given two products (A and B), the trial design would be A vs B vs AB vs vehicle, and the trial outcome would need to demonstrate

AB > A and AB > B. Unfortunately, the regulatory requirement to independently assess microbicide effectiveness for components of a combination product may result in significant delays in completing the clinical trials needed to license combination products (Coplan *et al.*, 2004).

Unprotected anal intercourse (AI) is the primary risk factor for HIV transmission in the MSM population, and drives much of the HIV epidemic in the Americas and Western Europe. However, the prevalence of AI in heterosexual couples is under-appreciated, and most commonly does not involve the use of condoms. A recent study of almost 13,000 heterosexual Americans documented 35–40 percent prevalence of ever having practiced AI (Mosher *et al.*, 2005). The rectal mucosa is extremely vulnerable to HIV infection, and product that may have a reasonable safety profile in the vagina may cause major problems in the rectum. However, once vaginal microbicides are available, it is clear they will be used in AI. As a consequence, there is a clear need to integrate rectal safety studies into vaginal microbicide development.

The rectal and vaginal compartments differ with respect to anatomy, histology, microbiology and physiology. The vagina is a relatively confined space, whereas the colon is an open-ended tube extending from the anus to the small intestine. Recent data suggest that semen simulants administered rectally may travel to the splenic flexure, a region some 60 cm from the anus (Hendrix *et al.*, 2008). The potential implication of this finding is that a rectal microbicide may need to provide colonic protection from the rectum to the splenic flexure.

The evaluation of candidate microbicides for rectal safety is a work in progress. Microbicides have been evaluated rectally in a number of animal model systems, including mice (Phillips and Zacharopoulos, 1998) and non-human primates (Patton *et al.*, 2002, 2004, 2006). Epithelial damage is the main criterion for toxicity, but Zeitlin and colleagues have also documented increased vulnerability to infections such as HSV-2 as another index of microbicide toxicity (Zeitlin *et al.*, 2001). Human intestinal explants provide a useful *ex vivo/in vitro* means to assess microbicide toxicity. Intestinal tissue is obtained from surgical resection specimens or endoscopic biopsies and set up in culture. The test agent is added to the explant and incubated for a variable period of time. Assessment of cytotoxicity can be performed on the basis of histology or the MTT assay. Abner and Fletcher have both recently published detailed descriptions of this approach to toxicity assessment (Abner *et al.*, 2005; Fletcher *et al.*, 2006).

Human rectal safety studies are extremely limited. Tabet and colleagues evaluated N-9 but, interestingly, saw little histological evidence of toxicity (Tabet *et al.*, 1999). One possible issue was that the collection of rectal tissue occurred 12 hours after exposure, and N-9 induced epithelial disruption might be a very early phenomenon. In contrast, Phillips and colleagues clearly documented epithelial damage using a combination of histology and collection of rectal lavage (Phillips *et al.*, 2000, 2004; Sudol and Phillips, 2004). It is unclear what techniques will be used in future rectal safety studies. Epithelial disruption may be a relatively crude index of microbicide toxicity. It is conceivable that products may induce immunological changes (increased cell activation, cell recruitment, upregulation of HIV-1 co-receptors) that increase the risk of HIV transmission but are not apparent using standard histological techniques. Formal Phase I rectal safety studies with the RT-inhibitor UC-781 are currently ongoing, with data expected in 2008.

The microbicide pipeline

It is estimated that there are approximately 60–80 candidate microbicides in development (McGowan, 2006). However, many of these candidates have only been evaluated in the test tube and have not progressed to animal or human studies. In addition, many candidates will have poor physicochemical properties for formulation or simply be too expensive to manufacture for a developing world market, and may never progress to preclinical or clinical studies. It has been estimated that 10,000 candidates are needed to develop a single licensed drug. The major candidate attrition occurs at the discovery level, and so even if the microbicide discovery pipeline has 50 candidates, this would be a fraction of the number that would be needed in a conventional pharmaceutical research platform. Pipeline expansion remains a major priority in the field, especially as combination approaches become more established.

Microbicides can be classified by their primary mechanism of action:

1. Vaginal defense enhancers help maintain the vaginal pH in an acid range, or facilitate colonization of vaginal flora with lactobacilli.
2. Surfactants or detergents disrupt microbial cell membranes.
3. Entry or fusion inhibitors target viral epitopes or cell receptors (CD4, CCR5, CXCR4) to prevent the sequence of viral binding, fusion, and entry that leads to cell infection.
4. Current replication inhibitors work by inhibiting HIV-1 reverse transcriptase and preventing viral replication. Some are already licensed for the treatment of chronic HIV infection; others have poor oral bioavailability and are being exclusively developed as topical microbicides.
5. Finally, some products such as Praneem, an Indian polyherbal microbicide, have no clear mechanism of action.

The current and potential microbicide pipeline has been reviewed extensively (McGowan, 2006; Klasse *et al.*, 2008), and is summarized in Table 4.1.

Initial formulation of microbicides was based on product availability, and the current range of microbicides in late-stage clinical development is restricted to gel formulations. Other options include foams, suppositories, films and vaginal rings (Garg *et al.*, 2003a, 2003b; Neurath *et al.*, 2003; Woolfson *et al.*, 2006).

91

Table 4.1 Microbicide pipeline by class of action

Stage of development	Membrane disruption	Defense enhancers	RT inhibitors	Entry/fusion inhibitors		Uncertain mechanism
Clinical*		MucoCept Acidform™ Buffergel™	Tenofovir TMC-120 UC-781 Carraguard™ -MIV-150	Carraguard™ VivaGel™	Cellulose acetate phthalate Invisible condom™ PRO-2000	Praneem™
Preclinical discovery	β-cyclodextrin Nisin Retrocyclins Octylglycerol Lactoferrin		MC1220 C-731, 988 PHI-236, PHI-346 PHI-443 S-Dabo	C85FL K5-N, OS(H) & K5OS(H) SAMMA Novaflux Porphyrins PSC RANTES & RANTES analogues BMS-806 BMS-378806 CMPD167 Cyclotriazadisulfonamide Aptamers ADS-J1 Zinc-carageenan Plant lectins Polystyrene sulphonate C52L	ICAM-1 B12, 2G12 2F5, 4E10 CD4 IgG2 T20 T-1249 SCH-C, D UK-427, 857 AMD3100 SFD-1 Bicyclams Zinc finger inhibitors siRNA Griffithsia Scytovirin Soluble DC-SIGN Nanobodies™	Sodium rutin sulfate SPM8CHAS BIL PSMA / PEHMB Magainin PVAS ZCS

Note: *Microbicide candidates with planned, ongoing, or completed human clinical trials.

To some extent the choice of formulation is driven by the chemical characteristics of the microbicide candidate, consumer preference, performance characteristics of the final product and economic considerations. Most microbicide studies now include detailed acceptability assessments of both female and male participants (Mantell *et al.*, 2005; Severy *et al.*, 2005). Microbicide development now often includes assessment of rheological properties of candidate formulations (Owen *et al.*, 2000) and *in vivo* imaging studies of gel coverage (Pretorius *et al.*, 2003; Barnhart *et al.*, 2004). Although the majority of studies have focused on vaginal microbicides, parallel rectal microbicide studies are now ongoing.

Challenges to microbicide development

Retention

Clinical trial retention can be defined as the proportion of patients who are evaluable for the primary study endpoint. Poor study retention can profoundly affect a study's ability to demonstrate product efficacy. As a general rule, the ratio of study endpoints to lost-to-follow-up rate (LFUP) should be greater than one. For example, in HPTN-035, an ongoing Phase IIB efficacy study of PRO 2000 and BufferGel sponsored by the NIH, it has been estimated that the study will need to observe 192 new HIV infections to demonstrate efficacy. Consequently, there should not be more than 192 participants lost to follow-up. With an enrollment base of 3220, retention rates will have to be >94 percent. This is clearly a major challenge for study execution. There can be difficulties in tracking down participants, and the exercise can be costly and/or time-consuming. However, ongoing studies have been able to meet these demanding retention rates.

Adherence

Adherence is another important aspect of clinical trial conduct. Adherence can be defined as the proportion of participants who comply with the clinical trial protocol in terms of using the study product. In a study where microbicides were meant to be used in a coital fashion, this would mean that the participant used study product for each sexual act. Non-adherence can be due to a number of causes; clinicians may ask the participant to stop taking product because of an adverse event or because she is pregnant, or the participant may decide herself to stop taking product. As the number of participants off product increases, the ability to demonstrate product effectiveness falls. Put more directly, when 25 percent of participants are off product, the sample size needs to double to allow the study to demonstrate effectiveness. This course of action is clearly not possible, and so the impact of non-adherence is to lead to study failure. Increasingly, trialists are being asked to develop more objective evidence of product adherence. In studies involving orally administered study products, it is possible to use electronic medication event monitoring systems (MEMS) (Santschi *et al.*, 2007). This approach is not possible with a gel product, but studies have been conducted in which an applicator-based dye test is used to provide information on whether the applicator has been inserted intravaginally (Wallace *et al.*, 2004, 2007). A more sophisticated approach might be to use vaginal rings manufactured with a microchip that could collect comprehensive information about insertion of the product and, possibly, sexual exposure. The concept of directly observed therapy (DOT) has been an important component of tuberculosis (TB) treatment programs (Saltini, 2006), and has been considered as a means of delivering antiretrovirals (Mitty *et al.*, 2003). A modification of DOT has been proposed as a means to monitor adherence in microbicide trials. This approach would probably involve daily contact with the participants and collection of used applicators. A recent Cochrane assessment of effectiveness of TB DOT has produced equivocal results (Volmink and Garner, 2006), and it is unclear how successful this approach might be in microbicide trials.

Pregnancy

Pregnancy is a major challenge in the execution and analysis of microbicide clinical trials. Microbicide efficacy studies require sexual activity, and pregnancy is one natural consequence of sexual activity. Unfortunately, pregnancy may affect participant retention and the power of the study to demonstrate an effect. Because microbicides are investigational agents and their effect on the human fetus is unknown, regulatory authorities mandate that once a woman is pregnant, she should not use the gel. An important issue that has recently emerged is the concept of a chemical pregnancy. This is an event in which a woman has a positive pregnancy test, in the absence of clinical symptoms, which does not progress to parturition. This is partly a consequence of the use of exquisitely sensitive human chorionic gonadotrophin (hCG) tests in many studies. A significant proportion of chemical pregnancies never proceed to delivery. A recent NIH-sponsored site preparedness study (HPTN-055) followed cohorts of women for 1 year; in this study there were 105 pregnancies but only 32 (30 percent) reached the third trimester. Again, these data suggest there is a natural attrition rate for women becoming pregnant, and overzealous identification of very early pregnancy (with the implied cessation of clinical trial participation) may be unnecessary and damaging to the integrity of microbicide clinical trials. The counter-argument is that the first trimester is the most vulnerable period for teratogenicity, and so it could be argued that this level of ascertainment is appropriate. One possible pathway through this challenging area is to ensure that reproductive toxicology is completed before efficacy studies are initiated. In this setting, it would be possible to continue study drug administration during pregnancy. A more practical approach is to optimize the use of contraception among trial participants. Contraceptive technology varies in the ideal and typical pregnancy rates. Implantable hormonal methods are the most effective means of avoiding pregnancy, with a 0.05 percent pregnancy rate in the first year of use. In contrast, condoms have an ideal rate of 2 percent but a typical rate of 15 percent. In HPTN-055, only 33 percent of the women were using implantable contraceptives at baseline, suggesting that significant improvements in contraceptive behaviors are needed. Pregnancy remains one of the most important challenges in microbicide development and, clearly, innovative approaches to addressing this issue are needed.

Viral resistance

Although RT microbicides have the potential to be highly potent and safe microbicides, their Achilles' heel is likely to be the potential for inducing viral resistance. HIV-1 is known to have an error-prone reverse transcriptase enzyme that leads to the production of 10^9–10^{10} potentially drug-resistant mutants each day. There is a significant theoretical potential for the selection of resistant mutants when antiretroviral microbicides are used in individuals with HIV infection. The public health concerns focus on whether introduction of RT microbicides to a community might lead to an increase in the prevalence of HIV resistance among HIV-positive individuals and potentially limit their therapeutic options. A second concern is that individuals who seroconvert while on an RT microbicide might acquire primary resistance. The breadth of virological resistance will depend on the RT microbicide. Tenofovir is associated with the K65R mutation, which has a variable effect on drug sensitivity, whereas TMC-120 (dapirivine) might induce the K103N mutation, which may render individuals resistant to a whole class of NNRTI drugs. The latter scenario is clearly more worrying than the former. A final concern is whether use of an RT microbicide would be effective if the donor virus carried the mutations listed above. Unfortunately, at this point in time we have very limited, if any, data regarding these questions. Mathematical models of both treatment of chronic infection (Blower *et al.*, 2005) and antiretroviral prophylaxis of HIV-negative populations in the developing world (Abbas *et al.*, 2007) give grounds for cautious optimism. Standard genotyping methods are unlikely to identify mutant genomes if they are present at low frequency, and more sensitive methods are prohibitively expensive for all but the most preliminary studies. The current consensus within the field is to proceed with caution. Trials should only be conducted on individuals who are known to be HIV negative, and whose serological status is confirmed every month during the study. Any participants who seroconvert should stop taking the study product as quickly as possible and be carefully followed up in prospective observational studies. This is clearly not a viable public health strategy, and if RT microbicides are shown to be safe and effective the logistical implications of confining their use to HIV-negative men and women will be challenging. RT microbicide-associated viral resistance is a concern, but the incidence of this problem is likely to be significantly less than viral resistance associated with the treatment of chronic HIV infection in the developing world setting.

Socio-cultural perspectives on trial conduct in the developing world

Early phase development of candidate microbicides is routinely conducted in North America or Europe. Phase I and II clinical studies are traditionally undertaken in populations at low risk of HIV infection, and these participants can be readily recruited in the developed world. In contrast, effectiveness studies (Phase IIB/III studies), where HIV seroconversion is the primary endpoint, need to be conducted in populations with an annual incidence of new HIV infections of at least 3 percent. Such populations can only be found in the developing world. As a consequence, the majority of Phase IIB/III research is conducted in sub-Saharan Africa. Microbicide research is focused on conducting clinical trials that will provide the basis for approval by regulatory agencies such as the Food and Drug Administration (FDA) in the US or the European Agency for the Evaluation of Medicinal Products (EMEA) in Europe. To meet this requirement, these studies have to be conducted at the same level of regulatory rigor demanded from pharmaceutical trials based in the US or Europe. This reality has significant implications for microbicide research conducted in the developing world, and is discussed in more detail below.

It is clear that appropriate populations for Phase IIB/III research can be identified in the developing world. However, often there is limited or absent clinical research infrastructure to support such microbicide research. Basic structural requirements include a clinical examination area, laboratory support, pharmacy, and storage area for regulatory documents. More importantly, there is a critical need for local clinical staff with adequate training to conduct regulatory-grade research. In addition, sites need to develop a mechanism to integrate the local community into the research program through the development of groups such as community advisory boards. The development model that has evolved is for a graduated mentorship program. New sites are identified and research centers are established through upgrading existing buildings or, in some cases, building new structures for clinical research. Staff are recruited and provided with appropriate training in research methodology, including the Good Clinical Practice (GCP) standards required by regulatory authorities. Sites are then ready to undertake preparedness studies that might include recruitment of the type of participants required for Phase IIB/III research, screening for STIs, pregnancy, and HIV seroincidence. Such studies are important in training staff and critical in providing contemporaneous estimates on HIV seroincidence that will ultimately determine whether a site has an appropriate profile for participation in late-stage microbicide trials.

Many of the endpoints in microbicide trials involve the use of laboratory techniques and/or shipping samples to centralized laboratories located in the US or Europe. This requires that sites receive training and demonstrate proficiency in a range of laboratory techniques. It also requires that sites have ongoing access to the reagents and equipment needed to perform these assays. Clearly, establishing and maintaining this level of functionality over time can be challenging. A major problem is retention of skilled staff for the duration of the study. These same issues relate to the clinical aspects of trial execution. Communication between sponsors and sites can be problematic in regions where Internet and telephone access may be limited or intermittent. Despite all of these potential pitfalls, many sites have been able to develop the required level of expertise, quality assurance and quality control to conduct regulatory grade microbicide research.

HIV prevalence rates of 30–40 percent are not uncommon in sub-Saharan Africa. A consequence of this is that when women are screened for Phase IIB/III microbicide studies, a significant number will be found to be HIV seropositive. In many cases, this will be the first time a woman finds out about her HIV status. This process places huge stress on clinic staff, who have to conduct pre- and post-test counseling and arrange referral of seropositive women to medical services.

Microbicide trials are designed to characterize the safety and efficacy of microbicides. However, when these trials are conducted in the developing world, the participants are potentially exposed to adverse events associated with both study product as well as diseases that are prevalent in the local community, such as malaria, TB and gastrointestinal infections. This can present a challenge for the trial sponsors in interpreting safety data, but also for the research clinicians who need to find mechanisms to provide treatment for women enrolled on the study. Participants bringing sick children to the trials site only amplify this problem further. In the developed world, participants can routinely access health care outside the clinical research facility; however, in the developing world the trial site may be the only source of health care. Unfortunately, trials sites may lack the staff or resources to provide generic health care, and often sponsors such as the National Institutes of Health (NIH) may not allow their research funds to be used to support this type of clinical activity.

As mentioned above, a critical component in the development of trial site capability is the identification, training and retention of skilled clinical trial staff. Unfortunately, once such individuals are fully trained they are very marketable, and may migrate to other clinical trial networks within the region or to more lucrative positions in the private sector. In some situations, there are simply very few in-country physicians and nurses. This gap can be transiently circumvented by the use of expatriate staff, but does little to develop the long-term viability of clinical sites.

Phase IIB/III microbicide trials enroll HIV-seronegative participants. However, HIV-seropositive individuals will be identified during the screening process. In addition, a number of participants will be anticipated to seroconvert during Phase IIB/III studies. It is the responsibility of the local trial site to identify local resources to provide clinical care for these people. The individuals who screen out prior to enrollment may have long-standing HIV infection and meet the requirements for initiation of ART. In contrast, individuals who seroconvert during the study may not require ART for many years after the initial infection. This presents the sponsors with two divergent problems. First, the sponsors need to have the ability to provide real-time access to ART to individuals with HIV infection who need prompt treatment. Secondly, they need to have a mechanism whereby individuals who seroconvert on study can be provided with ART at some point in the future (Kim *et al.*, 2006). This may be long after the study has been completed. Providing real-time access to ART is perhaps the easier of these two problems. Initiatives such as PEPFAR and the Global AIDS Fund have increased the availability of ART in the African sub-continent. More importantly, microbicide sponsors are using local availability to ART as a criterion for site selection. From an operational perspective, developing a mechanism to provide access to ART years after an individual seroconverts is more challenging. Some sponsors have created foundations and others have provided access to private health insurance programs to address this need. Optimistically, it might be hoped that access to ART in Africa will be even more widespread in 5–10 years time.

The era of "safari" or "parachute" research, when investigators conducted trials in the developing world for a brief period and left without establishing any ongoing infrastructure, has hopefully come to an end. In addition, the sense that clinical research in the developing world need not adhere to standards of practice in the developed world is clearly both unethical and unacceptable. Contemporary research should adhere to the same clinical and ethical standards irrespective of geographical location (Varmus and Satcher, 1997; Benatar and Singer, 2000; Shapiro and Meslin, 2001; Killen *et al.*, 2002). However, experiences from the past have left a sense of suspicion among participants and clinicians in the developing world. Also, recent high-profile termination of African pre-exposure prophylaxis (PREP) studies has highlighted the need for community education about the nature of clinical research. Community engagement needs to be broad, and to extend from clinical trial participants to Ministers of Health in the countries where trials will be conducted.

Most studies will provide participants with reimbursement for expenses associated with involvement in clinical studies. The intent is to cover items such as travel costs and childcare. The level of reimbursement is closely monitored by local Institutional Review Boards (IRBs) to avoid the possibility of financial inducements to participate in clinical research. Unfortunately, even very modest levels of reimbursement might be considered as providing inappropriate inducement for participants in the developing world. Financial inducements can be quantified and hopefully avoided, but more subtle pressures exist, such as the availability of health care to participants. Ultimately this is a difficult problem, and best addressed by local research staff who are optimally positioned to determine ethical equipoise.

Clinical research has not always been conducted under the most ethical circumstances. The extreme depravity of the experiments carried out in the Nazi concentration camps represents the most shocking example of medical malpractice in recent history, but other examples of abuse can be found in the Tuskegee syphilis studies. As a response to these activities, a series of reports, including the Nuremberg Code, the Belmont Report, and the Declarations of Geneva and Helsinki helped define a concept of Good Clinical Practice (GCP) whereby the conduct of clinical research placed the rights of human participants as the most important aspect of the research process. A key element of GCP is to ensure that study participants give informed consent to participate in the studies proposed. Informed consent is a process rather than a document, and should be reaffirmed throughout the study. The generic consent document used in the West can be a 15-page document, and clearly this has limited or no utility in a developing world setting where participants may have poor or no reading ability. Innovative responses to this problem include the development of other consent tools, and the use of quizzes to determine that participants understand the nature of the trial they are volunteering to participate in. Despite best efforts, there is always room for improving responses to the bioethical challenges raised by these studies (Moodley, 2007).

Despite these safeguards, local communities can often harbor suspicion about the nature of clinical research and the motivation of the scientists conducting the research. These suspicions are exacerbated by the reality that most research in the developing world is funded by governments and/or philanthropic foundations based in the developed world. Community doubts and suspicions can be significantly diminished by outreach activities into the local community, and focused education on the research process. Unfortunately, when problems arise in clinical research the local community can quickly revert to a position of hostility towards ongoing studies. Two recent studies illustrate this point, and both relate to HIV prevention research. In August 2004 a Cambodian PREP study of Tenofovir was stopped prior to enrollment of any participants, and in February 2005 a second PREP study was closed prematurely in Cameroon (Grant *et al.*, 2005; Mills *et al.*, 2005). In both cases, the premature termination probably resulted from community concerns about the adequacy of informed consent and the availability of ongoing treatment for study participants who seroconverted. These concerns subsequently led to political intervention and the closure of the studies. In February 2007, two Phase III studies of the vaginal microbicide cellulose sulfate (CS) were stopped. Both studies planned to enroll approximately 2000 women. The first study was based in Benin, India, South Africa and Uganda; the second was based in Nigeria. A scheduled interim analysis of the trial data from the first study demonstrated a disproportionate number of cases of HIV infection in the CS arm of the study, leading to premature closure of the study. Amongst the 1425 women enrolled in the study, there were 41 seroconversions; 25 were in the cellulose sulfate arm and 16 in the placebo arm. The difference was not statistically significant, but clearly the product was not working to prevent HIV infection. A subsequent analysis of the data from the second study did not show the same problem, but did not suggest that CS had any efficacy as a HIV microbicide and so the second study was also stopped. Subsequently, the local media characterized the studies as using South African women as "guinea pigs", suggested that hundreds of women had become infected on the study, and alleged that informed consent was inadequate and that no provision had been made to provide treatment for women who needed antiretrovirals (Ramjee *et al.*, 2007). These cases are interesting because they illustrate how quickly communities, the media, politicians and in-country regulatory agencies can move from enthusiastic support for research to a position of antipathy and suspicion. It is unfortunate, but necessary, that Phase IIB/III HIV prevention trials are conducted in countries with a high incidence of new HIV infections. This can lead to a misperception that the sponsors are conducting trials that facilitate HIV infection. In fact, the contrary is true. When HIV prevention studies are conducted, the study participants almost always have a decreased risk of HIV acquisition compared to community members not enrolled in the study. The impact of repeated safer-sex counseling, screening for sexually transmitted infections and the provision of condoms all lead to a lower HIV incidence rate in the study participants.

Microbicides as part of the broader prevention agenda

Despite more than 25 years of health education and the widespread availability of condoms, the AIDS pandemic continues. In this setting, a range of behavioral and biomedical interventions have been developed as possible approaches to HIV prevention. Behavioral interventions such as counseling have had limited success (Koblin *et al.*, 2004), and so attention is currently focused on biomedical approaches, including circumcision, pre-exposure prophylaxis (PREP), treatment of serodiscordant couples, and the use of acyclovir, vaccination and microbicides. It now appears that male circumcision is associated with a 50 percent reduction in the risk of acquiring HIV infection (Auvert *et al.*, 2005; Bailey *et al.*, 2007; Gray *et al.*, 2007; Sawires *et al.*, 2007). Pre-exposure prophylaxis, treatment of serodiscordant couples, and the use of acyclovir, vaccination and microbicides are still in the Phase IIB/III stage of development. However, preliminary results from these trials will start to emerge in late 2007. Any evidence of efficacy in these studies will have important implications for the design of future prevention trials. As an example, the observation that circumcision reduces the risk of HIV acquisition has led to the suggestion that all male subjects in prevention trials should be offered circumcision in addition to any other intervention. This will then likely decrease the incidence of new HIV infections in the study population, and the study sample size will have to be increased and/or participants studied for a longer period of time to determine whether an intervention works. The impact of circumcision on microbicide trials is less clear. Women enrolled in the study may have multiple male sexual partners, and the logistics of identifying, counseling, and providing circumcision would be very challenging.

A more likely research scenario is that an efficacy signal will be seen in another microbicide trial. At this time the degree of clinical equipoise that allows randomization to a placebo arm will have been lost or significantly reduced, and it may be necessary to consider redesigning ongoing studies to replace the placebo arm with an active compound. Again, as with circumcision, the net result will be to reduce the power of a study to detect a significant result. The next 5 years will prove to be a fascinating period in the design of intervention trials.

Microbicide research is undergoing a period of rapid evolution. It is hoped that proof of concept data for first-generation products will be available within the next 1–2 years. Meanwhile, research is ongoing to identify new pipeline candidates, improve formulations and develop combination products (Veazey *et al.*, 2005). A critical step will be to develop products that do not have to be used in a coitally dependent fashion. The RT-inhibitor group of products may be suitable for this indication, and the development of ring delivery systems may allow products to be administered on a monthly basis. Currently, microbicides and vaccines are viewed as two very different modalities of HIV prevention science. However, microbicides may be used to deliver immunogens topically, and effectively function as mucosal vaccines. Another, more intriguing, possibility is that microbicides may prevent viral infection but allow priming of the local immune system by live virus. This phenomenon has been seen in non-human primate studies (Shattock, 2006), and it is conceivable that it may occur in ongoing Phase IIB/III effectiveness studies. Whether such immune responses would protect against infection in the absence of a microbicide is an important scientific question, and warrants evaluation in animal models. A more mundane, but critical, issue is to find ways to reduce the cost of production for second- and third-generation microbicide candidates. One innovative approach is to use plants to generate microbicidal peptides such as cyanovirin (Sexton *et al.*, 2006). There is now breadth and depth to the development pipeline, although there is concern that some of the candidates may not be economically viable in the developing world because of the cost of production of these candidates.

First-generation products are still being evaluated in effectiveness trials. It is hoped that these studies will demonstrate a significant, albeit modest, efficacy against HIV-1. Results from the Population Council's Phase III study of Carraguard were announced in February 2008, and failed to show any evidence of efficacy although the product appeared very safe – an important consideration, as the Population Council is planning to evaluate a combination of Carraguard and MIV-150 (an RT-inhibitor). Data from the HPTN-035 study (PRO 2000 and BufferGel) is expected in 2009, and from the UK MRC PRO 2000 study in 2010. It is hoped that subsequent antiretroviral microbicide candidates will have increased potency. The CAPRISA 004 study will evaluate the efficacy of tenofovir gel when used in a coitally dependent fashion in a 980-participant study.

Until recently, the development of antiretroviral PREP and that of microbicides have been seen as quite distinct prevention strategies. However, the Microbicide Trials Network (MTN) will launch a study in 2008 that will combine both oral and topical delivery of antiretroviral agents. The MTN-003 or VOICE (*V*aginal and *O*ral *I*nterventions to *C*ontrol the *E*pidemic) study will randomize approximately 4200 women to receive a microbicide gel (1% tenofovir gel or placebo) or a tablet (Viread®, Truvada® or placebo). In contrast to the CAPRISA 004 study, the VOICE study will use gel on a daily basis. The VOICE study is powered to compare placebo versus active agent in both the oral and topical arms, but has more limited power to compare oral versus topical therapy. Topical and oral PREP have theoretical advantages and disadvantages that will be evaluated in the VOICE study. It might be imagined that oral PREP may be more potent than topical delivery, while topical delivery may result in less adverse events. In addition, women may prefer to take a pill rather than to use a gel. A more important consideration is whether either modality might be more or less likely to result in evolution of antiretroviral resistance. At this point in time, all of these important questions remain unanswered.

Providing access to microbicides

A common theme in HIV prevention is the question of what happens when an intervention works, or, more specifically, how will intervention X be rolled out to community Y? As with other approaches to HIV prevention, the microbicide research community and, especially, the advocacy community are beginning to raise these questions (Forbes and Engle, 2005). Much depends on the degree of effectiveness of the intervention. A microbicide with 30 percent efficacy will be viewed with limited enthusiasm, whereas one with 80 percent efficacy might become a priority for governments, international funding agencies and philanthropic foundations. A vaccine is probably the easiest intervention to roll out, but microbicides may be able to utilize available services such as family planning and STI clinics. Access will also depend on whether a product can be dispensed over the counter or requires medical supervision. Unfortunately, the RT microbicide family will, at least initially, need to be rolled out in a supervised fashion so that the HIV status of individuals and the community incidence of HIV resistance can be determined.

Microbicide development remains a critical component of HIV prevention research

The developing world desperately needs an HIV-1 vaccine. However, despite more than 20 years of research and a current annual research budget of close to $1 billion, there is no evidence to suggest that any of the current candidate vaccines are likely to provide sterilizing immunity against HIV-1. The closure of two Phase IIB studies of the Merck MRKAd5 HIV-1 gag/pol/nef trivalent adenovirus vaccine (HVTN 502 and 503) in October 2007 for lack of efficacy has been another major setback in the field. In addition, Padian and colleagues recently published the results of an innovative study to investigate whether the use of cervical diaphragms and lubricant gel might reduce HIV infection (Padian *et al.*, 2007). Unfortunately, this study, which enrolled almost 5000 women, also did not demonstrate efficacy. Early in 2008 Celum announced the results of HPTN-039, a study that evaluated the efficacy of HSV-2 suppression using acyclovir in preventing HIV acquisition. Sadly, this was another negative study.

Microbicides still have the potential to be a major component of the biomedical HIV-1 prevention portfolio. Despite a disappointing experience with N-9 and, more recently, cellulose sulfate, the microbicide research community continues to grow and mature. The focus is now on the development of candidates that target specific stages of viral infection that can be scaled up at an economically viable cost. The future probably lies in combination microbicides delivered in a coitally independent fashion. The critical challenge will be to find sufficient financial resources to accelerate microbicide development.

References

Abbas, U. L., Anderson, R. M. and Mellors, J. W. (2007). Potential impact of antiretroviral chemoprophylaxis on HIV-1 transmission in resource-limited settings. *PLoS ONE*, 2, e875.

Abner, S. R., Guenthner, P. C., Guarner, J. *et al.* (2005). A human colorectal explant culture to evaluate topical microbicides for the prevention of HIV infection. *J. Infect. Dis.*, 192, 1545–56.

Auvert, B., Taljaard, D., Lagarde, E. *et al.* (2005). Randomized, controlled intervention trial of male circumcision for reduction of HIV infection risk: the ANRS 1265 Trial. *PLoS Med.*, 2, e298.

Bailey, R. C., Moses, S., Parker, C. B. *et al.* (2007). Male circumcision for HIV prevention in young men in Kisumu, Kenya: a randomised controlled trial. *Lancet*, 369, 643–56.

Barnhart, K. T., Pretorius, E. S., Timbers, K. *et al.* (2004). In vivo distribution of a vaginal gel: MRI evaluation of the effects of gel volume, time and simulated intercourse. *Contraception*, 70, 498–505.

Benatar, S. R. and Singer, P. A. (2000). A new look at international research ethics. *Br. Med. J.*, 321, 824–6.

Blower, S., Bodine, E., Kahn, J. and McFarland, W. (2005). The antiretroviral rollout and drug-resistant HIV in Africa: insights from empirical data and theoretical models. *AIDS*, 19, 1–14.

Coplan, P. M., Mitchnick, M. and Rosenberg, Z. F. (2004). Public health. Regulatory challenges in microbicide development. *Science*, 304, 1911–12.

Corey, L., Wald, A., Celum, C. L. and Quinn, T. C. (2004). The effects of Herpes Simplex Virus-2 on HIV-1 acquisition and transmission: a review of two overlapping epidemics. *J. Acquir. Immune. Defic. Syndr.*, 35, 435–45.

Csete, J. (2004). Bangkok 2004. Not as simple as ABC: making real progress on women's rights and AIDS. *HIV AIDS Policy Law Rev.*, 9, 68–71.

D'Cruz, O. J. and Uckun, F. M. (2004). Clinical development of microbicides for the prevention of HIV infection. *Curr. Pharm. Des.*, 10, 315–36.

Duerr, A., Wasserheit, J. N. and Corey, L. (2006). HIV vaccines: new frontiers in vaccine development. *Clin. Infect. Dis.*, 43, 500–11.

Dworkin, S. L. and Ehrhardt, A. A. (2007). Going beyond "ABC" to include "GEM": critical reflections on progress in the HIV/AIDS epidemic. *Am. J. Public Health*, 97, 13–18.

Fletcher, P. S., Elliott, J., Grivel, J. C. *et al.* (2006). *Ex vivo* culture of human colorectal tissue for the evaluation of candidate microbicides. *AIDS*, 20, 1237–45.

Forbes, A. and Engle, N. (2005). Re-building distribution networks to assure future microbicide access. *AIDS Public Policy J.*, 20, 92–101.

Garg, S., Kandarapu, R., Vermani, K. *et al.* (2003a). Development pharmaceutics of microbicide formulations. Part I: preformulation considerations and challenges. *AIDS Patient Care STDS*, 17, 17–32.

Garg, S., Tambwekar, K. R., Vermani, K. *et al.* (2003b). Development pharmaceutics of microbicide formulations. Part II: formulation, evaluation, and challenges. *AIDS Patient Care STDS*, 17, 377–99.

Grant, R. M., Buchbinder, S., Cates, W. Jr. *et al.* (2005). AIDS. Promote HIV chemoprophylaxis research, don't prevent it. *Science*, 309, 2170–71.

Gray, R. H., Kigozi, G., Serwadda, D. *et al.* (2007). Male circumcision for HIV prevention in men in Rakai, Uganda: a randomised trial. *Lancet*, 369, 657–66.

Haghighi, D. B. and Lashner, B. A. (2004). Left-sided ulcerative colitis. *Gastroenterol. Clin. North Am.*, 33, 271–84, ix.

Hendrix, C. W., Fuchs, E. J., Macura, K. J. *et al.* (2008). Quantitative imaging and sigmoidoscopy to assess distribution of rectal microbicide surrogates. *Clin. Pharmacol. Ther.*, 83, 97–105.

Hillier, S. L., Moench, T., Shattock, R. *et al.* (2005). *In vitro* and *in vivo*: the story of nonoxynol 9. *J. Acquir. Immune. Defic. Syndr.*, 39, 1–8.

Hu, J., Gardner, M. B. and Miller, C. J. (2000). Simian immunodeficiency virus rapidly penetrates the cervicovaginal mucosa after intravaginal inoculation and infects intraepithelial dendritic cells. *J. Virol.*, 74, 6087–95.

Killen, J., Grady, C., Folkers, G. K. and Fauci, A. S. (2002). Ethics of clinical research in the developing world. *Nat. Rev. Immunol.*, 2, 210–15.

Kim, N. H., Tabet, S. R., Corey, L. and Celum, C. L. (2006). Antiretroviral therapy for HIV-1 vaccine efficacy trial participants who seroconvert. *Vaccine*, 24, 532–9.

Klasse, P. J., Shattock, R. and Moore, J. P. (2008). Antiretroviral drug-based microbicides to prevent HIV-1 sexual transmission. *Annu. Rev. Med.*, 59, 455–71.

Koblin, B., Chesney, M. and Coates, T. (2004). Effects of a behavioural intervention to reduce acquisition of HIV infection among men who have sex with men: the EXPLORE randomised controlled study. *Lancet*, 364, 41–50.

Lard-Whiteford, S. L. (2004). Recommendations for the nonclinical development of topical microbicides for prevention of HIV transmission: an update. *J. Acquir. Immune. Defic. Syndr.*, 36, 541–52.

Mantell, J. E., Myer, L., Carballo-Dieguez, A. *et al.* (2005). Microbicide acceptability research: current approaches and future directions. *Soc. Sci. Med.*, 60, 319–30.

Mauck, C., Rosenberg, Z. and Van Damme, L. (2001). Recommendations for the clinical development of topical microbicides: an update. *AIDS*, 15, 857–68.

McGowan, I. (2006). Microbicides: a new frontier in HIV prevention. *Biologicals*, 34, 241–55.

Mills, E., Rachlis, B., Wu, P. *et al.* (2005). Media reporting of tenofovir trials in Cambodia and Cameroon. *BMC Intl Health Hum. Rights*, 5, 6.

Mitty, J. A., Macalino, G., Taylor, L. *et al.* (2003). Directly observed therapy (DOT) for individuals with HIV: successes and challenges. *MedGenMed.*, 5, 30.

Moodley, K. (2007). Microbicide research in developing countries: have we given the ethical concerns due consideration? *BMC Med. Ethics*, 8, 10.

Mosher, W. D., Chandra, A. and Jones, J. (2005). Sexual Behavior and Selected Health Measures: Men and Women 15–44 Years of Age, United States, 2002. National Center for Health Statistics. Hyattsville, MD: 362.

Neurath, A. R., Strick, N. and Li, Y. Y. (2003). Water dispersible microbicidal cellulose acetate phthalate film. *BMC Infect. Dis.*, 3, 27.

Owen, D. H., Peters, J. J. and Katz, D. F. (2000). Rheological properties of contraceptive gels. *Contraception*, 62, 321–6.

Padian, N. S., van der Straten, A., Ramjee, G. *et al.* (2007). Diaphragm and lubricant gel for prevention of HIV acquisition in southern African women: a randomised controlled trial. *Lancet*, 370, 251–61.

Patton, D. L., Cosgrove Sweeney, Y. T., Rabe, L. K. and Hillier, S. L. (2002). Rectal applications of nonoxynol-9 cause tissue disruption in a monkey model. *Sex. Transm. Dis.*, 29, 581–7.

Patton, D. L., Sweeney, Y. C., Cummings, P. K. *et al.* (2004). Safety and efficacy evaluations for vaginal and rectal use of BufferGel in the macaque model. *Sex. Transm. Dis.*, 31, 290–6.

Patton, D. L., Sweeney, Y. T., Balkus, J. E. and Hillier, S. L. (2006). Vaginal and rectal topical microbicide development: safety and efficacy of 1.0 percent Savvy (C31G) in the pigtailed macaque. *Sex. Transm. Dis.*, 33, 91–5.

Phillips, D. M., Taylor, C. L., Zacharopoulos, V. R. and Maguire, R. A. (2000). Nonoxynol-9 causes rapid exfoliation of sheets of rectal epithelium. *Contraception*, 62, 149–54.

Phillips, D. M. and Zacharopoulos, V. R. (1998). Nonoxynol-9 enhances rectal infection by herpes simplex virus in mice. *Contraception*, 57, 341–8.

Phillips, D. M., Sudol, K. M., Taylor, C. L. *et al.* (2004). Lubricants containing N-9 may enhance rectal transmission of HIV and other STIs. *Contraception*, 70, 107–10.

Poles, M. A., Elliott, J., Taing, P. *et al.* (2001). A preponderance of CCR5(+) CXCR4(+) mononuclear cells enhances gastrointestinal mucosal susceptibility to human immunodeficiency virus type 1 infection. *J. Virol.*, 75, 8390–99.

Pretorius, A., Timbers, K., Malamud, D. and Barnhart, K. (2003). Magnetic resonance imaging to determine the distribution of vaginal gel: before, during, and after both stimulated and real intercourse. *Contraception*, 66, 443–51.

Quinn, T. C. and Overbaugh, J. (2005). HIV/AIDS in women: an expanding epidemic. *Science*, 308, 1582–3.

Ramjee, G., Govinden, R., Morar, N. S. and Mbewu, A. (2007). South Africa's experience of the closure of the cellulose sulphate microbicide trial. *PLoS Med.*, 4, e235.

Saltini, C. (2006). Chemotherapy and diagnosis of tuberculosis. *Respir. Med.*, 100, 2085–97.

Santschi, V., Wuerzner, G., Schneider, M. P. *et al.* (2007). Clinical evaluation of IDAS II, a new electronic device enabling drug adherence monitoring. *Eur. J. Clin. Pharmacol.*, 63, 1179–84.

Sawires, S. R., Dworkin, S. L., Fiamma, A. *et al.* (2007). Male circumcision and HIV/AIDS: challenges and opportunities. *Lancet*, 369, 708–13.

Severy, L. J., Tolley, E., Woodsong, C. and Guest, G. (2005). A framework for examining the sustained acceptability of microbicides. *AIDS Behav.*, 9, 121–31.

Sexton, A., Drake, P. M., Mahmood, N. *et al.* (2006). Transgenic plant production of Cyanovirin-N, an HIV microbicide. *FASEB J.*, 20, 356–8.

Shapiro, H. T. and Meslin, E. M. (2001). Ethical issues in the design and conduct of clinical trials in developing countries. *N. Engl. J. Med.*, 345, 139–42.

Shattock, R. J. (2006). Protection of macaques against rectal SIV challenge by mucosally-applied PMPA. Microbicides 2006, Cape Town, South Africa. Abstract OA15.

Shattock, R. J. and Moore, J. P. (2003). Inhibiting sexual transmission of HIV-1 infection. *Nat. Rev. Microbiol.*, 1, 25–34.

Stein, Z. A. (1990). HIV prevention: the need for methods women can use. *Am. J. Public Health*, 80, 460–2.

Sudol, K. M. and Phillips, D. M. (2004). Relative safety of sexual lubricants for rectal intercourse. *Sex. Transm. Dis.*, 31, 346–9.

Tabet, S. R., Surawicz, C., Horton, S. *et al.* (1999). Safety and toxicity of nonoxynol-9 gel as a rectal microbicide. *Sex. Transm. Infect.*, 26, 564–71.

Tsai, C. C., Emau, P., Jiang, Y. *et al.* (2003). Cyanovirin-N gel as a topical microbicide prevents rectal transmission of SHIV89.6P in macaques. *AIDS Res. Hum. Retroviruses*, 19, 535–41.

US Food and Drug Administration (2006). Additional resources regarding drug development of topical microbicides. Available at www.fda.gov/cder/ode4/preind/Top_Links.htm (accessed 05-29-08).

Van Damme, L., Ramjee, G., Alary, M. *et al.* (2002). Effectiveness of COL-1492, a nonoxynol-9 vaginal gel, on HIV-1 transmission in female sex workers: a randomised controlled trial. *Lancet*, 360, 971–7.

Varmus, H. and Satcher, D. (1997). Ethical complexities of conducting research in developing countries. *N. Engl. J. Med.*, 337, 1003–5.

Veazey, R. S., Shattock, R. J., Pope, M. *et al.* (2003). Prevention of virus transmission to macaque monkeys by a vaginally applied monoclonal antibody to HIV-1 gp120. *Nat. Med.*, 9, 343–6.

Veazey, R. S., Klasse, P. J., Schader, S. M. *et al.* (2005). Protection of macaques from vaginal SHIV challenge by vaginally delivered inhibitors of virus-cell fusion. *Nature*, 438, 99–102.

Volmink, J. and Garner, P. (2006). Directly observed therapy for treating tuberculosis. *Cochrane Database Syst. Rev.*, 2, CD003343.

Wallace, A., Thorn, M., Maguire, R. A. *et al.* (2004). Assay for establishing whether microbicide applicators have been exposed to the vagina. *Sex. Transm. Dis.*, 31, 465–8.

Wallace, A. R., Teitelbaum, A., Wan, L. *et al.* (2007). Determining the feasibility of utilizing the microbicide applicator compliance assay for use in clinical trials. *Contraception*, 76, 53–6.

Woolfson, A. D., Malcolm, R. K., Morrow, R. J. *et al.* (2006). Intravaginal ring delivery of the reverse transcriptase inhibitor TMC 120 as an HIV microbicide. *Intl J. Pharm.*, 325, 82–9.

Zeitlin, L., Hoen, T. E., Achilles, S. L. *et al.* (2001). Tests of buffer gel for contraception and prevention of sexually transmitted diseases in animal models. *Sex. Transm. Dis.*, 28, 417–23.

Using antiretrovirals to prevent HIV transmission

5

Cynthia L. Gay, Angela D. Kashuba and Myron S. Cohen

ART has drastically reduced HIV-associated mortality and improved the quality of lives of those living with HIV infection (Palella *et al.*, 1998; Mocroft *et al.*, 2003). More recently, global initiatives to expand the use of ART in resource poor countries have achieved some degree of success (UNAIDS, 2006, 2007). However, HIV prevention has not gone so well; for every person treated with ART four new people become infected (WHO, 2007). The purpose of this chapter is to describe the ways in which ART can be used for prevention as well as treatment.

HIV transmission

The biology of HIV transmission has been reviewed in Chapter 2. The transmission of HIV depends on the infectiousness of the host and the susceptibility of the partner (Royce, 1997). The likelihood of transmission can be described in terms of "efficiency", which varies greatly by the route of transmission and, for sexual transmission, the details of the sexual behavior. Transmission of HIV by all routes (Royce, 1997) has been most strongly associated with HIV viral load in the blood (Garcia *et al.*, 1999; Quinn *et al.*, 2000; Chakraborty *et al.*, 2001; Fideli *et al.*, 2001; Tovanabutra *et al.*, 2002; Richardson *et al.*, 2003; Rousseau *et al.*, 2003). The best empirical data related to sexual transmission of HIV comes from a study of Ugandan serodiscordant couples. In this study, the probability of HIV transmission correlated directly with increasing blood viral load, and no transmission events were observed when HIV RNA was less than 1500 copies/ml (Quinn *et al.*, 2000). This "concentration response" association between viral load and the risk of HIV transmission among serodiscordant couples was confirmed in subsequent studies in Zambia and Thailand (Fideli *et al.*, 2001; Tovanabutra *et al.*, 2002). In the latter study, no HIV transmission occurred when the viral load of the infected partners was less than 1094 copies/ml (Tovanabutra *et al.*, 2002). However, these studies are somewhat misleading because they were not designed to measure transmission events related to acute HIV infection, during which the blood viral burden is very high and transmission is very efficient (Pilcher *et al.*, 2001a).

HIV in genital secretions

The association between HIV viral load and sexual transmission must reflect a correlation between the concentration of HIV in genital secretions and in blood. While HIV concentrations in genital (Coombs *et al.*, 2003) and rectal secretions (Lampinen *et al.*, 2000) correlate with levels of virus in the blood, the correlation is imperfect and unpredictable from sample to sample. The concentration of HIV in semen is generally lower than in blood (Xu *et al.*, 1997; Coombs *et al.*, 2003), but can equal or exceed that in blood in the setting of inflammation as seen with STDs (Cohen *et al.*, 1997). The concentration of HIV in female genital secretions is greatly affected by the collection method used, with lower concentrations generally detected only intermittently (Brambilla *et al.*, 1999; Coombs *et al.*, 2003; Cu-Uvin *et al.*, 2006). Menses affects excretion of HIV, especially when cervico-vaginal lavage is used to collect specimens (Al-Harthi *et al.*, 2001). HIV excretion can be correlated with HSV-2 activation, a potentially important fact given the remarkable prevalence of co-infection with HIV and HSV (Wald and Link, 2002; Serwadda *et al.*, 2003).

Acute HIV infection

A significant proportion of sexual HIV transmission appears to be driven by acute HIV infection (Yerly *et al.*, 2001; Pilcher *et al.*, 2004a; Wawer *et al.*, 2005; Brenner *et al.*, 2007). Acute HIV infection represents the period following HIV acquisition during which HIV RNA and p24 antigen can be detected in the blood, when standard screening tests for HIV antibodies remain negative. Because HIV-specific immune responses, including HIV-specific antibodies, have not yet developed during acute HIV infection (Rosenberg *et al.*, 1997), viral replication continues unimpeded and results in very high levels of circulating virus (Pilcher *et al.*, 2001b, 2004a); greater than 1 million copies/ml in blood among acutely infected individuals in Malawi (Pilcher *et al.*, 2004b). In the Rakai study, Wawer and colleagues reported that nearly half of HIV transmission events among HIV discordant couples occurred during early HIV infection (Wawer *et al.*, 2005). Greatly increased shedding of HIV virus in the genital tract occurs during acute HIV infection (Figure 5.1) (Pilcher *et al.*, 2001b, 2007), and concurrent sexually

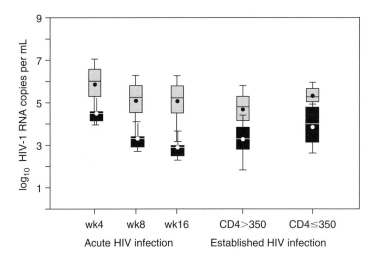

Figure 5.1 Comparison of HIV viral load in blood and semen in acute versus chronic HIV infection. Box-and-whisker plots depict HIV-1 RNA levels in blood (gray) and semen (black), and were estimated of individual subjects by disease stage. Reproduced from Pilcher *et al.* (2007), with permission.

transmitted infections (Cohen *et al.*, 1997; Ghys *et al.*, 1997; McClelland *et al.*, 2001) may play an important role.

ART pharmacology

The first antiviral drug, zidovudine (AZT), was approved in 1987, and quickly demonstrated the feasibility and importance of antiviral therapy. Currently, more than 20 drugs are available, divided into classes by mode of action and used in "optimized" combinations (Table 5.1). ART predictably reduces HIV RNA in blood to undetectable levels (Hogg *et al.*, 1998). ART concomitantly decreases HIV RNA concentrations in blood and seminal plasma (Vernazza *et al.*, 2000), female genital tract secretions (Cu-Uvin *et al.*, 2006) and rectal secretions (Kotler *et al.*, 1998). Such suppression can be sustained in blood and genital secretions over long periods of time in some patients (Pereira *et al.*, 1999; Vernazza *et al.*, 2000); however, breakthrough shedding despite ART suppression of plasma viral load has been demonstrated (Coombs *et al.*, 1998; Graham *et al.*, 2007).

In a recent prospective study of female sex workers in Kenya, HIV-1 RNA levels in plasma and in genital secretions declined rapidly and in parallel following initiation of an ART regimen containing non-nucleoside reverse transcriptase inhibitors (NNRTI) (Graham *et al.*, 2007). However, virus was still detectable in cervical and vaginal secretions in half the subjects after 28 days of ART, suggesting

Table 5.1   Preferred and alternative regimens for initial antiretroviral therapy per the Department of Health and Human Services (DHHS) and IAS-USA guidelines (Hammer *et al.*, 2006; Panel on Clinical Practices for the Treatment of HIV Infection and DHHS Guidelines, 2006)

Class	Medication	Recommendation
Nucleoside/nucleotide reverse transcriptase inhibitor (NRTI)	Tenofovir/emtricitabine[a,b] Zidovudine/lamivudine[a,b] Abacavir/lamivudine[a,b] Didanosine and lamivudine or emtricitabine	Preferred (IAS-USA and DHHS) Preferred (IAS-USA and DHHS) Preferred (IAS-USA) Alternative (DHHS) Alternative (DHHS)
Non-nucleoside reverse transcriptase inhibitor (NNRTI)	Efavirenz[c] Nevirapine[d]	Preferred (IAS-USA and DHHS) Preferred (IAS-USA) Alternative (DHHS)
Protease inhibitor (PI)	Atazanavir plus ritonavir Fos-amprenavir plus ritonavir (twice daily) Lopinavir/ritonavir[a] (twice daily) Saquinavir plus ritonavir Atazanavir[e] Fos-amprenavir Fos-amprenavir plus ritonavir (once daily) Lopinavir/ritonavir[a] (once daily)	Preferred (IAS-USA and DHHS) Preferred (IAS-USA and DHHS) Preferred (IAS-USA and DHHS) Preferred (IAS-USA and DHHS) Preferred (IAS-USA) Alternative (DHHS) Alternative (DHHS) Alternative (DHHS)

Adapted from Eron and Hirsch (2007).
Notes:
[a]Fixed dose combinations (FDC)
[b]emtricitabine may be used in place of lamivudine and vice versa (not available as FDC)
[c]efavirenz is not recommended for use in the first trimester of pregnancy or in sexually active women who are not using effective contraception
[d]nevirapine should not be initiated in women with CD4 cell counts $>250/mm^3$ or men with CD4 cell counts $>400/mm^3$
[e]atazanavir should be boosted with ritonavir when given with tenofovir.

persistent host infectiousness. In contrast, a cross-sectional study among men found that only 2 of 114 subjects with plasma HIV RNA <400 copies/ml had detectable HIV RNA in seminal plasma (Si-Mohamed *et al.*, 2000; Vernazza *et al.*, 2000). STDs can increase genital tract shedding of HIV even in subjects with ART suppression (Sadiq *et al.*, 2002). Furthermore, there is clearly compartmentalization of virus between blood and genital secretions, as indicated by discordant viral RNA levels (Coombs *et al.*, 1998; Kovacs *et al.*, 2001) and differences in viral phenotypes and genotypes between the two compartments (Coombs *et al.*, 1998; Ping *et al.*, 2000; Philpott *et al.*, 2005; Andreoletti *et al.*, 2007).

ART in the genital tract

Antiretroviral agents differ greatly in their ability to penetrate the genital tract (Figure 5.2) (Coombs *et al.*, 1998; Pereira *et al.*, 1999; Taylor *et al.*, 2000; van Praag *et al.*, 2001; Chaudry *et al.*, 2002; Reddy *et al.*, 2003; Solas *et al.*, 2003; Ghosn *et al.*, 2004; Min *et al.*, 2004; Dumond *et al.*, 2007; Vourvahis *et al.*, 2008). Since highly protein-bound antiretroviral agents achieve lower concentrations in genital tract than in blood plasma, affinity for albumin and a_1-acid glycoprotein may prevent penetration of antiviral agents into genital secretions. For example, with the exception of indinavir (which is only 60 percent protein bound), the concentrations of protease inhibitors in the male and female genital tracts are less than 10 percent and 50 percent, respectively, of those in blood plasma (Figure 5.2). The limited penetration of protease inhibitors into the genital tract may account for protease resistance in studies of HIV isolates from seminal plasma and vaginal lavage fluid (Eron *et al.*, 1998; Mayer *et al.*, 1999; Si-Mohamed *et al.*, 2000; Solas *et al.*, 2003). In contrast, most nucleoside/tide analogue reverse transcriptase inhibitors have a low degree of protein-binding and achieve genital tract concentrations two- to six-fold higher than in plasma (Figure 5.2) (Pereira *et al.*, 1999; Taylor *et al.*, 2000; van Praag *et al.*, 2001; Chaudry *et al.*, 2002; Reddy *et al.*, 2003; Solas *et al.*, 2003; Ghosn *et al.*, 2004; Min *et al.*, 2004; Dumond *et al.*, 2007; Vourvahis *et al.*, 2008). Most recently, compared to all other currently marketed antiretroviral agents, maraviroc has been found to have the highest exposure in cervicovaginal secretions relative to blood plasma (approximately 500 percent with steady-state dosing (Dumond *et al.*, 2008). Since maraviroc is 85 percent bound to plasma proteins, other physicochemical factors (e.g. lipophilicity) must also be important in dictating genital tract concentrations.

Phosphorylation of N(t)RTIs

The active component of nucleoside/tide analogue reverse transcriptase inhibitors is the intracellular phosphorylated form, rather than the parent drug present in blood plasma. Studies quantifying the concentrations of active, intracellular phosphorylated forms of tenofovir, zidovudine and lamivudine in seminal mononuclear cells have been reported (Reddy *et al.*, 2003; Vourvahis *et al.*, 2008). Intracellular tenofovir diphosphate concentrations are 5- to 10-fold higher in

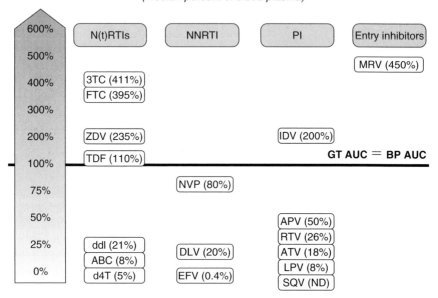

Figure 5.2 Antiretroviral drug levels in the male and female genital tract relative to blood plasma levels (ratio of genital to blood plasma levels). Data from Pereira *et al.* (1999), Taylor *et al.* (2000), van Praag *et al.* (2001), Chaudry *et al.*, 2002, Solas *et al.* (2003), Reddy *et al.* (2003), Ghosn *et al.* (2004), Dumond *et al.* (2007, 2008) and Vourvahis *et al.* (2008). 3TC, lamivudine; ABC, abacavir; APV, amprenavir; ATV, atazanavir; d4T, stavudine; ddI, didanosine; DLV, delavirdine; EFV, efavirenz; ENF, enfuvirtide; FTC, emtricitabine; IDV, indinavir; LPV, lopinavir; NFV, nelfinavir; NVP, nevirapine; RTV, ritonavir; SQV, saquinavir; TDF, tenofovir; ZDV, zidovudine.

seminal mononuclear cells than in peripheral blood mononuclear cells (PBMCs) (Vourvahis *et al.*, 2008). Intracellular lamivudine triphosphate concentrations in seminal mononuclear cells are similar to those found in PBMCs, and intracellular zidovudine triphosphate concentrations are approximately 50 percent of that found in PBMCs (Reddy *et al.*, 2003).

ART to Prevent Transmission of HIV

There are three ways ART can be used to prevent HIV transmission: by reducing HIV viral load in people who know they are infected, as post-exposure prophylaxis, and as pre-exposure prophylaxis both orally and as a topical microbicide (see Chapter 4). In each case, biological plausibility as well as clinical and epidemiological data are needed to guide therapy. Issues common to the application of ART for prevention include the potential changes in risk behaviors evoked by ART, fear of resistance, and public health relevance.

Effects of ART on Infectiousness

Without a doubt, the greatest potential for ART in HIV prevention lies in the ability of treatment to render the infected host less contagious. Nevertheless, the public health considerations of ART have been largely ignored by this generation of health workers.

There are three lines of evidence to suggest that ART reduces infectiousness of treated patients: retrospective analysis, prospective observational studies and ecological data. In a retrospective study of HIV transmission among 436 serodiscordant couples (Musicco *et al.*, 1994), the relative risk from a man to his female partner was decreased among the 15 percent of men who took zidovudine monotherapy with more advanced disease (odds ratio 0.5, 95% CI 0.1–0.9). Another retrospective study evaluated HIV transmission events in 393 serodiscordant couples in the pre-highly active antiretroviral therapy (HAART), early HAART, and post-HAART periods (1991 to 2002), and found an 80 percent reduction in HIV transmission following the introduction of HAART (OR 0.14, CI 0.03–0.66) (Castilla *et al.*, 2005).

Sexual transmission of HIV among couples lies at the root of the HIV epidemic. Recent massive household screening studies in Uganda demonstrated that a substantial number of unrecognized HIV infections and HIV-discordant couples could be detected. This is important, because ongoing transmission within discordant couples occurs at a rate of about 8–11 percent per year (Kamenga *et al.*, 1991; Allen *et al.*, 1992; Fideli *et al.*, 2001). Accordingly, prospective studies of the effects of ART in discordant couples are particularly important. In a recent observational study of 1034 discordant couples in Zambia and Rwanda, the index partners in 248 couples received ART due to CD4 cell counts less than 200 cells/µl (Kayitenkore *et al.*, 2006). Among the 42 partners who acquired HIV in this cohort since 2003, only two had an HIV-infected partner on ART. The risk of acquiring HIV in susceptible partners was decreased if their HIV-infected partner was receiving ART (OR 0.19, CI 0.05–0.80), and this finding persisted after adjusting for self-reported condom usage (adjusted OR 0.21, CI 0.05–0.80). A similar prospective observational study of Ugandan patients initiating ART reported a 98 percent reduction in the estimated risk of HIV transmission following the start of ART; a decrease from 45.7 to 1.0 transmission per 1000 person-years in 454 of 926 participants with 2 years of follow-up (Bunnell *et al.*, 2006). Although only 1 of 62 serodiscordant couples with 2 years of follow-up experienced a transmission event, interpretation of these results is limited because almost 50 percent of the initial cohort was excluded in the analysis.

It should be emphasized that the observational studies described above are susceptible to the effects of unexpected modifiers, including sexual behavior and condom use. In addition, the periods of observation are not long enough to determine long-term benefit, or to detect transmission of resistance viruses (see below). Perhaps most importantly, the studies only include index subjects who require ART for low CD4 counts or advancing HIV disease, whereas ART for prevention might be used at a much higher CD4 count, especially in people at greater risk for transmitting HIV. The latter issue emphasizes the importance of marrying the therapeutic and public health benefits of ART.

Several ecologic studies which focused on the preventative benefit of ART have been completed. In a large closed cohort of homosexual men in San Francisco, California, a 60 percent reduction in anticipated cases of HIV was attributed to availability of ART for infected sexual partners (Porco *et al.*, 2004). A study from Taiwan showed a 53 percent reduction in the expected cases of HIV following the free provision of ART in 1997, also accredited to the wider availability of ART in this country (Fang *et al.*, 2004). More recently, a study in British Columbia, Canada suggested that up to 50 percent of expected incident HIV cases were averted by ART (Montaner *et al.*, 2006). However, this conclusion may be limited, as a sharp decline in HIV among injection drug users was reported to offset an increase in HIV cases among Australian homosexual men during the study period (Grulich, 2006), implying that the decrease could have been due to changes in behavior among intravenous drug users and not the impact of ART on sexual transmission.

Ecological prevention benefits of ART have not been universal. In a city-wide analysis, no reduction in incident HIV infections among men who have sex with men (MSM) in San Francisco was observed despite widespread availability of ART (Katz *et al.*, 2002). Increases in HIV incidence were also found among homosexual men attending sexually transmitted disease clinics in Amsterdam, the Netherlands, from 1991 to 2001 (Dukers *et al.*, 2002), regardless of treatment roll-out. Increases in other sexually transmitted diseases in both settings (Dukers *et al.*, 2002; Katz *et al.*, 2002) demonstrated considerable ongoing risk-taking behaviors in these populations, raising the question of whether the increased availability of effective ART prompted high-risk sexual behavior in the respective populations (see below and Chapter 10). Ecologic studies are greatly limited by their inability to relate the patients who receive therapy to the actual incidence or prevalence of HIV in the community, and the accuracy of HIV prevalence and incidence data in these settings are largely unknown. In addition, ecologic evidence cannot directly correlate HIV transmission with HIV prevalence, since a small number of individuals with very high viral load (as seen in acute HIV infection) could account for a disproportionate number of incident cases (Wawer *et al.*, 2005; Brenner *et al.*, 2007). Furthermore, substantial HIV transmission occurs because infected persons remain unaware of their HIV status, including virtually all individuals with acute HIV infection (Pilcher *et al.*, 2001a, 2004a).

Given the limits of observational and ecologic studies, a randomized, controlled trial is needed to better define the impact of ART on HIV transmission. The HIV Prevention Trials Network (HPTN) is currently enrolling subjects into the HPTN 052 randomized trial to evaluate the effectiveness of two different treatment strategies to prevent the sexual transmission of HIV among 1750 serodiscordant couples (reviewed at www.hptn.org). HIV-infected partners with a CD4 count between 350 and 550 cells/mm^3 are randomly assigned to initiate ART at enrollment, or to delay ART until their CD4$^+$ T-cell count falls below 250 cells/ mm^3 or they develop an AIDS-defining illness. The study is designed to detect a 35 percent reduction in HIV transmission to sexual partners due to ART treatment of HIV-infected subjects. Subjects are being followed on study for 5 years to compare long-term safety and tolerability, patient adherence to therapy, and the patterns and prevalence of antiviral resistance between the two ART strategies. In addition, this study compares the benefits of early versus delayed ART (ACTG 5245) and examines some aspects of the biology of HIV transmission (CHAVI 007). The study is projected to finish in the year 2012.

Prevention of mother-to-child transmission

ART prevents HIV transmission from mother to child (Chapter 17). ART to prevent vertical transmission has primarily been utilized to decrease maternal viral load prior to infant delivery (Mofenson and McIntyre, 2000). However, an additional benefit is observed when ART is given to neonates (Guay *et al.*, 1999; Moodley *et al.*, 2003), demonstrating the benefits of post-exposure ART in infants. However, vertical and postnatal transmission of HIV is so unique as to preclude extrapolation of these results to other potential applications of PEP.

ART for post-exposure prophylaxis (PEP)

Three lines of evidence suggest ART can prevent HIV acquisition following exposure: the success of mother-to-child HIV transmission (Connor *et al.*, 1994; Cardo *et al.*, 1997; Wade *et al.*, 1998), animal studies (Tsai and Follis, 1995; Tsai *et al.*, 1998; Le Grand *et al.*, 2000; Otten *et al.*, 2000; Cohen *et al.*, 2002), and a case-control study of prophylaxis after needlestick injury in the health-care setting (Cardo *et al.*, 1997). There are no clinical trials of antiretroviral efficacy after sexual exposure to HIV.

Post-exposure ART in animal models

HIV-1 replicates within mucosal dendritic cells following initial HIV exposure and before spreading via lymphatic vessels to cause systemic infection (Spira

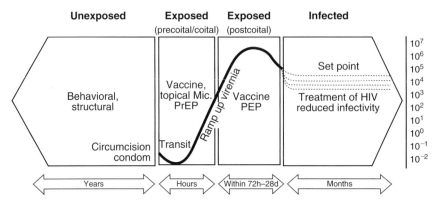

Figure 5.3   Dynamics of HIV-1 RNA levels in genital fluid or plasma during acute HIV infection.

et al., 1996; Hu *et al.*, 2000; CDC, 2001; Zhang *et al.*, 2004). Accordingly, the time between exposure and the systemic spread of HIV-1 (including ramp up viremia and seroconversion; Figure 5.3) provides a critical opportunity to intervene with antiretroviral therapy to prevent established HIV infection.

The rhesus macaque model (Pauza *et al.*, 1998) most often used to simulate and study HIV infection utilizes simian immunodeficiency virus (SIV) infection or SIV/HIV chimeric viruses (SHIV) (Sun *et al.*, 2007). The SIV strain used, the concentration of virus selected and the mode of exposure (intravenous, vaginal or rectal) must be carefully noted in interpreting and comparing the results from studies of post-exposure ART in macaques. Currently, SIV_{sm} or SIV_{mac} variants are felt to most closely simulate the natural progression of human HIV-1 infection. Infection of rhesus monkeys with SIV_{sm} or SIV_{mac} leads to a period of high peak and set-point blood viral RNA levels, followed by a subsequent, gradual $CD4^{+}$ T-cell decline, and the eventual development of opportunistic infections and neoplasms, similar to HIV-1 infection in humans. Chimeric SIV-HIV constructs which express the HIV-1 envelopes on a SIV backbone and reliably infect macaques have been developed; however, they have not simulated human infection as closely as SIV (see Chapter 3).

In early studies of post-exposure ART in macaques, a single, high-dose model of SIV or SHIV was employed (reviewed in Hosseinipour *et al.*, 2002), but such high inoculum concentrations could override the potential protective effects of ART. More recently, a repeated-low-dose rectal exposure model which utilizes multiple inoculations with lower virus titers (3.8×10^{5} viral particle equivalents) has been employed to more closely parallel high-risk human exposure and HIV concentrations in semen during acute HIV infection (Otten *et al.*, 2005; Garcia-Lerma *et al.*, 2006; Subbarao *et al.*, 2006, 2007).

Results from SIV and SIV-HIV macaque experiments suggest that appropriate antiretroviral agents given in sufficient concentration and for an appropriate duration after exposure can prevent HIV infection (Tsai and Follis, 1995; Tsai *et al.*, 1998; Le Grand *et al.*, 2000; Otten *et al.*, 2000).

Tenofovir has been used most extensively in the macaque model of post-exposure prophylaxis, due to its efficacy, tolerability and long half-life. Tenofovir prevented uncloned SIV_{mne} infection after intravenous challenge when given within 24 hours of exposure and continued for 10 days (Tsai *et al.*, 1998). In the same study, the maximal benefit of tenofovir was observed when animals were treated promptly and for a full 28 days (Tsai *et al.*, 1998). Tenofovir also provided protection to macaques after intravaginal exposure when initiated within 36 hours, but 1 failure resulted when therapy was delayed until 72 hours following exposure (Otten *et al.*, 2000). Despite such promising findings, other animal studies using different ART drug combinations, inoculum sizes, and timing of drug initiation following exposure have reported less reliable protection in animals (Le Grand *et al.*, 2000; Cohen *et al.*, 2002; Hosseinipour *et al.*, 2002; Heine, 2007; W. Heine, personal communication). It should be noted that the previous animal studies are limited by the small number of animals studied, and the failure to sacrifice the animals to prove that infection was averted.

Because HIV is very species tropic, other animal studies of HIV transmission have not been feasible to date. However, a novel SCID mice model recently demonstrated the ability to infect bone marrow/liver/thymus (BLT) humanized mice following one intrarectal exposure (Sun *et al.*, 2007). Using this model, Sun and colleagues demonstrated HIV-1 infection in BLT mice, CD4 T-cell depletion in gut-associated lymphoid tissue (GALT) and AIDS-associated pathology in the infected mice. The model shows great potential for further study on prevention strategies, including pre-exposure prophylaxis (Denton *et al.*, 2008).

Post-exposure prophylaxis following occupational exposure

Post-exposure prophylaxis following occupational exposure to HIV is considered the standard of care in the United States (CDC, 2001) and several other countries. However, evidence on the efficacy of post-exposure ART to prevent human HIV transmission is limited to a single, small, retrospective case-control study of ART prophylaxis in health-care workers following needlestick exposures in 1994 (Cardo *et al.*, 1997). In this study, 33 health-care workers who seroconverted following percutaneous exposure were retrospectively identified from national surveillance programs in the United States, Italy, France and the United Kingdom, and compared with controls selected from 679 individuals who did not seroconvert after post-exposure prophylaxis. Zidovudine (and other antiretrovirals in a few cases) given to individuals after percutaneous exposure to HIV led to an 81 percent risk reduction (CI 48–94%) in HIV seroconversion. Findings from this single study resulted in the publication of widely-adapted guidelines for the use of ART following occupational HIV exposure by the Centers for Diseases Control and Prevention (CDC) in 2001 (CDC, 2001), which were updated in 2005 (Panlilio *et al.*, 2005).

Non-occupational post-exposure prophylaxis

Conducting randomized, controlled, clinical trials of post-exposure prophylaxis to prevent the sexual transmission of HIV in humans is not feasible because of the inefficient transmission of HIV per sexual exposure, and the prohibitive cost of enrolling the very large number of subjects that would be needed to establish benefit. Such trials are also limited by the absence of information on, and the inability to confirm, the HIV status in the majority of source cases (Poynten *et al.*, 2007). Finally, the increasing worldwide adaptation of post-exposure prophylaxis guidelines after non-occupational HIV exposures likely precludes the possibility of a randomized, placebo-controlled trial for ethical reasons. Several institutions in the United States (Kunches *et al.*, 2001) and other countries (Bernasconi *et al.*, 2001; Almeda *et al.*, 2002; EURO-NONOPEP, 2002; Giele *et al.*, 2002; Rey *et al.*, 2002; Meel, 2005; Vitoria *et al.*, 2006; Poynten *et al.*, 2007; van Oosterhout *et al.*, 2007) have established guidelines for the implementation of non-occupational post-exposure prophylaxis after HIV exposure.

In 2001, 14 countries established the European Project on Non-occupational Post-Exposure Prophylaxis for HIV (EURO-NONOPEP) to formulate more unified guidelines for post-exposure prophylaxis and to establish a prospective registry of possible HIV non-occupational exposures (Martin *et al.*, 2005). The registry will be used to describe the use and safety of non-occupational post-exposure prophylaxis, and to evaluate its efficacy and sustainability. In addition, the CDC released the first US guidelines for non-occupational post-exposure prophylaxis in 2005 (Smith *et al.*, 2005). CDC guidelines recommend the use of 3 antiretroviral agents for 28 days following high-risk sexual exposure to a known or suspected HIV-infected partner. As in the case of occupational exposures, prophylaxis is recommended only if initiated within 72 hours of exposure, but more prompt initiation is recommended. It should be noted that the recommendation to start ART within 72 hours of sexual exposure is based upon decreased post-exposure efficacy in one trial of macaques with delayed ART (Otten *et al.*, 2000).

Clinical studies of non-occupational post-exposure prophylaxis

The use of ART prophylaxis following sexual exposures has met with cynicism due to cost (Pinkerton *et al.*, 1998; Low-Beer *et al.*, 2000), poor completion rates in certain settings (Wiebe *et al.*, 2000) as well as the potential impact on sexual disinhibition. In the absence of prospective controlled trials, several feasibility studies on non-occupational post-exposure prophylaxis have been undertaken to assess acceptability, associated toxicity, adherence and impact on sexual behavior (Table 5.2) (Wiebe *et al.*, 2000; Kahn *et al.*, 2001; Roland *et al.*, 2001; Cordes *et al.*, 2004; Martin *et al.*, 2004; Schechter *et al.*, 2004; Garcia *et al.*, 2005; Meel, 2005; Roland *et al.*, 2005; Winston *et al.*, 2005; Poynten *et al.*, 2007; Shoptaw *et al.*, 2008; Sonder *et al.*, 2007). In general, studies suggest that post-exposure ART is acceptable given completion rates for a 28 day ART course ranging from 64 to 100 percent.

The studies also revealed failures of non-occupational post-exposure prophylaxis to prevent HIV acquisition (Table 5.2) (Cordes *et al.*, 2004; Roland *et al.*, 2005). One study evaluated 702 individuals in a US non-occupational post-exposure prophylaxis registry who received either 2 or 3 drugs for 28 days with 12 weeks of follow-up (Roland *et al.*, 2005). Of seven participants who seroconverted, four reported 100 percent adherence to ART therapy, suggesting prophylactic failure. Failure was associated with anal intercourse exposure and delayed initiation of treatment. Of note, three seroconverters initiated treatment more than 55.5 hours after exposure, and none of the seven seroconverters received triple drug therapy. The finding of a delay to prophylactic initiation is particularly concerning, since the majority of those who view themselves at risk for HIV infection do not start therapy promptly (Kindrick *et al.*, 2006), yet the macaque model suggests that early therapy may be critical (Hu *et al.*, 2000; Otten *et al.*, 2000). Although no conclusions can be made regarding the impact of dual versus triple ART regimens for prophylaxis in the study, the fact that all seroconverters in the trial received dual regimens lends some support to CDC recommendations for triple therapy for high-risk exposures.

Antiretroviral selection for non-occupational post-exposure prophylaxis

With or without efficacy data, non-occupational post-exposure prophylaxis is increasingly being used for prevention (Poynten *et al.*, 2007), and the selection of ART agents must consider the pharmacology of specific agents as previously discussed, as well as cost, tolerability and the presence or likelihood of resistance in source partners.

Pinkerton and colleagues have argued that the only cost-effective use of non-occupational post-exposure prophylaxis might arise in the treatment of the highest-risk persons engaged in the most risky sexual act or unprotected anal intercourse (Pinkerton *et al.*, 2004). Obviously, such models of post-exposure prophylaxis are greatly limited by the absence of proof of benefit and are bounded by assumptions about the degree of benefit theoretically possible.

Table 5.2 Studies on antiretroviral non-occupational post-exposure prophylaxis

Reference	Study design	Study aim	Participants, n	Participants repeating course, %	Participants completing 4-week regimen, %	Seroconversions, n	Comments
		Sexual exposure, IV drug use exposure, or both					
Roland et al., 2001[a]	Cross-sectional study of source participants	Evaluate accuracy of source participants' report of HIV status and ART history	64	–	–	–	61% of source participants with ≥1 mutation on genotype testing
Kahn et al., 2001	Prospective cohort	Feasibility study	401	12[b]	78	0	Demonstrated safety and acceptability of 4-week nPEP regimen
Schechter et al., 2004	Prospective cohort	Feasibility and effect on sexual behavior	200	24.9[c]	89	1	No increase in high-risk sexual behavior
Martin et al., 2004	Prospective cohort	Effect on sexual behavior	397	17[d]	–	0	No increase in high-risk sexual behavior
Cordes et al., 2004	Case report	Report nPEP failure	1	0	100	1	Failure after anal intercourse exposure with triple, drug-sensitive regimen
Roland et al., 2005	Retrospective analysis	Characterize HIV seroconversion	702	–	–	7	4 of 7 non-responders reported 100% adherence; failure associated with unprotected anal intercourse and treatment delay
Winston et al., 2005	Retrospective cohort	Tolerability comparing 3 regimens	385	–	68–85	0	TDF–3TC–d4T regimen better tolerated than ZDV–3TC or ZDV–3TC–NFV

Study	Design	Purpose	N				Comments
Poynten et al., 2007	Prospective cohort	Characterize use and estimate HIV infections prevented	1552	–	80% of 1146 with 4-week data	0	Estimated that 0.9–9.2 HIV infections were prevented
Sonder et al., 2007	Retrospective analysis	Evaluate trends, compliance and outcomes	245	–	85	0	Used a single dose of nevirapine 200{ts}mg followed by a 28-day course of AZT/3TC/NFV
Shoptaw et al., 2007	Prospective cohort	Feasibility and safety study	98	–	64	0	No increase in high-risk sexual behavior in this community based nPEP program
Sexual assault							
Wiebe et al., 2000	Prospective cohort	Program evaluation	71	–	11	0	Adherence associated with higher-risk exposures
Garcia et al., 2005	Prospective cohort	Program evaluation	278	–	61	0	Adherence associated with dual regimens
Meel, 2005	Retrospective cohort	Evaluate acceptability and seroconversions	501	–	7	0	Dual therapy; only 1.4% of patients had repeated HIV testing at 12 weeks

Adapted from Cohen et al. (2007)[e]

Notes: Dashes indicate that the data were not collected in the study. 3TC, lamivudine; ART, antiretroviral therapy; d4T, stavudine; IV, intravenous; NFV, nelfinavir; nPEP, non-occupational post-exposure prophylaxis; ZDV, zidovudine.

[a]Substudy of Kahn et al. (7).

[b]Follow-up of 6 months.

[c]Median follow-up of 24.2 months.

[d]Follow-up of 12 months.

[e]The American College of Physicians is not responsible for the accuracy of this translation.

In a study comparing rates of toxicity in uninfected individuals to rates among HIV-infected patients on treatment, a six-fold higher rate of ART toxicity and an eight-fold higher rate of discontinuation was observed in those taking ART for non-occupational post-exposure prophylaxis (Quirino *et al.*, 2000). Using an evidence-based mathematical model, Bassett and colleagues suggested that two-drug post-exposure prophylaxis may be more effective than triple therapy in preventing HIV transmission because of its equivalent efficacy and higher completion rates (Bassett *et al.*, 2004). Additional studies have reported on the discontinuation of post-exposure prophylaxis due to adverse effects for occupational (Ippolito and Puro, 1997; Wang *et al.*, 2000) and non-occupational prophylaxis (Puro *et al.*, 2001; Sonder *et al.*, 2007). However, no difference was observed in rates of discontinuation when a triple drug regimen including a protease inhibitor was compared with a dual NRTI regimen for non-occupational ART prophylaxis (Puro *et al.*, 2001). Although rates of side-effects are high during non-occupational post-exposure prophylaxis, they are generally mild, do not lead to treatment discontinuation and have been reversible.

The number of drugs to initiate and the choice of agents for ART must also take into account the presence or risk of HIV-resistant variants and the pharmacology of antiviral agents. The prevalence of *de novo* resistance in individuals with incident HIV infection differs greatly by country and region (Vella and Palmisano, 2005), but should be incorporated into selection of ART prophylactic regimens. As previously discussed, recent findings on the pharmacology of antiretrovirals in the genital tract suggest that certain antiretroviral agents may be preferable for the prevention of HIV following sexual exposure (Figure 5.2) (Reddy *et al.*, 2003; Dumond *et al.*, 2007; Vourvahis *et al.*, 2008). First-dose genital tract drug-exposure data are available for men who received tenofovir (Dumond *et al.*, 2007; Vourvahis *et al.*, 2008) and for women who received a total of 12 different antiretroviral agents (Dumond *et al.*, 2007; Vourvahis *et al.*, 2008). In the two studies by Dumond and colleagues, lamivudine, emtricitabine, zidovudine, tenofovir and maraviroc concentrations in the female genital tract were higher than in blood plasma, and lopinavir and atazanavir achieved low to moderate genital tract concentrations (Dumond *et al.*, 2007). Efavirenz achieved female genital secretion concentrations that were <1 percent of those in blood plasma. In addition, many antiretrovirals are detected in genital secretions within 1–2 hours after the first dose of ART. Some (e.g. abacavir, tenofovir, didanosine) achieve higher genital tract concentrations after a single dose than during steady-state dosing conditions (Dumond *et al.*, 2007). These results suggest that the emergent and appropriate ART with some combination of lamivudine or emtricitabine, zidovudine, tenofovir, maraviroc, and possibly lopinavir or atazanavir, can be expected to rapidly concentrate in the target tissues. In contrast, the poor penetration of efavirenz into the genital compartment argues against its use for the prevention of sexual transmission. The poor penetration of efavirenz may explain the persistence of HIV RNA in cervical and vaginal secretions in the previously described study among

Kenyan female sex workers taking an NNRTI-containing regimen (Graham *et al.*, 2007). Data on newer antiretrovirals including integrase inhibitors are not yet available, but represent additional options following further study.

Ease of therapy is also an important consideration in choosing an ART regimen for non-occupational post-exposure prophylaxis. Two case-controlled studies of non-occupational post-exposure prophylaxis following high-risk sexual exposures were conducted using tenofovir DF and lamivudine in 44 subjects, and the combination of tenofovir DF and emtricitabine in an additional 68 subjects. Subjects in both studies with tenofovir-based dual regimens had higher completion rates of a 28-day post-exposure regimen than historical controls taking 2- or 3-drug regimens containing zidovudine ($P < 0.0001$) (Mayer *et al.*, 2008). Dropout rates during non-occupational post-exposure prophylaxis treatment are high (Kahn *et al.*, 2001; Winston *et al.*, 2005), particularly in the setting of sexual assault (Wiebe *et al.*, 2000; Garcia *et al.*, 2005; Meel, 2005; Sonder *et al.*, 2007). Although the reasons for discontinuation of therapy may include reassessment of risk exposure and/or intolerable side-effects, the lessons on increased adherence with simpler regimens in the case of ART for treatment should not be ignored.

Pre-exposure prophylaxis to prevent HIV transmission

Biological strategies to prevent an HIV transmission event are limited to barrier methods, an HIV vaccine and the use of topical or oral antiviral agents as pre-exposure prophylaxis. Barrier methods, including condoms and circumcision, clearly work; however, diaphragms failed in at least one trial (Padian *et al.*, 2007). An efficacious HIV vaccine will need to be capable of producing neutralizing antibodies, and this may or may not prove possible (Johnston and Fauci, 2007; see also Chapter 3).

Data to support the feasibility of oral therapy or topical microbicides (Van Rompay *et al.*, 2002; Lederman *et al.*, 2004) come from studies with rhesus macaques in which ART administration before exposure prevented infection (Tsai and Follis, 1995; Van Rompay *et al.*, 2001, 2002). As with post-exposure prophylaxis, success of the intervention in studies of macaques varied according to the challenge model and the doses and routes of antiretroviral agents employed. The "low-dose" (3.8×10^5 viral particle equivalents) vaginal (Kim *et al.*, 2006) and rectal (Subbarao *et al.*, 2006) mucosal challenge with a SIV-HIV construct ($SHIV_{SF162P3}$) has gained the greatest attention in studies of pre-exposure prophylaxis.

Animal models of pre-exposure prophylaxis

A series of studies from the CDC using the rectal mucosal challenge model demonstrate that oral tenofovir delays $SHIV_{SF162P3}$ infection; however, after repeated

Table 5.3 Ongoing and proposed clinical trials of pre-exposure prophylaxis

Study (sponsor)	Study and agent(s) (dose)	Population (target N)	Sites
US CDC-NCHSTP-4323	Phase II daily TDF or daily oral placebo	MSM Age: 18 to 60 (400)	USA
US CDC-NCHSTP-4370	Phase II/III daily TDF or daily oral placebo	IDU Age: 20 to 60 (2000)	Thailand
CDC-NCHSTP-4940; BOTUSA MB06	Phase III daily Truvada or daily oral placebo	Men and women Age: 18 to 29 (1200)	Botswana
iPrEX (NIAID/BMGF)	Phase III daily Truvada (tenofovir disoproxil fumarate 300 mg + emtricitabine 200 mg) or daily oral placebo	MSM Age: 18 and up (3000)	Peru, Ecuador, Brazil, Thailand, South Africa, USA
FHI (USAID)	Phase III daily Truvada or daily oral placebo	High-risk women Age: 18 to 35 (3900)	Kenya, Malawi, South Africa, Tanzania, Zimbabwe
Partners' study (BMGF)	Phase III daily TDF, daily Truvada, or daily oral placebo	Discordant heterosexual couples Age: 18 to 60 (3900)	Uganda, Kenya
VOICE/MTN 003 (NIAID)	Phase IIB safety and effectiveness of daily Tenofovir gel (1%) or placebo gel, or daily TDF (300 mg), Truvada (TDF 300 mg/FTC 200 mg), or oral placebo	Non-pregnant premenopausal women Age: 18 to 35 (2400 oral, 1600 gel)	South Africa, Zambia, Malawi, Uganda, Zimbabwe

Reprinted from Cohen and Kashuba (2008).

exposure, infection was prevented in only one of four animals studied (Subbarao *et al.*, 2006). Interestingly, in a follow-up comparison study using a single high-dose intrarectal inoculum, two of five tenofovir-treated macaques were protected, suggesting that the single high-dose model did not overwhelm the protective effects of oral tenofovir in all exposed macaques, and that the repeated-low-dose model was at least as stringent (Subbarao *et al.*, 2007). However, in another study in which the combination of high-dose tenofovir and emtricitabine was given subcutaneously, six of six macaques were completely protected from SIV-HIV despite repeated rectal exposures (Garcia-Lerma *et al.*, 2006). With daily therapy it remains unclear whether protection in this animal model is based on the effects of antiviral agents before or after exposure; whether the drugs are actually preventing infection, or treating and eliminating early infection and blocking seroconversion. Since the animals have not been sacrificed, absolute proof that they remained uninfected is lacking. Perhaps most intriguing, subcutaneous dosing of emtricitabine (20 mg/kg) with a supratherapeutic dose of tenofovir (22 mg/kg) given 2 hours before and 24 hours following intrarectal exposure prevented infection in 6 macaques exposed (Garcia-Lerma *et al.*, 2008). Such an approach is far closer to true pre-exposure prophylaxis than any other model, and therefore, of greatest therapeutic relevance. However, it should be noted that only supratherapeutic doses of tenofovir, achieving two-fold higher concentrations in the macaques than seen in humans, were uniformly protective.

Pre-exposure clinical trials

Based on the above data in animals, several trials of oral pre-exposure prophylaxis for uninfected high-risk individuals are under way in Peru, Ecuador, Thailand, Botswana and the United States using either tenofovir or the combination of tenofovir and emtricitabine (Bill and Melinda Gates Foundation, 2002; CDC, 2006; AIDS, 2007) (Table 5.3). Earlier attempts to launch similar pre-exposure prophylaxis trials in Cambodia, Cameroon, Malawi and Nigeria were aborted due to protests regarding trial design, the lack of perceived community participation, and concerns regarding the long-term benefit for the study populations (Grant *et al.*, 2005; Page-Shafer *et al.*, 2005). In addition, considerable concern arose regarding the risk of resistance with use of a single agent, such as tenofovir, based on data from animal studies (Magierowska *et al.*, 2004; Subbarao *et al.*, 2006). In response, mathematical modeling of 600 participants receiving prophylaxis in the Botswana tenofovir trial revealed that less than 1 percent of the predicted 45 seroconverters would acquire or develop a tenofovir-resistant strain (Smith *et al.*, 2006). However, the preventive advantages of the tenofovir–emtricitabine combination in the rectal transmission model in rhesus macaques (Garcia-Lerma *et al.*, 2006) led to reconsideration of mono versus dual therapy, and some trials replaced tenofovir with the tenofovir and emtricitabine combination.

A safety trial of pre-exposure tenofovir in 936 high-risk women in Ghana, Cameroon and Nigeria was completed, and showed no difference in adverse events or grade-3 or -4 laboratory abnormalities in subjects receiving tenofovir versus placebo (Peterson *et al.*, 2007). Although fewer seroconversion events occurred in participants receiving tenofovir versus placebo (two events versus six events), the study was not of sufficient size or duration to confirm the efficacy of tenofovir in preventing HIV infection. However, the absence of detectable tenofovir resistance among subjects who seroconverted and an increase in self-reported condom use during the study period are reassuring as other pre-exposure studies proceed.

Even prior to the findings from the above trial on pre-exposure prophylaxis, news reports emerged on the sale of antiretrovirals at clubs and self-administered use prior to high-risk sex (Costello, 2005; Cohen, 2006a). Seven percent of attendees of minority gay pride events in four cities in 2004 reported having ever used ART as pre-exposure prophylaxis (Kellerman *et al.*, 2006). These reports suggest that intermittent pre-exposure prophylaxis has begun in some communities.

Topical microbicides as pre-exposure prophylaxis

While studies of topical microbicides have been disappointing to date (Chapter 4), studies of topical antiretroviral agents are moving forward. The pharmacology of tenofovir has been studied (Deeks *et al.*, 1998; Ghosn *et al.*, 2004; Mayer *et al.*, 2006). Results from a Phase I study of tenofovir vaginal gel in HIV-infected and -uninfected women found twice-daily application to be well tolerated. Low-level, systemic absorption was observed, and raised concerns regarding the selection of tenofovir resistance, although none of the HIV-infected women with detectable plasma or cervicovaginal HIV RNA developed mutations associated with tenofovir resistance (Mayer *et al.*, 2006). More recently, pharmacokinetic evaluation of tenofovir cervicovaginal fluid, vaginal tissue and blood plasma concentrations was performed over 24 hours after a single 4-ml dose of 1% tenofovir gel in 21 healthy volunteers (Schwartz *et al.*, 2008). Using an assay sensitivity of 1 ng/ml, all women had detectable tenofovir concentrations in blood plasma: most were <5 ng/ml, although 20 percent of subjects had concentrations up to 19.5 ng/ml. At the end of 24 hours, tenofovir vaginal fluid and tissue exposures were as high: $4.5–47.1 \times 10^4$ ng/ml and 15×10^3 ng/g of tissue, respectively. These are similar to protective systemic exposures previously seen in macaques given subcutaneous dosing of tenofovir.

ART as public health prevention

The degree of public benefit of "therapeutic" ART for prevention will depend on (1) the proportion of HIV-infected individuals treated, (2) the ability to target

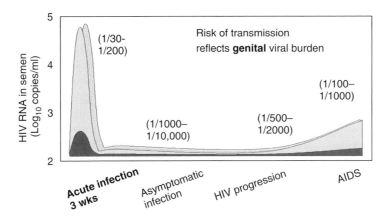

Figure 5.4 Sexual transmission of HIV. The numbers in parentheses are estimates of HIV transmission per episode of sexual intercourse. Prevention efforts such as ART that reduce viral burden in the semen below the required transmission threshold (dark gray) would be expected to be effective. HIV, Human Immunodeficiency Virus; AIDS, Acquired Immune Deficiency Syndrome; RNA, ribonucleic acid. Reproduced from Cohen and Pilcher (2005), with permission.

ART to those most likely to transmit HIV (Figure 5.4), (3) the efficacy of ART to reduce viral load in the genital tract, (4) the residual infectiousness of treated participants and the emergence and transmission of drug-resistant strains, and (5) behavioral disinhibition and the degree to which this can overwhelm the suppressive effects of ART.

Targeting ART preventative therapy to the most infectious persons or providing prophylactic medications to persons at the greatest risk for acquiring HIV is a major challenge. In addition, even if appropriate targeting of therapy could be accomplished, the prevention benefits of ART for the general population have been the subject of great debate. Several mathematical modeling studies have suggested that widespread administration of ART could substantially reduce HIV-related mortality and HIV incidence (Blower *et al.*, 2000; Law *et al.*, 2001), but that any possible preventative benefit of ART could be undermined by behavioral disinhibition (Blower *et al.*, 2000; Law *et al.*, 2001; Velasco-Hernandez *et al.*, 2002). Baggaley and colleagues published a review of modeling strategies and methods used to explore this question (Baggaley *et al.*, 2005). In another modeling paper, they concluded that ART is unlikely to reduce HIV transmission due to (1) limited magnitude of benefit of ART on HIV transmission, (2) widespread emergence of ART-resistant variants, and (3) considerable increases in HIV risk-taking behavior (Baggaley *et al.*, 2006). Mellors and colleagues concluded that 2.7–3.2 million incident HIV cases could be averted in sub-Saharan Africa by targeting pre-exposure prophylaxis with 90 percent efficacy to the highest-risk groups over a 10-year period (Abbas *et al.*, 2007), and by "preventing" disinhibition (Abbas *et al.*, 2007). Notably, in this model, the efficacy of ART was the most important determinant of preventative benefit, and other key factors included rates of discontinuation, target coverage, and sexual disinhibition.

Mathematical models are inherently limited by the assumptions that are used to construct the models, and are dependent on the availability and accuracy of surveillance, behavioral and resistance data in site-specific settings – which are often limited in resource-poor settings. In addition, modeling studies of HIV transmission can be limited by the assumption that sexual behavior is homogeneous within populations. As such, mathematical modeling offers unconfirmed predictions within a range of possibilities, but is not an alternative to evidence-based research. It can, however, be used to guide research and facilitate in the interpretation of research findings.

ART and sexual behaviors: non-occupational post-exposure prophylaxis

Several studies have evaluated the impact of post-exposure ART on sexual behavior. In one prospective study, 39 of 401 participants who received post-exposure prophylaxis (12 percent by Kaplan-Meier estimation) requested a second post-exposure prophylaxis course after a subsequent exposure (Kahn *et al.*, 2001). Schechter and colleagues supplied 200 high-risk, HIV-negative homosexual men with a 4-day non-occupational post-exposure prophylaxis starter pack to be initiated after an eligible exposure (Schechter *et al.*, 2004). Non-occupational post-exposure prophylaxis was started once by 68 participants, twice by 14 participants and 3 times by 2 participants. Overall, high-risk sexual behaviors decreased over a median follow-up of 24.2 months for participants who received non-occupational post-exposure prophylaxis and those who declined therapy, but a decrease in unprotected oral sex approached statistical significance only among those who took non-occupational post-exposure prophylaxis (23.5 percent at baseline vs 11.8 percent at last visit; = 0.06). In another study of 397 participants who received non-occupational post-exposure prophylaxis (Martin *et al.*, 2004), only 17 percent sought repeated post-exposure prophylaxis and 73 percent reported a decrease in high-risk sexual acts. More recently, a community-based program providing non-occupational exposure found that the 49 percent of participants who remained on study at week 26 reported a decrease in the number of sexual partners following a course of ART to prevent sexual transmission (Shoptaw *et al.*, 2008). Furthermore, despite the widespread availability of post-exposure prophylaxis in all hospital emergency departments and the Municipal Health Service in Amsterdam, the number of requests for post-exposure prophylaxis following sexual exposure increased very minimally between 2000 and 2004 (Sonder *et al.*, 2007).

In contrast, Australian men who had sex with men who received non-occupational post-exposure prophylaxis were more likely to report unprotected anal intercourse 1 year later compared with those who had not had non-occupational post-exposure prophylaxis (50 percent vs 36 percent; $P = 0.009$), and were at increased risk for acquiring HIV (incidence 2.37 cases per 100 person-years, relative risk 2.30, CI 1.05–5.06) (Grulich *et al.*, 2006). Roland and colleagues reported that 66 percent of sexual assault victims in South Africa who initiated non-occupational post-exposure prophylaxis reported unprotected intercourse at the 6-month follow-up, and that 2 of 4 seroconversions among 135 participants were probably due to ongoing high-risk exposures (Roland *et al.*, 2006a).

Based on concerns of sexual disinhibition, the post-exposure program in San Francisco incorporated five risk-reduction counseling sessions and found a reduction in unprotected intercourse (Martin *et al.*, 2004). A follow-up study of a randomized, controlled trial evaluating the impact of a two- versus five-session risk-reduction strategy has been completed, and preliminary analysis indicated no difference in unprotected sex or repeat post-exposure requests between the two arms (Roland, 2007). However, more recently, a concerning 21 percent of 89 subjects reported unprotected sex during their course of ART post-exposure prophylaxis, which in multivariate analysis was associated with a history of engagement with HIV/AIDS service organizations (Golub *et al.*, 2007).

Kalichman and colleagues evaluated attitudes toward post-exposure ART among a convenience sample of men who have sex with men (MSM) (Kalichman, 1998). Those expressing intent to use post-exposure prophylaxis in the future were more likely to state that ART could prevent HIV infection, and that newer antiretroviral treatments decreased their concern regarding unprotected sex and acquiring HIV infection. In addition, those intending to use ART as prophylaxis were more likely to report high-risk sex behaviors and substance abuse – thus the individuals who might benefit the most from additional prevention interventions.

In summary, the majority of data from several prospective studies of post-exposure prophylaxis after sexual exposure have failed to demonstrate misuse or abuse of this approach or an increase in risk-taking behavior (Martin *et al.*, 2004; Schechter *et al.*, 2004; Roland *et al.*, 2006b; Sonder *et al.*, 2007; Shoptaw *et al.*, 2008). It should be noted that repeat requests for non-occupational post-exposure prophylaxis and reports of subsequent high-risk behavior observed in these studies cannot be interpreted as an increase in behavior, but might simply represent the persistence of pre-existing sexual behaviors.

Pre-exposure prophylaxis

The trial of pre-exposure tenofovir in West Africa, as described above, provides the only available data on the impact of pre-exposure ART on sexual behavior (Peterson *et al.*, 2007). Subjects in this study reported an increase in condom use; from an average of 52 percent at last coital act prior to the screening visit to 95 percent at the 12-month visit (for coital acts in the previous 7 days). The effect of pre-exposure ART on sexual behaviors will be measured in the ongoing and planned trials of pre-exposure prophylaxis. Clearly, reports on the black-market availability of ART for use prior to high-risk sex acts (Costello, 2005; Cohen, 2006a; Kellerman *et al.*, 2006) raise the concern of whether even the unproven concept of pre-exposure prophylaxis could propagate or lead to high-risk sexual behavior in certain communities or sexual networks.

Antiretroviral therapy for HIV-infected individuals

Data regarding the impact of ART on sexual behavior among HIV-infected individuals have been inconsistent (Kalichman *et al.*, 2006). Crepaz and colleagues conducted a meta-analysis to examine the sexual behavior of persons who receive ART, and did not find an increase in high-risk sexual behavior between treated and untreated HIV-infected persons (Crepaz *et al.*, 2004). However, unprotected sex was reported more often by individuals who believed that therapy prevented transmission or who expressed a perceived lesser HIV threat given the availability of ART.

There is some evidence that advances in ART have resulted in sexual disinhibition (Kalichman, 1998; Kelly *et al.*, 1998; Ostrow *et al.*, 2002; Tun *et al.*, 2004). In a study of MSM, HIV-positive subjects with a decreased concern about transmitting HIV to sexual partners due to the availability of ART were three to six times more likely to engage in unprotected insertive anal intercourse (Ostrow *et al.*, 2002). Similarly, in a study of intravenous drug users, those on ART had a three-fold increase in unprotected sex following treatment initiation (Tun *et al.*, 2004). In contrast, another study of MSM found no change in risk behavior for those on ART (Remien *et al.*, 2005).

Conflicting data also exist regarding the association between undetectable viral load measurement and risky sexual behavior. Limited data have suggested that undetectable viral loads increase risk behaviors (Dukers *et al.*, 2001; Crepaz *et al.*, 2004). However, in another study, HIV-infected MSM with detectable viral loads were more likely to report unprotected anal intercourse (Vanable *et al.*, 2003).

Studies among HIV-negative MSM have also found a reduced concern about safe sex and acquiring HIV (Ostrow *et al.*, 2002) due to advances in ART, and found an association between a decreased concern of acquiring HIV and engaging in high-risk sex (Stolte *et al.*, 2004).

Antiretroviral therapy for acute HIV infection

To date, there are limited data on sexual behaviors following diagnosis with acute HIV infection. A recent study found a decline in unprotected sex acts following diagnosis with acute HIV infection, from 17.6 to 11.3 average unprotected sex acts per participant (Steward *et al.*, 2007). However, a total of 170 unprotected sex acts were reported following diagnosis with acute HIV infection, indicating ongoing risk behavior during the period of highest infectivity. Of the 170 unprotected sex acts, 97 percent reportedly took place with HIV-positive partners, indicating an adoption of serosorting behavior among acutes. Such data emphasize the need to elucidate the currently unclear impact of ART selective pressure on virus in acutely infected persons who may continue to engage in unprotected sex.

ART resistance

A priori antiretroviral resistance is a growing concern, and studies have demonstrated that substantial HIV resistance in newly diagnosed and treatment-naïve patients is common worldwide (Little *et al.*, 1999, 2002; Grant *et al.*, 2002). Such *de novo* resistance indicates that HIV transmission has occurred from individuals taking incompletely suppressive ART, or from individuals who have discontinued failing therapy. However, the prevalence of HIV resistance in Montreal, Canada (Routy *et al.*, 2004), decreased from 13 percent between 1997 and 2000 to 4 percent between 2001 and 2003 ($P = 0.04$). Nevertheless, resistant HIV variants have been detected in genital secretions (Eron *et al.*, 1998) and have been linked with sexual transmission of HIV (Angarano *et al.*, 1994; Imrie *et al.*, 1997). In addition, recent studies have suggested that ART-resistant HIV viral strains may be less readily transmissible (Turner *et al.*, 2004; Yerly *et al.*, 2004) – a finding which, if confirmed, could diminish controversy over ART for prevention due to concerns regarding development of resistance.

Possible future strategies for ART as prevention: acute HIV infection

Currently there are no established guidelines for the initiation of ART during acute HIV infection, primarily due to a lack of data confirming virologic or immunologic benefit following discontinuation of ART after some period of treatment (Kaufmann *et al.*, 2004; Jansen *et al.*, 2005; Hecht *et al.*, 2006). However, the public health benefit of treating individuals with AHI has largely been ignored in discussions on the risk versus benefit of this strategy. Since a significant amount of incident HIV is due to sexual transmission from acutely infected index partners (Yerly *et al.*, 2001; Pilcher *et al.*, 2004a; Wawer *et al.*, 2005; Brenner *et al.*, 2007) and ART has been shown to decrease infectiousness (Wawer *et al.*, 2005), targeted ART for this group could prevent a substantial number of new infections in areas with high HIV incidence. However, the public health benefit would depend greatly on the ability to detect acute HIV infection, currently diagnosed via the detection of HIV RNA in antibody-negative individuals – technology which is costly and largely unavailable in resource-poor settings.

Even in the presence of appropriate HIV diagnostic tests, acute HIV infection is largely missed on presentation, given protean symptoms of acute retroviral syndrome (Rosenberg *et al.*, 1999; Weintrob *et al.*, 2003). Although wide-scale screening of HIV antibody-negative specimens via HIV RNA pooling has proved feasible and effective in detecting acute HIV infection (Bollinger *et al.*, 1997; Rosenberg *et al.*, 1999; Pilcher *et al.*, 2004b, 2005; Fiscus *et al.*, 2007), its implementation remains limited. Nevertheless, the widespread and targeted implementation of this strategy to those at the highest risk of AHI, such as those presenting to STD clinics in sub-Saharan Africa, represents an important opportunity to prevent ongoing transmission and allow early intervention. In resource-poor settings, this strategy could be limited to temporary ART administration during the period of AHI. Since rebound viral load following discontinuation has been observed, combining ART with counseling on the high risk of transmission during acute infection and following treatment discontinuation would be crucial.

Intermittent post- or pre-exposure prophylaxis

The pre-exposure prophylaxis strategy utilized in the one completed study and ongoing pre-exposure clinical trials incorporates daily dosing of tenofovir or combination tenofovir/emtricitabine. However, intermittent ART dosing prior to unprotected or high-risk sex may represent a more practical and cost-effective, and thus more feasible, option for high-risk patients in high prevalence, resource-poor settings. Additional studies of intermittent or weekly tenofovir/emtricitabine in macaques, or utilizing the new SCID mice model with HIV-infected BLT humanized mice, will hopefully provide additional data to guide development of such pre-exposure dosing strategies. Currently, the CAPRISA 004 tenofovir gel trial is investigating coitally dependent dosing (12 hours prior and 12 hours after exposure). To date, studies of pre- and post-exposure prophylaxis in humans have focused on one strategy, and a broader approach combining oral and vaginal ART in high-risk individuals, such as commercial sex workers, may provide the greatest efficacy in ART for prevention.

Newer antiretroviral agents for HIV prevention

Novel compounds that prevent HIV cellular entry are attractive candidates for HIV prevention. Maraviroc, a novel chemokine receptor antagonist (CCR5), may be well suited for this purpose. Recently, a non-blinded pharmacokinetic study was performed in 12 healthy women receiving maraviroc (300 mg BID) for 7 days (Dumond *et al.*, 2008). Within 8 hours after a single dose, all women had greater cervicovaginal fluid concentrations of maraviroc compared to those in blood plasma, and concentrations remained greater than in blood plasma for all subsequent days. Additionally, maraviroc cervicovaginal fluid protein binding was 10-fold lower than in blood plasma, suggesting that >90 percent of the drug is available for pharmacologic activity. Vaginal tissue biopsy samples had two-fold higher exposure than blood plasma throughout the dosing interval. As maraviroc exhibits the greatest female genital tract penetration relative to blood plasma of all commercially-available antiretrovirals reported to date, these data show promise as an effective oral PrEP/PEP agent.

Treatment of adolescent and young women

Each of the above post-exposure and pre-exposure strategies must consider the risk versus benefit of ART in women of child-bearing age and its potential impact on pregnancy and a fetus. If ART is employed to prevent HIV infection, the choice of agents must consider the risk of resistance in relation to the potential need for ART to prevent mother-to-child transmission (MTCT), particularly given limited options in many settings. To date, nevirapine and/or zidovudine are the most widely used agents worldwide for prevention of vertical transmission. Nevirapine is not a good candidate for pre-exposure prophylaxis, given the likely unacceptable risk of severe toxicity in uninfected, immune-competent women with a CD4 cell count greater than $250\,cells/mm^3$. Limited tenofovir pharmacokinetic data in pregnancy are available, although it achieves cord blood concentrations 4- to 6-fold higher than maternal plasma concentrations, and amniotic fluid concentrations 1.6-fold higher than maternal plasma concentrations (Yem *et al.*, 2006).

Importantly, both oral and vaginal ART represent prevention methods which can be controlled by women and would not require partner consent for their use. Thus, in settings where women are unable to negotiate safe sex practices, both would provide a critically needed alternative to prevent HIV infection.

Conclusions

Now more than 25 years into the AIDS pandemic, we are just realizing the global expansion of ART (UNAIDS, 2006). Antiretroviral therapy will prolong survival, but at the same time will greatly increase the time available for viral transmission and the transmission of resistant isolates. The degree to which ART, provided as treatment and/or for prevention, will reduce infectiousness at the individual level remains unknown, yet the first results from pre-exposure trials are anticipated in 2008. At present, we are limited to important, but hypothetical, arguments about the benefit of ART at the population level (Montaner *et al.*, 2006). Confirming the degree to which ART lessens the infectiousness of HIV-infected individuals is of critical importance, and could potentially impact CD4 cell count guidelines for ART initiation in resource-poor countries to allow for earlier initiation.

The expanded use of ART for prevention in research, clinical and casual practice seems inevitable. Post-exposure prophylaxis following occupational exposures will likely remain the standard of care, with anticipated changes in the guidelines for the number and specific agents used based on accumulating evidence from large registries. Post-exposure prophylaxis following non-occupational exposure based on exposure risk will likely become the standard of care, and guidelines will evolve in parallel with findings on genital tract concentrations, with newer antiretroviral agents yet to be studied. Both strategies for post-exposure prophylaxis may prevent a small, but important, number of cases of HIV. Given recent data on the penetration of various antiretroviral agents into the genital tract, the selection of antiretroviral agents to prevent sexual transmission should likely include lamivudine or emtricitabine, zidovudine and/or tenofovir, and maraviroc, with consideration of the addition of boosted lopinavir or atazanavir, as tolerability and resistance patterns allow. Non-occupational post-exposure prophylaxis should be started as soon as possible after exposure, and within 72 hours following high-risk sexual exposures, and continued for 28 days.

Pre-exposure prophylaxis has attracted great attention as a potentially safe, time-limited prevention method for very high-risk people (Cohen, 2006b). Ongoing clinical trials will demonstrate the costs and benefits of pre-exposure prophylaxis, allowing evidence-based use of this approach. Future strategies could also recommend ART in acutely infected individuals to lower the very high viral load and risk of HIV transmission with unprotected sexual encounters during this phase of infection.

Acknowledgments

This work was supported by the University of North Carolina Center for AIDS Research (P30HD-37260 and R01AI041935). Dr Kashuba has been supported by AI54980. Dr Gay has been supported by a Centers for Disease Control and Prevention Association of Teachers of Preventive Medicine Fellowship and National Institutes of Health Training Grant (T32AI07151).

References

Abbas, U. L., Anderson, R. M. and Mellors, J. W. (2007). Potential impact of antiretroviral chemoprophylaxis on HIV-1 transmission in resource-limited settings. *PLoS ONE*, 2, e875.

AIDS, V. A. C. (2007). PrEP Watch.

Al-Harthi, L., Kovacs, A., Coombs, R. W. *et al.* (2001). A menstrual cycle pattern for cytokine levels exists in HIV-positive women: implication for HIV vaginal and plasma shedding. *AIDS*, 15, 1535–43.

Allen, S., Tice, J., Van de Perre, P. *et al.* (1992). Effect of serotesting with counselling on condom use and seroconversion among HIV discordant couples in Africa. *Br. Med. J.*, 304, 1605–9.

Almeda, J., Casabona, J., Allepuz, A. *et al.* (2002). [Recommendations for non-occupational post-exposure HIV prophylaxis. Spanish Working Group on Non-occupational Post-exposure HIV Prophylaxis of the Catalonian Center for Epidemiological Studies on AIDS and the AIDS Study Group]. *Enferm. Infecc. Microbiol. Clin.*, 20, 391–400.

Andreoletti, L., Skrabal, K., Perrin, V. *et al.* (2007). Genetic and phenotypic features of blood and genital viral populations of clinically asymptomatic and antiretroviral-treatment-naive clade a human immunodeficiency virus type 1-infected women. *J. Clin. Microbiol.*, 45, 1838–42.

Angarano, G., Monno, L., Appice, A. *et al.* (1994). Transmission of zidovudine-resistant HIV-1 through heterosexual contacts. *AIDS*, 8, 1013–14.

Baggaley, R. F., Ferguson, N. M. and Garnett, G. P. (2005). The epidemiological impact of antiretroviral use predicted by mathematical models: a review. *Emerg. Themes Epidemiol.*, 2, 9.

Baggaley, R. F., Garnett, G. P. and Ferguson, N. M. (2006). Modelling the impact of antiretroviral use in resource-poor settings. *PLoS Med.*, 3, e124.

Bassett, I. V., Freedberg, K. A. and Walensky, R. P. (2004). Two drugs or three? Balancing efficacy, toxicity, and resistance in postexposure prophylaxis for occupational exposure to HIV. *Clin. Infect. Dis.*, 39, 395–401.

Bernasconi, E., Jost, J., Ledergerber, B. *et al.* (2001). Antiretroviral prophylaxis for community exposure to the human immunodeficiency virus in Switzerland, 1997–2000. *Swiss Med. Wkly*, 131, 433–7.

Bill and Melinda Gates Foundation (2002). Family Health International Receives Grant to Evaluate Once-Daily Antiretroviral as a Potential Method of HIV Prevention. Available at http://www.gatesfoundation.org (accessed 24 August 2006).

Blower, S. M., Gershengorn, H. B. and Grant, R. M. (2000). A tale of two futures: HIV and antiretroviral therapy in San Francisco. *Science*, 287, 650–4.

Bollinger, R. C., Brookmeyer, R. S., Mehendale, S. M. *et al.* (1997). Risk factors and clinical presentation of acute primary HIV infection in India. *J. Am. Med. Assoc.*, 278, 2085–9.

Brambilla, D., Reichelderfer, P. S., Bremer, J. W. *et al.* (1999). The contribution of assay variation and biological variation to the total variability of plasma HIV-1 RNA measurements. The Women Infant Transmission Study Clinics. Virology Quality Assurance Program. *AIDS*, 13, 2269–79.

Brenner, B. G., Roger, M., Routy, J. P. *et al.* (2007). High rates of forward transmission events after acute/early HIV-1 infection. *J. Infect. Dis.*, 195, 951–9.

Bunnell, R., Ekwaru, J. P., Solberg, P. *et al.* (2006). Changes in sexual behavior and risk of HIV transmission after antiretroviral therapy and prevention interventions in rural Uganda. *AIDS*, 20, 85–92.

Cardo, D. M., Culver, D. H., Ciesielski, C. A. *et al.* (1997). A case-control study of HIV seroconversion in health care workers after percutaneous exposure. Centers for Disease Control and Prevention Needlestick Surveillance Group. *N. Engl. J. Med.*, 337, 1485–90.

Castilla, J., Del Romero, J., Hernando, V. *et al.* (2005). Effectiveness of highly active antiretroviral therapy in reducing heterosexual transmission of HIV. *J. Acquir. Immune Defic. Syndr.*, 40, 96–101.

CDC (2001). Updated US Public Health Service Guidelines for the Management of Occupational exposure to HBV, HCV, and HIV and recommendations for Postexposure Prophylaxis. *Morb. Mortal. Wkly Rep.*, 50, 1–42.

CDC (2006). CDC Trials of Pre-Exposure Prophylaxis for HIV Prevention. Atlanta, GA: CDC.

Chakraborty, H., Sen, P. K., Helms, R. W. *et al.* (2001). Viral burden in genital secretions determines male-to-female sexual transmission of HIV-1: a probabilistic empiric model. *AIDS*, 15, 621–7.

Chaudry, N. I., Eron, J. J., Naderer, O. J. *et al.* (2002). Effects of formulation and dosing strategy on amprenavir concentrations in the seminal plasma of human immunodeficiency virus type 1-infected men. *Clin. Infect. Dis.*, 35, 760–2.

Cohen, J. (2006a). Protect or disinhibit? New York, NY: New York Times.

Cohen, J. (2006b). Infectious disease. At International AIDS Conference, big names emphasize big gaps. *Science*, 313, 1030–1.

Cohen, M. S. and Pilcher, C. D. (2005). Amplified HIV transmission and new approaches to HIV prevention. *J. Infect. Dis.*, 191, 1391–3.

Cohen, M. S., Hoffman, I. F., Royce, R. A. *et al.* (1997). Reduction of concentration of HIV-1 in semen after treatment of urethritis: implications for prevention of exual transmission of HIV-1. AIDSCAP Malawi Research Group. *Lancet*, 349, 1868–73.

Cohen, M. S., Hosseinipour, M., Kashuba, A. and Butera, S. (2002). Use of antiretroviral drugs to prevent sexual transmission of HIV. *Curr. Clin. Top. Infect. Dis.*, 22, 214–51.

Cohen, M. S. and Kashuba, A. D. (2008). Antiretroviral therapy for prevention of HIV infection: new clues from an animal model. *PLoS Med.*, 5, e30.

Cohen, M. S., Gay, C., Kashuba, A. D. *et al.* (2007). Narrative review: antiretroviral therapy to prevent the sexual transmission of HIV-1. *Ann. Intern. Med.*, 146, 591–601.

Connor, E. M., Sperling, R. S., Gelber, R. *et al.* (1994). Reduction of maternal-infant transmission of human immunodeficiency virus type 1 with zidovudine treatment. Pediatric AIDS Clinical Trials Group Protocol 076 Study Group. *N. Engl. J. Med.*, 331, 1173–80.

Coombs, R. W., Speck, C. E., Hughes, J. P. *et al.* (1998). Association between culturable human immunodeficiency virus type 1 (HIV-1) in semen and HIV-1 RNA levels in semen and blood: evidence for compartmentalization of HIV-1 between semen and blood. *J. Infect. Dis.*, 177, 320–30.

Coombs, R. W., Reichelderfer, P. S. and Landay, A. L. (2003). Recent observations on HIV type-1 infection in the genital tract of men and women. *AIDS*, 17, 455–80.

Cordes, C., Moll, A., Kuecherer, C. and Marcus, U. (2004). HIV transmission despite HIV post-exposure prophylaxis after non-occupational exposure. *AIDS*, 18, 582–4.

Costello, D. (2005). AIDS pill as party drug? Los Angeles, CA: Los Angeles Times.

Crepaz, N., Hart, T. A. and Marks, G. (2004). Highly active antiretroviral therapy and sexual risk behavior: a meta-analytic review. *J. Am. Med. Assoc.*, 292, 224–36.

Cu-Uvin, S., Snyder, B., Harwell, J. I. *et al.* (2006). Association between paired plasma and cervicovaginal lavage fluid HIV-1 RNA levels during 36 months. *J. Acquir. Immune Defic. Syndr.*, 42, 584–7.

Deeks, S. G., Barditch-Crovo, P., Lietman, P. S. *et al.* (1998). Safety, pharmacokinetics, and antiretroviral activity of intravenous 9-[2-(R)-(Phosphonomethoxy)propyl]adenine, a novel anti-human immunodeficiency virus (HIV) therapy, in HIV-infected adults. *Antimicrob. Agents Chemother.*, 42, 2380–4.

Denton, P. W., Estes, J. D., Sun, Z. *et al.* (2008). Antiretroviral pre-exposure prophylaxis prevents vaginal transmission of HIV-1 in humanized BLT mice. *PLoS Med.*, 5, e16.

Dukers, N. H., Goudsmit, J., de Wit, J. B. *et al.* (2001). Sexual risk behaviour relates to the virological and immunological improvements during highly active antiretroviral therapy in HIV-1 infection. *AIDS*, 15, 369–78.

Dukers, N. H., Spaargaren, J., Geskus, R. B. *et al.* (2002). HIV incidence on the increase among homosexual men attending an Amsterdam sexually transmitted disease clinic: using a novel approach for detecting recent infections. *AIDS*, 16, F19–24.

Dumond, J. B., Yeh, R. F., Patterson, K. B. *et al.* (2007). Antiretroviral drug exposure in the female genital tract: implications for oral pre- and post-exposure prophylaxis. *AIDS*, 21, 1899–907.

Dumond, J., Patterson, K., Pecha, A. *et al.* (2008). Maraviroc (MRV) genital tract (GT) fluid and tissue pharmacokinetics (PK) in healthy female volunteers: implications for pre- or post-exposure prophylaxis (PrEP or PEP). Paper presented at the *Conference on Retroviruses and Opportunistic Infections, Boston, MA,* Abstract no. 135LB.

Eron, J. J. and Hirsch, M. (2007). Antiviral therapy of human immunodeficiency virus infection. In: K. K. Holmes, P.-A. Mårdh, F. P. Sparling *et al.* (eds), *Sexually Transmitted Diseases,* 4th edn. New York, NY: McGraw-Hill, pp. 1393–421.

Eron, J. J., Vernazza, P. L., Johnston, D. M. *et al.* (1998). Resistance of HIV-1 to antiretroviral agents in blood and seminal plasma: implications for transmission. *AIDS*, 12, F181–9.

EURO-NONOPEP (2002). Management of non-occupational post-exposure prophylaxis to HIV (NONOPEP): sexual, injecting drug user or other exposures. European Project on Non-occupational Exposure Prophylaxis.

Fang, C. T., Hsu, H. M., Twu, S. J. *et al.* (2004). Decreased HIV transmission after a policy of providing free access to highly active antiretroviral therapy in Taiwan. *J. Infect. Dis.*, 190, 879–85.

Fideli, U. S., Allen, S. A., Musonda, R. *et al.* (2001). Virologic and immunologic determinants of heterosexual transmission of human immunodeficiency virus type 1 in Africa. *AIDS Res. Hum. Retroviruses*, 17, 901–10.

Fiscus, S. A., Pilcher, C. D., Miller, W. C. *et al.* (2007). Rapid, real-time detection of acute HIV infection in patients in Africa. *J. Infect. Dis.*, 195, 416–24.

Garcia-Lerma, J., Otten, R., Qari, S. *et al.* (2006). Prevention of rectal SHIV transmission in macaques by tenofovir/FTC combination. Paper presented at the *13th Annual Conference on Retroviruses and Opportunistic Infections, February 5–8, Denver, CO,* Abstract no. 32LB.

Garcia-Lerma, J. G., Otten, R. A., Qari, S. H. *et al.* (2008). Prevention of rectal SHIV transmission in macaques by daily or intermittent prophylaxis with emtricitabine and tenofovir. *PLoS Med.*, 5, e28.

Garcia, M. T., Figueiredo, R. M., Moretti, M. L. *et al.* (2005). Postexposure prophylaxis after sexual assaults: a prospective cohort study. *Sex. Transm. Dis.*, 32, 214–19.

Garcia, P. M., Kalish, L. A., Pitt, J. *et al.* (1999). Maternal levels of plasma human immunodeficiency virus type 1 RNA and the risk of perinatal transmission. Women and Infants Transmission Study Group. *N. Engl. J. Med.*, 341, 394–402.

Ghosn, J., Chaix, M. L., Peytavin, G. *et al.* (2004). Penetration of enfuvirtide, tenofovir, efavirenz, and protease inhibitors in the genital tract of HIV-1-infected men. *AIDS*, 18, 1958–61.

Ghys, P. D., Fransen, K., Diallo, M. O. *et al.* (1997). The associations between cervicovaginal HIV shedding, sexually transmitted diseases and immunosuppression in female sex workers in Abidjan, Cote d'Ivoire. *AIDS*, 11, F85–93.

Giele, C. M., Maw, R., Carne, C. A. and Evans, B. G. (2002). Post-exposure prophylaxis for non-occupational exposure to HIV: current clinical practice and opinions in the UK. *Sex. Transm. Infect.*, 78, 130–2.

Golub, S. A., Rosenthal, L., Cohen, D. E. and Mayer, K. H. (2007). Determinants of high-risk sexual behavior during post-exposure prophylaxis to prevent HIV infection. *AIDS Behav.*, e-pub ahead of print.

Graham, S. M., Holte, S. E., Peshu, N. M. *et al.* (2007). Initiation of antiretroviral therapy leads to a rapid decline in cervical and vaginal HIV-1 shedding. *AIDS*, 21, 501–7.

Grant, R. M., Hecht, F. M., Warmerdam, M. *et al.* (2002). Time trends in primary HIV-1 drug resistance among recently infected persons. *J. Am. Med. Assoc.*, 288, 181–8.

Grant, R. M., Buchbinder, S., Cates, W. Jr. *et al.* (2005). AIDS. Promote HIV chemo-prophylaxis research, don't prevent it. *Science*, 309, 2170–1.

Grulich, A. E. (2006). Highly active antiretroviral therapy and HIV transmission. *Lancet*, 368, 1647, author reply 1647.

Grulich, A., Jin, F. Y., Prestage, G. *et al.* (2006). Previous use of non-occupational post exposure prophylaxis against HIV (NPEP) and subsequent HIV infection in homo-sexual men: data from the HIM cohort. Paper presented at the *XVIth International AIDS Conference, August 13–18, 2006, Toronto, Canada,* Abstract no. TUPE0434.

Guay, L. A., Musoke, P., Fleming, T. *et al.* (1999). Intrapartum and neonatal single-dose nevirapine compared with zidovudine for prevention of mother-to-child transmission of HIV-1 in Kampala, Uganda: HIVNET 012 randomised trial. *Lancet*, 354, 795–802.

Hammer, S. M., Saag, M. S., Schechter, M. *et al.* (2006). Treatment for adult HIV infection: 2006 recommendations of the International AIDS Society – USA panel. *J. Am. Med. Assoc.*, 296, 827–43.

Hecht, F. M., Wang, L., Collier, A. *et al.* (2006). A multicenter observational study of the potential benefits of initiating combination antiretroviral therapy during acute HIV infection. *J. Infect. Dis.*, 194, 725–33.

Hogg, R. S., Rhone, S. A., Yip, B. *et al.* (1998). Antiviral effect of double and triple drug combinations amongst HIV-infected adults: lessons from the implementation of viral load-driven antiretroviral therapy. *AIDS*, 12, 279–84.

Hosseinipour, M., Cohen, M. S., Vernazza, P. L. and Kashuba, A. D. (2002). Can antiret-roviral therapy be used to prevent sexual transmission of human immunodeficiency virus type 1? *Clin. Infect. Dis.*, 34, 1391–5.

Hu, J., Gardner, M. B. and Miller, C. J. (2000). Simian immunodeficiency virus rapidly penetrates the cervicovaginal mucosa after intravaginal inoculation and infects intraep-ithelial dendritic cells. *J. Virol.*, 74, 6087–95.

Imrie, A., Beveridge, A., Genn, W. *et al.* (1997). Transmission of human immunode-ficiency virus type 1 resistant to nevirapine and zidovudine. Sydney Primary HIV Infection Study Group. *J. Infect. Dis.*, 175, 1502–6.

Ippolito, G. and Puro, V. (1997). Zidovudine toxicity in uninfected healthcare workers. Italian Registry of Antiretroviral Prophylaxis. *Am. J. Med.*, 102, 58–62.

Jansen, C. A., De Cuyper, I. M., Steingrover, R. *et al.* (2005). Analysis of the effect of highly active antiretroviral therapy during acute HIV-1 infection on HIV-specific CD4T cell functions. *AIDS*, 19, 1145–54.

Johnston, M. I. and Fauci, A. S. (2007). An HIV vaccine – evolving concepts. *N. Engl. J. Med.*, 356, 2073–81.

Kahn, J. O., Martin, J. N., Roland, M. E. *et al.* (2001). Feasibility of postexposure prophy-laxis (PEP) against human immunodeficiency virus infection after sexual or injection drug use exposure: the San Francisco PEP Study. *J. Infect. Dis.*, 183, 707–14.

Kalichman, S. C. (1998). Post-exposure prophylaxis for HIV infection in gay and bisex-ual men. Implications for the future of HIV prevention. *Am. J. Prev. Med.*, 15, 120–7.

Kalichman, S. C., Eaton, L., Cain, D. *et al.* (2006). HIV treatment beliefs and sexual transmission risk behaviors among HIV positive men and women. *J. Behav. Med.*, 29, 401–10.

Kamenga, M., Ryder, R. W., Jingu, M. *et al.* (1991). Evidence of marked sexual behavior change associated with low HIV-1 seroconversion in 149 married couples with dis-cordant HIV-1 serostatus: experience at an HIV counselling center in Zaire. *AIDS*, 5, 61–7.

Katz, M. H., Schwarcz, S. K., Kellogg, T. A. *et al.* (2002). Impact of highly active antiret-roviral treatment on HIV seroincidence among men who have sex with men: San Francisco. *Am. J. Public Health*, 92, 388–94.

Kaufmann, D. E., Lichterfeld, M., Altfeld, M. *et al.* (2004). Limited durability of viral control following treated acute HIV infection. *PLoS Med.*, 1, e36.

Kayitenkore, K., Bekan, B., Rufagari, J. *et al.* (2006). The impact of ART on HIV transmission among HIV serodiscordant couples. Paper presented at the *XVIth International AIDS Conference, August 13–18, Toronto, Canada*, Abstract no. MOKC101.

Kellerman, S. E., Hutchinson, A. B., Begley, E. B. *et al.* (2006). Knowledge and use of HIV pre-exposure prophylaxis among attendees of minority gay pride events, 2004. *J. Acquir. Immune Defic. Syndr.*, 43, 376–7.

Kelly, J. A., Hoffman, R. G., Rompa, D. and Gray, M. (1998). Protease inhibitor combination therapies and perceptions of gay men regarding AIDS severity and the need to maintain safer sex. *AIDS*, 12, F91–5.

Kim, C. N., Adams, D. R., Bashirian, S. *et al.* (2006). Repetitive exposures with simian/human immunodeficiency viruses: strategy to study HIV pre-clinical interventions in non-human primates. *J. Med. Primatol.*, 35, 210–66.

Kindrick, A., Tang, H., Sterkenberg, C. *et al.* (2006). HIV post-exposure prophylaxis following sexual exposure is started too late for optimal benefit. Paper presented at the *13th Conference on Retroviruses and Opportunistic Infections, February 5–8, Denver, CO*, Abstract no. 190.

Kotler, D. P., Shimada, T., Snow, G. *et al.* (1998). Effect of combination antiretroviral therapy upon rectal mucosal HIV RNA burden and mononuclear cell apoptosis. *AIDS*, 12, 597–604.

Kovacs, A., Wasserman, S. S., Burns, D. *et al.* (2001). Determinants of HIV-1 shedding in the genital tract of women. *Lancet*, 358, 1593–601.

Kunches, L. M., Meehan, T. M., Boutwell, R. C. and McGuire, J. F. (2001). Survey of non-occupational HIV postexposure prophylaxis in hospital emergency departments. *J. Acquir. Immune Defic. Syndr.*, 26, 263–5.

Lampinen, T. M., Critchlow, C. W., Kuypers, J. M. *et al.* (2000). Association of antiretroviral therapy with detection of HIV-1 RNA and DNA in the anorectal mucosa of homosexual men. *AIDS*, 14, F69–75.

Law, M. G., Prestage, G., Grulich, A. *et al.* (2001). Modelling the effect of combination antiretroviral treatments on HIV incidence. *AIDS*, 15, 1287–94.

Lederman, M. M., Veazey, R. S., Offord, R. *et al.* (2004). Prevention of vaginal SHIV transmission in rhesus macaques through inhibition of CCR5. *Science*, 306, 485–7.

Le Grand, R., Vaslin, B., Larghero, J. *et al.* (2000). Post-exposure prophylaxis with highly active antiretroviral therapy could not protect macaques from infection with SIV/HIV chimera. *AIDS*, 14, 1864–6.

Little, S. J., Daar, E. S., D'Aquila, R. T. *et al.* (1999). Reduced antiretroviral drug susceptibility among patients with primary HIV infection. *J. Am. Med. Assoc.*, 282, 1142–9.

Little, S. J., Holte, S., Routy, J. P. *et al.* (2002). Antiretroviral-drug resistance among patients recently infected with HIV. *N. Engl. J. Med.*, 347, 385–94.

Low-Beer, S., Weber, A. E., Bartholomew, K. *et al.* (2000). A reality check: the cost of making post-exposure prophylaxis available to gay and bisexual men at high sexual risk. *AIDS*, 14, 325–6.

Magierowska, M., Bernardin, F., Garg, S. *et al.* (2004). Highly uneven distribution of tenofovir-selected simian immunodeficiency virus in different anatomical sites of rhesus macaques. *J. Virol.*, 78, 2434–44.

Martin, J. N., Roland, M. E., Torsten, B. N. *et al.* (2004). Use of postexposure prophylaxis against HIV infection following sexual exposure does not lead to increases in high-risk behavior. *AIDS*, 18, 787–92.

Martin, N. V., Almeda, J. and Casabona, J. (2005). Effectiveness and safety of HIV post-exposusre prophylaxis after sexual, injecting-drug-use or other non-occupational exposure. Cochrane Database of Systematic Reviews Chichester: John Wiley and Sons, Ltd.

Mayer, K. H., Boswell, S., Goldstein, R. *et al.* (1999). Persistence of human immunodeficiency virus in semen after adding indinavir to combination antiretroviral therapy. *Clin. Infect. Dis.*, 28, 1252–9.

Mayer, K. H., Maslankowski, L., Gai, F. *et al.* (2006). Safety and tolerability of tenofovir vaginal gel in abstinent and sexually active HIV-infected and uninfected women. *AIDS*, 20, 543–51.

Mayer, K. H., Mimiaga, M. J., Cohen, D. *et al.* (2008). Tenofovir DF plus lamivudine or emtricitabine for non-occupational postexposure prophylaxis (NPEP) in a Boston Community Health Center. *J. Acquir. Immune Defic. Syndr.*, 47, 494–9.

McClelland, R. S., Wang, C. C., Mandaliya, K. *et al.* (2001). Treatment of cervicitis is associated with decreased cervical shedding of HIV-1. *AIDS*, 15, 105–10.

Meel, B. L. (2005). HIV/AIDS post-exposure prophylaxis (PEP) for victims of sexual assault in South Africa. *Med. Sci. Law*, 45, 219–24.

Min, S. S., Corbett, A. H., Rezk, N. *et al.* (2004). Protease inhibitor and nonnucleoside reverse transcriptase inhibitor concentrations in the genital tract of HIV-1-infected women. *J. Acquir. Immune Defic. Syndr.*, 37, 1577–80.

Mocroft, A., Ledergerber, B., Katlama, C. *et al.* (2003). Decline in the AIDS and death rates in the EuroSIDA study: an observational study. *Lancet*, 362, 22–9.

Mofenson, L. M. and McIntyre, J. A. (2000). Advances and research directions in the prevention of mother-to-child HIV-1 transmission. *Lancet*, 355, 2237–44.

Montaner, J. S., Hogg, R., Wood, E. *et al.* (2006). The case for expanding access to highly active antiretroviral therapy to curb the growth of the HIV epidemic. *Lancet*, 368, 531–6.

Moodley, D., Moodley, J., Coovadia, H. *et al.* (2003). A multicenter randomized controlled trial of nevirapine versus a combination of zidovudine and lamivudine to reduce intrapartum and early postpartum mother-to-child transmission of human immunodeficiency virus type 1. *J. Infect. Dis.*, 187, 725–35.

Musicco, M., Lazzarin, A., Nicolosi, A. *et al.* (1994). Antiretroviral treatment of men infected with human immunodeficiency virus type 1 reduces the incidence of heterosexual transmission. Italian Study Group on HIV Heterosexual Transmission. *Arch. Intern. Med.*, 154, 1971–6.

Ostrow, D. E., Fox, K. J., Chmiel, J. S. *et al.* (2002). Attitudes towards highly active antiretroviral therapy are associated with sexual risk taking among HIV-infected and uninfected homosexual men. *AIDS*, 16, 775–80.

Otten, R. A., Smith, D. K., Adams, D. R. *et al.* (2000). Efficacy of postexposure prophylaxis after intravaginal exposure of pig-tailed macaques to a human-derived retrovirus (human immunodeficiency virus type 2). *J. Virol.*, 74, 9771–5.

Otten, R. A., Adams, D. R., Kim, C. N. *et al.* (2005). Multiple vaginal exposures to low doses of R5 simian-human immunodeficiency virus: strategy to study HIV preclinical interventions in nonhuman primates. *J. Infect. Dis.*, 191, 164–73.

Padian, N. S., van der Straten, A., Ramjee, G. *et al.* (2007). Diaphragm and lubricant gel for prevention of HIV acquisition in southern African women: a randomised controlled trial. *Lancet*, 370, 251–61.

Page-Shafer, K., Saphonn, V., Sun, L. P. *et al.* (2005). HIV prevention research in a resource-limited setting: the experience of planning a trial in Cambodia. *Lancet*, 366, 1499–503.

Palella, F. J. Jr., Delaney, K. M., Moorman, A. C. *et al.* (1998). Declining morbidity and mortality among patients with advanced human immunodeficiency virus infection. HIV Outpatient Study Investigators. *N. Engl. J. Med.*, 338, 853–60.

Panel on Clinical Practices for the Treatment of HIV Infection and DHHS Guidelines (2006). Guidelines for the use of antiretroviral agents in HIV-infected adults and adolescents. Rockville, MD: DHSS.

Panlilio, A. L., Cardo, D. M., Grohskopf, L. A. *et al.* (2005). Updated US Public Health Service guidelines for the management of occupational exposures to HIV and recommendations for postexposure prophylaxis. *Morb. Mortal. Wkly Rep. Recomm. Rep.*, 54, 1–17.

Pauza, C. D., Horejsh, D. and Wallace, M. (1998). Mucosal transmission of virulent and avirulent lentiviruses in macaques. *AIDS Res. Hum. Retroviruses*, 14 (Suppl. 1), S83–87.

Pereira, A. S., Kashuba, A. D., Fiscus, S. A. *et al.* (1999). Nucleoside analogues achieve high concentrations in seminal plasma: relationship between drug concentration and virus burden. *J. Infect. Dis.*, 180, 2039–43.

Peterson, L., Taylor, D., Roddy, R. *et al.* (2007). Tenofovir disoproxil fumarate for prevention of HIV infection in women: a phase 2, double-blind, randomized, placebo-controlled trial. *PLoS Clin. Trials*, 2, e27.

Philpott, S., Burger, H., Tsoukas, C. *et al.* (2005). Human immunodeficiency virus type 1 genomic RNA sequences in the female genital tract and blood: compartmentalization and intrapatient recombination. *J. Virol.*, 79, 353–63.

Pilcher, C. D., Eron, J. J. Jr., Vemazza, P. L. *et al.* (2001a). Sexual transmission during the incubation period of primary HIV infection. *J. Am. Med. Assoc.*, 286, 1713–14.

Pilcher, C. D., Shugars, D. C., Fiscus, S. A. *et al.* (2001b). HIV in body fluids during primary HIV infection: implications for pathogenesis, treatment and public health. *AIDS*, 15, 837–45.

Pilcher, C. D., Price, M. A., Hoffman, I. F. *et al.* (2004a). Frequent detection of acute primary HIV infection in men in Malawi. *AIDS*, 18, 517–24.

Pilcher, C. D., Tien, H. C., Eron, J. J. Jr. *et al.* (2004b). Brief but efficient: acute HIV infection and the sexual transmission of HIV. *J. Infect. Dis.*, 189, 1785–92.

Pilcher, C. D., Fiscus, S. A., Nguyen, T. Q. *et al.* (2005). Detection of acute infections during HIV testing in North Carolina. *N. Engl. J. Med.*, 352, 1873–83.

Pilcher, C. D., Joaki, G., Hoffman, I. F. *et al.* (2007). Amplified transmission of HIV-1: comparison of HIV-1 concentrations in semen and blood during acute and chronic infection. *AIDS*, 21, 1723–30.

Ping, L. H., Cohen, M. S., Hoffman, I. *et al.* (2000). Effects of genital tract inflammation on human immunodeficiency virus type 1 V3 populations in blood and semen. *J. Virol.*, 74, 8946–52.

Pinkerton, S. D., Holtgrave, D. R. and Bloom, F. R. (1998). Cost-effectiveness of postexposure prophylaxis following sexual exposure to HIV. *AIDS*, 12, 1067–78.

Pinkerton, S. D., Martin, J. N., Roland, M. E. *et al.* (2004). Cost-effectiveness of postexposure prophylaxis after sexual or injection-drug exposure to human immunodeficiency virus. *Arch. Intern. Med.*, 164, 46–54.

Porco, T. C., Martin, J. N., Page-Shafer, K. A. *et al.* (2004). Decline in HIV infectivity following the introduction of highly active antiretroviral therapy. *AIDS*, 18, 81–8.

Poynten, I. M., Smith, D. E., Cooper, D. A. *et al.* (2007). The public health impact of widespread availability of non-occupational postexposure prophylaxis against HIV. *HIV Med.*, 8, 374–81.

Puro, V., De Carli, G., Orchi, N. *et al.* (2001). Short-term adverse effects from and discontinuation of antiretroviral post-exposure prophylaxis. *J. Biol. Regul. Homeost. Agents*, 15, 238–42.

Quinn, T. C., Wawer, M. J., Sewankambo, N. *et al.* (2000). Viral load and heterosexual transmission of human immunodeficiency virus type 1. Rakai Project Study Group. *N. Engl. J. Med.*, 342, 921–9.

Quirino, T., Niero, F., Ricci, E. *et al.* (2000). HAART tolerability: post-exposure prophylaxis in healthcare workers versus treatment in HIV-infected patients. *Antivir. Ther.*, 5, 195–7.

Reddy, S., Troiani, L., Kim, J. *et al.* (2003). Differential phosphorylation of zidovudine and lamivudine between semen and blood mononuclear cells in HIV-1 infected men. Paper presented at the *10th Conference on Retroviruses and Opportunistic Infections, February 10–14, Boston, MA*, Abstract no. 530.

Remien, R. H., Halkitis, P. N., O'Leary, A. *et al.* (2005). Risk perception and sexual risk behaviors among HIV-positive men on antiretroviral therapy. *AIDS Behav.*, 9, 167–76.

Rey, D., Bendiane, M., Moatti, J. *et al.* (2002). Policy on non-occupational post-exposure prophylaxis for HIV in 14 European countries. Paper presented at the *XIVth International AIDS Conference, July 7–12, Barcelona, Spain*, Abstract no. TUPEF5382.

Richardson, B. A., John-Stewart, G. C., Hughes, J. P. *et al.* (2003). Breast-milk infectivity in human immunodeficiency virus type 1-infected mothers. *J. Infect. Dis.*, 187, 736–40.

Roland, M. E. (2007). Postexposure prophylaxis after sexual exposure to HIV. *Curr. Opin. Infect. Dis.*, 20, 39–46.

Roland, M., Martin, J., Grant, R. *et al.* (2001). Postexposure prophylaxis for human immunodeficiency virus infection after sexual or injection drug use exposure: identification and characterization of the source of exposure. *J. Infect. Dis.*, 184, 1608–12.

Roland, M. E., Neilands, T. B., Krone, M. R. *et al.* (2005). Seroconversion following non-occupational postexposure prophylaxis against HIV. *Clin. Infect. Dis.*, 41, 1507–13.

Roland, M., Myer, L., Chuunga, R. *et al.* (2006a). Post-exposure prophylaxis following sexual assault in Cape Town: adherence and HIV risk behavior. Paper presented at the *XVIth International AIDS Conference, August 13–18, Toronto, Canada*, Abstract no. MOPDCO3.

Roland 2006b – Abstract 902 at cited conference.

Roland, M., Neilands, T., Krone, M. *et al.* (2006b). A randomized trial of standard versus enhanced risk reduction counseling for individuals receiving post-exposure prophylaxis following sexual exposures to HIV. Paper presented at the *13th Annual Conference on Retroviruses and Opportunistic Infections, February 5–8, Denver, CO*, Abstract no. 902.

Rosenberg, E. S., Billingsley, J. M., Caliendo, A. M. *et al.* (1997). Vigorous HIV-1-specific CD4+ T cell responses associated with control of viremia. *Science*, 278, 1447–50.

Rosenberg, E. S., Caliendo, A. M. and Walker, B. D. (1999). Acute HIV infection among patients tested for mononucleosis. *N. Engl. J. Med.*, 340, 969.

Rousseau, C. M., Nduati, R. W., Richardson, B. A. *et al.* (2003). Longitudinal analysis of human immunodeficiency virus type 1 RNA in breast milk and of its relationship to infant infection and maternal disease. *J. Infect. Dis.*, 187, 741–7.

Routy, J. P., Machouf, N., Edwardes, M. D. *et al.* (2004). Factors associated with a decrease in the prevalence of drug resistance in newly HIV-1 infected individuals in Montreal. *AIDS*, 18, 2305–12.

Royce, R., Sena, A., Cates, W. J. and Cohen, M. S. (1997). Sexual transmission of HIV. *N. Engl. J. Med.*, 336, 1072–8.

Sadiq, S. T., Taylor, S., Kaye, S. *et al.* (2002). The effects of antiretroviral therapy on HIV-1 RNA loads in seminal plasma in HIV-positive patients with and without urethritis. *AIDS*, 16, 219–25.

Schechter, M., do Lago, R. F., Mendelsohn, A. B. *et al.* (2004). Behavioral impact, acceptability, and HIV incidence among homosexual men with access to postexposure chemoprophylaxis for HIV. *J. Acquir. Immune Defic. Syndr.*, 35, 519–25.

Schwartz, J., Kashuba, A., Rezk, N. *et al.* (2008). Preliminary results from a pharmacokinetic study of the candidate vaginal microbicide agent 1 percent tenofovir gel. Paper presented at the *Microbicides Conference 2008, New Delhi.*

Serwadda, D., Gray, R. H., Sewankambo, N. K. *et al.* (2003). Human immunodeficiency virus acquisition associated with genital ulcer disease and herpes simplex virus type 2 infection: a nested case-control study in Rakai, Uganda. *J. Infect. Dis.*, 188, 1492–7.

Shoptaw, S., Rotheram-Fuller, E., Landovitz, R. J. *et al.* (2008). Non-occupational post exposure prophylaxis as a biobehavioral HIV-prevention intervention. *AIDS Care*, 20, 376–81.

Si-Mohamed, A., Kazatchkine, M., Heard, I. *et al.* (2000). Selection of drug-resistant variants in the female genital tract of human immunodeficiency virus type 1-infected women receiving antiretroviral therapy. *J. Infect. Dis.*, 182, 112–22.

Smith, D., Kebaabetswe, P., Disasi, K. *et al.* (2006). Antiretroviral resistance is not an important risk of the oral tenofovir prophylaxis trial in Botswana: a simple mathematical modelling approach. Paper presented at the *XVIth International AIDS Conference, August 13–18, Toronto, Canada*, Abstract no. THAX0105.

Smith, D. K., Grohskopf, L. A., Black, R. J. *et al.* (2005). Antiretroviral postexposure prophylaxis after sexual, injection-drug use, or other non-occupational exposure to HIV in the United States: recommendations from the US Department of Health and Human Services. *Morb. Mortal. Wkly Rep. Recomm. Rep.*, 54, 1–20.

Solas, C., Lafeuillade, A., Halfon, P. *et al.* (2003). Discrepancies between protease inhibitor concentrations and viral load in reservoirs and sanctuary sites in human immunodeficiency virus-infected patients. *Antimicrob. Agents Chemother.*, 47, 238–43.

Sonder, G. J., van den Hoek, A., Regez, R. M. *et al.* (2007). Trends in HIV postexposure prophylaxis prescription and compliance after sexual exposure in Amsterdam, 2000–2004. *Sex. Transm. Dis.*, 34, 288–93.

Spira, A. I., Marx, P. A., Patterson, B. K. *et al.* (1996). Cellular targets of infection and route of viral dissemination after an intravaginal inoculation of simian immunodeficiency virus into rhesus macaques. *J. Exp. Med.*, 183, 215–25.

Steward, W. T., Remien, R. H., Truong, H. M. *et al.* (2007). A move toward serosorting following acute HIV diagnosis: part 1 of 4 on findings from the NIMH multisite acute HIV infection study. Paper presented at the *2007 National HIV Prevention Conference, Atlanta, GA*, Abstract no. C20-1.

Stolte, I. G., Dukers, N. H., Geskus, R. B., Coutinho, R. A. *et al.* (2004). Homosexual men change to risky sex when perceiving less threat of HIV/AIDS since availability of highly active antiretroviral therapy: a longitudinal study. *Aids*, 18, 303–9.

Subbarao, S., Otten, R. A., Ramos, A. *et al.* (2006). Chemoprophylaxis with tenofovir disoproxil fumarate provided partial protection against infection with simian human immunodeficiency virus in macaques given multiple virus challenges. *J. Infect. Dis.*, 194, 904–11.

Subbarao, S., Ramos, A., Kim, C. *et al.* (2007). Direct stringency comparison of two macaque models (single-high vs. repeat-low) for mucosal HIV transmission using an identical anti-HIV chemoprophylaxis intervention. *J. Med. Primatol.*, 36, 238–43.

Sun, Z., Denton, P. W., Estes, J. D. *et al.* (2007). Intrarectal transmission, systemic infection, and CD4+ T cell depletion in humanized mice infected with HIV-1. *J. Exp. Med.*, 204, 705–14.

Taylor, S., van Heeswijk, R. P., Hoetelmans, R. M. *et al.* (2000). Concentrations of nevirapine, lamivudine and stavudine in semen of HIV-1-infected men. *AIDS*, 14, 1979–84.

Tovanabutra, S., Robison, V., Wongtrakul, J. *et al.* (2002). Male viral load and heterosexual transmission of HIV-1 subtype E in northern Thailand. *J. Acquir. Immune Defic. Syndr.*, 29, 275–83.

Tsai, C.-C. and Follis, K. E. (1995). Prevention of SIV infection in macaques by (R)-9-(2-phosphonylmethoxypropyl)adenine. *Science*, 270, 1197–9.

Tsai, C. C., Emau, P., Follis, K. E. *et al.* (1998). Effectiveness of postinoculation (R)-9-(2-phosphonylmethoxypropyl). adenine treatment for prevention of persistent simian immunodeficiency virus SIVmne infection depends critically on timing of initiation and duration of treatment. *J. Virol.*, 72, 4265–73.

Tun, W., Gange, S. J., Vlahov, D. *et al.* (2004). Increase in sexual risk behavior associated with immunologic response to highly active antiretroviral therapy among HIV-infected injection drug users. *Clin. Infect. Dis.*, 38, 1167–74.

Turner, D., Brenner, B., Routy, J. P. *et al.* (2004). Diminished representation of HIV-1 variants containing select drug resistance-conferring mutations in primary HIV-1 infection. *J. Acquir. Immune Defic. Syndr.*, 37, 1627–31.

UNAIDS (2006). *2006 Report on the Global AIDS Epidemic* (available online at http://www.unaids.org/en/HIV_data/2006GlobalReport).

UNAIDS (2007). *AIDS Epidemic Update. World Health Organization* (available at http://data.unaids.org/pub/EPISlides/2007/2007_epiupdate_en.pdf).

Vanable, P. A., Ostrow, D. G. and McKirnan, D. J. (2003). Viral load and HIV treatment attitudes as correlates of sexual risk behavior among HIV-positive gay men. *J. Psychosom. Res.*, 54, 263–9.

van Oosterhout, J. J., Nyirenda, M., Beadsworth, M. B. *et al.* (2007). Challenges in HIV post-exposure prophylaxis for occupational injuries in a large teaching hospital in Malawi. *Trop. Doct.*, 37, 4–6.

van Praag, R. M., van Heeswijk, R. P., Jurriaans, S. *et al.* (2001). Penetration of the nucleoside analogue abacavir into the genital tract of men infected with human immunodeficiency virus type 1. *Clin. Infect. Dis.*, 33, e91–92.

Van Rompay, K. K., McChesney, M. B., Aguirre, N. L. *et al.* (2001). Two low doses of tenofovir protect newborn macaques against oral simian immunodeficiency virus infection. *J. Infect. Dis.*, 184, 429–38.

Van Rompay, K. K., Schmidt, K. A., Lawson, J. R. *et al.* (2002). Topical administration of low-dose tenofovir disoproxil fumarate to protect infant macaques against multiple oral exposures of low doses of simian immunodeficiency virus. *J. Infect. Dis.*, 186, 1508–13.

Velasco-Hernandez, J. X., Gershengorn, H. B. and Blower, S. M. (2002). Could widespread use of combination antiretroviral therapy eradicate HIV epidemics? *Lancet Infect. Dis.*, 2, 487–93.

Vella, S. and Palmisano, L. (2005). The global status of resistance to antiretroviral drugs. *Clin. Infect. Dis.*, 41(Suppl. 4), S239–46.

Vernazza, P. L., Troiani, L., Flepp, M. J. *et al.* (2000). Potent antiretroviral treatment of HIV-infection results in suppression of the seminal shedding of HIV. The Swiss HIV Cohort Study. *AIDS*, 14, 117–21.

Vitoria, M., Beck, E., Madalia, S. *et al.* (2006). Guidelines for post-exposure prophylaxis for HIV in developing countries. Paper presented at the *13th Conference on Retroviruses and Opportunistic Infections, February 5–8, Denver, CO*, Abstract no. 151.

Vourvahis, M., Tappouni, H. L., Patterson, K. B. *et al.* (2008). The pharmacokinetics and viral activity of tenofovir in the male genital tract. *J. Acquir. Immune Defic. Syndr.*, 47, 329–33.

Wade, N. A., Birkhead, G. S., Warren, B. L. *et al.* (1998). Abbreviated regimens of zido-vudine prophylaxis and perinatal transmission of the human immunodeficiency virus. *N. Engl. J. Med.*, 339, 1409–14.

Wald, A. and Link, K. (2002). Risk of human immunodeficiency virus infection in herpes simplex virus type 2-seropositive persons: a meta-analysis. *J. Infect. Dis.*, 185, 45–52.

Wang, S. A., Panlilio, A. L., Doi, P. A. *et al.* (2000). Experience of healthcare workers taking postexposure prophylaxis after occupational HIV exposures: findings of the HIV Postexposure Prophylaxis Registry. *Infect. Control Hosp. Epidemiol.*, 21, 780–5.

Wawer, M. J., Gray, R. H., Sewankambo, N. K. *et al.* (2005). Rates of HIV-1 transmission per coital act, by stage of HIV-1 infection, in Rakai, Uganda. *J. Infect. Dis.*, 191, 1403–9.

Weintrob, A. C., Anderson, A. M. L., Seshadri, C. *et al.* (2003). Prevalence of undiag-nosed HIV infection in febrile patients presenting to an emergency department in Southeastern United States. Paper presented at the *2nd IAS Pathogenesis Meeting, Paris, France*, Abstract no. 84.

WHO (2007). Towards universal access: scaling up priority HIV/AIDS inter-ventions in the health sector. *Progress Report* (available at http://www.who. int/hiv/mediacentre/universal_access_progress_report_en.pdf).

Wiebe, E. R., Comay, S. E., McGregor, M. and Ducceschi, S. (2000). Offering HIV prophylaxis to people who have been sexually assaulted: 16 months' experience in a sexual assault service. *Can. Med. Assoc. J.*, 162, 641–5.

Winston, A., McAllister, J., Amin, J. *et al.* (2005). The use of a triple nucleoside-nucle-otide regimen for non-occupational HIV post-exposure prophylaxis. *HIV Med.*, 6, 191–7.

Xu, C., Politch, J. A., Tucker, L. *et al.* (1997). Factors associated with increased levels of human immunodeficiency virus type 1 DNA in semen. *J. Infect. Dis.*, 176, 941–7.

Yem, R., Patterson, K., Dumond, J. *et al.* (2006). Genital Tract (GT), Cord Blood (CB), and Amniotic Fluid (AF) Exposures of 6 Antiretroviral (ARV) Drugs during Pregnancy and Postpartum (PP). San Francisco, CA: ICAAC.

Yerly, S., Vora, S., Rizzardi, P., Chave, J. P. *et al.* (2001). Acute HIV infection: impact on the spread of HIV and transmission of drug resistance. *AIDS*, 15, 2287–92.

Yerly, S., Jost, S., Telenti, A. *et al.* (2004). Infrequent transmission of HIV-1 drug-resistant variants. *Antivir. Ther.*, 9, 375–84.

Zhang, Z. Q., Wietgrefe, S. W., Li, Q. *et al.* (2004). Roles of substrate availability and infection of resting and activated CD4+ T cells in transmission and acute simian immunodeficiency virus infection. *Proc. Natl Acad. Sci. USA*, 101, 5640–5.

Male circumcision and HIV prevention

6

Ronald Gray, David Serwadda, Godfrey Kigozi and
Maria J. Wawer

The HIV epidemic in sub-Saharan Africa continues to grow despite prevention efforts, and there is an urgent need for preventive interventions to reduce transmission. Male circumcision has been shown to reduce HIV acquisition in men by over 50 percent in three randomized trials (Auvert *et al.*, 2005; Bailey *et al.*, 2007; Gray *et al.*, 2007a) and is now recommended by WHO/UNAIDS as a component of HIV prevention strategies, particularly in countries with a low prevalence of male circumcision and with generalized HIV epidemics (WHO & UNAIDS, 2007a). We review the observational and randomized trial evidence documenting the effects of male circumcision on HIV risk in men and women, and the mechanisms whereby circumcision might reduce HIV risk. We then consider models to estimate the impact of circumcision on the HIV epidemic, and the potential cost-effectiveness of the procedure.

Observational data on male circumcision and heterosexual HIV acquisition in men

Early in the sub-Saharan HIV epidemic, ecologic studies noted an inverse association between national HIV prevalence estimates and the prevalence of male circumcision (Bongaarts *et al.*, 1989; Moses *et al.*, 1990). The low prevalence of HIV in West Africa was associated with high rates of male circumcision, among both Muslims and non-Muslims, whereas in Eastern and southern Africa the high prevalence of HIV was associated with low rates of male circumcision. However, such ecologic analyses provide only a crude correlation, and cannot directly link circumcision status to HIV infection in individuals.

Numerous individual-level observational studies have also shown an association between male circumcision and lower risks of prevalent or incident heterosexually acquired HIV in Africa, Asia and the US (Bongaarts *et al.*, 1989;

146

Jessamine *et al.*, 1990; Moses *et al.*, 1990; de Vincenzi and Mertens, 1994; Seed *et al.*, 1995; Urassa *et al.*, 1997; Lavreys *et al.*, 1999; Bailey and Halperin, 2000; Buve, 2000; Gray *et al.*, 2000, 2006; Green, 2000; Harrison, 2000; Quinn *et al.*, 2000; Weiss *et al.*, 2000; Reynolds *et al.*, 2004; Baeten *et al.*, 2005; Siegfried *et al.*, 2005; Sateren *et al.*, 2006). However, these studies were not designed specifically to address the question of whether circumcision affects HIV risk; in many cases the circumcision-related findings were based on *post-hoc* analyses, and the findings are not consistent across studies. A meta-analysis of 21 cross-sectional and prospective studies published before 2000 (Weiss *et al.*, 2000) found a significantly lower relative risk (RR) of HIV in circumcised men (adj. RR = 0.52, 95% CI 0.40–0.68) overall, but there was significant variation between studies. In general populations, the adjusted RR of HIV associated with circumcision was 0.56 (95% CI 0.44–0.70), but there was significant heterogeneity between studies ($P = 0.008$), and three population-based studies found increases in HIV risk among circumcised as compared to uncircumcised men. In contrast, among men at high risk of HIV, such as those attending STD clinics, or truck drivers, the RR was 0.29 (95% CI 0.20–0.41); reduced HIV risk among the circumcised was observed in all studies and there was no significant heterogeneity between investigations ($P = 0.09$). The most extreme case of highly exposed men comes from a study of HIV-uninfected males in discordant relationships with HIV-positive female partners in Rakai, Uganda. In this study, no transmissions were observed in 50 HIV-negative circumcised men with infected partners, whereas the female-to-male transmission rate was 16.7/100 py (person-years) among uncircumcised men (Gray *et al.*, 2000). A more recent Cochrane review of 35 studies (Siegfried *et al.*, 2005) found an overall lowered risk of HIV among circumcised men, and this was particularly true of 19 studies conducted in high-risk populations where the odds ratios (OR) of HIV associated with circumcision ranged from 0.10 to 0.88. In general populations, the risks ranged from 0.28 to 1.55, and 7 of 14 studies showed no evidence of protection or increased risk. Because of the significant heterogeneity across studies the authors declined to estimate pooled risks, and judged many studies to be of poor quality and inherently limited by confounding factors (Siegfried *et al.*, 2005). Recent Demographic and Health Surveys (DHS) which included HIV testing (Mishra, 2006) have provided mixed findings. HIV prevalence was lower among circumcised than uncircumcised men in Uganda and Kenya, but in Malawi and Lesotho HIV prevalence was higher among circumcised men. In West African countries where circumcision is common (e.g., Burkina Faso, Ghana and Cameroon), HIV prevalence was low and was not associated with circumcision status. Thus, with the exception of high-risk men, for whom there are strong and consistent observational study findings suggesting a protective effect of circumcision on male HIV acquisition, the observational data do not provide consistent evidence for protective effects in general populations.

All the observational studies are vulnerable to confounding by reasons for circumcision and differences in behaviors between circumcised and uncircumcised men (de Vincenzi and Mertens, 1994; Bailey *et al.*, 1999; Kelly *et al.*, 1999; Gray *et al.*, 2000; Gray, 2004). Age at circumcision may also be a potential confounder, since surgery performed before puberty is more protective than surgery at later ages (Kelly *et al.*, 1999), and the age at circumcision varies with the reasons for performing the procedure. In most societies, particularly in sub-Saharan Africa, circumcision is performed for religious or cultural reasons (WHO & UNAIDS, 2007b). The majority of circumcised men are Muslims, among whom the procedure is almost universal and performed in infancy or childhood. In countries such as Kenya, tribal affiliation is the main determinant of circumcision status; the Luo are largely uncircumcised, but the Kikyu and other tribes practice the procedure. Thus, the possibility of confounding by correlated cultural behaviors which vary between tribal groups cannot be excluded (Lavreys *et al.*, 1999; Baeten *et al.*, 2005). In Rakai, Uganda, almost all pre-pubertal and adolescent circumcisions were among Muslims, and cultural characteristics such as non-use of alcohol, genital cleansing after intercourse and before prayer, and closed sexual networks are likely to reduce HIV risk (de Vincenzi and Mertens, 1994; Kelly *et al.*, 1999; Green, 2000). Thus, behaviors correlated with Islamic religion rather than a biological protective effect of circumcision on HIV susceptibility cannot be ruled out through observational studies alone. In Rakai, 75 percent of post-pubertal circumcisions were performed for health indications (e.g., phimosis, intractable GUD, etc.), and these medical conditions, particularly GUD, may reflect previous high-risk sexual behaviors. Furthermore, in southern Africa and Tanzania, where circumcision is part of a coming of age or puberty ritual, such ceremonies are often associated with sexual initiation which may place young men at risk of HIV, especially if intercourse occurs before full wound healing is completed (WHO & UNAIDS, 2007b). Traditional circumcisions, often performed in groups of boys or young men and using shared unsterilized instruments, may also result in HIV infection due to cross-contamination. Thus, the context of circumcision surgery may affect HIV risk and confound observational studies. Finally, misclassification bias could also affect observational studies due to misstatement of self-reported circumcision status (Diseker *et al.*, 2001) and variability in the amount of foreskin removed (Brown *et al.*, 2001).

In conclusion, the ecological and observational studies suggest that circumcision may reduce HIV risk particularly in men with high risks of HIV exposures, but no studies could resolve the possibility of complex confounding. Thus, randomized trials were needed before circumcision could be promoted as a means of HIV prevention.

Male circumcision and HIV acquisition in men who have sex with men

The possible effects of circumcision on HIV risk in men who have sex with men (MSM) are less well studied, and are complicated by varying practices of insertive or receptive anal intercourse. A study of 63 recently HIV-infected men obtained information on unprotected anal intercourse in the prior 6 months, and on the circumcision status of their partners. There was no association between infection and insertive anal intercourse (Grulich *et al.*, 2001). However, another study of 502 MSM reported a 50 percent reduced risk of prevalent HIV in circumcised MSM (Kreiss and Hopkins, 1993). Similar findings of protection were reported for incident HIV infections in 3257 initially uninfected MSM (Buchbinder *et al.*, 2005). However, cross-sectional studies in the US found that circumcision was not associated with prevalent HIV infection in Latino and Black MSM (Millett *et al.*, 2007). In Peru, circumcised men who practiced insertive intercourse had lower rates of prevalent HIV than did uncircumcised men, but circumcision did not appear to be protective against HIV acquisition with receptive anal intercourse (J. V. Guanira, personal communication, 2007). However, it has been argued that the association between circumcision and HIV among MSM is not observed in ecologic analyses of populations, since HIV prevalence is higher among MSM in North America where most men are circumcised, but prevalence is lower in Europe, where most men are uncircumcised.

Randomized trials of male circumcision for HIV prevention in men

Three trials of male circumcision for prevention of heterosexually acquired HIV infection in men have now been completed (Auvert *et al.*, 2005; Bailey *et al.*, 2007; Gray *et al.*, 2007a). All three trials employed a similar design whereby consenting HIV-negative men were enrolled and randomized to receive immediate circumcision (the intervention arm), or circumcision delayed for 21–24 months (the control arm). HIV incidence was then detected during follow-up for estimation of efficacy. The three trials were, of necessity, open-label, since circumcision status could not be concealed. The key features of the three trials are summarized in Table 6.1.

The South African trial enrolled 3128 HIV-negative men aged 18–24 in the semi-urban township of Orange Farms, and followed participants at 3, 12 and 21 months. Circumcision was performed by general practitioners using the forceps-guided method under local anesthesia. The Kenyan trial enrolled 2784 men aged 18–24 in urban Kisumu, and follow-up was at 1, 3, 6, 12, 18 and 24 months. Surgery used the forceps-guided procedure performed by study physicians in a single theater. The Ugandan trial enrolled 4996 men in rural Rakai District, and followed participants at 6, 12 and 24 months. The sleeve procedure was used for circumcision, which was performed by trained physicians in three operating theaters. All trials were designed to have the power to detect a 50 percent reduction in incident HIV, and all three trials were stopped prematurely because interim analyses showed evidence of significant efficacy of circumcision for HIV prevention in men.

Table 6.1 Characteristics of three randomized trials for HIV prevention in men

	South Africa	Kenya	Uganda
Setting	Peri-urban township	Urban	Rural
Age range	18–24	18–24	15–49
Numbers enrolled	3128	2784	4996
Intervention	1546	1391	2474
Control	1582	1393	2522
Surgical procedure	Forceps guided	Forceps guided	Sleeve
Follow-up schedule	3, 12, 21 months	1, 3, 6, 12, 18, 24 months	6, 12, 24 months
Follow-up rates	92%	91.4%	90%
Number of incident events	69	69	67
Intervention arm	20	22	22
Control arm	49	47	45
Cumulative HIV incidence			
Intervention arm	0.85	2.1	0.66
Control arm	2.1	4.2	1.33
Cumulative incidence rate ratio, 95% CI	0.40 (0.24–0.68)	0.47 (0.28–0.78)	0.43 (0.24–0.75)
Intent-to-treat efficacy based on survival analyses	60% (32%–76%)	53% (22%–72%)	57% (25%–76%)
Crossovers			
Intervention	6.5%	4.0%	5.6%
Control	10.1%	1.0%	1.4%
As treated efficacy (95% CI)	76% (56%–86%)	60% (32%–77%)	60% (30%–77%)
Moderate/severe adverse events related to surgery	3.6%	1.5%	3.6%

In all trials, enrollment characteristics and behaviors were similar in the intervention and control arms. Also, the three trials achieved comparably high retention rates ($\geq$90 percent) and, despite differences in sample size, all observed comparable numbers of incident HIV events (67–69) which were similarly distributed between the intervention (20–22) and control arms (45–49). The HIV incidence varied between trial populations, and was lowest in rural Rakai (Uganda) and highest in urban Kisumu (Kenya). Despite these differences, the estimates of efficacy based on an intent-to-treat survival analysis were similar, at 53 percent in Kenya, 57 percent in Uganda and 60 percent in South Africa.

The proportions of crossovers among men randomized to circumcision who failed to come for surgery was similar in the three studies (4.0–6.5 percent), but the proportions of controls who sought circumcision from non-trial sources differed markedly. In South Africa, 10.1 percent of controls were circumcised during the trial, whereas in Kenya and in Uganda control crossovers were infrequent (1.0 and 1.4 percent, respectively). Such crossovers are likely selective and thus to introduce bias into as-treated or per-protocol efficacy estimates. As a consequence, the as-treated efficacy was higher in South Africa (76 percent) compared to Kenya and Uganda (both 60 percent).

In the Ugandan trial, the efficacy of circumcision increased with duration of follow-up. The incidence rate ratio (IRR) of HIV acquisition in intervention compared to control men was 0.76 (95% CI 0.35–1.60) during the first 6 months, 0.35 (95% CI 0.10–1.04, $P = 0.04$) during the 6–12 months' follow-up interval and 0.25 (95% CI 0.05–0.94, $P = 0.02$) during the second year of follow-up. This increasing efficacy was due to a significant decline of incidence in the intervention arm over time (P for trend = 0.01), but no statistically significant decline of incidence was observed in the control arm (Gray *et al.*, 2007a). The South African trial did not report incidence by arm and follow-up duration, although model-based estimates suggested a similar trend (Auvert *et al.*, 2005). However, no temporal trends in intervention arm incidence were observed in the Kenyan trial (Bailey *et al.*, 2007). The differences between trials in the temporal trends of efficacy might be due to the differences in ages at enrollment. The South African and Kenyan trials enrolled young men aged 18–24 who are likely to increase their sexual activity over time as they progressively enter higher-risk age groups (e.g., older than 20). However, the Ugandan trial enrolled men aged 15–49, and thus is less affected by progression into higher-risk age groups during the follow-up period. Nevertheless, none of the trials can estimate efficacy beyond 2 years post-surgery. If circumcision efficacy increases over time, as observed in Uganda, the procedure might ultimately provide greater protection than that observed in the trials over 24 months' observation, and this has implications for the longer-term impact of circumcision, since the procedure is likely to afford life-long protection.

The Ugandan trial assessed circumcision efficacy by participant characteristics and behaviors. For example, the incidence rate ratios (IRR) of HIV acquisition in circumcised relative to uncircumcised men were 0.55 for men reporting one sexual partner and 0.3 for men reporting two or more partners. Similarly, among men who only reported marital partners the circumcision-associated IRR was 0.64, but among men with non-marital partners the IRR was 0.34. Among men with no GUD symptoms the IRR was 0.60, whereas among men with GUD the circumcision-associated IRR was 0.29. Thus, the trial showed evidence of greater circumcision efficacy among men at higher risk of HIV exposure, which is consistent with findings from observational studies. Unfortunately, the South African and Kenyan trials did not report circumcision efficacy by subgroups of risk exposure.

Safety of circumcision was a major consideration in all the trials, but the methods for ascertaining surgery-related adverse events varied between them and this may have affected comparability of reported complications. In South Africa, adverse events were ascertained by study personnel using participant self-reports at a follow-up interview 3 months postoperatively (Auvert *et al.*, 2005). In Kenya and Uganda, participants were seen within 24–48 hours and around 1 week and 4 weeks post-surgery (Bailey *et al.*, 2007; Gray *et al.*, 2007a). However, in Kenya recording of complications was performed by study physicians who performed the surgery, whereas in Uganda complications were recorded by independent clinical officers who did not perform the surgery. Thus, the timing of postoperative visits, mode of ascertainment and possible observer bias may have affected estimates of complications. Nevertheless, moderate to severe complications were uncommon in all trials, ranging from 1.5 percent to 3.6 percent. This suggests that circumcision performed by trained medical practitioners is safe, despite differences in the surgical procedures used (forceps-guided method in South Africa and Kenya, and sleeve circumcision in Uganda).

Assessment during follow-up showed reductions in risk behaviors over time in all three trials. However, in the South African and Kenyan trials, intervention-arm men reported higher-risk behaviors than control-arm men during follow-up, suggesting possible risk compensation or behavioral disinhibition (Auvert *et al.*, 2005; Bailey *et al.*, 2007). No behavioral disinhibition was observed in the Ugandan trial (Gray *et al.*, 2007a).

On the basis of these three trials, WHO/UNAIDS concluded that "The research evidence that male circumcision is efficacious in reducing sexual transmission of HIV from women to men is compelling" and that "The efficacy of male circumcision in reducing female to male transmission has been proven beyond reasonable doubt." WHO/UNAIDS recommended that male circumcision should now be recognized as an efficacious intervention for HIV prevention, and that promoting male circumcision should be recognized as an additional, important strategy for the prevention of heterosexually acquired HIV in men" (WHO & UNAIDS, 2007a).

Biological evidence for the protective effects of circumcision for HIV prevention in men

It is biologically plausible that circumcision can reduce male susceptibility to HIV. The foreskin is vulnerable to trauma during intercourse, and the presence of the prepuce increases the likelihood of genital ulceration and inflammatory conditions (such as balanitis) which may act as cofactors for HIV acquisition (Moses *et al.*, 1998; Szabo and Short, 2000; Reynolds *et al.*, 2004; M. Thoma and

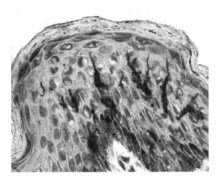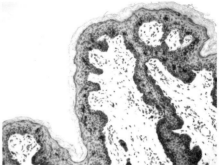

Figure 6.1 CD1a immunostain for dendritic cells within foreskin 40× and 10× Magnification.

R. Gray, 2007, personal communication). The penile epidermis is rich in Langerhans cells and subepithelial dendritic cells, macrophages and CD4 T cells expressing CD4 and CCR5/CXCR4 coreceptors (Bailey *et al.*, 2001; Patterson *et al.*, 2002; McCoombe and Short, 2006; Simon *et al.*, 2006). Figure 6.1 shows Langerhans cells in inner preputial mucosa from foreskins derived from the Ugandan trial. The density of Langerhans cells is greater in inner than outer foreskin mucosa, and the dendritic processes are particularly superficial on the inner surface of the foreskin and frenulum. Moreover, the density of Langerhans cells and macrophages is increased in men with a history of STDs (McCoombe and Short, 2006). Histopathology of foreskin tissues from the Ugandan trial shows an increase in Langerhans cell density with evidence of inflammation (Johnstone *et al.*, 2008). The outer surface of the foreskin, penile shaft and glans are heavily keratinized and relatively impermeable to HIV in the absence of lesions or trauma, whereas the inner mucosa is lightly keratinized and vulnerable to HIV infection (Szabo and Short, 2000; Patterson *et al.*, 2002; McCoombe and Short, 2006). During intercourse the foreskin is retracted, exposing the vulnerable inner mucosa to vaginal secretions (Szabo and Short, 2000). After circumcision, the only remaining vulnerable, unkeratinized mucosa is the urethral meatus, which is a much smaller surface area than the intact foreskin. In addition, the moistness of the subpreputial space may be conducive to HIV survival (Bailey *et al.*, 2001), and has been associated with an increased risk of HIV (O'Farrell *et al.*, 2006).

In the Ugandan trial, the foreskin area was measured following surgery, and secondary data analyses have shown that control participants who received circumcision after completing follow-up had higher risks of preoperative incident HIV if their foreskin surface area was in the largest quartile (area $> 45.9\,\mathrm{mm}^2$, incidence 7.8/100 py) compared with men whose foreskins were in the lower quartile (area $<27\,\mathrm{mm}^2$, incidence 3.2/100 py, IRR = 2.46, 95% CI 1.08–6.09). This clearly suggests that the size of the foreskin surface area is associated with higher HIV risk, presumably because of the larger number of HIV target cells (Kigozi *et al.*, 2008a).

Observational data and the Ugandan trial suggest that circumcision may be more highly protective against HIV acquisition among men who have repeated HIV exposures, compared to men with less frequent HIV exposures. This was noted in meta-analyses (Weiss *et al.*, 2000), in uninfected men in HIV-discordant relationships with HIV-positive women in Rakai (Gray *et al.*, 2000; Quinn *et al.*, 2000), STD clinic patients in Pune, India (Reynolds *et al.*, 2004) and Kenyan truck drivers (Lavreys *et al.*, 1999; Baeten *et al.*, 2005). We have hypothesized that following circumcision in high-risk men, the reduced surface area of unkeratinized mucosa may result in recurrent frequent exposures to low doses of HIV, which could provide antigenic stimulation via repeated subinfectious HIV inoculums resulting in a degree of mucosal immunity which enhances protection over and above the reduced risk afforded by removal of the foreskin *per se* (Wawer *et al.*, 2005). Such induced local immunity has been observed in highly exposed but uninfected members of HIV-discordant couples and in commercial sex workers, among whom there was evidence of enhanced mucosal $CD8^+$ T-cell responses (Fowke *et al.*, 2000; Kaul *et al.*, 2001) and HIV-1 specific IgA (Devito *et al.*, 2000), and increased α-defensin production and α-defensin-expressing CD8 lymphocytes (Trabattoni *et al.*, 2004). Irrespective of mechanism, the paradoxically higher protection in the highest-risk individuals might, in part, explain the heterogeneity of findings from observational studies, and could affect efficacy estimates from randomized trials.

Circumcision and STI acquisition in men

The effects of circumcision on STIs are difficult to study because of the transient nature of many genital tract infections and the effectiveness of treatment for curable conditions. Most information on circumcision and STIs is derived from incidental findings in observational studies designed for other purposes, and much of the data collection has been limited to STD clinic patients. A meta-analysis by Weiss *et al.* (2006) found that circumcision was associated with a reduced risk of HSV-2 (adj. summary RR = 0.85, CI 0.74–0.98) and serologic syphilis (adj. summary RR = 0.70, CI 0.51–0.96), but there was significant heterogeneity across studies ($P = 0.002$). In observational data from the Rakai Community Cohort Study, the adjusted RR of HSV-2 in circumcised versus uncircumcised men was 0.82 (CI 0.67–0.97) (Gray *et al.*, 2004). Rates of symptomatic genital ulcer disease are lower in circumcised than uncircumcised men, based on either self-reports or clinical observation (Lavreys *et al.*, 1999; M. Thoma and R. Gray, 2007, personal communication). Also, the protective effects of circumcision for GUD are more marked in high-risk men than in the general populations (M. Thoma and R. Gray, 2007, personal communication). In the Ugandan circumcision trial, the prevalence rate of GUD was significantly lower in the intervention arm

(RR = 0.0.53, CI 0.43–0.64, $P < 0.001$). Since HSV-2 is the commonest cause of GUD, it is possible that reductions in HSV-2 acquisition could explain the protective effects of circumcision against genital ulcers, although the South African and Kenyan trials did not find a reduction of HSV-2 seroincidence associated with circumcision using the Kalon HSV-2 assay with an index value of 1:1 (Bailey *et al.*, 2007). However, analyses of the Ugandan trial suggest a reduction in incident HSV-2 using higher Kalon index values which have been shown to be more specific in Ugandan sera (Tobian *et al.*, 2008). With a Kalon index value ≥ 1.5, the incidence of HSV-2 over 24 months was 4.1 per 100 person-years in the intervention group compared with 5.7/100 py in the control group by intention-to-treat analysis. The adjusted estimate of efficacy was 29 percent (95% CI 10–44%, $P = 0.005$). Since GUD is a risk factor for HIV (Rottingen *et al.*, 2001), the protective effects of circumcision for HIV acquisition may, in part, be mediated by reduced GUD and, possibly, HSV-2.

Findings with respect to gonorrhea are conflicting (Moses *et al.*, 1998). A US study (Diseker *et al.*, 2000) found lower rates of gonorrhea among circumcised men (adj. OR 0.6, CI 0.4–1.0), but this is not supported by findings from India (Reynolds *et al.*, 2004) and Uganda (Gray *et al.*, 2004). There is little evidence that circumcision affects the risks of chlamydia (Moses *et al.*, 1998; Diseker *et al.*, 2000; Gray *et al.*, 2004) or non-gonoccocal urethritis (Moses *et al.*, 1998; Lavreys *et al.*, 1999). There were no differences between study arms in urethral discharge or dysuria in the Ugandan circumcision trial. However, neonatal circumcision is associated with a reduced rate of infant urinary infections, and a New Zealand study found reduced rates of self-reported STIs associated with neonatal circumcision in men followed up at ages 18–25 (Fergusson *et al.*, 2006). It is possible that circumcision is not protective against urethral infections such as gonorrhea and chlamydia, whereas the foreskin may increase vulnerability to cutaneously acquired infections such as HSV-2, syphilis, *H. ducreyi* and HPV.

Male circumcision was associated with lower rates of penile HPV infection in a multinational case-control study (OR 0.37, CI 0.2–0.9) (Castellsague *et al.*, 2002), and in a US study of oncogenic HPV (OR 0.36, CI 0.2–0.7) and non-oncogenic HPV (OR 0.47, CI 0.3–0.9) (Baldwin *et al.*, 2004). It is possible that the moist subpreputial area may favor survival of HPV, or autoinfection (Castellsague *et al.*, 2002). In a prospective study, the incidence of penile HPV infection was lower in circumcised men, and genital warts were markedly less common in circumcised men (3 percent) than in uncircumcised men (26 percent) (Aynaud *et al.*, 1994). The majority of penile cancers are due to HPV infection, predominantly HPV 16, and the risk of penile cancer is markedly lower in circumcised than in uncircumcised men (OR 0.3–0.4) (Maden *et al.*, 1993; Dillner *et al.*, 2000). Penile cancer is among the most frequently diagnosed cancers in Ugandan men, the majority of whom are uncircumcised (Buonaguro *et al.*, 2000). There is, therefore, consistent evidence to suggest that circumcision may protect men from penile HPV infections and, possibly, penile neoplasia.

Ronald Gray, David Serwadda, Godfrey Kigozi and Maria J. Wawer

Male circumcision and HIV/STI infections in women

Cross-sectional studies in Kenyan women found a lower prevalence of HIV in women with circumcised husbands or partners, ranging from OR 0.34 (Hunter *et al.*, 1994) to 0.05 (Fonck *et al.*, 2000). However, one cannot infer that circumcision directly affects male infectivity, since circumcision is associated with lower HIV prevalence in Kenya, and this could reduce female exposures. A direct effect of circumcision on male-to-female transmission is suggested from studies of HIV-discordant couples with HIV-infected males and -uninfected female partners in Rakai, Uganda (Gray *et al.*, 2000). HIV incidence was 5.2/100 py in women with HIV-positive circumcised husbands, compared with 13.2/100 py in women with HIV-positive uncircumcised husbands, but this difference was not statistically significant (RR = 0.38, CI 0.13–1.22). However, among HIV-positive male partners with a viral load <50,000 cps ml, no transmissions were observed to the HIV-negative female partner if the man was circumcised, whereas the transmission rate was 9.6/100 py if the man was uncircumcised ($P = 0.02$) (Gray *et al.*, 2000). This suggests that circumcision may reduce male HIV infectivity at lower viremic levels, and that the intact foreskin may be an additional source of HIV shedding, over and above the seminal fluid. A prospective study in Uganda and Zambia found lower HIV incidence in women with circumcised partners (HR = 0.75), but this differential was attenuated after adjustment (Norris Turner *et al.*, 2006). A trial in Rakai, Uganda, assessed the effects of male circumcision in HIV-positive men on male-to-female HIV transmission among HIV-discordant couples (Wawer *et al.*, 2008). The trial was stopped after an interim analysis showed no evidence of efficacy; HIV incidence was 13.8/100 py in wives of circumcised men compared to 9.6/100 py in wives of uncircumcised men ($P = 0.4$). Among circumcised men, female HIV acquisition was highest if the couple resumed sexual intercourse prior to completed wound healing (27.8 percent), compared to those delaying resumption of sex until healing (9.5 percent).

There is limited information on the effects of male circumcision on STIs in women. In an IARC study, monogamous women whose male partners reported six or more lifetime sex partners were at reduced risk of cervical neoplasia if the man was circumcised (OR 0.37, CI 0.2–0.8) (Castellsague *et al.*, 2002). An ecologic study also suggested an inverse association between the prevalence of circumcision and rates of cervical cancer (Drain *et al.*, 2006). In observational studies from Rakai, Uganda, wives of circumcised men were at lower risk of BV (RR = 0.79), trichomonas (RR = 0.65), HSV-2 (RR = 0.82) and HPV (RR = 0.72), but there was no effect on gonorrhea (RR = 1.19) or chlamydia (RR = 1.06) (Gray *et al.*, 2006). Data from the Ugandan trial of male circumcision show lower rates of GUD (RR = 0.78, 95% CI 0.63–0.97), BV (RR = 0.60, 95% CI 0.38–0.94) and trichomonas (RR = 0.52, 95% CI 0.5–0.98) among wives of intervention-arm men relative to wives of control-arm men (Tobian *et al.*, 2008). It is plausible that the lower prevalence of GUD and vaginal STIs in female partners of circumcised men is due to reduced carriage or shedding of organisms in circumcised men.

The safety of male circumcision

Complication rates following circumcision are enormously variable, depending on the age at circumcision, the experience of the provider, the circumstances under which surgery is performed and the presence of pre-existent conditions which may be indications for the surgery. As previously noted, the three randomized trials reported surgery-related moderate/severe complications ranging from 1.5 percent to 3.6 percent. The most common complications were infections, followed by hematoma or bleeding and wound dehiscence. The frenulum is the most vulnerable site for these complications due to the friability of the tissue and possible contamination with urine. The Kenyan trial found that rates of complications decreased over time as surgeons acquired more experience (Krieger *et al.*, 2005). Resumption of intercourse before wound healing is associated with increased complication rates (Kigozi *et al.*, 2008b). However, the complication rates from the trials in which surgery was performed by well-trained physicians in properly equipped settings may be lower than in future programs where less experienced personnel may conduct the procedure under less optimal circumstances.

Complications following adult male circumcision in developed countries range from 2 to 10 percent, but adult circumcision is generally performed for medical indications, and these rates may not be comparable to those following elective surgery in otherwise healthy men. In Africa, much higher complication rates have been reported. For example, in Kenya a 11.1 percent rate of surgical complications was observed in public hospitals, 22.5 percent in private medical facilities and 34.3 percent when surgery was performed by traditional practitioners. High complication rates have also been reported with traditional pubertal circumcision in South Africa and Nigeria (Halperin and Bailey, 1999; Lavreys *et al.*, 1999). Moreover, many complications under these traditional circumstances are severe and may result in partial or complete loss of the penis, fistulas, meatal stenosis, and deaths. In Turkey, a mass hospital-based circumcision program in young boys observed a complication rate of 15.7 percent, which was substantially higher than the rate of 3.8 percent with routine procedures (Moses *et al.*, 1990). Surgical complications are much lower in infants. For example, in Israel, 19,478 neonatal circumcisions largely conducted in non-medical settings reported a complication rate of 0.3 percent. Similar low neonatal rates have been reported elsewhere.

Safety among HIV-infected men is of importance, since future programs are likely to include HIV-infected men. The WHO/UNAIDS guidelines recommend that circumcision be provided to HIV-positive men if medically indicated or if they request the procedure (WHO & UNAIDS, 2007a). The rationale is that if HIV-positive men were excluded from receipt of surgery it would be potentially stigmatizing. Moreover, if programs were restricted to HIV-negative men, circumcision could be perceived as a marker of uninfected status. Thus, HIV-infected men might seek surgery from unsafe sources in order to mask their status, and circumcised men (both HIV-positive and HIV-negative), might use their circumcision status to negotiate unsafe sex. The South African trial circumcised 73 HIV-positive men and reported a complication rate of 8.2 percent. In Rakai, Uganda, 420 HIV-positive men in latent stage disease were circumcised with a complication rate of 3.1 percent, which is comparable to that observed in uninfected men. At 30 days post-surgery, complete wound healing was found in 71.2 percent of HIV-positive and 80.7 percent of HIV-negative men, suggesting slower wound healing in those who were HIV-positive (Kigozi *et al.*, 2008b).

There are conflicting reports if the effects of circumcision on subsequent sexual satisfaction, penile sensitivity and sexual dysfunction (Bongaarts *et al.*, 1989; de Vincenzi and Mertens, 1994; Bailey and Halperin, 2000; Buve, 2000; Gray *et al.*, 2000, 2006; Green, 2000; Baeten *et al.*, 2005). Findings range from decreased satisfaction and function, to improvement following circumcision. Interpretation of these results is difficult, because adult circumcised men were highly selected due to medical indications for surgery; infant circumcision cannot provide before-and-after comparisons; sample sizes were small; and follow-up was short in most studies. Mild or moderate erectile dysfunction was reported by 0.4 percent of circumcised men in the South African trial (Auvert *et al.*, 2005) and 0.1 percent of circumcised men in the Kenyan trial (Bailey *et al.*, 2007). In the Ugandan trial, 0.3 percent of circumcised and 0.1 percent of uncircumcised men reported any sexual dysfunction, and this difference was statistically insignificant (Kigozi *et al.*, 2008b). It is therefore reasonable to conclude that circumcision does not adversely affect sexual function. Normal sexual satisfaction was reported by 98.4 percent of the circumcised and 99.9 percent of the uncircumcised men in the Ugandan trial after 2 years of follow-up and, although small, this difference was statistically significant (Kigozi *et al.*, 2008b). Thus, there may be a small reduction in sexual satisfaction among a minority of circumcised men.

The prevalence and acceptability of male circumcision

Worldwide, approximately 30 percent of men are circumcised, of whom around two-thirds are Muslim. Neonatal circumcision is common in predominantly Islamic countries of the Middle East, North Africa and Asia, as well as in Muslim and non-Muslim populations of West Africa. Circumcision is also common in North America, Australia, New Zealand and Israel. However, neonatal circumcision is uncommon in East and Southern Africa, and if men are circumcised in these countries it is performed at later ages, including adolescence or young adulthood (WHO & UNAIDS, 2007b).

Studies of acceptability of circumcision among uncircumcised men, conducted before the randomized trial results were known, suggest potentially high levels of acceptance. Acceptability among women is also high. In sub-Saharan African studies acceptability rates range from 45 to 85 percent (Nnko *et al.*, 2001; Bailey *et al.*, 2002; Halperin *et al.*, 2005; Quinn, 2006), and a high proportion of parents state that they would be willing to circumcise their infant sons (Quinn, 2006). The primary reasons stated for acceptability are improvement in hygiene, sexual pleasure, and potential protection from HIV and STIs. However, all these studies posed hypothetical questions, since circumcision was not readily available in any setting. Prior to the initiation of the randomized trials in Rakai, Uganda, surveys suggested that 60 percent of men would be willing to accept circumcision. The major barriers were concerns about pain and complications of surgery, and possible adverse effects on sexual pleasure or performance. Overall, 45 percent of eligible HIV-negative, uncircumcised males enrolled into the trial. Among control trial participants, 85 percent accepted circumcision on completion of 2 years of follow-up.

Male circumcision and behavioral disinhibition or risk compensation

It has been postulated that adoption of risk-reduction strategies may be partially offset by compensatory behaviors that increase risk because of a false sense of protection, or because people are comfortable with a certain level of risk and self-regulate their behaviors to maintain a perceived constant risk level (Pinkerton 2001; Rottingen *et al.*, 2001; Crepaz *et al.*, 2004). There is an extensive literature suggesting that newly introduced preventive measures may engender an exaggerated belief in protective efficacy which leads to increased risk behaviors (Rottingen *et al.*, 2001; Crepaz *et al.*, 2004; Cassell *et al.*, 2006). In the South African and Kenyan circumcision trials, the intervention-arm men had higher-risk behaviors during follow-up than controls, suggesting possible disinhibition (Auvert *et al.*, 2005). However, the differentials in risk behaviors were small. No disinhibition was observed in the Ugandan trial (Gray *et al.*, 2007a), and an observational study in Kenya reported no increase in risk behaviors following circumcision (Agot and Kiarie, 2006). Modeling suggests that even modest increases in the number of partners could negate the protective effects of circumcision (Gray *et al.*, 2007b), and it will be necessary to reinforce prevention messages of abstinence, fidelity and condom use in future circumcision programs.

Modeling of the effects of male circumcision on population HIV incidence, the number of surgeries and cost per HIV infection averted

A dynamic simulation model developed by Williams *et al.* (2006), using country-level prevalence data for HIV and male circumcision in sub-Saharan Africa, estimated that circumcision programs could potentially avert 2 million new HIV infections and 0.3 million deaths over 10 years, assuming an efficacy of 60 percent (as observed in the South African trial) and full program coverage of uncircumcised men within the next decade (Williams *et al.*, 2006). The impact of circumcision is likely to be most profound in East and Southern Africa, where HIV prevalence is high and male circumcision is infrequent. However, the authors acknowledged uncertainties in the data on HIV prevalence and the proportions of males who are circumcised, and that the possibility of behavioral disinhibition could offset the impact of circumcision programs. Another model, by Mesesan (2006), also suggested marked impact of circumcision on HIV incidence in South Africa over 20 years.

A stochastic simulation model was used to estimate the potential impact of male circumcision programs on HIV incidence in Rakai, Uganda (Gray, 2005; Gray *et al.*, 2007b). Under many scenarios, with a RR = 0.5, circumcision could potentially abort the HIV epidemic, but behavioral disinhibition could offset these benefits (Gray *et al.*, 2007b). It was estimated that approximately 35 surgeries were required to avert a single HIV infection over a period of 10 years, and that if the cost per surgery was approximately \$35, the cost per infection averted would be approximately \$1225. However, if the efficacy of circumcision increases with longer duration, as indicated in the Rakai trial, the costs could be substantially less. Kahn (2006) modeled the South African trial efficacy results and estimated a cost of \$181 per infection averted over 20 years. Although the cost per infection averted is likely to be substantially lower in settings with high HIV incidence, as in South Africa, it is likely that these latter cost estimates are unrealistically low.

Overall, these simulations suggest that male circumcision could have a substantial impact on the HIV epidemic, and is likely to be cost effective.

Scale-up of circumcision programs

The three randomized trials and the WHO recommendation that male circumcision be promoted for HIV prevention in men received wide publicity, and there has been anecdotal evidence of increased demand for circumcision in several African countries. If this demand is not met with safe services, men could resort to unsafe circumcision, with the attendant risks of complications. Thus, there is a compelling need to provide services as soon as possible. Increasingly, African Ministries of Health are including circumcision in their AIDS control strategies, and funding has been provided in selected countries under the President's Emergency Fund for AIDS Support. The Director of the Global Fund has also invited countries to apply for support. WHO/UNAIDS has prepared a manual for circumcision surgery (WHO, UNAIDS & JHPIEGO, 2006), and needs assessments are ongoing in several countries. Thus the scale-up of services is in process.

A major decision for policy-makers regards the relative investment in infant versus adolescent/adult circumcision. Surgery in infants or young boys is simpler, safer and less demanding in terms of personnel training, facilities and equipment compared to surgery at older ages. However, any protective effects against HIV conferred by infant circumcision would be deferred for one to two decades, until these young boys reach sexual maturity. Adolescent and adult surgery is more demanding and expensive, but potential HIV benefits are likely to accrue over the shorter term. Models suggest that high coverage of adolescent/adult circumcision could, in many settings, potentially abate the HIV epidemic over 10–20 years. Thus, placing emphasis on these older age groups could reduce HIV prevalence, and this would offset the potential benefits that might be realized from neonatal surgeries. Thus, each country will have to decide on a policy which weighs these two options.

Circumcision for HIV prevention presents a novel public health challenge, since this is the first time a surgical procedure has been used to prevent an infectious disease. The experience with mass sterilization in India provides a somewhat analogous example of the use of surgery for a public health goal, and it is noteworthy that these sterilization campaigns were a major catastrophe because of poor surgical management which resulted in unnecessary complications and deaths, poor health education, and a failure to fully inform participants of risks and benefits. Circumcision cannot stand alone; it is at best partially effective for HIV prevention in men and of unknown efficacy for HIV prevention in women. Circumcision must be integrated into established HIV prevention strategies such as ABC and VCT, both to prevent behavioral disinhibition and to ensure maximum effectiveness. HIV testing and counseling, although not mandatory, are highly desirable, and it will be necessary to provide treatment for STIs and penile infections before surgery to minimize risks of complications. Proper postoperative care and management of complications will also be needed. Programs will require major investments in training of personnel, construction or upgrading of facilities, purchase of equipment and supplies, and measures to ensure quality control of services. In addition, careful counseling will be needed for both men and their female partners to avoid early initiation of sex before complete wound healing, and to prevent potential behavioral disinhibition. Evaluation of the safety and long-term effectiveness of circumcision on HIV incidence in men and women, as well as monitoring sexual risk behaviors, will be needed in selected settings, and such evaluation is now being conducted in the Kenyan and Rakai trial sites.

In countries where circumcision has not traditionally been practiced, cultural and religious beliefs may affect the acceptability of services, and there is need for sensitive health education to avoid possible objections from religious or community authorities. In addition, where traditional male circumcision is currently practiced, it will be necessary to engage traditional practitioners in order to avoid their opposition to programs or their continued provision of potentially unsafe surgery. In such contexts, programs may have to include traditional practitioners

Ronald Gray, David Serwadda, Godfrey Kigozi and Maria J. Wawer

in a non-surgical role. Finally, the involvement of women is important, because acceptability studies suggest that women are more willing than men to accept circumcision, primarily for hygienic reasons, and because female support will be needed for postoperative wound care and sexual abstinence, as well as prevention of unsafe sexual practices. Male circumcision programs may also provide a venue for provision of male reproductive services, since current programs, with the exception of STI care, often fail to reach the high-risk groups of younger men.

References

Agot, K. and Kiarie, J. (2006). Male circumcision in Siaya and Bondo districts, Kenya: a prospective cohort study to assess behavioural disinhibition following circumcision. In: *Program and Abstracts of the XVIth International AIDS Conference, Toronto, Canada, 13–28 August*, Abstract No. TUAC0205.

Auvert, B., Taljaard, D., Lagarde, E. *et al.* (2005). Randomized, controlled intervention trial of male circumcision for reduction of HIV infection risk: the ANRS 1265 Trial. *PLoS Med.*, 2, e298.

Aynaud, O., Ionesco, M. and Barrasso, R. (1994). Penile intraepithelial neoplasia – specific clinical features correlate with histologic and virological findings. *Cancer*, 74, 1762–7.

Baeten, J. M., Richardson, B. A., Lavreys, L. *et al.* (2005). Female-to-male infectivity of HIV-1 among circumcised and uncircumcised Kenyan men. *J. Infect. Dis.*, 191, 546–53.

Bailey, R. C. and Halperin, D. T. (2000). Male circumcision and HIV infection – Reply. *Lancet*, 355, 927.

Bailey, R. C., Neema, S. and Othieno, R. (1999). Sexual behaviors and other HIV risk factors in circumcised and uncircumcised men in Uganda. *J. Acquir. Immune Defic. Syndr.*, 22, 294–301.

Bailey, R. C., Plummer, F. A. and Moses, S. (2001). Male circumcision and HIV prevention: current knowledge and future research direction. *Lancet Infect. Dis.*, 1, 223–31.

Bailey, R. C., Muga, R., Poulussen, R. and Abicht, H. (2002). The acceptability of male circumcision to reduce HIV infections in Nyanza Province, Kenya. *AIDS Care – Psychol. Socio-Med. Aspects AIDS/HIV*, 14, 27–40.

Bailey, R. C., Moses, S., Parker, C. B. *et al.* (2007). Male circumcision for HIV prevention in young men in Kisumu, Kenya: a randomised controlled trial. *Lancet*, 369, 643–56.

Baldwin, S. B., Wallace, D. R., Papenfuss, M. R. *et al.* (2004). Condom use and other factors affecting penile human papillomavirus detection in men attending a sexually transmitted disease clinic. *Sex. Transm. Dis.*, 31, 601–7.

Bongaarts, J., Reining, P., Way, P. and Conant, F. (1989). The relationship between male circumcision and HIV infection in African populations. *AIDS*, 3, 373–7.

Brown, J. E., Micheni, K. D., Grant, E. M. J. *et al.* (2001). Varieties of male circumcision – a study from Kenya. *Sex. Transm. Dis.*, 28, 608–12.

Buchbinder, S. P., Vittinghoff, E., Heagerty, P. J. *et al.* (2005). Sexual risk, nitrite inhalant use, and lack of circumcision associated with HIV seroconversion in men who have sex with men in the United States. *J. Acquir. Immune Defic. Syndr.*, 39, 82–9.

Buonaguro, F. M., Tornesello, M. L., Salatiello, I. *et al.* (2000). The Uganda study on HPV variants and genital cancers. *J. Clin. Virol.*, 19, 31–41.

Buve, A. (2000). HIV/AIDS in Africa: why so severe, why so heterogenous? Paper presented at the 7th Conference on Retroviruses and Opportunistic Infections, San Francisco, Abstract S28, Study Group on Heterogeneity of HIV Epidemics in African Cities.

Cassell, M. M., Halperin, D. T., Shelton, J. D. and Stanton, D. (2006). HIV and risk behaviour – risk compensation: the Achilles' heel of innovations in HIV prevention? *Br. Med. J.*, 332, 605–7.

Castellsague, X., Bosch, F. X., Munoz, N. *et al.* (2002). Male circumcision, penile human papillomavirus infection, and cervical cancer in female partners. *N. Engl. J. Med.*, 346, 1105–12.

Crepaz, N., Hart, T. A. and Marks, G. (2004). Highly active antiretroviral therapy and sexual risk behavior – A meta-analytic review. *J. Am. Med. Assoc.*, 292, 224–36.

de Vincenzi, I. and Mertens, T. (1994). Male circumcision – a role in HIV prevention. *AIDS*, 8, 153–60.

Devito, C., Hinkula, J., Kaul, R. *et al.* (2000). Mucosal and plasma IgA from HIV-exposed seronegative individuals neutralize a primary HIV-1 isolate. *AIDS*, 14, 1917–20.

Dillner, J., von Krogh, G., Horenblas, S. and Meijer, C. J. L. M. (2000). Etiology of squamous cell carcinoma of the penis. *Scand. J. Urol. Nephrol.*, 34, 189–93.

Diseker, R. A. III, Peterman, T. A., Kamb, M. L. *et al.* (2000). Circumcision and STD in the United States: cross-sectional and cohort analyses. *Sex. Transm. Infect.*, 76, 474–9.

Diseker, R. A., Lin, L. S., Kamb, M. L. *et al.* (2001). Fleeting foreskins: the misclassification of male circumcision status. *Sex. Transm. Dis.*, 28, 330–5.

Drain, P., Halperin, D., Hughes, J. *et al.* (2006). Male circumcision, religion, and infectious diseases: an ecologic analysis of 118 developing countries. In: *Program and Abstracts of the XVIth International AIDS Conference, Toronto, Canada, 13–28 August*, Abstract no. TUPE0400.

Fergusson, D. M., Boden, J. M. and Horwood, L. J. (2006). Circumcision status and risk of sexually transmitted infection in young adult males: an analysis of a longitudinal birth cohort. *Pediatrics*, 118, 1971–7.

Fonck, K., Kidula, N., Kirui, P. *et al.* (2000). Pattern of sexually transmitted diseases and risk factors among women attending an STD referral clinic in Nairobi, Kenya. *Sex. Transm. Dis.*, 27, 417–23.

Fowke, K. R., Kaul, R., Rosenthal, K. L. *et al.* (2000). HIV-1-specific cellular immune responses among HIV-1-resistant sex workers. *Immunol. Cell Biol.*, 78, 586–95.

Gray, P. B. (2004). HIV and Islam: is HIV prevalence lower among Muslims? *Social Sci. Med.*, 58, 1751–6.

Gray, R., Azire, J., Serwadda, D. *et al.* (2004). Male circumcision and the risk of sexually transmitted infections and HIV in Rakai, Uganda. *AIDS*, 18, 2428–30.

Gray, R. H. (2005). What Will it Take to Control the Epidemic? Reducing HIV Transmission: Lessons from Rakai and Other African Studies. Rio de Janeiro: IAS.

Gray, R. H., Kiwanuka, N., Quinn, T. C. *et al.* (2000). Male circumcision and HIV acquisition and transmission: cohort studies in Rakai, Uganda. *AIDS*, 14, 2371–81.

Gray, R. H., Wawer, M., Thoma, M. *et al.* (2006). Male circumcision and risks of female HIV and STI acquisition in Rakai, Uganda. In: *13th Conference on Retroviruses and Opportunistic Infections, Denver, Colorado*.

Gray, R. H., Kigozi, G., Serwadda, D. *et al.* (2007a). Male circumcision for HIV prevention in men in Rakai, Uganda: a randomised trial. *Lancet*, 369, 657–66.

Gray, R. H., Li, X. B., Kigozi, G. *et al.* (2007b). The impact of male circumcision on HIV incidence and cost per infection prevented: a stochastic simulation model from Rakai, Uganda. *AIDS*, 21, 845–50.

Green, E. C. (2000). Male circumcision and HIV infection – Reply. *Lancet*, 355, 927.

Grulich, A. E., Hendry, O., Clark, E. *et al.* (2001). Circumcision and male-to-male sexual transmission of HIV. *AIDS*, 15, 1188–9.

Halperin, D. T. and Bailey, R. C. (1999). Male circumcision and HIV infection: 10 years and counting. *Lancet*, 354, 1813–15.

Halperin, D. T., Fritz, K., McFarland, W. and Woelk, G. (2005). Acceptability of adult male circumcision for sexually transmitted disease and HIV prevention in Zimbabwe. *Sex. Transm. Dis.*, 32, 238–9.

Harrison, D. C. (2000). Male circumcision and HIV infection. *Lancet*, 355, 926.

Hunter, D. J., Maggwa, B. N., Mati, J. K. G. *et al.* (1994). Sexual behavior, sexually transmitted diseases, male circumcision and risk of HIV infection among women in Nairobi, Kenya. *AIDS*, 8, 93–9.

Jessamine, P. G., Plummer, F. A., Achola, J. O. N. *et al.* (1990). Human Immunodeficiency-Virus, genital ulcers and the male foreskin – synergism in HIV-1 transmission. *Scand. J. Infect. Dis.*, 69, 181–6.

Kahn, J. G. (2006). Cost-effectiveness of male circumcision in sub-Saharan Africa. In: *Program and Abstracts of the XVIth International AIDS Conference, Toronto, Canada, 13–28 August*, Abstract No. TUAC0204.

Kaul, R., Rowland-Jones, S. L., Kimani, J. *et al.* (2001). Late seroconversion in HIV-resistant Nairobi prostitutes despite pre-existing HIV-specific CD8(+) responses. *J. Clin. Inv.*, 107, 341–9.

Kelly, R., Kiwanuka, N., Wawer, M. J. *et al.* (1999). Age of male circumcision and risk of prevalent HIV infection in rural Uganda. *AIDS*, 13, 399–405.

Kigozi, G., Gray, R., Settuba, A. *et al.* (2008a). Foreskin surface area and HIV acquisition in Rakai, Uganda (size matters). Poster accepted for the forthcoming Mexico City Conference.

Kigozi, G., Watya, S., Polis, C. B. *et al.* (2008b). The effect of male circumcision on sexual satisfaction and function, results from a randomized trial of male circumcision for human immunodeficiency virus prevention, Rakai, Uganda. *Br. J. Urol. Intl*, 101, 65–70.

Kreiss, J. K. and Hopkins, S. G. (1993). The association between circumcision status and human immunodeficiency virus infection among homosexual men. *J. Infect. Dis.*, 168, 1404–8.

Krieger, J. N., Bailey, R. C., Opeya, J. *et al.* (2005). Adult male circumcision: results of a standardized procedure in Kisumu District, Kenya. *Br. J. Urol. Intl*, 96, 1109–13.

Lavreys, L., Rakwar, J. P., Thompson, M. L. *et al.* (1999). Effect of circumcision on incidence of human immunodeficiency virus type 1 and other sexually transmitted diseases: a prospective cohort study of trucking company employees in Kenya. *J. Infect. Dis.*, 180, 330–6.

Maden, C., Sherman, K. J., Beckmann, A. M. *et al.* (1993). History of circumcision, medical conditions, and sexual activity and risk of penile cancer. *J. Natl Cancer Inst.*, 85, 19–24.

McCoombe, S. G. and Short, R. V. (2006). Potential HIV-1 target cells in the human penis. *AIDS*, 20, 1491–5.

Mesesan, K. (2006). The potential benefits of expanded male circumcision programs in Africa: predicting the population-level impact on heterosexual HIV transmission in Soweto. In: *Program and Abstracts of the XVI International AIDS Conference, Toronto, Canada, 13–28 August*, Abstract No. TUAC0203.

Millett, G. A., Ding, H., Lauby, J. *et al.* (2007). Circumcision status and HIV infection among Black and Latino men who have sex with men in 3 US cities. *J. Acquir. Immune Defic. Syndr.*, 46, 643–50.

Mishra, V. (2006). Patterns of HIV seroprevalence and associated risk factors. PEPFAR Annual Meeting, Durban, South Africa.

Moses, S., Bailey, R. C. and Ronald, A. R. (1998). Male circumcision: assessment of health benefits and risks. *Sex. Transm. Infect.*, 74, 368–73.

Moses, S., Bradley, J. E., Nagelkerke, N. J. D. *et al.* (1990). Geographical patterns of male circumcision practices in Africa – association with HIV seroprevalence. *Intl J. Epidemiol.*, 19, 693–7.

Nnko, S., Washija, R., Urassa, M. and Boerma, J. T. (2001). Dynamics of male circumcision practices in northwest Tanzania. *Sex. Transm. Dis.*, 28, 214–18.

Norris Turner, A., Morrison, C., Padian, N. *et al.* (2006). Men's circumcision status and women's risk of HIV: a longitudinal analysis of Zimbabwean and Ugandan women. In: *Program and Abstracts of the XVIth International AIDS Conference, Toronto, Canada, 13–28 August*, Abstract No. TUPE0457.

O'Farrell, N., Morison, L., Moodley, P. *et al.* (2006). Association between HIV and subpreputial penile wetness in uncircumcised men in South Africa. *J. Acquir. Defic. Syndr.*, 43, 69–77.

Patterson, B. K., Landay, A., Siegel, J. N. *et al.* (2002). Susceptibility to human immunodeficiency virus-1 infection of human foreskin and cervical tissue grown in explant culture. *Am. J. Pathol.*, 161, 867–73.

Pinkerton, S. D. (2001). Sexual risk compensation and HIV/STD transmission: Empirical evidence and theoretical considerations. *Risk Analysis*, 21, 727–36.

Quinn, T. C. (2006). Circumcision and HIV transmission: the cutting edge. Paper #120. In: *13th Conference on Retroviruses and Opportunistic Infections, Denver, Colorado.*

Quinn, T. C., Wawer, M. J., Sewankambo, N. *et al.* (2000). Viral load and heterosexual transmission of human immunodeficiency virus type 1. *N. Engl. J. Med.*, 342, 921–9.

Reynolds, S. J., Shepherd, M. E., Risbud, A. R. *et al.* (2004). Male circumcision and risk of HIV-1 and other sexually transmitted infections in India. *Lancet*, 363, 1039–40.

Rottingen, J. A., Cameron, D. W. and Garnett, G. P. (2001). A systematic review of the epidemiologic interactions between classic sexually transmitted diseases and HIV – how much really is known? *Sex. Transm. Dis.*, 28, 579–97.

Sateren, W. S., Bautista, C. T., Shaffer, D. N. *et al.* (2006). Male circumcision and HIV infection risk among tea plantation residents in Kericho, Kenya: incidence results after 1.5 years of follow up. In: *Program and Abstracts of the XVIth International AIDS Conference, Toronto, Canada, 13–28 August*, Abstract No. TUAC0202.

Seed, J., Allen, S., Mertens, T. *et al.* (1995). Male circumcision, sexually transmitted disease, and risk of HIV. *J. Acquir. Immune Defic. Syndr. Hum. Retrovirol.*, 8, 83–90.

Siegfried, N., Muller, M., Deeks, J. *et al.* (2005). HIV and male circumcision – a systematic review with assessment of the quality of studies. *Lancet Infect. Dis.*, 5, 165–73.

Simon, V., Ho, D. D. and Karim, Q. A. (2006). HIV/AIDS epidemiology, pathogenesis, prevention, and treatment. *Lancet*, 368, 489–504.

Szabo, R. and Short, R. V. (2000). How does male circumcision protect against HIV infection? *Br. Med. J.*, 320, 1592–4.

Tobian, A., Gray, R., Serwadda, D. *et al.* (2008). Late breaker abstract: trial of male circumcision: prevention of HSV-2 in men and vaginal infections in female partners, Rakai, Uganda. CROI. Boston.

Trabattoni, D., Lo Caputo, S., Maffeis, G. *et al.* (2004). Human alpha defensin in HIV-exposed but uninfected individuals. *J. Acquir. Immune Defic. Syndr.*, 35, 455–63.

Urassa, M., Todd, J., Boerma, J. T. *et al.* (1997). Male circumcision and susceptibility to HIV infection among men in Tanzania. *AIDS*, 11, 73–9.

Wawer, M. J., Reynolds, S. J., Serwadda, D. *et al.* (2005). Might male circumcision be more protective against HIV in the highly exposed? An immunological hypothesis. *AIDS*, 19, 2181–2.

Wawer, M. J., Kigozi, G., Serwadda, D. *et al.* (2008). Late breaker abstract: trial of male circumcision in HIV+ men, Rakai, Uganda: effects in HIV+ men and women partners. CROI. Boston.

Weiss, H. A., Quigley, M. A. and Hayes, R. J. (2000). Male circumcision and risk of HIV infection in sub-Saharan Africa: a systematic review and meta-analysis. *AIDS*, 14, 2361–70.

Weiss, H. A., Thomas, S. L., Munabi, S. K. and Hayes, R. J. (2006). Male circumcision and risk of syphilis, chancroid, and genital herpes: a systematic review and meta-analysis. *Sex. Transm. Infect.*, 82, 101–9.

WHO & UNAIDS (2007a). New Data on Male Circumcision and HIV Prevention: Policy and Programme Implications. Montreaux: WHO/UNAIDS.

WHO & UNAIDS (2007b). Male Circumcision: Global Trends and Determinants of Prevalence, Safety and Acceptability. Geneva: WHO/UNAIDS.

WHO, UNAIDS & JHPIEGO (2006). Manual for Male Circumcision under Local Anaesthesia, Version 2.2. Geneva: WHO, UNAIDS, JHPIEGO.

Williams, B. G., Lloyd-Smith, J. O., Gouws, E. *et al.* (2006). The potential impact of male circumcision on HIV in sub-Saharan Africa. *PLoS Med.*, 3, 1032–40.

Behavioral issues in HIV prevention

PART

II

Payoff from AIDS behavioral prevention research

7

Willo Pequegnat and Ellen Stover

The reality of AIDS is human. Human choice and the limits of our repertoire of human behaviors primarily determine who will contract HIV and will get AIDS. Ultimately, the human mind and spirit will determine the end of the epidemic. I hope it will be within my lifetime.

Walter F. Batchelor

We put absolutely too much emphasis on treatment and too little on prevention. ...A trillion dollars a year is spent on treatment and 1% of that is for population based prevention. We depend on biotechnology. We believe that we can solve everything with it. And its just not true and becomes less true every day that we live.

David Satcher

On current trends, AIDS will kill tens of millions of people over the next 20 years. But this need not happen. We know prevention works.

Peter Piot

Despite the remarkable improvement in HIV treatment, there is still no cure or preventive vaccine. For every person in sub-Saharan Africa that starts on HIV Highly Active Antiretroviral Therapy (HAART), conservatively four persons become HIV infected (UNAIDS, 2007). Given these odds, it is not possible to treat our way out of this epidemic. However, HIV infection is preventable, primarily through behavior change (NIH, 1997; Coutinho and Cates, 2000). In the 25 years during which HIV/STD prevention research has been conducted in the US, much has been learned about the theoretical variables that need to be considered when predicting and changing high-risk HIV-related behaviors (Fishbein *et al.*, 1992). The model of the reproductive rate of sexually transmitted diseases includes: (1) infectivity or transmissibility, (2) interaction rates between susceptibles and infectors, and (3) duration of infectiousness (May and Anderson, 1987). The former two factors can be influenced by interventions that decrease the number of partners, increase abstinence or mutual monogamy between HIV seronegatives, increase condom use, or delay sexual initiation. The first can be influenced by adherence to HAART, the second can be moderated by minimizing interactions among social networks (e.g., truck stops and CSWs), and the third can be changed by interventions that increase the uptake of HIV counseling and testing to identify newly infected individuals and provide treatment while they are most infectious.

HIV/STD prevention research tests ways to change behaviors that place persons at risk for HIV/STD infection, while HIV/STD operational research identifies methods to translate effective programs to public health settings as rapidly as possible. Several themes conceptually unify these research areas:

1. Utilization of theory-based interventions
2. Utilization of knowledge of determinants and distribution of high-risk HIV/ STD risk behaviors
3. Identification of co-factors, such as mental health, STDs, alcohol and substance use, and developmental stage as targets for interventions
4. Identification of moderators and mediators of behavior change
5. Utilization of similar methods to assess HIV/STD knowledge and risk behaviors
6. A commitment to the assessment of both behavioral and biomedical outcomes to ensure convergent validity of findings.

Prevention and operations research are conducted at multiple levels: individual, couple, family, institutional, community and societal. Targeted outcomes are self-report of safer sexual behavior (correct and consistent condom use, fewer partners, monogamy by both partners, moderate alcohol use, adoption of safer drug-using practices) and lower incidence of STDs and HIV. The involvement of the community in the development and review of the conduct of HIV/STD research has been a hallmark of this work.

The National Institutes of Health (NIH) and the Centers for Disease Control and Prevention (CDC) have been at the forefront of AIDS research in the US since the beginning of the epidemic, developing research knowledge essential for understanding and preventing the spread of disease. The AIDS epidemic is a public health emergency that requires a coordinated and interdisciplinary federal response. Mobilizing the best scientific minds in both behavioral and medical research, federal support continues to play a pivotal role in developing new ways to understand and prevent HIV, and to treat its consequences. Prevention research efforts have been designed for vulnerable groups who are at higher risk for both HIV and co-occurring medical conditions that are mediated by health behaviors (e.g., diet, exercise, stress reduction, drug and alcohol use, mental illness).

While the payoff from AIDS medical research has been significant, this review is focused on behavioral research. Medical breakthroughs are cited only if they led to the development of different behavioral strategies. While many AIDS researchers might disagree with the division of AIDS research into these five periods, there would be agreement that there has been unprecedented progress in AIDS prevention research during these 25 years (Pequegnat, 2005). In some cases, an area of research may have begun in one period but is discussed in another period because that is where the payoff occurred. Because NIMH funds the majority of NIH-supported behavioral prevention, much of the focus is on its initiatives and findings from investigations that it supports.

Periods in AIDS prevention research

Period I (1983–1985): Identification of risk factors (knowledge, attitudes and behaviors (KAB)

In 1983 the federal government began mobilizing a response to this new public health threat, even before the routes of transmission were fully understood. Public health officials suspected that it was a social disease associated with sexual behavior (CDC, 1983). Federal support for sexual behavioral research had been terminated in the 1980s because it was controversial; consequently, there were no good prevalence data on HIV-related risk behaviors of Americans, nor proven models for prevention of high-risk HIV-related sexual behaviors. The first generation of AIDS researchers were drawn primarily from health and behavior research which had adapted social psychological processes and methods to study health outcomes. Their work focused on cardiovascular and other chronic diseases (e.g., stress, coping). Investigators who had been using methods from classical and operant conditioning to develop smoking-cessation and other lifestyle programs adapted these methods for stopping HIV-associated risk behaviors. Experimental social psychologists (e.g., attitudes, impression formation, loss–gain messages) contributed well-articulated theories and well-controlled methods. Because it is essential to have a better understanding of what people know, think and do in order to develop population-specific prevention programs, early studies collected data on knowledge, attitudes and behavior (KAB) (McKusick *et al.*, 1985; Catania *et al.*, 1990; Coates, 1990; Fisher and Fisher, 1992).

Studies were conducted with people who were exhibiting symptoms of the illness and those who were not, in order to identify KAB patterns that differentiated them. Studies also identified antecedent behaviors (e.g., drinking alcohol, sharing needles) and settings (e.g., bars, bathhouses, shooting galleries) that placed people at risk for HIV/STDs. The initial period of HIV/AIDS research was characterized by multiple KAB studies and the legitimatization of AIDS as a behavioral research area.

In 1985, when an HIV test was licensed for screening blood supplies, it was possible to identify reliably those who were HIV seropositive. Investigators then intensified their efforts to develop interventions to keep at-risk populations from seroconverting.

Payoff from this period

- Investigators identified risk groups: gay men, IDUs, partners of IDUs, children of seropositive women
- Investigators described KABs of these at-risk groups and put out public health messages that it was these behaviors not membership in a group that placed individuals at risk
- Investigators established risk factors for newly emerging at-risk groups, especially those transitioning to adolescence.

Period II (1986–1991): Test of concept of HIV interventions

The second period of payoff from AIDS research was from 1986 to 1991, and was characterized by the activism of ACT UP and other groups in the HIV-infected and -affected communities that adopted the slogan "Silence = Death". Zidovudine (AZT), a nucleoside reverse transcriptase inhibitor, was the first drug to treat HIV approved by the FDA in 1987. Shortly thereafter the first candidate AIDS vaccine began testing, and there was hope that an effective AIDS vaccine would be available in 5 years. The first comprehensive needle-exchange program was established. The Americans with Disabilities Act was expanded to include people living with HIV/AIDS.

Despite these breakthroughs, early in the epidemic, surveys of public opinion about AIDS revealed widespread fear, lack of accurate information about its transmission, and willingness to support policies that would restrict the civil liberties of persons living with HIV (Herek, 1990). HIV/AIDS is a major barrier to persons seeking testing and counseling services, and to health workers providing care (Valdiserri, 2001).

A significant minority of people in the US have consistently expressed negative attitudes toward persons living with HIV, although they have fluctuated in the measures that they would support: quarantine, universal mandatory testing, and even tattooing of infected individuals (Herek and Capitanio, 1993). AIDS-related discrimination in employment, health care, insurance and education, and physical attacks, have been reported. People living with AIDS have been fired from their jobs, evicted from their homes, and denied services (Gostin, 1989, 1990; Hunter and Rubenstein, 1992). Multiple social, psychological and demographic variables have been correlated with AIDS-related attitudes: age, education, ethnicity, personal contact with persons living with AIDS, knowledge about HIV transmission, and attitudes toward homosexuality. Younger and better-educated respondents consistently manifest lower levels of negative attitudes toward persons living with AIDS. Persons who know someone living with AIDS tend to be more favorably inclined and to have better attitudes toward AIDS-related policies. Some data suggest that racial and ethnic differences exist in AIDS stigma, and that African-Americans and Hispanics tend to overestimate the risks of HIV transmission through casual contact and to favor policies that would separate persons living with HIV from the public (Alcalay *et al.*, 1988–1989). Despite major public health education campaigns, people with HIV throughout the world are still stigmatized in various ways. There appears to be a greater degree of stigma expressed toward individuals with AIDS than toward individuals with other diseases (Crawford, 1996).

The Centers for Disease Control (CDC) promulgated guidelines in 1985, encouraging people to participate in HIV counseling and testing programs and suggesting measures to help persons living with AIDS to avoid opportunistic infections, and these guidelines were later updated (CDC, 1986). The CDC recommended that community health education programs be aimed at members of high-risk groups to: (1) increase knowledge about AIDS, (2) facilitate behavioral changes to reduce risk of infection, and (3) encourage voluntary counseling and testing (VCT). They laid out a plan to triage persons who came in for tests and either tested negative or positive. Pre- and post-test counseling was touted by the CDC as an effective preventive intervention. In 1992, Allen and colleagues demonstrated that HIV serotesting and counseling led to a large increase in condom use and was associated with a lower rate of new infections (Allen *et al.*, 1992).

A Request for Applications (RFA) soliciting AIDS research centers to be state-of-the-art resources was released by NIMH (NIMH, 1985). One of the innovative aspects of the RFA was the requirement that the community and populations with which the research was conducted should participate in every stage of the research and be given briefings about the findings from the studies. Each center developed a Community Advisory Board (CAB) to facilitate communication between the Center and the community. Initially AIDS Research Centers were funded as P50s, which is a funding mechanism that requires both core (measurement, statistical analysis, laboratory) and at least three research projects. Beginning in 1995, Centers were funded using the P30 mechanism, which supports core research projects. Investigators then propose individual research grants (RO1s).

During this period six centers were awarded under the original solicitation. The HIV Center for Clinical and Behavioral Studies at Columbia University and the Center for Biopsychosocial Study of AIDS at the University of Miami were funded in 1986. The Center for AIDS Prevention Studies (CAPS) at the University of California at San Francisco was begun in 1986, and the Center at the University of Michigan (Midwest AIDS Biobehavioral Research Center) was initiated in 1988. The HIV Neurobehavioral Research Center (HNRC) at the University of California at San Diego was awarded in 1989, and the Center for NeuroAIDS Preclinical Studies at the Scripps Institute was awarded in 1990.

NIMH AIDS Research Centers provided an environment in which investigators pursued interdisciplinary behavioral, clinical and treatment research on prevention of transmission and neuropsychiatric consequences of HIV/AIDS. They were viewed as national and international resources. AIDS investigators at these Centers collaborated with other AIDS researchers and community service organizations. While the range of their activities was multidisciplinary and interdisciplinary, each Center focused its activities on a major theme: behavioral prevention, NeuroAIDS, or HIV policy research.

There was a scarcity of reliable data on sexual behavior available to design effective behavioral prevention research. During this period, several sexual behavior surveys were conducted that provided rich data sets. The National AIDS Behavioral Survey (NABS) was a behavioral epidemiological survey using a national probability population-based sample (Catania *et al.*, 1992).

Two samples were recruited by random-digit dialing: (1) 23 high-risk cities (individuals 18–49 years of age) that at time one had had large number of AIDS cases, and (2) the national sample (individuals 18–75 years of age). The majority of people (93 percent) who were surveyed at Time 1 (June 1990–February 1991) volunteered to be re-surveyed at Time 2 (January–August 1992), and 73 percent were re-interviewed. This survey was designed to describe changes in (1) risk factors relevant to HIV and other sexually transmitted disease, (2) condom use, and (3) HIV antibody testing. Nationally, none of the risk estimates changed significantly from collection of data at Time 1 to collection at Time 2. However, the proportion of respondents reporting multiple sexual partners increased significantly in high-risk cities (a 4 percent increase). Respondents who appear to have been monogamous may have been serially monogamous rather than being with the same partners.

The other survey, the National Health and Social Life Survey (NORC), was designed to capture data on adult sexual behavior in the US, and there were 3432 respondents. The goals were to describe systematic variation in the timing and sequence of individual sexual activity in response to life-course events and changes in social and cultural environment. This survey also collected data on variations in sexual relationships based on the assumption that the characteristics of the couple, rather than the individual, contributed to the spread of disease. To assess these aspects, social network techniques of data collection and analysis were used. In this survey, the investigators looked at two major indicators of whether the respondents had changed behavior in response to acquiring HIV/AIDS (Laumann *et al.*, 1994); the first was the number of HIV tests that they had had to determine whether they were HIV seropositive, and the second was direct question asking if they had changed behavior. The results indicated that 26.6 percent of all respondents had been tested for HIV/AIDS (slightly more men than women, substantially more younger than older persons, more blacks than whites or Hispanics, more educated, and more unmarried, and more in larger cities). Those at greater risk for HIV (who had had many partners, same-gender partners, or had paid or been paid for sex) were more likely to have been tested: There appeared to be a negligible relationship between having been tested and condom use. Approximately 30 percent of the respondents indicated that they had changed their sexual behavior as a result of the risk of HIV/AIDS. This was more pronounced in men, younger people, blacks, the unmarried, and those in larger cities.

Because of the public health imperative, HIV behavioral interventions were designed as rapidly as possible for seronegative at-risk persons to reduce their sexual and drug-use behaviors associated with high risk of HIV infection (e.g., unprotected sex or needle-sharing). To accomplish this, investigators used results from the two surveys and adapted ethnographic methods to identify social and cultural norms, mores, values and beliefs of groups of at-risk people (Herdt and Boxer, 1991; Pequegnat *et al.*, 1995). Ethnography traditionally required immersion in cultures and long periods of field-based research, but in the 1980s models for rapid or focused ethnographic assessment were developed and applied in various areas of research (Scrimshaw and Hurtado, 1987; Bentley *et al.*, 1988). To understand the risk behaviors of these populations, participant observational data were combined with open-ended qualitative interview data that permitted in-depth explorations of sensitive topics, such as substance abuse and sexual behavior. These focused ethnographic assessments accelerated conduct of formative research, which is essential when designing, tailoring and implementing interventions in a reasonable timeframe. Formative research also facilitated community involvement in the design of prevention and clinical trial protocols, making them more acceptable to the community (Cornell *et al.*, 2007).

This strategy led to single-site, single-population studies that tested proof of concept for prevention programs with seronegative persons. The interventions tested in this period were based on social cognitive theory and delivered at the individual level. Because individuals who were seropositive could be distinguished from those who were seronegative, comparative studies to identify predictors of seroconversion were possible.

One of the major reasons that there have been major payoffs from these behavioral research studies is the fact that the interventions are based on theories (Ajzen and Fishbein, 1980; Fishbein *et al.*, 1992; Herek, 1995). All empirical research is based on cause-and-effect assumptions; even formative or ethnographic studies involve choices about what will be observed and at what depth. A major challenge has been to make these assumptions explicit so that intervention research can evaluate and, if necessary, modify them. Theory development and evaluation is more than an intellectual exercise because it can operationalize every aspect of the study, from describing the causal path, to the hypotheses, selection of measures, development of the targets in the intervention and, finally, interpretation of the data (Herek, 1995). If the overall intervention is efficacious, a theory that has posited moderators and mediators can identify which subgroups are best matched for different interventions and even which aspects of an intervention works best with subgroups (West *et al.*, 1993).

Studies demonstrated that interventions based on social cognitive theory were effective in reducing high-risk sexual practices in gay and bisexual men (Kelly *et al.*, 1989, 1991, 1995, 1997; Stall *et al.*, 1988; Valdiserri *et al.*, 1989; Osmond *et al.*, 1994). As other risk groups emerged, those interventions were adapted for them: adolescents and runaways (Kirby *et al.*, 1991; Rotheram-Borus *et al.*, 1991; Jemmott *et al.*, 1992; Rotheram-Borus, 1993); minority women (Ehrhardt and Exner, 1991; Exner *et al.*, 1997; Ehrhardt *et al.*, 2002; El-Bassell *et al.*, 2003); and the mentally ill (Kelly *et al.*, 1992; Carey *et al.*, 1995).

During this period there was also a program of research that examined the relationships among stress, immunity, lifestyle changes and secondary prevention of persons living with HIV/AIDS. A study of severe environmental stress and declining social integration in a lower social support structure and feelings of loneliness were associated with higher HHV-6 antibody titers (Dixon *et al.*, 2001). Reactivation of HHV-6 has been related to poorer immune status in HIV-seropositive persons. However, feelings of social integration mediated the relationship. These findings are consistent with research by Leserman and colleagues (Leserman *et al.*, 1999) showing that diminished social support is associated with lower immune status. In one study, 81 percent of HIV-seropositive people began eating healthier food and taking vitamins when they found out they were HIV-positive (Namir *et al.*, 1990). They also reported reducing alcohol and drug use, and increasing sleep (Siegel and Krauss, 1991). These changes in lifestyle empowered individuals, imparted a sense of control over their health and increased their quality of life. A study examined aerobic exercise training as a buffer against affective distress and immune decrements which are associated with learning about HIV antibody status and initiating treatment (LaPerriere *et al.*, 1990; Antoni *et al.*, 1991). Research subjects received 5 weeks of training and, at a point 72 hours before serostatus notification, researchers collected psychometric, fitness and immunologic data on all subjects. Control subjects that tested positive showed significant increases in anxiety and depression, as well as decrements in natural killer cell numbers, following notification, whereas those who exercised did not exhibit negative consequences.

In addition to examining the physical health of seropositive persons, investigators studied the neurological and neuropsychiatric manifestations of HIV infection. Those early reports suggested that even asymptomatic HIV-positive persons have impaired neurocognitive skills, such as executive functions (planning, initiation, solving problems, engaging in purposeful behavior); motor and perceptual-motor speed; speed of information-processing, including reaction time; attention; and working memory (Grant *et al.*, 1987). Because these skills are essential to perform many jobs safely (e.g., air traffic controller, commercial aircraft pilot, bus driver and train operator, dangerous equipment operator), these revelations stimulated the development of employment policies. However, the final suggestions were that HIV-positive individuals should not be excluded from employment simply on the basis of HIV infection.

Payoff from this period

- Stigma was recognized as a major concern which must be addressed to ensure the uptake of prevention and treatment opportunities
- AIDS research centers scaled up quickly and became state-of-the-art resources for prevention and mental health research
- Community Advisory Boards (CAB) demonstrated that community input is essential to ensure the development of a culturally relevant and sustainable prevention programs
- Ethnographic approaches were recognized as essential for collecting cultural and gender data to design or tailor HIV/STD interventions for different populations
- Efficacious prevention programs for at-risk groups at the individual-level were developed
- Prevention programs were based on theories which permitted examination of which interventions worked for subgroups of at-risk populations
- Investigators recognized that HIV testing and counseling could change HIV/STD risk behaviors
- Despite controversy, investigators successfully conducted sexual behavior surveys in successive waves and provided significant databases for AIDS prevention work
- Investigators identified the cognitive domains that were impaired in HIV infection and developed standardized HIV neuropsychological assessment batteries
- Investigators began conducting secondary prevention programs enhancing quality of life, associated with slowing disease progression in seropositive persons.

Period III (1992–1997): Multisite, multipopulation randomized controlled trials

Despite the advancements being made in behavioral and medical research, by 1994 AIDS was the leading cause of death among Americans between the ages of 25 and 44. However, in 1994 the study results from ACTG 076 gave hope that early treatment of pregnant women with the drug zidovudine (AZT) could reduce or prevent all mother-to-child transmission of HIV (Connor *et al.*, 1994). In 1995 highly active antiretroviral therapy (HAART) was introduced, with the first protease inhibitor, saquinavir. The following year, the FDA approved the first non-nucleoside reverse transcriptase inhibitor, nevirapine, and viral load test to measure levels of HIV. These developments revived many patients who were near death (the "Lazarus effect"), and led to a 70 percent reduction in AIDS-related deaths.

Replications of single-site, single-population studies in different contexts and emerging populations increased. Having determined that an intervention would work, large randomized, controlled trials (RCTs) were conducted in multiple populations and sites using both behavioral and biological outcomes in order to increase the generalizability of the results (Fishbein and Pequegnat, 2000).

While AIDS research was only initiated in 1983, strong federal support for public health research and committed HIV investigators (many of whom were HIV-positive) had generated such a large amount of data from randomized controlled trials (RCTs) by 1997 that the National Institutes of Health held a Consensus Development Conference (CDC) on Interventions to Prevent HIV Risk Behaviors. This was the first behavioral intervention selected for this format. The Consensus Conference is a mechanism established by Congress to evaluate behavioral and biomedical research to determine whether drugs, technologies or interventions are ready to be implemented in public health agencies (NIH, 1997; Coutinho and Cates, 2000). The Consensus conference is the closest mechanism that the US has to a Supreme Court for science.

The impartial expert consensus panel declared that behavioral prevention efforts were effective in stopping the spread of HIV in at-risk populations. They specifically stated, "Prevention interventions are effective for reducing behavioral risk for HIV/AIDS and must be widely disseminated" (NIH, 1997; Coutinho and Cates, 2000). They suggested to the field that prevention work should be initiated with HIV-seropositive individuals, which led to the development of two trials conducted in a later period: Project EXPLORE and the Healthy Living Project (HLP). Furthermore, they weighed in on a controversial issue and stated that "an impressive body of evidence suggests powerful effects from needle exchange programs".

Over the years these opinions have been corroborated by the CDC, which has established a three-part technology transfer program to ensure that tested prevention programs are adopted in public health agencies. The first is the Prevention Research Synthesis (PRS) project, which conducts systematic reviews based on efficacy criteria to select HIV prevention behavioral interventions (Sogolow *et al.*, 2002). The second is called Replicating Effective Programs Plus (REP+), which is dedicated to packaging evidence-based prevention programs that meet the rigorous criteria established by the PRS (CDC's HIV/AIDS Prevention Research Synthesis Project, 2001). Research protocols are packaged as user-friendly prevention programs which can guide providers in replicating evidence-based risk-reduction programs in their own settings and communities.

The third part of this process is the Diffusion of Effective Behavioral Interventions (DEBI) project (see also Chapter 22), which moves effective HIV interventions into program practice (Collins *et al.*, 2006). (Table 7.1 lists the evidence-based programs and the funders.) The DEBI project coordinates the dissemination of packaged interventions and provides training and technical assistance. In summary, interventions that meet the PRS efficacy criteria are packaged by REP+ and are then disseminated by the DEBI project.

Table 7.1 Sample of evidence-based interventions selected for REP+

Name of intervention	Principal investigator	Funding institution	Targeted population	Year RCT completed
Popular Opinion Leader (POL)	Jeffrey A. Kelly	National Institute of Mental Health (NIMH)	MSM	1991
Reducing the Risk	Doug Kirby	NIH & Private Foundation	Youth	1991
Be Proud! Be Responsible!	John and Loretta Jemmott	NIMH	Adolescents	1992
Healthy Relationships	Seth Kalichman	NIMH	African American men and women	1995
Becoming a Responsible Teen (BART)	Janet St Lawrence	NIMH	Youth	1995
Focus on Kids (FOK)	Bonita Stanton	NIMH	Youth	1996
Mpowerment Project	Susan Kegeles	NIMH	Young MSM	1996
Street Smart	Mary Jane Rotheram-Borus	NIMH	Adolescents	1997
Project Respect	Mary Kamb, Martin Fishbein	Centers for Disease Control (CDC)	Heterosexual adults	1998
Project Light	David Celentano, Tyler Hartwell, Jeff Kelly, Raul Magana, Willo Pequegnat, Mary Jane Rotheram, Robert Shilling	NIMH	Heterosexual adults	1998
Real AIDS Prevention Project	Jennifer L. Lauby	CDC	Heterosexual adults	1998
VOICES/VOCES	L.H. O'Donnell	CDC	Heterosexual adults	1998
Community PROMISE	CDC AIDS Community Demonstration Projects	CDC	IDUs & female sex partners, sex workers, non-Gay identified MSM, youth, residents of high STD-prevalence areas	1999
Together Learning Choices	Mary Jane Rotheram-Borus	NIDA	Young HIV+ persons (13–29)	2001
Project Connect	Nabila El-Bassel	NIMH	Women and partners	2004
Partnerships for Health	Jean Richardson	NIMH	HIV+ persons over 18	2004

While important work was being conducted at the individual level during this period, there was recognition that prevention could be implemented at different levels. AIDS was not solely an individual disease, because of the ways that it could affect the entire family: multiple members and generations could be at risk or infected. NIMH convened a group of investigators in 1989 to discuss the role of families in preventing and adapting to HIV/AIDS. Having identified family-level research questions, NIMH issued two RFAs. The first, in 1992, was for 3 years and called for studies examining process in families heavily impacted by HIV, as a basis for developing prevention programs for families. The second, in 1995, called for developing and testing family-based interventions; these are described in the book *Working with Families in the Era of AIDS* (Pequegnat and Szapocznik, 2000).

During this period, there was a strong payoff from a program of community-level prevention research developed by Kelly and colleagues (Kelly *et al.*, 1991, 1992, 1997, 2006; Kelly, 2004). This work was based on the theory of diffusion of innovation, which was a model to explain how new technological and behavioral innovations are initiated and become adopted, accepted and normative with community populations (Rogers, 1983). Innovations often begin with "trusted leaders", whose actions and opinions lead other people to change their behaviors and norms. This alters what are acceptable normative behaviors in a community. Kelly and colleagues were able to design interventions that changed the norms about safe sexual behaviors in different community settings (e.g., bars, public housing), which resulted in significant behavior changes (Sikkema *et al.*, 2000). The Center for AIDS Interventions Research (CAIR), established by Kelly and his colleagues at the Medical College of Wisconsin in Milwaukee, was funded in 1994. CAIR is a multidisciplinary HIV prevention research center that develops and evaluates innovative new interventions to prevent HIV among persons most vulnerable to HIV. It also focuses on prevention of adverse health and mental health outcomes among persons living with HIV infection, and their friends and families. Its international research focuses on prevention work in Russia and Eastern Europe.

Instruments, assessment batteries and survey methods were standardized during this period so that data could be utilized by multiple research teams. Important methodological issues were identified, such as recruitment and sampling issues, appropriate control or comparison groups for RCTs, interviewing techniques for sensitive topics, criteria for establishing change in behaviors, ethical issues, quasi-experimental design, combining qualitative and quantitative techniques, response bias in surveys, validity of self-report, modeling of AIDS epidemic, and cultural issues with ethnic and racial minority populations. These issues were explored in a 1993 book, edited by Ostrow and Kessler, entitled *Methodological Issues in AIDS Behavioral Research* (Ostrow and Kessler, 1993). Investigators from the AIDS Research Centers also formed working groups to identify instruments that provided valid and reliable data across populations and settings. Some groups developed batteries that could be used in assessing multiple HIV-related problems (Butters *et al.*, 1990).

When introduced in 1993, the female condom was hailed as the first woman-initiated (dual protection) method, capable of preventing both unwanted pregnancy and infection (Worley, 2005). This opened up a potential new line of research of behavior change in women; women were taught to avoid risky situations and triggers for their risk behaviors, and to negotiate female condom use (Caso *et al.*, 2002). International and US experience with the female condom has shown that the device empowers diverse populations of women, helping them to negotiate protection with their partners, promoting healthy behaviors, and increasing self-efficacy and sexual confidence and autonomy (Golub, 2000). However, the use of the female condom still depends on relationship dynamics and partner negotiation (Artz *et al.*, 2000), and male condoms tend to be more widely used (Ehrhardt *et al.*, 2002). Evidence suggests that the male partner's objection is the main reason for non-use of the female condom (Artz *et al.*, 2000). A series of research projects was initiated to examine relationships in couples in order to address these communication and power imbalances (Exner *et al.*, 1997; El-Bassell *et al.*, 2003).

In previous years Voluntary Counseling and Testing (VCT) had been viewed as an intervention to reduce HIV infection (see also Chapter 19), but, with the advent of treatment, there was emphasis on early screening of at-risk populations, to identify and treat people (Rotheram-Borus, 1997). Rotheram and colleagues developed an AIDS Research Center – the Center for HIV Identification, Prevention, and Treatment Services (CHIPTS) – funded in 1997. This Center is dedicated to developing a new model of counseling and testing, promoting early detection, and engaging people into systems of care as early as possible. To implement these goals, investigators at CHIPTS collaborate with other research programs and service providers in the Los Angeles area and hold an annual policy forum to discuss research that informs health policy. Studies have suggested that people with STDs have a two- to five-fold increased risk of HIV acquisition, and have suggested multiple biological mechanisms by which they may increase HIV susceptibility or infectiousness. In this period, two major RCTs were conducted to determine if treating STDs could decrease HIV incidence. The first trial, conducted in Mwanza, Tanzania, demonstrated a decrease of 40 percent in heterosexually transmitted HIV infections in communities with continuous access to treatment of symptomatic STDs as compared to the control communities with minimal services, where the incidence remained the same (Grosskuth, 1995). However, in the second study, in Rakai, Uganda, there was no reduction in HIV incidence when community-wide mass treatment was administered to both asymptomatic and symptomatic individuals intermittently (Wawer, 1998). In explaining these mixed results, they identified the following reasons: Mwanza has an HIV prevalence of only 4 percent, versus 16 percent in Rakai; Mwanza adopted a continue treatment strategy for only symptomatic persons, versus intermittent treatment in Rakai for both symptomatic and asymptomatic persons; and Mwanza's prevalence of curable STDs was slightly higher than that in Rakai. This work stimulated an HIV prevention program to treat STDs in symptomatic persons, and to reduce risky sexual behaviors to prevent acquisition of STDs.

In 1995, another treatment to prevent HIV infection – post-exposure prophylaxis (PEP) – became the standard of care for health-care workers who had occupational exposure to HIV (see also Chapter 6). Post-exposure use of zidovudine (AZT) was associated with an approximate 81 percent reduction in becoming HIV seropositive after work-related exposures (CDC, 1995; Cardo *et al.*, 1997). Studies were later initiated that examined the efficacy of PEP for non-occupational (i.e., sexual and drug use) exposures to HIV. Studies have demonstrated that delivery of PEP for non-occupational exposure is feasible and cost-effective (Kahn *et al.*, 2001; Schechter *et al.*, 2002; Pinkerton *et al.*, 2004). Despite initial concern that it would provide a rationale for being more risky, PEP was primarily used by individuals who usually engaged in safe sexual or injection behaviors but experienced a lapse or an accident and used PEP as a last resort (Kahn *et al.*, 2001). Martin and colleagues (Kahn *et al.*, 2001) followed 397 adults with a high-risk sexual or drug-use exposure and found, after 12 months of followup, that the majority of participants did not request a repeat course of PEP (83 percent). Compared with baseline, 73 percent of participants reported a decrease in the number of times they had performed high-risk sexual acts, and 13 percent reported no change. However, Roland and colleagues found a substantial number of seroconversions among those using PEP for non-occupational exposure (Kahn *et al.*, 2001). Although some guidelines exist for non-occupational PEP, the safe and effective delivery of PEP is complicated and efficacy data on the use of PEP are limited.

While PEP did not appear to be associated with behavioral disinhibition, there were widespread reports of increased sexual risk-taking and sexually transmitted infections and HIV among MSM associated with the advent of highly active antiretroviral therapy (HAART) (Schechter *et al.*, 2002). HAART is a treatment for human immunodeficiency virus (HIV) infection that uses a combination of several antiretroviral drugs. The goals of HAART are to suppress HIV replication to a level sufficient to prevent the development of drug-resistant mutations, to prevent the progression of HIV disease, and to reconstitute the immune system (Friedland and Williams, 1999). Patients begin to feel better and, when the virus is not detectable, they feel that unprotected sex is safe behavior.

Patients who are able to suppress HIV replication below the level of detection have a substantially greater chance to achieve long-term viral suppression on HAART. However, it is necessary to obtain consistent and adequate drug levels in order to prevent development of drug-resistant mutations. This has led to a vigorous field of research on adherence to medications.

Treatment adherence to any medical regime is complicated, but HAART is particularly complex and requires a high level of adherence to be effective. When

HAART is used properly and paired with reduced HIV-related risky behavior, it is effective and is a major payoff for public health.

The factors associated with adherence have been divided into five categories in major reviews of the topic (Reiter *et al.*, 2000; Ickovics and Meade, 2002; Simoni *et al.*, 2003). Common predictors of non-adherence include depression/psychiatric morbidity, active drug or alcohol use, stressful life events, lack of social support, and inability to identify the drug regimen or describe the relationship between adherence and drug resistance (Gordillo *et al.*, 1999; Golin *et al.*, 2002). The complexity of the regimen and side-effects are associated with non-adherence. The types of pills prescribed are not associated with adherence (Golin *et al.*, 2002). Some studies report that if the regimen fits into the individual's daily routine, there is a greater likelihood of adherence (Chesney, 2000). The stage and duration of HIV infection and the associated opportunistic infections and HIV-related symptoms can increase adherence because persons are more aware of the need for treatment. Dissatisfaction with previous health-care systems has been associated with non-adherence (Chesney, 2000). A patient's trust in the physician has been shown to improve adherence, especially if the relationship is longstanding (Stone *et al.*, 1998; Altice *et al.*, 2001).

Interventions to promote adherence have focused on identifying adherence barriers, developing problem-solving skills, integrating the treatment regimen into daily life, increasing adherence self-efficacy, increasing motivation to adhere, and enhancing patients' readiness to take HAART before initiation of the treatment. Some indirect methods that have been tested in interventions are pill counts, self-reporting, MEMS caps, pharmacy records, metabolic drug levels, pharmacy refill records, service utilization records, patient self reports and provider reporting. Effective direct methods are directly observed therapy (DOT), biomedical assay of blood, and urine analysis.

In 1997 the Center for Interdisciplinary Research on AIDS (CIRA) was established at Yale University, and brought together collaborating scientists from 20 different disciplines. These investigators focused on basic social and behavioral research aimed at identifying the determinants of HIV-related risk in different populations. They conducted prevention interventions and policy and modeling-based research that examines the cost-effectiveness of interventions and models the impact on public health. This research is undertaken at multiple levels of analysis, and combines social/behavior with biomedical/laboratory research. This Center is unique because it was co-funded by the National Institute of Drug Abuse (NIDA), which was the first time an AIDS Center had been a joint venture.

Payoff from this period

- The expert panel from the NIH Consensus Development Conference stated that HIV preventive interventions were ready to be implemented in public health agencies

- Teams of investigators developed methods to handle the problems of conducting research in real settings where control was a challenge
- Investigators identified components of effective preventive interventions and the process of behavior change:
  - before a person can change, he or she must perceive personal risk for HIV;
  - a person must have complete and accurate knowledge of HIV, which is necessary but not sufficient;
  - a person must recognize personal triggers to engage in high-risk HIV-related behaviors (e.g., drinking alcohol, going to a certain bar, etc.);
  - a person must learn to problem-solve in order to avoid risky situations and triggers;
  - a person must learn the skills to practice safer sex (correct and consistent condom use, ability to negotiate safer sex) *in vitro* and *in vivo*, and receive feedback from the facilitator and other group members; and
  - a person must practice these skills so that he or she can engage in them when required
- Investigators began analyzing moderators and mediators, which permitted identification of factors that must be changed to be effective and identification of which interventions worked best for subgroups
- Investigators determined that behavioral prevention must be titrated to the risk behaviors and life context of at-risk individuals (some people may benefit from brief interventions but others may require a more sustained intervention)
- Investigators recognized that the intervention did not need to be delivered to individuals to change their risk behaviors and they began exploring intervening at multiple levels to prevent HIV transmission
- Researchers established the role of the family in preventing and adapting to HIV/AIDS
- Investigators demonstrated that community-level interventions can change social norms and prevent HIV/STD risk behaviors in a community
- It was found that HAART could be effective when HIV seropositive persons adhere
- Availability of treatment led investigators to develop interventions to engage at-risk populations in early screening and testing and in adherence to complicated medical strategies; and
- NIMH funded three new AIDS Research Centers.

Period IV (1998–2004): Technology transfer and cost-effectiveness of programs

This period of HIV/AIDS research began in 1998, when the Centers for Disease Control published results from Project Respect in the *Journal of the American*

Medical Association (Kamb *et al.*, 1998) and the NIMH Multisite HIV Prevention Trial (1998) published outcome data from Project LIGHT in *Science*.

These two significant RCTs demonstrated HIV-risk reduction with vulnerable, at-risk ethnic minority individuals. Project Respect was a multisite randomized controlled trial that evaluated the relative efficacy of three interventions based on social cognitive theory with 5872 HIV-negative participants recruited from clinics (Kamb *et al.*, 1998). The research participants were randomly assigned to receive one of the following: (1) two HIV education sessions; (2) two sessions aimed at increasing risk perception and condom use; or (3) four enhanced, skill-focused sessions. At the 3-month follow-up, 44 percent of participants who had received the brief two-session counseling and 46 percent of those who had received the enhanced four-session intervention were more likely to report no unprotected intercourse than those who received the risk education alone (38 percent). These results were corroborated by the 30 percent lower incidence of new STDs among brief- and enhanced-counseling participants compared to controls.

The efficacy of a behavioral intervention – Project LIGHT (Living in Good Health Together) – to reduce human immunodeficiency virus (HIV) risk behaviors was tested in a randomized, controlled trial with three high-risk populations at 37 clinics from 7 sites across the United States (National Institute of Mental Health, 1998). Compared with the 1855 individuals in the control condition, the 1851 participants assigned to a small-group, 7-session HIV risk-reduction program reported fewer unprotected sexual acts, had higher levels of condom use, and were more likely to use condoms consistently over a 12-month follow-up period. On the basis of clinical record review, no difference in overall sexually transmitted disease (STD) re-infection rate was found between intervention and control condition participants. However, among men recruited from STD clinics, those assigned to the intervention condition had a gonorrhea incidence rate one-half that of those in the control condition. Intervention-condition participants also reported fewer STD symptoms over the 12-month follow-up period. Study outcomes suggest that behavioral interventions can reduce HIV-related sexual risk behavior among low-income women and men served in public health settings.

Cost-effectiveness analysis refers to the economic analysis of intervention, such as the number of HIV infections averted. The purpose is to quantify the impact of the intervention and to determine its overall value to public health decision-makers. Kahn (1998) calculated that needle exchange programs were effective (typically $4000–$40,000 saved per HIV infection averted), HIV testing and counseling ($5000 to $10,000 per HIV infection averted), and drug treatment ($40,000 per HIV infection averted, which may include other important benefits such as reduction in crime). A cost-effectiveness analysis of Project LIGHT demonstrated that both brief and intensive sexual risk-reduction interventions for high-risk populations can be cost-effective (Pinkerton *et al.*, 2002). AIDS has a high cost to society because it primarily impacts young people during their most productive years. In this period, awareness increased that the lifetime cost of care and treatment for just one HIV seropositive person can be high because of the excellent treatment options.

Another prevention issue that stunned the investigators was the fact that 25 percent of men and 27 percent of women in Project LIGHT endorsed one question about experiencing unwanted touching when they were children (DiIorio *et al.*, 2002). Because of the ubiquitous role of CSA in HIV prevalence, the NIMH and CDC held a conference in 1998 on Child Sexual Abuse and HIV, and in 2004 published a book entitled *From Child Sexual Abuse to Adult Sexual Risk: Trauma, Revictimization and Intervention* (Koenig *et al.*, 2004). There was mounting research evidence suggesting that both men and women with a history of unwanted sexual activity during childhood are more likely to experience sexual assault and be HIV-seropositive as adults (Carballo-Dieguez and Dolezal, 1995; Coid *et al.*, 2001; Wyatt, 2002). In a study of men who experienced early unwanted sexual advances, they were more likely to abuse drugs, engage in unprotected intercourse and assume the riskiest positions with partners (Carballo-Dieguez and Dolezal, 1995).

Another at-risk group of children was found to be orphans. There had been growing concern about orphans since 1991, when it was estimated that there were less than a million. However, orphans became an international issue when UNAIDS issued a sobering report on the orphan crisis at the International AIDS Conference at Barcelona in 2002. In sub-Saharan Africa, a larger proportion of orphans have lost a parent to AIDS than to any other cause. Although the number of orphans varies among countries, the age of orphans is fairly constant across countries: 15 percent of orphans are 0–4 years old, 35 percent are 5–9 years old and 50 percent are 10–14 years old (Monasch and Boerma, 2004). Children whose parents are living with HIV experience negative changes in their lives, and often become the *de facto* adults in the household. As a result, they can experience exploitation, premature adult-level stress, emotional trauma and mental health problems (Stein, 2003; Subbarao and Coury, 2004). Orphans can encounter problems accessing the basic necessities, such as shelter, food, clothing, health and education, and even greater difficulty accessing mental health treatment. While still mourning their parents, they often experience stigma, shame, fear and isolation, which has negative psychological consequences. In a study in rural Uganda, high levels of psychological distress were identified in children whose parents had died from AIDS. Anxiety, depression and anger were more prevalent in AIDS orphans than in other children. In fact, 12 percent of AIDS orphans revealed that they hoped for death, as compared to 3 percent of other children (Atwine *et al.*, 2005). Additionally, in Zimbabwe, maternal orphans have been found to be at greater risk for HIV when compared to non-orphans (Kang *et al.*, 2008). To address these psychosocial problems, NIH issued a joint RFA in 2004 entitled "Psychosocial Needs of Children Affected by AIDS in Low Resource Countries", which has led to the development of a program of research that is currently being conducted.

Orphans often live in large female-headed or child-headed households with limited income. To address this problem, Ssewamala and colleagues tested a novel economic intervention to reduce HIV risks among AIDS-orphaned adolescents (Ssewamala *et al.*, 2008). Adolescents were randomly assigned to receive the intervention or usual care for orphans in Uganda. Data obtained at baseline and 12-month follow-up revealed significant differences between the treatment and control groups in HIV prevention attitudes and educational planning. This is a very promising approach to help the plight of orphans and other vulnerable children.

The Global Campaign for Microbicides was officially launched in July 1998 at the XIIth International AIDS Conference in Geneva. This effort had been initiated by members of the women's health and HIV community, and became a worldwide force as the Global Campaign for Microbicides. Microbicides – gels, films or other substances that can be used by the woman vaginally – have shown great potential, but as yet there has not been a successful candidate. The microbicide is designed to kill or inactivate HIV, stop the virus entering human cells, enhance the body's normal defense mechanisms against HIV, or inhibit HIV replication. There have been multiple studies, but the results have not been promising. Two trials were halted because of adverse events associated with non-nonoxynol 9 and cellulose sulfate, and a third trial found that Carraguard had no protective effect against becoming HIV seropositive. However, there are over 50 candidates in the pipeline that appear to have promising characteristics. In addition to being a successful barrier, its effectiveness will be determined by the willingness of the woman and her partner to use the product correctly and consistently. This has led to a vigorous research program on acceptability by social and behavioral scientists. This work has included questions about motivations for prevention activities, past and current condom use, and their attitudes about the characteristics of the actual or potential microbicide candidate (Mantell *et al.*, 2005) (see also Chapter 4).

Recreational drug use had been a known but underground co-factor in becoming HIV-positive. While originally associated with the West Coast, crystal meth is a stimulant commonly abused in many parts of the US. The link between crystal meth use and high-risk sexual practices has been documented: multiple partners, casual partners, high rates of STIs, low rates of condom use, increased desire for anal sex and fisting, and prolonged sexual activity (Molitor *et al.*, 1998). Most meth users are White men between the ages of 18 and 25, but the highest usage rates have been found in native Hawaiians, persons of more than one race, Native Americans, and men who have sex with men (MSM) (Winslow *et al.*, 2007). Increased numbers of HIV/AIDS cases among MSM have been linked to the use of methamphetamine (Semple *et al.*, 2002, 2003). Meth produces a rapid, pleasurable rush that reduces inhibitions and increases sexual drive, but prolonged use can lead to chronic depression and anxiety, an inability to enjoy activities, loss of friends, criminal involvement, and even myocardial infarctions, strokes, seizures, psychosis and death.

Investigators have been developing prevention programs for this problem (Mausbach *et al.*, 2007). A total of 341 men were randomly assigned to receive either a safer-sex behavior intervention (nicknamed EDGE) or a time equivalent diet-and-exercise attention-control condition. Participants in EDGE intervention engaged in significantly more protected sex acts at the 8-month and 12-month assessments. Participants in EDGE demonstrated a greater increase in self-efficacy over time. Results suggest that it is possible to reduce high-risk sexual behaviors in the context of ongoing meth use among MSM. In another study, participants felt that websites were the major starting point for crystal meth sexual "hook ups", and the Internet would be an excellent site for preventive interventions for meth use and risky sexual behavior (Matthew *et al.*, 2008) (see also Chapter 13).

Recent reports have linked the use of phosphodiesterase type 5 (PDE-5) inhibitors with increased rates of high-risk sexual behavior and HIV transmission in some individuals (Rosen *et al.*, 2006). A NIMH-funded multidisciplinary conference was convened to evaluate scientific research, clinical and ethical considerations, and public policy implications of this topic. PDE-5 inhibitors are selective and highly effective peripheral vasodilator drugs that treat male erectile dysfunction (ED). Three agents in this class (sildenafil, tadalafil, vardenafil) are in use worldwide. The use of PDE-5 inhibitors has had a tremendous impact on treatment of ED, and has solved an age-old problem. Despite the benefits, there are reports of misuse of PDE-5 inhibitors as recreational drugs, often in association with meth, ecstasy, cocaine and other stimulant drugs (Catania *et al.*, 2001; Ostrow *et al.*, 2006). This is most pronounced among MSM, and is associated with unprotected sex with multiple partners in a short period of time – which has been associated with an increased prevalence of HIV.

Prevention efforts have largely focused on keeping the HIV-negative persons from becoming infected. The steady rate of new HIV infections in the US for over 10 years (CDC, 2001), as well as alarming increases in many other countries, however, led to a realization that prevention also needs to target HIV-positive persons, particularly those in treatment (Merson *et al.*, 2000; Jansen *et al.*, 2001) (see also Chapter 11).

Initially, HIV-infected persons who were aware of their HIV serostatus tended to reduce behaviors that might transmit HIV to others (Allen *et al.*, 1992; Weinhardt *et al.*, 1999). However, more recent studies showed that some persons in treatment across a range of cultural backgrounds and geographical locations were increasing their risk behavior and acquiring other STDs (Collis and Celum, 2001; Crepaz and Marks, 2002). Several studies suggested that optimism about HAART effectiveness might be contributing to relaxed attitudes toward safer sex practices and increased sexual risk-taking by some HIV-positive persons. Media coverage and marketing campaigns for medications may have had an unwanted impact on risk behavior. It became apparent that behavioral interventions must accompany medical treatment for HIV in order to curtail sexual risk-taking during treatment.

A prospective clinical trial comparing the impact of a clinician-delivered intervention versus the standard-of-care on unprotected sexual behavior of HIV-infected patients was conducted (Catania *et al.*, 2001). Medical providers delivered brief client-centered interventions to a total of 497 HIV-infected patients as part of each clinical appointment. When the patients received a prescription for their medications, they also received a prescription for their behavioral risk reduction. The HIV-infected patients who received the clinician-delivered intervention showed significantly reduced unprotected intercourse over a follow-up period of 18 months. However, these behaviors increased across the study for HIV patients in the standard of care. As clinician-delivered HIV prevention intervention targeting HIV-infected patients resulted in reductions in unprotected sex, interventions of this kind can be integrated into routine clinical care.

Another test of a brief safer-sex counseling by medical providers of HIV-positive patients during medical visits was conducted in clinics (Richardson *et al.*, 2004). The clinics were randomized to intervention and control conditions. A total of 485 HIV-positive persons, sexually active prior to enrollment, were recruited for this study. Prevention counseling from medical providers was supplemented with written information. Two clinics used a gain-framed approach where the messages stressed the positive consequences of safer sex. A second group of medical providers delivered messages that used a loss-frame approach where the negative consequences of unsafe sex were highlighted. In the attention control condition, medical providers delivered messages about medication adherence. The intervention was given to all patients who came to the clinics. Of the patients who had two or more sex partners at baseline, unprotected sex was reduced by 38 percent among those who received the loss-frame intervention, and at follow-up they were significantly lower in the loss-frame condition as compared to the control. Similar results were obtained in participants with casual partners at baseline. No effects were observed in participants with only one partner, or only a main partner at baseline. No significant changes were seen in the gain-frame condition. Brief provider counseling emphasizing the negative consequences of unsafe sex can reduce HIV transmission behaviors in HIV-positive patients presenting with risky behavioral profiles.

Two important studies were conducted that addressed prevention with positives. EXPLORE was a multisite two-ground randomized controlled trial testing the efficacy of a behavioral intervention in preventing HIV infection among 4295 US men who have sex with men (Koblin and The Explore Study Team, 2004). The experimental intervention consisted of 10 one-on-one counseling sessions followed by maintenance sessions. The standard condition was twice yearly Project RESPECT individual counseling. The rate of acquisition of HIV was 18.2 percent lower in the intervention group than in the standard group, but the adjustment for baseline covariates lowered this to 15.7 percent. This fell short of the 35 percent that had been hypothesized. Contrary to expectation, more men in the intervention group than in the control group experienced a new STD. The behavioral outcomes did not significantly differ between the groups. While these were not the hoped-for results, the trial did demonstrate that it is possible successfully to conduct a prevention trial for positives.

The NIMH Healthy Living Project (HLP), a randomized behavioral intervention trial for people living with HIV, enrolled 943 individuals, including women, heterosexual men, injection drug users and men who have sex with men, from Los Angeles, Milwaukee, New York and San Francisco (NIMH Healthy Living Project Team, 2007). The intervention, which is based on qualitative formative research and Ewart's Social Action Theory, addresses three interrelated aspects of living with HIV: stress and coping, transmission risk behavior, and medication adherence. Fifteen 90-minute structured sessions, divided into three modules of five sessions each, are delivered to individuals: (1) stress, coping, and adjustment; (2) risk behaviors; and (3) health behaviors. Sessions are tailored to individuals within a structure that uses role-plays, problem-solving and goal-setting techniques. A "Life Project" – or overarching goal related to personal striving – provides continuity throughout the sessions. Overall, a significant difference in mean transmission risk acts was shown between the intervention and control arms over 50 to 25 months. The greatest reduction occurred at the 20-month follow-up, with a 36 percent reduction in the intervention group compared with the control group. Cognitive behavioral intervention programs can effectively reduce the potential of HIV transmission to others among HIV-seropositive person who report significant transmission risk behavior. The intervention was designed to be useful for prevention case-management settings where repeated one-on-one contact is possible, and where a structured but highly individualized intervention approach is desired.

In an early study of the Internet as a risky environment, Klausner and colleagues found that an outbreak of syphilis among 22 MSM could be traced back to meeting partners in the same chat room (Klauser *et al.*, 2000). This suggested that the Internet was an important tool for promoting disease awareness, prevention and control, and accessing partners of individuals acquiring STDs, because of its reach. Populations that seem to be engaging in the most risky behaviors appear to be the early adopters of the Internet as an alternative to the bar as a place to meet new partners. Often individuals who are engaging in the riskiest behaviors are the most difficult to identify and engage in prevention programs, but they can be accessed via the Internet because of the privacy that it offers: MSM, sensation-seekers, transgenders, married men engaging in same-gender sex, and other deeply closeted persons. Online surveys provide an opportunity to collect data quickely from large samples in response to unexpected events (Butler *et al.*, 2008).

NIMH put out a solicitation for "Communications and HIV/STD Prevention" in 2000 that specifically requested exploratory research studies examining the utility of the Internet in HIV prevention. While the CDC earlier had funded Internet research that demonstrated its utility, this was the first concerted effort to develop a program of research to harness the potential of the Internet for HIV prevention and treatment programs.

Based on the early experience with this research, a paper that provides guidelines on how to conduct Internet research was developed (Pequegnat *et al.*, 2007). Because the Internet is a unique environment, this paper explores the adaptations that must be made in research design and methods. Questions and response categories must be designed differently. If the banner does not appeal to and engage the target population, they will immediately move on.

Internet-based HIV prevention research is increasingly viewed as a powerful public health tool to prevent HIV transmission and the consequences of HIV (Bowen *et al.*, 2006; Rosser *et al.*, 2008) because of the ability to tailor messages for different demographic or behavioral risk characteristics based on information that users provide. Different role models can also be tailored to the participants (race, ethnicity, age, gender and sexual orientation) (Bull *et al.*, 2004). The Internet permits multiple research and prevention activities, such as chat-room recruitment and outreach, self-administered risk assessments with tailored messages, advertising using banners to recruit subjects, online partner notification, and surveys to collect rapid data on natural disasters affecting HIV risk behaviors.

During this period there was increased interest in technology transfer, and novel means beyond DEBI were tried. Dissemination approaches are more successful when providers gain skills and receive ongoing support for the use of a new method. A randomized trial evaluating technology transfer approaches using distance training approaches through the Internet (Kelly *et al.*, 2004) involved 86 leading AIDS non-governmental organizations (NGOs) in 78 countries (29 countries in Africa, 25 in Central/Eastern Europe and Central Asia, and 24 in Latin America and the Caribbean) as study participants. Two-part baseline interviews were carried out, by telephone, with NGO directors. The first interview asked about the general organization, while the second probed for information to determine if the POL intervention would be successful. NGOs were then randomized within their region into control and experimental conditions. Both conditions received training, a computer, subsidized Internet service, access to a study website to network with other NGOs, and briefing papers on program evaluation, needs assessment and organizational management. Only the experimental condition received a behavioral science consultant in who helped them to tailor the POL intervention to the site. At follow-up, 18 of the 42 experimental condition (43 percent) but only 7 of the 41 controls (17 percent) had developed a new HIV prevention program based on the mode that was disseminated. This appears to be a promising method to transfer programs.

Payoff from this period

- Researchers unblinded two important multisite clinical trials that demonstrated prevention works with ethnic minority populations (Project Respect and Project LIGHT)

- Researchers demonstrated that people can change their HIV-related risk behaviors and sustain this, making the interventions cost-effective
- Researchers analyzed components of effective intervention and identified processes of behavior change
- Investigators recognized the importance of integrating HIV/STD behavioral preventive interventions into treatment settings
- "Prevention for positives" interventions were demonstrated to be efficacious (OPTIONS, Brief Counseling, Healthy Living Project)
- Researchers analyzed evidence-based preventive interventions to assess cost effectiveness
- Investigators developed prevention programs that addressed the psychosocial and prevention issues of orphans
- Investigators recognized the role of childhood sexual abuse in the prevalence of HIV infection, and designed interventions to address this concern
- Behavioral scientists joined with biomedical scientists to evaluate the acceptability of different candidate microbicides
- Investigators demonstrated that interventions can be designed to prevent HIV risk behaviors among active meth users
- Investigators recognized that guidelines (toolkits) for technology transfer/information dissemination must be developed if NGOs can adapt them for local conditions
- Investigators realized that the seeds of technology transfer must be built into the original design of the study
- The Internet began to fulfill its promise as an important environment in which individually tailored prevention programs and important information about AIDS treatment and adherence can be effectively delivered.

Period V (2005–2008): Biomedical prevention and international research

While women had only accounted for 8 percent of new HIV cases in 1985, by 2005 they accounted for 27 percent of new cases. Most of these women were ethnic-minority women and, along with black men, they disproportionately constitute over half of new HIV infections while the incidence among whites has begun to decrease. There are currently 29 drugs on the FDA's list of drugs approved for the treatment of HIV/AIDS, and a vigorous adherence program has developed as effective strategies and treatment regimens have been simplified.

While behavioral interventions have been demonstrated to be efficacious in preventing HIV infection, the protection relies on the decisions made jointly by the couple, and must be consistent and correct. Investigators are exploring other prevention methods that are more technology based. This period of research has been characterized by a shift to biomedical prevention techniques that are more technology based. It is likely that successful HIV prevention efforts will rely on multiple techniques and strategies in combination.

While international research had been conducted in the past, much of the prevention work was formative work or Phase I or Phase II studies. In this period there have been some significant Phase III trials with both behavioral and biological outcomes that are more feasible and cost-effective to conduct in international settings.

One extremely successful example of this is the evaluation of male circumcision as a biomedical prevention strategy. For years, anthropological studies and a meta-analysis of observational studies had suggested that populations where male circumcision was the norm were experiencing lower rates of HIV transmission (Weiss *et al.*, 2000). In 2005, three studies reported on their results of male circumcision from randomized, controlled intervention trials. In the study conducted in South Africa, 3274 uncircumcised men aged 18–24 years were randomized to a group to be circumcised immediately and to a control group that was offered circumcision at the end of the follow-up period (Bongaarts *et al.*, 1989). The trial was so successful that the Data and Safety Monitoring Board (DSMB) recommended stopping the trial at the interim analysis. The protective rate was 60 percent. The protective effect of male circumcision was unchanged when controlling for sexual behavior, including condom use, seeking treatment for STIs, and abstinence during recovery from surgery. The investigators speculate that the direct or indirect factors that explain the results may be keratinization of the glans when not protected by the foreskin; rapid drying after sexual contact, which reduces the life expectancy of HIV; and a reduction of target cells, which are numerous on the foreskin.

Investigators conducted a second randomized control trial of 2784 men between the ages of 18 to 24 in Kenya (Shaffer *et al.*, 2007). Again this trial was stopped at the third interim analysis when the results were evaluated as being extremely beneficial by the DSMB. Male circumcision provided a 53 percent protective effect against HIV acquisition as compared to the control, and a 60 percent protective effect after adjustments for non-adherence and for those individuals who were found to be HIV-positive at baseline. There was little difference between circumcised and uncircumcised men on measures of change in sexual behavior across the follow-up visits, with the exception of a reduction in the number of men who had had two or more partners in the previous 6 months.

In another male circumcision study conducted in Uganda, 4996 uncircumcised, HIV-negative men between the ages of 15 and 40 agreed to HIV testing and counseling (Gray *et al.*, 2007; see also Chapter 6). They were randomized to receive immediate or delayed circumcision. The efficacy of circumcision for prevention of incident HIV was 51 percent and the protective efficacy of circumcision increased progressively during later follow-up periods, while the rate of HIV infection remained constant in the control group. This trial did not find evidence that men in the intervention group engaged in higher sexual risk behaviors. *Time* magazine announced that circumcision was the number one medical breakthrough in 2007. *The Lancet* declared that it ushered in a "new era for HIV prevention" (Guthrie, 2007).

Another study explored the possibility that circumcision of men, which reduces their risk of infection from HIV, would also protect their female sexual partners. Wawer and colleagues (2008) found that 1015 HIV-positive men selected at random to be circumcised conferred no indirect risk reduction for their female partners. In fact, circumcision increased the risk if the couples resumed sex before the surgical wound was fully healed – which takes 4–6 weeks. Among the women, the rates of condom use, bacterial vaginal infections, vaginal discharge, painful urination and urinary tracts infections remained the same.

Acute HIV infection refers to the first stage of infection, which is the period during which a person is most infectious and antibodies have not developed. The person frequently does not have symptoms, and the widely available tests cannot detect HIV. Nucleic acid amplification testing (NAAT) can detect acute HIV infection, but it is expensive. Some labs have developed a NAAT pooling strategy to screen in areas with high prevalence. Identifying persons with acute HIV infection is a high prevention priority because they may transmit HIV infection to multiple persons because they are unaware of their status, and treatment at this point may boost the immune system and slow the progression of HIV disease. Persons who initiate HAART during the acute infection stage had a much better viral load and CD4 counts compared to those who initiated it at a later stage (Hecht *et al.*, 2005). This suggests a major behavioral intervention in order to identify people in this early infectious stage. More HIV testing and counseling sites need to test for acute infection in high-prevalence areas and high-risk settings, and individuals should be alerted to practice safer sex after they have engaged in high-risk behaviors.

Pre-exposure prophylaxis (PrEP) is a treatment approach to prevention of HIV transmission in which antiretroviral drugs (ARVs) are used by an individual prior to potential HIV exposure (see also Chapter 5). PrEP should be distinguished from post-exposure prophylaxis (PEP), in which an individual takes ARVs soon after a potential HIV exposure with the goal of reducing the likelihood of infection. There are multiple studies in the field internationally with high-risk populations. The CDC is conducting a randomized, double-blinded, placebo-controlled study of PrEP using TDF in high-risk, HIV-negative MSM in three cities in the US (CDC, 2007). They will recruit a diverse sample of 400 MSM in Atlanta, Boston and San Francisco over a 9-month period, with 2 years of follow-up (CDC, 2008). Adherence to the treatment program will be extremely important in ensuring efficacy.

The NIMH Collaborative HIV/STD Prevention Trial is the first multicountry randomized trial of a community-level HIV prevention intervention in five countries (China, India, Peru, Russia and Zimbabwe) (NIMH Collaborative HIV/STD Prevention Trial Group, 2007). The sites and populations selected for study are among those most imminently threatened by HIV infection. The trial is evaluating the efficacy of the community popular opinion leader (C-POL) community-level intervention. Based on the diffusion of innovation theory, the intervention has identified, trained and engaged C-POLs within a high-risk community population to endorse the importance of safer behaviors to other members of the same population. The C-POL intervention taps into community strengths, altruism, and people's desire to do something to help fight AIDS. The results of this international community-level trial will be available shortly (see also Chapter 20).

Payoff from this period

- International research has provided an important scientific opportunity to investigate the moderators and mediators of intervention programs with a biological outcome
- The program of international behavioral research has contributed to the development of infrastructure and resource development in international settings
- Investigators have begun developing behavioral studies to address ethical and disinhibition issues related to circumcision
- NIMH currently supports seven AIDS Research Centers which continue to be state-of-the-art resources both nationally and internationally.

Conclusions

There is a direct interaction between the public's perception of a crisis, the availability of resources, and the development of science. Opportunities are a negotiation between what a field is prepared to contribute and the perceived need of society for certain kinds of information and activity. The continued development of scientific issues in AIDS research requires an adequate and stable level of support for well-trained investigators, and an ultimate public health payoff for that support.

Some converging trends in public health policy highlight the richness of the area of AIDS research for continued research contributions to the entire spectrum of public health problems. For example, findings from AIDS research are contributing to the knowledge of CNS aspects of multiple diseases and to behavior change strategies for other public health problems.

As part of this planning for the future, alternatives need to be laid out so that the choice points are clearly articulated and the cost/benefit ratio or other qualitative heuristics can be computed in advance. More than 30 years ago, Toffler (1970) remarked in *Future Shock* that finding the non-zero-sum solution to society's pressing problems requires infinite imagination. The collective imagination of interdisciplinary teams conducting research will result in continued payoff for public health from AIDS behavioral prevention research.

Willo Pequegnat and Ellen Stover

Acknowledgments

We would like to thank Dr Martin Fishbein for reading and comment on an earlier manuscript, Rayford Kytle for his extensive literature review and comments, and Cheryl Reese for her skill in preparing the manuscript correctly.

References

Ajzen, I. and Fishbein, M. (1980). Understanding Attitudes and Predicting Social Behavior. Englewood Cliffs, NJ: Prentice-Hall.

Alcalay, R., Sniderman, P. M., Mitchell, J. and Griffin, R. (1988–1989). Ethnic differences on knowledge of AIDS transmission and attitudes toward gays and people with AIDS. *Intl Q. Comm. Health Educ.*, 10, 213–22.

Allen, S., Tice, J., van de Perre, P. *et al.* (1992). Effect of serotesting and counseling on condom use and seroconversion among HIV discordant couples in Africa. *Br. Med. J.*, 204, 1605–19.

Altice, F. L., Mostashari, F. and Friedland, G. H. (2001). Trust and the acceptance of and adherence to antiretroviral therapy. *J. Acquir. Immune Defic. Syndr.*, 28, 47–58.

Antoni, M. H., Baggett, L., Ironson, G. *et al.* (1991). Cognitive behavioral stress management intervention buffers stress responses and immunologic changes following notification of HIV-1 seropositivity. *J. Consult. Clin. Psychol.*, 264, 62–9.

Artz, L., Macaluso, M., Brill, I. *et al.* (2000). Effectiveness of an intervention promoting the female condom to patients at sexually transmitted disease clinics. *Am. J. Public Health*, 90, 237–44.

Atwine, B., Cantor-Graae, E. and Banjunirwe, F. (2005). Psychological distress among AIDS orphans in rural Uganda. *Social Sci. Med.*, 61, 555–64.

Bentley, M. E., Pelto, C. H., Staus, W. L. *et al.* (1988). Rapid ethnographic assessment: application in a diarrhea management program. *Social Sci. Med.*, 27, 107–16.

Bongaarts, J., Reining, P. and Conant, F. (1989). The relationship between male circumcision and HIV infection in African populations. *AIDS*, 3, 373–7.

Bowen, A., Horvath, K. and Williams, M. (2006). Randomized control trial of an Internet delivered HIV knowledge intervention with MSM. *Health Educ. Res.*, 22, 120–7.

Bull, S. S., Mcfarlane, M., Lloyd, L. and Rietmeijer, C. (2004). The process of seeking sex partners online an implications for STD/HIV prevention. *AIDS Care*, 16, 1012–20.

Butler, L. D., Blasey, C. M., Azarow, J. *et al.* (2008). Posttraumatic growth following the terrorist attacks of September 11, 2001: cognitive, coping, and trauma symptom predictors of posttraumatic growth in an Internet convenience sample. *J. Nerv. Mental Dis.*, in press.

Butters, G. R., Grant, I., Haxby, J. *et al.* (1990). Assessment of AIDS-related cognitive changes: Recommendations of the NIMH workshop on neuropsychological assessment approaches. *J. Clin. Exp. Neuropsychol.*, 12, 963–78.

Carballo-Dieguez, A. and Dolezal, C. (1995). Association between history of childhood sexual abuse and adult HIV-risk sexual behavior in Puerto Rican men who have sex with men. *Child Abuse Neglect*, 19, 595–605.

Cardo, D. M., Culver, D. H., Ciesielski, C. A. *et al.* (1997). A case-control study of HIV seroconversion in health care workers after percutaneous exposure. *N. Engl. J. Med.*, 337, 1485–90.

Carey, M. P., Weinhardt, L. S. and Carey, K. B. (1995). Prevalence of infection with HIV among the seriously mentally ill: review of research and implications for practice. *Profess. Psychol.*, 26, 1–7.

Caso, L. D., Egremy, G., Uribe, P. *et al.* (2002). The use of female condom as a possible alternative with clients who reject male condom use and are under the effects of alcohol and/or use of drugs. Paper presented at the *XIVth International AIDS Conference, July 7–12, Barcelona, Spain*, Abstract no. TUPEF5469.

Catania, J., Coates, T., Stall, R. *et al.* (1992). Prevalence of AIDS-related risk factors and condom use in the United States. *Science*, 258, 1101–6.

Catania, J., Osmond, D., Stall, R. *et al.* (2001). The continuing HIV epidemic among men who have sex with men. *Am. J. Public Health*, 1, 907–14.

Catania, J. A., Kegeles, S. M. and Coates, T. J. (1990). Towards an understanding of risk behaviors: AIDS risk reduction model (ARRM). *Health Educ. Q.*, 17, 53–72.

CDC (1983). Epidemiological notes and report immunodeficiency among female sexual partners of males with AIDS. *Morbid. Mortal. Wkly Rep.*, 31, 697–8.

CDC (1986). Current trends: additional recommendations to reduce sexual and drug abuse-related transmission of Human T-Lymphotropic Virus Type III Lymphadenopathy-Associated Virus. *Morbid. Mortal. Wkly Rep.*, 35, 152–5.

CDC (1995). Case-control study of HIV seroconversion in health care workers after percutaneous exposure to HIV-infected blood – France, United Kingdom, and United States, January 1988–August 1994. *Morbid. Mortal. Wkly Rep.*, 44, 929–33.

CDC (2001). Updated US Public Health Service guidelines for the management of occupational exposures to HBC, HCV, and HIV and recommendations for postexposure prophylaxis. *Morbid. Mortal. Wkly Rep.*, 50, 1–42.

CDC (2008). CDC Trials of Pre-exposure Prophylaxis. CDC Fact Sheet. Atlanta, GA: CDC.

CDC's HIV/AIDS Prevention Research Synthesis Project (2001). Compendium of HIV Prevention Interventions with Evidence of Effectiveness. Atlanta, GA: CDC.

Chesney, M. A. (2000). Factors affecting adherence to antiretroviral therapy. *Clin. Infect. Dis.*, 30(Suppl. 2), 171–6.

Coates, T. J. (1990). Strategies for modifying sexual behavior for primary and secondary prevention of HIV disease. *J. Consult. Clin. Psychol.*, 58, 57–69.

Coid, J., Petruckevitch, A., Feder, G. *et al.* (2001). Relation between childhood sexual and physical abuse and risk of revictimization in women: a cross-sectional survey. *Lancet*, 385, 440–54.

Collins, C., Harshbarger, C., Sawyer, R. and Hamdallah, M. (2006). The Diffusion of Effective Behavioral Interventions Project: development, implementation, and lessons learned. *AIDS Educ. Prev.*, 18, 5–20.

Collis, T. K. and Celum, C. L. (2001). The clinical manifestations and treatment of diseases in Human Immunodeficiency Virus. *Clin. Infect. Dis.*, 32, 611–22.

Connor, E., Sperling, R., Gelber, R. *et al.* (1994). Reduction of maternal-infant transmission of Human Immunodeficiency Virus Type 1 with zidovudine treatment. *N. Engl. J. Med.*, 331, 1173–80.

Cornell, A. L., Piwoz, E., Bentley, M. E. *et al.* (2007). Involving communities in the design of clinical trial protocols: the BAN Study in Linongwe, Malawi. *Contemp. Clin. Trials*, 28, 59–67.

Coutinho, R. and Cates, W. E. (2000). Interventions to prevent HIV risk behaviors. *AIDS*, 14(Suppl. 2), 19–6.

Crawford, A. M. (1996). Stigma associated with AIDS: a meta-analysis. *J. Appl. Social Psychol.*, 29, 398–416.

Crepaz, N. and Marks, G. (2002). Toward an understanding of sexual risk behavior in people living with HIV: a review of social, psychological, and medical findings. *AIDS*, 16, 135–49.

DiIorio, C. T., Hartwell, T. and Hansen, N. (2002). NIMH Multisite HIV Prevention Trial Group: childhood sexual abuse and risk behaviors among men at high risk for HIV infection. *Am. J. Public Health*, 92, 214–19.

Dixon, D., Cruess, S., Kilbourn, K. *et al.* (2001). Social support mediates loneliness and human herpesvirus type 6 (HHV6) antibody titers. *J. Appl. Social Psychol.*, 31, 1111–32.

Ehrhardt, A. and Exner, T. (1991). The Impact of HIV Infection on Women's Sexuality and Gender Role. Rockville, MD: National Technical Information Services, NIMH/ NIDA Monograph.

Ehrhardt, A. A., Exner, T., Hoffman, S. *et al.* (2002). HIV/STD risk and sexual strategies among women family planning clients in New York: Project FIO. *AIDS Behav.*, 6, 1–13.

El-Bassell, N., Witte, S. S., Gilbert, L. *et al.* (2003). The efficacy of a relationship-based HIV/STD prevention program for heterosexual couples. *Am. Public Health Assoc.*, 93, 963–9.

Exner, T., Seal, D. W. and Ehrhardt, A. (1997). A review of HIV interventions for at-risk women. *AIDS Behav.*, 1, 93–124.

Fishbein, M. and Pequegnat, W. (2000). Evaluating AIDS prevention intervention using behavioral and biological outcome measures. *Sex. Transm. Dis.*, 27, 101–10.

Fishbein, M., Bandura, A. and Triandis, H. C. (1992). Factors influencing behavior and behavior change: Final Report – Theorist's Workshop. Workshop, Washington, DC, 3–5 October 1991

Fisher, J. D. and Fisher, W. A. (1992). Changing AIDS risk behavior. *Psychological Bull.*, 111, 455–74.

Friedland, G. H. and Williams, A. (1999). Attaining higher goals in HIV treatment: the central importance of adherence. *AIDS*, 13(Suppl. 1), 61–72.

Golin, C. E., Liu, H., Hays, R. D. *et al.* (2002). A prospective study of predictors of adherence to combination antiretroviral medication. *J. Gen. Int. Med.*, 17, 756–65.

Golub, E. L. (2000). The female condom: tool for women's empowerment. *Am. J. Public Health*, 90, 377–81.

Gordillo, V., del Amo, J., Soriano, V. and Gonzalez-Lahoz, J. (1999). Sociodemographic and psychological variables influencing adherence to antiretroviral therapy. *AIDS*, 13, 1763–9.

Gostin, L. O. (1989). Public health strategies for confronting AIDS: legislative and regulatory policy in the United States. *J. Am. Med. Assoc.*, 261, 1621–30.

Gostin, L. O. (1990). The AIDS litigation project: a national review of court and human rights commission decisions, Part II: Discrimination. *J. Am. Med. Assoc.*, 263, 2086–93.

Grant, I., Atkinson, J. H., Hesselink, J. R. *et al.* (1987). Evidence for early central nervous system involvement in the acquired immunodeficiency syndrome (AIDS) and other human immunodeficiency virus (HIV) infections. Studies with neuropsychologic testing and magnetic resonance imaging. *Ann. Int. Med.*, 107, 828–36.

Gray, R. H., Kigozi, G., Serwadda, D. *et al.* (2007). Male circumcision for HIV prevention in men in Rakai, Uganda: a randomised trial. *Lancet*, 369, 657–66.

Grosskuth, H. (1995). Impact of improved treatment of sexually transmitted diseases on HIV in rural Tanzania: randomized controlled trial. *Lancet*, 346, 530–6.

Guthrie, C. (2007). Top 10 Medical Breakthroughs. *Time Magazine*, 24 December.

Herdt, G. and Boxer, A. (1991). Ethnographic issues in the study of AIDS. *J. Sexual Res.*, 28, 171–87.

Herek, G. H. (1990). Illness, stigma, and AIDS. *Am. Psychological Assoc.*, 103–50.

Herek, G. H. (1995). Developing a theoretical framework and rationale for a research proposal. In: W. Pequegnat and E. Stover (eds), *How To Write A Successful Research*

Grant Application: A Guide For Social And Behavioral Scientists. New York, NY: Plenum Press, pp. 85–92.

Herek, G. H. and Capitanio, J. P. (1993). Public reactions to AIDS in the United States: a second decade of stigma. *Am. J. Public Health*, 83, 574–7.

Hunter, N. D. and Rubenstein, W. B. E. (1992). AIDS Agenda: Emerging Issues in Civil Rights. New York, NY: New Press.

Ickovics, J. R. and Meade, C. S. (2002). Adherence to antiretroviral therapy among patients with HIV: a critical link between behavioral and biomedical sciences. *J. Acquir. Immune Defic. Syndr.*, 31(Suppl. 3), 98–102.

Jansen, R. S., Holtgrave, D. R., Valdiserri, R. O. *et al.* (2001). The serostatus approach to fighting the HIV epidemic: prevention strategies for infected individuals. *Am. J. Public Health*, 91, 1019–24.

Jemmott, J. B., Jemmott, L. S. and Fong, G. T. (1992). Reductions in HIV risk-associated sexual behaviors among Black adolescents: effects of AIDS prevention. *Am. J. Public Health*, 82, 372–7.

Kahn, J. G. (1998). Economic evaluation of primary HIV prevention in travenous drug users. In: D. R. Holtgrave (ed.), *Handbook of Economic Evaluation of HIV Prevention Programs.* New York, NY: Plenum Press.

Kahn, J. O., Martin, J. N., Roland, M. E. *et al.* (2001). Feasibility of postexposure prophylaxis (PEP) against human immunodeficiency virus infection after sexual or injection drug exposure: The San Francisco PEP Study. *J. Infect. Dis.*, 183, 707–14.

Kamb, M. L., Fishbein, M., Douglas, J. M. *et al.* (1998). Efficacy of risk-reduction counseling to prevent human immunodeficiency virus and sexually transmitted diseases. *J. Am. Med. Assoc.*, 280, 1161–7.

Kang, M., Dunbar, M., Laver, S. and Padian, N. (2008). Maternal versus paternal orphans and HIV/STI risk among adolescent girls in Zimbabwe. *AIDS Care*, 20, 221–4.

Kelly, J. A. (2004). Popular opinion leaders and HIV prevention peer education: resolving discrepant findings, and implications for the development of effective community programmes. *AIDS Care*, 16, 139–50.

Kelly, J. A., Lawrence, J. S., Hood, H. V. and Brasfield, T. L. (1989). Behavioral intervention to reduce AIDS risk activities. *J. Consult. Clin. Psychol.*, 57, 60–7.

Kelly, J. A., St Lawrence, J. S., Diaz, Y. E. *et al.* (1991). HIV risk behavior reduction following intervention with key opinion leaders of a population: an experimental analysis. *Am. J. Public Health*, 81, 168–71.

Kelly, J. A., Murphy, D. A., Bahr, G. R. *et al.* (1992). AIDS/HIV risk behavior among the chronic mentally ill. *J. Am. Psychiatry*, 149, 886–9.

Kelly, J. A., St Lawrence, J. S., Stevenson, L. Y. *et al.* (1995). Community AIDS/HIV risk reduction: the effects of endorsements by popular people in three cities. *Am. J. Public Health*, 82, 1483–9.

Kelly, J. A., Murphy, D. A., Sikkema, K. J. *et al.* (1997). Randomized, controlled, community-level HIV prevention intervention for sexual risk behaviour among homosexual men in US cities. *Lancet*, 350, 1500–5.

Kelly, J. A., Somlai, A. M., Benotsch, E. G. *et al.* (2004). Distance communication transfer of HIV prevention interventions to service providers. *Science*, 305, 1953–5.

Kelly, J. A., Somlai, A. M., Benotsch, E. G. *et al.* (2006). Programmes, resources, and needs of HIV-prevention nongovernmental organizations (NGOs) in Africa, Central/Eastern Europe and Central Asia, Latin America and the Caribbean. *AIDS Care*, 18, 12–21.

Kirby, D., Barth, R. P., Leland, N. and Vetro, J. V. (1991). Reducing the risk: impact of a new curriculum on sexual risk taking. *Family Plan. Persp.*, 23, 253–63.

Klauser, J., Wolf, W., Fischer-Ponce, L. *et al.* (2000). Tracing a syphilis epidemic through cyberspace. *J. Am. Med. Assoc.*, 284, 484–7.

Koblin, B. A. and The EXPLORE Study Team (2004). Effects of a behavioral intervention to reduce acquisition of HIV infection among men who have sex with men: The EXPLORE randomised controlled study. *Lancet*, 364, 41–50.

Koenig, L. J., Doll, L. S., O'Leary, A. and Pequegnat, W. (2004). From Child Sexual Abuse to Adult Sexual Risk: Trauma, Revictimization, and Intervention, eds. Washington, DC: American Psychological Association.

LaPerriere, A., Antoni, M. H., Schneiderman, N. *et al.* (1990). Exercise intervention attenuates emotional distress and natural killer cell decrements following notification of positive serologic status for HIV-1. *Biofeedback Self Reg.*, 15, 229–42.

Laumann, E. O., Gagnon, J. H., Michael, R. T. and Michaels, S. (1994). The Social Organization of Sexuality: Sexual Practices in United States. Chicago, IL: University of Chicago Press.

Leserman, J., Jackson, E. D., Petito, J. M. *et al.* (1999). Progression to AIDS: the effects of stress, depressive symptoms, and social support. *Psychosom. Med.*, 61, 397–406.

Mantell, J. E., Myer, L., Carballo-Dieguez, A. *et al.* (2005). Microbicide acceptability research: current approaches and future directions. *Social Sci. Med.*, 60, 319–30.

Matthew, M. J., Fair, A. D., Mayer, K. H. *et al.* (2008). Experiences and sexual behaviors of HIV-infected MSM who acquired HIV in the context of crystal methamphetamine use. *AIDS Educ.*, 20, 30–41.

Mausbach, B. T., Semple, S. J., Strathdee, S. A. *et al.* (2007). Efficacy of a behavioral intervention for increasing safer sex behaviors in HIV-positive MSM methamphetamine users: results from the EDGE study. *Drug Alcohol Dep.*, 87, 249–57.

May, R. M. and Anderson, R. M. (1987). Transmission dynamics and HIV infection. *Nature*, 326, 137–42.

McKusick, L., Horstman, W. and Coates, T. J. (1985). AIDS and sexual behavior reported by gay men in San Francisco. *Am. J. Public Health*, 75, 493–6.

Merson, M. H., Dayton, J. M. and O'Reilly, K. (2000). Effectiveness of HIV prevention interventions in developing countries. *AIDS*, 14 (Suppl. 2), 85–96.

Molitor, F., Truax, S., Ruiz, J. and Sun, R. (1998). Association of methamphetamine use during sex with risky sexual behavior and HIV infection among non-injection drug users. *W. J. Med.*, 168, 93–7.

Monasch, R. and Boerma, J. T. (2004). Orphanhood and childcare patterns in Sub-Saharan Africa: An analysis of national surveys from 40 countries. *AIDS*, 18 (Suppl. 2), 55–65.

Namir, S., Wolcott, D., Fawzy, F. and Alumbaugh, M. (1990). Implications of Different Strategies for Coping with AIDS. Hillsdale, NJ: Erlbaum.

National Institute of Mental Health (1998). The NIMH Multisite HIV Prevention Trial: reducing HIV sexual risk behavior. *Science*, 280, 1889–94.

National Institute of Mental Health (1985). RFA for AIDS Research Centers. Bethesda, MD: NIMH.

National Institute of Mental Health Healthy Living Project Team (2007). Effects of a behavioral intervention to reduce risk of transmission among people living with HIV: the Healthy Living Project randomized controlled study. *J. Acquir. Immune Defic. Syndr.*, 44, 213–21.

National Institutes of Health (1997). Interventions to Prevent HIV Risk Behaviors. NIH Consensus Statement. Bethesda, MD: NIH.

Osmond, D. H., Page, K., Wiley, J. *et al.* (1994). HIV infection in homosexual and bisexual men 18 to 29 years of age: The San Francisco Young Men's Health Study. *Am. J. Public Health*, 84, 1933–7.

Ostrow, D. and Kessler, R. (1993). Methodological Issues in AIDS Behavioral Research. Philadelphia, PA: Springer Press.

Ostrow, D. G., Plankey, M. W., Li, X. *et al.* (2006). Methamphetamine and combination of drugs and HIV seroincidence in the MACS study. Paper presented at the *XVIth International AIDS Conference, Toronto, Canada, 13–28 August.*

Pequegnat, W. (2005). AIDS behavioral prevention: unprecedented progress and emerging challenges. In: K. H. Mayer and H. F. Pizer (eds), *The AIDS Pandemic: Impact on Science and Society.* San Diego, CA: Elsevier Academic Press, pp. 236–60.

Pequegnat, W. and Szapocznik, J. (2000). Working with Families in the Era of AIDS. New York, NY: Sage.

Pequegnat, W., Page, J. B., Stauss, A. *et al.* (1995). Qualitative Inquiry: An Underutilized Approach in AIDS Research. New York, NY: Plenum.

Pequegnat, W., Rosser, B. R. S., Bowen, A. M. *et al.* (2007). Conducting Internet-based HIV/STD prevention survey research: considerations in design and evaluation. *AIDS Behav.,* 11, 505–21.

Pinkerton, S. D., Holtgrave, D. R., Johnson, M. A. P. *et al.* (2002). Cost-effectiveness of the NIMH Multisite HIV Prevention Intervention. *AIDS Behav.,* 6, 83–96.

Pinkerton, S. D., Martin, J. N., Roland, M. E. *et al.* (2004). Cost-effectiveness of HIV postexposure prophylaxis following sexual or injection drug exposure in 96 metropolitan areas in the United States. *AIDS,* 18, 2065–73.

Reiter, G. S., Wojtusik, L., Hewitt, R. *et al.* (2000). Elements of success in HIV clinical care. *Topics HIV Med.,* 8, 67.

Richardson, J. L., McCutchan, A., Stoyanoff, S. *et al.* (2004). Effect of brief provider safer-sex counseling of HIV-1 seropositive patients: a multi-clinic assessment. *AIDS,* 18, 1179–86.

Rogers, E. M. (1983). Diffusion of Innovations. New York, NY: The Free Press.

Rosen, R. C., Catania, J. A., Ehrhardt, A. *et al.* (2006). The Bolger Conference on PDE–5 inhibition and HIV risk: implications for health policy and prevention. *Intl Soc. Sexual Med.,* 3, 960–75.

Rosser, B. R. S., Miner, M. H., Bockting, W. O. *et al.* (2008). HIV risk and the Internet: results of the Men's INTernet Study (MINTS). *AIDS Behav.,* in press.

Rotheram-Borus, M. J. (1993). Suicidal behavior and risk factors among runaway youths. *Am. J. Psych.,* 150, 103–7.

Rotheram-Borus, M. J. (1997). HIV prevention challenges – realistic strategies and early detection programs. *Am. J. Public Health,* 87, 544–6.

Rotheram-Borus, M. J., Koopman, C., Haignere, C. and Davies, M. (1991). Reducing HIV sexual risk behaviors among runaway adolescents. *J. Am. Med. Assoc.,* 266, 1237–41.

Schechter, M., Lago, R. F., Ismerio, R. *et al.* (2002). Acceptability, behavioral impact, and possible efficacy of post-sexual-exposure chemoprophylaxis (PEP) for HIV. Paper presented at the *9th Conference on Retroviruses and Opportunistic Infections, Seattle, Washington, 24–28 February,* Abstract no. 15.

Scrimshaw, S. M. and Hurtado, E. (1987). Rapid Assessment Procedures for Nutrition and Primary Health Care: Anthropological Approaches to Improving Program Effectiveness. References Series. Los Angeles, CA: UCLA.

Semple, S. H., Patterson, T. L. and Grant, I. (2002). Motivations associated with methamphetamine use among HIV+ men who have sex with men. *J. Substance Abuse Treat.,* 22, 149–56.

Semple, S. H., Patterson, T. L. and Grant, I. (2003). Binge use of methamphetamine among HIV-positive men who have sex with men: pilot data and HIV prevention implications. *AIDS Educ. Prev.,* 15, 133–47.

Shaffer, D. N., Bautista, C. T., Sateren, W. B. *et al.* (2007). The protective effect of circumcision on HIV incidence in rural low-risk men circumcised predominantly by traditional circumcisers in Kenya: two-year follow-up of the Kericho HIV cohort study. *J. Acquir. Immune Defic. Syndr.*, 45, 371–9.

Siegel, K. and Krauss, B. (1991). Living with HIV infection: adaptive tasks of seropositive gay men. *J. Health Social Behav.*, 32, 17–32.

Sikkema, K. J., Kelly, J. A., Winett, R. A. *et al.* (2000). Outcomes of a randomized community-level HIV prevention intervention for women living in 18 low-income housing developments. *Am. J. Public Health*, 90, 57–63.

Simoni, J. M., Frick, P. A., Pantalone, D. W. and Turner, B. J. (2003). Antiretroviral adherence interventions: A review of current literature and ongoing studies. *Topics HIV Med.*, 11, 185–98.

Sogolow, E., Peersman, G., Semaan, S. *et al.* HIV/AIDS Prevention Research Project Team (2002). The HIV/AIDS Prevention Research Synthesis Project: scope, methods, and study classification results. *J. Acquir. Immune Defic. Syndr.*, 1, 15–29.

Ssewamala, F. M., Alicea, S., Bannon, W. M. and Ismayilova, L. (2008). A novel economic intervention to reduce HIV risks among school-going AIDS orphans in rural Uganda. *J. Adolesc. Health*, 42, 102–4.

Stall, R. D., Coates, T. J. and Hoff, C. (1988). Behavioral risk reduction for HIV infection among gay and bisexual men: a review of results from the United States. *Am. Psychologist*, 43, 878–85.

Stein, J. (2003). Sorrow makes children of us all: a literature review of the psycho-social impact of HIV/AIDS on children. CSSR Working Paper No. 47.

Stone, V. E., Clarke, J., Lovell, J. *et al.* (1998). HIV/AIDS patients (perspectives on adhering to regimens containing protease inhibitors). *Ann. Int. Med.*, 124, 968–77.

Subbarao, K. and Coury, D. (2004). *Reaching out to Africa's Orphans: A Framework for Public Action.* UNAIDS Report, The World Bank.

Toffler, A. (1970). Future Shock. New York, NY: Random House.

UNAIDS (2007). AIDS Epidemic Update. Joint United Nations Program on HIV/AIDS. Montreaux: UNAIDS.

Valdiserri, R. O. (2001). HIV/AIDS stigma: an impediment to public health. *Am. J. Public Health*, 93, 341–2.

Valdiserri, R. O., Lyter, D., Leviton, I. *et al.* (1989). AIDS prevention in homosexual and bisexual men: results of a randomized trial evaluating two-risk reduction interventions. *AIDS*, 3, 21–6.

Wawer, M. J. (1998). The Radai randomized, community based trial of STD control for AIDS prevention: no effect on HIV incidence despite reductions in STDs. Paper presented at the *12th World AIDS Conference, Geneva, 28 June–3 July.*

Weinhardt, L. S., Carey, M. P., Johnson, B. T. and Bickham, N. L. (1999). Effects of HIV counseling and testing on sexual risk behavior: Meta-analysis of published research, 1985–1997. *Am. J. Public Health*, 89, 1397–405.

West, S. G., Aiken, L. S. and Todd, M. (1993). Probing the effects of individual components in multiple component prevention programs. *Am. J. Community Psychol.*, 21, 571–605.

Winslow, B. T., Voorhees, K. I. and Pehl, K. A. (2007). Battling meth addiction in the gay community. *Am. Family Physician*, 76, 1169–74.

Worley, H. (2005). Obstacles remain to wide adoption of female condom. Population Reference Bureau, 1–6 (available at http://www.prb.org/Articles/2005/ObstaclesRem aintoWideAdoptionofFemaleCondom.aspx)

Wyatt, G. E., Myers, H. F., Williams, J. K. *et al.* (2002). Does a history of trauma contribute to HIV for risk for women of color? Implications for prevention and policy. *Am. J. Public Health*, 92, 660–5.

Individual interventions

8

Matthew J. Mimiaga, Sari L. Reisner, Laura Reilly,
Nafisseh Soroudi and Steven A. Safren

Interventions delivered at the individual level measure outcomes on an individual basis, and seek to influence and modify HIV risk behavior one individual at a time (Cohen and Scribner, 2000). This chapter describes individual interventions for HIV prevention from the perspective of a three-stage model of behavioral therapy development. This is highly applicable to individual interventions for HIV prevention, due to the sequential approach to designing and refining culturally appropriate interventions. First, we review the most common theoretical models used to design behavioral interventions, and how successful they have been when applied to individual interventions for HIV risk behaviors. We emphasize current theories regarding how people make decisions to engage in HIV risk behavior, such as sexual risk and concomitant substance use. Second, we describe examples of published, randomized, controlled efficacy trials of individualized interventions which are generally based on psychosocial theories of health behavior. Also discussed are the differences in addressing risk-related issues in various at-risk populations, such as men who have sex with men (MSM), injection drug users, women, and ethnic and racial minorities such as African-Americans. Third, we draw conclusions, with an emphasis on conducting effectiveness research on individualized interventions, and the challenges that exist within this stage.

In biomedical intervention development, clinical researchers use the regulatory standards set forth by the Food and Drug Administration (FDA) for pharmacotherapy and other biomedical clinical intervention trials. Similarly, emerging standards exist for behavioral research, where the goal is to develop, test and refine behavioral interventions to be used in real-life settings. These standards are required in order not only to be efficient, but also to rigorously develop and test behavioral interventions so that they meet similar standards of evidence-based research as biomedical interventions. Onken and colleagues (1997) cogently

203

articulated a three-stage model of evidence-based research, progressing from feasibility and acceptability of intervention development to formative conceptualization, from efficacy and refinement of intervention to effectiveness research, and, finally, from larger generalizability to dissemination.

Accordingly, the principal aim in the first stage of treatment (see also Rounsaville *et al.*, 2001) development is to incorporate basic behavioral research into studies to develop new interventions, as well as to improve understanding of the behavioral change process. The main objectives of this stage are to develop the components needed for an efficacy trial, and to begin to establish a basis for the ability to change behavior. Hence, Stage 1 includes activities such as establishing a firm conceptual base behind the intervention, conducting necessary formative research with the target population, and using these data to generate materials needed for a pilot trial. These materials include training manuals and protocols that describe the techniques, interventionist training and process measures to assess interventionist fidelity and competence. Additionally, this stage involves pilot/feasibility testing to gain information about patient acceptability, and logistics such as the ability to recruit and the feasibility of implementation with the target population. The initial section of the current chapter reviews conceptual models that are at the base of many individual interventions for HIV prevention, and have been tested and used across a wide range of at-risk populations.

In Stage 2 of Onken's model, full-scale randomized controlled trials (RCTs) are conducted to evaluate the treatment efficacy of interventions that have undergone some pilot testing. Mechanisms of action, as per the conceptual model, and effective treatment components are also addressed to further elucidate the hypothesized salient intervention effects. These RCTs are conducted in order to establish efficacy of intervention effects, and therefore set the groundwork to disseminate an intervention that can be used in real-world settings. The second part of this chapter reviews Stage 2 studies: efficacy studies of individualized behavioral interventions for HIV risk reduction among various at-risk populations.

Stage 3 of this model therefore entails "effectiveness" testing. "Effectiveness" refers to the process by which research findings are connected to clinical "real-world" applications. This phase addresses any issues with generalizability (i.e., treatment effectiveness across health-care professionals carrying out the intervention, patients, varied populations and settings); implementation (e.g., identifying the qualities of trainers, trainees, and the training necessary to carry out the treatment); acceptability outside of research in real-world settings; and cost-effectiveness (i.e., does the treatment compare with existing ones on the costs versus savings incurred?). The third component of this chapter provides evidence for the continued need for more effectiveness research in HIV prevention interventions.

Stage 1: Common theoretical models applicable to individual interventions for HIV prevention

Familiarity with the conceptual models that underlie HIV prevention behaviors is important because it allows clinicians and researchers to develop and refine intervention efforts. Typically, the reviewers of grant applications, with the aim to develop and test interventions, heavily weigh the conceptual and theoretical models in their evaluations. Most of the theoretical models that underlie HIV prevention interventions come from the fields of social and health psychology. Each model posits various factors that are thought to influence HIV risk-taking, which then become intervention targets in Stage 2 and 3 of the research process.

The models that are described include the biomedical model, the behavioral model (learning), communication, the health belief model, social cognitive theory, the theory of reasoned action, the theory of planned behavior, protection motivation theory, information–motivation–behavioral skills, self-regulation, the transtheoretical model, relapse prevention, the health-decision model of behavior change and the AIDS risk-reduction model. For each, we describe the major tenets and strengths of the model as well as any frequent criticisms or drawbacks in predicting HIV risk behaviors.

The biomedical model

The biomedical model focuses on biological factors in attempting to understand a person's medical illness, disorder or behavior, excluding psychological and social factors as relevant determinants (Porter, 1997; Wade, 2004). In the biomedical model, individuals are seen as passive agents of their health, with little or no personal responsibility for the presence of, or ability to change, the behavior (Wade, 2004). Although patient characteristics such as age and gender may influence the consequences of illness and play a role in HIV risk behaviors, they are not deemed related to the etiology, development or manifestations of illness or behavior. For example, a biological factor associated with male-to-male sexual behavior that may increase vulnerability to HIV is anal intercourse. Both vaginal and anal intercourse have been shown to be efficient routes for HIV transmission, yet, relative to the vagina, rectal tissue is much more vulnerable to tearing during intercourse, and the larger surface area of the rectum/colon provides more opportunity for viral penetration and infection, placing men who have sex with men at greater risk of HIV acquisition and transmission (Roehr *et al.*, 2001; Shattock and Moore, 2003; amFAR AIDS Research, 2005). HIV prevention interventions based on the biomedical model include microbicides and vaccines to address such biological factors associated with susceptibility to infection.

Although the biomedical model has dominated health care for the past century, it has been critiqued for its inability to fully explain many forms of illness and behavior due to three assumptions: all illness has a single underlying cause, pathology is always the single cause of illness, and removal or attenuation of the disease will result in a return to health (i.e., health is equivalent to lack of disease) (Wade, 2004). Importantly, the biomedical model has been criticized in HIV prevention because it is viewed as overlooking important factors such as patients' views of risk behaviors, as well as psychosocial and economic factors that contribute to risk (Munro *et al.*, 2007). Moreover, because some patient characteristics and factors are marked as unamenable to intervention in the biomedical model, this model may exclude certain patient groups from treatment.

The behavioral learning model

The behavioral learning model emphasizes the role of antecedents (what comes before acting – i.e. thoughts, environmental cues) and consequences (what comes after acting – rewards/punishments) in shaping overt health behaviors. An example of behavioral learning theory is *operant conditioning*. According to operant conditioning, behaviors are performed in response to stimuli, and the frequency of occurrence of the behavior post-stimuli (response) increases if the behavior is reinforced (Skinner, 1938). For example, if a woman finds condom negotiation with a partner to be difficult, and every time condom negotiation is initiated she experiences anxiety and negative resistance from her sexual partner, she may have greater frequency of unprotected sex (the overt behavior of unprotected sex is being reinforced by the reduction of anxiety and lack of argument that results from no condom use). Reinforcers that have a stronger effect on behavior are considered to have more reinforcing value (Skinner, 1938).

Behavior change in this model focuses on either gaining control of the stimuli and reinforcers and reinforcing only desired behaviors, or presenting only the stimuli which are already linked to desired behaviors. Learning theories emphasize that learning a new pattern of behavior, such as reducing sexual risk, requires modifying many of the small behaviors that compose the overall complex behavior (e.g., buying condoms, negotiating condom use, etc.). Principles of behavior modification suggest that a complex-pattern behavior can be learned by first breaking it down into smaller segments. Incremental increases are then made as the complex pattern of behavior is "shaped" toward the targeted behavioral outcome goal in successive approximations. The drawbacks of the behavioral learning perspective are that there is little attention to cognition or thought as explanations of behavior, and it does not account for the fact that the relevance of rewards may be individualized and may differ depending on time after the stimulus-response chain.

206

The communication perspective

The communication perspective is a client-centered approach that emphasizes the key role played by good provider–patient relationship and clear communication in determining patient health behaviors, including reducing risk for HIV. The most successful approach related to the communication perspective is *motivational interviewing* (MI), a client-centered, participatory communication technique that elicits behavioral change by helping clients explore and work through ambivalence about changing their behavior (Miller and Rollnick, 1991; Emmons and Rollnick, 2001; Lewis *et al.*, 2002). From the MI perspective, the client is the expert in evaluation of his or her own behavior and generates potential solutions to problems; the provider or practitioner uses Rogerian empathetic and reflective listening (based on the work of American psychologist Carl Rogers, 1902–1987, researcher and founder of client-centered humanistic therapy) to focus on ambivalence associated with behavior change, encouraging the client to examine his or her own behavior (Lewis *et al.*, 2002). This model, however, has been argued to overlook attitudes, motivation, skills, and interpersonal factors that may affect the way messages are received and translated into health behaviors.

The cognitive perspective

The cognitive perspective considers the effects of attitudes, beliefs and expectations of outcomes on engaging in health behaviors (Janz *et al.*, 2002). Mental processes, including thinking, reasoning, hypothesizing and expecting, play an integral role in cognitive theories of behavior. In this perspective, behavior is a function of the subjective value of an outcome and of a subjective expectation (Janz *et al.*, 2002). Cognitive theorists view reinforcement or consequence of behavior as being important in so far as it influences expectations about a situation, but generally do not consider the role of reinforcement in directly influencing behavior (Bandura, 1977a). Several models that follow this premise include the health belief model, social cognitive theory, the theory of reasoned action and the protection motivation theory, which are described in more detail below. Several criticisms have been levied against the cognitive perspective. First, it does not take into account the impact of non-voluntary factors; second, it does not address behavioral skills associated with sexual risk behaviors directly; and third, the theory gives a paucity of attention to the origin of beliefs and how these beliefs may impact behavior (Munro *et al.*, 2007).

The health belief model

The health belief model (HBM) is a value-expectancy theory, and assumes that an individual's behavior is guided by expectations of consequences of adopting new practices (Janz *et al.*, 2002). The model has four key concepts (Hornik, 1991; Fisher and Fisher, 1992):

1. Susceptibility: does the person perceive vulnerability to the specific disease?
2. Severity: does the individual perceive that getting the disease has negative consequences?
3. Benefits minus costs: what are the positive and negative effects of adopting a new practice?
4. Health motive: does the person have concern about the consequences of contracting the disease?

In addition, self-efficacy, a sense of competence as a cogent agent of long-term behavior change, has recently been integrated into HBM. Thus, increased sexual risk-taking or unprotected sex may be explained and addressed by HMB as follows: one's beliefs about the benefits of condoms (protection from HIV or STDs) do not outweigh the costs of condom use (pleasure reduction due to reduced sensation, partner-related concerns such as creation of distrust in a relationship or reduction of spontaneity); interventions would focus on shifting the benefit–cost. A criticism of this model is that it lacks clear definitions of components and the relationship between them; thus the model has been critiqued for inconsistent measurement in both descriptive and intervention research. HBM has been further critiqued for not fully addressing several behavioral determinants, including socio-cultural factors, and assuming that health is a high priority for most individuals (thus, it may not be applicable to those who do not place as high a value on health).

Social cognitive theory

Social cognitive theory (SCT) is one of the most frequently applied theories of health behavior (Baranowski *et al.*, 2002). SCT posits a reciprocal deterministic relationship between the individual, his or her environment, and behavior; all three elements dynamically and reciprocally interact with and upon one another to form the basis for behavior, as well as potential interventions to change behaviors (Bandura, 1977a, 1986, 2001). Social cognitive theory has often been called a bridge between behavioral and cognitive learning theories, because it focuses on the interaction between internal factors such as thinking and symbolic processing (e.g., attention, memory, motivation) and external determinants (e.g., rewards and punishments) in determining behavior.

A central tenet of social cognitive theory is the concept of self-efficacy – individuals' belief in their capability to perform a behavior (Bandura, 1977b). Behaviors are determined by the interaction of outcome expectations (the extent to which people believe their behavior will lead to certain outcomes) and efficacy expectations (the extent to which they believe they can bring about the particular outcome) (Bandura, 1977b, 1997). For example, individuals may hold the outcome expectation that if they consistently use condoms, they will significantly reduce risk of becoming HIV-infected; however, they must also hold the efficacy expectation that they are incapable of such consistent behavioral practice. Behavior change would necessitate bringing outcome and efficacy expectations in alignment with one another. SCT emphasizes predictors of health behaviors, such as motivation and self-efficacy, perception of barriers to and benefits of behavior, perception of control over outcome, and personal sources of behavioral control (self-regulation) (Bandura, 1977a, 1977b). Another important tenet with respect to behavioral and learning is SCT's emphasis that individuals learn from one another via observation, imitation and modeling; effective models evoke trust, admiration and respect from the observer, and they do not appear to represent a level of behavior that observers are unable to visualize attaining for themselves. Thus, a change in efficacy expectations through vicarious experience may be effected by encouraging an individual to believe something akin to the following: "if she can do it, so can I". SCT has been critiqued for being too comprehensive in its formulation, making for difficulty in operationalizing and evaluating the theory in its entirety (Munro *et al.*, 2007). Moreover, some researchers using SCT as a theoretical basis have been criticized for using only one or two concepts from the theory to explain behavioral outcomes (Baranowski *et al.*, 2002).

The theory of reasoned action

The theory of reasoned action (TRA; Ajzen and Fishbein, 1980) maintains that volition and intention predict behavior. According to TRA, if people evaluate the suggested behavior as positive (attitude) and if they think others want them to perform the behavior (subjective norm), this results in a higher intention (motivation) and they are more likely to perform the behavior. A high correlation of attitudes and subjective norms to behavioral intention and to behavior has been confirmed in many studies (Sheppard *et al.*, 1998). However, results of some studies gesture to a limitation of this theory: behavioral intention does not always lead to actual behavior. A counter-argument against the strong relationship between behavioral intention and actual behavior led to the evolution of the theory of planned behavior, a model which includes the impact of non-volitional factors on behavior.

The theory of planned behavior

The theory of planned behavior (TPB; Ajzen, 1985, 1987, 1991) was developed from the theory of reasoned action, and is more applicable when the probability of success and actual control over performance of a behavior are suboptimal. In addition to attitudes and subjective norms which comprise the theory of reasoned action, the TPB's key contribution is the concept of perceived behavioral control, defined as an individual's perception of the ease or difficulty of performing the particular behavior (Ajzen, 1987). How strong an attempt the individual makes to engage in the behavior and how much control that individual has over the behavior (behavioral control) are influential in whether he or she engages in the behavior. Behavioral intention is produced from a combination of attitude toward the behavior, subjective norm, and perceived behavioral control (Ajzen, 2002). Behavioral control is similar to self-efficacy, and depends on the individual's perception of how difficult it is going to be to engage in the behavior. The more favorable a person's attitude is toward behavior and subjective norms, and the greater the perceived behavioral control, the stronger that person's intention will be to perform the behavior in question. Moreover, given a sufficient degree of actual control over the behavior, people will be expected to carry out their intentions when the opportunity arises (Ajzen, 2002). Thus, an individual with positive attitudes about always using condoms during vaginal or anal intercourse, who perceives social support for these behaviors from key referent others and who has the conviction that he or she can carry out these behaviors effectively, will likely take consistent HIV preventive actions (Fisher, 1997). The model emphasizes the roles played by knowledge regarding necessary skills for performing the behavior, environmental factors, and past experience with the behavior (Ajzen and Madden, 1986). Critics have argued that these models would benefit from a more clear and explicit definition of behavior control. Others have suggested adding the role of beliefs and moral and religious norms would help improve predictive ability of the models (Godkin and Koh, 1996).

The protection motivation model

The protection motivation model (Rogers, 1975, 1983) describes processes of adaptive and maladaptive coping with a health threat, emphasizing two appraisal processes: (1) a process of threat appraisal; and (2) a process of coping appraisal, in which the behavioral options to diminish the threat are evaluated. The appraisal of the health threat and the appraisal of the coping responses result in either the intention to perform adaptive responses (protection motivation) or to maladaptive responses (responses that place the individual at high risk and lead to negative consequences, such as unsafe sex) (Boer and Seydel, 1996). Central to the decision-making process is fear, which is thought to increase both the motivation and the likelihood to engage in protective action. Intention to protect oneself depends upon perceptions associated with four factors: severity (expected harmfulness), probability (or vulnerability that the event will occur), response efficacy and self-efficacy. The interaction of these four components is in turn responsible for arousing fear. Criticisms of this model, however, are that it does not comprehensively identify socio-cultural, environmental and cognitive variables that impact motivation.

The information–motivation–behavioral skills model

The information–motivation–behavioral (IMB) skills model (Fisher and Fisher, 1992) was originally developed as a simple model to address HIV risk behaviors. Three factors constitute important foci for intervention efforts in this model: basic knowledge and information about the medical condition, motivation based on attitudes about and social support for the behavior, and behavioral skills and self-efficacy specific for the behavior. Moreover, five skills are identified by the model as necessary for the practice of HIV prevention: self-acceptance of sexuality, acquisition of behaviorally relevant information, negotiation of preventive behavior with partner, performance of public prevention acts (such as condom purchase), and consistent performance of prevention behavior (Fisher and Fisher, 1992). For example, an IMB-based intervention addressing women's use of female condoms would focus on ensuring that a woman has basic knowledge about how to use a female condom, is motivated to use it, has the technical skills to properly insert it, and possesses the negotiation skills to get her partner to agree to use it. The IMB model also considers economic and environmental factors, such as living conditions and whether the individual has access to good health care. Moderators of the IMB model, however, include problems such as depression, substance abuse, and other mental health concerns (Starace *et al.*, 2006).

The self-regulation theory

The self-regulation theory appreciates the individual as an active agent who engages in a dynamic process of first assessing health threats and then using problem-solving strategies to address them (Leventhal *et al.*, 1980). This model focuses on individuals' cognitive representation of a health threat, which is based on past experience and newly acquired information, to guide their health behaviors. The self-regulation model captures the influence of the complex interaction between individual and socio-cultural factors on health behaviors. However, this model has not provided much information to help develop interventions to promote health behaviors.

The transtheoretical model

The transtheoretical model (TTM) (Prochaska *et al.*, 1994, 2002; Prochaska and Velicer, 1997) is a dynamic theory of change based on the assumption that there is a common set of change processes that can be applied across a broad range of health behaviors. TTM conceptualizes behavior change as a process involving a series of six distinct stages: precontemplation, contemplation, preparation, action, maintenance and termination. These stages are transtheoretical, and integrate principles of change from across a variety of theories of intervention. In the early stages of change, individuals apply cognitive, affective and evaluation processes to progress forward; during the later stages, commitments, conditioning, contingencies, environmental controls and support to move toward maintenance and termination (Procheska *et al.*, 2002). Each stage brings an individual closer to making or sustaining behavioral changes. Unique variables, processes and benefits versus costs of behavior change define each stage, and interventions based on this model are meant to increase motivation to change and to resolve ambivalence about change. At times individuals may move back to earlier stages (relapse), but movement through the stages recommences the process of change. TTM is one of the most widely cited and utilized models for interventions regarding health behavior changes. A criticism of TTM is that such distinct stages can not capture the complexity of human behavior; the stages may be more properly understood as mere points on a larger continuum of the process of change. Motivational interviewing (described above) is a technique that is also consistent with the transtheoretical model.

The relapse prevention model

The relapse prevention (RP) model is based on the principles of social learning theory. As described by Marlatt and George (1984), the RP model integrates skills training, cognitive therapy and lifestyle change. It aims to increase awareness and begin to change habits in an effort to rebalance lifestyle and improve ability to cope with stressors. Principles of relapse prevention include identifying high-risk situations for relapse (e.g., drug/alcohol use during sex) and developing appropriate solutions (e.g., abstaining or moderating drug/alcohol use during sex). Helping individuals to distinguish between a slip (e.g., an instance of not participating in the planned activity) and a relapse (e.g., an extended period of not participating) is thought to improve adherence in the relapse prevention model.

The health decision model

The health decision model (HDM), combining the health belief model and individual preferences, defers to patient preferences in making health decisions with respect to weighing the benefits and risks of behaviors (McNeil *et al.*, 1975; Weinstein and Fineberg, 1980). In addition to the health belief model's variables (perceived severity, susceptibility, evaluation of action, motivations), the health decision model also incorporates social variables (including knowledge, past experience, socio-demographic and socio-cultural factors, and patient-provider factors) (Eraker *et al.*, 1985), and recognizes that behaviors and health decisions are often made in an interactional context with other people (Fan *et al.*, 2004). For example, using a condom to prevent HIV requires two people to jointly take action, and necessitates the consideration of other people's views – either implicitly or explicitly – to successfully negotiate behavior implementation. Importantly, HDM also considers the impact of cultural values on behavior; for example, in Latino/Hispanic culture, condoms have a number of negative associations (e.g., machismo, religious prohibitions, association with prostitution, etc.) that may interfere with condom use. As with the health belief model, critics have argued that the health decision model envisions health attitudes and behavior change in a static and linear way (i.e., progressing through a series of stages that lead to the outcome of decision to change) and does not allow for the fluidity, dynamism and flexibility needed to account for or encourage complex behavior change (Fan *et al.*, 2004).

The AIDS risk reduction model

The AIDS risk reduction model (ARRM) is one of several stages of change models that posit behavior change to be a process in which individuals move from one step to the next as a result of a given stimulus (Catania *et al.*, 1990). The ARRM combines aspects of the health belief model, the diffusion of innovation theory, and social cognitive theory. In the ARRM, an individual must pass through three stages: behavior labeling, commitment to change, and taking action. Consequently, interventions using this model focus on conducting an individual risk assessment, influencing the decision to reduce risk through perceptions of enjoyment or self-efficacy, and assisting the individual with support to enact the change (e.g., access to condoms, social support).

Stage 2: Selected efficacy trials of HIV prevention interventions based on the conceptual models

Efficacy studies of individualized behavioral interventions for HIV risk reduction have been conducted among various at-risk populations. This section of the chapter reviews selected efficacy studies with populations at high risk of HIV, namely men who have sex with men (MSM), injection drug users (IDUs), women, HIV-infected individuals and racial/ethnic minority populations (African-American and Latino). Table 8.1 offers a summary of several meta-analyses examining overall effectiveness of HIV behavioral interventions. Table 8.2 provides an overview of selected individual interventions, their theoretical basis, description of implementation, and key efficacy findings.

Men who have sex with men (MSM)

Because men who have sex with men are the risk group with the highest incidence and prevalence of HIV in the US, there have been a good number of studies of individual interventions for this group. Below we report on two meta-analyses

213

Table 8.1 Summary of meta-analyses: Stage II individual HIV behavioral intervention research

Author(s) & Date	Specified population	Study description and other information	Effect measure	Results for individual interventions	Overall results
Herbst et al. (2007a)	MSM	20 studies with 4689 participants were included	Any unprotected anal intercourse	OR = 0.57(0.37–0.87), 95% CI	HIV behavioral interventions reduce odds of UAI ranging from 27% to 43%
		4 individual-level interventions	Receptive UAI	OR = 0.77(0.65–0.92), 95% CI	
			Incident HIV	OR = 0.62(0.36–1.06), 95% CI	
Johnson et al. (2005)	MSM	54 interventions with 16,224 participants were evaluated in 40 either randomized trials or controlled observational studies	Unprotected anal intercourse	Individual interventions versus minimal to no HIV prevention (RR = 0.87; CI = 0.60, 1.26)	Contrasted to minimal or no HIV prevention interventions: reduced unprotected sex by 27%. RR = 0.73, 95% CI, 0.63–0.85
		18 individual-level interventions		Individual interventions versus standard or other HIV prevention interventions (RR = 0.88; CI = 0.95, 1.04)	Contrasted to standard or other HIV prevention interventions: RR = 0.83, 95% CI, 0.73–0.95
				Among HIV MSM, individual interventions versus standing or other HIV prevention interventions (RR = 0.91; CI = 0.62, 1.34)	HIV-positive contrasted to standard or other HIV prevention interventions: RR = 0.79, 95% CI, 0.61–1.02

Neumann et al. (2002)	Heterosexuals	14 studies with 4354 participants (10 studies for the behavioral analysis, 6 for the biologic portion; 2 were included in both behavioral and biologic)	Unprotected sex	Analysis did not stratify effect size by intervention type (e.g., individual vs. group)	Reduced sex-related risk: OR = 0.81, 95% CI, 0.53–0.90
		4 of 10 behavioral studies were individual-level intervention studies; 6 of 10 behavioral studies focused on women	STD incidence		Decreased STDs: OR = 0.74, 95% CI, 0.62–0.89
Herbst et al. (2007b)	Latinos/ Hispanics	20 randomized and nonrandomized trials with 6173 participants	Unprotected sex	Sex risk behavior OR = 0.77, 95% CI, 0.49, 1.21	Sex risk behavior (unprotected sex): OR = 0.75, 95% CI (0.66–0.85)
		2 studies were individual-level interventions			Incident STD infections (chlamydia or gonorrhea): OR = 0.69, 95% CI (0.54–0.88)
					Drug risk behavior (incl. drug use freq., needle-sharing, and cotton/cooker sharing): OR = 0.83, 95% CI (0.72–0.96)

(continued)

Table 8.1 (continued)

Author(s) & Date	Specified population	Study description and other information	Effect measure	Results for individual interventions	Overall results
Semaan et al. (2002)	Drug users	33 studies with 7146 participants (94% IDU; 21% crack users) 30 studies were individual and community level interventions (analysis calculated only combined effect)	Unprotected sex	OR = 0.89(0.78–1.01), 95% CI (includes individual or community interventions with a corrected variance for effect size)	OR = 0.60(0.43–0.85), 95% CI
Crepaz et al. (2007)	African-Americans/ Blacks and Latinos/ Hispanics	18 randomized controlled trials. 14 trials evaluated unprotected sex; $n = 11,590$; 13 trials evaluated incident STDs; $n = 16,172$ 7 studies were individual level interventions	Unprotected sex Incident STDs	OR = 0.77(0.63–0.95); 95% CI (7 trials) OR = 0.91(0.70–1.18); 95% CI (7 trials)	OR = 0.77(0.68–0.87); 95% CI; (14 trials; $n = 11,590$) OR = 0.85(0.73–0.99); (13 trials; $n = 16,172$)

Table 8.2 Selected Stage II individual HIV behavioral intervention effectiveness studies

Author(s) & Date	Specified population; guiding theory	Intervention description and other information	Effect measure	Results	Follow-up
Rosser (1990)	MSM; theory not reported	Intervention: Individual HIV prevention counseling (1 session, 20–30 min, 1 day), 15 min safer sex video. Comparison: Wait-list. Sample size: 57	Inverse of % safe sex (number UAI + CU + monogamous relationship)	OR = 0.83(0.15–4.57); 95% CI	6 mo
Picciano et al. (2001)	MSM; motivational enhancement	Intervention: Telephone-based motivational enhancement including immediate counseling by telephone (1 session, 90–120 min, 1 day). Comparison: Wait-list. Sample size: 89	Mean # UAI	OR = 0.60(0.28–1.27); 95% CI	6 wk
			Mean # partners	OR = 0.96(0.45–2.06); 95% CI	
			Mean CU during anal intercourse	OR = 1.54(0.73–3.33); 95% CI	
			Mean unprotected oral intercourse	OR = 0.58(0.27–1.24); 95% CI	
Rotheram-Borus et al. (2004)	HIV-infected young people (69% MSM); social action model, cognitive-behavioral therapy, social learning theory	RCT. Intervention: 3 modules; 18 weekly 2 hr sessions via telephone, in-person, or a delayed-intervention condition. Comparison: Non-intervention. Sample size: 175	Unprotected sex acts with HIV-negative partners	OR = 0.26, 95% CI, 0.12–0.59 Significant difference in unprotected sex acts in past three months at follow-up	3 mo

(continued)

Table 8.2 (continued)

Author(s) & Date	Specified population; guiding theory	Intervention description and other information	Effect measure	Results	Follow-up
Dilley et al. (2002)	MSM; Gold's theory of online versus offline thinking, cognitive theory	Intervention: Single-session cognitive-behavioral intervention counseling and sex diary. Comparison: Standard HIV C&T only. Sample size: 124	% unprotected anal intercourse with nonprimary partner of unknown HIV status	OR = 0.36(0.15–0.86); 95% CI	6, 12 mo
Dilley et al. (2002)	MSM; Gold's theory of online versus offline thinking, cognitive theory	Intervention: Single-session cognitive-behavioral intervention counseling only. Comparison: Standard HIV C&T only. Sample size: 124	% unprotected anal intercourse with nonprimary partner of unknown HIV status	OR = 0.24(0.10–0.56); 95% CI	6, 12 mo
Dilley et al. (2002)	MSM; Gold's theory of online versus offline thinking, cognitive theory	Intervention: Sex diary only. Comparison: standard HIV C&T only. Sample size: 124	% unprotected anal intercourse with nonprimary partner of unknown HIV status	OR = 0.74(0.34–1.60); 95% CI	6, 12 mo
Koblin et al. (2004)	MSM; information-motivation-behavioral skills model (IMB), social	Intervention: 10 one-on-one counseling sessions, maintenance sessions every 3 mo and HIV C&T. Comparison: Treatment (2 HIV C&T sessions per year with	% UAI % UAI with serodiscordant partners	OR = 0.81(0.71–0.93); 95% CI OR = 0.81(0.71–0.93); 95% CI	12, 18 mo

Study	Theory/Focus	Intervention	Outcome	Results	Follow-up
	learning theory, motivational enhancement	Project RESPECT individual counseling). Sample size: 4295	% receptive UAI Incident HIV infection	OR = 0.77(0.65–0.92); 95% CI OR = 0.62(0.36–1.06); 95% CI	30 days, 12 mo
Shoptaw et al. (2005)	MSM with methamphetamine use; cognitive behavioral	Intervention: (1) standard CBT (the Matrix Model, n = 40), (2) contingency management (n = 42), (3) combined CBT and contingency management (n = 40), and (4) an adapted gay-specific CBT (GCBT, n = 40) that incorporated standard CBT with referents of gay culture for cultural specificity. Comparison: none. Sample size: 162	UAI with other than primary partner in the past 30	Reduced from an average of 3 episodes in past 30 days to an average of less than 1 episode at 12-month follow-up	
Gold and Rosenthal (1995)	MSM; relapse prevention	Intervention: (1) Sexual behavior diary plus one individual brief intervention group focused on evaluating self-justifications for risk behaviors; (2) sex diary plus standard AIDS education posters. Comparison: sex diaries only. Sample size: 109	# UAI slips	Group 1 which included individual brief intervention were less likely to have had multiple UAI slip-ups than the other two groups	6 mo

(continued)

Table 8.2 (continued)

Author(s) & Date	Specified population; guiding theory	Intervention description and other information	Effect measure	Results	Follow-up
Robles (2004)	IDU in treatment; motivational interviewing	Intervention: Six weekly interventions sessions, case management for 1.5 months, and 2 sessions of HIV counseling and testing. Comparison: none. Sample size: 557	Drug use and injection-related HIV risk	Intervention group significantly less likely to continue drug injection and less likely to share needles	6 mo
Stephens et al. (1993)	IDU; psychoeducation and behavioral skills training	Intervention: One-on-one format; basic information on HIV transmission using film; discuss sexual risk reduction and condom use; ways to reduce risk due to injection drug use; and provide information on HIV testing. Comparison: Wait-list. Sample Size: 322	% reporting injecting drugs % reporting sharing syringes	decreased from 92.2% to 70.5% decreased from 67.4% to 24.3%	3 mo
Belcher et al. (1998)	Women; motivational interviewing	Intervention: Single-session skill-based intervention adopting Miller's motivational enhancement interviewing model. Comparison: AIDS-education session among women. Sample size: 72	Instances of condom use	Significantly higher instances of condom use among women in the skill-based intervention at follow-ups	1, 3 mo

			% condom use		
Eldridge et al. (1997)	Women court-ordered to an inpatient drug treatment program; psychoeducation, behavioral skills training	Intervention: Brief individual behavioral skills training intervention. Comparison: Psychoeducational intervention. Sample size: 117		Women in the behavioral skills training intervention had increased condom use at follow-up (from 35.7% to 49.5% of vaginal intercourse); women in psychoeducation intervention evidenced a decrease in frequency of condom use (28.8% to 15.8%)	2 mo
St. Lawrence et al. (1997)	Incarcerated women; social cognitive theory, theory of gender and power	Intervention: Intervention based on social cognitive theory. Comparison: Intervention based on the theory of gender and power. Sample size: 90		Both interventions produced increased self-esteem, self-efficacy, Attitudes Toward Prevention Scale scores, AIDS knowledge, communication skills, and condom application skills that were maintained at follow-up. Women in the social cognitive theory group showed greater improvement in condom application skills; women in the theory of gender and power group displayed greater commitment to behavior change	6 mo

(continued)

Table 8.2 (continued)

Author(s) & Date	Specified population; guiding theory	Intervention description and other information	Effect measure	Results	Follow-up
Sterk et al. (2003)	Women crack abusers not in treatment; theory not specified	Intervention: Culturally appropriate, gender-tailored HIV intervention for crack abusers, a revised National Institute on Drug Abuse standard intervention. Comparison: Wait-list. Sample size: 265		Significant ($P < .05$) decreases in the frequency of crack use, the number of paying partners, the number of times sex (vaginal, oral, or anal) was had with a paying partner, and sexual risks, such as trading sex for drugs. Significant ($p < .05$) increases in male condom use with sex partners were observed	3, 6mo
El-Bassel et al. (2003)	Heterosexual couples; relationship-based HIV prevention	Intervention: (1) 6 sessions provided to couples together ($n = 81$), (2) the same intervention provided to the woman alone ($n = 73$), or (3) a 1-session control condition provided to the woman alone ($n = 63$). Sample size: 217		The intervention was effective in reducing the proportion of unprotected sex and increasing the proportion of protected sex. No significant differences in effects were observed between couples receiving the intervention together and those in which the woman received it alone	

Kamb et al. (1998)	Heterosexuals (84% ethnic/racial minority: 59% Black, 19% Hispanic, 6% other race/ethnicity); theory of reasoned action, social cognitive theory	RCT Intervention: Individual-level counseling (4 sessions, 3 hours and 20 min duration, over 3–4 weeks) focused on condom use behavioral goal setting; discussion and reinforcement of positive condom use attitudes, self-efficacy, and perceived norms; and setting a risk reduction plan. Comparison: Educational messages only (two 5 minute sessions). Sample size: 5758 men (56.7%) and women (43.2%).	Unprotected sex Incident STDs	$OR = 0.80(0.67–0.96)$; 95% CI (at 6 mo) $OR = 0.76(0.61–0.94)$; 95% CI (at 12 mo)	6, 12 mo
		reductions in self-reported at and in incident STDs			
Orr et al. (1996)	Ethnic/racial minority women; health belief model	Intervention: Women received one individual counseling session (10–20 min duration) to increase perception of vulnerability, decrease condom use barriers by fostering positive attitudes about condoms and negotiation skills, and rehearsing condom negotiation via role-playing. Comparison: Women received one standard counseling session (10–20 min duration). Sample size: 209	Condom use Incident STDs	For STDs (OR = 2.4; $p = 0.02$) and vaginal intercourse (OR = 3.1; $p = 0.005$). No significant change	6 mos

of these studies, followed by selected studies that either were not used in the meta-analysis, or that cogently illustrate specific findings for the theories reviewed above.

Meta-analyses

Individual HIV behavioral interventions with adult MSM have been shown to result in significant reductions in self-reported sexual risk behaviors (Herbst *et al.*, 2005, 2007a; Johnson *et al.*, 2005). A meta-analytic review by Herbst *et al.* (2007b), synthesizing evidence and evaluating the effectiveness of four HIV individual interventions based in a variety of conceptual models, found an average 43 percent reduced odds of engaging in unprotected anal intercourse among individual intervention-group members relative to the comparison group (aggregate effect size OR 0.57, 95% CI 0.27–0.87; $n = 4689$) (Herbst *et al.*, 2007a). Similarly, when comparing the effects of 16 individual interventions versus minimal to no HIV prevention (RR = 0.87; CI 0.60–1.26) or versus standard or other HIV prevention interventions (RR = 0.88; CI 0.95–1.04), Johnson *et al.* (2005) found significant reductions in unprotected anal intercourse supporting the effectiveness of individual interventions (Johnson *et al.*, 2005). Moreover, relative to standard or other HIV prevention interventions, individual interventions were shown to reduce the proportion of MSM reporting unprotected sex by 5 percent (OR 0.95; CI 0.62–1.34) (Johnson *et al.*, 2005).

Gold's theory of online versus off-line thinking, cognitive theory

Dilley *et al.* (2002) compared the effects of personalized single-session cognitive behavioral therapy (CBT) in addition to standard HIV test counseling, to HIV test counseling and keeping a sexual diary, demonstrating the effective application of cognitive theory to an individual intervention with MSM. Participants were 248 MSM with a previous HIV-negative test result and self-reported unprotected anal intercourse (UAI) with partners of unknown or discordant HIV status. The CBT focused on using self-justifications for HIV high-risk behaviors (thoughts, attitudes or beliefs that allow participant to engage in high-risk behaviors). Counselors delivered the intervention to individuals, and follow-up evaluation occurred at 6 and 12 months. Compared to controls, intervention participants reported significant decreased UAI with non-primary partners of unknown or discordant HIV status at 6 and 12 months (from 66 percent, to 21 percent at 6 months and 26 percent at 12 months; OR 0.24, 95% CI 0.10–0.56).

Motivational interviewing, social learning theory, social cognitive theory

Employing motivational interviewing based in social cognitive theory, Picciano and colleagues (2001) evaluated the effects of a telephone-based intervention to reduce sexual risk-taking in MSM. Participants were 89 MSM who reported engaging in three or more recent episodes of oral or anal sex without a condom. They were assigned either to an immediate 90-minute motivational interview (MI) counseling group or to a control group for delayed MI counseling (wait-list). MI involved encouraging readiness for change to increase condom use and safer sex practices. Results showed that those who received the one-time counseling intervention significantly increased safer sex practices and intention to use condoms and decreased UAI at 6 weeks' follow-up, in comparison to those in the control group (OR 0.60, 95% CI 0.28–1.27).

Social action model, cognitive behavioral therapy, social learning theory

Rotheram-Borus *et al.* (2004) conducted a randomized controlled trial with a three-module preventive intervention. Young, HIV-infected substance users (*n* = 175, who were 69 percent MSM) were assigned to 18 weekly 2-hour sessions delivered via telephone, or delivered in person, or to a delayed-intervention condition. The intervention focused on improving physical health; coping with HIV status; maintaining drug regimens and making health-care decisions; identifying life goals; reducing distress; anticipating situations that raise anxiety, depression, fear or anger; and recognizing and controlling negative emotion with relaxation, self-instruction, and meditation. The in-person intervention group showed significantly higher condom use, as well as higher condom use with HIV-infected partners, while participants in the delayed-intervention control reported fewer sexual partners, a decrease in drug use and emotional distress, as well as a decrease in antiretroviral therapy. Significant differences in unprotected sex acts in the past 3 months were observed at 3-month follow-up (OR 0.26, 95% CI 0.12–0.59) (Crepaz *et al.*, 2006).

Theoretical model not reported

Rosser (1990) conducted a trial involving gay men (mean age 34) in Auckland, New Zealand, to examine the effects of five different experimental conditions on creating safer sex behaviors. The 159 participants were randomized; each individual either watched a video on AIDS, received individual counseling for HIV, took part in a group program on AIDS with safer sex guidelines, took part in a group program on eroticizing safer sex, or was assigned to a control condition (wait-list). Health counselors or program facilitators delivered the interventions to individual participants. At baseline, only 74.7 percent of the sample reported safer sex practices during the previous 2 months; at 6-month follow-up this had increased to 82.7 percent. No significant differences were found between the effectiveness of different interventions when safer sex (outcome) was measured globally; however, a trend emerged toward individual counseling as a more effective intervention ($P < 0.10$). Those receiving the individualized intervention ($n = 57$) were more likely to reduce unprotected sexual encounters at 6-month follow-up (OR 0.83, 95% CI) (Herbst *et al.*, 2007a).

Motivational interviewing, social learning theory, social cognitive theory

The EXPLORE study was a large-scale randomized HIV-prevention trial among MSM conducted in 6 US cities, with 4295 participants total (Chesney *et al.*, 2003; Koblin *et al.*, 2003, 2004). Inclusion criteria for the EXPLORE study were men who were HIV-uninfected, 16 years or older, had had anal sex with another man during the past year, and had not been involved in a mutually monogamous relationship in the past 2 years with a male partner who was HIV-uninfected. Men were randomized to receive a behavioral intervention versus standard risk-reduction counseling. The experimental intervention, described in detail by Chesney *et al.* (2003), consisted of 19 core counseling modules delivered at one-on-one counseling sessions. In the main EXPLORE trial, participants in both arms had HIV testing every 6 months and completed a sexual risk behavior and psychosocial assessment battery using audio-computer-assisted self-interviewing (ACASI).

While the intervention did not achieve a targeted level of efficacy (35 percent) in preventing new HIV infections compared to semi-annual HIV voluntary counseling and testing, the study results suggest a possible modest benefit of the intervention in reducing new HIV infections (EXPLORE Study Team, 2004). Further, the reporting of unprotected receptive anal sex with HIV-positive or unknown-status partners was significantly lower in the intervention group compared with the standard group (Koblin *et al.*, 2004).

Cognitive behavioral therapy (CBT)

In another study, by Shoptaw *et al.* (2005), 162 MSM with methamphetamine use were randomized to one of four interventions to address methamphetamine use and sexual risk behavior. The intervention groups included: (1) standard CBT (the matrix model, $n = 40$), (2) contingency management ($n = 42$), (3) combined CBT and contingency management ($n = 40$), and (4) an adapted gay-specific CBT (GCBT, $n = 40$) that incorporated standard CBT with referents of gay culture for cultural specifity. Reductions from baseline were significant across all groups. Results indicated that participants in all the intervention groups greatly decreased both sexual risk behaviors as measured by frequency of UAIs, and methamphetamine use as assessed by drug screening. Episodes of UAI with other than the primary partner in the past 30 days were reduced from an average of 3 episodes to an average of less than 1 episode at 12-month follow-up. Although promising, this study lacked a non-intervention control group, making it impossible to control for potential cohort effects.

Relapse prevention model

In a study by Gold and Rosenthal (1995) based on the relapse prevention model, 109 MSM who had "slipped up" and had unprotected anal intercourse kept diaries of their sexual behavior for 16 weeks. After 4 weeks, participants were randomized to three different conditions: one individual brief intervention group focused on evaluating self-justifications for risk behaviors (self-justifications), the second intervention group used standard AIDS education posters (standards), while the third condition kept diaries only (control). The outcome variable was the number of UAI slips. Although the three groups did not differ in the incidence of sexual activity or in the proportion that slipped up at least once, the self-justifications group members were less likely to have had multiple UAI slip-ups than the other two groups; however, results were not significant.

A subsequent study by Gold and Rosenthal (1998), also based on the relapse prevention model, further evaluated the effects of use of self-justification with posters versus vivid recalls of events to reduce risk of STDs. Participants were again asked to keep diaries of their sexual behavior for 16 weeks. After 4 weeks, they were randomized into a control group without any intervention, an intervention group with instructions to recall a detailed description of a safe-sex slip-up without self-justifying behaviors, or a second intervention group with instructions to examine posters designed to focus on promoting evaluation of one's self-justifications. Post-intervention evaluations showed no differences in the frequency of UAI slip-ups among the three groups. The two studies did not provide cohesive support for use of the self-justification relapse prevention model.

Injection drug users (IDUs)

In the US, injection drug users (IDUs) represent a risk group with a high incidence and prevalence of HIV. There have been a number of studies of individual interventions among IDUs. Below we report on a meta-analysis of these studies, followed by selected studies that either were not used in the meta-analysis or that exemplify specific findings for theories reviewed above.

Meta-analyses

Semaan *et al.* (2002) examined the efficacy of 33 intervention studies with drug users (94 percent injection drug users, 21 percent crack users). Relative to no HIV intervention, drug users in intervention conditions significantly reduced sexual risk behaviors among drug users (OR 0.60, 95% CI 0.43–0.85). In other words, when extrapolating results to a population with a 72 percent prevalence of risk behaviors, the proportion of drug users who reduced their risk behaviors was 12.6 percent greater in the intervention groups than in comparison groups (Semaan *et al.*, 2002). Although the effect size for individual interventions was combined with community interventions (30 studies in total), analyses indicated that drug users in the intervention groups were significantly more likely to reduce unsafe sexual behaviors relative to participants in the comparison groups (OR 0.89, 95% CI 0.78–1.01) (Semaan *et al.*, 2002).

Motivational interviewing, social learning theory, social cognitive theory

Robles (2004) examined the effectiveness of a combined counseling and case-management behavioral intervention, using motivational interviewing strategies, in engaging Hispanic injection drug users in treatment and reducing drug use and injection-related HIV risk behaviors. The intervention group received six weekly intervention sessions, case management for one-and-a-half months, and two sessions of HIV counseling and testing. Follow-up data were collected from 440 (79.0 percent) of 557 randomized participants at 6 months after the initial interview. Subjects in the experimental arm were significantly less likely to continue drug injection independent of entering drug treatment, and were also more likely to enter drug treatment. Subjects in both arms who entered drug treatment were less likely to continue drug injection. Among subjects who continued drug injection, those in the experimental arm were significantly less likely to share needles.

Psychoeducation, behavioral learning, behavioral skills training

Stephens *et al.* (1993) conducted a psychoeducationally-based intervention with mostly heterosexual street addicts not in treatment (*n* = 322). The intervention was conducted by a health educator in a one-on-one format, and provided basic information on HIV transmission using a segment of a film, discussed sexual risk reduction and condom use, discussed ways to reduce risk due to injection drug use, and provided information on HIV testing. Risk was assessed prior to the intervention, and was compared to a follow-up assessment obtained approximately 3 months later. The percentage who reported injecting drugs decreased from 92.2 to 70.5, and the percentage who reported sharing syringes dropped from 67.4 to 24.3. Reductions of these and other high-risk behaviors were detected across various demographic subgroups, and analyses show that the impact of the intervention endured at 3-month follow-up. The research did not focus on sexual risk.

Women

Among US women, a high incidence and prevalence of HIV have been reported; consequently, reviewing studies of individual interventions with women provides important information about effective HIV prevention strategies. Below we report on a meta-analysis of these studies, followed by selected studies either that were not used in the meta-analysis or that fruitfully illustrate specific findings for some of the theoretical models reviewed previously.

Meta-analyses

Behavioral HIV interventions with women have been shown to be effective in reducing overall HIV risk (Mize *et al.*, 2002; Neumann *et al.*, 2002). Neumann *et al.* (2002) conducted a meta-analysis of HIV behavioral interventions among heterosexuals and found statistically significant effects in reducing sex-related risk (OR 0.81, 95% CI 0.53–0.90) and decreased incidence of STDs (OR 0.74, 95% CI 0.62–0.89). Of the 10 studies, 4 (40%) included in the behavioral analyses were individual interventions and 6 (60%) focused exclusively on women; however, the analysis did not stratify effect size by intervention type (e.g., individual vs group) or gender (women vs men).

Motivational interviewing, social learning theory, social cognitive theory

Belcher *et al.* (1998) compared the effects of a single-session skill-based intervention adopting Miller's motivational enhancement interviewing model to an AIDS-education session among women. Participants were 72 women from a low-income housing project. Outcome variables included AIDS knowledge, behavioral intentions, self-efficacy, and sexual risk behaviors. Results showed significantly higher instances of condom use among women in the skill-based intervention at 1- and 3-month follow-up, providing evidence for the effectiveness of an individual intervention based on a combination of motivational interviewing, social learning theory and social cognitive theory.

Psychoeducation, behavioral learning, behavioral skills training

Eldridge *et al.* (1997) compared the effects of a psychoeducation intervention relative to a brief individual behavioral skills training intervention geared towards reducing sexual risk behaviors in women with substance addiction. Participants were 117 women who had been court-ordered to an inpatient drug treatment program. They were evaluated for sexual risk behavior, and attitudes toward HIV prevention and condom use at baseline, post-intervention, and 2 months after discharge from the drug program. Women in both conditions had reduced drug use and drug-related high-risk sex activities at follow-up. Results indicated that both interventions were helpful in improving positive attitudes toward HIV prevention, and reported greater partner agreement with condom use at the post intervention assessment. However, at 2-month follow-up women in the behavioral skills training group continued to show improvement in communication and condom application skills, whereas women in the psychoeducation group did not. Moreover, women in the behavioral skills group had increased their condom use at follow-up (from 35.7 percent to 49.5 percent of incidences of vaginal intercourse) while women in the psychoeducation group evidenced a decrease in frequency of condom use (28.8 percent to 15.8 percent).

Social cognitive theory

St Lawrence *et al.* (1997) evaluated two HIV risk reduction interventions for incarcerated women, comparing an intervention based on social cognitive theory against a comparison condition based on the theory of gender and power. Incarcerated women ($n = 90$) were assessed at baseline, post-intervention, and at 6-month follow-up. Both interventions produced increased self-efficacy, self-esteem, Attitudes Toward Prevention Scale scores, AIDS knowledge, communication skills, and condom application skills that were maintained at 6-month follow-up. Women in the social cognitive theory group showed greater improvement in condom application skills; women in the theory of gender and power group displayed greater commitment to behavior change.

Theoretical model not reported

El-Bassel *et al.* (2003) examined the efficacy of a relationship-based HIV prevention program for heterosexual couples, as well as to determine whether the intervention was more effective when delivered to the couple or to the woman alone. Couples ($n = 217$) were recruited and randomized to (1) six sessions provided to couples together ($n = 81$), (2) the same intervention provided to the woman alone ($n = 73$), or (3) a single-session control condition provided to the woman alone ($n = 63$). The intervention was effective in reducing the proportion of unprotected sex and increasing the proportion of protected sex. No significant differences in effects were observed between couples receiving the intervention together and those in which the woman received it alone.

Sterk *et al.* (2003) evaluated the effectiveness of a culturally appropriate, gender-tailored HIV intervention with 265 African-American women who use crack cocaine (Sterk *et al.*, 2003). A substantial proportion of women reported no past 30-day crack use at 6-month follow-up (100 percent to 61 percent, $P < 0.001$). Significant ($P < 0.05$) decreases were reported in the frequency of crack use, the number of paying partners, the number of times sex (vaginal, oral or anal) was had with a paying partner, and sexual risks, such as trading sex for drugs. Significant ($P < 0.05$) increases in male condom use with sex partners were observed. Findings suggest that including culturally appropriate, gender-tailored components in individual interventions may be critical to enhancing the effectiveness of HIV preventive behaviors.

Ethnic/racial minorities

Ethnic and racial minorities, in particular African-American and Latino/Hispanic populations, bear a disproportionate burden of new HIV diagnoses in the United States. Few rigorous efficacy trials have been conducted with these populations to date. A review of meta-analyses focusing on HIV interventions with ethnic and racial minorities is presented below.

Meta-analyses

Reviews and meta-analyses summarizing results of interventions with ethnic and racial minority communities have suggested that HIV interventions are effective in reducing HIV risk (Crepaz *et al.*, 2007; Herbst *et al.*, 2007b; Lyles *et al.*, 2007). Herbst *et al.* (2007) conducted a meta-analysis of randomized and non-randomized trials of Latinos/Hispanics in the US and Puerto Rico, and found that interventions with sex risk behavior data yielded a significant 56 percent increased odds of condom use, 25 percent reduced odds of unprotected sex, 25 percent reduced odds of multiple sex partners and 31 percent reduced odds of acquiring an incident STD (Herbst *et al.*, 2007b). However, only 2 of the 20 studies included in the review were individual-level interventions, suggesting a need for additional research on individual interventions with Hispanic/Latino populations.

Crepaz *et al.* (2007) evaluated the efficacy of behavioral interventions in reducing unprotected sex and incident STD among African-American/Black and Latino/Hispanic STD clinic patients. Individual interventions were effective in reducing unprotected sex (OR 0.77, 95% CI 0.63–0.95, seven trials) and incidents of STDs (OR 0.91, 95% CI 0.70–1.18), seven trials). Interestingly, significantly greater efficacy ($P < 0.05$) was found among interventions that used ethnically matched individuals to deliver the intervention. Interventions using an ethnically matched deliverer were more likely to reduce unprotected sex (OR 0.59, 95% CI 0.45–0.78) and incident STDs (OR 0.65, 95% CI 0.51–0.82) relative to those interventions that did not use an ethnically matched deliverer (Crepaz *et al.*, 2007). Moreover, theory-based interventions were more likely to affect behavioral change among ethnic/racial minority participants ($P < 0.05$).

Mize *et al.* (2002) evaluated HIV prevention interventions on three HIV-related sexual outcome variables (HIV/AIDS knowledge, self-efficacy, and sexual risk-reduction behavior), stratifying results by five ethnic groupings (all ethnicities combined, African-American, White, Hispanic, and a Mixed Ethnicity group). Results indicate that interventions appear effective at improving knowledge about HIV/AIDS and increasing sexual risk-reduction behaviors for all ethnicities examined at follow-up, with one exception: self-efficacy findings. Interventions were less consistently effective for African-American women.

Theory of reasoned action, social cognitive theory

Kamb *et al.* (1998) conducted a randomized controlled trial with 5758 men (56.8 percent) and women (43.2 percent); 59 percent black, 19 percent Hispanic, 6 percent other race/ethnicity; 16 percent white; 100 percent heterosexual. The intervention arm received individual-level counseling (4 sessions, 3 hours and 20 minutes duration, over 3–4 weeks) focused on condom use behavioral goal-setting; discussion and reinforcement of positive condom use attitudes, self-efficacy, and perceived norms; and setting a risk-reduction plan. The control arm received educational messages only (two 5-minute sessions). The intervention led to significant reductions in self-reported unprotected sex at 6-month follow-up (OR 0.80, 95% CI 0.67–0.96) and in incident STDs at 12-month follow-up (OR 0.76, 95% CI 0.61–0.94) (Kamb *et al.*, 1998).

Health belief model

In formative research on a culturally tailored intervention based on the health belief model, Orr *et al.* (1996) implemented a randomized controlled trial with 209 women diagnosed with an STD (55 percent black; 45 percent race/ethnicity not reported). Women in the intervention group received one individual counseling session (10–20 minutes) to increase perception of vulnerability, decrease condom use barriers by fostering positive attitudes about condoms and negotiation skills, and rehearse condom negotiation via role-playing. Controls received one standard counseling session (10–20 minutes). Women in the intervention group were much more likely to use condoms for protection against STDs at follow-up than at baseline (OR 2.4, $P = 0.02$) and for vaginal intercourse (OR 3.1, $P = 0.005$). However, the rate of reinfection with *C. trachomatis* in the intervention group was not significantly different than the control group (26 percent vs 17 percent; $P = 0.30$).

Stage 3: Summary and conclusions

This chapter has reviewed selected HIV risk-reduction interventions that have undergone randomized efficacy trials. Several studies have found evidence for the efficacy of interventions, but support for their effectiveness and generalizability in different geographical locations with various at-risk populations remains insufficient. Data suggest that higher-risk clients may be best served by individual-level HIV interventions (Johnson *et al.*, 2005). In particular, as measured by self-report, interventions delivered on a one-to-one basis have been found to reduce unprotected sex significantly among HIV-infected individuals (Crepaz *et al.*, 2006), who may be more complex because of associated potential co-morbidities, and MSM (Herbst *et al.* 2007a). Studies suggest that individual approaches may be especially appropriate for MSM who are difficult to reach with group- or community-level interventions (such as those who wish to remain anonymous) (Herbst, 2007a). The strength of individual interventions and its potential to effect individual-level behavior change lies in the ability to tailor interventions to meet the specific needs of the individual client (Herbst *et al.*, 2007a). However, a significant limitation is that, although affecting individual behavior change, individual interventions may be limited in reach and ability to impact the broader population (e.g., by reducing overall rates of HIV infection).

Yet the literature remains contradictory and suggests that interventions may be differentially effective depending on the targeted at-risk population. For example, Neumann *et al.* (2002) report that although individual interventions were effective in reducing HIV risk among heterosexuals, group interventions showed more favorable effects than interventions delivered to individuals. Other reviews of interventions with heterosexuals have also found that individual-level counseling was less likely to be a good primary prevention strategy (Ickovics and Yoshikawa, 1998) relative to group interventions (Exner *et al.*, 1999; Rotheram-Borus *et al.*, 2000). Similarly, the literature suggests that individual interventions are less effective in improving self-efficacy among African-American women, relative to women of other races and ethnicities (Mize *et al.*, 2002).

These differences have a significant impact on how we conceptualize the theory driving effective interventions. Generally, the existing theoretical models of HIV risk-reduction for individual interventions are based in disciplines of social and health psychology and can be applied across various at-risk groups. Using these models to inform the development of interventions to reduce HIV risk behaviors and promote health behaviors has several benefits and some drawbacks. Because there are multiple theories available to aid mental and medical health care and public health researchers and providers, the insufficient and at times contradictory support for the various models does not allow for developing clearly defined interventions to promote health behaviors (Munro *et al.*, 2007).

Another limitation of existing models is that that they often do not fully take into account psychosocial factors, such as clinically significant mental health diagnoses or the treatment of concomitant substance abuse or dependence, which often potentiate sexual risk-taking or can moderate the degree to which an individual can benefit from an intervention that does not address these problems. While some studies have directly demonstrated the predictive value of psychosocial comorbidities with respect to increased HIV-related risk (Stall *et al.*, 2001; Koblin *et al.*, 2003), other research has not supported this conclusion (Crepaz *et al.*, 2002, 2004). However, even if psychosocial variables are not directly correlated with HIV risk, they may represent indirect moderators that must be considered in the design and implementation of interventions, both practically and theoretically. To date, individual interventions have been based in social psychology theory, with little attention to factors such as depression and anxiety. Depression, for example, can affect information acquisition (poor concentration, apathy), motivation (loss of interest, sadness, apathy) and skills training in IMB. It can also impact variables like perceived norms and attitudes, because people with depression view a negative sense of self, others, world and future. Targeting individuals at high risk for additional psychosocial problems may necessitate addressing psychosocial problems such as depression and anxiety in order for an HIV intervention to be beneficial to the individual. Existing efficacy trials of interventions that target individuals at risk for HIV have not sufficiently addressed the significant co-morbidities and multifaceted problems that may exist. Thus, although some interventions have been successful in reducing sexual risk-taking among populations at greater risk for HIV, they have not addressed the "intertwined syndemics" (Singer and Snipes, 1992; Stall and Purcell, 2000) – the complex and multifaceted issues associated with risk behavior.

Applicability of existing interventions to other more complex populations thus has limited generalizability. In order for interventions to be of use to practitioners treating at-risk populations, successful efficacy trials tested with specific groups need to be tested across a wide variety of populations, across time and place. There is a paucity of efficacy research on heterosexual men who do not inject drugs and on racial/ethnic minorities (Neumann *et al.*, 2002). Moreover, few studies have examined longer-term effects of interventions in various real-world settings. That may be because effectiveness trials are somewhat more difficult to carry out due to the time, expense and involvement of multiple sites, and hence there are fewer of them. Further, with the great importance of implementing interventions that are effective in efficacy trials, it is important to translate interventions appropriately to different populations and settings, instead of implementing them exactly as they were developed or tested in a research setting. The latter step would be necessary to optimize the intervention's ability to be cost-effective and to be implemented in, accepted by and generalized to a variety of populations and real-world settings.

Munro *et al.* (2007) suggest that future studies may help by answering several key queries related to whether theory-based interventions can be effective and reliably tested. It may be possible to address the incongruence and limitations of the various models by identifying factors that are present in most of the models and that best predict behavior change, to take out redundant concepts, and to integrate them into a more cohesive approach. Therefore, future research that is informed by existing models rather than new ones can be helpful to clarify key components and provide a deeper understanding of and support for their effective application to behavior change across populations and settings. It is important to create a structure to account for the interaction of the various models' constructs at each level of change (Onken *et al.*, 1997). It is also recommended that future studies identify and address factors that can be modified, that are not fully under the control of the individual, and that are external. Incorporating component analyses to specify components responsible for effectiveness of behavior change is also warranted (Exner *et al.*, 1999). In summary, to improve the efficacy and effectiveness of HIV risk behavior interventions, it is recommended that interventions comprehensively consider factors within the social, psychological, economical and environmental domains.

Acknowledgment

Some of the investigator time for preparation of this manuscript was supported by grant number DA 5R01DA018603 to Steven Safren.

References

Ajzen, I. (1985). From intentions to actions: a theory of planned behavior. In: J. Kuhl and J. Beckman (eds), *Action-control: From Cognition to Behavior*. Heidelberg: Springer, pp. 11–39.

Ajzen, I. (1987). Attitudes, traits, and actions: dispositional prediction of behavior in personality and social psychology. In: L. Berkowitz (ed.), *Advances in Experimental Social Psychology*, Vol. 20. New York, NY: Academic Press, pp. 1–63.

Ajzen, I. (1991). The theory of planned behavior. *Org. Behav. Human Dec. Process.*, 50, 179–211.

Ajzen, I. (2002). Residual effects of past on later behavior: habituation and reasoned action perspectives. *Pers. Social Psychol. Rev.*, 6, 107–22.

Ajzen, I. and Fishbein, M. (1980). *Understanding Attitudes and Predicting Social Behavior*. Englewood Cliffs, NJ: Prentice Hall.

Ajzen, I. and Madden, T. J. (1986). Prediction of goal-directed behavior: attitudes, intentions, and perceived behavioral control. *J. Exp. Social Psychol.*, 22, 453–74.

amFAR AIDS Research. (2006). Issue Brief: HIV prevention for men who have sex with men. June 2006. Available at http://www.amfar.org/binary-data/AMFAR_PUBLICATION/download_file/46.pdf (accessed 10-10-2007).

Bandura, A. (1977a). *Social Learning Theory*. Englewood Cliffs, NJ: Prentice Hall.

Bandura, A. (1977b). Self-efficacy: toward a unifying theory of behavioral change. *Psychological Rev.*, 84, 191–215.

Bandura, A. (1986). *Social Functions of Thought and Action*. Englewood Cliffs, NJ: Prentice Hall.

Bandura, A. (1997). *Self-efficacy: The Exercise of Control*. New York, NY: Freeman.

Bandura, A. (2001). Social cognitive theory: An agentic perspective. *Ann. Rev. Psychol.*, 52, 1–26.

Baranowski, T., Perry, C. L. and Parcel, G. S. (2002). How individuals, environments, and health behavior interact: social cognitive theory. In: K. Glanz, B. K. Rimer and R. M. Lewis (eds), *Health Behavior and Health Education: Theory, Research, and Practice*, 3rd edn. San Francisco, CA: John Wiley & Sons, pp. 165–84.

Belcher, L., Kalichman, S., Topping, M. *et al.* (1998). A randomized trial of a brief HIV risk reduction counseling intervention for women. *J. Consult. Clin. Psychol.*, 66, 856–61.

Boer, H. and Seydel, E. R. (1996). Protection motivation theory. In: M. Connor and P. Norman (eds), *Predicting Health Behavior*. Buckingham: Open University Press, pp. 95–120.

Catania, J. A., Kegeles, S. and Coates, T. J. (1990). Towards an understanding of risk behavior: an AIDS risk reduction model (AARM). *Health Educ. Q.*, 17, 53–72.

Chesney, M. A., Koblin, M. A., Barresi, P. J. *et al.* (2003). An individually tailored intervention for HIV prevention: baseline data from the EXPLORE study. *Am. J. Public Health*, 93, 933–8.

Cohen, D. A. and Scribner, R. (2000). An STD/HIV prevention intervention framework. *AIDS Patient Care STDs*, 14, 37–45.

Crepaz, N. and Marks, G. (2002). Towards an understanding of sexual risk behavior in people living with HIV: a review of social, psychological, and medical findings. *AIDS*, 16, 135–49.

Crepaz, N., Hart, T. A. and Marks, G. (2004). Highly active antiretroviral therapy and sexual risk behavior: a meta-analytic review. *J. Am. Med. Assoc.*, 292, 224–36.

Crepaz, N., Lyles, C. M., Wolitski, R. J. *et al.* (2006). Do prevention interventions reduce HIV risk behaviours among people with HIV? A meta-analytic review of controlled trials. *AIDS*, 20, 143–57.

Crepaz, N., Horn, A. K., Rama, S. M. *et al.* (2007). The efficacy of behavioral interventions in reducing HIV risk sex behaviors and incident sexually transmitted disease in black and hispanic sexually transmitted disease clinic patients in the United States: a meta-analytic review. *Sex. Transm. Dis.*, 34, 319–32.

Dilley, J. W., Woods, W. J., Sabatino, J. *et al.* (2002). Changing sexual behavior among gay male repeat tester for HIV: a randomized, controlled trial of a single-session intervention. *J. Acquir. Immune Defic. Syndr.*, 30, 177–86.

El-Bassel, N., Witte, S. S., Gilbert, L. *et al.* (2003). The efficacy of a relationship-based HIV/STD prevention program for heterosexual couples. *Am. J. Public Health*, 93, 963–9.

Eldridge, G. D., St Lawrence, J. S., Little, C. E. *et al.* (1997). Evaluation of an HIV risk reduction invention for women entering inpatient substance abuse treatment. *AIDS Educ. Prev.*, 9, S62–sS76.

Emmonds, K. M. and Rollnick, S. (2001). Motivational interviewing in health care settings: opportunities and limitations. *Am. J. Prev. Med.*, 20, 68–74.

Eraker, S. A., Becker, M. H., Strecher, V. J. *et al.* (1985). Smoking behavior, cessation techniques, and the health decision model. *Am. J. Med.*, 78, 817–25.

Exner, T. M., Gardos, P. S., Seal, D. W. *et al.* (1999). HIV sexual risk reduction interventions with heterosexual men: the forgotten group. *AIDS Behav.*, 3, 347–58.

EXPLORE Study Team (2004). Effects of a behavioural intervention to reduce acquisition of HIV infection among men who have sex with men: the EXPLORE randomised controlled study. *Lancet*, 364, 41–50.

Fan, H. Y., Conner, R. F. and Villarreal, L. P. (2004). *AIDS: Science and Society*, 4th edn. Boston, MA: Jones and Bartlett Publishers, Inc.

Fisher, J. D. and Fisher, W. A. (1992). Changing AIDS-risk behavior. *Psychological Bull.*, 111, 455–74.

Fisher, W. A. (1997). A theory-based framework for intervention and evaluation in STD/HIV prevention. *Can. J. Human Sexuality*, 6, e-pub only.

Godkin, G. and Koh, G. (1996). The theory of planned behavior: a review of its application to health-related behaviours. *Am. J. Health Promotion*, 11, 87–8.

Gold, R. S. and Rosenthal, D. A. (1995). Preventing unprotected anal intercourse in gay men: a comparison of two intervention techniques. *Intl J. STD AIDS*, 6, 89–94.

Gold, R. S. and Rosenthal, D. A. (1998). Examining self-justifications for unsafe sex as a technique of AIDS education: the importance of personal relevance. *Intl J. STD AIDS*, 9, 208–13.

Herbst, J., Sherba, T., Crepaz, N. *et al.* (2005). A meta-analytic review of HIV behavioral interventions for reducing sexual risk behavior of men who have sex with men. *J. Acquir. Immune Defic. Syndr.*, 39, 228–41.

Herbst, J. H., Beeker, C. and Mathew, A. (2007a). The effectiveness of individual-, group-, and community-level HIV behavioral risk-reduction interventions for adult men who have sex with men: a systematic review. *Am. J. Prev. Med.*, 32, S38–67.

Herbst, J. H., Kay, L. S., Passin, W. F. *et al.* (2007b). A systematic review and meta-analysis of behavioral interventions to reduce HIV risk behaviors of Hispanics in the United States and Puerto Rico. *AIDS Behav.*, 11, 25–47.

Hornik, R. (1991). Alternative models of behavior change. In: J. N. Wasserheit, S. O. Aral and K. K. Holmes (eds), *Research Issues in Human Behavior and Sexually Transmitted Diseases in the AIDS Era*. Washington, DC: American Society for Microbiology, pp. 201–18.

Ickovics, J. R. and Yoshikawa, H. (1998). Preventive interventions to reduce heterosexual HIV risk for women: current perspectives, future directions. *AIDS*, 12, 197–208.

Janz, N. K., Champion, V. L. and Strecher, V. J. (2002). The health belief model. In: K. Glanz, B. K. Rimer and F. M. Lewis (eds), *Health Behavior and Health Education: Theory, Research, and Practice*, 3rd edn. San Francisco, CA: John Wiley & Sons, pp. 45–66.

Johnson, W., Holtgrave, W., Flanders, D. *et al.* (2005). HIV intervention research for men who have sex with men: a 7-year update. *AIDS Educ. Prev.*, 17, 568–89.

Kamb, M. L., Fishbein, M., Douglas, J. M. Jr. *et al.* (1998). Efficacy of risk-reduction counseling to prevent human immunodeficiency virus and sexually transmitted diseases: a randomized controlled trial. *J. Am. Med. Assoc.*, 280, 1161–7.

Koblin, B. A., Chesney, M. A., Jusnik, M. H. *et al.* (2003). High risk behaviors among men who have sex with men in 6 US cities: baseline data from the EXPLORE study. *Am. J. Public Health*, 93, 926–32.

Koblin, B., Chesney, M., Coates, T. *et al.* (2004). Effects of a behavioral intervention to reduce acquisition if HIV infection among men who have sex with men: the EXPLORE randomized controlled study. *Lancet*, 364, 41–50.

Leventhal, H., Meyer, D. and Nerenz, D. (1980). The commonsense representation of illness danger. In: S. Rachman (ed.), *Medical Psychology,* Vol. 2. New York, NY: Pergamon, pp. 7–30.

Lewis, M. A., DeVellis, B. M. and Sleath, B. (2002). Social influence and interpersonal communication in health behavior. In: K. Glanz, B. K. Rimer and F. M. Lewis (eds), *Health Behavior and Health Education: Theory, Research, and Practice*, 3rd edn. San Francisco, CA: John Wiley & Sons, pp. 240–64.

Lyles, C. M., Kay, L. S., Crepaz, N. *et al.* (2007). Best-evidence interventions: findings from a systematic review of HIV behavioral interventions for US populations at high risk, 2000–2004. *Am. J. Public Health*, 97, 133–43.

Marlatt, G. A. and George, W. H. (1984). Relapse prevention: introduction and overview of the model. *Br. J. Addictions*, 79, 261–73.

McNeil, B. J., Keeler, E. and Adelstein, S. J. (1975). Primer on certain elements on medical decision making. *N. Engl. J. Med.*, 293, 211–15.

Miller, W. R. and Rollnick, S. (1991). *Motivational Interviewing: Preparing People to Change Addictive Behavior*. New York, NY: Guilford Press.

Mize, S. J. S., Robinson, B. E., Bochting, W. O. *et al.* (2002). Meta-analysis of the effectiveness of HIV prevention interventions for women. *AIDS Care*, 14, 163–80.

Munro, S., Lewin, S., Swart, T. *et al.* (2007). A review of health behaviour theories: How useful are these for developing interventions to promote long-term medication adherence for TB and HIV/AIDS? *BMC Public Health*, 7, 104.

Neumann, M. S., Johnson, W. D., Semaan, S. *et al.* (2002). Review and meta-analysis of HIV prevention intervention research for heterosexual adult populations in the United States. *J. Acquir. Immune Defic. Syndr.*, 30, S106–17.

Onken, L. S., Blaine, J. D. and Battjes, R. J. (1997). Behavioral therapy research: a conceptualization of a process. In: S. W. Henggeler and A. B. Santos (eds), *Innovative Approaches for Difficult-to-Treat Populations*. Washington, DC: American Psychiatric Press, pp. 477–85.

Orr, D. P., Langefelt, C. D., Katz, B. P. *et al.* (1996). Behavioral intervention to increase condom use among high-risk female adolescents. *J. Pediatrics*, 128, 288–95.

Picciano, J., Roffman, R., Kalichman, S. *et al.* (2001). A telephone based brief intervention using motivational enhancement to facilitate HIV risk reduction among MSM: a pilot study. *AIDS Behav.*, 5, 251–62.

Porter, R. (1997). *The Greatest Benefit to Mankind. A Medical History of Humanity from Antiquity to the Present*. London: HarperCollins.

Prochaska, J. O. and Velicer, W. F. (1997). The transtheoretical model of health behavior change. *Am. J. Health Promot.*, 12, 38–48.

Prochaska, J. O., Redding, C. A., Harlow, L. L. *et al.* (1994). The transtheoretical model of change and HIV prevention: a review. *Health Educ. Q.*, 21, 471–86.

Prochaska, J. O., Redding, C. A. and Evers, K. E. (2002). The transtheoretical model and stages of change. In: K. Glanz, B. K. Rimer and F. M. Lewis (eds), *Health Behavior and Health Education: Theory, Research, and Practice*, 3rd edn. San Francisco, CA: John Wiley & Sons, pp. 99–120.

Robles, R. R., Reyes, J. C., Colon, H. M. *et al.* (2004). Effects of combined counseling and case management to reduce HIV risk behaviors among Hispanic drug injectors in Puerto Rico: a randomized controlled study. *J. Subst. Abuse Treat.*, 27, 145–52.

Roehr, B., Gross, M. and Mayer, K. (2001). Creating a research and development agenda for microbicides that protect against HIV infection. *amfAR AIDS Research*, 6, 7–8, Jun.

Rogers, R. W. (1975). A protection motivation theory of fear appeals and attitude change. *J. Psychol.*, 91, 93–114.

Rogers, R. W. (1983). Cognitive and physiological processes in fear appeals and attitude change: a revised theory of protection motivation. In: J. Cacioppo and R. Petty (eds), *Social Psychophysiology*. New York, NY: Guilford Press, pp. 153–76.

Rosser, B. R. S. (1990). Evaluation of the efficacy of AIDS education interventions for homosexually active men. *Health Educ. Res.*, 5, 299–308.

Rotheram-Borus, M. J., Cantwell, S. and Newman, P. A. (2000). HIV prevention programs with heterosexuals. *AIDS*, 14, 59–67.

Rotheram-Borus, M. J., Swendeman, D., Comulada, W. S. *et al.* (2004). Prevention for substance-using HIV-positive young people: telephone and in-person delivery. *J. Acquir. Immune Defic. Syndr.*, 37, S68–77.

Rounsaville, B. J., Carroll, K. M. and Onken, L. S. (2001). NIDA's stage model of behavioral therapies research: getting started and moving on from I. *Clin. Psychol. Sci. Practice*, 8, 133–42.

Semaan, S., Des Jarlais, D. C., Sogolow, E. *et al.* (2002). A meta-analysis of the effect of HIV prevention interventions on the sex behaviors of drug users in the United States. *J. Acquir. Immune Defic. Syndr.*, 30, S73–93.

Shattock, R. J. and Moore, J. P. (2003). Inhibiting sexual transmission of HIV-1 infection. *Nat. Rev. Microbiol.*, 1, 25–34.

Sheppard, B. H., Hartwick, J. and Warshaw, P. R. (1998). The theory of reasoned action: a meta-analysis with recommendations for modifications and future research. *J. Consumer Res.*, 15, 25–343.

Shoptaw, S., Reback, C. J., Peck, J. A. *et al.* (2005). Behavioral treatment approaches for methamphetamine dependence and HIV-related sexual risk behaviors among urban gay and bisexual men. *Drug Alcohol Depend.*, 78, 125–34.

Singer, M. and Snipes, C. (1992). Generations of suffering: experiences of a treatment program for substance abuse during pregnancy. *J. Health Care Poor Underserved*, 3, 222–34.

Skinner, B. F. (1938). *The Behavior of Organisms: An Experimental Analysis*. Englewood Cliffs, NJ: Appleton Century Crofts.

Stall, R. and Purcell, D. W. (2000). Intertwining epidemics: a review of research on substance use among men who have sex with men and its connection to the AIDS epidemic. *AIDS Behav.*, 4, 181–92.

Stall, R., Paul, J. P., Greenwood, G. *et al.* (2001). Alcohol use, drug use and alcohol-related problems among men who have sex with men: the Urban Men's Health Study. *Addiction*, 96, 1589–601.

Starace, F., Massa, A., Amico, K. R. *et al.* (2006). Adherence to antiretroviral therapy: An empirical test of the information-motivation-behavioral skills model. *Health Psychol.*, 25, 153–62.

Stephens, R. C., Feucht, T. E. and Roman, S. W. (1991). Effects of an intervention program on AIDS-related drug and needle behavior among intravenous drug users. *Am. J. Public Health*, 81, 568–71.

Sterk, C. E., Theall, K. P. and Elifson, K. W. (2003). Effectiveness of a risk reduction intervention among African-American women who use crack cocaine. *AIDS Educ. Prev.*, 15, 15–32.

St Lawrence, J. S., Eldrige, G. D., Shelby, M. C. *et al.* (1997). HIV risk reduction for incarcerated women: a comparison of brief interventions based on two theoretical models. *J. Consult. Clin. Psychol.*, 65, 504–9.

Wade, D. T. (2004). Do biomedical models of illness make for good healthcare systems? *Br. Med. J.*, 329, 1398–401.

Weinstein, M. and Fineberg, H. (1980). *Clinical Decision Analysis*. W.B. Philadelphia, PA: Saunders.

Couples' voluntary counseling and testing

9

Kathy Hageman, Amanda Tichacek and Susan Allen

Until a preventive vaccine is found, strategies that attempt to modify behavior remain the cornerstone of HIV prevention efforts. Globally, most HIV transmissions occur among heterosexual cohabiting couples (Trask *et al.*, 2002; Dunkle *et al.*, 2008). Although cohabitation has been identified as a risk factor for HIV almost since the start of the epidemic among gay (DeGruttola *et al.*, 1989; Doll *et al.*, 1991; Remien *et al.*, 1995) and hemophiliac (Pitchenik *et al.*, 1984; Kreiss *et al.*, 1985; Goedert *et al.*, 1987) couples, couple-level prevention programs have not been widely incorporated into the global prevention campaigns (Carael *et al.*, 1988; Allen *et al.*, 1991a, 1992a; Hunter *et al.*, 1994). Couples remain a largely overlooked population for HIV prevention strategies (Painter, 2001). Although efficacy for voluntary counseling and testing (VCT) has been established (Kamenga *et al.*, 1991; Padian *et al.*, 1993; The Voluntary HIV-1 Counseling and Testing Efficacy Study Group, 2000) and individual-level serostatus knowledge has been promoted globally, the expansion and large-scale uptake of more effective and efficient prevention VCT models for higher-risk populations has yet to be translated into public health practice. With less than 1 percent of African couples having been jointly tested for HIV (Chomba *et al.*, 2007), awareness of one's own and one's partner's serostatus has not been strongly emphasized, although HIV-discordant stable relationships remain one of the world's most vulnerable populations (De Cock *et al.*, 2002).

Prevention through behavior change remains the best tool to control the epidemic

Behavior change at the partnership level is critical to the prevention of HIV among sexually active couples, and cannot be achieved without mutual serostatus

240

knowledge (Allen *et al.*, 1999). At the individual level of VCT, uncertainty remains regarding the success of risk reduction and behavior change (Cates and Handsfield, 1988; Coates *et al.*, 1988; Higgins *et al.*, 1991; Wolitski *et al.*, 1997; Ickovics *et al.*, 1998; Weinhardt *et al.*, 1999), including limited disclosure to sexual partner (17–87 percent) (Keogh *et al.*, 1994; van der Straten *et al.*, 1995; Antelman *et al.*, 2001; Maman *et al.*, 2003). The most desired outcomes for VCT – risk reduction and behavior change – occurred among HIV-positive participants and HIV-discordant couples (where one partner is HIV-positive and the other is HIV-negative), while HIV-negative individuals did not modify their behavior (Wolitski *et al.*, 1997).

As the largest risk group for HIV (Allen *et al.*, 1991a; McKenna *et al.*, 1997; Allen, 2005a; Bunnell *et al.*, 2005) and unintended pregnancies (Allen, 2005b), cohabiting couples in high-HIV-prevalence areas experience the most daily infections (Allen *et al.*, 1991a, 2003; Fylkesnes *et al.*, 1997; McKenna *et al.*, 1997; Painter, 2001; Malamba *et al.*, 2005), contributing 56–93 percent of incident heterosexual infections in Africa (Dunkle *et al.*, 2008). Each year, it is estimated that 20–25 percent of African serodiscordant couples who do not know their HIV-discordant status transmit the virus to the negative partner (Allen *et al.*, 1992b; Hira *et al.*, 1990).

Considered a "high leverage HIV prevention intervention" (Painter, 2001), couples' voluntary counseling and testing (CVCT) is the most effective behavioral intervention to prevent HIV transmission in this at-risk population (Allen *et al.*, 2003), with the potential to avert more than two-thirds of new HIV infections among urban African men and women (Dunkle *et al.*, 2008). Although grounded in 20 years of evidence-based research establishing efficacy and acceptability of CVCT within diverse cultures and populations, CVCT continues to be one of the most greatly underutilized prevention strategies globally, particularly in high-prevalence areas, despite the initial publications regarding discordant couples in several African countries in the early 1990s (Chomba *et al.*, 2007).

CVCT and correct and consistent condom use: what is known to work at the dyad level

The established HIV prevention paradigm of abstinence, monogamy and condom use ("ABC") has limited benefit to couples who are unaware of their mutual HIV serostatus. Abstinence is not an appropriate counseling message for sexually active couples, and there is no protective benefit to monogamy among discordant couples (Allen *et al.*, 1992a; Central Statistical Office Zimbabwe and Macro International Inc., 2000; Newmann *et al.*, 2000; Agha *et al.*, 2002; Chatterjee and Hosain, 2006). It is vitally important that HIV prevention messages acknowledge that being faithful to one's spouse is only protective once both partners are aware that they have the same serostatus and remain mutually monogamous (Chomba *et al.*, 2007). The promotion of monogamy without mutual serostatus knowledge is misleading, and an insufficient strategy for HIV prevention. Furthermore without mutual serostatus knowledge, condom use is negligible within stable relationships (Allen *et al.*, 1992b; Adentunji, 2000; Ali *et al.*, 2004; Maharaj and Cleland, 2004; Bunnell *et al.*, 2005).

Why CVCT is important and why it is not happening

The success of CVCT is driven by the involvement of both partners in learning each other's serostatus, removing the primary challenge of serostatus disclosure for couples, and by the couple jointly receiving serostatus-specific risk-reduction counseling (concordant HIV-negative, concordant HIV-positive, or HIV-discordant) by a trained CVCT counselor who is also able to provide psychosocial support, (Painter, 2001; Glick, 2005). As the appropriate intervention message is dependent upon the couples' mutual serostatus, counseling a couple and their respective relationship is different from counseling separate individuals. It is crucial for the counseling process to include the adoption of a risk-reduction strategy that *both* individuals in a couple are comfortable with and able to implement with limited difficulty. Concordant HIV-negative couples are advised to remain monogamous with each other, and to use condoms with any outside partners. Concordant HIV-positive couples are advised to use condoms with each other to prevent exposure to different strains of HIV, as well as with outside partners to prevent transmission. For HIV-discordant couples, correct and consistent condom use becomes the primary prevention strategy to protect the HIV-negative partner from infection. Both concordant HIV-positive and discordant couples are counseled on the benefits of dual-method use, with long-acting contraceptives providing additional protection against unplanned pregnancies.

Globally, couples are rarely tested together. The incorporation of CVCT as an integral part of the ABC prevention model would greatly strengthen the protective benefits of the model, and also provide a critically-needed and essential *first* step for all sexually active couples prior to any decision-making regarding a couple's strategy for HIV prevention within the partnership.

In addition to providing a non-judgmental and power-balanced environment for dialogue between partners, CVCT also:

- supports open communication between partners;
- removes the barrier of partner notification;
- increases condom use and condom skills;
- encourages couples to take steps to maintain their health as a couple;

- prevents transmission of HIV and sexually transmitted infections (STIs), with the correct and consistent use of condoms;
- provides positive living counseling, emotional support and social service referral for couples with positive partners;
- reduces stigma associated with HIV/AIDS;
- allows a couple to plan for their own and their family's future (Mendenhall *et al.*, 1997; Stephenson *et al.*, 2008);
- acknowledges the desire for additional children and fosters discussion of prevention of mother-to-child transmission (PMTCT); and
- promotes the identification of pregnant women and HIV-positive persons in need of PMTCT and antiretroviral (ARV) treatment and care.

The development of CVCT: listening to the public's needs

In Rwanda, one of the first countries in Africa to implement voluntary HIV testing, CVCT began in 1986 with the recruitment and enrollment of women into an HIV prospective study among those attending the antenatal clinic at the Centre Hospitalier Kigali in the capital city of Rwanda (Carael *et al.*, 1988). At post-test counseling, many women inquired whether their steady sexual partners could also attend the clinic for HIV voluntary counseling and testing and to see the AIDS education video. With no incentives or benefits offered, approximately one-third of the male partners came for VCT. Although evaluating the effectiveness of voluntary counseling and testing was not the original intent of the prospective study, the changes in reported behavior were striking (Allen *et al.*, 1992a). Two years after the women had learned of their serostatus and received counseling, those who were HIV-negative and whose partners had not participated in VCT had a small reduction in seroconversion, from 4.1/100 person-years (py) to an initial 3.4/100-py seroconversion rate. In comparison, the seroincidence rate among women whose partners had participated dropped by more than half (from 4.1/100 py to 1.8/100 py, $p < 0.04$). Although the women's participation in the program led to increased condom use, participation of the men prompted far more drastic changes. Results from this non-randomized assessment of the impact of men's participation in counseling and testing confirm that male participation is important in reducing a couple's HIV risk (Allen *et al.*, 1992b, 1993; Seed *et al.*,1995). Overall, 31 percent of couples in which both partners were tested for HIV were using condoms 1 year after VCT, in comparison to 16 percent if only the woman was tested. The impact was greatest among discordant couples: 3–9 percent of the discordant couples seroconverted annually after VCT, compared to an estimated 20–25 percent transmission rate in uncounseled discordant couples (Allen *et al.*, 1992b; Fideli *et al.*, 2001). There were other benefits as well. Gonorrhea declined by more than 50 percent in HIV-positive women (Allen *et al.*, 1992a), and the incidence of HIV in concordant negative couples dropped to less than 0.5 percent per year (Allen *et al.*, 2003; Roth *et al.*, 2001). Combining CVCT and family planning messages with HIV/STI prevention has also been shown to reduce unplanned pregnancies by half (Allen *et al.*, 1992b; King *et al.*, 1995). All couples – regardless of HIV test results – may benefit from the counseling and testing, as the desired behavior depends on their mutual HIV test results which is promoted by the process of CVCT (Allen *et al.*, 1992a, 2003).

Between 1988 and 1991, 60 HIV-discordant couples were identified during VCT, of whom 53 were followed for an average of 2.2 years. The proportion of discordant couples using condoms increased from 4 percent to 57 percent after 1 year of follow-up. During follow-up, 2 of the 23 HIV-negative men and 6 of the 30 HIV-negative women seroconverted (seroconversion rates of 4 and 9 per 100 py). The rate among women was less than half that estimated for similar women in discordant couples whose partners had not been serotested. Condom use was less common among those who seroconverted (0/2 seroconvertors vs 20/21 non-converting men, $p = 0.01$; 2/6 seroconvertors vs 18/24 non-converting women, $p = 0.14$).

Based upon these results, couples' VCT centers in Kigali, Rwanda (1992), Lusaka, Zambia (1994) and Copperbelt, Zambia (2003) have been established under the umbrella organization of the Rwanda Zambia HIV Research Group (RZHRG) to research the best ways to deliver and disseminate sustainable CVCT. Among couples living in Kigali, 10 percent have either one or both partners living with HIV (Institut National de la Statistique du Rwanda (INSR) and ORC Macro, 2006). In Lusaka and the Copperbelt area of Zambia, HIV-discordant couples make up approximately 20 percent of cohabiting couples (Allen *et al.*, 2003; Kempf *et al.*, 2008; Lingappa *et al.*, 2008), resulting in an even greater need for dyad-level interventions. RZHRG has also contributed to the development of best practices in CVCT as experienced counselors from the Rwanda and Zambia sites participated in the development of a detailed procedure manual and training program with joint sponsorship from the Centers for Disease Control (CDC), the National Institutes of Mental Health (NIMH) and the Liverpool School of Tropical Medicine (www.cdc.gov/nchstp/od/gap/CHCTintervention). As a result of the growing recognition of the importance of couple-level HIV prevention strategies, this past year has seen the addition of couples' counseling to the President's Emergency Plan for AIDS Relief (PEPFAR) priorities for VCT (United States President's Emergency Plan for AIDS Relief, 2007).

Towards sustainable HIV prevention: structural and economic aspects, psychosocial elements, and social norms

The main obstacles to couples' VCT are identifiable and can be overcome. Twenty years of research and implementation in a research setting has confirmed that

CVCT and correct and consistent condom use provide a sustainable and successful method of prevention of HIV infection. Unfortunately, although technical advances in rapid HIV-testing technology and the expansion of PMTCT and ARV programs have increased access to HIV testing, most clients are managed as individuals. The likelihood of testing both partners is greatly increased when CVCT is offered (Weinhardt *et al.*, 1999), but couples remain largely overlooked in the global expansion of VCT.

Structural barriers: lack of perceived risk at multiple levels resulting in low demand and low supply

Most adults, including many health-care personnel and policy-makers, do not realize that one partner in a long-term relationship may have HIV while the other does not (Bakari *et al.*, 2000). Many are also not aware that (1) the majority of new HIV infections occur in married couples (McKenna *et al.*, 1997; Dunkle *et al.*, 2008), (2) these transmissions could be prevented by couples' VCT (Higgins *et al.*, 1991), and (3) the negative consequences of couples' VCT are uncommon and far outweigh the benefits of mutual serostatus knowledge (Grinstead *et al.*, 2000, 2001; Kilewo *et al.*, 2001; Maman *et al.*, 2001; Varghese *et al.*, 2001). Although knowledge of the HIV blood test has been widespread for over a decade, and individuals may be aware of their increased risk of HIV specifically due to their partnership, few VCT and health centers are trained to deliver CVCT. This is a greatly missed opportunity for couples' VCT globally (de Zoysa *et al.*, 1995), particularly among pregnant women seeking antenatal care. Over the past 5 years, there has been successful global promotion and service provision of antenatal care and clinics. On average, African women of reproductive age become pregnant once every 3–4 years; therefore, with the incorporation of CVCT into antenatal care and clinics, every African woman who has sought antenatal care in the past 5 years could also have participated in CVCT. This is, and continues to be, a vitally missed opportunity to prevent the transmission of HIV to both parents through risk-reduction counseling.

Low knowledge regarding HIV discordance within partnerships, perception that marriage and/or cohabitation is "safe" from HIV transmission, stigma, gender inequality, and lack of promotion of available CVCT services has resulted in a low demand for CVCT services paired with low supply due to minimal global funding and promotion, thus resulting in underutilization of this highly effective prevention strategy (Allen *et al.*, 2007a). The investment in and training of CVCT personnel is an addressable obstacle at the national and local levels as funding agencies, policy-makers and local health-care professionals become more aware of the increased risk for HIV that cohabitation brings to an individual, as well as the success of CVCT implementation. This cycle of missed opportunities due to low demand and low supply must be broken in order to successfully address the global needs of couples, to protect their individual health and to end the risk of unknowingly having unprotected sexual contact with an HIV-positive partner in the context of a stable relationship.

Promotions-based supply and demand

A variety of CVCT promotion strategies have been utilized in the past 20 years in Rwanda and Zambia. These include mass media campaigns (newspaper ads, radio announcements, billboards and posters), door-to-door invitations delivered by community workers and, the most recent and successful approach, recruiting members of the community to promote the benefits of CVCT and available services. We found paid community workers to be a successful short-term promotion model for the duration of employment, but once this was discontinued, the attendance of couples at the testing sites dropped significantly, resulting in a cost-prohibitive and unsustainable community-level promotion model (Chomba *et al.*, 2007). To ensure ongoing community awareness and financial sustainability, an approach was designed that identified Influential Network Agents (INAs) and Influential Network Leaders (INLs). INAs and INLs come from four distinct sectors in the community: health, private, non-governmental (NGO) or other community-based (CBO) organizations, and religious (Allen *et al.*, 2007a). The INA/INL model identifies influential community members to make public endorsements about CVCT, to speak with couples in groups and/or one-on-one, and to invite couples for testing. INAs are working-age men and women who are trained to invite cohabiting couples for CVCT using their established networks, such as at the workplace, church, market or neighborhood. INLs have greater influence as community leaders, and may have a larger sphere of influence and the ability to reach larger groups of people. INLs include physicians, religious leaders, teachers, business leaders and senior members of NGO/CBOs. INLs help with the recruitment of INAs, and offer their promotional support and endorsement to INAs during CVCT promotion activities. Our research has also found that after 4 months of active promotions by an INA and approximately a year for INLs, saturation of their community networks occurs and the number of couples they invite for CVCT declines. To maintain attendance at or near the capacity of the CVCT centers, we actively recruit and train a new cohort of INAs and INLs as needed to replace the cohort members that have exhausted the network of people they may influence. In order to train the INAs and INLs appropriately, RZHRG has developed a 4-day training curriculum on how to reach out and speak to couples, answer questions about HIV and invite couples to CVCT.

CVCT invitations distributed by INAs provide information about the location and times of the fixed and mobile CVCT sites and CVCT services at the district hospital. INAs are reimbursed according to the number of couples who attend CVCT, regardless of whether a couple decides to test. To supplement the individual-level promotions of the INAs and INLs, community-level promotions of billboards, posters, theater performances and radio announcements are utilized.

Participatory approaches were used for the development of the INA/INL recruitment model; focus groups were conducted to elicit better understanding of issues within communities and to revise the model accordingly. Ongoing monitoring is provided for each cohort of INAs through regular group meetings to provide feedback on their experiences. This feedback is incorporated into the next training sessions, and into the structure of the promotions program. By using information from the influential community members, RZHRG has been able to identify successes and barriers to promotions, and tailor activities to meet the specific needs of each community.

Promotion success has differed greatly by city, most likely due to differences in language (there is one primary language in Kigali and Copperbelt, but many languages in Lusaka), transport infrastructure and city size. In Kigali, Rwanda, 30 percent of couples tested were invited by INAs and 70 percent were walk-ins (Allen *et al.*, 2007a). The primary ways in which Kigali couples heard of CVCT were from a previously tested couple (56.8 percent), radio (47.7 percent), INA (33.5 percent), posters (32.5 percent) and friends (12.8 percent). During the same period in Lusaka, the majority (72.5 percent) of couples heard of CVCT through an INA, followed by radio (26.5 percent), TV (17.2 percent), friends (14.0 percent), posters (14 percent) and previously tested couples (10.3 percent). In general, predictors of successful invitations included inviting couples together and/or involving the man in the invitation process, inviting couples known to the INA, issuing invitations after public announcements, and delivering invitations at the INAs home or at the workplace. There is a need for widespread community and NGO advocacy to successfully target couples for testing (Chomba *et al.*, 2007).

Due to the significant role INAs contribute to CVCT promotion, we investigated the effectiveness of each of the four sectors. In Zambia, health-related INAs had the greatest success of recruitment, with attendance of 34 percent of couples, followed by NGO/CBO-related INAs, who recruited 29 percent of couples (Allen *et al.*, 2007a). Religious-sector and private-sector related INAs had similar rates, 19 percent and 18 percent respectively. Systematic efforts to use church networks to promote CVCT have not been highly successful, although 80 percent of the Zambian population self-identifies as Christian, and churches are a place that couples attend together. We hypothesize that hearing about couples'-testing in a church may provoke a paradoxical negative response, and denial that HIV risk may be present due to conservative viewpoints about abstinence and fidelity. Likewise, we found a very low number of religious INLs interested in participating in CVCT (8 percent) themselves; it is possible that the lack of "leading by example" may also be contributing to the low recruitment success rate from this sector. More research is needed regarding religious dynamics and their effects on public perception of CVCT to strengthen prevention efforts for couples.

Financial and logistical barriers

In Zambia and Rwanda, we have been able to successfully overcome financial and logistical barriers to CVCT through the implementation of same-day services, facilitation of transport, and provision of childcare at the CVCT center. Rapid HIV testing allows clients to avoid a second trip, and prolonged anxiety about test results (Mashu *et al.*, 1996; McKenna *et al.*, 1997). For couples, organizing simultaneous free time, childcare, and transport costs is not a trivial exercise, and doing so twice for pre- and post-test counseling is daunting. Unfortunately, despite the availability of rapid HIV tests, same-day VCT services are not often provided (Plourde *et al.*, 1998). To reach a larger number of couples, it is vitally important for CVCT to become integrated into the medical care infrastructure, including health clinics, hospitals, and PMTCT, ARV, STI treatment and care services.

Psychological factors

In addition to logistical and financial obstacles, numerous psychosocial factors impact a couple's decision to participate in CVCT. These include: a couple's communication and negotiation abilities; self-efficacy to keep the negative partner(s) free of the virus; the perceived risk of infection versus the perceived safety of being in a cohabiting relationship; and the lack of personal sexual control. Providing a venue in which gender and power dynamics in the relationship can be addressed is particularly important in environments where women are not accustomed to participating in sexual decision-making and/or men generally control both sexual and economic decision-making (Larson, 1989; Armstrong, 1992; Bizimungu, 1992; van der Straten *et al.*, 1995; WHO/UNFPA & World Bank, 1995). HIV-negative women with untested or seronegative partners are the least likely to use condoms, or to discuss or attempt to negotiate condom use (van der Straten *et al.*, 1995). Couple communication has been associated with condom use, but only when the discussion was specific to sexual behavior, STIs and using condoms (van der Straten *et al.*, 1998). CVCT's facilitation of effective risk-reduction strategies helps to mediate lack of communication and/or agreement within the partnership.

Social norms: alcohol use and desire for children

Alcohol use in men (Coldiron *et al.*, 2008) and the desire to conceive (Mark *et al.*, 2007) are two predictors of unprotected sex and HIV transmission risk in couples. A crucial step in HIV prevention for couples is the opportunity to discuss HIV-related issues and sexual behavior decisions based upon their dyad-specific HIV status, as well as to develop risk-reduction strategies for situations relating to alcohol use (Kalichman *et al.*, 2007), having outside partners, and/or the desire to have children (Allen *et al.*, 1991a; van der Straten *et al.*, 1995; Painter, 2001). CVCT provides an opportunity for couples to discuss their specific situation, while the counselor is able to specifically address these concerns/problems and help develop additional coping strategies.

Although being HIV-positive is largely perceived by the medical community as a barrier to reproduction, and HIV-positive persons are advised not to become pregnant (Thornton *et al.*, 2004), numerous studies have found that the desire for children is comparable to that in the general population (McGrath *et al.*, 1993; VanDevanter *et al.*, 1999; Ryder *et al.*, 2000; Chen *et al.*, 2001; Klein *et al.*, 2003; Panozzo *et al.*, 2003; Thornton *et al.*, 2004), particularly within sub-Saharan Africa, where a high value is put upon childbearing (Allen *et al.*, 1993, 2003, 2005b; King *et al.*, 1995; Mark *et al.*, 2007). Social norms may inhibit decisions to limit fertility among HIV-discordant couples, particularly in cultures where a woman's role largely centers on childbearing. Much needed recognition of ongoing fertility desires and intention to become pregnant further demonstrates the limited research and efforts to address the risks and needs of HIV-discordant couples at a global level.

Controversies: past and present

The development of CVCT has not been without challenges. At the most basic level, what is a couple? Definitions range from any two sexually active individuals, to those who are married and/or cohabiting. Understanding and being prepared to handle the variety of relationship types is crucial to creating a supportive, non-judgmental environment, and providing counseling to address successfully the couple's emotional and decision-making needs following CVCT. RZHRG's model of mutual disclosure developed from the initial experience of providing married men and women with VCT as individuals, who were then encouraged to share their HIV results with their partners at home. Over time, it was realized that some couples were unable to achieve mutual and accurate serostatus disclosure; therefore, mutual disclosure and couple-specific risk reduction were integrated into the CVCT model. As sexual activity involves two people, it is necessary to ensure that both partners have the relevant information and skills to negotiate HIV prevention techniques successfully.

As the importance of addressing the couple-unit became a crucial component in the success of long-term prevention for couples, questions arose as to whether to address couples coming for CVCT as a group, or each couple-unit alone. We have found that a group environment for the HIV education and question-and-answer sessions has been highly beneficial to couples, as it provides an opportunity to hear each other's questions and concerns and an opportunity to dispel myths and rumors that are circulating in the neighborhood regarding voluntary counseling and testing. In addition to reducing the amount of time necessary for pre-test counseling, we have also found that the group education and discussion session significantly contributes to the comfort of a couple's decision to seek CVCT services. Although Rwanda and Zambia are culturally distinct, in both settings the group session provides an active and supportive discussion among the participants. Risky behaviors and alternative options are discussed as a group, and allow couples to benefit from the questions and responses posed by the other participants. As a group, couples also discuss the implications of each combination of possible serostatus results, again benefiting from other's responses.

Initially, the idea of joint couples counseling and testing was met with apprehension at many levels. Concerns included confidentiality regarding the sharing of results, perceived lack of interest among couples in sharing their own serostatus results and/or knowing a partner's results, and the possibility of marriage dissolution and domestic violence based upon a couple's results. Although valid concerns, these must be weighed against the benefits of mutual serostatus knowledge. CVCT (at our sites as well as others) has been administered in two ways: (1) the couple completes all steps together, including the disclosure of the test results and risk-reduction counseling; or (2) not all steps are completed together as a couple, particularly the disclosure of the test results. As mentioned earlier, we found that when couples were unaware of their mutual status and did not participate in joint risk-reduction counseling, inaccurate or no serostatus disclosure would occur at times. Similar to other studies, we have not found an increased risk of HIV infection, domestic violence, marriage dissolution, or sexual coercion due to having participated in CVCT (van der Straten *et al.*, 1998; Maman *et al.*, 2003).

Logistical aspects also need to be considered when working with couples. Having both individuals available at the same time requires organization of their schedules. Many couples prefer to attend CVCT in a discreet manner, so that family and friends are not aware of their intentions. Other logistical obstacles include time, transport and childcare. Time and transport difficulties can be minimized through same-day pre and post-test counseling, which eliminates the need for a second trip. Initially, resistance to same-day results centered on concerns about the accuracy of rapid tests, and about individuals not being emotionally prepared to receive their test results immediately. However, we have found that by the time couples come to our sites, they have made the mutual emotional commitment to know their serostatus and are ready to receive the results. Delaying this process not only compounds the emotional burden of not knowing the results, but also contributes to an increased number of couples who do not return at a later date to receive their mutual results.

Whereas the provision of childcare onsite is becoming a more common practice at research sites that cater to couples and/or women, this service has yet to be routinely provided by service-based providers. Likewise the provision of incentives such as transport reimbursement to couples to participate in CVCT is desirable until communities understand the importance of joint testing and the services become integrated into routine health care.

Best practices: a day in the life of a same-day CVCT clinic

RZHRG's main activities are conducted at stand-alone CVCT clinics offering couples a full-day program of HIV education, and mutual serostatus knowledge. We recognize the importance of and greatly promote the integration of CVCT into district health clinics, antenatal clinics, and other VCT and medical facilities. Therefore, the adoption of CVCT into other services will require changes to the best practices described below.

CVCT timetable

9.00–9.30   Check-in
9.30–10.30   Group HIV education and discussion of pre-test issues
10.30–12.30   Pre-test counseling

- Each couple meets with a counselor in a private room to discuss questions remaining from earlier group discussion, and how they as a couple will handle the possible results.
- The counselor performs a risk assessment and explains the testing process.
- The couple chooses whether to be tested, and blood is collected.

12.30–13:30   Couples are offered free lunch, regardless of whether they have chosen to be tested. Rapid HIV tests are run for couples who chose to test.
13.30–16.00   Post-test counseling

- Couples are given their individual results in separate envelopes in the presence of each other and are asked to share their results with each other and the counselor.
- The counselor facilitates a discussion regarding the most appropriate prevention strategies, based upon the couple's mutual serostatus.
- Couples sign a form to show that they have received their results, and are treated for syphilis, if needed.

- Couples who have doubtful/indeterminate test results are asked to return in 1 month for a follow-up test.
- Couples are given transportation money.

Group HIV education and discussion

Group education refers to the initial HIV/AIDS information session facilitated by a couples' counselor, and offers basic education about HIV, modes of transmission and risk-reduction behaviors, presents the possible test results, and offers couples the opportunity to ask questions about HIV/AIDS. Many of the people who come for CVCT have limited accurate information about HIV, and the provision of basic information is critical before any fruitful counseling can take place. In short, the group discussion is only one of many ways to help couples to "know" and "decide", as well as to prepare for their potential results.

The group education and discussion session is usually modeled on the following agenda:

- Introduction and welcome
- What is HIV and AIDS?
  - Modes of transmission (unprotected sex, blood-to-blood contact, mother-to-child transmission)
  - Prevention (ABC – abstinence, be faithful to sexual partner, condom use)
  - ARVs and the importance of nevirapine use in pregnant women
- What is CVCT?
  - The importance of HIV testing as a couple, as well as why a couple should stay together regardless of results
  - Types of HIV results and their benefits and implications: concordant negative $(-/-)$, concordant positive $(+/+)$, discordant $(+/-)$, indeterminate/doubtful/discrepant
- Couple's question and answer session
- Male and female condom demonstration
- Summary and close discussion.

Pre-test counseling and risk assessment

The aim of pre-test counseling is to assist couples to make an informed decision about whether to test. The pre-test counseling session also allows the counselor to conduct a couple-specific risk assessment. Most couples have already started considering and identifying their own risks, and are quite prepared to discuss them with the counselors. Regardless, this needs to be done sensitively, and does not necessarily require deep probing for details. Instead, the pre-test risk assessment should address the specific risks that the couple are willing to discuss at that point in time. A counselor may also choose, using his or her own discretion, to separate a particularly problematic couple during pre-test counseling. However, this should only be done as a last resort, as couples' VCT ultimately aims to empower the couple together.

During the pre-test counseling session, the counselor seeks to:

- Assess a couple's reasons for seeking testing
- Review each partner's basic understanding of HIV and modes of transmission
- Identify both partners' history of HIV testing
- Discuss and clarify the couple's understanding of different HIV test results
- Discuss how they, as a couple, will handle receiving each type of result, and what they can do to move forward as a couple
- Explore the couple's feelings about taking an HIV test and receiving results together
- Discuss the advantages/disadvantages of knowing their serostatus
- Help the couple to identify sources of support
- Discuss disclosure issues, including the importance of keeping one's own and one's partner's results confidential, unless the couple mutually agrees to disclose to other parties
- Assess the couple's readiness to be tested and to receive results together
- Describe the HIV test process.

Pre-test counseling also allows for the assessment of couple dynamics to ensure mutual readiness and intentions. For example, if a counselor suspects conflicts with one or both partners, that one or both partners are trying to use HIV testing as an excuse to leave the relationship, or suspects that there will be potential problems after leaving the counseling session, the counselor may defer testing and suggest instead that the couple think things over and come back at another time.

Post-test counseling (results session)

The post-test counseling session is when the couple receives their HIV test results. The couple exchanges their results slips, received in a sealed envelope, after each partner has viewed his or her individual test results. It is important to inquire with all couples if they understand their mutual results and the respective implications. To facilitate the communication process, it is also helpful to inquire whether they are surprised by the results, and for the counselor to mediate any blame or anger regarding past sexual behaviors. Likewise, the counselor is able to draw upon the support and disclosure discussion that occurred during pre-test counseling to help the couple to develop joint coping and risk-reduction strategies. It is crucially important that a counselor *never rushes* post-test counseling, and ensures enough time for the couple to absorb their results while offering support and advice when appropriate. Although the three possible mutual serostatus results require different counseling messages, inclusive in all messages is risk-reduction counseling.

Risk-reduction counseling: concordant negative serostatus (−/−)

Couples who share HIV-negative test results are counseled to remain monogamous or use condoms with any outside partners. If they have been mutually monogamous for 3 months prior to testing, they are informed that they do not need to use condoms when they are sexually active with each other. If either partner has had unprotected sex with an outside person in the last 3 months, then condoms should be used until 3 months have passed and retesting can occur. Couples are also informed that being HIV-negative does not mean immunization or protection from the virus, regardless of their past sexual practices, and are counseled on how to avoid future high-risk contacts.

Concordant positive serostatus (+/+)

Couples who share HIV-positive test results are counseled to use condoms with each other to prevent additional exposure to the virus, as well as with any outside relations to prevent transmission of HIV and acquisition of an STI. The counselor also discusses the importance of supporting each other, remaining committed to the relationship, coping strategies, and positive living. Positive living includes good nutrition, adequate social and emotional support, seeking health care immediately if an individual starts to feel ill, and the importance of remaining healthy for the family and children.

Discordant serostatus (male+/female− or male−/female+)

As with concordant positive couples, discordant couples are counseled regarding coping strategies, positive living and emotional support. Emphasis is placed upon the fact that the relationship does not need to end, and the HIV-positive partner's ability to remain healthy with proper medical, social and nutritional support. Prevention of transmission to the negative partner is promoted through proper precautions, such as condoms, monogamy and/or abstinence. It is important to clarify that beginning a sexual relationship with an outside partner is not safe, as that partner might also be HIV-positive. Both partners must also understand that the HIV negative partner is not immune or protected from the virus regardless of how long the couple has been having unprotected sex. Learning of a discordant status is a very difficult and emotion-filled issue, and requires delicate handling in order to defuse blame, anger, shame and confusion. Each individual and couple's reaction is unique, and should be handled appropriately.

Monitoring and evaluation of CVCT services

Due to the complexity of CVCT delivery, the administration of CVCT at stand-alone testing sites or integrated with other health-care services, and cultural variations, evaluation is a vital aspect to ensure quality assurance and satisfaction.

UNAIDS and UNFPA are just two agencies that have evaluation kits available online that include monitoring and evaluation (M&E) forms and protocols. *Tools for Evaluating HIV Voluntary Counseling and Testing* (UNAIDS, 2000) covers eight areas for M&E, ranging from set-up of services to site and logistical considerations to training and skills assessment. *Integrating HIV Voluntary Counseling and Testing Services into Reproductive Health Settings* (UNFPA & International Planned Parenthood Federation, 2004) includes a chapter on monitoring and evaluation that includes assessing mobilization efforts, counselling and referral skills, and management-related issues. It is important that each testing site identifies its specific needs for assessment derived from its established protocols that can be replicated over time to provide a longitudinal perspective of the sites strengths, weaknesses, gains and lapses. M&E can be conducted on a designated routine basis (annually, bi-annually, etc.) or be ongoing through the establishment of an M&E team.

The remainder of this section will discuss the following four suggested monitoring and evaluation components (Shah *et al.*, 2008): client interviews, counselor written and oral HIV-knowledge and skills exams, skill-based counselor evaluations, and counselor feedback interviews.

Client interviews

To ensure high client satisfaction, exit interviews are conducted to assess the clients' understanding of HIV prevention and transmission; the counseling and risk-reduction messages received; and satisfaction with the CVCT services, confidentiality, and the counselor assigned to their visit. Although couples are counseled together, client exit interviews should occur separately to ensure an equal opportunity for each individual to share his or her knowledge and opinions. Selection of the facilitator for client interviews is another important consideration, to ensure non-biased client reporting. Suggestions include a senior counselor or a designated M&E member of the community trained in CVCT procedures but seen as separate from the service provider (Taegtmeyer and Doyle, 2003).

To evaluate and monitor the quality and consistency of CVCT counselors' work, three counselor assessments are suggested. At RZHRG, these assessments are based upon our standard operating procedures (SOPs).

Counselor written and oral HIV knowledge and skills exams

The written exam assesses knowledge of HIV transmission and prevention, CVCT counseling protocols, living positively, end-of-life issues, and client confidentiality. For research-based sites, or facilities that have integrated services with a research site, examination of the knowledge of research concepts and participant consent procedures must also be assessed for all employees on a routine basis.

The oral exam requires counselors to recall the specific steps to each of the counseling protocols, including group discussion, pre-test counseling, post-test counseling for all serostatus combinations, and appropriate referrals.

Skill-based counselor observations

To ensure the highest quality and consistency of counseling protocols, all counselors undergo skill-based evaluations. Developed from our SOPs, observation checklists are completed by senior counselors based upon each counselor's delivery of pre- and post-test counseling sessions.

Counselor feedback sessions

To provide suggestions regarding protocols and work-related needs (e.g. training, workload and work environment), individual and/or group feedback sessions are held. At the individual level, feedback sessions provide counselors with an opportunity to reflect upon their performance and ability to provide client-centered counseling for couples. For integrated CVCT sites, it is equally important to provide feedback sessions for the hosting health-care provider, such as antenatal or district clinic personnel, to assess the success of the integration of services, the understanding of CVCT procedures and referral processes, as well as identifying potential areas for improvement.

Topics for consideration based upon RZHRG's evaluation findings

After CVCT, couples were able to answer questions accurately regarding HIV knowledge and the counseling messages that they have received. Couples also felt comfortable with their counselor, trusted the confidentiality of the program, appreciated the amount of time that the counselor spent with them, and felt that their questions were answered clearly. Although some couples stated that they would have preferred an older counselor as he or she would have had more experience with marriage, or a female counselor as women are more sensitive, in general, clients reported being pleased with their counselor and counseling experience.

Regarding the skill-based written and oral counselor assessments and counselor observations, counselors had a substantial grasp of counseling theory and CVCT protocols. Areas for continued improvement included knowledge regarding country-available antiretrovirals and their side-effects, disease progression, and explaining inconclusive or indeterminate serostatus results (Au *et al.*, 2006). Examination results also identified the critical importance of ensuring that sufficient time is provided during the counseling sessions to prepare the couple psychologically for their serostatus results during pre-test counseling, as well as for the condom demonstration by the counselor and practice by both partners.

Counselor feedback has included requests for informational sheets for service referrals for ARVs, PMTCT, and living-positively support groups. Counselors have also acknowledged their lack of comfort when counseling a couple with inconclusive serostatus results, and have requested additional training and refresher courses for the abovementioned areas, as well as the latest information on HIV-related treatment and counseling techniques. Similar to others' findings, counselors acknowledged the emotional strain experienced in facilitating serostatus disclosure among couples, particularly with discordant and concordant positive couples, and suggested that scheduled meetings and/or debriefing sessions to discuss difficult cases would aid them professionally and personally (Ginwalla *et al.*, 2002; Taegtmeyer and Doyle, 2003). The establishment of a mentoring system between experienced and new counselors has also been suggested.

Twenty years of barriers and progress/concrete gains as CVCT evolves

Optimism remains regarding advancements in microbiocides and vaccines to reduce risk of sexual transmission, yet until a viable universal solution is available, prevention through behavior change and harm reduction is the only option. CVCT reduces transmission more than male circumcision does, and has a beneficial impact on transmission from men to women as well as from women to men. The cost per infection prevented is also lower, and there are no surgical complications. Even with such success, CVCT remains a significantly underutilized HIV prevention method globally.

Bringing couples together for mutual serostatus knowledge and serostatus-specific harm-reduction counseling should be the first step in the global approach to HIV prevention, and is a logical pairing with additional HIV and reproductive-related services, treatment and care. Service providers must come together to collectively target the people most at risk of transmission of HIV – individuals having sex with an HIV-positive partner, and couples who are having unprotected sex (Chomba *et al.*, 2007). The integration of CVCT with other health-care and treatment services, such as STI diagnosis, ARV treatment, prevention of mother-to-child treatment, and family planning, is vitally important, as these programs currently prioritize the individual rather than the couple (Were *et al.*, 2006; Guthrie *et al.*, 2007; PEPFAR, annual publication). In areas with a high prevalence of HIV/STI in heterosexual populations, the target audiences for HIV/STI and family planning services overlap broadly, and can benefit most from joint services (Sweeney *et al.*, 1992; Peterman *et al.*, 1996; Stover, 1996). Unfortunately, historical, philosophical and structural differences in the fields of family planning, mother–child health and HIV/STI pose obstacles to integration both in the US and in developing countries (Cates, 1992). Staff tend to see these categories as distinctly different (Daley, 1994). Service delivery styles also differ: family planning clinics often rely on a fact-giving approach, while HIV-testing services tend to emphasize client-centered counseling approaches (Becker and Ureno, 1996). Furthermore, with the scaling up of male circumcision in many countries as a HIV prevention method, all eligible men who have a steady sex partner should be encouraged to participate in CVCT to identify whether the male is HIV-negative and the steady sexual partner is HIV-positive prior to accepting the risk of the medical procedure.

In the capital cities of Kigali (Rwanda) and Lusaka (Zambia), same-day couples' VCT technology has successfully been transferred to antenatal care (Bakari *et al.*, 2000) and mobile weekend clinics, clearly showing that obstacles at the individual- and service-provider level can be overcome. Due to the high volume of antenatal clients during the week, the existing burden of ANC procedures, and limited clinic rooms and staff, the ANC environment did not allow for the inclusion of husbands in VCT procedures. Therefore, weekend CVCT sessions for couples were added through the use of mobile CVCT units, and invitations to attend CVCT on Saturdays and Sundays were distributed to ANC clients during the week. This system worked extremely well, as clients and their husbands did not have to take additional time off from work during the week, and space is readily available in clinics on weekends. Building upon research that found that VCT is feasible and acceptable at antenatal clinics in Africa (Bakari *et al.*, 2000; Stringer *et al.*, 2003; Etiebet *et al.*, 2004), we found the same to be true for the provision of CVCT services. Furthermore, the involvement of the male partner in VCT increased compliance with the single-dose nevirapine regimen to prevent mother-to-child transmission (Shutes *et al.*, 2008) and provided an opportunity to counsel couples on post-pregnancy contraception choices. In addition, leveraging the success of CVCT with family planning and antenatal clinics reduces not only pediatric AIDS cases but also the number of children who would become AIDS orphans and the adverse family consequences of both parents becoming ill or dying of HIV. Both capital cities continue to work with government leaders, funding agencies, service providers and community leaders to ensure self-sustaining couples' VCT programs within the antenatal system. Unfortunately, VCT is not being provided in the most effective manner. The onset of rapid testing nearly 8 years ago and the scaling up of antenatal testing, along with the natural cycling of pregnancies occurring every few years in international settings and particularly in Africa, has provided the opportunity to incorporate husbands into the antenatal mechanism to participate in couples' testing. This has been one of the greatest missed opportunities to date in the field of VCT.

Integral to the incorporation of CVCT into the larger medical infrastructure is the collaboration with local organizations and agencies to identify where ARVs and PMTCT services are being offered, and what is the best way to proceed with these referrals.

Vital to the restructuring of the global approach to HIV to include CVCT as the first step in prevention for couples is an extension of the UNAIDS & WHO recommendation for provider-initiated VCT (UNAIDS & WHO, 2007) to include CVCT for persons in cohabiting or steady relationships. Equally important is the need for greater understanding by couples themselves regarding the susceptible risk of infection associated with cohabitation and/or marriage so that CVCT becomes *couple- and provider-initiated*. Ideally, with appropriate education, successful promotion to couples, efforts to reduce stigma, and open discussion regarding ongoing fertility desires, CVCT can become a "standard request" for couples and the "standard of care" for medical professionals, particularly for medical professionals who treat antenatal couples and/or couples in which STIs, tuberculosis or an opportunistic infection has been diagnosed.

To achieve a societal norm that supports couple- and provider-initiated CVCT, tangible support for and endorsement of CVCT by governmental officials, funding and development agencies, policy-makers, non-governmental agencies, and community and faith-based leaders must occur (Chomba *et al.*, 2007). Unfortunately a gap remains between research, practice and public health policy: policy-makers are sometimes reluctant to support couples' VCT even when the prevention impact is understood (Allen *et al.*, 2007a). Reducing stigma and fear of repercussions among local policy-makers is crucial to the global advancement of CVCT as international and bilateral funding agencies take their cues from in-country policy-makers. Without active support and request for support for CVCT by such policy-makers, government and community leaders, little to no resources will be allocated by international agencies for couples' VCT.

Conclusion

With a shift in social and political norms, long-term sustainability is possible for CVCT. With greater emphasis on training and the incorporation of CVCT in routine VCT centers, hospitals and clinics, plus additional hours on weekends to cater to the working population, the goal of integrating CVCT into national medical schemes is achievable. Thus, making CVCT a standard of care service that contributes to capacity-building at the local and national level, and ensures sustainable feasibility.

We must seize this opportunity to ensure that couples' VCT becomes the standard of care in high-prevalence areas. To make this happen, several changes are needed: (1) a global shift that recognizes that couples are a high-risk population; (2) incorporation of couple-level prevention into the current ABC prevention model; (3) integration of CVCT into the larger medical schemes so that it serves as a "point of entry" for additional services. Once such changes occur in collaboration with greater support from policy-makers and commitment from donor agencies, CVCT can provide a cost-effective and sustainable model for global HIV prevention.

References

Adentunji, J. (2000). Condom use in marital and nonmarital relationships in Zimbabwe. *Intl Family Plan. Persp.*, 26, 196–200.

Agha, S., Kusanthan, T., Longfield, K. *et al.* (2002). *Reasons for Non-use of Condoms in Eight Countries in sub-Saharan Africa*. Washington, DC: Population Services International.

Ali, M. M., Cleland, J. and Shah, I. H. (2004). Condom use within marriage: a neglected HIV intervention. *Bull. WHO*, 82, 180–86.

Allen, S. (2005a). International data. *J. Acquir. Immune Defic. Syndr.*, 38(Suppl. 1), S7–8.

Allen, S. (2005b). Why is fertility an issue for HIV-infected and at-risk women? *J. Acquir. Immune Defic. Syndr.*, 38(Suppl. 1), S1–3.

Allen, S., Lindan, C., Serufilira, A. *et al.* (1991a). Human immunodeficiency virus infection in urban Rwanda. Demographic and behavioral correlates in a representative sample of childbearing women. *J. Am. Med. Assoc.*, 266, 1657–63.

Allen, S., Serufilira, A., Bogaerts, J. *et al.* (1992a). Confidential HIV testing and condom promotion in Africa. Impact on HIV and gonorrhea rates. *J. Am. Med. Assoc.*, 268, 3338–43.

Allen, S., Tice, J., van de Perre, P. *et al.* (1992b). Effect of serotesting with counselling on condom use and seroconversion among HIV discordant couples in Africa. *Br. Med. J.*, 304, 1605–9.

Allen, S., Serufilira, A., Gruber, V. *et al.* (1993). Pregnancy and contraception use among urban Rwandan women after HIV testing and counseling. *Am. J. Public Health*, 83, 705–10.

Allen, S., Karita, E., Ng'andu, N. and Tichacek, A. (1999). The evolution of voluntary testing and counseling as an HIV prevention strategy. In: L. Gibney, R. J. Diclemente and S. H. Vermund (eds), *Preventing HIV in Developing Countries: Biomedical and Behavioral Approaches*. New York, NY: Plenum Press, pp. 87–105.

Allen, S., Meinzen-Derr, J., Kautzman, M. *et al.* (2003). Sexual behavior of HIV discordant couples after HIV counseling and testing: validation of self-report using biological markers in discordant couples from Lusaka, Zambia. *AIDS*, 17, 733–40.

Allen, S., Karita, E., Chomba, E. *et al.* (2007a). Promotion of couples voluntary counselling and testing for HIV through influential networks in two African capital cities. *BMC Public Health*, 7, 349.

Antelman, G., Smith Fawzi, M. C., Kaaya, S. *et al.* (2001). Predictors of HIV-1 serostatus disclosure: a prospective study among HIV-infected pregnant women in Dar es Salaam, Tanzania. *AIDS*, 15, 1865–74.

Armstrong, A. (1992). Maintenance payments for child support in southern Africa: using law to promote family planning. *Studies Family Plan.*, 23, 217–18.

Au, J. T., Kayitenkore, K., Shutes, E. *et al.* (2006). Access to adequate nutrition is a major potential obstacle to antiretroviral adherence among HIV-infected individuals in Rwanda. *AIDS*, 20, 2116–18.

Bakari, J. P., McKenna, S., Myrick, A. *et al.* (2000). Rapid voluntary testing and counseling for HIV. Acceptability and feasibility in Zambian antenatal care clinics. *Ann. NY Acad. Sci.*, 918, 64–76.

Becker, J. and Ureno, M. (1996). How HIV has helped family planning: sexuality, the essential link. In: Futures Group International (ed.), *XIth International Conference on AIDS, Vancouver, 7–12 July*, Vol. 2 Abstracts.

Bizimungu, C. (1992). Instruction Ministerielle No 779 du 3 Mars 1988 relative a la promotion du programme de sante familiale dans les etablissements de sante du Rwanda. *Imbonezamuryango*, 25, 20–3.

Bunnell, R. E., Nassozi, J., Marum, E. *et al.* (2005). Living with discordance: knowledge, challenges, and prevention strategies of HIV-discordant couples in Uganda. *AIDS Care*, 17, 999–1012.

Carael, M., Nkurunziza, J., Allen, S. and Almedal, C. (1988). Knowledge about AIDS in a Central African city. *AIDS Health Prom Ex.*, 1, 6–8.

Cates, W. Jr. and Handsfield, H. H. (1988). HIV counseling and testing: does it work? [editorial]. *Am. J. Public Health*, 78, 1533–4.

Cates, W. Jr. and Stone, K. M. (1992). Family planning, sexually transmitted disease, and contraceptive choice: a literature update Part II. *Family Plan. Persp.*, 24, 122–8.

Central Statistical Office Zimbabwe and Macro International Inc. (2000). *Zimbabwe Demographic and Health Survey*. Calverton, MD: Central Statistical Office and Macro International Inc.

Chatterjee, N. and Hosain, G. M. M. (2006). Perceptions of risk and behaviour change for prevention of HIV among married women in Mumbai, India. *J. Health Pop. Nutr.*, 24, 81–8.

Chen, J. L., Philips, K. A., Kanouse, D. E. *et al.* (2001). Fertility desires and intentions of HIV-positive men and women. [see comment] *Family Plan. Persp.*, 33, 144–52.

Chomba, E., Allen, S., Kanweka, W. *et al.* and Rwanda Zambia HIV Research Group (2007). Evolution of couples voluntary counseling and testing for HIV in Lusaka, Zambia. *J. Acquir. Immune Defic. Syndr.*, 47, 108–15.

Coates, T. J., Stall, R. D., Kegeles, S. M. *et al.* (1988). AIDS antibody testing. Will it stop the AIDS epidemic? Will it help people infected with HIV? *Am. Psychologist*, 43, 859–64.

Coldiron, M. E., Stephenson, R., Chomba, E., *et al.* (2008). The relationship between alcohol consumption and unprotected sex among known HIV-discordant couples in Rwanda and Zambia. *AIDS and Behavior*, 12, 594–603.

Daley, D. (1994). Reproductive health and AIDS-related services for women: how well are they integrated? *Family Plan. Persp.*, 26, 264–9.

De Cock, K. M., Mbori-Ngacha, D. and Marum, E. (2002). Shadow on the continent: public health and HIV/AIDS in Africa in the 21st century. [see comment] *Lancet*, 360, 67–72.

Degruttola, V., Seage, G. R. III, Mayer, K. H. and Horsburgh, C. R. Jr. (1989). Infectiousness of HIV between male homosexual partners. *J. Clin. Epidemiol.*, 42, 849–56.

De Zoysa, I., Phillips, K. A., Kamenga, M. C. *et al.* (1995). Role of HIV counseling and testing in changing risk behavior in developing countries. *AIDS*, 9, S95–101.

Doll, L. S., Byers, R. H., Bolan, G. *et al.* (1991). Homosexual men who engage in high-risk sexual behavior. A multicenter comparison. *Sex. Transm. Dis.*, 18, 170–5.

Dunkle, K. L., Stephenson, R., Karita, E. *et al.* (2008). New heterosexually transmitted HIV infections in married and cohabiting couples in urban Zambia and Rwanda: an analysis of survey and clinical data. *Lancet*, 371, 2183–91.

Etiebet, M. A., Fransman, D., Forsyth, B. *et al.* (2004). Integrating prevention of mother-to-child HIV transmission into antenatal care: learning from the experiences of women in South Africa. *AIDS Care*, 16, 37–46.

Fideli, U. S., Allen, S. A., Musonda, R. *et al.* (2001). Virologic and immunologic determinants of heterosexual transmission of human immunodeficiency virus type 1 in Africa. *AIDS Res. Hum. Retroviruses*, 17, 901–10.

Fylkesnes, K., Musonda, R. M., Kasumba, K. *et al.* (1997). The HIV epidemic in Zambia: socio-demographic prevalence patterns and indications of trends among childbearing women. *AIDS*, 11, 339–45.

Ginwalla, S. K., Grant, A. D., Day, J. H. *et al.* (2002). Use of UNAIDS tools to evaluate HIV voluntary counselling and testing services for mineworkers in South Africa. *AIDS Care*, 14, 707–26.

Glick, P. (2005). Scaling up HIV voluntary counseling and testing in Africa. what can evaluation studies tell us about potential prevention impacts? *Evaluation Rev.*, 29, 331–57.

Goedert, J. J., Eyster, M. E., Biggar, R. J. and Blattner, W. A. (1987). Heterosexual transmission of human immunodeficiency virus: association with severe depletion of T-helper lymphocytes in men with hemophilia. *AIDS Res. Hum. Retroviruses*, 3, 355–61.

Grinstead, O. and van der Straten, A. and Voluntary HIV-1 Counseling and Testing Efficacy Group (2000). Counselors' perspectives on the experience of providing HIV Counselling and testing Efficacy Study. *AIDS Care*, 12, 625–42.

Grinstead, O. A., Gregorich, S. E., Choi, K. H. and Coates, T. J. (2001). Positive and negative life events after counselling and testing: the Voluntary HIV-1 Counselling and Testing Efficacy Study. *AIDS*, 15, 1045–52.

Guthrie, B. L., De Bruyn, G. and Farquhar, C. (2007). HIV-1-discordant couples in sub-Saharan Africa: explanations and implications for high rates of discordancy. *Curr. HIV Res.*, 5, 416–29.

Higgins, D. L., Galavotti, C., O'Reilly, K. R. *et al.* (1991). Evidence for the effects of HIV antibody counseling and testing on risk behaviors. *J. Am. Med. Assoc.*, 266, 2419–29.

Hira, S. K., Nkowane, B. M., Kamanga, J. *et al.* (1990). Epidemiology of human immunodeficiency virus in families in Lusaka, Zambia. *J. Acquir. Immune Defic. Syndr.*, 3, 83–6.

Hunter, D. J., Maggwa, B. N., Mati, J. K. *et al.* (1994). Sexual behavior, sexually transmitted diseases, male circumcision and risk of HIV infection among women in Nairobi, Kenya. *AIDS*, 8, 93–9.

Ickovics, J. R., Druley, J. A., Grigorenko, E. L. *et al.* (1998). Long-term effects of HIV counseling and testing for women: behavioral and psychological consequences are limited at 18 months posttest. *Health Psychol.*, 17, 395–402.

Institut National de la Statistique du Rwanda (INSR) and ORC Macro (2006). Rwanda Demographic and Health Survey 2005. Calverton, MD: INSR and ORC Macro.

Kalichman, S. C., Simbayi, L. C., Kaufman, M. *et al.* (2007). Alcohol use and sexual risks for HIV/AIDS in sub-Saharan Africa: systematic review of empirical findings. *Prevent. Sci.*, 8, 141–51.

Kamenga, M., Ryder, R. W., Jingu, M. *et al.* (1991). Evidence of marked sexual behavior change associated with low HIV-1 seroconversion in 149 married couples with discordant HIV-1 serostatus: experience at an HIV counselling center in Zaire. *AIDS*, 5, 61–7.

Kempf, M. C., Allen, S., Zulu, I. *et al.* (2008). Enrollment and retention of HIV discordant couples in Lusaka, Zambia. *J. Acquir. Immune Defic. Syndr.*, 47, 116–25.

Keogh, P., Allen, S., Almedal, C. and Temahagili, B. (1994). The social impact of HIV infection on women in Kigali, Rwanda: a prospective study. *Social Sci. Med.*, 38, 1047–53.

Kilewo, C., Massawe, A., Lyamuya, E. *et al.* (2001). HIV counseling and testing of pregnant women in sub-Saharan Africa: experiences from a study on prevention of mother-to-child HIV-1 transmission in Dar Es Salaam, Tanzania. *J. Acquir. Immune Defic. Syndr.*, 28, 458–62.

King, R., Estey, J., Allen, S. *et al.* (1995). A family planning intervention to reduce vertical transmission of HIV in Rwanda. *AIDS*, 9(Suppl. 1), S45–51.

Klein, J., Pena, J. E., Thornton, M. H. and Sauer, M. V. (2003). Understanding the motivations, concerns, and desires of human immunodeficiency virus 1-serodiscordant couples wishing to have children through assisted reproduction. *Obstet. Gynecol.*, 101, 987–94.

Kreiss, J. K., Kitchen, L. W., Prince, H. E. *et al.* (1985). Antibody to human T-lymphotropic virus type III in wives of hemophiliacs. Evidence for heterosexual transmission. *Ann. Int. Med.*, 102, 623–6.

Larson, A. (1989). Social context of HIV transmission in Africa: historical and cultural bases of East and Central African sexual relations. *Rev. Infect. Dis.*, 11, 716–31.

Lingappa, J. R., Lambdin, B., Bukusi, E. A. *et al.* (2008). Regional differences in prevalence of HIV-1 discordance in Africa and enrollment of HIV-1 discordant couples into an HIV-1 prevention trial. *PLoS ONE*, 3, e1411.

Maharaj, P. and Cleland, J. (2004). Condom use within marital and cohabiting partnerships in KwaZulu-Natal, South Africa. *Studies Family Plan.*, 35, 116–24.

Malamba, S. S., Mermin, J. H., Bunnell, R. *et al.* (2005). Couples at risk: HIV-1 concordance and discordance among sexual partners receiving voluntary counseling and testing in Uganda. *J. Acquir. Immune Defic. Syndr.*, 39, 576–80.

Maman, S. M., Mbwambo, J. K., Hogan, N. M. *et al.* (2001). Women's barriers to HIV-1 testing and disclosure: challenges for HIV-1 voluntary counselling and testing. *AIDS Care*, 13, 595–603.

Maman, S. M., Mbwambo, J. K., Hogan, N. *et al.* (2003). High rates and positive outcomes of HIV-serostatus disclosure to sexual partners: reasons for cautious optimism from a voluntary counseling and testing clinic in Dar es Salaam, Tanzania. *AIDS Behav.*, 7, 373–81.

Mark, K. E., Meinzen-Derr, J., Stephenson, R. *et al.* (2007). Contraception among HIV concordant and discordant couples in Zambia: a randomized controlled trial. *J. Women's Health (Larchmt)*, 16, 1200–10.

Mashu, A., Mbizvo, M., Makura, E. *et al.* (1996). Evaluation of rapid on-site clinic HIV test (Capillus) combined with counseling. In: Futures Group International (ed.), *XIth International AIDS Conference, Vancouver, 7–12 July*, 11, p. 445.

McGrath, J. W., Rwabukwali, C. B., Schumann, D. A. *et al.* (1993). Anthropology and AIDS: the cultural context of sexual risk behavior among urban Baganda women in Kampala, Uganda. *Social Sci. Med.*, 36, 429–39.

McKenna, S. L., Muyinda, G. K., Roth, D. *et al.* (1997). Rapid HIV testing and counseling for voluntary testing centers in Africa. *AIDS*, 11(Suppl. 1), S103–110.

Mendenhall, E., Muzizi, L., Stephenson, R. *et al.* (2007). Property grabbing and will writing in Lusaka, Zambia: an examination of wills of HIV-infected cohabiting couples. *AIDS Care*, 19, 369–74.

Newmann, S., Sarin, P., Kumarasamy, N. *et al.* (2000). Marriage, monogamy and HIV: a profile of HIV-infected women in south India. *Intl J. STD AIDS*, 11, 250–3.

Padian, N. S., O'Brien, T. R., Chang, Y. *et al.* (1993). Prevention of heterosexual transmission of human immunodeficiency virus through couple counseling. *J. Acquir. Immune Defic. Syndr.*, 6, 1043–8.

Painter, T. M. (2001). Voluntary counseling and testing for couples: a high-leverage intervention for HIV/AIDS prevention in sub-Saharan Africa. *Social Sci. Med.*, 53, 1397–411.

Panozzo, L., Battegay, M., Friedl, A. *et al.* (2003). High risk behaviour and fertility desires among heterosexual HIV-positive patients with a serodiscordant partner–two challenging issues. *Swiss Medical Wkly*, 133, 124–7.

President's Emergency Plan for AIDS Relief (PEPFAR) (annual publication). Country Profile: Zambia. Available from: www.pepfar.gov/pepfar/press/81694.htm.

Peterman, T. A., Todd, K. A. and Mupanduki, I. (1996). Opportunities for targeting publicly funded human immunodeficiency virus counseling and testing. *J. Acquir. Immune Defic. Syndr.*, 12, 69–74.

Pitchenik, A. E., Shafron, R. D., Glasser, R. M. and Spira, T. J. (1984). The acquired immunodeficiency syndrome in the wife of a hemophiliac. *Ann. Int. Med.*, 100, 62–5.

Plourde, P. J., Mphuka, S., Muyinda, G. K. *et al.* (1998). Accuracy and costs of rapid human immunodeficiency virus testing technologies in rural hospitals in Zambia. *Sex. Transm. Dis.*, 25, 254–9.

Remien, R. H., Carballo-Dieguez, A. and Wagner, G. (1995). Intimacy and sexual risk behaviour in serodiscordant male couples. *AIDS Care*, 7, 429–38.

Roth, D. L., Stewart, K. E., Clay, O. J. *et al.* (2001). Sexual practices of HIV discordant and concordant couples in Rwanda: effects of a testing and counselling programme for men. *Intl J. STD AIDS*, 12, 181–8.

Ryder, R. W., Kamenga, C., Jingu, M. *et al.* (2000). Pregnancy and HIV-1 incidence in 178 married couples with discordant HIV-1 serostatus: additional experience at an HIV-1 counselling centre in the Democratic Republic of the Congo. *Trop. Med. Intl Health*, 5, 482–7.

Seed, J., Allen, S., Mertens, T. *et al.* (1995). Male circumcision, sexually transmitted disease, and risk of HIV. *J. Acquir. Immune Defic. Syndr.*, 8, 83–90.

Shah, H., Vwalika, C. and Haworth, A. (2008). Process evaluation of couples voluntary counseling and testing (CVCT): a comprehensive approach. Working Paper, Rwanda–Zambia HIV Research Group.

Shutes, E., Iwanowski, M., Karita, E. *et al.* (2008). Couples voluntary counseling and testing and nevirapine use in antenatal clinics in two African capitals. Working Paper, Rwanda–Zambia HIV Research Group.

Stephenson, R., Mendenhall, E., Muzizi, L. *et al.* (2008). The influence of motivational messages on future planning behaviors among HIV concordant positive and discordant couples in Lusaka, Zambia. *AIDS Care*, 20, 150–60.

Stover, J. (1996). Reducing the costs of effective AIDS control programmes through appropriate targeting of interventions. In: Futures Group International (ed.), *XIth International Conference on AIDS, Vancouver, 7–12 July,* 11, p. 229.

Stringer, E. M., Sinkala, M., Stringer, J. S. *et al.* (2003). Prevention of mother-to-child transmission of HIV in Africa: successes and challenges in scaling-up a nevirapine-based program in Lusaka, Zambia. *AIDS*, 17, 1377–82.

Sweeney, P. A., Onorato, I. M., Allen, D. M. and Byers, R. H. (1992). Sentinel surveillance of human immunodeficiency virus infection in women seeking reproductive health services in the United States, 1988–1989. The Field Services Branch. *Obstet. Gynecol.*, 79, 503–10.

Taegtmeyer, M. and Doyle, V. (2003). *Quality Assurance Resource Pack for Voluntary Counselling and Testing Service Providers.* Nairobi: Liverpool VCT Centre.

The Voluntary HIV-1 Counseling and Testing Efficacy Study Group (2000). Efficacy of voluntary HIV-1 counselling and testing in individuals and couples in Kenya, Tanzania, and Trinidad: a randomised trial. *Lancet*, 356, 103–12.

Thornton, A. C., Romanelli, F. and Collins, J. D. (2004). Reproduction decision making for couples affected by HIV: a review of the literature. *Topics HIV Med.*, 12, 61–7.

Trask, S. A., Derdeyn, C. A., Fideli, U. *et al.* (2002). Molecular epidemiology of Human Immunodeficiency Virus type 1 transmission in a heterosexual cohort of discordant couples in Zambia. *J. Virol.*, 76, 397–405.

UNAIDS (2000). *Tools for Evaluting HIV Voluntary Counseling and Testing.* Geneva: UNAIDS. (available at http://data.unaids.org/Publications/IRC.pub02/ JC685-Tools%20for%20Eval_en.pdf).

UNAIDS & WHO (2007). *Guidance on Provider-Initiated HIV Testing and Counseling in Health Facilities.* Geneva: UNAIDS & WHO (available at http://libdoc.wbo.int/ publications/2007/9789241595568_eng.pdf).

UNFPA & International Planned Parenthood Federation (2004). *Integrating HIV Voluntary Counseling and Testing Services into Reproductive Health Settings* (available at http:// www.unfpa.org/upload/lib_pub_file/245_filename_hiv_publication.pdf).

United States Centers for Disease Control and Prevention (2007). *Couples HIV Counseling and Testing Intervention and Curriculum.* Atlanta, GA: CDC (available at http://www.cdc.gov/nchstp/od/gap/CHCTintervention/page2.html).

United States President's Emergency Plan for AIDS Relief (2007). *HIV Counseling and Testing* (updated November 2007) (available at http://www.pepfar.gov/pepfar/ press/76382.htm).

Van der Straten, A., King, R., Grinstead, O. *et al.* (1995). Couple communication, sexual coercion and HIV risk reduction in Kigali, Rwanda. *AIDS*, 9, 935–44.

Van der Straten, A., King, R., Grinstead, O. *et al.* (1998). Sexual coercion, physical violence, and HIV infection among women in steady relationships in Kigali, Rwanda. *AIDS Behav.*, 2, 61–73.

VanDevanter, N., Thacker, A. S., Bass, G. and Arnold, M. (1999). Heterosexual couples confronting the challenges of HIV infection. *AIDS Care*, 11, 181–93.

Varghese, B., Peterman, T. A. and Mugalla, C. (2001). Voluntary counselling and testing for HIV-1. *Lancet*, 357, 144–5.

Weinhardt, L. S., Carey, M. P., Johnson, B. T. and Bickham, N. L. (1999). Effects of HIV counseling and testing on sexual risk behavior: a meta-analytic review of published research, 1985–1997. *Am. J. Public Health*, 89, 1397–405.

Were, W. A., Mermin, J. H., Wamai, N. *et al.* (2006). Undiagnosed HIV infection and couple HIV discordance among household members of HIV-infected people receiving antiretroviral therapy in Uganda. *J. Acquir. Immune Defic. Syndr.*, 43, 91–5.

WHO/UNFPA & World Bank (1995). An assessment of the need for contraceptive introduction in Zambia. Lusaka, Special Programme of Research, Development, and

Research Training in Human Reproduction. Geneva: WHO/UNPFA & World Bank (available at http://www.who.int/reproductive-health/publications/HRP_ITT_95_4/HRP_ITT_95_4_contents_en.html).

Wolitski, R. J., MacGowan, R. J., Higgins, D. L. and Jorgensen, C. M. (1997). The effects of HIV counseling and testing on risk-related practices and help-seeking behavior. *AIDS Ed. Prev.*, 9, 52–67.

Updating HIV prevention with gay men: current challenges and opportunities to advance health among gay men

10

Ron Stall, Amy Herrick, Thomas E. Guadamuz and
Mark S. Friedman

At the close of the first quarter century of AIDS, HIV prevention among gay men seems paralyzed. Gay men commonly report weariness with HIV prevention strategies first defined when AIDS was understood to be a crisis and not an ongoing burden whose toll would be extracted across decades of the life course. Younger cohorts of gay men, raised in the post-HAART era, have been spared witnessing the worst onslaughts of the AIDS epidemic, and so are not motivated to respond to prevention messages with firsthand knowledge of the devastating effects of AIDS. While HIV prevention programs continue among gay men, much of the day-to-day prevention work is being funded by government agencies who operate under regulations designed first and foremost to cater to the sensibilities of religious conservatives, even when these sensibilities conflict with principles of effective public health practice. It is also fair to point out that HIV prevention for gay men has never been funded at a level necessary to have maximum effectiveness (Holtgrave, 2002). Instead, HIV prevention for gay men has been fielded in practice by a heroic network of overworked and underfunded community-based organizations that often must adhere to a very specific set of government-imposed guidelines. Hence, HIV prevention work can sometimes lack a cutting-edge, street-smart quality that directly addresses the specific prevention needs of gay men. Prevention strategies for gay men must be designed so that they are attractive to men of widely varying class, racial, educational and sexual-identity backgrounds, who, with the exception of sexual practice, may perceive that they hold very little else in common. It should also be noted that AIDS prevention continues to be challenged by all the features that have always made it a difficult topic to discuss. In the end, AIDS prevention among gay men requires a clear-eyed discussion of the specifics of how a sexual minority has sex, uses drugs, and experiences disease and death – all in the contexts of social stigma, racism and homophobia. It is hard to imagine how one could possibly raise a more explosive combination of topics in current American life. It is no wonder that keeping HIV prevention fresh, exciting and effective for decades on end has proven to be such a difficult challenge.

Given these challenges, the question of whether or not ongoing HIV prevention efforts are having any effect among gay men should be raised. Reports in the scientific literature continue to appear, describing high prevalence rates of sexual risk-taking behaviors among gay men (Valleroy *et al.*, 2000), high rates of HIV seroprevalence even among young gay men who came of age after AIDS prevention efforts had been fielded (CDC, 2005), high HIV incidence rates (CDC, 2001) and outbreaks of sexually transmitted diseases (CDC, 2004). Current HIV prevalence rates among African-American men exceed those found even in many sub-Saharan countries (CDC, 2002; UNAIDS, 2007), with rates of HIV infection found among substance-abusing men also approaching these levels (Catania *et al.*, 2001). Rates in other subpopulations of gay men or men who have sex with men are also quite elevated, especially when compared to heterosexual counterparts in the US. This body of ongoing reports has led some commentators to question whether HIV prevention has had any effect among gay men, and to call for a cessation of their characterization of AIDS prevention messages, at least in the short term (Rofes, 2007). However, it should be noted that the data sets on which these arguments are based are ecological – not experimental – data, and raise the question of what the rates of HIV infection would have been had HIV prevention efforts not been attempted at all among gay men in the United States. To answer whether HIV prevention has a direct effect in reducing HIV risk behaviors, experimental data are required.

The purpose of this chapter is to review the evidence base for the efficacy of AIDS prevention programs among gay men in the United States, and to describe a set of agenda items that might be expected to increase the effectiveness of AIDS prevention efforts among American gay men. Regarding terminology, the phrase "gay men" will generally be used here to refer to all men who have sex with men, on the grounds that the majority of such men described in the HIV prevention research literature identify themselves as gay. Other groups of homosexually active men who may be less likely to identify themselves as gay will be referred to in the chapter as "men who have sex with men". The chapter will end with an argument that current HIV prevention strategies that rely on efforts to manipulate individual behaviors alone will not be very effective as a long-term strategy. As an alternative, a model is proposed to use multiple mechanisms of support for HIV prevention behaviors – mimicking the strategy behind HAART medication regimens – to constitute an AIDS prevention cocktail.

What is the evidence base for efficacy of HIV prevention efforts among gay men?

The ability briefly to summarize the efficacy of current HIV prevention efforts among gay men has been enhanced by the publication of a set of meta-analyses (Johnson *et al.*, 2002, 2005; Herbst *et al.*, 2005, 2007), each of which have reported quite consistent and positive findings for the effects of these model programs. For example, Herbst *et al.* (2005) extracted an odds ratio of 0.77 overall decline in the prevalence of unprotected anal sex, and a significant increase in condom use during anal sex (odds ratio of 1.61) in their meta-analytic review of this literature. Furthermore, they identified a set of variables that were associated with greater intervention efficacy in their review, among them interpersonal skills training, the use of several methods to deliver messages, and the use of multiple intervention sessions over a time period of at least 3 weeks. A follow-up systematic review conducted by the Task Force on Community Prevention Services concluded that not only were the effect sizes for risk reduction among gay men likely to be cost-effective; they could also be cost-saving (Herbst *et al.*, 2007). Together, these meta-analyses, and the literature on which they are based, provide strong empirical evidence that AIDS prevention among gay men can reduce sexual risk-taking behaviors.

To summarize, we now have a scientific basis to demonstrate that HIV prevention programs for gay men can reduce risk – a goal that some denied was possible early in the AIDS epidemic. We can field a set of behavioral intervention risk-reduction programs that have been shown, through randomized controlled trial data, to lower sexual risk-taking among gay men. The effect sizes for these programs are sufficiently large so that they are not only cost-effective, but also potentially cost-saving.

The question should then be raised: if we have interventions that yield impressive results in terms of behavioral risk reduction, why are we still detecting ongoing HIV infections among gay men? One answer to that question is that the current HIV prevention interventions, while efficacious, do not address all of the prevention challenges that are facing gay men. The remainder of this chapter will attempt to list at least some of these prevention challenges, with the goal of suggesting strategies that might be adopted to raise even further the levels of HIV prevention program efficacy for gay men.

Ron Stall, Amy Herrick, Thomas Guadamuz and Mark S. Friedman

What are the current challenges in HIV prevention work among gay men?

What exactly is sexual risk?

Although knowledge of health risks is not sufficient in and of itself to induce behavior change, it is generally necessary to maintain long-term behavioral change. As sexual risk-reduction strategies have evolved over time within the gay community and with the advent of the HAART era, the informational needs of gay men have become more sophisticated. Sexually active gay men are now asking questions regarding HIV transmission risks concerning very specific sexual practices, to inform their long-term risk-reduction strategies. Unfortunately, sound epidemiological research to answer these questions has yet to be published, thereby leaving gay men in the position of needing to test epidemiological hypotheses regarding the risks of HIV transmission of very specific sexual acts with their own bodies.

For example, we still have no carefully conducted epidemiological estimates to measure the risks of unprotected positive-on-positive sex, HIV transmission risks during unprotected sex from HIV-positive men who have undetectable viral loads, HIV transmission risks within serodiscordant sexual relationships in which the HIV-positive partner is receptive during unprotected sex, HIV transmission risks of insertion by an HIV-positive partner without ejaculation (i.e. "dipping"), or HIV transmission risks that derive in practice within relationships that practice negotiated safety. The strategies listed here are only among the most common strategies incorporating unprotected anal sex that are being utilized by gay men as "risk-reduction" strategies. Given the fact that these strategies are not necessarily mutually exclusive (e.g. dipping by an HIV-positive man with an undetectable viral load), this means that the combined number of possible strategies being considered by gay men is quite large and complex.

Notably, all of the strategies mentioned in the previous paragraph are being attempted to allow unprotected anal sex, albeit in ways that are believed to reduce risk of HIV transmission. It should also be noted that these strategies are being attempted by men who (with some variation in exposure as a function of area of residence) have been the target of ongoing HIV prevention campaigns since the onset of the AIDS epidemic that have emphasized the efficacy of condoms in preventing HIV transmission. It is therefore reasonable to assume that these strategies are being attempted by men who understand that condom use effectively reduces risk, but also find that consistent condom use is difficult or impossible to achieve over long periods of time in their own particular case. This situation replicates that of many gay men earlier in the AIDS epidemic with regard to the risks of unprotected oral sex. Most men refused to attempt a risk-reduction strategy as invasive as the use of condoms during oral sex, and chose instead to monitor possible harm associated with unprotected oral sex through ongoing HIV testing.

Accordingly, it is also quite possible that many HIV-negative men who are attempting the current set of "risk-reduction" strategies are undergoing regular HIV testing, so that they can form an opinion as to the overall risks of their particular strategy (or sets of strategies) and to ensure that they have not seroconverted. To the extent that given strategies reduce but do not eliminate the risk of HIV seroconversion, men could come to an overly optimistic conclusion regarding the efficacy of a particular strategy over the short term, and put off ongoing HIV monitoring. If men continue to engage in a flawed risk-reduction strategy after seroconversion, a cascading string of new HIV infections could occur within a sexual network.

Hence, finding ways to measure the long-term risks of specific sexual risk-reduction strategies that include unprotected anal sex may well give men the tools that they need to make better decisions to inform their risk-reduction strategies. There are probably very few at-risk populations that would read reports of epidemiological findings regarding the risks of HIV transmission to guide their risk-reduction strategies as carefully as will gay men.

Who bears the responsibility for HIV prevention?

At present, many HIV-negative gay men feel that HIV-positives should always ensure that their sex partners cannot be infected, as HIV transmission must by definition involve an HIV-positive person. On the other hand, many HIV-positive men feel that HIV-negatives should always be vigilant to ensure that they cannot be infected, as they are the men who will bear the burden of a new infection. Additional groups of men believe that HIV prevention agencies should be in charge of maintaining prevention standards, as they are being funded, after all, to do this work. This means, in practice, that substantial proportions of gay men believe that someone else should bear the responsibility for HIV prevention. This raises the question of whether we've recreated a public health version of the "tragedy of the commons", in which public spaces are abused by allowing everyone use of a public resource while no one is charged with the responsibility of caring for it.

It is clear that community-based approaches to HIV prevention can only work if all members of an at-risk community take responsibility for it. How, then, can responsibility for HIV prevention be supported among all gay men? Perhaps the first step in the process would be to recognize the considerable assets that exist within many gay communities in terms of supporting responsibility for HIV prevention. Among HIV-positive men, for example, a sense of responsibility for HIV prevention has been identified as a predictor of long-term sexual safety in an intervention trial (O'Leary *et al.*, 2005), and risk reductions tend to follow notification of HIV seroconversion among HIV-positive men (Higgins *et al.*, 1991; Wolitski *et al.*, 1997; Weinhardt *et al.*, 1999). It therefore follows that interventions that seek to support responsibility to prevent HIV transmission among

HIV-positives may well yield useful public health results. Regarding barriers, it should be recognized that efforts to promote disclosure by HIV-seropositive individuals requires men with the greatest to lose as a result of HIV stigma (i.e. HIV seropositives) to bear the responsibility for HIV prevention. One wonders what the result of a campaign that asks HIV-negatives to disclose their status would yield in terms of supporting safe sex among gay men. An "I'm negative" campaign would allow men who have relatively little to lose in terms of AIDS stigma to start a conversation about HIV prevention without necessarily requiring their sexual partners to disclose their status. Even if HIV-positive men do not disclose their status to HIV-negative sexual partners, men can still agree to have safe sex as a means of protecting the HIV-negative partner.

Creating self-knowledge as a basis for HIV prevention

HIV prevention practice among gay men is based on the assumption that all gay men want to avoid HIV transmission every time that they have sex. While this might seem a reasonable assumption at face value, absolute sexual safety may not be the most overriding concern of gay men every single time that they have sex. Although this assertion may seem shocking to some, the strategies that some gay men have adopted when it comes to sexual safety may resemble the strategies that many drivers take when operating automobiles. That is, while no one wants to be involved in a traffic accident, many drivers take the chance of occasionally driving without seatbelts, after having had a few drinks, a bit too fast, or on slippery winter roads, even though they know full well that these conditions increase the risks of long-term disability or even death.

Accordingly, it stands to reason that it is important to help gay men to understand more about the specific needs that they want to fill by having sex, such as understanding clearly what they like and don't like in terms of sex, why they pick the sex partners that they do and, perhaps most importantly, distinguishing between the times that they have safe or unsafe sex. These insights may be as useful in terms of maintaining HIV prevention strategies as memorizing a list of sexual practices that are understood to be safe and unsafe. If men are able to understand the conditions under which they are most likely to cut prevention corners while having sex, they might well be able to form opinions as to which conditions seem reasonable bets (analogous to a driver who dispenses with wearing a seatbelt to drive to the nearest supermarket for some milk) and which seem less reasonable (driving too fast on a wintry road at night). In the end, men are free to choose whether to have safe or unsafe sex, and are the individuals who will suffer the consequences of having made a poor choice. Helping men to identify the conditions during which risky sex is most likely to occur and to therefore make informed judgments as to the times when foregoing condoms might reasonably seem to be a good bet may help men to maintain lower levels of risk over longer periods of time. Thus, incorporating strategies to help men achieve self-knowledge and to help them design the strategies that will serve them best to avoid HIV transmission may be more effective in maintaining lower levels of risk than asking men to adhere to a memorized list of safe and unsafe practices.

Incorporating the demographic diversity of gay men as part of prevention strategies

Perhaps the single most unique quality of gay male communities is their demographic diversity: gay men are drawn from every religious tradition, every age group, every race, every economic class and every educational level found in the societies in which gay communities are located. Given this fact, it should be self-evident that demographic diversity will be an important variable to consider in the design of public health and HIV prevention programs designed for gay men.

One of the most consistent findings of the epidemiology of HIV is the strength of demographic and behavioral variables in shaping risk for HIV infection among gay men. The disparities in terms of HIV seroprevalence suffered by African-American men who have sex with men are striking (CDC, 2002, 2005; Millett *et al.*, 2007), and the added burden of disease found among substance-abusing gay men is only marginally smaller (Catania *et al.*, 2001) than that found among African-American men. Although the general strength and direction of these differences have been reported for decades (Samuel and Winkelstein, 1987), interventions specifically developed to lower risk among African-American or substance-abusing MSM have yet to be developed with demonstrated efficacy. Additional arguments can also be made in terms of life-course issues among gay men that work to shape risk, particularly to address risk levels among men in their 30s. Other work could be designed to convey prevention messages to rural and suburban men who may not have had access to prevention programs designed to operate in large urban settings (Preston *et al.*, 2002; Bowen *et al.*, 2007). Although the challenge that demographic diversity poses for HIV prevention is well recognized in the field, the fact that we still have yet to develop efficacious interventions for subgroups of gay men at increased risk strongly suggests that we have a great deal to learn in terms of intervention development for demographic or behaviorally distinct groups. AIDS prevention cannot work within gay communities as a whole unless the public health needs of all subgroups of gay men that constitute the larger community are addressed.

Addressing the multiple psychosocial health problems that drive risk among gay men

HIV is only the best known of several life-threatening epidemics that disproportionately affect gay men (Wolitski *et al.*, 2008). Depression, substance abuse and

273

violence victimization, among others, have a higher prevalence rate in gay male communities than in the general population of men in the United States. Worse, two separate analyses have shown that these individual epidemics intertwine in ways to amplify each others' deleterious effects and to amplify HIV risk among gay men (Stall *et al.*, 2003; Mustanski *et al.*, 2007). This intimate intertwining of epidemics is referred to collectively as a *syndemic* (Singer, 1996), and is an epidemiological phenomenon found in many other marginalized populations than gay men.

One of the implications of the recognition of the importance of syndemic processes in driving HIV risk among gay male populations is to emphasize the importance of raising health levels across a broad range of health problems. By creating a broader approach to supporting health programs among gay men than that found in most HIV prevention work, one might expect the effects of HIV prevention programs to be enhanced. That is, the theory that underlies most HIV prevention work assumes that men are free actors to respond to HIV prevention messages regarding risk. However, if men are mired in the effects of depression, substance abuse and partner violence, the assumption underlying HIV prevention programs that men are free to act may well be flawed. To the extent, then, that men are freed from the confines created by high prevalence rates of coexisting psychosocial problems, their ability to respond successfully to HIV prevention efforts may well be enhanced. Thus, by raising health levels of gay men across multiple fronts, the effectiveness of HIV prevention as well as other health promotion efforts may well be improved.

Community viral load approaches to HIV prevention: reducing risk by changing context

Risk of HIV transmission is governed by more than individual behavioral risk levels; it is also governed by the prevalence of HIV infection within the group in which men find sexual partners, by the proportion of individuals in that group who know that they are infected and so are less likely to engage in high-risk behaviors, and by the proportion of individuals in that group who have accessed HAART treatment and are no longer efficient HIV transmitters. Thus, risk for HIV transmission for gay men can be thought of as an interaction between risks taken at the individual level and risk that is driven by the "community viral load" (i.e., the proportion of a community who have high HIV viral loads) in which they meet new sexual partners.

To date, nearly all HIV prevention efforts have been devoted to reducing risk at the level of the individual, with less attention being devoted to reducing risk by reducing the community viral loads within the social groups in which men find new sexual partners. This raises the possibility of increasing the effectiveness of HIV prevention programs by working to reduce the proportion of an at-risk population who know that they are HIV infected, helping them to access medical care for HIV infection, lowering their viral loads to the point of being undetectable and promoting sexual safety among HIV seropositive individuals. Substantial proportions of HIV-seropositive gay men remain unaware of their HIV status (CDC, 2002, 2005); other men know that they are infected, but are not accessing HIV care. Addressing the prevention and care needs of both groups of men would lower community viral load levels. Prevention work that seeks to identify unknown HIV-positives and help them to access effective medical care that will lower their viral loads so that they cannot transmit HIV infection as efficiently, and which aims to lower risk levels among them, would prove to be an important addition to current HIV prevention efforts. This work would also, of course, have the additional benefit of helping HIV-seropositives to remain healthy for longer periods of time.

How can we translate efficacy into effectiveness?

Current meta-analyses have shown that HIV prevention programs can reduce risk among gay men (Herbst *et al.*, 2005, 2007). However, proof of concept is only one part of mounting an effective public health response to the AIDS epidemic. One must also find ways to field interventions with evidence of efficacy so that they can be widely used in the field, and so become effective HIV prevention tools. Although ramping up efficacy trials for broad public health use may sound simple, the challenges should not be underestimated. Translating efficacy to effectiveness will require substantial pre-testing and development of proven interventions so that they can be used widely by public health agencies, as well as substantial training of public health workers and community-based organization staff so that they can field interventions with fidelity. Additionally, it will necessitate ongoing evaluation to ensure that interventions are yielding expected results in the field, and the adaptation of prevention interventions so that they can be used in populations in which the original intervention had not been tested but which could be expected to yield similar results in new settings (e.g., across racial groups of gay men). It should also be noted that the success of scaling up interventions so that they are widely accessible to at-risk populations will require substantial resources to be devoted to the institutional development of community-based partners that will field such interventions. Although the CDC has begun important work to address this need through its REP/DEBI process, it is clear that a much larger effort will be needed to ensure that all gay men in the United States, from men who live in rural areas to men who reside in America's largest cities, can access prevention programs proven to reduce risk. Although this will be an expensive and time-consuming effort, there is probably no other way for gay men to benefit fully from the investments that have already been made in HIV prevention science.

Ron Stall, Amy Herrick, Thomas Guadamuz and Mark S. Friedman

Towards a prevention cocktail: strategies to move HIV prevention among gay men forward

The bulk of current HIV prevention approaches have been designed to help individual gay men to avoid the threat of HIV infection and, if HIV-positive, HIV transmission. Drawing on dominant cognitive behavioral theories, these prevention models emphasize to varying degrees interventions that were designed to change knowledge, self-efficacy, peer-group norms, safe-sex negotiation and condom use skills in order to increase rates of safe-sex practice among gay men. These interventions are designed to manipulate individual-level variables, and have become the dominant behavioral intervention approach to reducing HIV transmission among gay men.

Evolutions in the response of gay men to the AIDS epidemic over the past quarter of a century, however, may have rendered obsolete a reliance on interventions based on the single mechanism of cognitive behavioral approaches designed to operate at the level of the individual. That is, the list of current challenges in HIV prevention practice listed above cannot be reduced to a set of variables that only operate at the level of the individual. The list of current prevention challenges described previously can be grouped according to varying levels of mechanism of action, as follows:

Individual	Knowledge of risk levels
	Self-knowledge of risk goals
Interpersonal	Negotiating prevention responsibility
Community	Community norms regarding responsibility
	Addressing demographic diversity
	Addressing syndemic processes
	Lowering community viral load
Public health infrastructure	Programs to lower community viral loads
	Programs to address syndemic conditions
	Efficacy into effectiveness
Governmental policy	Funding and scientific leadership to address issues surrounding efficacy into effectiveness, syndemic conditions and lowering of community viral loads

Thus, given the wide range of levels of mechanism of action at which each of these current prevention challenges operate, it should be clear that reliance on a single mechanism of action that operates at the level of the individual alone is likely to doom HIV prevention efforts among gay men to low levels of effectiveness. The list of challenges in current prevention practice has grown far beyond those that operate at the level of the individual, hence making the use of interventions based on cognitive behavioral theory alone very problematic. Attempts to modify this list to incorporate additional challenges may well serve to strengthen this argument even further. For example, it would be difficult to modify the list of current challenges so that it is especially appropriate for African-American men without listing the conjoined effects of homophobia and racism that hobble HIV prevention efforts for this population. Dealing with either homophobia or racism requires intervention work at the level of cultural understandings of a marginalized minority, and thereby the addition of cultural-level mechanisms of action to the above list. If we are to achieve the ambitious goals of HIV prevention, multiple mechanisms for risk reduction will need to be marshaled to promote successful HIV prevention work with gay men.

Perhaps it is time to consider one of the lessons learned from a success in the fight against AIDS, i.e., the development of HAART medications. Understanding the mechanisms of viral replication led to the development of drugs that interfere with viral replication at multiple levels. This insight allowed the combined use of drugs, that when used alone had limited effectiveness but in combination constituted a treatment cocktail of highly effective treatment action. If we are to translate this principle to the field of HIV prevention, we would need to ask a central question: what would a prevention cocktail strategy look like?

HIV prevention among gay men that is based on the principle that multiple mechanisms of prevention action are likely to be mutually reinforcing would clearly move beyond prevention practice that seeks to manipulate variables at the level of the individual alone. It would also seek to address multiple STI and psychosocial epidemics among gay men as a way of raising the level of HIV prevention effectiveness. It would work closely with HIV treatment programs, not only to lower rates of high-risk behaviors among positives, but also to lower the prevalence of individuals with high HIV viral loads in the community at large. In particular, it would work to raise levels of treatment access among populations that currently have very high incidence rates of HIV transmission (i.e. African-American MSM and substance-abusing gay men), with a view to lower rates of community viral loads within the high-risk sexual networks where these men meet. It would work to field community- and structural-level interventions, where possible, that have proven so effective in other aspects of AIDS prevention practice (i.e., testing the blood supply, interventions to prevent vertical HIV transmission). It would work to identify through rigorous epidemiological research the strategies used by gay men that seem to protect against HIV transmission and those that convey a high risk for seroconversion. Over time, as specific prevention activities wane in effectiveness, different combinations would be attempted to maintain potency of the prevention cocktail. And, finally, it would work to support a positive policy environment so that the interventions that have been proven to work can be fielded with the financial backing that they need to succeed.

Ron Stall, Amy Herrick, Thomas Guadamuz and Mark S. Friedman

Steps toward the creation of a prevention cocktail

As has been observed many times before, the AIDS epidemic arrived at a time when medical and public health responses were poorly equipped to deal with the emergency. The knowledge base necessary to inform a response to the epidemic was lacking on many fronts, among them immunology, virology, pharmacology and the behavioral epidemiology of risk behaviors. To this list should be added basic knowledge about how best to construct sexual risk-reduction public health programs among gay men so that they yield strong outcomes that could be sustained over time. Had the need to respond to AIDS not been so dire, few public health workers would have fielded sexual risk-reduction programs at such an early stage of development with the hope of yielding effective outcomes over long periods of time.

The past quarter-century of public health practice has provided the field with a wealth of experience from which lessons should be learned. Among these is that a strong, concerted effort in medical and public health can yield results: we are far better equipped to deal with the epidemic now than when it first emerged. This is even the case in terms of prevention among gay men, which has not enjoyed the funding base or political support that other aspects of AIDS prevention and treatment research have been given. We have strong data to demonstrate that HIV prevention is efficacious among gay men, which raises the possibility that bringing efficacious programs up to scale would yield effective outcomes among this group.

That said, it is also clear that even behavioral interventions with evidence of efficacy have limited effects among gay men. In some cases, HIV prevention can be fairly judged to have failed some gay male communities – most notably African-American men who have sex with men and substance-abusing gay men. This raises the question of how we can best raise the efficacy levels of the interventions that are available for becoming widely used as standard public health practice. This chapter has argued that a strategy that works to employ multiple mechanisms of prevention practice will likely yield far more effective outcomes than will a strategy that relies on the single mechanism of manipulating variables at the level of the individual. Developing ways to field and conduct efficacy trials of "prevention cocktail" strategies should become a priority in the field.

Ultimately, one truism has held constant since the beginning of the epidemic: the most effective tool that we have to confront the epidemic is prevention. This has held true despite the promise of biomedical interventions to either cure AIDS or prevent HIV transmission. Thus, despite the potential promise of vaccines, microbicides and/or treatments for HIV infection, behavioral interventions remain among our most effective tools to manage the AIDS epidemic. While we do not wish to suggest that the efforts to develop biomedical interventions to prevent HIV transmission be diminished, we do point out that ongoing scientific and public health attention should be devoted to making sure that the tools that we do have in hand remain effective. We cannot claim that prevention has a high priority if we continue to field interventions that are theoretically and operationally static, even as the adaptations of gay men to the epidemic are moving forward. To make sure that prevention is meeting the needs of gay male communities, we need to continue to refine interventions with evidence of efficacy so that they are even more effective, and to look for new methods of prevention activity that enhance the effects of the interventions that are already in the field. This work should not proceed as a stop-gap until such time as biomedical interventions appear; it should be designed, evaluated, fielded and studied as if it were an important public health tool in its own right. If prevention remains the most effective tool that we have to manage the AIDS epidemic among gay men, the time to get even more serious about it is long overdue.

References

Bowen, A., Horvath, K. and Williams, M. (2007). A randomized control trial of internet-delivered HIV prevention targeting rural MSM. *Health Educ. Res.*, 22, 120–7.

Catania, J. A., Osmond, D., Stall, R. D. *et al.* (2001). The continuing HIV epidemic among men who have sex with men. *Am. J. Public Health*, 91, 907–14.

CDC (2001). HIV incidence among young men who have sex with men – seven US cities, 1994–2000. *Morbid. Mortal. Wkly Rep.*, 50, 440–4.

CDC (2002). Unrecognized HIV infection, risk behaviors, and perceptions of risk among young black men who have sex with men – six US cities, 1994–1998. *Morbid. Mortal. Wkly Rep.*, 51, 733–6.

CDC (2004). Trends in primary and secondary syphilis and HIV infections in men who have sex with men – San Francisco and Los Angeles, California, 1998–2002. *Morbid. Mortal. Wkly Rep.*, 53, 575–8.

CDC (2005). HIV prevalence, unrecognized infection, and HIV testing among men who have sex with men – five U.S. cities, June 2004–April 2005. *Morbid. Mortal. Wkly Rep.*, 54, 597–601.

Herbst, J., Sherba, R., Crepaz, N. *et al.* (2005). The HIV/AIDS Prevention Research Synthesis Team: a meta-analytic review of HIV behavioral interventions for reducing sexual risk behavior of men who have sex with men. *J. Acquir. Immune Defic. Syndr.*, 39, 228–41.

Herbst, J., Beeker, C., Mathew, A. *et al.* (2007). The effectiveness of individual-, group-, and community-level HIV behavioral risk-reduction interventions for adult men who have sex with men: a systematic review. *Am. J. Prev. Med.*, 32, S38–67.

Higgins, D. L., Galavotti, C., O'Reilly, K. R. *et al.* (1991). Evidence for the effects of HIV antibody testing on risk behaviors. *J. Am. Med. Assoc.*, 266, 2419–29.

Holtgrave, D. R. (2002). Estimating the effectiveness and efficiency of HIV prevention efforts in the U.S. using scenario and cost-effectiveness analysis. *AIDS*, 16, 2347–9.

Johnson, W., Hedges., L., Ramirez, G. *et al.* (2002). HIV prevention research for men who have sex with men: a systematic review and meta-analysis. *J. Acquir. Immune Defic. Syndr.*, 30, S118–129.

Johnson, W., Holtgrave, D., McClellan, W. *et al.* (2005). HIV intervention research for men who have sex with men: a 7-year update. *AIDS Educ. Prev.*, 17, 568–89.

Millett, G. A., Flores, S. A., Peterson, J. L. and Bakeman, R. (2007). Explaining disparities in HIV infection among black and white men who have sex with men: a meta-analysis of HIV risk behaviors. *AIDS*, 21, 2083–91.

Mustanski, B., Garofalo, R., Herrick, A. and Donenberg, G. (2007). Psychosocial health problems increase risk for HIV among urban young men who have sex with men: preliminary evidence of a syndemic in need of attention. *Ann. Behav. Med.*, 34, 37–45.

O'Leary, A., Hoff, C. C., Purcell, D. W. *et al.* (2005). What happened in the SUMIT trial? Mediators and responders. *AIDS*, 19, S111–21.

Preston, D., D'Augelli, A., Cain, R. and Schulze, F. (2002). Issues in the development of HIV-prevention interventions for men who have sex with men (MSM) in rural areas. *J. Prim. Prev.*, 23, 199–214.

Rofes, E. (2007). The condom backlash. Available at Planet Out (http://www.planetout.com/health/hiv/?sernum=3162).

Samuel, M. and Winkelstein, W. (1987). Prevalence of HIV in ethnic minority homosexual/bisexual men. *J. Am. Med. Assoc.*, 257, 1901–2.

Singer, M. (1996). A dose of drugs, a touch of violence and case of AIDS: conceptualizing the SAVA syndemic. *Free Inq. Creat. Sociol.*, 24, 99–110.

Stall, R., Mills, T. C., Williamson, J. *et al.* (2003). Association of co-occurring psychosocial health problems and increased vulnerability to HIV/AIDS among urban men who have sex with men. *Am. J. Public Health*, 93, 939–42.

UNAIDS (2007). *AIDS Epidemic Update: December 2007*. Geneva: UNAIDS & WHO.

Valleroy, L. A., Mackellar, D. A., Karon, J. M. *et al.* (2000). HIV prevalence and associated risks in young men who have sex with men. *J. Am. Med. Assoc.*, 284, 198–204.

Weinhardt, L. S., Carey, M. P., Johnson, B. T. and Bickman, N. L. (1999). Effects of HIV counseling and testing on sexual risk behavior: a meta-analytic review of published research, 1985–97. *Am. J. Public Health*, 89, 1397–405.

Wolitski, R. J., MacGowan, R. J., Higgins, D. L. and Jorgensen, C. M. (1997). The effects of HIV counseling and testing on risk-related practices and help-seeking behavior. *AIDS Educ. Prev.*, 9, 52–67.

Wolitski, R. J., Stall, R. and Valdiserri, R. O. (eds) (2008). *Unequal Opportunity: Health Disparities Affecting Gay and Bisexual Men in the United States*. New York, NY: Oxford University Press.

Reducing sexual risk behavior among men and women with HIV infection

11

Jean L. Richardson and Tracey E. Wilson

There is little question that the control of HIV/AIDS presents one of the most difficult public health problems in the United States and around the world. The challenge of HIV prevention has increased even as HIV therapy has greatly improved survival. In the US, the annual number of AIDS cases reported decreased after the introduction of more-effective HIV therapy in 1996 (notably protease inhibitors), leveled from 2001 to 2002, and has increased since that time. Similarly, the number of cases of HIV infection in those states with names-based reporting has continued to increase. While nobody knows how many people are infected with HIV (whether AIDS or not), recent CDC surveillance reports suggest that the number could be over a million in the US (CDC, 2006).

It is estimated that 75 percent of persons in the United States with HIV are aware of their infection (CDC, 2006). Promotion of HIV testing has been an important component of efforts to control the spread of infection. Efforts aimed at promoting positive intentions to have an HIV test and toward increasing access to testing have been important strategies, because it has been estimated that acts of unprotected sex decrease by 50 percent following an HIV diagnosis (Marks *et al.*, 2005). Prior to the development of the current more-effective therapies, most of the other efforts to control HIV were focused on targeting reduction of HIV transmission risk among those who are HIV-negative. Although new HIV infections are significantly more likely to occur in situations where the infected partner is unaware of his or her HIV status, a great many new infections occur in the context in which the HIV-infected partner is aware of his or her status (Marks *et al.*, 2006). In recent years, therefore, the CDC and others have strongly supported an increase in prevention efforts which include those who are

281

HIV-positive and who are aware of their infection (Janssen *et al.*, 2001; CDC, 2003a, 2003b). In order to control an infectious disease effectively, it is essential to address both those who have the disease and might transmit it as well as those who do not have the disease and are at risk of acquiring it. While this may seem to be a simple paradigm, the disease characteristics that are unique to HIV make this a daunting task. These include the fact that because HIV is often transmitted through sexual contact, controlling it means intervening with behaviors related to a natural biological drive. In addition, because the infection currently cannot be eradicated, persons living with HIV/AIDS (PLWHA) are potentially infectious in any encounter and for their entire lifetime.

Since the early 1980s, many studies have been conducted to examine risk factors for sexual risk behavior among PLWHA. Far fewer studies have empirically tested interventions to reduce high-risk sexual behavior among PLWHA. These interventions have typically been targeted at particular risk groups, such as serodiscordant couples, injection drug users, men who have sex with men (MSM), adolescents and women. Studies have tested small-group interventions, individual counseling and clinic-based interventions. Since the advent of effective therapies for HIV, many more PLWHA receive ongoing medical care and medication prescriptions that bring them into regular contact with the health-care system. This provides an opportunity for health-care providers to evaluate, educate, problem-solve, counsel and provide referrals for their patients in order to help them reduce their high-risk behavior.

This chapter will review the literature on the prevalence of sexual risk behavior on the part of PLWHA, and on factors that are associated with increased risk for engaging in these behaviors. It will also review interventions that have been tested to reduce sexual risk behaviors among PLWHA. This chapter concludes with a discussion of the challenges in this approach to HIV prevention, suggestions for additional areas of investigation, and recommendations for approaches to incorporating these programs, particularly in clinical settings.

Sexual behavior among PLWHA

Recent reviews and subsequently published data from current and well-characterized cohorts of PLWHA are important for understanding risk behaviors within specific subpopulations, and for identifying the context in which these behaviors occur. In general, these studies suggest that at any given time, approximately one-third of PLWHA report no sexual behavior of any kind, one-third report being sexually active but always using condoms, and the remainder report engaging in unprotected intercourse (Kalichman, 1999; Erbelding *et al.*, 2000; Richardson *et al.*, 2004a; Wilson *et al.*, 2004; Purcell *et al.*, 2006; Courtenay-Quirk *et al.*, 2007). Of note, the 30 percent who report no sexual risk behavior may not be reflective of a stable group with a long-term commitment to abstinence. Data from the

HIV Cost Services Utilization Study (HCSUS), for instance, reveal that approximately 50 percent of those reporting no anal, vaginal or oral sex in the prior 6 months said that their abstinence was deliberate (Bogart *et al.*, 2006); for some of the other half of respondents, lack of sexual activity may reflect unavailability of sexual partners at any given time.

Many studies have described the sexual risk behavior of PLWHA who are aware of their HIV-seropositive status, and have described factors associated with transmission risk behaviors. These studies vary considerably in their definition of sexual risk behavior (condom-use consistency among sexually active PLWHA, prevalence of unprotected vaginal and/or anal sex, the inclusion of oral sex as a transmission risk behavior), in the extent to which they associate risk behavior directly with sexual partners at risk for HIV acquisition (partners of unknown or negative serostatus or who are known to be HIV-seropositive versus any sexual partners), and in the extent to which PLWHA are at risk for transmitting HIV (viral resistance, whether viral load is detectable). Estimates of 30 percent unprotected anal or vaginal intercourse (UAVI), therefore, may create a perception that a greater proportion of PLWHA are putting HIV-seronegative partners at risk for HIV acquisition than may actually be the case. In fact, much of the unprotected sex described in these studies is taking place within the context of HIV-seroconcordant partnerships in which disclosure had previously occurred (Weinhardt *et al.*, 2004; Duru *et al.*, 2006). Although the degree of variability in regard to these issues makes it difficult to establish precise estimates of transmission risk, they have been useful for identifying factors associated with sexual risk behavior and for identifying the prevention needs of specific subpopulations. In fact, for many PLWHA, decision-making regarding acceptable levels of risk behavior varies as a function of sex partner characteristics (e.g., partner serostatus, relationship status), characteristics of the individual (e.g., beliefs regarding HIV transmission risk given current disease status, intentions toward consistent condom use, alcohol and drug use, mental health issues), and by contextual factors, including community norms regarding sexual behavior, venues that may offer opportunities for risk behaviors and, alternatively, access to prevention services. In a review of sexual risk behavior among PLWHA, Crepaz and Marks (2002) reported that a number of intrapersonal factors were associated with increased risk for UAVI. These include lower self-efficacy/perceived behavioral control related to condom use, lower intentions/commitment to use condoms, and lower knowledge regarding HIV/AIDS transmission and health risks.

Sex-partner characteristics and transmission risk behaviors

Sex-partner characteristics are important determinants of high-risk behavior among PLWHA. Seroselection, or "serosorting", in which sexual partners use information regarding the actual or perceived HIV serostatus of their partners as a tool in gauging the level of acceptable risk in regard to condom use consistency, may be very important in determining sexual behaviors. Early studies of HIV-positive MSM found that unprotected insertive anal sex was less likely to have occurred with HIV-negative partners than with HIV-positive partners or with partners whose HIV status was not known (Wiktor *et al.*, 1990; Marks *et al.*, 1994; Wenger *et al.*, 1994). These findings generally have been confirmed by subsequent studies. Among young MSM on the West Coast, 67 percent of concordant HIV-positive couples engaged in unprotected anal intercourse, compared to 36 percent of HIV-discordant couples (Hays *et al.*, 1997). Similarly, Klitzman *et al.* (2007) found 36 percent of 1828 HIV-positive MSM had unprotected anal sex with partners who were negative or unknown. This is similar to the findings among Latino HIV-positive men and women; UAVI was most likely to occur with sex partners whose HIV status was positive or unknown than with partners who were HIV-negative (Marks *et al.*, 1998). Hoff and colleagues (1997) found that unprotected anal intercourse (UAI) was less likely to occur among men in discordant relationships, and also less likely to occur with non-primary partners. Another study of 1268 MSM found that 66 percent were sexually active, 63 percent with an HIV-positive partner; however, among those active with a negative- or unknown-status partner, 13 percent of the acts were unprotected and these were largely with casual partners (Morin *et al.*, 2004). These results suggest that while PLWHA appear to use more protection with HIV-negative or unknown-serostatus partners than with HIV-positive partners, this is not unexpected and does not mitigate the fact that a substantial number – again approaching one-third – engage in UAVI with HIV-negative or unknown partners.

Studies have shown that the number and type of sexual partners is another critical factor in determining whether a PLWHA engages in unprotected sexual behavior. One study (Richardson *et al.*, 2004a) of 840 sexually active PLWHA attending outpatient clinics showed that 34 percent engaged in UAVI. However, among the 64 percent that had had 1 sex partner in the prior 3 months, 26 percent engaged in UAVI; among the 36 percent who had had 2 or more partners, 50 percent engaged in UAVI. For those with one partner the proportions engaging in unprotected sex did not differ across sexual orientation, suggesting that, for those with one partner, relationship status tends to be the primary determinant of unsafe sex. However, sexual orientation was important in identifying those with more partners. Fifty-four percent of MSM reported 1 partner and 46 percent reported 2 or more; however, 95 percent of those with 2 or more partners were MSM. Among those with 2 or more partners the relationships were more likely to be casual, of short duration, and higher in UAI and non-disclosure. Another large study of 982 HIV-positive persons showed that 27 percent had engaged in UAVI with serodiscordant partners, and among gay and heterosexual men this was associated with having two or more partners – although this was not the case among women (Courtenay-Quirk *et al.*, 2007). In a recent examination of MSM in San Francisco, Schwarcz and colleagues (2007) reported that men in committed relationships were less likely to engage in UAVI than were men who reported casual relationships. Primary or "main partner" relationships, however, are not always exclusive. In a study of HIV-discordant MSM in primary relationships, 67 percent reported that one or both partners had had sex outside the relationship during the previous year but 75 percent reported always using condoms (Wagner *et al.*, 1998). In a study of young MSM with a new HIV diagnosis, Davidovich and colleagues (2001) found that earlier in the epidemic these men were more likely to contract HIV from a casual partner, but later in the epidemic from a steady partner. A qualitative study of 28 discordant heterosexual couples who were part of the California Partner Study (van der Straten *et al.*, 2000) found that HIV-negative partners were often the ones willing to engage in unsafe sexual behaviors.

Disclosure of HIV status and its relationship to sexual behavior has also been investigated. The type of relationship (main or casual) and its duration have important correlations with disclosure as well as with sexual risk behavior. A study of HIV-positive men and women found that 40 percent had not disclosed their HIV status to all sex partners (Stein *et al.*, 1998); 21 percent of those with one partner and 58 percent of those with two or more partners had not disclosed to all of their partners. Of those who did not disclose, 43 percent used condoms all of the time. While some studies suggest that those who had not disclosed their HIV status to all of their partners were more likely to use safer sex as a strategy for protecting their partner (Kalichman, 1999), other studies show the opposite. DeRosa and Marks (1998) found the prevalence of unsafe sex was higher among men who had not disclosed their HIV status to their HIV-negative partners than among men who did. Among 840 PLWHA, Richardson *et al.* (2004a) found non-disclosure was lower for those with one partner (20 percent) than for those with two or more partners (60 percent), although, surprisingly, non-disclosure and UAVI were not correlated. In a study of HIV-positive men whose most recent partner was HIV-negative or of unknown status, 40 percent disclosed their HIV-positive status and had safer sex; 35 percent did not disclose and had safer sex; 12 percent disclosed and had unsafe sex; and 13 percent did not disclose and had unsafe sex. Among those who disclosed, 78 percent had safer sex; and among those who did not disclose, 73 percent had safer sex (Marks and Crepaz, 2001) – again suggesting that disclosure and unsafe sex are not correlated. Duru *et al.* (2006) found that of 875 sexually active PLWHA, 41 percent did not disclose, and this was more frequent among MSM than among women or MSW. Among women living with HIV, mutual disclosure is related to condom use, but these relationships differ depending on whether sexual partners are newer or more established. As compared with an established partner, sexual partnerships that are more recently initiated appear more likely to involve consistent condom use and less likely to involve mutual HIV disclosure. More recently, Klitzman *et al.* (2007) found that among 1828 HIV-positive MSM, 46 percent disclosed to all partners and those with more partners were less likely to disclose; for those with casual partners, 21.5 percent did not disclose to any. Established partnerships are also less likely to involve a partner whose HIV serostatus is unknown, but not more or less likely to involve a partner with an HIV-negative serostatus (Wilson *et al.*, 2007). The reasons for non-disclosure and the relationship with unsafe sex suggest that some PLWHA do not disclose, but engage in safer sexual practices to mitigate risk to the partner without risking rejection; others do not disclose and engage in high-risk sexual behavior because of lower perceived responsibility, perceptions that HAART makes HIV less serious, or substance use (Duru *et al.*, 2006).

Relationship between HAART and sexual risk behavior

An important outstanding question that must be asked is whether unprotected high-risk sexual behavior has increased in recent years, and especially since the introduction of HAART. Mathematical transmission models developed by Blower *et al.* (2000) suggest that even with HAART an increase in unsafe sexual activity could lead to an increase in HIV incidence, and so the impact of HAART on sexual behavior is particularly important to consider.

Several studies suggest that being on HAART is not associated with increasing risky sexual behavior. Van der Straten *et al.* (2000) found that over two-thirds of 104 heterosexual HIV discordant couples reported any UAVI during the prior 6 months, but those who were not on HAART were more likely to report unprotected sex. More than one-third of HIV-negative partners and about 15 percent of HIV-positive partners stated they had taken a chance with unprotected sex because of HAART, with HIV-negative partners more likely to report decreased concerns about infectivity. Diamond *et al.* (2005) also found that, among 874 clinic patients who were all sexually active, less UAVI was found for both ART use and adherence to ART, and for those with undetectable HIV RNA viral load. These results persisted when stratified by many patient characteristics, suggesting that the finding is robust across various groups. A meta-analysis of 25 studies conducted at the CDC (Crepaz *et al.*, 2004) concluded that the prevalence of unprotected sex was not higher among HIV-positive persons on HAART (33 percent overall) as compared to those not receiving HAART (44 percent overall). It was also not higher among those with an undetectable viral load (39 percent overall) as compared to those with a detectable viral load (42 percent overall). However, unprotected sex was 80 percent higher among those who believed that receiving HAART or having an undetectable viral load was protective against transmitting HIV to sexual partners.

However, there are many studies that suggest that HAART is related to increased risk-taking. Data from several studies have attempted to address the issue of changes in sexual behavior since the introduction of HAART. Several studies have found that between 6 percent and 15 percent stated that new medical treatments made them more willing to engage in riskier sexual behaviors

(Dilley *et al.*, 1997; Kelly *et al.*, 1998; Remien *et al.*, 1998). HAART use was an independent predictor of unprotected anal intercourse in a sample of HIV-positive MSM in Chicago (Vanable *et al.*, 2000). In a study of HIV-positive MSM in Amsterdam, undetectable viral loads and increased CD4 counts resulting from HAART were associated with higher levels of unprotected sex with casual partners (Dukers *et al.*, 2001). Another study (Stolte *et al.*, 2004) from the same group followed 57 HIV-positive patients and found that, between 2000 and 2003, risky sex with casual partners increased from 10.5 percent to 28 percent and that with steady partners of negative or unknown HIV status from 5 percent to 11 percent, and this was associated with a more favorable perception of their viral load. A study of MSM on protease inhibitor therapy in France found that after initiating PI therapy, patients were more likely to engage in unprotected sex with partners whose HIV status was unknown to them (Miller *et al.*, 2000). This is consistent with the more recent results suggesting that beliefs about transmissibility while on HAART are critically important to sexual decision-making (Kalichman *et al.*, 2006; Wilson *et al.*, 2007; Ostrow *et al.*, 2007).

Although previous studies suggest that HAART is not independently associated with increases in sexual risk behavior among PLWHA, data from several subsequent studies suggest that these relationships may exist. A study conducted in San Francisco found that an increase in sexually transmitted infections (STI) from 1995 to 1998 among PLWHA was associated with having ever been on HAART (Scheer *et al.*, 2001). Williamson *et al.* (2006) found an increase in sexual risk behavior among more than 8000 MSM in London and Glasgow between 1996 and 2002 including more UAI overall and with seronegative or unknown partners, although the link with HAART use was inferred. Data from the Women's Interagency HIV Study found that sexually active women were more likely to engage in unprotected sex during a 6-month period after HAART initiation than before HAART initiation (Wilson *et al.*, 2004). Chen *et al.* (2003) found that UAI with at least two serodiscordant sex partners increased from 11 percent in 1999 to 16 percent in 2001 among 10,000 MSM. Truong *et al.* (2006) suggested that recent trends lead to a somewhat different conclusion. While STI among MSM in San Francisco rose from 1998 to 2004 and UAI increased as well, UAI on the part of HIV-positive MSM with serodiscordant partners decreased from 31 percent to 21 percent. However, a careful review of 53 cross-sectional and longitudinal studies conducted since 2000 suggests that risk behaviors, particularly UAI among HIV-positive MSM, may be increasing (van Kesteren *et al.*, 2007). This review suggests that in half of the studies, over 40 percent of HIV-positive MSM continue to engage in UAI, often with partners whose serostatus is negative or unknown. Additionally, UAI among HIV-positive MSM exceeds that among negative MSM. Further, most of the longitudinal studies reviewed from the US and Europe indicate that UAI is increasing among HIV-positive MSM, and this is consistent with increasing rates of STI's among MSM.

Thus, the data are not completely consistent with regard to change in risky sexual behavior since the introduction of HAART or whether any increase in risk is attributable to HAART. While there is certainly much error in self-reported sexual behavior and self presentation biases that would consistently lead to an underestimation of HIV risk behaviors, both the data from surveillance of HIV and STI's and the self report data suggest at least a modest increase in risky sexual behaviors. Certainly, risky sexual behavior on the part of PLWHA does not seem to have decreased since the start of the epidemic and since the introduction of HAART.

Intervention research addressing reduction in HIV transmission risk among PLWHA

Interventions to decrease HIV risky sexual behaviors on the part of PLWHA have been developed since the beginning of the epidemic. Early on, several of these programs addressed risky sexual behavior among couples in which one partner was HIV-positive – often due to transfusions for the treatment of hemophilia. Subsequent interventions addressed risky sexual behavior on the part of PLWHA in community settings, through social service organizations and at HIV treatment clinics. It is extremely important to address the issue of HIV prevention among PLWHA who are in main-partner relationships. These relationships are more likely to be stable and of longer duration, and in most cases the unaffected partner is aware of their partner's HIV infection. An early study by Padian *et al.* (1993) addressed serodiscordant couples in main-partner relationships. This study suggested that a 1-hour couple-counseling session had a significant impact on increasing abstinence (0 percent to 17 percent) and on increasing condom use among those not abstinent (49 percent to 88 percent). Similarly, Parsons *et al.*'s (2000) study of hemophiliacs and their partners found that those reporting only one sexual partner were less likely to engage in unsafe sexual behavior. After three hour-long sessions, unsafe sexual behavior decreased, communication about safer sex increased, and condom use self-efficacy and actual use increased. In a study of HIV-positive women with a main partner, Gielen *et al.* (2001) also showed more consistent condom use after weekly meetings with a peer advocate. One unique study in Brazil promoted condom use among the male partners of 340 HIV-positive women being seen at an outpatient clinic (Silveira and Santos, 2006). The intervention, which included education from physicians and unlimited distribution of free condoms, resulted in a 14 percent increase in condom use on the part of the male partner, but condom use in the control group also increased by 10 percent, and at 2 months 67 percent were using condoms in both groups.

Most other interventions targeting sexual risk reduction have targeted PLWHA independent of their relationship status. Kelly *et al.* (1993) conducted one of the earliest efforts to reduce unprotected sexual practices among PLWHA who were depressed. This study of 68 men tested a cognitive behavioral arm, a social support arm, and a control condition, that used eight sessions each. All groups showed reductions in UAI from the baseline to 3-month follow-up. However, the social support group was most effective in reducing unprotected receptive anal intercourse. Cleary *et al.* (1995) tested a combined individual and group intervention among 271 newly diagnosed HIV-positive patients comparing information and counseling with information and stress reduction. The proportion who were sexually active decreased (88 percent to 64 percent), and the proportion who engaged in unsafe sex in the prior week (57 percent to 31 percent) also decreased at a 4-week follow-up. Similarly, DiScenza *et al.* (1996) tested a nurse counseling intervention with 20 newly-diagnosed HIV-positive patients and found a decrease in high-risk behavior, although it did not drop below 50 percent. Patterson *et al.* (2003) tested a 90-minute counseling intervention with HIV-positive patients who had engaged in unsafe sex with a seronegative or unknown-status partner. The session focused on condom use, safer sex and disclosure, and the control focused on diet and exercise. At follow-up, all groups, including the control group, had a decrease in the total number of unprotected sex acts; however, the effects were strongest for the group that received booster sessions. In the Healthy Relationships study, Kalichman *et al.* (2001) evaluated the impact of five 120-minute group sessions that were developed based on social cognitive theories. This intervention resulted in less anal or vaginal intercourse overall, less unprotected sexual behavior, and greater condom-use consistency, particularly with partners who were HIV-negative or unknown. The WiLLOW study (Wingood *et al.*, 2004) tested an intensive intervention consisting of four 4-hour group sessions, facilitated by an educator and a peer leader, to reduce HIV transmission risk behavior among 366 women with HIV. Intervention effects at follow-up indicated reductions in unprotected vaginal intercourse and bacterial infections (chlamydia and gonorrhea), and increased knowledge, condom acceptability and condom usage.

The SUMIT study (Wolitski *et al.*, 2005) evaluated the effectiveness of a peer-led intervention to reduce sexual risk behavior and to increase disclosure among 811 HIV-positive MSM and bisexual men. The intervention group received six sessions including information and discussion led by an HIV-positive peer, while the control group received one information session. At 3-month follow-up, but not at 6-month follow-up, those who had received six sessions reported less engagement in unprotected receptive anal intercourse with a negative or unknown serostatus partner (21 percent as compared to the control group, 26 percent). Similarly, there was a lower level of UAI among the intervention group which approached significance. This intervention may have been impeded by group attendees who did not believe it was their responsibility to protect a partner's health, and may have led a group consensus away from safer sexual behaviors. Further, the intervention did not address self-protective motivations for safer sex, which may have been more salient than partner-protection motivations.

Several studies have sought to change sexual risk behavior of PLWHA who are at particularly high risk, including substance users and recent prison inmates. Mausbach *et al.* (2007) provided eight sessions of individual therapy on a weekly basis addressing either safer sex or a control condition of diet and exercise for methamphetamine users. Both groups declined in the total unprotected sex acts over time, but there was no difference between conditions in unprotected sex at any time point. Grinstead *et al.* (2001) tested an 8-week intervention delivered at the prison by community service providers for HIV-positive inmates. The goal of this program was to decrease sexual risk behavior and to increase use of social services after release. After release, those who attended the sessions showed less risky behavior than those who signed up but were unable to attend; 72 percent vs 80 percent had engaged in sex since release, and 81 percent versus 69 percent had used a condom at their first sexual encounter after release.

Other interventions have sought to reduce risks for sexual risk behavior among persons with a history of childhood sexual abuse. Wyatt and colleagues (2004) tested an intervention designed to impact both sexual risk behavior and medication adherence for HIV-positive women who had histories of sexual abuse during childhood. Women randomized to an 11-session intervention reported greater risk reduction than did those in an attention-control group. More recently, an intervention developed by Sikkema and colleagues (2008) sought to compare the effects of a 15-session therapeutic support group versus a 15-session coping group intervention among 247 men and women with HIV infection and a history of sexual abuse. While both groups exhibited reduced UAVI, participants randomly assigned to the coping group had reduced UAVI by 54 percent as compared with the support group intervention at a 12-month assessment.

Sexual activity is a natural part of human existence – and while this is true of all age groups, it is especially so for young persons. Many people acquire the disease while they are young, although there are few interventions developed to address developmentally appropriate approaches to help reduce risk among youth with HIV infection. One such approach, the TLC: Together Learning Choices program (Rotheram-Borus *et al.*, 2001), examined the impact of a 3-module, 31-session intervention on the health behaviors of HIV-positive youth aged 13 to 24. Intervention sessions focused not only on sexual risk behavior, but also on the importance of retention in HIV care, healthy lifestyle, and methods to promote improved quality of life. There was a reduced risk for unprotected sex among youth who engaged in intervention sessions, as compared to those who were assigned to the control group and to those who were assigned to, but did not attend, intervention sessions. Studies among younger PLWHA are still evolving, but will need to keep pace with changes in the HIV epidemic. While those who were infected earlier in the epidemic and are now older survivors saw the devastation, pain and suffering that HIV/AIDS can cause, younger persons who are newly infected have never had the experience of high mortality, frequent funerals, stigma and hopelessness. Their experience is of a disease that can be controlled. The new and emerging HIV cohorts may have a very different lifetime experience, and the impact of that changing experience may have a large effect on their sexual behavior.

In response to the substantial number of studies that have shown changes in both the intervention and the control groups, Lightfoot *et al.* (2007) suggested that the process of responding to interview questions about sexual behavior was a form of self-assessment for PLWHA, and that the process of program evaluation was in part responsible for changes in behavior. In line with this hypothesis, she tested an intervention that asked clinic patients to respond to a computer-assisted self-interview. Results showed that PLWHA reported fewer partners and higher rates of abstinence, as well as greater condom-protected sexual interactions with known HIV-negative or unknown partners and changes in attitudes more conducive to prevention, and this was related to the increased number of self-assessments. If we are able to discount the idea that there may be a greater propensity to respond in a socially desirable manner following repeated assessments, then it may be that a relatively low-cost intervention could have benefit for patients.

Sexual risk reduction approaches within the HIV primary-care setting

There are many reasons to integrate HIV prevention interventions into the clinical care settings at which HIV care is provided. Health-care providers in these settings are knowledgeable about the disease, how to prevent transmission, what the patient's health status is, why prevention is important for the patient as well as the partner, what treatments the patient is on, and how to discuss the possibility of transmission even with a low viral load. Besides the obvious expectation that the provider will have uniquely accurate information to share with the patient, the provider will see the patient repeatedly for an extended period of time, often discussing intimate matters as well as conducting physical exams; the provider may be the only individual who is fully aware of the patient's health status. Health-care providers are generally respected and, in most cases, patients believe it is in their best interest to comply with the provider recommendations (although this may be more true regarding clinical management rather than sexual behavior). Discussion of ways to avoid transmitting the virus is more acceptable and expected in the health-care setting, and generally consistent with the physician's code of ethics. Clinic settings provide an optimal location for training large numbers of physicians efficiently. Thus, both the feasibility and potential benefit of counseling in clinical settings is compelling.

The integration of HIV-prevention counseling into HIV care has been suggested by public health leaders and agencies both before and since the advent of HIV therapy (Francis *et al.*, 1989; Francis, 1996; Janssen *et al.*, 2001; CDC,

2003b). Despite this, there is a strong suggestion that counseling is rarely done in the absence of specific interventions to encourage this behavior, and certainly there are many barriers that providers face in this regard. Margolis *et al.* (2001) reported that 23 percent of seropositive men indicated that their health-care provider had never talked with them about safer sex. Marks *et al.* (2002) similarly reported that 29 percent of HIV-positive patients had not discussed safer sex and 50 percent had not discussed disclosure with their provider. Metsch *et al.* (2004) found that 90 percent of physicians spent more than 15 minutes with each patient and 60 percent counseled almost all of their new patients about HIV risk reduction, but only 14 percent counseled almost all of their established patients. Morin *et al.* (2004) found that only 25 percent had discussed safer sex and only 6 percent had discussed specific sexual activities at the review visit, while 53 percent and 24 percent respectively had had these discussions in the prior 6 months. In clinics with written procedures, counseling was as much as three times higher than in clinics without such procedures (Myers *et al.*, 2004).

Two studies tested interventions that involved brief counseling sessions with health-care providers in clinic settings during routine HIV visits (Richardson *et al.*, 2004b; Fisher *et al.*, 2006). Both of these relied on integrating counseling within a clinic setting, training providers to deliver brief, individually-delivered risk reduction messages, and implemented procedures that helped ensure that counseling was conducted regularly. The theoretical basis for the programs was different; however, the actual process for the intervention interactions by providers with patients was very similar.

The first of these, the Partnership for Health (PfH) study, tested the efficacy of brief, safer-sex counseling delivered by HIV primary-care providers to patients during routine medical examinations (Richardson *et al.*, 2004b). Three compatible behavioral theories were incorporated into the PfH model. Message framing was formally tested, and was supported by the mutual participation and the stages of change models. This study emphasized the patient–provider relationship to encourage safer behaviors. The intervention incorporated elements of the transtheoretical model of behavior change by suggesting that patients may go through the "stages of change" to maintaining safer behavior. This study tested a communication strategy known as message framing, which uses motivational messages that emphasize the benefits of safer behaviors (gain frame) or the costs of unsafe behavior (loss frame). The study was conducted in six clinics in California, based on 585 patients present at baseline and follow-up. Two clinics used gain-frame prevention messages/counseling, two clinics used loss-frame messages/counseling, and two clinics implemented an attention-control protocol (adherence to antiretroviral therapy). Among participants who had two or more sex partners at baseline (almost all were gay or bisexual men), there was a 38 percent reduction ($P < 0.001$) in the prevalence of UAVI among those who received the loss-frame intervention. No significant changes were observed in gain-frame or attention-control clinics. In participants with multiple partners at baseline, the likelihood of UAVI at follow-up was significantly lower in the loss-frame arm (OR 0.42, 95% CI 0.19–0.91) compared with the control arm after adjusting for baseline differences. Adjustment for clustering did not change the conclusions. Similar results were also obtained among participants with casual partners at baseline. No effects were seen in participants who reported only one partner or only a main partner at baseline. In addition, no effects were seen for disclosure of HIV status to sex partners, regardless of the frame of the intervention and regardless of number of partners at baseline (although non-disclosure was primarily a concern for those with two or more partners and those with casual partners, and less of an issue for those with one main partner).

The OPTIONS project was designed and tested by Fisher and colleagues (2004, 2006). The study was based on the information–motivation–behavioral skills model of prevention developed by Fisher and his colleagues. The intervention used techniques drawn from motivational interviewing which emphasized a patient-centered strategy to promote risk reduction. This study was carried out in two clinics in Connecticut, with clinics assigned to either the intervention or the control condition. This study assessed the number of unprotected vaginal, anal and insertive oral sexual acts with HIV-negative or status-unknown partners, and a similar measure without the inclusion of oral sex. The study recruited 490 patients, and followed many of them for 4 time points. The study results showed that the unprotected vaginal, anal and insertive oral sex decreased significantly over time among those who received the intervention program, but in the control arm these behaviors increased significantly. The intervention effect was also present when analysis was conducted only for UAVI. The analysis also suggests that these results hold when examined for risk behaviors engaged in with HIV-negative or unknown-serostatus partners.

Prevention case management, or comprehensive counseling, responds to a number of potential needs experienced by PLWHA that may interfere with their ability to engage in safer behavior. These might include medical needs, drug dependency, social service needs, housing needs, emotional needs such as coping or depression, and others. These types of approaches help patients to deal with the contextual factors that may be associated with lower skills, self-efficacy, or motivation to engage in safer sexual behaviors, and can be implemented within the clinical setting. They help to promote and maintain a consistent presence in HIV primary care, which would increase exposure to risk-reduction interventions. Several studies have examined the impact of prevention case management models on the sexual risk behavior of PLWHA. In an examination of the impact of a prevention case management approach on sexual risk behaviors of a group of men and women with HIV infection in Wisconsin, Gasiorowicz and colleagues (2005) reported that men and women reported lower levels of risk behaviors following individual risk-reduction counseling paired with prevention case management. Sorensen *et al.* (2003) compared 12 months of case management with a brief contact intervention for PLWHA who were substance abusers.

Both the intervention and control groups reported a decrease in problems and in transmission risk behavior from baseline to the 6-month assessment, but no change at 12 and 18 months. However, there did not appear to be a dose–effect relationship between the amount of case management services received and shifts in sexual risk behavior. A similar study of IDU in methadone maintenance compared a program that included 6 months of individual case management with twice-weekly harm-reduction group sessions. Again both groups decreased sexual risk behaviors, but the effect was significantly stronger in the harm-reduction program (Margolin *et al.*, 2003). The Healthy Living Project was an intervention that included 15 sessions, each of 90 minutes, that were individually delivered and focused on goal-setting and problem-solving. The study randomized patients to prevention case management and to a control group. At follow-up, both groups reduced the number of acts of unprotected sex with HIV-negative or unknown partners; however, the reduction was greater among the intervention group by 22 percent, 23 percent and 36 percent at 10, 15 and 20 months respectively (Gore-Felton *et al.*, 2005; Healthy Living Project Team, 2007).

Effectiveness of sexual risk-reduction programs among PLWHA

The studies described here often encountered problems with drop-outs, inconsistent attendance and excessive burden that would negatively impact the probability of diffusion as a large-scale intervention. In addition, as with all self-report assessments, the possibility of self-presentation bias in reporting as well as demand effects, especially for those in the intervention groups, cannot be ruled out. Despite these limitations, however, it does appear that these programs are effective at reducing sexual risk behavior among PLWHA. Two recent meta-analyses have summarized the results of many of these programs. Johnson *et al.* (2006) examined 15 studies that included follow-up to 47 weeks. Overall, there was a statistically significant increase in condom use, with greater effects seen among younger participants and when the programs included motivational and behavioral skill-training elements. Crepaz and colleagues (2006) examined the results of 12 studies which included follow-up periods from 3 months to 1 year. Overall, the reviewers found statistically significant reductions in HIV risk behaviors (OR 0.57) and a decreased risk for acquisition of STI (OR 0.20) among those receiving intervention activities. Interventions were found to have greater effectiveness if they were based on behavioral theory, included skills-building components, if they were delivered in intensive sessions and delivered by health-care providers or counselors, and when they were delivered in HIV primary-care settings.

In sum, the ongoing risky sexual behavior of at least one-third of PLWHA, the increasing prevalence of HIV, the few studies that have tested interventions, the difficulty of conducting this type of intervention research, the possible changes in attitudes toward preventive behavior among PLWHA since the advent of HAART, and the age cohort effect all make this a very sobering point in the epidemic. Yet, evidence exists to support the adoption and adaptation of many of the approaches to risk-reduction described here, particularly in regard to prevention programs administered in HIV clinical-care settings.

Integration of HIV prevention programs into the clinical setting

As with any new program, there is a gradual process of integration of a program within a clinic. In order for these programs to be effective, they need to be adopted by the primary-care providers at the clinic. However, there are many barriers to HIV prevention counseling in the clinical setting (Mayer *et al.*, 2004; Grodensky *et al.*, 2007). The first barrier, and the most difficult to overcome, is the issue of limited time with each patient. Providers must offer clinical care and assessment as well as counseling about the importance of medication adherence. Counseling about prevention is often considered to be another burden – and one that should be or could be better handled by somebody else. In some cases providers may feel that there is little probability of patients changing their high-risk sexual behavior, and therefore there does not seem to be any benefit in spending precious time counseling them (Steward *et al.*, 2006). In fact, this provider "fatalism" may be highest in clinics where there is the highest level of UAI and where the patients need this counseling the most. Additional barriers include feelings of discomfort about discussing risky sexual behavior with their patients (Mayer *et al.*, 2004). This could be because of lack of training, the provider's own religious views, beliefs about the patient's resistance, or concerns that the provider may appear judgmental. Providers may not believe that prevention is their responsibility, and that it is not necessarily a part of providing good medical care to the patient – who is, after all, their primary responsibility. Providers may also believe that patients are not truthful with them, and that an honest discussion cannot take place. Providers may feel that other issues, such as drug and alcohol use or mental illness, may interfere with the ability to communicate, reason and discuss important prevention issues with some patients. Finally, providers may feel that there are legal issues, such as reporting policies, that can be raised in these sessions that they would not want, or be able, to address. Research has shown that adoption is more likely to occur when organizational factors such as strong morale, active and general support of administrators, a good fit between program and organization/local needs and the presence of a program champion to take leadership are all in place. Individual providers are more likely to adopt a program when they have positive attitudes toward the program, when they are comfortable with the approach and their involvement, when the instructional methods are clear and straightforward, and when they have high self-efficacy to implement the program (Grodensky *et al.*, 2007).

Brief training programs to address clinician-perceived individual and organizational barriers to counseling, however, can address many of these barriers by improving provider self-efficacy for risk-reduction counseling (Bluespruce *et al.*, 2001). Evidence for the efficacy and feasibility of implementation of these types of programs was demonstrated in the Positive STEPs demonstration project (Gardner *et al.*, 2008). This CDC-funded initiative was designed to evaluate risk-reduction interventions delivered by medical providers to HIV patients during routine HIV primary care. The program, implemented at seven HIV clinics, included screening patients for behavioral risks, administration to patients at primary-care visits of brief, provider-initiated messages regarding safer sex during routine HIV care visits, provision of a brochure with prevention messages and strategies, and posters located in waiting areas, exam rooms and staff common areas. For some clinics, in-depth approaches to risk reduction were provided by on-site prevention case managers for those patients reporting HIV transmission risk behaviors. Each participating clinic received a standardized training for program activities, followed by a "booster" training session once program activities had been implemented. Trained study staff at each clinic enrolled a measurement cohort of sexually active patients, administered a baseline questionnaire, and collected baseline medical record data for each cohort participant. Overall, the 3-month prevalence of UAVI among the 767 measurement cohort participants declined significantly from baseline (42 percent) to follow-up (26 percent at 6-month follow-up, 23 percent at 12-month follow-up). In addition, the reduction in the prevalence of UAVI was associated with patients' self-reported receipt of safer-sex counseling. Reports of counseling at all, some, or no visits (from baseline to first follow-up) showed a dose–response relationship with a decline in UAVI in that interval, with relative reductions of 45 percent, 35 percent and 19 percent, respectively. The Positive STEPs program demonstrated that prevention programs for PLWHA can be successfully integrated into HIV care, and can significantly impact the risk behaviors of patients.

While several interventions have successfully reduced UAVI among HIV-positive patients through implementation of different approaches such as prevention case management and provider-administered counseling, there are few available models for the full integration of all of these prevention activities. One such approach is depicted in Figure 11.1. This figure suggests that the clinical-care setting can be used as a triage point for HIV risk-reduction counseling and other prevention services. As the use of HAART has become more available, effective and sought out by PLWHA, the clinical setting has become the most likely point that PLWHA will continue to access with regularity. The clinical setting can act as a location for assessment of risky sexual behavior, as well as to provide accurate information and brief counseling. Within this setting, a program such as Partnership for Health, OPTIONS, or another approach could be implemented in a way to match patients with the intensity of service most appropriate to their needs. In this way, those who do not exhibit risky behavior primarily

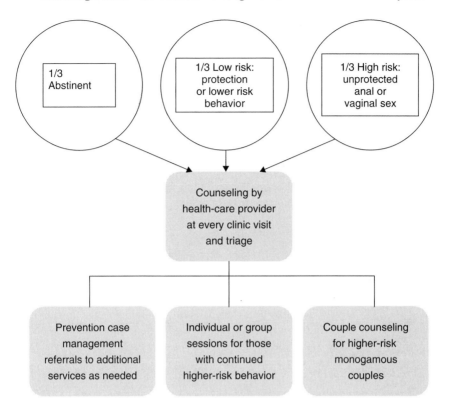

Figure 11.1   An integrated program of HIV prevention for PLWHA.

because they are abstinent (approximately one-third of PLWHA) or because they have one main partner and do not have risky sexual interactions with that person (approximately half of those who are sexually active or about one-third of PLWHA) can be briefly counseled in that setting. This leaves the one-third of PLWHA who are engaging in risky sexual behavior. This group can be identified in the clinic setting, and provided with brief counseling. While brief discussions in the clinical setting by a health-care professional – preferably the primary-care provider but possibly another health professional – will have an impact on a portion of those who engage in high-risk sexual interactions, it is likely that many will need more intensive services. If the PLWHA is in a serodiscordant relationship and is continuing to engage in high-risk sexual behavior, then both people in the relationship may need to be counseled, either alone or with other similar serodiscordant couples. On the other hand, a sizeable number of those who are engaging in high-risk sexual behavior are primarily involved in casual relationships, and perhaps one of the previously tested individual- or group-counseling approaches might be helpful. Finally, some of the patients may need help with other demands of living, and prevention case management may be necessary. In order to integrate this multicomponent prevention program for PLWHA, careful planning, expanded resources and integration with other AIDS service organizations need to take place.

Conclusions

There is little evidence that there has been an overall reduction in high-risk sexual activity on the part of PLWHA since the studies conducted in the 1980s. To some extent this is to be expected. Some studies of sexual behavior suggest that there may have been an increase in risky sexual behavior since the advent of HAART. This may be due in part to assumptions about reduced infectivity on HAART, especially when the viral load is undetectable. In addition, few symptoms of infection, slowed disease progression, and fewer acute episodes may be linked to higher levels of sexual functioning; PLWHA feel better, and are likely to become more sexually active as a result. As HIV therapeutic regimens become simplified and easier to tolerate, the need to visit the physician or clinic may decrease and there would then be fewer patient and provider encounters. Thus patients would come into contact with counseling sessions less often, potentially lessening the impact of clinic-based interventions. So, while good models exist, they continue to operate in a changing environment. Those interventions that were tested and in the field in the 1980s and early 1990s are very different from those possible in the post-HAART period.

What is the model for integrating prevention programs for PLWHA in the post-HAART era? There are many theoretical and practical guides for developing and integrating prevention programs for PLWHA, and the experience of existing studies provides a guide for program development. However, in addition, a thoughtful look at what makes a program diffusible at the public health level of implementation also needs to be examined. Issues of cost, training, resources, complexity and fidelity to the original plan need to be considered. As was discussed early in this chapter, PLWHA form a heterogeneous group. Interventions will need to be available for those with different needs, and the ability to match individuals with interventions needs to be clarified and efficient strategies to be developed to link persons with the services that fit their needs.

This chapter has reviewed the level of sexual activity among PLWHA, and the prevention interventions that have been designed and tested to reduce HIV transmission risk behaviors among PLWHA. It is important to note at the outset that, from the studies that have been conducted, it appears that the majority of PLWHA are either abstinent or engage in safer sexual behaviors. More than half are in one-partner relationships, and for the most part disclosure has taken place in these relationships, although at times these couples also engage in unprotected sexual behavior. A majority of PLWHA who are aware of their HIV status are sexually active; however, a significant minority engage in unprotected sexual intercourse. PLWHA represent the full spectrum of sexual behavior, from abstinence to multipartner sexually compulsive behavior and from no partners to many partners, from monogamous to serial monogamy to multiple casual partners and exchange partners. They are young and old, educated and uneducated, male and female, self-interested and other-interested, wise and unwise – in fact, they are as heterogeneous as the population. From this viewpoint, PLWHA form a diverse group, cannot be easily considered as homogeneous, have different reasons prompting their sexual behavior, and will respond to different persuasive appeals and interventions to maintain lower-risk sexual behaviors. To further complicate the issue, the entire public perception of HIV/AIDS continues to change, and is undergoing a transformation from a news-dominating fearsome epidemic among the American people to an often ignored chronic disease – although in reality the stigma has not been eliminated. This change in public perception must have an impact on the behavior of those with HIV– after all, they are part of the public as well.

It is clear that there are many studies that have attempted to measure the extent of risky sexual behavior and to identify the myriad intrapersonal and contextual factors that are related to these behaviors. Almost all of them end with a suggestion that interventions be designed for PLWHA. Despite this long-held view, there is a remarkably small number of studies that have implemented and evaluated approaches to reducing HIV transmission behaviors among PLWHA. These are difficult studies to conduct, and are fraught with methodological problems; however, this cannot be regarded as reason for the paucity of research on this topic. Nevertheless, the studies that have been done suggest that these interventions can have an effect on reducing high-risk sexual behavior. One note of caution in this regard, however, is that many of the studies seem to have an influence on the control group as well as on the intervention group, and this may be due to self-presentation bias, contamination of intervention activities, historical events, or the effect of measurement itself on behaviors. Further, there is much to learn about how new behavioral strategies can be readily adapted and maintained widely, about how to ensure that these strategies stay current with the evolving intrapersonal and interpersonal factors that influence risk, and how to address gaps in our knowledge regarding understudied populations such as heterosexual and MSM minority populations of PLWHA. Thus, it is certainly not a time to become complacent. The development of feasible, acceptable and effective approaches to reducing sexual risk behavior among PLWHA needs to be a high priority for public health practitioners at the clinical, community-based, academic and/or federal levels.

References

Blower, S. M., Gershengorn, H. B. and Grant, R. M. (2000). A tale of two futures: HIV and antiretroviral therapy in San Francisco. *Science*, 287, 650–4.

Bluespruce, J., Dodge, W. T., Grothaus, L. *et al.* (2001). HIV prevention in primary care: impact of a clinical intervention. *AIDS Patient Care STDs*, 15, 243–53.

Bogart, L. M., Collins, R. L., Kanouse, D. E. *et al.* (2006). Patterns and correlates of deliberate abstinence among men and women with HIV/AIDS. *Am. J. Public Health*, 96, 1078–84.

CDC (2003a). Advancing HIV prevention: new strategies for a changing epidemic – United States. *Morbid. Mortal. Wkly Rep.*, 52, 329–32.

CDC (2003b). Incorporating HIV prevention into the medical care of persons living with HIV. *Morbid. Mortal. Wkly Rep.*, 52, 1–23.

CDC (2006). *HIV/AIDS Surveillance Report. (2005).* Atlanta GA: US Department of Health and Human Services, pp. 1–54.

Chen, S. Y., Gobson, S., Weide, D. and McFarland, W. (2003). Unprotected anal intercourse between potentially HIV-serodiscordant men who have sex with men, San Francisco. *J. Acquir. Immune Defic. Syndr.*, 33, 166–70.

Cleary, P. D., Devanter, N., Steilen, M. *et al.* (1995). A randomized trial of an education and support program for HIV-infected individuals. *AIDS*, 9, 1271–8.

Courtenay-Quirk, C., Pals, S. L., Colfax, G. *et al.* (2007). Factors associated with sexual risk behavior among persons living with HIV: gender and sexual identity group differences. *AIDS Behav.*, in press (e-pub ahead of print, 26 June).

Crepaz, N. and Marks, G. (2002). Towards an understanding of sexual risk behavior in people living with HIV: a review of social, psychological, and medical findings. *AIDS*, 16, 135–49.

Crepaz, N., Hart, T. A. and Marks, G. (2004). Highly active antiretroviral therapy and sexual risk behavior. *J. Am. Med. Assoc.*, 292, 224–36.

Crepaz, N., Lyles, C. M., Wolitski, R. J. *et al.* (2006). Do prevention interventions reduce HIV risk behaviours among people living with HIV? A meta-analytic review of controlled trials. *AIDS*, 20, 143–57.

Davidovich, U., de Wit, J., Albrecht, N. *et al.* (2001). Increase in the share of steady partners as a source of HIV infection: a 17-year study of seroconversion among gay men. *AIDS*, 15, 1303–8.

DeRosa, C. J. and Marks, G. (1998). Preventive counseling of HIV-positive men and self-disclosure of serostatus to sex partners: new opportunities for prevention. *Health Psychology*, 17, 224–31.

Diamond, C., Richardson, J. L., Milam, J. *et al.* (2005). Use of and adherence to antiretroviral therapy is associated with decreased sexual risk behavior in HIV clinic patients. *J. Acquir. Immune Defic. Syndr.*, 39, 211–18.

Dilley, J. W., Woods, W. J. and McFarland, W. (1997). Are advances in treatment changing views about high-risk sex? *N. Engl. J. Med.*, 337, 501–2.

DiScenza, S., Nies, M. and Jordan, C. (1996). Effectiveness of counseling in the health promotion of HIV-positive clients in the community. *Public Health Nursing*, 13, 209–16.

Dukers, N. H. T. M., Goudsmit, J., deWit, J. B. F. *et al.* (2001). Sexual risk behavior related to the virological and immunological improvements during highly active antiretroviral therapy in HIV-1 infection. *AIDS*, 15, 369–78.

Duru, O. K., Collins, R. L., Ciccarone, D. H. *et al.* (2006). Correlates of sex without serostatus disclosure among a national probability sample of HIV patients. *AIDS Behav.*, 10, 495–507.

Erbelding, E. J., Stanton, D., Quinn, T. C. and Rompalo, A. (2000). Behavioral and biologic evidence of persistent high-risk behavior in an HIV primary care population. *AIDS*, 14, 297–301.

Fisher, J. H., Cornman, D. H., Osborn, C. Y. *et al.* (2004). Clinician initiated HIV risk reduction intervention for HIV-positive persons: formative research, acceptability, and fidelity of the OPTIONS project. *J. Acquir. Immune Defic. Syndr.*, 37, S78–87.

Fisher, J. H., Fisher, W. A., Cornman, D. H. *et al.* (2006). Clinician-delivered intervention during routine clinical care reduces unprotected sexual behavior among HIV-infected patients. *J. Acquir. Immune Defic. Syndr.*, 41, 44–52.

Francis, D. P. (1996). Every person infected with HIV-1 should be in a lifelong early intervention program. *Sex. Transm. Dis.*, 23, 351–2.

Francis, D. P., Anderson, R. E., Gorman, M. E. *et al.* (1989). Targeting AIDS prevention and treatment toward HIV-1 infected persons: the concept of early intervention. *J. Am. Med. Assoc.*, 262, 2572–6.

Gardner, L., Marks, G., O'Daniels, C. *et al.* (2008). Implementation and evaluation of a clinic-based behavioral intervention: Positive STEPs for HIV patients. *AIDS Patient Care STDs* (in press).

Gasiorowicz, M., Llanas, M. R., DiFranceisco, W. *et al.* (2005). Reductions in transmission risk behaviors in HIV-positive clients receiving prevention case management services: findings from a community demonstration project. *AIDS Educ. Prev.*, 17(1 Suppl. A), 40–52.

Gielen, A. C., Fogarty, L. A., Armstrong, K. *et al.* (2001). Promoting condom use with main partners: a behavioral intervention trial for women. *AIDS Behav.*, 5, 193–204.

Gore-Felton, C., Rotheram-Borus, M. J., Weinhardt, L. S. *et al.* (2005). The Healthy Living Project: an individually tailored, multidimensional intervention for HIV-infected persons. *AIDS Educ. Prev.*, 17(1 Suppl. A), 21–39.

Grinstead, O., Zack, B. and Faigeles, R. (2001). Reducing postrelease risk behavior among HIV seropositve prison inmates: the health promotion program. *AIDS Educ. Prev.*, 13, 109–19.

Grodensky, C. A., Golin, C. E., Boland, M. S. *et al.* (2007). Translating concern into action: HIV care providers' views on counseling patients about HIV prevention in the clinical setting. *AIDS Behav.*, in press (e-pub ahead of print available at http://www.springerlink.com/content/10j47u2118672mw6/).

Hays, R. B., Kegeles, S. M. and Coates, T. J. (1997). Unprotected sex and HIV risk taking among young gay men within boyfriend relationships. *AIDS Educ. Prev.*, 9, 14–329.

Healthy Living Project Team (2007). Effects of a behavioral intervention to reduce risk of transmission among people living with HIV: the healthy living project randomized controlled study. *J. Acquir. Immune Defic. Syndr.*, 44, 213–21.

Hoff, C. C., Stall, R., Paul, J. *et al.* (1997). Differences in sexual behavior among HIV discordant and concordant gay men in primary relationships. *J. Acquir. Immune Defic. Syndr.*, 14, 72–8.

Janssen, R. S., Holtgrave, D. R., Valdiserri, R. O. *et al.* (2001). The serostatus approach to fighting the HIV epidemic: prevention strategies for infected individuals. *Am. J. Public Health*, 91, 1019–24.

Johnson, B. T., Carey, M. P., Chaudoir, S. R. and Reid, A. E. (2006). Sexual risk reduction for persons living with HIV: research synthesis of randomized controlled trials 1993 to 2004. *J. Acquir. Immune Defic. Syndr.*, 41, 642–50.

Kalichman, S. C. (1999). Psychological and social correlates of high-risk sexual behavior among men and women living with HIV/AIDS. *AIDS Care*, 11, 415–28.

Kalichman, S. C., Rompa, D., Cage, M. *et al.* (2001). Effectiveness of an intervention to reduce HIV transmission risks in HIV-positive people. *Am. J. Prev. Med.*, 21, 84–92.

Kalichman, S. E., Eaton, L., Cain, D. *et al.* (2006). HIV treatment beliefs and sexual transmission risk behaviors among HIV positive men and women. *J. Behav. Med.*, 29, 401–10.

Kelly, J. A., Murphy, D. A., Bahr, R. *et al.* (1993). Outcome of cognitive-behavioral and support group brief therapies of depressed, HIV-infected persons. *Am. J. Psychiatry*, 150, 1679–86.

Kelly, J. A., Hoffman, R. G., Rompa, D. and Gray, M. (1998). Protease inhibitor combination therapies and perceptions of gay men regarding AIDS severity and the need to maintain safer sex. *AIDS*, 12, F91–95.

Klitzman, R., Exner, T., Correale, J. *et al.* (2007). It's not just what you say: relationships of HIV disclosure and risk reduction among MSM in the post-HAART era. *AIDS Care*, 19, 749–56.

Lightfoot, M., Rotheram-Borus, M. J., Comulada, S. *et al.* (2007). Self-monitoring of behaviour as a risk reduction strategy for persons living with HIV. *AIDS Care*, 19, 757–63.

Maragolis, A. D., Wolitski, R. J., Parsons, J. R. and Gomez, C. A. (2001). Are healthcare providers talking to HIV-seropositive patients about safer sex? *AIDS*, 23, 2335–7.

Margolin, A., Avants, S. K., Warburton, L. A. *et al.* (2003). A randomized clinical trial of a manual guided risk reduction intervention for HIV-positive injection drug users. *Health Psychol.*, 22, 223–8.

Marks, G. and Crepaz, N. (2001). HIV-positive men's sexual practices in the context of self disclosure of HIV status. *J. Acquir. Immune Defic. Syndr.*, 27, 79–85.

Marks, G., Ruiz, M. S., Richardson, J. L. *et al.* (1994). Anal intercourse and disclosure of HIV infection among seropositive gay and bisexual men. *J. Acquir. Immune Defic. Syndr.*, 7, 866–9.

Marks, G., Cantero, P. J. and Simoni, J. M. (1998). Is acculturation associated with sexual risk behaviours? An investigation of HIV-positive Latino men and women. *AIDS Care*, 10, 283–95.

Marks, G., Richardson, J. L., Crepaz, N. *et al.* (2002). Are HIV care providers talking with patients about safer sex and disclosure? A multi-clinic assessment. *AIDS*, 16, 1953–7.

Marks, G., Crepaz, N., Senterfitt, J. W. and Janssen, R. S. (2005). Meta-analysis of high risk sexual behavior in persons aware and unaware they are infected with HIV in the United States: implications for HIV prevention programs. *J. Acquir. Immune Defic. Syndr.*, 39, 446–53.

Marks, G., Crepaz, N. and Janssen, R. S. (2006). Estimating sexual transmission of HIV from persons aware and unaware that they are infected with the virus in the USA. *AIDS*, 20, 1447–50.

Mausbach, B. T., Semple, S. J., Strathdee, S. A. *et al.* (2007). Efficacy of a behavioral intervention for increasing safer sex behaviors in HIV-positive MSM methamphetamine users: results from the EDGE study. *Drug Alcohol Depend.*, 87, 249–57.

Mayer, K. H., Safren, S. A. and Gordon, C. M. (2004). HIV care providers and prevention: opportunities and challenges. *J. Acquir. Immune Defic. Syndr.*, 3, S130–133.

Metsch, L. R., Pereyra, M., del Rio, C. *et al.* (2004). Delivery of HIV prevention counseling by physicians at HIV medical care settings in 4 US cities. *Am. J. Public Health*, 94, 1186–92.

Miller, M., Meyer, L., Boufassa, F. *et al.* (2000). Sexual behavior changes and protease inhibitor therapy. *AIDS*, 14, F33–39.

Morin, S. F., Koester, K. A., Steward, W. T. *et al.* (2004). Missed opportunities: prevention with HIV-infected patients in clinical care settings. *J. Acquir. Immune Defic. Syndr.*, 36, 960–6.

Myers, J. J., Steward, W. T., Charlebois, E. *et al.* (2004). Written clinic procedures enhance delivery of HIV prevention with positives counseling in primary health care settings. *J. Acquir. Immune Defic. Syndr.*, 37, S95–100.

Ostrow, D. G., Silverberg, M. J., Cook, R. L. *et al.* (2008). Prospective study of attitudinal and relationship predictors of sexual risk in the Multicenter AIDS Cohort Study. *AIDS Behav.*, 12, 127–38.

Padian, N. S., O'Brien, T. R., Change, Y. C. *et al.* (1993). Prevention of heterosexual transmission of human immunodeficiency virus through couple counseling. *J. Acquir. Immune Defic. Syndr.*, 6, 1043–8.

Parsons, J. R., Huszti, H. C., Crudder, S. O. *et al.* (2000). Maintenance of safer sexual behaviors: evaluation of a theory-based intervention for HIV seropositive men with haemophilia and their female partners. *Haemophilia*, 6, 181–90.

Patterson, T. L., Shaw, W. S. and Semple, S. J. (2003). Reducing the sexual risk behaviors of HIV+ individuals: Outcome of a randomized controlled trial. *Ann. Behav. Med.*, 25, 137–45.

Purcell, D. W., Mizuno, Y., Metsch, L. R. *et al.* (2006). Unprotected sexual behavior among heterosexual HIV-positive injection drug using men: Associations by partner type and partner serostatus. *J. Urban Health*, 83, 656–68.

Remien, R. H., Wagner, G., Carballo-Dieguez, A. and Dolezal, C. (1998). Who may be engaging in high-risk sex due to medical treatment advances? *AIDS*, 12, 1560–1.

Richardson, J. L., Milam, J., Stoyanoff, S. *et al.* (2004a). Using patient risk indicators to plan prevention strategies in the clinical care setting. *J. Acquir. Immune Defic. Syndr.*, 37, S88–94.

Richardson, J. L., Milam, J., McCutchan, A. *et al.* (2004b). Effect of brief provider safer-sex counseling of HIV-1 seropositive patients: A multi-clinic assessment. *AIDS*, 18, 1179–86.

Rotheram-Borus, M. J., Lee, M. B., Murphy, D. A. *et al.* (2001). Efficacy of a preventive intervention for youths living with HIV. *Am. J. Public Health*, 91, 400–5.

Scheer, S., Chu, P. L., Klausner, J. D. *et al.* (2001). Effect of highly active antiretroviral therapy on diagnosis of sexually transmitted diseases in people with AIDS. *Lancet*, 357, 432–5.

Schwarcz, S., Scheer, S., McFarland, W. *et al.* (2007). Prevalence of HIV infection and predictors of high-transmission sexual risk behaviors among men who have sex with men. *Am. J. Public Health*, 97, 1067–75.

Sikkema, K., Wilson, P., Hansen, N. *et al.* (2008). Effects of a coping intervention on transmission risk behavior among people living with HIV/AIDS and a history of childhood sexual abuse. *J. Acquir. Immune Defic. Syndr.*, in press (e-pub ahead of print 12 July, available at: http://www.springerlink.com/content/r01j024p28134513/).

Silveira, M. F. and Santos, I. S. (2006). Impact of an educational intervention to promote condom use among the male partners of HIV positive women. *J. Eval. Clin. Pract.*, 12, 102–11.

Sorensen, J. L., Dilley, J., London, J. *et al.* (2003). Case management for substance abusers with HIV/AIDS: a randomized clinical trial. *Am. J. Drug Alcohol Abuse*, 29, 133–50.

Stein, M. D., Freedberg, K. A., Sullivan, L. M. *et al.* (1998). Sexual ethics: disclosure of HIV-positive status to partners. *Arch. Int. Med.*, 158, 253–7.

Steward, W. T., Koester, K. A., Myers, J. J. and Morin, S. F. (2006). Provider fatalism reduces the likelihood of HIV-prevention counseling in primary care settings. *AIDS Behav.*, 10, 3–12.

Stolte, I. G., deWit, J. B. F., van Eeden, A. *et al.* (2004). Perceived viral load, but not actual HIV-1-RNA load is associated with sexual risk behavior among HIV infected homosexual men. *AIDS*, 18, 1943–9.

Truong, H. H. M., Kellogg, T., Klausner, J. D. *et al.* (2006). Increases in sexually transmitted infections and sexual risk behaviour without a concurrent increase in HIV incidence among men who have sex with men in San Francisco: a suggestion of HIV serostorting? *Sex. Transm. Infect.*, 82, 461–6.

Vanable, P. A., Ostrow, D. G., McKirnan, D. J. *et al.* (2000). Impact of combination therapies on HIV risk perceptions and sexual risk among HIV-positive and HIV-negative gay and bisexual men. *Health Psychol.*, 19, 134–45.

Van Kesteren, N. M. C., Hospers, J. H. and Kok, G. (2007). Sexual risk behavior among HIV-positive men who have sex with men: a literature review. *Patient Educ. Couns.*, 65, 5–20.

Van der Straten, A., Gomez, C., Saul, J. *et al.* (2000). Sexual risk behaviors among heterosexual HIV serodiscordant couples in the era of post-exposure prevention and viral suppressive therapy. *AIDS*, 14, F47–54.

Wagner, G. J., Remien, R. H. and Carballo-Dieguez, A. (1998). Extramarital sex: is there an increased risk for HIV transmission? A study of male couples of mixed HIV status. *AIDS Educ. Prev.*, 10, 245–56.

Weinhardt, L. S., Kelly, J. A., Brondino, M. J. *et al.* (2004). HIV transmission risk behavior among men and women living with HIV in 4 cities in the United States. *J. Acquir. Immune Defic. Syndr.*, 36, 1057–66.

Wenger, N. S., Kusseling, F. S., Beck, K. and Shapiro, M. F. (1994). Sexual behavior of individuals infected with the human immunodeficiency virus. *Arch. Int. Med.*, 154, 1849–54.

Wiktor, S. Z., Biggar, R. J., Melbye, M. *et al.* (1990). Effect of knowledge of human immunodeficiency virus infection status on sexual activity among homosexual men. *J. Acquir. Immune Defic. Syndr.*, 3, 62–8.

Williamson, L. M., Dodds, J. P., Mercey, D. E. *et al.* (2006). Increases in HIV-related sexual risk behavior among community samples of gay men in London and Glasgow: how do they compare? *J. Acquir. Immune Defic. Syndr.*, 42, 238–41.

Wilson, T. E., Gore, M. E., Greenblatt, R. *et al.* (2004). Changes in sexual behavior among HIV-infected women after initiation of HAART. *Am. J. Public Health*, 94, 1141–6.

Wilson, T. E., Feldman, J., Vega, M. Y. *et al.* (2007). Acquisition of new sexual partners among women with HIV infection: patterns of disclosure and sexual behavior within new partnerships. *AIDS Educ. Prev.*, 19, 151–9.

Wingood, G. M., DiClemente, R. J., Mikhail, I. *et al.* (2004). A randomized controlled trial to reduce HIV transmission risk behaviors and sexually transmitted diseases among women living with HIV. *J. Acquir. Immune Defic. Syndr.*, 37, S58–67.

Wolitski, R. J., Gomez, C. A., Parsons, J. T. and The SUMIT Study Group (2005). Effects of a peer-led behavioral intervention to reduce HIV transmission and promote serostatus disclosure among HIV-seropositive gay and bisexual men. *AIDS*, 19, S99–109.

Wyatt, G. E., Longshore, D., Chin, D. *et al.* (2004). The efficacy of an integrated risk reduction intervention for HIV-positive women with child sexual abuse histories. *AIDS Behav.*, 8, 453–62.

Injection drug use and HIV: past and future considerations for HIV prevention and interventions

12

Crystal M. Fuller, Chandra Ford and Abby Rudolph

In this chapter we will discuss HIV prevention and intervention efforts among injection drug users (IDUs) in the United States and internationally. We will begin with the political and social context of injection drug use followed by a brief overview of the HIV epidemic among IDUs, and factors that have contributed to HIV acquisition and transmission. These factors will be described in the context of how they have been able to inform HIV prevention and intervention strategies throughout the epidemic. Finally, we will provide a detailed account of successful and unsuccessful prevention and intervention strategies, and suggestions for future research to reduce HIV among specific IDU subgroups who continue to be at high risk.

Political and social context of injection drug use

According to Galea and colleagues (2003), political and social factors fundamentally shape the nature of substance-abuse behaviors and HIV prevention and intervention approaches. Policies and regulations can influence neighborhood socioeconomic conditions by determining the availability, accessibility and nature of health and social services, as well as the types of interventions that are amenable to specific populations (e.g., IDUs). Although drug dependence is a mental illness which should in turn be dealt with from a public health and medical perspective, injection drug use is unlawful and is treated as a criminal offense in many settings (see "Anti-drug legislations" section below).

It is important to consider the social and political context of the HIV epidemic when designing and implementing HIV prevention programs for IDUs. In New York City, for example, the HIV epidemic among IDUs is estimated to have begun in the mid-1970s. The epidemic grew rapidly among IDUs since it was illegal to purchase and possess drug-use paraphernalia, which largely contributed to the reuse and sharing of syringes. Shooting galleries and clandestine locations, or abandoned apartment buildings, served as private sites for injecting drugs (mainly heroin early on in the epidemic) and for selling/renting drug paraphernalia. It was not until the second decade of the HIV epidemic, in 1992, that syringe exchange programs (SEPs) were legally (or quasi-legally) implemented and expanded. SEPs, programs that exchange used syringes for sterile syringes, now operate in many major metropolitan cities and in some rural communities as well, although not without contentious social and political opposition, which has dampened their potential (see "Increasing syringe access", below).

The availability and accessibility of drug treatment has been used as an HIV prevention strategy, and in some settings it has been the only strategy. Unfortunately, many drug users face barriers to quitting injection drug use because of other social and psychological problems, difficulty gaining entry into drug treatment, a lack of desire to cease drug use, or an inability to find one of the limited available treatment options agreeable (see "Substance-abuse treatment", below).

Extreme poverty and disordered social environments are two key contextual factors that also contribute to injection drug use (Maas *et al.*, 2007). Some have asserted that, despite the IDU and HIV-transmission risk associated with sex tourism, governments in some countries have not regulated the industry because of the economic benefits it brings to the country or region. A convergence of other contextual factors also helps to create high-risk environments that promote injection drug use and HIV risk behaviors. Some of these factors include limited opportunities for gainful employment, income inequality, pervasive crime, availability of drugs, and social norms that either permit or do not discourage drug use (Boardman *et al.*, 2001).

In the US, patterns of racial segregation of blacks from whites and Hispanics from whites apparent in the broader society have been historically linked with injection drug use. Within racially segregated areas, black neighborhoods are typically characterized by lower rates of employment, education, and access to services and resources than white neighborhoods. Even after removing the effect of neighborhood socioeconomic status (SES), neighborhoods characterized by predominantly black residents still remain worse off than those that are predominantly white (Fuller *et al.*, 2005a; Mays *et al.*, 2007). Higher prevalence rates of

HIV among black IDUs relative to white IDUs may stem from the fact that the HIV burden is higher in black communities compared with white communities, and risk behaviors within higher prevalence networks are more likely to result in seroconversion (Kottiri *et al.*, 2002). Research has often highlighted race/ethnicity as one of the strongest risk correlates of HIV, even when risk behaviors are also considered. This race effect does not suggest a biological mechanism, but a social construct that contributes to the racial disparities in HIV (Williams and Collins, 2001). Thus, looking deeper into the social context of the environment that likely influences behavior is a natural step in attempting to explain and remedy persistent racial disparities in HIV. It is important to note that segregated neighborhoods may not tell the complete story with respect to injection drug use. In a US metropolitan city, racial segregation coupled with low educational attainment of a neighborhood was found to create a "concentration effect" of neighborhood disadvantage, which in turn significantly contributed to black drug users initiating injection during young adolescence as opposed to young adulthood. However, neither neighborhood segregation nor low educational attainment alone contributed to racial differences in initiation of injection drug use (Fuller *et al.*, 2005a). Thus, by taking neighborhood characteristics into account, a high-risk subgroup of black drug users can be identified and more appropriately targeted with prevention and intervention programs (see "Demographic characteristics", below).

Epidemiology of HIV infection among IDUs

Epidemiologic trends in HIV among IDUs in the United States

In the United States, injection drug use has been a driving force of the HIV epidemic since the mid-1980s. At the start of the epidemic, close to half of all new infections were among IDUs in the northeastern cities; however Miami (48 percent) and San Juan, Puerto Rico (45 percent), also had high rates of HIV (Holmberg, 1996). In the late 1980s, the HIV prevalence in the northeast region of the United States remained high, ranging from 61 percent in New York City to 43 percent in Asbury Park (Lange *et al.*, 1988; Battjes *et al.*, 1991), and the prevalence on the West Coast (1.5 percent prevalence in Portland; 2.7 percent prevalence in Seattle, 1.8 percent prevalence in Los Angeles) and in the Midwest (10.1 percent prevalence in Chicago, 10.3 percent prevalence in Detroit, 5.0 percent prevalence in Denver) was much lower (Lange *et al.*, 1988; Battjes *et al.*, 1991; Kral *et al.*, 1998). The epidemic peaked in the early 1990s with rates highest in the Northeast and lowest in the West (Santibanez *et al.*, 2006). Based on HIV surveillance data from 25 US states, the incidence of HIV among IDUs dropped by 42 percent between 1994 and 2000 (Santibanez *et al.*, 2006). Similar to the trends in HIV prevalence, HIV incidence estimates along the East Coast ranged from 2.7–10.7 per 100 person-years, while the estimates were much lower

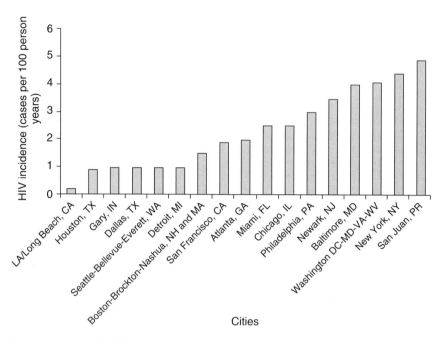

Figure 12.1 HIV incidence among IDUs in 17 US cities. Reproduced under BioMed Central Open Access license agreement; originally appeared in Holmberg (1996).

in cities on the West Coast (~0.3 per 100 py) (Santibanez *et al.*, 2006; Figure 12.1). The geographic distribution of HIV prevalence and incidence has closely followed the heroin and cocaine epidemics across the US, with the northeastern cities representing the heaviest drug markets (NIDA, 2000). Rates of HIV among IDUs have continued to drop across the US, and most drastically in cities with well-established and maintained SEPs (Des Jarlais *et al.*, 2000).

Given the concentrated drug markets that have historically targeted many urban, minority communities, it is not surprising that throughout the epidemic there have been significant differences in HIV incidence by race/ethnicity and by region. Among males, injection drug use accounted for a greater proportion of the AIDS cases in blacks and Hispanics (29.2 percent and 28.3 percent respectively). However, in females, injection drug use accounted for the greatest percentage of AIDS cases in whites (39.7 percent), followed closely by American Indian/Alaska Natives (38.8 percent), blacks (32.2 percent), Hispanics (32.6 percent) and Asian/ Pacific Islanders (15.8 percent) (CDC, 2004). These differential effects suggest that characteristics of disadvantaged neighborhoods may impact race and gender differently due to differing social circumstances faced by black men in particular (e.g., differential rates of incarceration, lack of employment opportunities, etc).

As seen in Figure 12.2a–d, the prevalence of HIV among IDUs entering drug treatment centers in the northeast region of the US was particularly high among

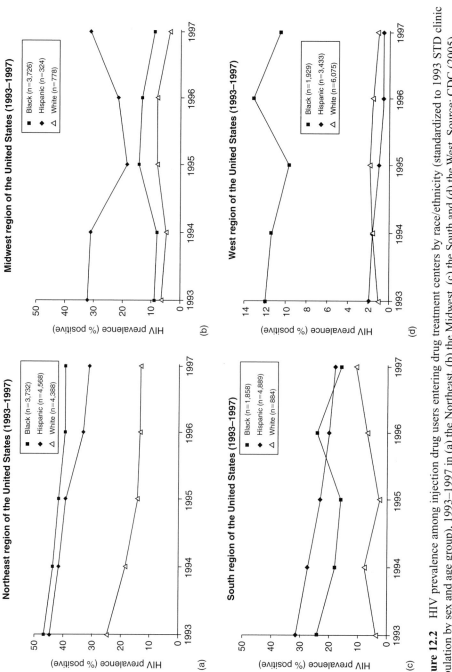

Figure 12.2 HIV prevalence among injection drug users entering drug treatment centers by race/ethnicity (standardized to 1993 STD clinic population by sex and age group), 1993–1997 in (a) the Northeast, (b) the Midwest, (c) the South and (d) the West. Source: CDC (2005).

blacks and Hispanics. Regarding the HIV prevalence among minority IDUs (with the exception of Hispanics in the Midwest Region of the US) and white IDUs between 1993 and 1997, there was an increase among white IDUs in the South Region of the US; however, prevalence among white IDUs has never surpassed those of black and Hispanics in the same region.

According to the 2004 CDC report, injection drug use accounted for 21.5 percent of all new AIDS cases (19.2 percent of men and 27.8 percent of women) in the United States, and at the end of that year 21.1 percent of the men and 33.6 percent of the women currently living with HIV had been infected through injection drug use (CDC, 2004). In many areas the number of newly diagnosed cases of HIV that can be attributed to injection drug use is on the decline (NYSDOH, 2000; NYC DHMH, 2003); however, HIV remains a public health burden for IDUs and their sexual partners (Klein *et al.*, 2000), particularly among minority communities and newly initiated IDUs (Nelson *et al.*, 1995; Neaigus *et al.*, 1996; Fuller *et al.*, 2003), and intervention efforts should target the IDU community and their sex partners.

Global epidemiologic trends in HIV among IDUs

Globally, approximately 10 percent of HIV/AIDS cases can be attributed to injection drug use (UNAIDS, 2003). By the end of 1999, over 10 million people worldwide reported injecting drugs, and of the 136 countries reporting injection drug use, 114 reported HIV infections associated with it (UNAIDS, 2002). While the HIV prevalence among IDUs in Australia has been relatively low for almost the entirety of the epidemic, injection drug use has substantially influenced the HIV epidemic in many countries in Europe, Asia, the Middle East, the Southern tip of Latin America and the United States, accounting for 30–90 percent of all reported infections (UNAIDS, 1997).

Figure 12.3 shows the trends in HIV prevalence among IDUs in 12 cities from the United States, the United Kingdom, Thailand, Belarus, Canada, Ukraine, Vietnam, India and Australia. While the US epidemic among IDUs began to peak in the mid-1980s and then gradually declined, the epidemic started much later in many other countries. In Glasgow and other cities in Western Europe, HIV prevalence among IDUs heightened in the late 1980s and then rapidly declined. The epidemic among IDUs in Eastern Europe peaked much later than the epidemic in the United States, and by the mid-1990s countries in Central, South and Southeast Asia and Eastern Europe had HIV seroprevalence rates of 80–90 percent (Department of Health Union of Myanmar AIDS Prevention and Control Programme, 1993; Hien, 1995; CDC, 2001). Between 1987 and 1995 injection drug use accounted for only a small portion of the HIV infections in the Russian Federation (0.3 percent); however, by the late 1990s it accounted for over 60 percent (Pokrovski, 1998). Similarly, the HIV prevalence in Ukraine

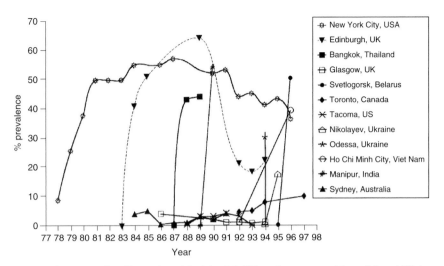

Figure 12.3   Trends in HIV prevalence among IDUs in selected cities. Reproduced under BioMed Central Open Access license agreement; originally appeared in Ball *et al.* (1998).

spiked in the mid-1990s (Figure 12.3). Injection drug use is now considered to be the primary mode of transmission in Eastern Europe and the newly independent states of the former Soviet Union (UNAIDS, 1997).

In Switzerland, the prevalence among IDUs started out high in the mid-1980s (~40 percent) but had dropped below 10 percent by the 1990s. Several cities in Spain and Italy initially reported explosive rates of HIV at the beginning of the epidemic, with rates approaching 76 percent in a community sample and 52 percent in a treatment sample (Hamers *et al.*, 1997). On the contrary, several cities in Canada have had a relatively low prevalence of HIV (e.g., 10.7 percent in Montreal and 23 percent in Vancouver) (Lamothe *et al.*, 1997; Strathdee *et al.*, 1997).

While the epidemic started much later in Asia, injection drug use quickly emerged as the main mode of transmission, accounting for over 80 percent of HIV cases in Kazakhstan, 75 percent in Malaysia, 75 percent in Vietnam and 50 percent in China (WHO 1997). Figure 12.4 looks more closely at the epidemic among IDUs in China, Indonesia and Vietnam between 1994 and 2003. The epidemic began to peak at the turn of the century, and by 2002 nearly 65 percent of IDUs in China, 50 percent in Indonesia and 30 percent in Vietnam were HIV-positive.

While the recent scale-up in HIV voluntary counseling and testing (VCT) funded by the President's Emergency Plan for AIDS Relief (PEPFAR) has resulted in decreased risk behaviors among those learning of their positive status, many IDUs continue to engage in unsafe sexual and injection risk behaviors,

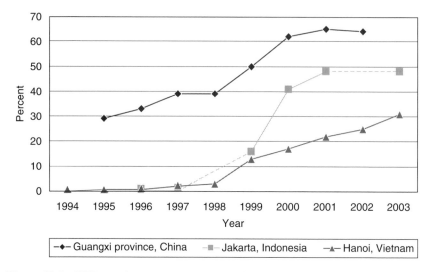

Figure 12.4 HIV prevalence among IDUs at selected sentinel sites in three countries, 1994–2003. Source: UNAIDS (2004).

suggesting that VCT programs should be supplemented with behavioral and structural interventions (Marks *et al.*, 1999; Wolitski *et al.*, 1997; Weinhardt, 1999; White House Press Release, 2003). Recent interventions have focused on improving the care of HIV-infected individuals to prevent further disease transmission (CDC, 2003; Gordon *et al.*, 2005). Others have focused on reducing sexual and injecting risk behaviors, to in turn reduce HIV transmission by (1) improving HIV knowledge, (2) facilitating voluntary disclosure of HIV status, and (3) strengthening social support among HIV-positive persons.

Injection drug use has also emerged as the primary mode for HIV transmission in North Africa, the Middle East, and West and Latin America (UNAIDS, 1997). Similarly, in Kenya, injection drug use accounts for 68–88 percent of the HIV prevalence (Ndetei, 2004). Based on current epidemiologic trends among IDUs, there is a need to focus more attention on the HIV epidemic among IDUs in many regions of the world, particularly Asia.

Factors that have informed prevention and intervention strategies

Based on decades of HIV research describing the changes in the epidemic, it has become evident that HIV prevention strategies among IDUs should encompass individual, group-level and structural-level factors. While proximal factors such as sexual and drug-use behaviors have been the primary focus of most interventions to date, researchers and advocates have begun to address ways in which social and contextual factors can better inform prevention approaches. In this section, we provide an overview of the most salient factors that have either contributed to or been suggested as having the potential to contribute to HIV prevention and intervention planning.

Demographic characteristics

United States' prevention and intervention strategies primarily have targeted four personal demographic characteristics: race/ethnicity, age, gender, and sexual orientation. Despite the convenience of using demographic categories to identify groups to target for interventions, doing so should be in ways that do not further stigmatize these particular communities (Ford *et al.*, 2007).

Race/ethnicity

Complex relationships exist between race/ethnicity, drug use and HIV infection (Cooper *et al.*, 2005). While lifetime prevalence of injection drug use is similar across US racial/ethnic groups, racial/ethnic differences have also been published (National Institute on Drug Abuse, 2002; Sherman, 2005). In a recent study conducted across 94 large US metropolitan areas, an estimated 55–66 percent of IDUs were non-Hispanic white, 13–30 percent were black and 8–10 percent were Hispanic/Latino. While these rates roughly reflect the proportion of each race/ethnicity in the US population (Cooper *et al.*, 2005), blacks account for 26 percent of HIV infections (CDC, 2002). To add to the complexity of the persistent epidemic among black IDUs, younger IDUs have an increased risk for HIV relative to older IDUs – which, again, is counterintuitive given that black IDUs tend to initiate injection drug use at an older age compared with whites and other racial/ethnic subgroups (van Ameijden, 1992; Nicolosi *et al.*, 1992; Nelson *et al.*, 1995; Des Jarlais *et al.*, 1999; Fuller *et al.*, 2001). When planning interventions, attention must be given to the phenomenon of race and age of initiation.

Racial/ethnic disparities exist in other countries as well. For instance, Canadian aboriginal populations are overrepresented among commercial sex workers and incarcerated persons, and account for nearly one-third of those utilizing syringe exchange services (National Task Force on HIV/AIDS and Injection Drug Use, 2001). As observed in the US, young and recent-onset IDUs were also found to have an increased risk for HIV in Amsterdam, with ethnic subgroups carrying an increased HIV risk (van Ameijden, 1992). Thus, more research needs to be done to explain and remedy racial differences in drug abuse and HIV so that more appropriate interventions can be implemented.

Age

Age of initiating injection drug use is associated with HIV transmission because (1) the highest risk for HIV seroconversion is the period immediately following injection initiation (Nicolosi *et al.*, 1992; Nelson *et al.*, 1995; Garfein *et al.*, 1996; Doherty *et al.*, 2000), and (2) newly initiated IDUs are an extremely hidden population that is difficult to identify for research and public health programs. For these reasons, multilevel studies may point public health practitioners in a new and more useful direction to help reduce the risk of HIV transmission among those who have been historically difficult to reach and whose behavior has been difficult to modify through large-scale, randomized controlled behavioral intervention trials (Purcell *et al.*, 2007).

Younger users also have unique treatment considerations. Data have consistently shown younger drug users as being less likely to seek methadone maintenance programs (Fennema *et al.*, 1997), and when seeking treatment younger IDUs tend to participate in short-term detoxification programs (Shin, 2007). Certain drug treatment venues may or may not reach those at highest risk and, given the different structure of detoxification (i.e., 24–48 hour stay) versus methadone maintenance (outpatient drop-in), interventions need to be carefully planned such that IDUs will be willing and available to participate.

Gender

Patterns of injection drug use, risk-behaviors and IDU-related HIV infection vary by gender. While the majority of IDUs are men, women become drug-dependent more easily and experience more severe addiction-related social problems (Ompad and Fuller, 2005; Institute of Medicine of the National Academies, 2007). Among incarcerated persons, women use drugs at levels equal to or higher than those of men (Henderson, 1998; CDC, 2001). In terms of HIV, women experience disproportionate levels of IDU-related HIV infection (CDC, 2002). Risk of acquiring HIV among female IDUs is double that of male IDUs (Nelson *et al.*, 1995). For men, IDU-related HIV risk derives primarily from direct exposure to HIV-infected blood via syringe sharing. For women, however, a considerable proportion of their risk stems from unprotected sex with male IDU sexual partners (CDC, 2002; Evans *et al.*, 2003; Frye *et al.*, 2007). Some research suggests that women may progress to AIDS at lower levels of viral load compared with men; however, further research is needed in this area (Farzadegan *et al.*, 1998).

Considerations for prevention and intervention strategies for women substance abusers should bear in mind that women are more likely than men to have children in their custody (Henderson, 1998). The presence of children may motivate women to obtain treatment; however, it also may serve as a barrier if women fear losing their children because of their drug dependence (Henderson, 1998).

In addition, public health strategies should also consider the high possibility of violence victimization among women. Women are more likely to experience intimate partner violence, which in turn is associated with homelessness, substance abuse and HIV infection (El-Bassel *et al*., 2005; Panchanadeswaran *et al*., 2008). A recent four-city study found that 40 percent of male IDUs reported perpetrating physical violence against their female partners (Frye *et al*., 2007). Given this increased risk for HIV and other drug-related consequences, few interventions have solely targeted females IDUs, and the need to target female IDUs remains critical. However, many female IDUs have been captured in interventions that have targeted all types of drug-using, high-risk women (Solomon, 1998).

Sexual orientation

Only one exposure category, men who have sex with men (MSM) and also inject drugs, reflects compounded HIV risk due to both a demographic characteristic and risk behavior. The category MSM encompasses several different groups, such as gay-identified men as well as heterosexual-identified men or boys, including sex workers, who exchange sex with other men solely for economic or drug-use purposes (Rietmeijer *et al*., 1998). Differences in sexual identity influence which intervention strategies are most appropriate. For instance, HIV prevention education delivered through known gay-identified organizations, venues or networks is not likely to reach non-gay identified MSM (Goldbaum *et al*., 1998). On the other hand, concerns about homophobia, psychosocial barriers and proclivity for higher-risk behaviors (such as "barebacking") may be key targets for HIV prevention in some segments of this population (Ibanez *et al*., 2005).

It is important to note that some state and national HIV and AIDS reporting agencies indicate that belonging to two high-risk groups (injection drug use and MSM) does not necessarily translate into the highest-risk category for HIV burden and risk of new infection. *Current* HIV infection/AIDS diagnoses and *newly*-diagnosed HIV/AIDS represent less than 3 percent for New York City (NYSDOH, 2007). However, this is based on reporting from HIV testing centers and clinics, and it has been suggested that some of the highest-risk individuals may not be reached by HIV testing services in some states, while other states may be better utilized by IDU-MSM (2000). A recent national surveillance report combining data from 33 US states between 2001 and 2005 indicated this combined-risk group (IDU-MSM) as having either the highest or nearly the highest percentage of newly-diagnosed HIV infection, more than either IDU or MSM categories alone (CDC, 2007). While under-reporting is likely to exist in some US states, it should be noted that the New York data (presented above) represent data from the tail end of the HIV epidemic in IDUs, and are likely to have some degree of accuracy with respect to the declining incidence in HIV (Des Jarlais *et al*., 2000), with rates among IDU-MSM demonstrating similar declines in New York since 1993. Even in the face of a declining HIV epidemic among IDUs in some US cities,

HIV prevention and intervention strategies should not ignore IDU-MSM, given the persistent racial/ethnic disparities that characterizes the epidemic in the US. Targeting this group with sexual risk-reduction strategies has been consistently and strongly suggested (Des Jarlais *et al.*, 2000; Fuller, 2005b).

Both country of origin and prevalence of IDU-related HIV infection influence HIV risk among IDU-MSM (Baral *et al.*, 2007). Homosexuality is a crime in many countries, which contributes to under-reporting of homosexual modes of HIV transmission (Baral *et al.*, 2007). In some Asian and African countries, MSM are a hard-to-reach population. As has been shown in China, HIV/AIDS stigma may be compounded by the stigma of being an IDU or homosexual (Qian *et al.*, 2006), which may in turn prevent access to needed care. MSM who reside in low-income countries carry a higher prevalence of HIV infection than do MSM in middle- or high-income countries (Baral *et al.*, 2007). Further, the odds of HIV infection among MSM are higher in countries in which IDU-related transmission accounts for a substantial proportion of national HIV infections (Baral *et al.*, 2007).

In terms of risk among women who have sex with women (WSW), data have consistently shown this group to be the least risky subgroup with respect to HIV when compared with heterosexual men and/or women. However, drug-using WSW have been one of the highest risk groups identified to date and carry the highest burden of HIV disease among women (Friedman *et al.*, 2003; Young *et al.*, 2005). Recent data collected among heroin, crack and cocaine injecting and non-injecting drug users revealed that young adult WSW (age 18–40) were 17 percent HIV-seropositive compared with only 13 percent among heterosexual drug-using women; the rate among WSW who also have sex with men was even higher, at 20 percent (Absalon *et al.*, 2006). It is important to note that these data highlighted injection drug use as not being a significant contributor to HIV-seropositive status, which in turn suggests that WSW are likely to have elevated behavioral risk resulting more from risky sexual behaviors with men as opposed to risky drug-use behaviors (Diaz *et al.*, 2001; Bell *et al.*, 2006). Thus, similar to IDU-MSM, WSW who also inject should be targeted with sexual risk-reduction prevention and intervention strategies with similar attention to the persistent racial disparities that exist among this group as well (Harawa and Adimora, 2008).

Drug-use behaviors

Drug-use behaviors are a natural focal point of HIV interventions targeting IDUs. Sharing of syringes and other injection equipment has traditionally been the most common mode of HIV transmission among IDUs, and exacts a heavy toll on IDUs in Eastern Europe and Central Asia, where an estimated 80 percent or more of IDU-related HIV infections result from sharing of needles or works (Institute of Medicine of the National Academies, 2007). Interventions targeting syringe-sharing have the potential to reduce individuals' risk of acquiring HIV (see "Increasing syringe access", below). However, the epidemic has not been restricted to syringe-sharing. Other drug-use risk has included injecting cocaine or other forms of "speed", such as methamphetamines. This is due to the increased frequency in injection episodes, which increases the likelihood of using previously used and/or contaminated syringes (Anthony *et al.*, 1991). As safe and sterile syringe-access has increased, risk among cocaine/speed IDUs has decreased. Other types of drug-use risk has elevated the risk of sexual transmission among IDUs, namely smoking crack cocaine (Booth *et al.*, 1993; Astemborski *et al.*, 1994; Compton *et al.*, 2000), which is highly correlated with trading sex for drugs, money or survival (Fullilove *et al.*, 1990; Edlin *et al.*, 1994). Sexual risk-reduction interventions have been recently conducted in the US; however, they have been unsuccessful in reducing sexual risk behavior (Purcell *et al.*, 2007). Design and implementation of social network-oriented as well as structural interventions may have better success in reducing sexual risk behavior (see below).

Sexual behaviors

Sexual risk behaviors also represent an important mode of HIV transmission among IDUs, particularly young or newly initiated IDUs. For example, young or newly initiated IDUs continue to be at highest risk for HIV among the general population of IDUs, with the most recent incidence of HIV infection being 6.6 per 100 py reported in 2003 among a sample of IDUs injecting for less than 3 years between the ages of 17 and 30 (Fuller *et al.*, 2003). Several studies have indicated that not only sexual risk but also, and in some cases more significantly, injection risk is important among young or new IDUs (Garfein *et al.*, 1996; Des Jarlais *et al.*, 1999; Doherty *et al.*, 2000). Sexual behaviors associated with injection drug use include unprotected intercourse, exchanging sex for drugs or money, sex with high-risk partners and sex with multiple partners (Institute of Medicine of the National Academies, 2007).

Characteristics of IDU sexual dyads are also important targets for sexual risk behaviors. Among IDUs in New York, odds of engaging in unprotected sex were higher in couples that smoked crack together, shared needles together, or in which one partner was known to have other sexual partners (Gyarmathy and Neaigus, 2007). Cumulative risk of inadvertent sexual transmission of HIV may be higher in main rather than casual or other sexual partnerships, because main partners tend not to use safer sex options (e.g., condoms) consistently (Iguchi *et al.*, 2001).

IDU and HIV risk correlate with commercial sex work. In many societies, women and some men turn to commercial sex work when economic opportunities are sparse. Up to 80 percent of women commercial sex workers (CSWs) in the Russian Federation, 50 percent in Eastern Europe and 25 percent in Central Asia are IDUs (Institute of Medicine of the National Academies, 2007). CSWs who inject drugs tend to engage in riskier sexual behaviors than other CSWs. Analyses of surveillance data of nearly 16,000 CSWs in China found that 3.2 percent were IDUs, and that IDU CSWs had lower odds of owning a condom, using a condom with their last client or consistently using condoms (Lau *et al.*, 2007).

Clearly, HIV prevention and intervention strategies should include sexual behavioral risk-reduction to further stem the epidemic among IDUs, particularly those subgroups who continue to carry high risk of HIV even after full implementation of SEPs.

Psychosocial factors

A number of psychosocial factors are correlated with injection drug use and HIV infection, and mental health co-morbidities are common (Institute of Medicine of the National Academies, 2007). It is unclear, however, whether they are more prevalent in some countries than others. Antisocial personality disorder (ASPD), for example, is common among drug users, and is also associated with high-risk sex behaviors and injection practices and, consequently, an increased rate of HIV infection (Brooner and Bigelow, 1990). Psychiatric status should be taken into account when designing and evaluating prevention and treatment interventions, since drug users with ASPD may not respond as well to HIV prevention interventions as those without ASPD (Compton and Cottler, 2000). Fear of being involuntarily institutionalized, medicated or otherwise penalized may drive IDU behavior underground and create other obstacles to HIV prevention (Henderson, 1998). For these reasons, concurrent treatment of mental illness is recommended in order comprehensively to address HIV prevention among IDUs (National Institute on Drug Abuse, 2006).

Addressing stigma is central to HIV prevention among IDUs. HIV/AIDS stigma is compounded by stigma associated with being a member of certain socially marginalized groups, or with modes of HIV transmission (Takahashi, 1997; Berger, 2004; Ahern *et al.*, 2007; Chan *et al.*, 2007). For example, drug policies and legal environments often constrain the success of harm-reduction interventions that target IDUs. Stigma and discrimination of IDUs is associated with decreased access to health services, especially for HIV and HCV treatment (Sarang *et al.*, 2007). Furthermore, in many settings a societal view of IDUs as outcasts may influence users' decision to participate in treatment and prevention programs, since participation labels them as substance abusers. Therefore, failure to participate in HIV prevention programs is often driven by a fear of being perceived negatively by society (Bobrova *et al.*, 2006).

Social environmental factors

An emerging body of research has sought to identify social, environmental and structural determinants of HIV and risk behaviors. Given that the persistence of the HIV epidemic over the past few decades has been closely correlated with low SES, it is important to consider how features of the environment can inform prevention and intervention programs.

Socioeconomic status

Low socioeconomic status (SES) connotes limited resources (e.g., education, income, employment opportunities) with which to improve one's social standing and quality of life (Krieger *et al.*, 1997). It can increase exposure to risk conditions (e.g., incarceration) and introduce individuals to risky drug-use behaviors, thereby increasing HIV risk (Blankenship *et al.*, 2005). Although prevention and treatment services may be physically located in poor communities, access may be limited by factors such as long waits for enrollment into free or low-cost treatment programs (Wyatt *et al.*, 2005). Inattention to other social circumstances that may foster continued drug abuse can also deter treatment entry. Socio-ecologic models posit that the SES of individuals, of their social network and of their community may all influence their behavior, and should therefore be considered when designing an intervention program. For example, neighborhoods characterized by low SES have been targeted with community outreach and other risk- and harm-reduction services to help reduce transmission of HIV. However, given the persistent racial disparities in HIV, more attention to social problems and going beyond simply targeting poor communities is necessary (e.g., considering characteristics of social networks) (Schoenbaum *et al.*, 1989; Kottiri *et al.*, 2002; CDC, 2007).

Social networks

Interventions targeting social networks are developed on the basis of network characteristics. Typically, network-oriented approaches identify core group members who have the greatest risk of transmitting HIV, determine high-risk links within a network, or strive to understand network characteristics (e.g., density of the network) that increase HIV risk (Latkin *et al.*, 1995). A widely-used approach draws on the potential of influential peer leaders to disseminate prevention messages or to model recommended behavior (Latkin, 1998). Peer leaders are members of a network who are held in high regard by other network members, and whose opinion they consider important. Peer leader outreach is particularly useful among IDUs, because it addresses IDUs' distrust of persons who are not members of their social networks (Ford *et al.*, 2007). Peer leaders can educate network members (e.g., about local SEPs), provide resources (e.g., condoms) to network members, and model desired lower HIV risk behaviors (Latkin, 1998; Latkin *et al.*, 2003). Not only have social network-oriented interventions been successful among IDUs (Latkin and Knowlton, 2005), but a body of research demonstrating how social networks are linked to the environment

319

is developing which may further enhance social network-oriented interventions among IDUs (Williams and Latkin, 2007).

Environment/neighborhood

The environment in which individuals inject drugs as well as the neighborhoods in which they live influence their risk for HIV infection and access to preventive services (Galea *et al.*, 2003; Maas *et al.*, 2007). Poor neighborhood conditions, such as neighborhood social and physical disorder, have shown an independent association with the onset and prevalence of injection drug use, even when individual behaviors are taken into account, which emphasizes the importance of the environment. However, it has been the exploration of how social networks are related to the macro-environment, and how this relationship impacts individual risk behavior among IDUs, that has been suggested as the "next step" in the development of better targeted prevention and intervention programs (Fuller *et al.*, 2005a). Both characteristics of the environment (e.g. drug-related arrests) and of a person's social network (e.g., drug-using networks) have shown to have an independent effect on continued heroin and cocaine use and sexual risk behaviors (Schroeder, 2001; Williams and Latkin, 2007); however, it is unclear whether positive network characteristics, such as social support and having ties to positive network members, are able to buffer against poor neighborhood conditions (Williams and Latkin, 2007). While the exact mechanism driving the relationship between social networks, neighborhoods and HIV risk are not entirely understood, researchers have suggested that interventions at the network and community levels are needed to address the HIV epidemic (Williams and Latkin, 2007).

Other social and contextual factors

Many other macro-level factors may inform intervention development, including culture, policies, political structure, conflict, economics and religion. These factors fundamentally determine the types of interventions that can be implemented in specific regions. Attempted interventions that challenge cultural beliefs or norms are unlikely to succeed, and may damage relations between interventionists and communities. In addition, as indicated by data from various conflict-affected regions, war can increase risk behaviors, promote drug-use economies and completely disrupt access to usual preventive services (Hankins *et al.*, 2000; Strathdee, 2003; Culbert *et al.*, 2007). Interventions in conflict-affected regions must be adapted to make them implementable in war-torn regions.

Finally, in the US, researchers have suggested further exploration into macro-level factors (e.g., neighborhood characteristics) that may help inform the design of HIV prevention and intervention strategies aimed at reducing the racial disparities in HIV. Neighborhood differences may reflect disparate access to care, services and other resources, which in turn may contribute to high-risk behaviors that will persist in the face of behavioral intervention strategies that do not take the social environment within which the behaviors occur into account. For example, relative to blacks and Hispanics, whites have better access to treatment services, which in turn helps reduce the severity of drug-related consequences in this population (Godette *et al.*, 2006). Prevention strategies that have been based upon race or ethnicity have typically included developing culturally competent strategies and matching intervention staff to the race or ethnicity of program participants (Airhihenbuwa *et al.*, 1992; Ford *et al.*, 2007); however, it is clear that a more comprehensive intervention strategy that goes above and beyond a person's individual race/ethnicity will be needed to combat the HIV epidemic among racial/ethnic minorities.

Successful and unsuccessful interventions

If all IDUs received treatment and were able to stop using drugs, IDU-related transmission of HIV/AIDS could be eliminated. However, many are not ready to enter treatment, some are unwilling or unable to abstain from drug use, and others face barriers to drug treatment programs and other social and medical services (e.g., drug treatment centers often have long waiting lists, rules for participation are strict, and relapses are common). Numerous intervention strategies have been utilized to prevent HIV transmission among IDUs. HIV prevention interventions targeting IDUs have included, but are not limited to, the following:

- peer education programs and outreach to IDUs to reduce high-risk sex and injection behaviors;
- social network interventions;
- increasing access to sterile syringes through SEPs, pharmacies and health-care providers;
- drug treatment;
- opioid substitution pharmacotherapy;
- buprenorphine; and
- anti-drug legislations.

Peer education programs and outreach to IDUs

At the beginning of the epidemic in the United States, HIV prevention interventions that targeted IDUs consisted mainly of community-based outreach programs that distributed condoms and bleach kits to IDUs and encouraged them to change their behaviors. These efforts to reduce the risk of HIV transmission among IDUs were very successful, especially at a time when syringe possession and distribution were prohibited. Outreach workers provided IDUs with a hierarchical message for behavioral risk reduction: first, stop using and/or injecting drugs; if you cannot stop, use your own equipment and do not share it with others; if you cannot avoid reusing or sharing equipment, disinfect it with bleach to reduce the risk of HIV transmission. In addition to providing IDUs with several behavioral strategies to reduce their risk, they were also given educational materials and condoms, and referrals to community-based programs for HIV testing and counseling, drug treatment, and medical and social services (Needle *et al.*, 1998).

Strategies that promote cleaning injection equipment have worked well, particularly in areas where sterile syringes are expensive and not readily available, or where restrictions on syringe distribution exist (Broadhead, 1991). For example, programs that provide injection safety information and distribute bleach and condoms have been implemented in Manipur, India (Chatterjee *et al.*, 1996; Hangzo *et al.*, 1997), Malaysia, Vietnam, Thailand and Nepal (Ball, 1996).

Social network interventions

While outreach has utilized social networks to reach hidden IDU populations with HIV education and prevention materials for many years, HIV prevention interventions have only recently been designed to affect and influence behavior at the group level. Individual-level interventions rely primarily on education to promote rational decision-making; however, structural-level interventions take the social and ecological context of behaviors into consideration, and try to identify influential figures within the target population to motivate changes in behavioral norms in ways that are socially acceptable. There are several challenges to designing a successful and sustainable peer-led behavior change intervention. First, HIV risk-reduction practices need to be socially acceptable, or linked in some way to a behavior that is socially rewarded. The behavior modification must also be meaningful, and linked to the priorities of those in the target community (e.g. sex and drug use are highly meaning-endowed behaviors) in order to create motivation for behavior modification. Since group concepts are often incorporated into an individual's concept of self, promoting a group or collective social identity may also aid in behavior change. Creating effective diffusion of the intervention is also important, and relies on the selection of influential and connected "opinion leaders" to disseminate intervention messages to those in the target network who are not directly involved in the intervention. Furthermore, to enhance intervention stability, it is essential that the promoted behavior be valued and reinforced by the target community (Latkin and Knowlton, 2005).

Social network members have a remarkable influence on the behavior of their peers, which is thought to result from norm formation and maintenance (Latkin and Knowlton, 2005), and peer outreach programs are designed based on cognitive consistency, social identity, social influence, and active learning theories (Latkin *et al.*, 2003). From the theoretical perspective, personal, behavioral and environmental factors interact in a dynamic way to produce human behavior, so interventions that promote behavior change cannot be implemented without considering all of these factors. However, because these factors are interrelated, it is possible to change human behavior by altering individuals' social conditions or improving their behavioral competencies. Therefore, successful interventions that utilize social cognitive theory can be designed to change attitudes, norms and values that exist in the larger community, and the social and environmental context of these risk behaviors in the community (Pajares, 2002).

The success of these interventions is typically evaluated by examining group interactions and the interactions among individuals in a group (i.e., sexual risk-taking behaviors and how people acquire, prepare, mix and share drugs) rather than through examination of individual risk behaviors. For example, successful network-oriented interventions have initiated behavior changes, reduced high-risk behaviors, increased treatment readiness, and prevented HIV transmission among network members (Needle *et al.*, 1998).

Latkin and colleagues (2003) have demonstrated that social–cognitive–behavioral approaches to intervention that target inner-city drug users are most successful when they incorporate both social identity and peer outreach components. Network-oriented interventions often identify key network members to relay HIV prevention messages and needle-cleaning materials and supplies to their sex and drug partners, family and friends, and other community members. These individuals, or "peer networkers", are usually members of the population or community at risk for HIV infection, or individuals who frequently interact with those at risk (i.e. current or former IDUs). They are selected because they share beliefs, attitudes, norms, behaviors or some other attribute with those at risk, and are recruited to help distribute health promotion information and materials in the community and/or to serve as positive social role models for the community (Guenther-Grey *et al.*, 1996). By changing their own behaviors and encouraging others in their social network to adopt safer-sex and -injection practices, peer networkers can initiate changes in the perceived social norms in the community, which may eventually lead to changes in actual community norms. According to social cognitive theory, when peer networkers distribute prevention and education materials to members of their drug-using and sexual risk networks, they provide not only encouragement and positive social reinforcement, but also the means necessary to make behavior changes (Guenther-Grey *et al.*, 1996). Furthermore, in an effort to maintain their credibility as a health educator, peer educators often reduce their own risk behaviors as well (Latkin and Knowlton, 2005).

It is also important for outreach workers and peer educators to incorporate risk-reduction education for high-risk sex behaviors because (1) HIV-positive IDUs often engage in risky sexual behaviors that put their uninfected partners at risk for infection and (2) several studies have demonstrated a strong relationship between unprotected sex and HIV seroconversion among IDUs (Kral *et al.*, 1998; Strathdee, 2003). According to a meta-analysis by Copenhaven and colleagues (2006), interventions that focused on drug and sexual behaviors equally were much more successful than those that focused on only one.

Developing countries like India (Chatterjee *et al.*, 1996; Dorabjee *et al.*, 1996) and Nepal (Maharajan and Singh, 1996) have also used former IDUs as peer educators in intervention programs. In some communities, peer educators function more as outreach workers than as peer educators. In other settings, most notably Australia, India and several European countries, drug users' organizations have formed to advocate on the behalf of IDUs. These unions are comprised of active drug users who work together to implement HIV prevention programs, including peer outreach, education, and needle-syringe exchange (Jose *et al.*, 1996; Sharan Society for Helping the Urban Poor, 1997).

Increasing syringe access

SEPs

Since unsafe injecting practices is the primary risk factor for transmission among IDUs, and access to sterile injection equipment is associated with safer injection practices, improved access to sterile syringes and injection equipment is required to reduce the risk of infection in this population (Vlahov and Junge, 1998; Rudolph *et al.*, 2006). In 1984, the first SEP was started by the Junky Union, a league of drug addicts, in response to an outbreak of HBV among Dutch IDUs (Needle *et al.*, 1998). While SEPs began as an experiment, they have become critically important to combat the spread of HIV and AIDS. Not only developed countries but also developing and transitional countries, such as Brazil, the Czech Republic, Hungary, India, Nepal, the Philippines, Poland, Russia, Thailand and Vietnam (Needle *et al.*, 1998), have implemented successful SEPs.

Most countries that were concerned about the spread of HIV among IDUs introduced SEPs to their already existing community-based outreach programs in the mid-1980s. However, it was not until the late 1980s that the United States established its first SEP in Tacoma, Washington. Point Defiance was established by an independent agent, and was probably the most successful SEP ever established in the US. One of the factors that contributed to its success was its ability to establish trust between IDUs and the social service system. This trust and the rapport that developed between customers and clients often created a window of opportunity for IDUs to seek treatment. Anonymity was another extremely important component of Point Defiance; IDUs were not required to provide their name or any form of identification to participate in syringe exchange. Point Defiance was also able to convince city officials that SEPs did not directly contradict abstinence-based drug treatment programs – in fact, in the year following the program's start, there was a 53 percent increase in the number of IDUs seeking drug treatment compared to the year before. The success of this project also demonstrates the importance of community and political support (Sherman and Purchase, 2001).

Other SEPs were established in Portland, San Francisco and New York City around the same time, but were not nearly as successful as Point Defiance. New York City, for example, had one of the highest rates of HIV in the country, but struggled to establish an exchange large enough to meet the needs of its IDUs. In 1988 it attempted to implement its first SEP but, due to the lack of political and community support for harm-reduction strategies, New York City's first pilot needle exchange program was funded as a controlled clinical trial. The trial was designed to recruit a limited number of IDUs into a treatment group that would be permitted to exchange used needles and syringes for sterile ones, and to compare their progress with that of a control group not given the same access to clean paraphernalia. To be eligible to participate in this pilot program, IDUs had to (1) be over 18 years of age, (2) have previously applied to a drug rehabilitation program but been turned away because it was full, and (3) register at the health department. The program was unsuccessful, and its failure can be attributed to several factors – lack of political and community support; a single distribution center with limited hours of operation; lack of anonymity (IDUs were required to identify themselves to a government agency); and caps on enrollment (Anderson, 1991).

Overall, SEPs have successfully increased access to sterile syringes and provided a means for disposal. Access to SEPs has also been associated with reduced drug use (Walters *et al.*, 1994; Vlahov and Junge, 1998) decreased incidence and prevalence of HIV and other blood-borne infections (Des Jarlais *et al.*, 2000; MacDonald *et al.*, 2003), improved access to HIV prevention programs, lower rates of criminal activity, less needle-sharing and other high-risk injection behaviors (Groseclose *et al.*, 1995; Des Jarlais *et al.*, 1996), and greater entry into and retention in drug treatment programs (Strathdee *et al.*, 2006). Many SEPs also provide additional services, such as health education; alcohol swabs to prevent abscesses and other bacterial infections; condoms to prevent the transmission of HIV and other sexually transmitted infections (STIs); on-site medical services; counseling and screening for tuberculosis, hepatitis B, hepatitis C, HIV and other infections; and referrals to substance-abuse treatment and other medical and social services. These additional services are important, especially since many drug users do not have health insurance (Stancliff, 2000) or fear being mistreated in the health-care system because of their drug use (Miller *et al.*, 2001). SEPs can therefore link IDUs to health and social services to which they might not otherwise have access. This harm-reduction approach has successfully reduced many of the negative consequences associated with drug use, in a non-judgmental way.

Laws and regulations affecting SEPs

In some settings, laws and regulations that permit syringe possession are absent and therefore prevent these programs from meeting local needs (Feldman and Biernacki, 1988; Zule, 1992; Booth *et al.*, 1993; Koester, 1994; Bluthenthal and Watters, 1995). For example, the majority of US states have drug paraphernalia laws that make the possession or sale of anything intended for drug use illegal. By 1997, the possession of drug paraphernalia was criminalized in 47 states for this reason (Gostin and Lazzarini, 1997). Qualitative studies consistently show that fear of being arrested for violating state laws that prohibit syringe possession deters IDUs from carrying their own syringes (Feldman and Biernacki, 1988; Zule, 1992; Booth *et al.*, 1993; Koester, 1994; Bluthenthal and Watters, 1995). Police harassment at SEP sites is also associated with decreased use of SEPs by IDUs (Bluthenthal *et al.*, 1997). Even where SEPs are legal, the criminalization of syringe possession creates a conflict between public health recommendations and law enforcement practices. In addition, since 1998 Congress has prohibited federal funding for programs that provide IDUs with access to sterile syringes, so SEPs rely on financial support from local governments and private funders. Due to limited funding, SEPs have restricted hours of operation, and too few are located in urban centers. Because of their sparse locations, long travel distances present another barrier to syringe access. Although many states and municipalities in the United States have taken steps towards improving access to sterile syringes, the possession, distribution and sale of syringes remains a criminal offense throughout much of the country. These legal and funding impediments restrict widespread utilization of these programs in the United States.

Pharmacies and health-care providers

Alternatively, IDUs can access syringes through pharmacies. Pharmacists are health-care professionals who can potentially provide discrete and confidential information regarding disease prevention and safe disposal of syringes to all syringe customers. Pharmacies have access to needles, are already established in most communities, have longer and more convenient hours of operation than SEPs, can provide greater anonymity for syringe customers, and may attract those uncomfortable with SEPs or those who require syringes outside the hours of SEP operation. Additionally, pharmacies have the potential to build partnerships with community and local agencies that provide health services, promote safer injection and disposal practices, and offer information about drug treatment options.

In the UK, Australia and the Netherlands, pharmacies play an integral role in providing HIV prevention services to IDUs. Pharmacies in these countries have expanded their services and currently counsel IDU customers on safer sex and injection practices, furnish both syringes and injection equipment, provide a means for the disposal of used syringes, and dispense oral methadone for treating opiate dependence. In Quebec, some pharmacies act as syringe exchange sites, and all pharmacies are required to indicate whether their injection equipment is for sale or exchange by displaying a logo. Some community pharmacists in Quebec also sit on advisory boards for community outreach programs, and others provide IDU customers with referrals to these programs.

In the United States, 23 states have pharmacy regulations or practice guidelines that limit pharmacy sales of sterile syringes to IDUs, and several other states have regulations that limit or prohibit the sale, distribution and dispensing of syringes without a prescription (Burris *et al.*, 2002). Even in the states that have repealed laws and regulations banning the sale of sterile syringes to IDUs, syringe sales may be hindered by individual pharmacy policies and/or pharmacy managers or pharmacists who are reluctant to sell syringes to IDUs. For example, some pharmacies require syringe customers to show identification, sign a register, or provide a reason for the purchase (Des Jarlais *et al.*, 1996). Collectively, these regulations deter IDUs from accessing sterile syringes and make it difficult to provide a sterile syringe for each injection, as is recommended by public health agencies in order to reduce the risk of transmitting HIV/AIDS and other infectious diseases. In an effort to counter the contribution that injection drug use has had on the HIV/AIDS epidemic, 11 states (Connecticut, Hawaii, Maine, Minnesota, New Hampshire, New Mexico, New York, Oregon, Rhode Island, Washington and Wisconsin) have implemented legislation to permit pharmacy sales of syringes without a prescription (Pouget *et al.*, 2005).

In 1996, the New York State Department of Health AIDS Advisory Council unanimously adopted a recommendation made by its Subcommittee on Harm Reduction to revise the Public Health Law to (1) expand syringe access by allowing non-prescription pharmacy sales of syringes and syringe distribution by health-care facilities and health-care practitioners, and (2) require hospitals and nursing homes in New York State to accept used syringes. This public health law was referred to as the New York State Expanded Syringe Access Demonstration Program (ESAP), and was enacted by the legislatures in August 2000, implemented in January 2001, and required an independent evaluation to be conducted to determine its continuance beyond 2003. According to the 2003 Final Evaluation Report to the Governor and State Legislature and other published studies:

- program utilization by pharmacies was initially slow, but increased over time (NYAM, 2003);
- syringe-sharing rates among IDUs declined (Pouget *et al.*, 2005);
- needle-stick injuries decreased (Fuller *et al.*, 2002);
- criminal activity dropped; and
- there was no increase in drug abuse (NYAM, 2003).

Additionally, New York City pharmacists' positive attitudes towards disease prevention increased and negative attitudes decreased over the evaluation period (Rudolph *et al.*, 2006).

States with similar legislations have reported consistent findings. In Washington State, for example, it is legal to purchase syringes in pharmacies. It was found that IDUs who obtain most of their syringes from pharmacies are less likely to share injection equipment than those with other primary syringe sources (Calsyn *et al.*, 1995). Groseclose and colleagues (1995) also showed that when legal restrictions on both possession and purchase of syringes are removed, IDUs change their syringe purchasing practices and their syringe-sharing behaviors in ways that reduce HIV transmission. Similar to strategies that have been underway in Australia and the UK, US pharmacies may also be able to incorporate various harm-reduction strategies such as those that have proven successful in SEP venues. Currently, evaluation of expanded pharmacy services targeted to IDU syringe customers is underway in New York City (e.g., to include provision of referrals, and safe injection and disposal information).

Substance-abuse treatment

Substance-abuse treatment has also been pursued as a means to prevent HIV transmission, because drug users in treatment are less likely to inject drugs. Furthermore, those who continue with treatment tend also to reduce their high-risk drug-related behaviors (i.e. syringe-sharing and high-risk sex), and are therefore less likely to acquire or transmit HIV (Needle *et al.*, 1998). Similarly, Friedman showed that participants in the National AIDS Demonstration Research treatment program were less likely to seroconvert than those out of treatment (Friedman *et al.*, 1995). Below, we provide summaries of pharmacologic treatment options available to IDUs.

Opioid substitution pharmacotherapy

Methadone maintenance programs were first implemented outside of the United States in the late 1980s in Australia and Europe (Ward, 1994; Farrell *et al.*, 1995). Since then, programs have been established in Nepal, Vietnam, Thailand, Latvia, Lithuania, Poland, Slovenia, the Slovak Republic, Hungary, Bulgaria, Hong Kong and the Former Yugoslav Republic of Macedonia. Other small-scale and pilot studies have been implemented in regions, such as Asia, Latin America, Eastern Europe, New Delhi, and the Newly Independent States (Ball *et al.*, 1998). Even governments who initially had reservations about opioid substitution programs are now considering implementation of these programs to tackle the high rates of HIV transmission in their IDU populations.

Methadone maintenance programs do, however, have several limitations. First, agonist pharmacotherapy programs are not always appropriate, feasible or affordable in many developing countries, and in the United States only a limited number of spots are available. On any given day, approximately 85 percent of

IDUs in the United States are not enrolled in a drug treatment program (Needle *et al.*, 1998). Second, individuals dependent on both heroin and cocaine present a unique challenge for HIV prevention strategies that promote drug treatment, since these individuals are much less likely to have successful treatment outcomes (owing to high dropout rates, involuntary dismissal from treatment, and high rates of relapse) (Needle *et al.*, 1998). Cocaine users in methadone maintenance treatment also tend to have higher-risk injecting behaviors (Reynaud-Maurupt *et al.*, 2000). Finally, there is no available drug treatment for cocaine injectors and crack smokers, who make up a high-risk group for HIV transmission. Therefore, many IDUs remain without viable treatment options.

Buprenorphine

While morphine and methadone are complete agonists, buprenorphine is a partial mu-opioid agonist, which may make it more acceptable in countries with political and/or philosophical reservations about prescribing opioid agonist medications. Buprenorphine has been shown to be an efficacious alternative to methadone for the treatment of opioid dependence, and has demonstrated its ability to decrease opioid use in IDUs in several studies. As a partial mu-opioid agonist, treatment with buprenorphine has several advantages over treatment with methadone. First, intravenous administration of buprenorphine, when combined with naloxone, causes symptoms similar to opioid withdrawal, making it an unlikely target for abuse among recovering addicts (Sullivan *et al.*, 2005). Additionally, while its effects increase with the dosage, they do so within a limited range, creating a greater margin of safety for its administration than methadone.

Its use was first approved by France in 1996. Several years later, in 2000, the Drug Addiction Treatment Act permitted office-based treatment with buprenorphine in the United States. In 2002, the Food and Drug Administration allowed buprenorphine to be prescribed in physicians' offices (e.g., addiction specialists, HIV medicine specialists, psychiatrists and primary-care physicians). It is becoming increasingly available internationally, and as of May 2004 it was approved for the treatment of opioid dependence in 37 countries (Sullivan and Fiellin, 2005).

While few studies report changes in HIV risk behavior, such as in the frequency of injection drug use, sharing of injection equipment, sexual behavior, overall HIV risk, or rates of HIV seroconversion, there is some evidence that HIV-positive IDUs taking buprenorphine are more adherent to their HIV medications than those who are not receiving buprenorphine (Sullivan and Fiellin, 2005). Just as with methadone maintenance, co-dependence on heroin and cocaine presents a challenge for HIV prevention, since these individuals are more likely to relapse and also tend to have higher-risk injecting behaviors. Similar to methadone, buprenorphine is not an effective HIV prevention strategy for those who inject cocaine or smoke crack. It is also important to note that many IDUs have difficulty in accessing medical care, do not have a primary care provider, and are unaware of buprenorphine providers in particular. Thus, drug treatment has not been an optimal HIV prevention strategy for high-risk IDUs (e.g., those who inject cocaine and who have difficulty accessing buprenorphine providers), and therefore better treatment options are needed.

Anti-drug legislations

Many other countries have implemented abstinence-based treatment approaches to prevent HIV transmission. Although the rate of relapse is high, IDUs may benefit from the HIV education, other prevention interventions and access to primary health care that they receive while in treatment (Ball *et al.*, 1998). Recognizing that risk reduction and improving an individual's health status and social functioning are important steps along the way to abstinence, several community-based drug treatment programs in India, Sri Lanka and Thailand promote rehabilitation before detoxification (Ball *et al.*, 1998).

Other countries, such as China, have adopted a zero tolerance policy on drug use. Both drug trafficking and drug abuse are illegal in China, and can lead to the death penalty. In China, there are approximately 1.14 million documented drug users, and about half of them inject drugs. Because many of these IDUs share needles, injection drug use is responsible for 42 percent of all HIV/AIDS cases reported thus far (Qian *et al.*, 2006). The Chinese government is currently targeting the drug abuse problem at several different levels. They have taken measures to (1) eliminate drug smuggling activities across borders by collaborating with neighboring countries, (2) use anti-drug education campaigns to discourage drug use, and (3) provide current drug users with drug detoxification and harm-reduction services. In an era of globalization, it seems unlikely that China will eradicate its drug abuse problem solely by enforcing strict laws to prohibit drug use and smuggling. Successful HIV prevention interventions in other countries have demonstrated that governmental support, harm-reduction programs, and aid from non-governmental organizations are also needed.

Future directions

To date, most HIV prevention efforts have targeted individual-level factors such as risk behaviors; future research must aim to understand social, political and other contextual determinants of IDU and HIV infection among IDUs. Epidemiologic methods are expanding in ways that make the measurement of these types of social and contextual factors feasible within epidemiologic research. In addition, the increasing globalization of commerce must be an important dimension of future research and intervention efforts, as highlighted by international drug-smuggling rings and the challenges facing China, discussed earlier. IDUs are considered a socially marginalized population, and comprise of severely marginalized subpopulations (e.g., young IDUs, sexual and racial minorities) who are not likely to seek preventive services and may be systematically missed in research and interventions. Thus, future research and intervention efforts, should aim (1) to ensure the inclusion of these specific groups, (2) to utilize methods that are social network-oriented, and (3) to include features of the larger social environment which influence individual behavior. Pharmacologic advances should also be a priority for future intervention efforts. Currently, pharmacologic interventions typically accessed by IDUs are limited primarily to methadone, a synthetic narcotic. Finally, the National Institute on Drug Abuse recommends that future research strives to incorporate drug users into the broader diagnosis-based strategies for *early* detection of HIV infection and reduction of transmission risk (NIDA, 2006). This may include strategies that target IDUs prior to onset of injection drug use by identifying drug users at risk of becoming injectors, such as those who use heroin intranasally and/or smoke crack cocaine and reside in high-risk environments (Fuller *et al.*, 2004; Vlahov *et al.*, 2004).

Acknowledgment

We would like to thank Rachel Stern for her meticulous review of this manuscript.

References

Absalon, J., Fuller, C. M., Ompad, D. *et al.* (2006). Gender differences in sexual behaviors, sexual partnerships, and HIV among drug users in New York City. *AIDS Behav.*, 10(6), 707–15.

Ahern, J., Stuber, J. and Galea, S. (2007). Stigma, discrimination and the health of illicit drug users. *Drug Alcohol Depend.*, 88(2–3), 188–96.

Airhihenbuwa, C. O., DiClemente, R. J., Wingood, G. M. and Lowe, A. (1992). HIV/AIDS education and prevention among African-Americans: a focus on culture. *AIDS Educ. Prev.*, 4, 267–76.

Anderson, W. (1991). The New York Needle Trial: the politics of public health in the age of AIDS. *Am. J. Public Health*, 81, 1506–17.

Anthony, J. C., Vlahov, D., Nelson, K. E. *et al.* (1991). New evidence on intravenous cocaine use and the risk of infection with human immunodeficiency virus type 1. *Am. J. Epidemiol.*, 134, 1175–89.

Astemborski, J., Vlahov, D., Warren, D. *et al.* (1994). The trading of sex for drugs or money and HIV seropositivity among female intravenous drug users. *Am. J. Public Health*, 84, 382–7.

Ball, A. L. (1996). Averting a global epidemic. *Addiction*, 91, 1095–8.

Ball, A. L., Rana, S. and Dehne, K. L. (1998). HIV prevention among injecting drug users, responses in developing and transitional countries. *Public Health Rep.*, 113(Suppl. 1), 170–81.

Baral, S., Sifakis, F., Cleghorn, F. and Beyrer, C. (2007). Elevated risk for HIV infection among men who have sex with men in low- and middle-income countries 2000–2006: a systematic review. *PLoS Med.*, 4, 1901–11.

Battjes, R. J., Pickens, R. W. and Amsel, Z. (1991). HIV infection and AIDS risk behaviors among intravenous drug users entering methadone treatment in selected US cities. *J. Acquir. Immune Defic. Syndr.*, 4, 1148–54.

Bell, A. V., Ompad, D. and Sherman, S. G. (2006). Sexual and drug risk behaviors among women who have sex with women. *Am. J. Public Health*, 96, 1066–72.

Berger, M. T. (2004). *Workable Sisterhood: The Political Journey of Stigmatized Women with HIV/AIDS*. Princeton, NJ: Princeton University Press.

Blankenship, K. M., Smoyer, A. B., Bray, S. J. and Mattocks, K. (2005). Black–white disparities in HIV/AIDS: the role of drug policy and the corrections system. *J. Health Care Poor Underserved*, 16(Suppl. B), 140–56.

Bluthenthal, R. N. and Watters, J. K. (1995). Multimethod research from targeted sampling to HIV environments. *NIDA Research Monographs*, 157, 212–30.

Bluthenthal, R. N., Kral, A. H., Lorvick, J. and Watters, J. K. (1997). Impact of law enforcement on syringe exchange programs, a look at Oakland and San Francisco. *Med. Anthropol.*, 18, 61–83.

Boardman, J. D., Finch, B. K., Ellison, C. G. *et al.* (2001). Neighborhood disadvantage, stress, and drug use among adults. *J. Health Social Behav.*, 42, 151–65.

Bobrova, N., Rhodes, T., Powers, R. *et al.* (2006). Barriers to accessing drug treatment in Russia, a qualitative study among injecting drug users in two cities. *Drug Alcohol Depend.*, 82(Suppl. 1), S57–63.

Booth, R. E., Watters, J. K. and Chitwood, D. D. (1993). HIV risk-related sex behaviors among injection drug users, crack smokers, and injection drug users who smoke crack. *Am. J. Public Health*, 83, 1144–8.

Broadhead, R. S. (1991). Social construction of bleach in combating AIDS among injection drug users. *J. Drug Issues*, 21, 713–37.

Brooner, R. K., Bigelow, G. E., Strain, E. and Schmidt, C. W. (1990). Intravenous drug abusers with antisocial personality disorder: increased HIV risk behavior. *Drug Alcohol Depend.*, 26, 39–44.

Burris, S., Welsh, J., Mitzi, N. G. *et al.* (2002). State syringe and drug possession laws potentially influencing safe syringe disposal by injection drug users. *J. Am. Pharm. Assoc.*, 42, S94–8.

Calsyn, D. A., Saxon, A. J., Freeman, G. and Wittaker, S. (1991). Needle-use practices among intravenous drug users in an area where needle purchase is legal. *AIDS*, 5, 187–93.

Carceres, W., Blaney, S., Lewis, E. *et al.* (2006). Pharmacy syringe access program for injection drug users in New York City: policy recommendation for targeted advertisement to increase sterile syringe access through pharmacies. *Annual Meeting of the American Public Health Association, Boston, MA*.

CDC (2001). Women, injection drug use and the criminal justice system. Available at http,//www.cdc.gov/idu/facts/cj-women.pdf (accessed 02-29-08).

CDC (2002). *Drug-associated HIV Transmission Continues in the United States*. Atlanta, GA: Department of Health and Human Services, Centers for Disease Control and Prevention.

CDC (2003). Advancing HIV prevention, new strategies for a changing epidemic – United States. *Morbid. Mortal. Wkly Rep.*, 52, 329–32.

CDC (2005). HIV prevalence among selected populations: high-risk populations. Available at http://www.cdc.gov/hiv/topics/testing/resources/reports/hiv_prevalence/high-risk.htm#figure11 (accessed 06-06-2008).

CDC (2004). *HIV/AIDS Surveillance Report.* Atlanta, GA: Department of Health and Human Services, Centers for Disease Control and Prevention, 16(21), Table 11.

CDC (2007). Racial/ethnic disparities in diagnoses of HIV/AIDS – 33 states, 2001–2005. *Morbid. Mortal. Wkly Rep.*, 56, 189–93.

Chan, K. Y., Yang, Y., Zhang, K. and Reidpath, D. D. (2007). Disentangling the stigma of HIV/AIDS from the stigmas of drugs use, commercial sex and commercial blood donation – a factorial survey of medical students in China. *BMC Public Health*, 7, 280.

Chatterjee, A., Uprety, L., Chapagain, M. and Kafle, K. (1996). Drug abuse in Nepal, a rapid assessment study. *Bull. Narcotics*, 48(1–2), 11–33.

Compton, W. M. and Cottler, L. B. (2000). The effects of psychiatric comorbidity on response to an HIV prevention intervention. *Drug Alcohol Depend.*, 58, 247–57.

Compton, W. M., Cottler, L. B., Ben-Abdallah, A. *et al.* (2000). The effects of psychiatric comorbidity on response to an HIV prevention intervention. *Drug Alcohol Depend.*, 58, 247–57.

Cooper, H., Friedman, S. R., Tempalski, B. *et al.* (2005). Racial/ethnic disparities in injection drug use in large US metropolitan areas. *Ann. Epidemiol.*, 15, 326–34.

Copenhaver, M., Johnson, B. T., Lee, I. E. *et al.* (2006). Behavioral HIV risk reduction among people who inject drugs: meta-analytic evidence of efficacy. *J. Subst. Abuse*, 31, 63–171.

Culbert, H., Tu, D., O'Brien, D. P. *et al.* (2007). HIV treatment in a conflict setting, outcomes and experiences from Bukavu, Democratic Republic of the Congo. *PLoS Med.*, 4, e129.

Department of Health Union of Myanmar (1993). *AIDS Prevention and Control Programme 1993. Sentinel Surveillance Data.* Yango: Ministry of Health.

Des Jarlais, D. C., Marmor, M., Panone, D. *et al.* HIV incidence among injecting drug users in NYC syringe-exchange programmes. *Lancet*, 348, 987–91.

Des Jarlais, D. C., Friedman, S. R., Perlis, T. *et al.* (1999). Risk behavior and HIV infection among new drug injectors in the era of AIDS in New York City. *J. Acquir. Immune Defic. Syndr.*, 20, 67–72.

Des Jarlais, D. C., Marmor, M., Paone, D. *et al.* (2000). HIV incidence among injection drug users in New York City, 1992–1997, evidence for a declining epidemic. *Am. J. Public Health*, 90, 352–9.

Diaz, T., Vlahov, D., Greenberg, B. *et al.* (2001). Sexual orientation and HIV infection prevalence among young Latino injection drug users in Harlem. *J. Women's Health Gender-based Med.*, 10, 371–80.

Doherty, M. C., Garfein, R. S., Monterroso, E. *et al.* (2000). Correlates of HIV infection among young adult short-term injection drug users. *AIDS*, 14, 717–26.

Dorabjee, J., Samson, L. and Dyalchand, R. (1996). A community-based intervention for injecting drug users (IDUs) in New Delhi slums. Paper presented at the *XIth International Conference on AIDS, 7–12 July, Vancouver, Canada.*

Edlin, B. R., Irwin, K. L., Faruque, S. *et al.* (1994). Intersecting epidemics – crack cocaine use and HIV infection among inner-city young adults. Multicenter Crack Cocaine and HIV Infection Study Team. *N. Engl. J. Med.*, 331, 1422–7.

El-Bassel, N. L., Gilbert, E., Wu, G. *et al.* (2005). Relationship between drug abuse and intimate partner violence, a longitudinal study among women receiving methadone. *Am. J. Public Health*, 95, 465–70.

Evans, J. L., Hahn, J. A., Page-Shafer, K. and Lum, P. J. (2003). Gender differences in sexual and injection risk behavior among active young injection drug users in San Francisco (The UFO Study). *J. Urban Health*, 80, 137–46.

Farrell, M., Neeleman, J., Gossop, M. *et al.* (1995). Methadone provision in the European Union. *Intl J. Drug Policy*, 6, 168–72.

Farzadegan, H., Hoover, D. R., Astemborski, J. *et al.* (1998). Sex differences in HIV-1 viral load and progression to AIDS. *Lancet*, 352, 1510–14.

Feldman, H. W. and Biernacki, P. (1988). The ethnography of needle sharing among intravenous drug users and implications for public policies and intervention strategies. *NIDA Res. Monogr.*, 80, 28–39.

Fennema, J. S., van Ameijden, E. J., van den Hoek, A. and Coutinho, R. A. (1997). Young and recent-onset injection drug users are at higher risk for HIV. *Addiction*, 92, 1457–65.

Ford, C. L., Miller, W. C., Smurzynski, M. and Leone, P. A. (2007). Key components of a theory-guided HIV prevention outreach model: pre-outreach preparation, community assessment, and a network of key informants. *AIDS Educ. Prev.*, 19, 173–86.

Friedman, S. R., Jose, B., Deren, S. *et al.* (1995). Risk factors for human immunodeficiency virus seroconversion among out-of-treatment injectors in high and low seroprevalence cities. The National AISA Research Consortium. *Am. J. Epidemiol.*, 142, 864–74.

Friedman, S. R., Ompad, D. C., Maslow, C. *et al.* (2003). HIV prevalence, risk behaviors, and high-risk sexual and injection networks among young women injectors who have sex with women. *Am. J. Public Health*, 93, 902–6.

Frye, V., Latka, M. H., Wu, Y. *et al.* (2007). Intimate partner violence perpetration against main female partners among HIV-positive male injection drug users. *J. Acquir. Immune Defic. Syndr.*, 46(Suppl. 2), S101–9.

Fuller, C. M., Vlahov, D., Arria, A. M. *et al.* (2001). Factors associated with adolescent initiation of injection drug use. *Public Health Rep.*, 116(Suppl. 1), 136–45.

Fuller, C. M., Ahern, J., Vadnai, L. *et al.* Impact of increased syringe access, preliminary findings on injection drug user syringe source, disposal, and pharmacy sales in Harlem, New York. *J. Am. Pharm. Assoc.*, 42 (Suppl. 2), S77–82.

Fuller, C. M., Vlahov, D., Latkin, C. A. *et al.* (2003). Social circumstances of initiation of injection drug use and early shooting gallery attendance, implications for HIV intervention among adolescent and young adult injection drug users. *J. Acquir. Immune Defic. Syndr.*, 32, 86–93.

Fuller, C. M., Ompad, D. C., Galea, S. *et al.* (2004). Hepatitis C incidence – a comparison between injection and non-injection drug users in New York City. *J. Urban Health*, 81, 20–4.

Fuller, C. M., Borrell, L. N., Latkin, C. A. *et al.* (2005a). Effects of race, neighborhood, and social network on age at initiation of injection drug use. *Am. J. Public Health*, 95, 689–95.

Fuller, C. M., Absalon, J., Ompad, D. C. *et al.* (2005b). A comparison of HIV seropositive and seronegative young adult heroin- and cocaine-using men who have sex with men in New York City, 2000–2003. *J. Urban Health*, 82(Suppl. 1), i51–61.

Fullilove, R. E., Fullilove, M. T., Bowser, B. P. and Gross, S. A. (1990). Risk of sexually transmitted disease among black adolescent crack users in Oakland and San Francisco, California. *J. Am. Med. Assoc.*, 263, 851–5.

Galea, S., Ahern, J. and Valhov, C. (2003). Contextual determinants of drug use risk behavior: a theoretic framework. *J. Urban Health*, 80(Suppl. 3), iii50–8.

Garfein, R. S., Vlahov, D., Galai, N. *et al.* (1996). Viral infections in short-term injection drug users, the prevalence of the hepatitis C, hepatitis B, human immunodeficiency, and human T-lymphotropic viruses. *Am. J. Public Health*, 86, 655–61.

Godette, D. C., Headen, S. and Ford, C. (2006). Windows of opportunity, fundamental concepts for understanding alcohol-related disparities experienced by young Blacks in the United States. *Prevent. Sci.*, 7, 377–87.

Goldbaum, G., Perdue, T., Wolitski, R. *et al.* (1998). Differences in risk behavior and sources of AIDS information among gay, bisexual, and straight-identified men who have sex with men. *AIDS Behav.*, 2, 13–21.

Gordon, C. M., Forsyth, A. D., Stall, R. and Cheever, L. W. (2005). Prevention interventions with persons living with HIV/AIDS, state of the science and future directions. *AIDS Educ. Prev.*, 17, 6–20.

Gostin, L. O. and Lazzarini, Z. (1997). Prevention of HIV/AIDS among injection drug users: the theory and science of public health and criminal justice approaches to disease prevention. *Emory Law J.*, 46, 587–696.

Groseclose, S. L., Weinstein, B., Jones, T. S. *et al.* (1995). Impact of increased legal access to needles and syringes on practices of injecting-drug users and police officers – Connecticut, 1992–1993. *J. Acquir. Immune Defic. Syndr.*, 10, 71–2.

Guenther-Grey, C., Noroian, D., Foneska, J. and Higgins, D. (1996). Developing community networks to deliver HIV prevention interventions. *Public Health Rep.*, 111(Suppl. 1), 41–9.

Gyarmathy, V. A. and Neaigus, A. (2007). The relationship of sexual dyad and personal network characteristics and individual attributes to unprotected sex among young injecting drug users. *AIDS Behav.*, (e-pub ahead of print).

Hamers, F. F., Batter, V., Downs, A. *et al.* (1997). The HIV epidemic associated with injecting drug use in Europe, geographic and time trends. *AIDS*, 11, 1365–74.

Hangzo, C., Chatterjee, A., Sarkar, S. *et al.* (1997). Reaching out beyond the hills, HIV prevention among injecting drug users in Manipur, India. *Addiction*, 92, 813–20.

Hankins, C. A., Friedman, S. R., Zafar, T. and Strathdee, S. A. (2002). Transmission and prevention of HIV and sexually transmitted infections in war settings: implications for current and future armed conflicts. *AIDS*, 16, 2245–52.

Harawa, N. and Adimora, A. (2008). Incarceration, African Americans and HIV: advancing a research agenda. *J. Natl Med. Assoc.*, 100, 57–62.

Hein, N. (1995) *Drug Use and HIV Infection in Viet Nam.* Report of the WHO Drug Injecting Project Planning Meeting, Phase II. Bangkok: WHO.

Henderson, D. J. (1998). Drug abuse and incarcerated women. *J. Subst. Abuse Treat.*, 15, 579–87.

Holmberg, S. D. (1996). The estimated prevalence and incidence of HIV in 96 large US metropolitan areas. *Am. J. Public Health*, 86, 642–54.

Ibanez, G. E., Purcell, D. W., Stall, R. *et al.* (2005). Sexual risk, substance use, and psychological distress in HIV-positive gay and bisexual men who also inject drugs. *AIDS*, 19(Suppl. 1), S49–55.

Iguchi, M. Y., Bux, D. A. Jr., Kushner, H. and Victor, L. (2001). Correlates of HIV risk among female sex partners of injecting drug users in a high-seroprevalence area. *Eval. Prog. Plan.*, 24, 175–85.

Institute of Medicine of the National Academies (2007). *Preventing HIV Infection among Injecting Drug Users in High-risk Countries: An assessment of the Evidence.* Washington, DC: National Academy of Sciences.

Jose, B., Friedman, S. R., Neaigus, A. (1996). Collective organization of injecting drug users and the struggle against AIDS. In: T. Rhodes and R. Hartnoll *et al.*, (eds) *AIDS, Drugs and Prevention: Perspectives on Individual and Community Action.* London: Routledge, pp. 216–33.

Klein, S. J., Guthrie, S. and Candelas, A. R. (2000). Expanded syringe access demonstration program in New York State: an intervention to prevent HIV/AIDS transmission. *J. Urban Health*, 77, 762–7.

Koester, S. K. (1994). The context of risk, ethnographic contributions to the study of drug use and HIV. *NIDA Res. Monogr.*, 143, 202–17.

Kottiri, B. J., Friedman, S. R., Neaigus, A. *et al.* (2002). Risk networks and racial/ethnic differences in the prevalence of HIV infection among injection drug users. *J. Acquir. Immune Defic. Syndr.*, 30, 95–104.

Kral, A. H., Bluthenthal, R. N., Booth, R. E. and Watters, J. K. (1998). HIV seroprevalence among street-recruited injection drug and crack cocaine users in 16 US municipalities. *Am. J. Public Health*, 88, 108–13.

Krieger, N., Williams, D. R. and Moss, N. E. (1997). Measuring social class in US public health research: concepts, methodologies, and guidelines. *Annu. Rev. Public Health*, 18, 341–78.

Lamothe, F., Bruneau, J., Coates, R. *et al.* (1993). Seroprevalence of and risk factors for HIV-1 infection in injection drug users in Montreal and Toronto, a collaborative study. *Can. Med. Assoc. J.*, 149, 945–51.

Lange, W. R., Snyder, F. R., Lozovsky, D. *et al.* (1988). Geographic distribution of human immunodeficiency virus markers in parenteral drug abusers. *Am. J. Public Health*, 78, 443–6.

Latkin, C. A. (1998). Outreach in natural settings, the use of peer leaders for HIV prevention among injecting drug users' networks. *Public Health Rep.*, 113(Suppl. 1), 151–9.

Latkin, C. A. and Knowlton, A. R. (2005). Micro-social structural approaches to HIV prevention, a social ecological perspective. *AIDS Care*, 17(Suppl. 1), S102–13.

Latkin, C. A., Mandell, W., Oziemkowska, M. *et al.* (1995). Using social network analysis to study patterns of drug use among urban drug users at high risk for HIV/AIDS. *Drug Alcohol Depend.*, 38, 1–9.

Latkin, C. A., Forman, V., Knowlton, A. and Sherman, S. (2003). Norms, social networks, and HIV-related risk behaviors among urban disadvantaged drug users. *Social Sci. Med.*, 56, 465–76.

Lau, J. T., Zhang, J., Zhang, L. *et al.* (2007). Comparing prevalence of condom use among 15,379 female sex workers injecting or not injecting drugs in China. *Sex. Transm. Dis.*, 34, 908–16.

Maas, B., Fairbairn, N., Thomas, K. *et al.* (2007). Neighborhood and HIV infection among IDU: place of residence independently predicts HIV infection among a cohort of injection drug users. *Health Place*, 13, 432–9.

MacDonald, M., Law, M., Kaldor, J. *et al.* (2003). Effectiveness of needle and syringe programmes for preventing HIV transmission. *Intl J. Drug Policy*, 14, 353–7.

Maharjan, S. H. and Singh, M. (1996). Street-based outreach program for injecting drug users. Paper presented at the *XIth International Conference on AIDS, July 7–12, Vancouver, Canada*, Abstract no. Mo.D.244.

Marks, G., Burris, S. and Peterman, T. A. (1999). Reducing sexual transmission of HIV from those who know they are infected, the need for personal and collective responsibility. *AIDS*, 13, 297–306.

Mays, V. M., Cochran, S. D. and Barnes, N. W. (2007). Race, race-based discrimination, and health outcomes among African Americans. *Ann. Rev. Psychol.*, 58, 201–25.

Miller, N. S., Sheppard, L. M., Colenda, C. C. and Magen, J. (2001). Why physicians are unprepared to treat patients who have alcohol- and drug-related disorders. *Acad. Med.*, 76, 410–18.

National Institute on Drug Abuse (2000). *The Science of Drug Abuse and Addiction.* Advance Report, Community Epidemiology Work Group. Available at http://www.drugabuse.gov/CEWG/AdvancedRep/6_20ADV/0600adv.html.

National Institute on Drug Abuse (2002). *Principles of HIV Prevention in Drug-using Populations.* Bethesda, MD: National Institute on Drug Abuse, Department of Health and Human Development.

National Institute on Drug Abuse (2006). *Research Report, HIV/AIDS.* Rockville, MD: National Clearinghouse on Alcohol and Drug Information, US Department of Health and Human Services, National Institutes of Health.

National Task Force on HIV/AIDS and Injection Drug Use (2001). HIV/AIDS and Injection Drug Use: A National Action Plan. Available at http,//www.cfdp.ca/hivaids. html (accessed 02-29-08).

Ndetei, D. (2004). *UNODC Study on the Linkages between Drug Use, Injecting Drug Use and HIV/AIDS in Kenya.* Nairobi: University of Nairobi.

Neaigus, A., Friedman, S. R., Jose, B. *et al.* (1996). High-risk personal networks and syringe sharing as risk factors for HIV infection among new drug injectors. *J. Acquir. Immune Defic. Syndr.,* 11, 499–509.

Needle, R. H., Coyle, S. L., Norman, J. *et al.* (1998). HIV prevention with drug-using populations–current status and future prospects, introduction and overview. *Public Health Rep.,* 113(Suppl. 1), 4–18.

Nelson, K. E., Vlahov, D., Solomon, S. *et al.* (1995). Temporal trends of incident human immunodeficiency virus infection in a cohort of injecting drug users in Baltimore, MD. *Arch. Intern. Med.,* 155, 1305–11.

New York Academy of Medicine (2003). *New York State Expanded Syringe Access Demonstration Program Evaluation.* Evaluation Report to the Governor and the New York State Legislatures. New York, NY: New York Academy of Medicine.

New York City Department of Health and Mental Hygiene (2003). *HIV Surveillance and Epidemiology Program. Quarterly Report.* New York, NY: New York City Department of Health and Mental Hygiene.

New York State Department of Health, Bureau of HIV/AIDS Epidemiology (2000). *AIDS Surveillance Quarterly Update for Cases Reported through March 2000.* Albany, NY: New York State Department of Health, Bureau of HIV/AIDS Epidemiology.

New York State Department of Health, Bureau of HIV/AIDS Epidemiology (2007). *New York State HIV/AIDS Surveillance Semiannual Report for Cases Diagnosed through December 2005.* Available at http,//www.health.state.ny.us/diseases/aids/statistics/ semiannual/2005/surveillance_semiannual_report_2005-12.pdf.

Nicolosi, A., Leite, M. L., Molinari, S. *et al.* (1992). Incidence and prevalence trends of HIV infection in intravenous drug users attending treatment centers in Milan and northern Italy, 1986–1990. *J. Acquir. Immune Defic. Syndr.,* 5, 365–73.

Ompad, D. and Fuller, C. M. (2005). The urban environment, drug use, and health. In: S. Galea and D. Vlahov (eds), *Handbook of Urban Health.* New York, NY: Springer, pp. 127–54.

Pajares, F. (2002). Overview of social cognitive theory and of self efficacy. Available at http://www.emory.edu/EDUCATION/mfp/eff.html (accessed 04-28-08).

Panchanadeswaran, S., Johnson, S., Sivaram, C. *et al.* (2008). Intimate partner violence is as important as client violence in increasing street-based female sex workers' vulnerability to HIV in India. *Intl J. Drug Policy,* 19, 106–12.

Pokrovski, V. (1998). *HIV Infections by Modes of Transmission in the Russian Federation.* Report to UNAIDS. Russia: AIDS Center.

Pouget, E. R., Deren, S., Fuller, C. *et al.* (2005). Receptive syringe sharing among injection drug users in Harlem and the Bronx during the New York State Expanded Syringe Access Demonstration Program. *J. Acquir. Immune Defic. Syndr.,* 39, 471–7.

Purcell, D. W., Garfein, R. S., Latka, M. H. *et al.* (2007). Development, description, and acceptability of a small-group, behavioral intervention to prevent HIV and hepatitis C virus infections among young adult injection drug users. *Drug Alcohol Depend.,* 91(Suppl. 1), S73–80.

Qian, H.-Z., Schumacher, J. E., Chen, H. T. and Ruan, Y. (2006). Injection drug use and HIV/AIDS in China, Review of current situation, prevention and policy implications. *Harm Reduction J.,* 1, 4.

Reynaud-Maurupt, C., Carrieri, N. P., Gastaud, J. A. *et al.* (2000). Impact of drug maintenance treatment on injection practices among French HIV-infected IDUs. *AIDS Care*, 12, 461–70.

Rietmeijer, C. A., Wolitski, R. J., Fishbein, M. *et al.* (1998). Sex hustling, injection drug use, and non-gay identification by men who have sex with men. Associations with high-risk sexual behaviors and condom use. *Sex. Transm. Dis.*, 25, 353–60.

Santibanez, S. S., Garfein, R. S., Swartzendruber, A. *et al.* (2006). Update and overview of practical epidemiologic aspects of HIV/AIDS among injection drug users in the United States. *J. Urban Health*, 83, 86–100.

Sarang, A., Stuikyte, A. and Bykoe, R. (2007). Implementation of harm reduction in Central and Eastern Europe and Central Asia. *Intl J. Drug Policy*, 18, 129–35.

Schoenbaum, G. M., Martin, R. J. and Roane, D. S. (1989). Relationships between sustained sucrose-feeding and opioid tolerance and withdrawal. *Pharmacol. Biochem. Behav.*, 34, 911–14.

Schroeder, J. R. (2001). Illicit drug use in one's social network and in one's neighborhood predicts individual heroin and cocaine use. *Ann. Epidemiol.*, 11, 389–94.

Sharan Society for Helping the Urban Poor (1997). *Beyond Appearances*. A monthly newsletter for the benefit of drug users and ex-drug users, August–September.

Sherman, S. G. (2005). Correlates of initiation of injection drug use among young drug users in Baltimore, Maryland: the need for early intervention. *J. Psychoactive Drugs*, 37, 437–43.

Sherman, S. G. and Purchase, D. (2001). Point Defiance: a case study of the United States' first public needle exchange in Tacoma, Washington. *Intl J. Drug Policy*, 12, 45–57.

Shin, S. H., Lundgren, L. and Chassler, D. (2007). Examining drug treatment entry patterns among young injection drug users. *Am. J. Drug Alcohol Abuse*, 33, 217–25.

Solomon, L., Stein, M., Flynn, C. *et al.* (1998). Health services use by urban women with or at risk for HIV-1 infection, the HIV Epidemiology Research Study (HERS). *J. Acquir. Immune Defic. Syndr.*, 17, 253–61.

Stancliff, S., Salomon, N., Perlman, D. C. and Russell, P. C. (2000). Provision of influenza and pneumococcal vaccines to injection drug users at a syringe exchange. *J. Subst. Abuse. Treat.*, 18, 263–5.

Strathdee, S. A. (2003). Rise in needle sharing among injection drug users in Pakistan during the Afghanistan war. *Drug Alcohol Depend.*, 71, 17–24.

Strathdee, S. A., Patrick, D. M., Currie, S. L. *et al.* (1997). Needle exchange is not enough, lessons from the Vancouver injecting drug use study. *AIDS*, 11, F59–65.

Strathdee, S. A., Zafar, T., Brahmbhatt, H. *et al.* (2003). Rise in needle sharing among injection drug users in Pakistan during the Afghanistan war. *Drug Alcohol Depend.*, 71, 17–24.

Strathdee, S. A., Ricketts, E. P. and Huettner, S. (2006). Facilitating entry into drug treatment among injection drug users referred from a needle exchange program: results from a community-based behavioral intervention trial. *Drug Alcohol Depend.*, 83, 225–32.

Sullivan, L. E. and Fiellin, D. A. (2005). Buprenorphine: its role in preventing HIV transmission and improving the care of HIV patients with opioid dependence. *HIV/AIDS, September*, 891–6.

Sullivan, L. E., Metzger, D. S., Fudala, P. J. *et al.* (2005). Decreasing international HIV transmission, the role of expanding access to opioid agonist therapies for injection drug users. *Addiction*, 100, 150–8.

Takahashi, L. M. (1997). Stigmatization, HIV/AIDS, and communities of color: exploring response to human service facilities. *Health Place*, 3, 187–99.

UNAIDS (1997). *Report on the Global HIV/AIDS Epidemic, 1997.* Geneva: United Nations Programme on HIV/AIDS.

UNAIDS (2002). *Report on the Global HIV/AIDS Epidemic, 2002.* Geneva: United Nations Programme on HIV/AIDS.

UNAIDS (2003). Health Canada: The Open Society Institute & The Canadian International Development Agency. The Warsaw Declaration: A Framework for Effective Action on HIV/AIDS and Injecting Drug Use. Second International Policy Dialogue on HIV/AIDS, Warsaw (Poland), November 12–14.

UNAIDS (2004). National Surveillance Report from China, Indonesia, and Vietnam. In: AIDS in Asia, Face the Facts, 2004 MAP Report, Figure 1. Geneva: United Nations Programme on HIV/AIDS.

Van Ameijden, E. J., Van den Hoek, J. A. R., Van Haastrecht, H. J. A. and Coutinho, R. A. (1992). The harm reduction approach and risk factors for human immunodeficiency virus (HIV) seroconversion in injecting drug users, Amsterdam. *Am. J. Epidemiol.*, 136, 236–43.

Vlahov, D. and Junge, B. (1998). The role of needle exchange programs in HIV prevention. *Public Health Rep.*, 113(Suppl. 1), 75–80.

Vlahov, D., Fuller, C. M., Ompad, D. C. *et al.* (2004). Updating the infection risk reduction hierarchy, preventing transition into injection. *J. Urban Health*, 81, 14–19.

Walters, J. K., Estilo, M. J., Clark, G. L. and Lorvick, J. K. (1994). Syringe and needle exchange as HIV/AIDS prevention for injection drug users. *J. Am. Med. Assoc.*, 271, 115–20.

Ward, J. R., Mattick, R. and Hall, W. (1994). The effectiveness of methadone maintenance treatment, an overview. *Drug Alcohol Rev.*, 13, 327–36.

Weinhardt, L. S., Carey, M. P., Johnson, B. T. and Bickham, N. L. (1999). Effects of HIV counseling and testing on sexual risk behavior, a meta-analytic review of published research, 1985–1997. *Am, J. Public Health*, 89, 1397–405.

White House Press Release (2003). President Delivers State of the Union. Available at http://www.whitehouse.gov/news/releases/2003/01/20030128-19.html (accessed 12-28-06).

Williams, C. T. and Latkin, C. A. (2007). Neighborhood socioeconomic status, personal network attributes, and use of heroin and cocaine. *Am. J. Prev. Med.*, 32(Suppl.), S203–10.

Williams, D. R. and Collins, C. (2001). Racial residential segregation, a fundamental cause of racial disparities in health. *Public Health Rep.*, 116, 404–16.

Wolitski, R. J., MacGowan, R. J., Higgins, D. L. and Jorgensen, C. M. (1997). The effects of HIV counseling and testing on risk-related practices and help-seeking behavior. *AIDS Educ. Prev.*, 9, 52–67.

World Health Organization (1997). *STD/HIV/AIDS Surveillance Report, Special Edition.* Manila: WHO Regional Office for the Western Pacific, Issue No. 10.

Wyatt, G. E., Carmona, J. V., Loeb, T. B. and Williams, J. K. (2005). HIV-positive black women with histories of childhood sexual abuse, patterns of substance use and barriers to health care. *J. Health Care Poor Underserved*, 16(Suppl. B), 9–23.

Young, R. M., Friedman, S. R. and Case, P. (2005). Exploring an HIV paradox, an ethnography of sexual minority women injectors. *J. Lesbian Studies*, 9, 103–16.

Zule, W. A. (1992). Risk and reciprocity, HIV and the injection drug user. *J. Psychoactive Drugs*, 24, 243–9.

HIV risk and prevention for non-injection substance users

13

Lydia N. Drumright and Grant N. Colfax

Approximately 2 billion people worldwide consume alcohol (WHO, 2004) and 200 million use illicit drugs annually (UN, 2007), corresponding to about one-third and 5 percent of the world's population respectively. Alcohol use disorders are estimated to affect 76.3 million people (WHO, 2004). According to United Nations estimates, 79.4 percent of those reporting drug use annually worldwide use cannabis, 12.5 percent amphetamine, 7.2 percent cocaine and 4.3 percent MDMA (UN, 2007). Approximately 8 percent report using opiates, including heroin (5.6 percent) (UN, 2007); however, most studies have not distinguished non-injection opiate use from injection, and therefore opiate use will not be discussed in the context of non-injection substance use (NISU).

It is important to note that the substances included in this chapter can be administered in a number of ways, and that not all studies query participants regarding how drugs are administered. There are many reports of cocaine (De *et al.*, 2007; Rhodes T. *et al.*, 2007b) and methamphetamine (Semple *et al.*, 2004; Pilowsky *et al.*, 2007; Rawstorne *et al.*, 2007) injection and, to a lesser extent, ketamine (Lankenau and Clatts, 2005) and GHB (Teter and Guthrie, 2001). Individuals may also change their pattern of drug use over time, with non-injectors taking up injection use, and injectors discontinuing injection but maintaining drug use through other forms of administration (Des Jarlais *et al.*, 2007). Therefore, associations between HIV prevalence and current or past NISU may be confounded by changes in drug-administration behavior.

Although injection drug use came to the forefront of the HIV epidemic as a risk factor for HIV transmission due to parenteral transmission, NISU is far more common than injection drug use, and has been studied as a factor in increased risk for HIV infection in a number of studies, either directly or through increased risky sexual behavior. In addition, there is growing recognition that non-injection substance users are heterogeneous in terms of their patterns of substance use, with respect to frequency of use, types of drugs combined and motivations for use. Prevention of HIV through reduction and cessation of NISU has been less straightforward than with injection drug use, because direct parenteral HIV transmission is not present with NISU as it is in the setting of high-risk injection behaviors.

Concern over establishing causation has been expressed for studies examining associations between substance use and HIV risk. Leigh and Stall (1993) addressed the methodological issues related to establishing causal associations between HIV risk and alcohol and drug use. Four types of studies examining associations between substance use and HIV risk were presented. Global association studies demonstrate associations between drug or alcohol use in general, and HIV/STI incidence, prevalence or risk behaviors. Situational association studies demonstrate associations between use of alcohol and drugs at the time of sexual activity and sexual risk or disease outcome in general. While these types of studies do demonstrate association, they provide the weakest evidence for a causal association, because factors such as personality could confound the associations between drug use and risky sexual behaviors. In contrast, event-level analyses examine substance use and sexual risk-taking in the same event, demonstrating a direct link between the two behaviors, but still not ruling out psychosocial factors such as personality. A case-crossover design was also suggested, where events in which risky sexual behavior do and do not occur are compared for substance use in the same individual. Such analyses essentially control for all factors not measured on an individual, and are likely to be less susceptible to confounding by unmeasured individual personality factors. Personality factors suggested to be the cause of both substance use and sexual risk, or to modify the association between substance use and HIV risk behaviors, include risky personality type (Kokkevi *et al.*, 1998; Semple *et al.*, 2000; Irwin *et al.*, 2006), sensation-seeking and lack of impulse control (McCoul and Haslam, 2001; Ross *et al.*, 2003; Lejuez *et al.*, 2005; Patterson *et al.*, 2005), an excuse for unsafe sexual activity (Rhodes, 1996; Rhodes and Stimson, 1994) or an "escape" from self-monitoring of sexual activity (McKirnan *et al.*, 1996, 2001; Ostrow, 1997).

In addition to personal characteristics, effects of substances on the user may also explain more direct associations between substances use and HIV/STI risk. A conceptual model of how drug use could lead to increased sexual risk explained risk in terms of drug effects on the user, including altered mental state, reduced sensation of pain, enhancement of sexual functioning, vasodilation, and administration (Drumright *et al.*, 2006a). A number of qualitative and quantitative studies examining the effects of drugs on their users support the hypothesis that these effects could explain how drug use could be directly associated with HIV/STI risk and acquisition; however, different substances will have more biological plausibility than others within this conceptual framework (Drumright *et al.*, 2006a).

While the current literature on substance use treatment for HIV prevention suggests that this may be an effective means of reducing HIV incidence, this area of research is still developing and many questions remain. In this chapter we examine current modalities of substance use treatment; the extent of the problem of alcohol and substance use, focusing mainly on studies done in the US; and evidence for associations between substance use and HIV/STI risk; and discuss the evidence that drug treatment could lead to HIV/STI prevention.

Types of interventions used to treat substance use problems

To better understand how substance use treatment may contribute to a reduction in HIV risk, a brief review of the most common substance use (including abuse) treatment approaches is indicated. Importantly, few programs employ just one of these modalities, and often combine multiple approaches in an effort to improve success. Programs also typically modify the approaches to individualize treatment plans, or to make the programs culturally appropriate for the target population.

Interventions to decrease substance use can be divided into pharmacotherapies and behavioral interventions, which may be used alone or in combination. Pharmacotherapies are medications that can be used for detoxification, as maintenance to prevent individuals from relapsing to a particular substance, or to treat co-occurring health disorders that may be related to substance use. Pharmacotherapies may involve the use of aversive agents such as disulfiram for alcohol disorders, agents of substitution such as methadone and buprenorphine for opiate disorders, and anti-craving agents such as naltrexone or acamprosate for alcohol use disorders. Behavioral interventions focus on modification of substance use behaviors through a variety of methods, including personalized and group therapies, positive behavior reinforcement, recognition of events and feelings that may lead to substance use, and multiple other methods. Numerous behavioral interventions have been applied to reduce substance use, and some of the more frequently utilized applications for substance use treatment in HIV prevention are summarized below.

Cognitive behavioral therapy (CBT) is a psychotherapy based on modifying cognitions, assumptions, beliefs and behaviors, with the aim of influencing disturbed emotions. It is widely used to treat various kinds of neurosis and psychopathology, including mood disorders and anxiety disorders (Gould *et al.*, 1997). In the context of treatment of substance use, CBT can be used to address the motivations for substance use, and to help clients practice behavioral and cognitive skills to cope with or avoid substance use. In this way, CBT is designed to help relapse triggered by both internal factors (e.g., craving, depression, anxiety) and external factors (e.g., exposure to an environment where substance use is taking place). CBT may also have the ability to assist in changing expectancies related to use of particular substances. Variations of CBT are widely used in alcoholism treatment under the label "relapse prevention". The matrix model is an enhanced form of CBT, with an approach that integrates individual, group, family and community support modalities to assist with substance use behavior change (Obert *et al.*, 2000).

Motivational enhancement therapy (MET), also referred to as motivational interviewing (MI), is a patient-oriented counseling approach focusing on resolving ambivalence in order to treat substance abuse (Miller, 1996, 2002). The focus of MET is to encourage patients to develop a negative view of their abuse, along with a desire to change their behavior. In MET, the therapist does not guide clients step-by-step through recovery, but instead strives to motivate them to use their own resources, through empathy, avoiding contradiction with the clients, and supporting self-efficacy. Such strategies could also assist in changing outcome expectancies related to substance use.

A twelve-step program is a set of guiding principles for recovery from addictive, compulsive or other behavioral problems (Chappel and DuPont, 1999). The twelve-step program was originally developed by the fellowship of Alcoholics Anonymous (AA) for recovery from alcoholism, but has been adapted as the foundation of other twelve-step programs such as Narcotics Anonymous. As summarized by the American Psychological Association, the Twelve Steps involve the following: admitting that one cannot control one's addiction or compulsion; recognizing a greater power that can give strength; examining past errors with the help of a sponsor (experienced member); making amends for these errors; learning to live a new life with a new code of behavior; and helping others that suffer from the same addictions or compulsions.

Contingency management (CM) is a type of treatment in which clients are rewarded (or, less often, punished) for their behavior, generally, for adherence to (or failure to adhere) to program rules and regulations or their treatment plan (Petry, 2000). CM is derived from an understanding that substance abuse disorders represent reinforced operant behaviors learned over time, and that these operant behaviors can be altered using environmental consequences. There is an extensive literature demonstrating use of multiple stimuli in CM to effectively reshape behavior (Bigelow and Silverman, 1999). CM for drug users, based on reward, operates under the premises that there is a normative response to positive effects of drug use, drug use is the failure of successful competition from other non-drug reinforcers to compete in the environment, and non-drug reinforcers can be introduced into the environment to successfully change behaviors.

Comparisons of substance use treatment interventions do not provide consistent evidence of superior efficacy of one type of intervention over another (Quimette, 1997; Petry and Martin, 2002; Prendergast *et al.*, 2006), with the possible exception of some evidence supporting CM over CBT (Rawson *et al.*, 2002a; Shoptaw *et al.*, 2005). Additionally, a meta-analysis of intervention-control studies demonstrated that drug treatment interventions were always more effective than no intervention or limited intervention, with very little difference in effect size by treatment modality (Prendergast *et al.*, 2002).

Given that substance treatment interventions in general tend to result in reduction or cessation of NISU for some individuals, and that some types of substance use have been associated with HIV risk behaviors, it is important to understand how substance use treatment could be used to prevent HIV transmission and acquisition among non-injection substance users. Associations between use of particular substances and HIV risk, and results of studies examining the effects of NISU treatment on HIV risk behaviors, are summarized below.

Alcohol

Alcohol is consumed in most countries in the world by a large proportion of adults (WHO, 2007), with the exception of some Muslim countries (WHO, 2004). Adult *per capita* consumption of alcohol is greatest in the European region, followed by the Americas, the African region, the Western Pacific, South East Asia, and the Eastern Mediterranean region (WHO, 2004). In the United States (US), the National Survey on Drug Use and Health (NSDUH) indicated that more than half of the US population aged 12 and older reported current alcohol consumption (i.e., drinking alcohol in the past month), with 23 percent engaging in binge drinking in the previous month (SAMHSA, 2006). Binge drinking in the past 30 days was common among US adolescents, with 8.9 percent of 14- to 15-year-olds, 20 percent of 16- to 17-year-olds, 36.2 percent of 18- to 20-year-olds and 46.1 percent of 21- to 25-year-olds reporting this drinking behavior. Following the peak in late adolescents, the proportion of people reporting binge drinking dropped.

Alcohol use and HIV risk

There is a vast scientific literature on alcohol use and sexual behavior. However, in the mid-1980s there was a shift in the literature from examining alcohol use and sexual behavior to examining the relation between drinking and specific HIV risk behaviors (Hendershot and George, 2007). Effects of alcohol proposed to contribute to increased HIV risk behavior are disinhibition and impaired judgment, alcohol myopia (i.e., narrowing of perception and cognition), and increased sexual arousal (Bryant, 2006). Additionally, learned behavioral scripts (i.e., beliefs, behaviors and expectations leading up to sexual activity) and settings that combine sexually charged environments with alcohol consumption have also been suggested to contribute to increased HIV risk behaviors (Bryant, 2006).

Controversy regarding the strength of the association between alcohol use and HIV risk behavior still exists. A recent review of the literature on alcohol use and sexually transmitted infections (STIs) suggests that over half of studies demonstrate a positive association between alcohol use and STI (Cook *et al.*, 2006). However, an earlier meta-analysis, examining event-level associations from studies of alcohol use and condom use, suggested that drinking may reduce condom use at sexual debut, but not in recent sexual encounters or with new partners (Leigh, 2002). Notably, the amount of alcohol consumed was not always included in these studies, although this may be an important factor for increasing risk behavior.

A number of studies examining sexual behavioral risk (i.e., condom use) and alcohol consumption have been conducted with a variety of study designs. Cross-sectional studies demonstrate associations between unprotected vaginal intercourse (UVI) and alcohol consumption (Barthlow *et al.*, 1995; Baskin-Sommers and Sommers, 2006), UVI and heavy or binge drinking (Graves and Leigh, 1995; Malow *et al.*, 2006; Weiser *et al.*, 2006; Coldiron, 2007), and unprotected anal intercourse (UAI) and heavy drinking or binge drinking (Ekstrand and Coates, 1990; Woody *et al.*, 1999; Bouhnik *et al.*, 2007; Theall *et al.*, 2007) among MSM. Studies examining alcohol use at the time of sexual activity also demonstrate associations between consumption and unprotected intercourse over the same time period among both heterosexuals (Graves and Hines, 1997; Myer *et al.*, 2002; Simbayi *et al.*, 2004; Brown and Vanable, 2007; Kalichman *et al.*, 2007) and MSM (Colfax *et al.*, 2004; Hirshfield *et al.*, 2004a; Rusch *et al.*, 2004; Celentano *et al.*, 2006; Irwin *et al.*, 2006). Although a number of studies have demonstrated associations between alcohol consumption and unprotected sexual activity, cross-sectional (Klitzman *et al.*, 2000; Shrier *et al.*, 2001; Guo *et al.*, 2002; Choi *et al.*, 2005; Raj *et al.*, 2007) and event-level (Kingree *et al.*, 2000; Clutterbuck *et al.*, 2001; Leigh, 2002; Kingree and Betz, 2003; Morrison *et al.*, 2003; Morin *et al.*, 2005; Bailey *et al.*, 2006; Springer *et al.*, 2007) studies have also demonstrated a lack of association.

Studies have also examined associations between alcohol consumption and incident or prevalent HIV infection. Prevalent HIV infection has been associated with alcohol consumption among heterosexuals in different African countries (Ayisi *et al.*, 2000; Hargreaves *et al.*, 2002; Talbot *et al.*, 2002; Zuma *et al.*, 2003; Lewis *et al.*, 2005; Mmbaga *et al.*, 2007). Far fewer studies have reported on associations between alcohol use and HIV incidence. Among heterosexuals in Uganda, HIV seroconversion has been associated with alcohol use just prior to sexual activity (Zablotska *et al.*, 2006). Among 4295 MSM in 6 US cities, heavy alcohol use and use of drugs or alcohol during sex were associated with HIV incidence (Koblin *et al.*, 2006). However, a number of longitudinal studies among MSM have also demonstrated a lack of association between heavy alcohol use (e.g., 60 or more drinks per month, 5 or more drinks on occasion weekly) or binge use (i.e., 5 or more drinks on one occasion) and HIV seroconversion (Ostrow *et al.*, 1995; Chesney *et al.*, 1998; Weber *et al.*, 2003; Plankey *et al.*, 2007).

A plethora of studies have also been published examining associations between STI incidence and prevalence and alcohol use. Studies have demonstrated global

345

associations between alcohol use and prevalent STI (Chetwynd *et al.*, 1992; Bairati *et al.*, 1994; Ericksen and Trocki, 1994; Zenilman *et al.*, 1994; Gwati, 1995; Rakwar *et al.*, 1997; Cu-Uvin *et al.*, 1999; Ebrahim *et al.*, 1999; Bjekic *et al.*, 2000; Feldblum *et al.*, 2000; Miranda, 2000; Cook *et al.*, 2002; Lewis *et al.*, 2005), heavy or binge drinking and prevalent STI (Shafer *et al.*, 1993; Miller *et al.*, 2001; Thomas *et al.*, 2001; Hallfors *et al.*, 2007), and current use of alcohol and incident STI (Yadav *et al.*, 2005) among heterosexual populations. However, many of these studies did not examine alcohol use over the same time period as STI risk. Fewer studies have demonstrated significant situational or event-level associations between alcohol use and STI (Morrison *et al.*, 1997; Molitor *et al.*, 1998; DiClemente *et al.*, 2002; Zachariah *et al.*, 2003). A number of studies have also demonstrated a lack of association between alcohol use and STI, including prevalent STI and alcohol consumption (Cotch *et al.*, 1991; Jamison *et al.*, 1995; Celentano *et al.*, 1996; Bjekic, 1997; Estebanez, 1997; Van den Eeden, 1998; Austin *et al.*, 1999; Boyer *et al.*, 1999; Cu-Uvin *et al.*, 1999; Ebrahim *et al.*, 1999; Vuylsteke *et al.*, 1999; Kalichman *et al.*, 2000; Lewis, 2000; Radcliffe *et al.*, 2001; Shrier *et al.*, 2001), binge drinking (Ericksen and Trocki, 1994; Ellen *et al.*, 1996; Chokephaibulkit *et al.*, 1997; Boyer *et al.*, 2006), and alcohol use just prior to sexual activity (Kraft and Rise, 1994; Sanchez, 1996; Molitor *et al.*, 1998; Noell *et al.*, 2001) among heterosexuals, and alcohol use and STI among MSM (Kim *et al.*, 2003; Hirshfield *et al.*, 2004b).

The vast number of studies of different design reporting differing results regarding the associations between alcohol use and HIV risk behaviors, seroconversion and STI make it difficult to interpret the potential effects of alcohol treatment on HIV prevention. Reasons for discrepancies may include differences in how alcohol use was measured, variations in data collection or analyses with regard to the timing of alcohol use and sexual behavior, and differences in how the amount of alcohol consumed was measured or analyzed. The ambiguity of these studies supports the need for randomized controlled trials of alcohol treatment with HIV risk endpoints.

HIV prevention and alcohol treatment

Few studies have examined alcohol cessation as an intervention for HIV/STI prevention. Although many studies have demonstrated high-risk sexual behaviors and high HIV prevalence among people in alcohol treatment (Windle, 1989; Avins *et al.*, 1994; Scheidt and Windle, 1995; Woods *et al.*, 2000) and among heavy alcohol users (Fisher *et al.*, 2008), suggesting the need for alcohol treatment, few published studies have reported on changes in HIV risk behaviors after alcohol treatment. A recent randomized control trial of an intervention comparing four sessions of MI to 12 sessions of MI plus CBT to test efficacy for reductions in alcohol use and sexual risk behavior among HIV-uninfected MSM revealed that those in the MI group had a greater reduction in drinking than those in the MI and CBT groups during intervention; however, both groups had equivalent declines in drinking at 12-month follow-up (Morgenstern *et al.*, 2007). Additionally, both MI and MI + CBT groups demonstrated a greater reduction in alcohol consumption than MSM who declined treatment. Results of the efficacy of this intervention on reducing HIV risk behaviors were not published in this report but will be available in a future manuscript, according to the authors. Additionally, in observational studies of people in alcohol treatment, reductions in risk behaviors have occurred from baseline to follow-up. Among 700 self-identified alcoholics recruited from 5 public alcohol treatment centers, all of which included HIV risk-reduction counseling, Avins *et al.* (1997) found an overall 26 percent reduction in having sex with an injection-drug-using partner, a 58 percent reduction in the use of injection drugs, and a 77 percent improvement in consistent condom use with multiple sexual partners after a mean of 13 months' follow-up compared to baseline. Similarly, in an observational study of alcoholics in treatment, among those reporting sobriety after treatment there was a significant reduction in HIV risk behaviors (Scheidt, 1999). These studies suggest that alcohol treatment in general may decrease sexual risk behaviors.

Reduction in sexual risk behaviors, but not alcohol use, following intervention has also been observed. In a group randomized intervention trial of adolescent schoolchildren in the Kwa-Zulu Natal region of South Africa, a combination HIV and alcohol use intervention revealed a significant reduction in alcohol use concurrent with sexual intercourse, and an increase in female students' refusal to engage in intercourse even in the absence of a reduction in alcohol use (Karnell *et al.*, 2006).

It is important to note that individuals in many of these studies were classified as alcoholics and most were seeking treatment, whereas studies of sexual risk may have included drinking in general, binge drinking or alcoholism, so might not be directly comparable. It is also important to consider that interventions may be more effective if tailored to specific needs of different groups. For example, among 921 patients entering alcohol treatment, prevalent HIV infection was associated with greater reports of alcohol impairment among heterosexual men and women, but was associated with greater bar socializing among MSM (Boscarino *et al.*, 1995) – suggesting the need to examine contextual factors associated with alcohol consumption and sexual risk behavior, and incorporate the findings into treatment programs.

Studies demonstrating reductions in sexual risk behaviors when treating alcohol dependence using pharmacotherapy are lacking. Currently, three pharmacotherapies that are approved for treatment of alcohol disorders appear to be effective in assisting in alcohol treatment and relapse prevention (Garbutt *et al.*, 1999; Kiefer and Mann, 2005; Tambour and Quertemont, 2007). Disulfiram is an alcohol deterrent medication, which acts through inhibiting alcohol dehydrogenase. Naltrexone has been used successfully to reduce alcohol craving, amount of alcohol used and relapse. Acamprosate has been effective in maintaining alcohol abstinence (Mann *et al.*, 2004). In addition to these medications, many other pharmacotherapies are currently being tested. Studies aimed at reducing HIV risk through alcohol cessation that consider evaluating the use of pharmacotherapies either alone or in conjunction with behavioral therapies are currently in progress.

Non-alcohol substance use

About 5 percent of the world's population uses illicit substances other than alcohol (UN, 2007). In the US in 2006, 8.3 percent of the population reported current illicit substance use (i.e., use in the past 30 days) and 9.2 percent reported substance dependence within the last year (SAMHSA, 2006). Non-alcohol substance use differs from alcohol use in that most non-alcohol substances are illegal worldwide and are used illicitly by more well-defined groups of individuals. Patterns of use and populations using differ across substances.

Studies examining associations between substance use and HIV/STI risk can be divided among those that examine specific substances, substance use in general, or broad categories of substances (such as club drugs or stimulants). While studies examining illicit substance use in general are not ideal, as different substances may have a lesser or greater association with HIV risk than others, it is important to understand how treatment in general may affect one's HIV risk. As with alcohol use, establishing both a direct association between substance use and HIV risk and demonstrating risk behavior changes after treatment are critical to understanding the effect that NISU treatment may have on preventing HIV.

Most studies examining unspecified or mixed (e.g., combination) drug use and HIV risk have examined global associations. In cross-sectional studies, mixed or general substance use has been associated with increased risk of UAI among MSM (Woody *et al.*, 1999; Fernandez *et al.*, 2005b; Operario *et al.*, 2006), HIV serodiscordant UAI (Chesney *et al.*, 2003; Morin *et al.*, 2005) and youth, even exchanging sex for money or drugs (Edwards *et al.*, 2006b). Unprotected intercourse has also been associated with general substance use at the time of sexual activity among both MSM (Purcell *et al.*, 2001; Darrow *et al.*, 2005) and adolescents (Hallfors *et al.*, 2007) and in a case-crossover analysis comparing episodes of sexual activity with and without condom use among MSM (Colfax *et al.*, 2005). Associations between substance use and HIV acquisition have also been reported. In a nested case–control analysis, use of nitrites or cocaine was associated with an increased risk of HIV seroconversion in adjusted analyses (DiFranceisco *et al.*, 1996). Similarly, among 4295 HIV-uninfected MSM from 6 cities in the US, after 48 months of follow-up HIV seroconversion was associated with use of drugs or alcohol before sex (Koblin *et al.*, 2006). Fewer studies have reported a lack of association between general substance use and HIV risk, including those examining substance use and UAI among MSM (Klitzman *et al.*, 2000), HIV prevalence among men in Peru (Lama *et al.*, 2006), UVI among adolescents (Guo *et al.*, 2002) and condom use at last sexual encounter (Palen *et al.*, 2006). Additionally, among 134 15- to 21-year-olds with substance abuse disorders in Pittsburgh, taking drugs just prior to sexual activity was not associated with condom use in an event-level analysis (Bailey *et al.*, 2006).

Studies also demonstrate reductions in sexual risk behaviors associated with substance use interventions that are not necessarily focused on a specific substance. A three-arm intervention trial with MET-like qualities reported significantly less substance use in general (i.e., type of substance was unspecified), increased condom use, and increased abstinence from sexual activity (St Lawrence *et al.*, 2002). Similarly, a non-randomized intervention based on using problem-solving therapy among adolescent male substance users in jail revealed that, compared to no intervention, those who participated in facilitator-guided discussion of substance use and sexual risk were more likely to report increased condom use; however, drug use behaviors did not change (Magura *et al.*, 1994). In a non-controlled multisite study of adolescents, substance use treatment in general also resulted in half of patients reporting a reduction in sexual risk behaviors (Joshi *et al.*, 2001). Among women in South Africa, a randomized trial of an adapted evidence-based intervention revealed reductions in substance use and sexual risk behaviors 1 month post-intervention regardless of treatment arm (Wechsberg *et al.*, 2008), suggesting that interventions that reduce substance use may also have an effect on sexual risk behaviors.

Lasting effects of sexual risk reduction have varied between studies. In a pilot intervention to reduce sexual risk behaviors and substance among HIV infected youth (16- to 25-years-old) using MET, and a medical care control group with no intervention, results immediately following the intervention demonstrated that there were no significant differences between groups by substance use; however, the MET group showed a significant reduction in unprotected sex and viral load, due to medication adherence, over the control group (Naar-King *et al.*, 2006). Reports at 6 and 9 months post-intervention revealed sustained reductions in viral load and reductions in alcohol consumption in the MET group over the controls; however, the reduction in sexual risk behavior was not sustained (Naar-King, 2007).

Amphetamine and methamphetamine

Amphetamine and methamphetamine (referred to collectively as methamphetamine for the remainder of this chapter) are derived from ephedrine and pseudo-ephedrine. The greatest producer of methamphetamine worldwide has historically been the US (UN, 2007); however, restrictions on the precursor substances in the US and Canada have more recently shifted production to Mexico

(UN, 2007), which is estimated to control 70–90 percent of US methamphetamine production and distribution (Brouwer *et al.*, 2006).

Methamphetamine use is reported by 5.8 percent of the US population 12 years of age and older, and 0.3 percent have reported use in the last 30 days (SAMHSA, 2006). More men in the US report recent methamphetamine use than women (0.4 percent versus 0.2 percent respectively) (SAMHSA, 2006). While use has increased throughout the US, methamphetamine is still more common in the West compared with other US regions (SAMHSA, 2006), and more common among specific populations, including MSM (Koblin *et al.*, 2003; Thiede *et al.*, 2003).

Methamphetamine use and HIV risk

Methamphetamine has been reported to increase sexual confidence (Semple *et al.*, 2002; Green, 2003; Halkitis *et al.*, 2005a) and libido (Buffum, 1982; Gorman *et al.*, 1997; Semple *et al.*, 2002); enhance sexual function (Diaz *et al.*, 2005; Halkitis *et al.*, 2005a); reduce the experience of pain (Green, 2003; Halkitis *et al.*, 2005a); prolong sexual encounters and increase the number of encounters over the intoxication period (Green, 2003); dry the genital and rectal mucosa and decrease physical sensitivity (Shoptaw and Reback, 2007); and alter the user's mental state, including judgment and decision-making (McKim, 2003). Hypersexuality has also been ascribed as a prominent effect of methamphetamine use (Green, 2003; Green and Halkitis, 2006), and stimulant users frequently report that sex and substance use always or often go together (Gorman *et al.*, 1997; Shoptaw *et al.*, 1998; Semple *et al.*, 2002; Parsons *et al.*, 2004).

Of all NISU, use of methamphetamine has been most consistently associated with risky sexual behavior and HIV seroconversion. Due to the high prevalence of both methamphetamine use (Stall *et al.*, 2001; Koblin *et al.*, 2003; Thiede *et al.*, 2003; SAMHSA, 2006) and HIV among MSM (CDC, 2007), most studies of methamphetamine use and HIV risk have been conducted within this population. Most cross-sectional studies examining sexual risk behavior and methamphetamine use among MSM have demonstrated at least a two-fold increase in risk behavior associated with use, including elevated risk of UAI in general (Mattison *et al.*, 2001; Buchacz *et al.*, 2005; Celentano *et al.*, 2006; Chiasson *et al.*, 2007; Fernandez *et al.*, 2007; Spindler *et al.*, 2007; Rhodes S. D. *et al.*, 2007) and elevated risk of UAI when used at the time of sexual intercourse (Rusch *et al.*, 2004). Studies of methamphetamine use and serodiscordant sexual activity have demonstrated associations with both general use of methamphetamine within the time period of UAI (Colfax *et al.*, 2001; Bolding *et al.*, 2006; Brewer *et al.*, 2006) and use just prior to sexual activity (Mansergh *et al.*, 2006; Vaudrey *et al.*, 2007). Conditional analyses, in which sexual activity with and without drug use is examined, thereby controlling for unmeasured individual level factors, have also found increased risk associated with methamphetamine use (Colfax *et al.*, 2004; Drumright *et al.*, 2006b; Koblin *et al.*, 2007).

Methamphetamine has also been associated with HIV seroconversion, including in studies that controlled for sexual risk behaviors and other potential confounders. Causes for this association remain unknown, but may include potential unmeasured behavioral cofactors (more prolonged or traumatic sex), misreporting of risk behavior, or possible direct effects of methamphetamine that increase risk of infection (such as effects on the immune system). At least seven longitudinal studies have demonstrated associations between methamphetamine use and HIV seroconversion (Burcham *et al.*, 1989; Page-Shafer, 1997; Chesney *et al.*, 1998; Weber *et al.*, 2003; Buchbinder *et al.*, 2005; Koblin *et al.*, 2006; Plankey *et al.*, 2007). Methamphetamine use among MSM has also been associated with early syphilis infection (Wong *et al.*, 2005), lifetime STI history (Rhodes S. D. *et al.*, 2007), self-reported incident STI (Hirshfield *et al.*, 2004b) and biologically confirmed STI (Kim *et al.*, 2003).

In addition to numerous studies among MSM that consistently demonstrate associations between HIV risk or acquisition and methamphetamine use, studies also demonstrate associations between methamphetamine use and increased viral load (Ellis *et al.*, 2003), decreased adherence to antiretroviral therapy (Reback *et al.*, 2003), and higher likelihood of acquiring a drug-resistant strain of HIV (Colfax *et al.*, 2007; Gorbach *et al.*, 2008). Additionally, studies have demonstrated that methamphetamine use in association with sexual activity may be more common among HIV-infected MSM than among those who are uninfected (Semple *et al.*, 2002; Halkitis *et al.*, 2005b). These studies suggest that methamphetamine may be having an effect on the HIV epidemic over and above its role in contributing directly to HIV risk behaviors.

Although few studies of methamphetamine use and sexual risk have been conducted among heterosexual populations, the majority demonstrate significant associations. Among adolescents, methamphetamine use has been associated with a decreased likelihood in reporting condom use in national studies, including the Youth Risk Behavior Survey (Springer *et al.*, 2007) and Add Health (Iritani *et al.*, 2007). Among 258,567 adult non-IDU HIV testers in California (Molitor *et al.*, 1998), men who had sex with women and reported using methamphetamine during sexual activity were significantly less likely to report condom use during vaginal or anal intercourse. Similarly, in the same study, women who used methamphetamine during sexual activity were significantly less likely to report condom use during vaginal or anal intercourse with a man. Additionally, in a national US study of heterosexual men, methamphetamine was also associated with a greater likelihood of trading sex for money and having had anal sex with female partners (CDC, 2006).

Some studies have reported no association between methamphetamine use and HIV risk behaviors (Purcell *et al.*, 2001; Harawa *et al.*, 2004; Choi *et al.*, 2005). However, not only do the majority of published studies that include methamphetamine as a risk factor for HIV or HIV-related behaviors demonstrate significant associations, but direct association between methamphetamine use and increased sexual risk behavior among MSM is also supported by qualitative studies in which participants report a direct effect between use of methamphetamine, sexual desire, and acting upon that desire (Green, 2003; Reback *et al.*, 2004; Green and Halkitis, 2006). Thus, the preponderance of evidence suggests the existence of a very strong direct association between methamphetamine use and HIV risk.

HIV prevention and methamphetamine treatment

Most studies demonstrate that methamphetamine treatment is associated with decreased HIV risk behaviors. In a non-randomized study of 58 MSM entering a 12-step program for methamphetamine use, Crystal Meth Anonymous, participants reported a reduction in UAI from 70 percent at baseline to 24 percent after 3 months' participation (Lyons *et al.*, 2006). Similarly, a study of 147 drug-using MSM in San Francisco who were non-randomly assigned to a group counseling and a 12-step program, or a group counseling and 12-step program with a safer-sex intervention, demonstrated a reduction in UAI from baseline through 1 year of follow-up regardless of treatment group (Stall *et al.*, 1999).

Among MSM randomized to four different interventions, focusing on contingency management (CM), cognitive behavioral therapy (CBT), culturally sensitive CBT for MSM, or a combination of CM and CBT, sexual risk behavior was significantly decreased a year following initiation of the intervention among participants reporting a reduction in methamphetamine use (Shoptaw *et al.*, 2005). Jaffe *et al.* (2007) assessed temporal changes in methamphetamine use, depression symptoms and unprotected anal intercourse in this same sample of MSM, demonstrating that participants with the greatest downward trajectory in methamphetamine use, as verified through urine testing, reported the greatest and most rapid decreases in depressive symptoms and sexual risk behaviors. Tailoring these interventions to be more socially appropriate for MSM demonstrated the greatest reduction in sexual risk behaviors (Shoptaw *et al.*, 2005). The CBT control group reported the most methamphetamine use over the 16 weeks; the tailored gay-specific group reported a more rapid decrease in methamphetamine use than the other participants. Additionally, in qualitative interviews with a subset of the same men, individuals reported changes in attitudes toward the role of methamphetamine use in their sex lives (Reback *et al.*, 2004). Prior to receiving drug-use interventions, methamphetamine was described as a sex drug and the participants described an inability to control their sexual needs. However, after completing the intervention, methamphetamine use was reduced and participants viewed themselves as having control over their sexual choices. Similarly, among 784 methamphetamine users participating in a trial comparing two interventions – a matrix model versus treatment as usual – completion of the treatment program, regardless of the intervention, was significantly associated with reductions in HIV-related sexual risk behaviors (Rawson *et al.*, 2008). Additionally after 3 years follow-up, participants reported decreased sexual risk, which was proportionately associated with decreased methamphetamine use. These studies suggest that high-risk behaviors often decreased with decreasing use of methamphetamine.

In contrast to these studies, a randomized clinical trial of CBT for meth-amphetamine use among 214 regular users in the Greater Brisbane Region of Queensland and Newcastle, New South Wales, had no effect on HIV risk-taking, despite improvements in stage of change and substance use, benzodiazepine use, tobacco smoking, polydrug use, injecting risk-taking behavior, criminal activity level, and psychiatric distress and depression level (Baker *et al.*, 2005). However, baseline sexual risk behavior was not reported, making it unclear whether the study was sufficiently powered to detect a change in this variable.

In addition to studies examining methamphetamine-use reduction or cessation for HIV prevention, interventions to prevent HIV risk behaviors among active methamphetamine users, which do not focus on drug cessation, have also been conducted. A randomized attention control trial containing three arms, interven-tion based on motivational interviewing, intervention plus follow-up booster and attention control, enrolled 451 HIV-uninfected heterosexual methamphetamine users to determine the effects of an intervention focusing on sexual risk reduc-tion, but not on cessation of substance use (Mausbach *et al.*, 2007a). Participants in the intervention groups reported significant decreases in the number of unpro-tected sexual acts and significant increases in the number of protected sexual acts as compared to the attention control group over the study period. Similarly, in a randomized control trial among HIV-infected methamphetamine-using MSM, those receiving the behavioral intervention reported more protected sexual acts than did those in the attention control group 6 and 12 months post-intervention (Mausbach *et al.*, 2007b). In both studies, participants in the intervention groups reported greater self-efficacy of condom use than did attention controls. While less than optimal retention rates limit drawing firm conclusions from these data, these important results suggest that, even in the absence of cessation, metham-phetamine users can reduce sexual risk behaviors following intervention.

While there are no approved pharmacotherapies for methamphetamine depend-ence (Srisurapanont *et al.*, 1999), testing of potential agents, including a number of medications used to treat a variety of psychiatric conditions, is an area of active research (Galloway *et al.*, 1996; Shoptaw *et al.*, 2006). Pharmacotherapies may have potential for advancing methamphetamine treatment, including minimizing severe withdrawal symptoms (McGregor *et al.*, 2004) and preventing the high relapse rates observed in behavioral interventions (Brecht *et al.*, 2000). There are several current pharmacotherapy trials underway examining sexual risk behavior change in addi-tion to reductions in methamphetamine use (Colfax, unpublished data).

Cocaine

Approximately 14 million people worldwide use cocaine annually. Although a strong decline in coca cultivation was observed between 2000 and 2006, produc-tion of cocaine has remained stable (UN, 2007). In the US, approximately 1 percent of the population reported using cocaine in 2006 30 percent of cocaine users reported crack cocaine use (SAMHSA, 2006). Men (1.4 percent) were more likely to report cocaine use than women (0.6 percent).

Cocaine use and HIV risk

As with methamphetamine, cocaine has been reported to give the user increased feelings of self-confidence, increase sexual desires and libido, and alter the user's mental state (Rawson *et al.*, 2002b). Additionally, higher prevalence of cocaine use has been reported among MSM than among heterosexual men (Cochran *et al.*, 2004). In the NSDUH (SAMHSA, 2006), the general population reported 1 percent prevalence of cocaine use in the previous year compared to 37 percent 1-year prevalence among MSM in the venue-sampled National HIV Behavioral Surveillance (NHBS) (Sanchez, 2006). Similar high prevalence has been reported in other studies of MSM as well (Stall *et al.*, 2001; Koblin *et al.*, 2003; Thiede *et al.*, 2003).

Associations between cocaine use and high-risk behaviors for HIV infection or HIV incidence and prevalence have been reported in numerous studies of MSM and heterosexuals. In studies examining risk behaviors among MSM who were both HIV-infected and -uninfected, use of cocaine in general (Molitor *et al.*, 1999; Celentano *et al.*, 2006), and just prior to or during sexual activity (Celentano *et al.*, 2006), has been associated with increased risk of UAI. Additionally, a study examining risk behaviors among HIV-infected MSM from across the US found that UAI with an HIV-uninfected or serostatus unknown partner was associated with crack cocaine use (Denning and Campsmith, 2005). Among heterosexuals, cocaine use has also been associated with decreased condom use (Braithwaite and Stephens, 2005; Szwarcwald *et al.*, 2005; Springer *et al.*, 2007). Additionally, buying and selling sex for money or drugs has been associated with cocaine use among adult heterosexuals (Edwards *et al.*, 2006a; Weiser *et al.*, 2006; Raj *et al.*, 2007), adolescents (Fullilove *et al.*, 1993) and MSM (Newman *et al.*, 2004).

Event-level analyses have also demonstrated associations between cocaine use and sexual risk behaviors. Among 327 homeless or runaway youth in Washington, DC, crack cocaine use at their last sexual encounter was significantly associated with unprotected intercourse (Bailey *et al.*, 1998). In a case-crossover study of MSM, conditional analyses using the participant as his own control demonstrated that use of cocaine and other substances just prior to sexual activity was significantly associated with serodiscordant UAI, and such risk was decreased in the absence of cocaine use (Colfax *et al.*, 2004).

Prevalent HIV infection among MSM has also been associated with cocaine use (Bautista *et al.*, 2004), crack cocaine use (Edlin *et al.*, 1994; Harawa *et al.*, 2004; McCoy *et al.*, 2004; Pechansky *et al.*, 2006), and cocaine use before or during sex (Lama *et al.*, 2006). Studies also report significant associations between cocaine use and STI prevalence among heterosexual populations (DeHovitz *et al.*, 1994; Molitor *et al.*, 1998; Buchacz *et al.*, 2000). As in the case with methamphetamine, while the causal factors remain unknown, longitudinal studies among MSM have demonstrated significant associations between HIV seroconversion and cocaine use (Chesney *et al.*, 1998; Weber *et al.*, 2003; Plankey *et al.*, 2007). Additionally, in a nested case-control study of 76 MSM who seroconverted and 380 matched controls from a longitudinal HIV seroconversion study in Chicago, use of cocaine during the follow-up period was associated with increased risk of HIV seroconversion (Ostrow *et al.*, 1995). Qualitative reports also indicate that cocaine use may lead to increased HIV risk behaviors (Lichtenstein, 1997; Djumalieva *et al.*, 2002; Essien *et al.*, 2005).

Some studies also demonstrate a lack of association between cocaine use and HIV or HIV risk behaviors. In cross-sectional studies, cocaine use just prior to sexual activity was not associated with UAI among MSM (Mattison *et al.*, 2001), general cocaine use was not associated with condom use among drug recovery patients (Raj *et al.*, 2007), and cocaine use was not associated with UAI during the same time period MSM (Purcell *et al.*, 2001; Rusch *et al.*, 2004). Additionally, cocaine use during sexual activity was not associated with condom use in a large cohort of men and women receiving HIV testing in California (Molitor *et al.*, 1998). In case-control studies, lack of association between cocaine use and self-reported STI incidence (Hirshfield *et al.*, 2004b) or clinically diagnosed rectal gonorrhea (Kim *et al.*, 2003) among MSM has been reported. Thus, while the evidence is not quite as compelling as it is for methamphetamine, many well-designed studies demonstrate an association between cocaine use and sexual risk.

HIV prevention and cocaine treatment

Studies among cocaine-using populations suggest that cessation of cocaine use may be associated with a reduction in risk behaviors. Among 487 cocaine-dependent individuals assigned to four different treatment groups – group counseling, individual and group counseling, CBT, or support expressive therapy – reduction in cocaine use from an average of 10 days per month to 1 day per month was associated with a 40 percent decrease in sexual risk behaviors (Woody *et al.*, 2003). Reductions in cocaine use and sexual risk behaviors were similar regardless of treatment group. Similarly, among 620 African-American women who used crack cocaine, all treatment groups reported reduced substance and reduced sexual risk behaviors at follow-up (Wechsberg *et al.*, 2004). In a non-randomized study of sexual risk behavior change following cocaine treatment, completion of a matrix model-based intervention for cocaine cessation that did not contain HIV counseling resulted in significant reduction of sexual risk (Shoptaw *et al.*, 1997). Among individuals entering or interested in substance use treatment who reported cocaine use, a study comparing the effects of substance use treatment on those with antisocial personality disorder to those without revealed a significant reduction in cocaine use and sexual risk behaviors in both groups from baseline (Compton *et al.*, 1998). These studies suggest that cocaine use treatment in general may have positive effects on reducing sexual risk behaviors of those seeking treatment. Studies of cocaine-using individuals undergoing methadone treatment for heroin use (Cottler *et al.*, 1998; Magura *et al.*, 1998; Robles *et al.*, 1998; Broome *et al.*, 1999; Latka *et al.*, 2005) and of injection cocaine users (Schroeder *et al.*, 2006) also report sexual risk behavior reduction after treatment; however, these studies are difficult to interpret in the context of NISU, as injection drug use is also reported.

As is true for methamphetamine, no effective pharmacotherapy agents have been identified for cocaine treatment, but this is an active area of research (de Lima *et al.*, 2002; Vocci and Ling, 2005; Sofuoglu and Kosten, 2006). As pharmacotherapies for cocaine become available, it will be important to study the effects of cocaine cessation on sexual risk behavior in studies employing these pharmacotherapies.

Polydrug use

Polydrug use (i.e., use of more than one drug in a given time period) is important when considering substance use treatment for HIV prevention. Perceived efficacy of intervention trials may be confounded if cessation of one substance is replaced by use of another. Additionally, polydrug users may require specifically tailored interventions for greater efficacy.

Polydrug use is frequently reported among various populations of substance users, including substance-using MSM (Mansergh *et al.*, 2001; Lee *et al.*, 2003; Banta-Green *et al.*, 2005; Clatts *et al.*, 2005; Fernandez *et al.*, 2005a; Bolding *et al.*, 2006; Drumright *et al.*, 2006b, 2007; Halkitis *et al.*, 2007; Lampinen *et al.*, 2007) and rave and circuit party attendees (Boys *et al.*, 1997; Lenton *et al.*, 1997; Colfax *et al.*, 2001; Gross *et al.*, 2002). Additionally, cocaine users (Prinzleve *et al.*, 2004) and methamphetamine users (Degenhardt *et al.*, 2004, 2007; Bolding *et al.*, 2006; Martin *et al.*, 2006; Halkitis *et al.*, 2007; Herman-Stahl *et al.*, 2007) often report polydrug use. Simultaneous polydrug use (i.e., use of more than one substance on the same occasion) has also been reported among a number of drug-using groups, including alcoholics (Martin *et al.*, 1996; Staines *et al.*, 2001), rave attendees (Boys *et al.*, 1997; Tossmann *et al.*, 2001; Barrett *et al.*, 2005) and adolescents (Martin, *et al.*, 1993; Martin *et al.*, 1996; Collins, *et al.*, 1998).

Simultaneous polydrug use or concurrent drug administration and order of drug administration may be important in regulating the amount of a particular drug used. For example, among college students alcohol was used in greater quantities when used in combination with cocaine than when used alone (Barrett *et al.*, 2006). Additionally, specific combinations of drugs, such as methamphetamine and erectile dysfunction medications (e.g., Viagra, Levitra, Cialis), may increase the probability of risky sexual activity. Similarly, polydrug users of any type may be more likely to practice higher-risk behaviors for HIV insection than single-drug users. Among adolescents, polydrug use has been associated with inconsistent condom use and trading sex for money or drugs (Castrucci and Martin, 2002). Risk behaviors identified among polydrug-using MSM include a greater number of visits to a bathhouse or sex-club (Stall *et al.*, 2001), increased risk of UAI (Operario *et al.*, 2006), and increased likelihood of UAI with partners after HIV diagnosis (Patterson, *et al.*, 2005). In addition to associations between high-risk behavior and polydrug use, studies suggest that polydrug-using MSM may practice riskier sexual behavior than single-drug using MSM (Patterson *et al.*, 2005; Drumright *et al.*, 2007). Among Hispanic MSM in Miami, polydrug users were more likely than those reporting use of a single drug to report sex under the influence of drugs (Fernandez *et al.*, 2005a). Additionally, among 261 HIV-infected MSM who reported methamphetamine use, those who were heavy polydrug users (used drugs other than marijuana and nitrites in combination with methamphetamine) reported more HIV-uninfected and serostatus-unknown partners than those who were light or single-drug users (Patterson *et al.*, 2005).

Use of other substances and HIV risk

Many substances that have been studied to determine associations between use and HIV risk behaviors have not been well studied in treatment settings with regard to HIV prevention, and therefore data on changes in sexual risk behaviors after treatment are not available. However, it is important to assess studies of use and risk behaviors to determine whether cessation or prevention of use of these substances could be effective for HIV prevention.

Popper or nitrite use and HIV risk

While most national surveys do not measure "popper" or volatile nitrite inhalant use, research studies report high use among some MSM populations (Mansergh *et al.*, 2001; Koblin *et al.*, 2003; Thiede *et al.*, 2003). Additionally, multiple studies among MSM over the past 25 years have demonstrated significant independent associations between nitrite use and HIV risk. Poppers have been reported to increase sexual desire (Newell *et al.*, 1985; Haverkos and Dougherty, 1988) and, as vasodilators, relax anal sphincter muscles (French and Power, 2001), potentially facilitating sexual activity – particularly, receptive anal sex. In a variety of studies, nitrite use has been associated with UAI among MSM (Strathdee *et al.*, 1998; Ekstrand *et al.*, 1999; Woody *et al.*, 1999; Colfax *et al.*, 2001, 2004; Mattison *et al.*, 2001; Purcell *et al.*, 2001, 2005; Parsons *et al.*, 2003; Choi *et al.*, 2005; Brewer *et al.*, 2006).

Some studies have also quantified risk by increasing use of nitrites. In a case-control analysis of 495 MSM in San Francisco, where cases were AIDS-diagnosed patients and controls were HIV-uninfected MSM from the same neighborhood, heavy nitrite inhalant use (>65 "hits" per month) was associated with AIDS diagnosis, but mild nitrite use (≤65 "hits" per month) was not (Moss *et al.*, 1987). In a longitudinal study of 337 MSM in San Francisco, HIV seroconversion was associated with consistent nitrite inhalant use over the course of the study period, but not with recent adoption of nitrite use (Chesney *et al.*, 1998). Additionally, studies have also demonstrated significant associations between nitrite use and HIV seroconversion (Ostrow *et al.*, 1995; Weber *et al.*, 2003; Koblin *et al.*, 2006; Plankey *et al.*, 2007; Macdonald *et al.*, 2008).

Some studies report no significant association with nitrite use (Kim *et al.*, 2003; Harawa *et al.*, 2004; Lampinen *et al.*, 2007). Taken as a whole, however, the literature strongly indicates that use of nitrites is an independent cofactor for HIV risk. Nevertheless, no known studies have examined the association between cessation of nitrite use and reduction in HIV risk behaviors. Twenty-five years into the HIV epidemic, this remains cause for concern.

Marijuana use and HIV risk

After alcohol, marijuana is the most widely used drug, with approximately 160 million cannabis users worldwide (UN, 2007). According to NSDUH reports of marijuana use in the US in 2006, 6.0 percent of the population aged 12 or older had used it in the last 30 days (SAMHSA, 2006).

Marijuana may both have mind-altering properties (which may reduce judgment) and enhance sexual pleasure (Green *et al.*, 2003). Studies of marijuana use and HIV risk vary by population studied. A large number of studies on marijuana use and HIV risk have been conducted among adolescents, revealing independent associations between marijuana use and prevalent STI (Boyer *et al.*, 1999, 2000; Kingree and Phan, 2001; Liau *et al.*, 2002; Mertz *et al.*, 2000; De Genna *et al.*, 2007), unprotected intercourse (Barthlow *et al.*, 1995; Kingree and Phan, 2001; Guo *et al.*, 2002; Liau *et al.*, 2002; Springer *et al.*, 2007), and unprotected intercourse when used at the time of the sexual event (Kingree *et al.*, 2000; Kingree and Betz, 2003). However, a number of studies among adolescents have also demonstrated a lack of association between marijuana use and HIV risk, including no association with condom use over the same time period (Bailey *et al.*, 1998; Crosby *et al.*, 2003) or condom use at a particular sexual event (Bailey *et al.*, 1998).

While marijuana use is reported by a high proportion of MSM (Stall *et al.*, 2001; Koblin *et al.*, 2003; Thiede *et al.*, 2003), few studies report an association between use of marijuana and HIV risk, especially after controlling for other substance use and other potential confounders. In cross-sectional studies among

MSM, use of marijuana just prior to sexual activity has been associated with receptive UAI (Celentano *et al.*, 2006), and UAI with serodiscordant or status unknown partners (Clutterbuck *et al.*, 2001). However, lack of association between marijuana use and HIV risk among MSM has been reported in studies examining marijuana use and UAI (Woody *et al.*, 1999; Mattison *et al.*, 2001; Colfax *et al.*, 2001; Purcell *et al.*, 2001; Choi *et al.*, 2005) and HIV seroconversion (Ostrow *et al.*, 1995; Koblin *et al.*, 2006).

Most studies of marijuana use and HIV risk behaviors among MSM demonstrate a lack of association, but mixed results from studies of adolescents suggest this substance may play a different role in HIV risk behaviors in these overlapping populations. Studies of adolescents and MSM also differ by controlling for use of multiple substances; studies demonstrating significant associations between marijuana use and HIV risk among adolescents often did not report on substance use other than marijuana and alcohol use. Future studies of marijuana use and sexual risk among adolescents should control for all types of substance use, to assist in developing a better understanding of the true associations between marijuana use and HIV risk.

Other substance use and HIV risk

In 2006, about 0.1 percent of the US population aged 12 years and older reported LSD use and 0.2 percent reported methylenedioxymethamphetamine (MDMA) or ecstasy use in the last 30 days (SAMHSA, 2006). National data were not available on gamma-hydroxybutrate (GHB) or ketamine.

Among adolescents MDMA use has been associated with decreased condom use (Springer *et al.*, 2007), and among MSM MDMA use has been associated with UAI (Klitzman *et al.*, 2000; Rusch *et al.*, 2004; Choi *et al.*, 2005). MDMA use has also been associated with HIV seroconversion among MSM (Burcham *et al.*, 1989; Plankey *et al.*, 2007). In contrast, lack of association has been demonstrated between MDMA use and self-reported incident STI (Kim *et al.*, 2003; Hirshfield *et al.*, 2004b) and HIV seroconversion among MSM (Weber *et al.*, 2003).

Fewer data are available on HIV risks related to use of ketamine, GHB and LSD, and most studies have demonstrated a lack of association with HIV risk, including studies examining UAI and ketamine (Mattison *et al.*, 2001) or GHB use (Ostrow *et al.*, 1995; Colfax *et al.*, 2001; Mattison *et al.*, 2001; Choi *et al.*, 2005) and HIV seroconversion and hallucinogen use among MSM (Chesney *et al.*, 1998). However, among young MSM in Vancouver, use of ketamine and GHB at the time of sexual activity was associated with UAI over the same time period (Rusch *et al.*, 2004).

Given the limited number of studies presenting specific results on MDMA, GHB, ketamine and LSD use and HIV risk, it is difficult to determine the effect treatment may have on HIV transmission. It appears that most studies that control for multiple other-substance use, especially alcohol, stimulants and poppers, find no significant associations between use of these other substances and HIV risk. In addition, we are aware of no interventions that have specifically examined whether treating these substances reduces HIV risk.

Conclusions

Given the high prevalence of substance use reported within populations at risk for HIV infection and the number of studies demonstrating significant associations between high-risk sexual behavior and substance use – particularly use of meth-amphetamine, cocaine and poppers, and perhaps alcohol – substance-use treatment appears to be a logical intervention for HIV prevention. Studies examining the association between substance-use treatment and sexual risk-taking behavior indicate that substance-use treatment is linked with reductions in HIV risk.

There is a need, however, to focus more research on prevention and treatment of substance use in populations at risk for HIV. Work on determining effective primary substance-use prevention efforts remains in its infancy. The limited research on the role NISU plays in the heterosexual HIV epidemic, particularly among African-Americans, is striking, and cause for concern. However, even among populations as extensively studied as MSM substance users, very few people are accessing substance-use treatment services. In the (NHBS) System, for instance, only 16 percent of MSM substance users reported ever access-ing treatment services (Sanchez *et al.*, 2006), demonstrating the great need to increase availability and access to treatment in this population. In addition, while most interventions have focused on heavy substance users, there is increasing recognition of the need to develop interventions for persons who use substances only episodically, and for whom episodic use is associated with increased HIV risk. Reports from current substance users suggest that more interventions are needed and would be welcomed (Menza *et al.*, 2006; Brems and Dewane, 2007).

Twenty-five years into the HIV epidemic, we continue to observe HIV trans-mission among MSM and increases in transmission among heterosexual popu-lations in the US (CDC, 2007). Given the compelling evidence that NISU can result in increased HIV risk, and treatment can diminish these risks, greater efforts to identify the most effective NISU treatment programs for HIV preven-tion must be made. HIV prevention efforts should prioritize research to identify effective treatment interventions, and make effective substance-use treatment available to those most in need.

References

Austin, H., Macaluso, M., Nahmias, A. *et al.* (1999). Correlates of herpes simplex virus seroprevalence among women attending a sexually transmitted disease clinic. *Sex. Trans. Dis.*, 26, 329–34.

Avins, A. L., Woods, W. J., Lindan, C. P. *et al.* (1994). HIV infection and risk behaviors among heterosexuals in alcohol treatment programs. *J. Am. Med. Assoc.*, 271, 515–18.

Avins, A. L., Lindan, C. P., Woods, W. J. *et al.* (1997). Changes in HIV-related behaviors among heterosexual alcoholics following addiction treatment. *Drug Alcohol Depend.*, 44, 47–55.

Ayisi, J. G., van Eijk, A. M., ter Kuile, F. O. *et al.* (2000). Risk factors for HIV infection among asymptomatic pregnant women attending an antenatal clinic in western Kenya. *Intl J. STD AIDS*, 11, 393.

Bailey, S. L., Gao, W. and Clark, D. B. (2006). Diary study of substance use and unsafe sex among adolescents with substance use disorders. *J. Adolesc. Health.*, 38, 297e13–e297e20.

Bailey, S. L., Camlin, C. S. and Ennett, S. T. (1998). Substance use and risky sexual behavior among homeless and runaway youth. *J. Adolesc. Health*, 23, 378–88.

Bairati, I., Sherman, K. J., McKnight, B. *et al.* (1994). Diet and genital warts – a case–control study. *Sex. Trans. Dis.*, 21, 149–54.

Baker, A., Lee, N. K., Claire, M. *et al.* (2005). Brief cognitive behavioural interventions for regular amphetamine users: a step in the right direction. *Addiction*, 100, 367–78.

Banta-Green, C., Goldbaum, G., Kingston, S. *et al.* (2005). Epidemiology of MDMA and associated club drugs in the Seattle area. *Subst. Use Misuse*, 40, 1295–315.

Barrett, S. P., Gross, S. R., Garand, I. and Pihl, R. O. (2005). Patterns of simultaneous polysubstance use in Canadian rave attendees. *Subst. Use Misuse*, 40, 1525–37.

Barrett, S. P., Darredeau, C. and Pihl, R. O. (2006). Patterns of simultaneous polysubstance use in drug using university students. *Hum. Psychopharmacol. Clin. Exp.*, 21, 255–63.

Barthlow, D. J., Horan, P. F., DiClemente, R. J. and Lanier, M. M. (1995). Correlates of condom use among incarcerated adolescents in a rural state. *Criminal Justice Behav.*, 22, 295–306.

Baskin-Sommers, A. and Sommers, I. (2006). The co-occurrence of substance use and high-risk behaviors. *J. Adolesc. Health*, 38, 609–11.

Bautista, C. T., Sanchez, J. L., Montano, S. M. *et al.* (2004). Seroprevalence of and risk factors for HIV-1 infection among South American men who have sex with men. *Sex. Transm. Infect.*, 80, 498–504.

Bigelow, G. E. and Silverman, K. (1999). Theoretical and empirical foundations of contingency management treatments for drug abuse. In: S. T. Higgins and K. Silverman (eds), *Motivating Behavior Change among Illicit-drug Abusers: Research on Contingency Management Interventions*. Washington, DC: American Psychology Association, pp. 15–31.

Bjekic, M., Vlajinac, H., Sipetic, S. and Marinkovic, J. (1997). Risk factors for gonorrhea: case–control study. *Genitourinary Med.*, 73, 518–21.

Bjekic, M., Vlajinac, H. and Marinkovic, J. (2000). Behavioural and social characteristics of subjects with repeated sexually transmitted diseases. *Acta Dermato-Venereol.*, 80, 44–7.

Bolding, G., Hart, G., Sherr, L. and Elford, J. (2006). Use of crystal methamphetamine among gay men in London. *Addiction*, 101, 1622–30.

Boscarino, J. A., Avins, A. L., Woods, W. J. *et al.* (1995). Alcohol-related risk-factors associated with HIV-infection among patients entering alcoholism-treatment – implications for prevention. *J. Stud. Alcohol*, 56, 642–53.

Bouhnik, A. D., Preau, M., Lert, F. *et al.* (2007). Unsafe sex in regular partnerships among heterosexual persons living with HIV: evidence from a large representative sample of individuals attending outpatient services in France (ANRS-EN12-VESPA Study). *AIDS*, 21, S57–62.

Boyer, C. B., Shafer, M. A., Teitle, E. *et al.* (1999). Sexually transmitted diseases in a health maintenance organization teen clinic – associations of race, partner's age, and marijuana use. *Arch. Ped. Adolesc. Med.*, 153, 838–44.

Boyer, C. B., Shafer, M. A., Wibbelsman, C. J. *et al.* (2000). Associations of sociodemographic, psychosocial, and behavioral factors with sexual risk and sexually transmitted diseases in teen clinic patients. *J. Adolesc. Health*, 27, 102–11.

Boyer, C. B., Sebro, N. S., Wibbelsman, C. and Shafer, M. A. (2006). Acquisition of sexually transmitted infections in adolescents attending an urban, general HMO teen clinic. *J. Adolesc. Health*, 39, 287–90.

Boys, A., Lenton, S. and Norcross, K. (1997). Polydrug use at raves by a Western Australian sample. *Drug Alcohol Rev.*, 16, 227–34.

Braithwaite, R. and Stephens, T. (2005). Use of protective barriers and unprotected sex among adult male prison inmates prior to incarceration. *Intl J. STD AIDS*, 16, 224–6.

Brecht, M. L., von Mayrhauser, C. and Anglin, M. D. (2000). Predictors of relapse after treatment for methamphetamine use. *J. Psychoactive Drugs*, 32, 211–20.

Brems, C. and Dewane, S. (2007). Hearing consumer voices: planning HIV/sexually transmitted infection prevention in alcohol detoxification. *J. Assoc. Nurses AIDS Care*, 18, 12–24.

Brewer, D. D., Golden, M. R. and Handsfield, H. H. (2006). Unsafe sexual behavior and correlates of risk in a probability sample of men who have sex with men in the era of highly active antiretroviral therapy. *Sex. Trans. Dis.*, 33, 250–5.

Broome, K. M., Joe, G. W. and Simpson, D. D. (1999). HIV risk reduction in outpatient drug abuse treatment: individual and geographic differences. *AIDS Educ. Prev.*, 11, 293–306.

Brouwer, K. C., Case, P., Ramos, R. *et al.* (2006). Trends in production, trafficking, and consumption of methamphetamine and cocaine in Mexico. *Subst. Use Misuse*, 41, 707–27.

Brown, J. L. and Vanable, P. A. (2007). Alcohol use, partner type, and risky sexual behavior among college students: findings from and event-level study. *Addictive Behav.*, 32, 2940–52.

Bryant, K. J. (2006). Expanding research on the role of alcohol consumption and related risks in the prevention and treatment of HIV/AIDS. *Subst. Use Misuse*, 41, 1465–507.

Buchacz, K., McFarland, W., Hernandez, M. *et al.* (2000). Prevalence and correlates of herpes simplex virus type 2 infection in a population-based survey of young women in low-income neighborhoods of Northern California. *Sex. Trans. Dis.*, 27, 393–400.

Buchacz, K., McFarland, W., Kellogg, T. A. *et al.* (2005). Amphetamine use is associated with increased HIV incidence among men who have sex with men in San Francisco. *AIDS*, 19, 1423–4.

Buchbinder, S. P., Vittinghoff, E., Heagerty, P. J. *et al.* (2005). Sexual risk, nitrite inhalant use, and lack of circumcision associated with HIV seroconversion in men who have sex with men in the United States. *J. Acquir. Immune Defic. Syndr.*, 39, 82–9.

Buffum, J. (1982). Pharmaco-sexology – the effects of drugs on sexual function – a review. *J. Psychoactive Drugs*, 14, 5–44.

Burcham, J. L., Tindall, B., Marmor, M. *et al.* (1989). Incidence and risk-factors for human immunodeficiency virus seroconversion in a cohort of Sydney homosexual men. *Med. J. Aust.*, 150, 634–7.

Castrucci, B. C. and Martin, S. L. (2002). The association between substance use and risky sexual behaviors among incarcerated adolescents. *Maternal Child Health J.*, 6, 43–7.

CDC (2006). Methamphetamine use and HIV risk behaviors among heterosexual men – preliminary results from five Northern California counties, December 2001–November 2003. *Morbid. Mortal. Wkly Rep.*, 55, 273–7.

CDC (2007). *HIV/AIDS Surveillance Report*, 2005. Atlanta, GA: CDC.

Celentano, D. D., Nelson, K. E., Suprasert, S. *et al.* (1996). Epidemiologic risk factors for incident sexually transmitted diseases in young Thai men. *Sex. Trans. Dis.*, 23, 198–205.

Celentano, D. D., Valleroy, L. A., Sifakis, F. *et al.* (2006). Associations between substance use and sexual risk among very young men who have sex with men. *Sex. Trans. Dis.*, 33, 265–71.

Chappel, J. N. and Dupont, R. L. (1999). Twelve-step and mutual-help programs for addictive disorders. *Psychiatr. Clin. North Am.*, 22, 425–46.

Chesney, M. A., Barrett, D. C. and Stall, R. (1998). Histories of substance use and risk behavior: Precursors to HIV seroconversion in homosexual men. *Am. J. Public Health*, 88, 113–16.

Chesney, M. A., Koblin, B. A., Barresi, P. J. *et al.* (2003). An individually tailored intervention for HIV prevention: Baseline data from the EXPLORE study. *Am. J. Public Health*, 93, 933–8.

Chetwynd, J., Chambers, A. and Hughes, A. J. (1992). Sexually-transmitted diseases amongst a sample of people seeking HIV testing. *NZ Med. J.*, 105, 444–6.

Chiasson, M. A., Hirshfield, S., Remien, R. H. *et al.* (2007). A comparison of on-line and off-line sexual risk in men who have sex with men – an event-based on-line survey. *J. Acquir. Immune Defic. Syndr.*, 44, 235–43.

Choi, K. H., Operario, D., Gregorich, S. E. *et al.* (2005). Substance use, substance choice, and unprotected anal intercourse among young Asian American and Pacific Islander men who have sex with men. *Aids Educ. Prev.*, 17, 418–29.

Chokephaibulkit, K., Patmasucon, P., List, M. *et al.* (1997). Genital chlamydia trachomatis infection in pregnant adolescents in east Tenessee: a 7-year case-control study. *J. Ped. Adolesc. Gynecol.*, 10, 95–100.

Clatts, M. C., Goldsamt, L. A. and Yi, H. (2005). Club drug use among young men who have sex with men in NYC: a preliminary epidemiological profile. *Subst. Use Misuse*, 40, 1317–30.

Clutterbuck, D. J., Gorman, D., McMillan, A. *et al.* (2001). Substance use and unsafe sex amongst homosexual men in Edinburgh. *AIDS Care*, 13, 527–35.

Cochran, S. D., Ackerman, D., Mays, V. M. and Ross, M. W. (2004). Prevalence of non-medical drug use and dependence among homosexually active men and women in the US population. *Addiction*, 99, 989–98.

Coldiron, M. E., Stephenson, R., Chomba, E. *et al.* (2007). The relationship between alcohol consumption and unprotected sex among known HIV-discordant couples in Rwanda and Zambia. *AIDS Behav.*, e-pub ahead of print.

Colfax, G., Mansergh, G., Guzman, R. *et al.* (2001). Drug use and sexual risk behavior among gay and bisexual men who attend circuit parties: a venue-based comparison. *J. Acquir. Immune Defic. Syndr.*, 28, 373–9.

Colfax, G., Vittinghoff, E., Husnik, M. J. *et al.* (2004). Substance use and sexual risk: a participant- and episode-level analysis among a cohort of men who have sex with men. *Am. J. Epidemiol.*, 159, 1002–12.

Colfax, G., Coates, T. J., Husnik, M. J. *et al.* (2005). Longitudinal patterns of methamphetamine, popper (Amyl nitrite), and cocaine use and high-risk sexual behavior among a cohort of San Francisco men who have sex with men. *J. Urban Health*, 82, 162–70.

Colfax, G. N., Vittinghoff, E., Grant, R. *et al.* (2007). Frequent methamphetamine use is associated with primary non-nucleoside reverse transcriptase inhibitor resistance. *AIDS*, 21, 239–41.

Collins, R. L., Ellickson, P. L., Bell, R. M. (1998). Simultaneous polydrug use among teens: Prevalence and predictors. *Journal of Substance Abuse*, 10(3), 233–253.

Compton, W., Cottler, L. B., Spitznagel, E. L. *et al.* (1998). Cocaine users with antisocial personality improve HIV risk behaviors as much as those without antisocial personality. *Drug Alcohol Depend.*, 49, 239–47.

Cook, R. L., Pollock, N. K., Rao, A. K. and Clark, D. B. (2002). Increased prevalence of herpes simplex virus type 2 among adolescent women with alcohol use disorders. *J. Adolesc. Health*, 30, 169–74.

Cook, R. L., Comer, D. M., Wiesenfeld, H. C. *et al.* (2006). Alcohol and drug use and related disorders: An underrecognized health issue among adolescents and young adults attending sexually transmitted disease clinics. *Sex. Trans. Dis.*, 33, 565–70.

Cotch, M. F., Pastorek, J. G., Nugent, R. P. *et al.* (1991). Demographic and behavioral predictors of *Trichomonas vaginalis* infection among pregnant women. *Obstet. Gynecol.*, 78, 1087–92.

Cottler, L. B., Compton, W. M., Ben Abdallah, A. *et al.* (1998). Peer-delivered interventions reduce HIV risk behaviors among out-of-treatment drug abusers. *Public Health Rep.*, 113, 31–41.

Crosby, R. A., Diclemente, R. J., Wingood, G. M. *et al.* (2003). Correlates of continued risky sex among pregnant African-American teens – implications for STD prevention. *Sex. Trans. Dis.*, 30, 57–63.

Cu-Uvin, S., Hogan, J. W., Warren, D. *et al.* (1999). Prevalence of lower genital tract infections among human immunodeficiency virus (HIV)-seropositive and high-risk HIV-seronegative women. *Clin. Infect. Dis.*, 29, 1145–50.

Darrow, W. W., Biersteker, S., Geiss, T. *et al.* (2005). Risky sexual behaviors associated with recreational drug use among men who have sex with men in an international resort area: Challenges and opportunities. *J. Urban Health*, 82, 601–9.

De, P., Cox, J., Boivin, J. F. *et al.* (2007). Rethinking approaches to risk reduction for injection drug users – differences in drug type affect risk for HIV and hepatitis C virus infection through drug-injecting networks. *J. Acquir. Immune Defic. Syndr.*, 46, 355–61.

Degenhardt, L., Barker, B. and Topp, L. (2004). Patterns of ecstasy use in Australia: findings from a national household survey. *Addiction*, 99, 187–95.

Degenhardt, L., Coffey, C., Carlin, J. B. *et al.* (2007). Who are the new amphetamine users? A 10-year prospective study of young Australians. *Addiction*, 102, 1269–79.

De Genna, N. M., Cornelius, M. D. and Cook, R. L. (2007). Marijuana use and sexually transmitted infections in young women who were teenage mothers. *Womens Health Issues*, 17, 300–9.

Dehovitz, J. A., Kelly, P., Feldman, J. *et al.* (1994). Sexually transmitted diseases, sexual behavior, and cocaine use in inner-city women. *Am. J. Epidemiol.*, 140, 1125–34.

De Lima, M. S., de Oliveira Soares, B. G., Reisser, A. A. and Farrell, M. (2002). Pharmacological treatment of cocaine dependence: a systematic review. *Addiction*, 97, 9321–49.

Denning, P. H. and Campsmith, M. L. (2005). Unprotected anal intercourse among HIV-positive men who have a steady male sex partner with negative or unknown HIV serostatus. *Am. J. Public Health*, 95, 152–8.

Des Jarlais, D. C., Arasteh, K., Perlis, T. *et al.* (2007). The transition from injection to non-injection drug use: long-term outcomes among heroin and cocaine users in New York City. *Addiction*, 102, 778–85.

Diaz, R. M., Heckert, A. L. and Sanchez, J. (2005). Reasons for stimulant use among Latino gay men in San Francisco: a comparison between methamphetamine and cocaine users. *J. Urban Health*, 82, 171–8.

DiClemente, R. J., Wingood, G. M., Sionean, C. *et al.* (2002). Association of adolescents' history of sexually transmitted disease (STD). and their current high-risk behavior and STD status-A case for intensifying clinic-based prevention efforts. *Sex. Trans. Dis.*, 29, 503–9.

DiFranceisco, W., Ostrow, D. G. and Chmiel, J. S. (1996). Sexual adventurism, high-risk behavior and human immunodeficiency virus-1 seroconversion among the Chicago MACS-CCS cohort, 1984 to 1992 – a case–control study. *Sex. Trans. Dis.*, 23, 453–60.

Djumalieva, D., Imamshah, W., Wagner, U. and Razum, O. (2002). Drug use and HIV risk in Trinidad and Tobago: qualitative study. *Intl J. STD AIDS*, 13, 633–9.

Drumright, L. N., Patterson, T. L. and Strathdee, S. A. (2006a). Club drugs as causal risk factors for HIV acquisition among men who have sex with men: a review. *Subst. Use Misuse*, 41, 1551–601.

Drumright, L. N., Little, S. J., Strathdee, S. A. *et al.* (2006b). Unprotected anal intercourse and substance use among men who have sex with men with recent HIV infection. *J. Acquir. Immune Defic. Syndr.*, 43, 344–50.

Drumright, L. N., Strathdee, S. A., Little, S. J. *et al.* (2007). Unprotected anal intercourse and substance use before and after HIV diagnosis among recently HIV-infected men who have sex with men. *Sex. Trans. Dis.*, 34, 401–7.

Ebrahim, S. H., Andrews, W. W., Zaidi, A. A. *et al.* (1999). Syphilis, gonorrhoea, and drug abuse among pregnant women in Jefferson County, Alabama, US, 1980–94: monitoring trends through systematically collected health services data. *Sex. Transm. Infect.*, 75, 300–5.

Edlin, B. R., Irwin, K. L., Faruque, S. *et al.* (1994). Intersecting epidemics – crack cocaine use and HIV infection among inner-city young adults. *N. Engl. J. Med.*, 331, 1422–7.

Edwards, J. M., Halpern, C. T. and Wechsberg, W. M. (2006a). Correlates of exchanging sex for drugs or money among women who use crack cocaine. *AIDS Educ. Prev.*, 18, 420–9.

Edwards, J. M., Iritani, B. J. and Hallfors, D. D. (2006b). Prevalence and correlates of exchanging sex for drugs or money among adolescents in the United States. *Sex. Transm. Infect.*, 82, 354–8.

Ekstrand, M. L. and Coates, T. J. (1990). Maintenance of safer sexual behaviors and predictors of risky sex – the San Francisco Men's Health Study. *Am. J. Public Health*, 80, 973–7.

Ekstrand, M. L., Stall, R. D., Paul, J. P. *et al.* (1999). Gay men report high rates of unprotected anal sex with partners of unknown or discordant HIV status. *AIDS*, 13, 1525–33.

Ellen, J. M., Langer, L. M., Zimmerman, R. S. *et al.* (1996). The link between the use of crack cocaine and the sexually transmitted diseases of a clinic population. A comparison of adolescents with adults. *Sex Transm. Dis.*, 23, 511–16.

Ellis, R. J., Childers, M. E., Cherner, M. *et al.* (2003). Increased human immunodeficiency virus loads in active methamphetamine users are explained by reduced effectiveness of antiretroviral therapy. *J. Infect. Dis.*, 188, 1820–6.

Ericksen, K. P. and Trocki, K. F. (1994). Sex, alcohol and sexually-transmitted diseases – a national survey. *Family Plan. Persp.*, 26, 257–63.

Essien, E. J., Meshack, A. F., Peters, R. J. *et al.* (2005). Strategies to prevent HIV transmission among heterosexual African-American men. *BMC Public Health*, 5.

Estebanez, P., Zunzunegui, M. V. and Grant, J. (1997). The prevalence of serological markers for syphilis amongst a sample of Spanish prostitutes. *Intl J. STD AIDS*, 8, 675–80.

Feldblum, P. J., Kuyoh, M., Omari, M. *et al.* (2000). Baseline STD prevalence in a community intervention trial of the female condom in Kenya. *Sex. Transm. Infect.*, 76, 454–6.

Fernandez, M. I., Bowen, G. S., Varga, L. M. *et al.* (2005a). High rates of club drug use and risky sexual practices among Hispanic men who have sex with men in Miami, Florida. *Subst. Use Misuse*, 40, 1347–62.

Fernandez, M. I., Perrino, T., Collazo, J. B. *et al.* (2005b). Surfing new territory: club-drug use and risky sex among hispanic men who have sex with men recruited on the internet. *J. Urban Health*, 82, 179–88.

Fernandez, M. I., Bowen, G. S., Warren, J. C. *et al.* (2007). Crystal methamphetamine: A source of added sexual risk for Hispanic men who have sex with men? *Drug Alcohol Depend.*, 86, 245–52.

Fisher, J. C., Cook, P. A., Sam, N. E. and Kapiga, S. H. (2008). Patterns of alcohol use, problem drinking, and HIV infection among high-risk African women. *Sex. Transm. Dis.*, e-pub ahead of print.

French, R. and Power, R. (2001). Self-reported effects of alkyl nitrate use: a qualitative study amongst targeted groups. *Addiction Res.*, 5, 519–48.

Fullilove, M. T., Golden, E., Fullilove, R. E. *et al.* (1993). Crack cocaine use and high-risk behaviors among sexually active black adolescents. *J. Adolesc. Health*, 14, 295–300.

Galloway, G. P., Newmeyer, J., Knapp, T. *et al.* (1996). A controlled trial of imipramine for the treatment of methamphetamine dependence. *J. Subst. Abuse Treat.*, 13, 493–7.

Garbutt, J. C., West, S. L., Carey, T. S. *et al.* (1999). Pharmacological treatment of alcohol dependence: a review of the evidence. *J. Am. Med. Assoc.*, 281, 1318–25.

Gorbach, P. M., Drumright, L. N., Javanbakht, M. *et al.* (2008). Antiretroviral drug resistance and risk behavior among recently HIV-infected men who have sex with men. *J. Acquir. Immune Defic. Syndr.*, 47, 639–43.

Gorman, E. M., Barr, B. D., Hansen, A. *et al.* (1997). Speed, sex, gay men, and HIV: ecological and community perspectives. *Med. Anthropol. Q.*, 11, 505–15.

Gould, R. A., Otto, M. W., Pollack, M. H. and Yap, L. (1997). Cognitive behavioral and pharmacological treatment of generalized anxiety disorder: a preliminary meta-analysis. *Behav. Ther.*, 28, 285–305.

Graves, K. L. and Hines, A. M. (1997). Ethnic differences in the association between alcohol and risky sexual behavior with a new partner: an event-based analysis. *AIDS Educ. Prev.*, 9, 219–37.

Graves, K. L. and Leigh, B. C. (1995). The relationship of substance use to sexual activity among young adults in the United States. *Family Plan. Persp.*, 27, 18.

Green, A. I. (2003). "Chem friendly": the institutional basis of "club-drug" use in a sample of urban gay men. *Deviant Behav.*, 24, 427–47.

Green, A. I. and Halkitis, P. N. (2006). Crystal methamphetamine and sexual sociality in an urban gay subculture: an elective affinity. *Culture Health Sexuality*, 8, 317–33.

Green, B., Kavanagh, D. and Young, R. (2003). Being stoned: a review of self-reported cannabis effects. *Drug Alcohol Rev.*, 22, 453–60.

Gross, S. R., Barrett, S. P., Shestowsky, J. S. and Pihl, R. O. (2002). Ecstasy and drug consumption patterns: a Canadian rave population study. *Can. J. Psychiatry*, 47, 546–51.

Guo, J., Chung, I. J., Hill, K. G. *et al.* (2002). Developmental relationships between adolescent substance use and risky sexual behavior in young adulthood. *J. Adolesc. Health*, 31, 354–62.

Gwati, B., Guli, A. and Todd, C. H. (1995). Risk factors for sexually transmitted disease amongst men in Harare, Zimbabwe. *C. Afr. J. Med.*, 41, 179–81.

Halkitis, P. N., Fischgrund, B. N. and Parsons, J. T. (2005a). Explanations for methamphetamine use among gay and bisexual men in New York City. *Subst. Use Misuse*, 40, 1331–45.

Halkitis, P. N., Shrem, M. T. and Martin, F. W. (2005b). Sexual behavior patterns of methamphetamine-using gay and bisexual men. *Subst. Use Misuse*, 40, 703–19.

Halkitis, P. N., Palamar, J. J. and Mukherjee, P. P. (2007). Poly-club-drug use among gay and bisexual men: a longitudinal analysis. *Drug Alcohol Depend.*, 89, 153–60.

Hallfors, D. D., Iritani, B. J., Miller, W. C. and Bauer, D. J. (2007). Sexual and drug behavior patterns and HIV and STD racial disparities: the need for new directions. *Am. J. Public Health*, 97, 125–32.

Harawa, N. T., Greenland, S., Bingham, T. A. *et al.* (2004). Associations of race/ethnicity with HIV prevalence and HIV-related behaviors among young men who have sex with men in 7 urban centers in the United States. *J. Acquir. Immune Defic. Syndr.*, 35, 526–36.

Hargreaves, J. R., Morison, L. A., Chege, J. *et al.* (2002). Socioeconomic status and risk of HIV infection in an urban population in Kenya. *Tropic. Med. Intl Health*, 7, 793–802.

Haverkos, H. W. and Dougherty, J. (1988). Health hazards of nitrite inhalants. *Am. J. Med.*, 84, 479–82.

Hendershot, C. S. and George, W. H. (2007). Alcohol and sexuality research in the AIDS era: trends in publication activity, target populations and research design. *AIDS Behav.*, 11, 217–26.

Herman-Stahl, M. A., Krebs, C. P., Kroutil, L. A. and Heller, D. C. (2007). Risk and protective factors for methamphetamine use and nonmedical use of prescription stimulants among young adults aged 18 to 25. *Addictive Behav.*, 32, 1003–15.

Hirshfield, S., Remien, R. H., Humberstone, M. *et al.* (2004a). Substance use and high-risk sex among men who have sex with men: a national online study in the USA. *AIDS Care*, 16, 1036–47.

Hirshfield, S., Remien, R. H., Walavalkar, I. and Chiasson, M. A. (2004b). Crystal methamphetamine use predicts incident STD infection among men who have sex with men recruited online: a nested case–control study. *J. Med. Internet Res.*, 6, 42–9.

Iritani, B. J., Hallfors, D. D. and Bauer, D. J. (2007). Crystal methamphetamine use among young adults in the USA. *Addiction*, 102, 1102–13.

Irwin, T. W., Morgenstern, J., Parsons, J. T. *et al.* (2006). Alcohol and sexual HIV risk behavior among problem drinking men who have sex with men: An event level analysis of timeline followback data. *AIDS Behav.*, 10, 299–307.

Jaffe, A., Shoptaw, S., Stein, J. A. *et al.* (2007). Depression ratings, reported sexual risk behaviors, and methamphetamine use: latent growth curve models of positive change among gay and bisexual men in an outpatient treatment program. *Exp. Clin. Psychopharmacol.*, 15, 301–7.

Jamison, J. H., Kaplan, D. W., Hamman, R. *et al.* (1995). Spectrum of genital human papillomavirus infection in a female adolescent population. *Sex. Trans. Dis.*, 22, 236–43.

Joshi, V., Hser, Y., Grella, C. E. and Houlton, R. (2001). Sex-related HIV risk reduction behavior among adolescents in DATOS-A. *J. Adolesc. Res.*, 16, 642–60.

Kalichman, S. C., Rompa, D. and Cage, M. (2000). Sexually transmitted infections among HIV seropositive men and women. *Sex Transm. Infect.*, 76, 350–4.

Kalichman, S. C., Simbayi, L. C., Vermaak, R. *et al.* (2007). HIV/AIDS risk reduction counseling for alcohol using sexually transmitted infections clinic patients in Cape Town, South Africa. *J. Acquir. Immune Defic. Syndr.*, 44, 594–600.

Karnell, A. P., Cupp, P. K., Zimmerman, R. S. *et al.* (2006). Efficacy of an American alcohol and HIV prevention curriculum adapted for use in South Africa: results of a pilot study in five township schools. *AIDS Educ. Prev.*, 18, 295–310.

Kiefer, F. and Mann, K. (2005). New achievements and pharmacotherapeutic approaches in the treatment of alcohol dependence. *Eur. J. Pharmacol.*, 526, 163–71.

Kim, A. A., Kent, C. K. and Klausner, J. D. (2003). Risk factors for rectal gonococcal infection amidst resurgence in HIV transmission. *Sex. Trans. Dis.*, 30, 813–17.

Kingree, J. B. and Betz, H. (2003). Risky sexual behavior in relation to marijuana and alcohol use among African-American, male adolescent detainees and their female partners. *Drug Alcohol Depend.*, 72, 197–203.

Kingree, J. B. and Phan, D. L. (2001). Marijuana use and HIV risk among adolescent offenders: the moderating effect of age. *J. Subst. Abuse*, 13, 59–71.

Kingree, J. B., Braithwaite, R. and Woodring, T. (2000). Unprotected sex as a function of alcohol and marijuana use among adolescent detainees. *J. Adolesc. Health*, 27, 179–85.

Klitzman, R. L., Pope, H. G. and Hudson, J. I. (2000). MDMA ("Ecstasy") abuse and high-risk sexual behaviors among 169 gay and bisexual men. *Am. J. Psychiatry*, 157, 1162–4.

Koblin, B. A., Chesney, M. A., Husnik, M. J. *et al.* (2003). High-risk behaviors among men who have sex with men in 6 US cities: Baseline data from the EXPLORE study. *Am. J. Public Health*, 93, 926–32.

Koblin, B. A., Husnik, M. J., Colfax, G. *et al.* (2006). Risk factors for HIV infection among men who have sex with men. *AIDS*, 20, 731–9.

Koblin, B. A., Murrill, C., Camacho, M. *et al.* (2007). Amphetamine use and sexual risk among men who have sex with men: results from the National HIV Behavioral Surveillance study – New York City. *Subst. Use Misuse*, 42, 1613–28.

Kokkevi, A., Stefanis, N., Anastasopoulou, E. and Kostogianni, C. (1998). Personality disorders in drug abusers: prevalence and their association with AXIS I disorders as predictors of treatment retention. *Addictive Behav.*, 23, 841–53.

Kraft, P. and Rise, J. (1994). The relationship between sensation seeking and smoking, alcohol-consumption and sexual-behavior among Norwegian adolescents. *Health Educ. Res.*, 9, 193–200.

Lama, J. R., Lucchetti, A., Suarez, L. *et al.* (2006). Association of herpes simplex virus type 2 infection and syphilis with human immunodeficiency virus infection among men who have sex with men in Peru. *J. Infect. Dis.*, 194, 1459–66.

Lampinen, T. M., Mattheis, K., Chan, K. and Hogg, R. S. (2007). Nitrite inhalant use among young gay and bisexual men in Vancouver during a period of increasing HIV incidence. *BMC Public Health*, 7.

Lankenau, S. E. and Clatts, M. C. (2005). Patterns of polydrug use among ketamine injectors in New York City. *Subst. Use Misuse*, 40, 1381–97.

Latka, M. H., Wilson, T. E., Cook, J. A. *et al.* (2005). Impact of drug treatment on subsequent sexual risk behavior in a multisite cohort of drug-using women – a report from the Women's Interagency HIV Study. *J. Subst. Abuse Treat.*, 29, 329–37.

Lee, S. J., Galanter, M., Dermatis, H. and Mcdowell, D. (2003). Circuit parties and patterns of drug use in a subset of gay men. *J. Addictive Dis.*, 22, 47–60.

Leigh, B. C. (2002). Alcohol and condom use – a meta-analysis of event-level studies. *Sex. Transm. Dis.*, 29, 476–82.

Leigh, B. C. and Stall, R. (1993). Substance use and risky sexual behavior for exposure to HIV – issues in methodology, interpretation, and prevention. *Am. Psychologist*, 48, 1035–45.

Lejuez, C. W., Bornovalova, M. A., Daughters, S. B. and Curtin, J. J. (2005). Differences in impulsivity and sexual risk behavior among inner-city crack/cocaine users and heroin users. *Drug Alcohol Depend.*, 77, 169–75.

Lenton, S., Boys, A. and Norcross, K. (1997). Raves, drugs and experience: drug use by a sample of people who attend raves in Western Australia. *Addiction*, 92, 1327–37.

Lewis, J. J. C., Garnett, G. P., Mhlanga, S. *et al.* (2005). Beer halls as a focus for HIV prevention activities in rural Zimbabwe. *Sex. Transm. Dis.*, 32, 364–9.

Lewis, L. M., Melton, R. S., Succop, P. A. and Rosenthal, S. L. (2000). Factors influencing condom use and STD acquisition among African-American college women. *J. Am. College Health*, 49, 19–23.

Liau, A., DiClemente, R. J., Wingood, G. M. *et al.* (2002). Associations between biologically confirmed marijuana use and laboratory-confirmed sexually transmitted diseases among African-American adolescent females. *Sex. Transm. Dis.*, 29, 387–90.

Lichtenstein, B. (1997). Women and crack-cocaine use: a study of social networks and HIV risk in an Alabama jail sample. *Addiction Res.*, 5, 279–96.

Lyons, T., Chandra, G. and Goldstein, J. (2006). Stimulant use and HIV risk behavior: the influence of peer support group participation. *AIDS Educ. Prev.*, 18, 461–73.

Macdonald, N., Elam, G., Hickson, F. *et al.* (2008). Factors associated with HIV seroconversion in gay men in England at the start of the 21st century. *Sex. Transm. Infect.*, 84, 8–13.

Magura, S., Kang, S. Y. and Shapiro, J. L. (1994). Outcomes of intensive AIDS education for male adolescent drug-users in jail. *J. Adolesc. Health*, 15, 457–63.

Magura, S., Rosenblum, A. and Rodriguez, E. M. (1998). Changes in HIV risk behaviors among cocaine-using methadone patients. *J. Addictive Dis.*, 17, 71–90.

Malow, R. M., Devieux, J. G., Rosenberg, R. *et al.* (2006). Alcohol use severity and HIV sexual risk among juvenile offenders. *Subst. Use Misuse*, 41, 1769–88.

Mann, K., Lehert, P. and Morgan, M. Y. (2004). The efficacy of acamprosate in the maintenance of abstinence in alcohol-dependent individuals: results of a meta-analysis. *Alcoholism Clin. Exp. Res.*, 28, 51–63.

Mansergh, G., Colfax, G. N., Marks, G. *et al.* (2001). The Circuit Party Men's Health Survey: findings and implications for gay and bisexual men. *Am. J. Public Health*, 91, 953–8.

Mansergh, G., Shouse, R. L., Marks, G. *et al.* (2006). Methamphetamine and sildenafil (Viagra) use are linked to unprotected receptive and insertive anal sex, respectively, in a sample of men who have sex with men. *Sex. Transm. Infect.*, 82, 131–4.

Martin, C. S., Arria, A. M., Mezzich, A. C., Bukstein, O. G. (1993). Patterns of polydrug use in adolescent alcohol abusers. *American Journal of Drug and Alcohol Abuse.*, 19(4), 511–521.

Martin, C. S., Clifford, P. R., Maisto, S. A. *et al.* (1996). Polydrug use in an inpatient treatment sample of problem drinkers. *Alcoholism Clin. Exp. Res.*, 20, 413–17.

Martin, I., Lampinen, T. M. and McGhee, D. (2006). Methamphetamine use among marginalized youth in British Columbia. *Can. J. Public Health*, 97, 320–4.

Mattison, A. M., Ross, M. W., Wolfson, T. and Franklin, D. (2001). Circuit party attendance, club drug use, and unsafe sex in gay men. *J. Subst. Abuse*, 13, 119–26.

Mausbach, B. T., Semple, S. J., Strathdee, S. A. *et al.* (2007a). Efficacy of a behavioral intervention for increasing safer sex behaviors in HIV-negative, heterosexual methamphetamine users: results from the Fast-Lane Study. *Ann. Behav. Med.*, 34, 263–74.

Mausbach, B. T., Semple, S. J., Strathdee, S. A. *et al.* (2007b). Efficacy of a behavioral intervention for increasing safer sex behaviors in HIV-positive MSM methamphetamine users: results from the EDGE study. *Drug Alcohol Depend.*, 87, 249–57.

McCoul, M. D. and Haslam, N. (2001). Predicting high risk sexual behaviour in heterosexual and homosexual men: the roles of impulsivity and sensation seeking. *Pers. Indiv. Diff.*, 31, 03–1310.

McCoy, C. B., Lai, S., Metsch, L. R. *et al.* (2004). Injection drug use and crack cocaine smoking: independent and dual risk behaviors for HIV infection. *Ann. Epidemiol.*, 14, 535–42.

McGregor, C., Srisurapanont, M., Jittiwutikarn, J. *et al.* (2004). The nature, time course and severity of methamphetamine withdrawal. *Addiction*, 100, 1320–9.

McKim, W. A. (2003). Upper Saddle River, NJ: Drugs and Behavior Prentice Hall.

McKirnan, D. J., Ostrow, D. G. and Hope, B. (1996). Sex, drugs and escape: a psychological model of HIV-risk sexual behaviours. *Aids Care*, 8, 655–69.

McKirnan, D. J., Vanable, P. A., Ostrow, D. G. and Hope, B. (2001). Expectancies of sexual "escape" and sexual risk among drug and alcohol-involved gay and bisexual men. *J. Subst. Abuse*, 13, 137–54.

Menza, T., Colfax, G., Shoptaw, S. *et al.* (2006). Interest in a methamphetamine intervention among men who have sex with men. *Sex. Transm. Dis.*, 34, 209–14.

Mertz, K. J., Finelli, L., Levine, W. C. *et al.* (2000). Gonorrhea in male adolescents and young adults in Newark, New Jersey – implications of risk factors and patient preferences for prevention strategies. *Sex. Trans. Dis.*, 27, 201–7.

Miller, P. J., Law, M., Torzillo, P. J. and Kaldor, J. (2001). Incident sexually transmitted infections and their risk factors in an Aboriginal community in Australia: a population based cohort study. *Sex. Transm. Infect.*, 77, 21–35.

Miller, W. R. (1996). Motivational Interviewing: research, practice, and puzzles *Addict. Behav.*, 21, 835–42.

Miller, W. R. and Rollnick, S. (2002). *Motivational Interviewing: Preparing People for Change.* New York, NY: Guilord Press.

Miranda, A., Vargas, P., St Louis, M. and Viana, M. (2000). Sexually transmitted diseases among female prisoners in Brazil. *Sex. Transm. Dis.*, 27, 491–5.

Mmbaga, E. J., Hussain, A., Leyna, G. H. *et al.* (2007). Prevalence and risk factors for HIV-1 infection in rural Kilimanjaro region of Tanzania: implications for prevention and treatment. *BMC Public Health*, 7.

Molitor, F., Truax, S. R., Ruiz, J. D. and Sun, R. K. (1998). Association of methamphetamine use during sex with risky sexual behaviors and HIV infection among non-injection drug users. *W. J. Med.*, 168, 93–7.

Molitor, F., Facer, M. and Ruiz, J. D. (1999). Safer sex communication and unsafe sexual behavior among young men who have sex with men in California. *Arch. Sex. Behav.*, 28, 335–43.

Morgenstern, J., Irwin, T. W., Wainberg, M. L. *et al.* (2007). A randomized controlled trial of goal choice interventions for alcohol use disorders among men who have sex with men. *J. Consult. Clin. Psychol.*, 75, 72–84.

Morin, S. F., Steward, W. T., Charlebois, E. D. *et al.* (2005). Predicting HIV transmission risk among HIV-Infected men who have sex with men – findings from the healthy living project. *J. Acquir. Immune Defic. Syndr.*, 40, 226–35.

Morrison, C. S., Sunkutu, M. R., Musaba, E. and Glover, L. H. (1997). Sexually transmitted disease among married Zambian women: the role of male and female sexual behaviour in prevention and management. *Genitourinary Med.*, 73, 555–7.

Morrison, D. M., Gillmore, M. R., Hoppe, M. J. *et al.* (2003). Adolescent drinking and sex: Findings from a daily diary study. *Persp. Sex. Reprod. Health*, 35, 162–8.

Moss, A. R., Osmond, D., Bacchetti, P. *et al.* (1987). Risk factors for AIDS and HIV seropositivity in homosexual men. *Am. J. Epidemiol.*, 125, 1035–47.

Myer, L., Mathews, C. and Little, F. (2002). Condom use and sexual behaviors among individuals procuring free male condoms in South Africa – a prospective study. *Sex. Transm. Dis.*, 29, 239–41.

Naar-King, S., Wright, K., Parsons, J. T. *et al.* (2006). Healthy choices: motivational enhancement therapy for health risk behaviors in HIV-positive youth. *AIDS Educ. Prev.*, 18, 1–11.

Naar-King, S., Lam, P., Wang, B. *et al.* (2007). Maintenance of effects of motivational enhancement therapy to improve risk behaviors and HIV-related health in a randomized controlled trial of youth living with HIV. *J. Ped. Psychol.*, advance access publication.

Newell, G. R., Mansell, P. W. A., Spitz, M. R. and Hersh, E. M. (1985). Volatile nitrites – use and adverse effects related to the current epidemic of the acquired immune deficiency syndrome. *Am. J. Med.*, 78, 811–16.

Newman, P. A., Rhodes, F. and Weiss, R. E. (2004). Correlates of sex trading among drug-using men who have sex with men. *Am. J. Public Health*, 94, 1998–2003.

Noell, J., Rohde, P., Ochs, L. *et al.* (2001). Incidence and prevalence of chlamydia, herpes, and viral hepatitis in a homeless adolescent population. *Sex. Transm. Dis.*, 28, 4–10.

Obert, J. L., McCann, M. J., Marinelli-Casey, P. *et al.* (2000). The matrix model of out-patient stimulant abuse treatment: history and description. *J. Psychoactive Drugs*, 32, 157–64.

Operario, D., Choi, K. H., Chu, P. L. *et al.* (2006). Prevalence and correlates of substance use among young Asian pacific islander men who have sex with men. *Prev. Sci.*, 7, 19–29.

Ostrow, D. and McKirnan, D. (1997). Prevention of substance-related high-risk sexual behavior among gay men: critical review of the literature and proposed harm reduction approach. *J. Gay Lesbian Med. Assoc.*, 1, 97–110.

Ostrow, D. G., DiFranceisco, W. J., Chmiel, J. S. and Wesch, J. (1995). A case–control study of human immunodeficiency virus type 1 seroconversion and risk-related behaviors in the Chicago MACS/CCS cohort, 1984–1992. *Am. J. Epidemiol.*, 142, 875–83.

Page-Shafer, K. (1997). Sexual risk behavior and risk factors for HIV-1 seroconversion in homosexual men participating in the Tricontinental Seroconverter Study, 1982–1994. *Am. J. Epidemiol.*, 146, 1076 (Vol. 146, p. 531, 1997).

Palen, L. A., Smith, E. A., Flisher, A. J. *et al.* (2006). Substance use and sexual risk behavior among South African eighth grade students. *J. Adolesc. Health*, 39, 761–3.

Parsons, J. T., Halkitis, P. N., Wolitski, R. J. and Gomez, C. A. (2003). Correlates of sexual risk behaviors among HIV-positive men who have sex with men. *AIDS Educ. Prev.*, 15, 383–400.

Parsons, J. T., Vicioso, K. J., Punzalan, J. C. *et al.* (2004). The impact of alcohol use on the sexual scripts of HIV-positive men who have sex with men. *J. Sex Res.*, 41, 160–72.

Patterson, T. L., Semple, S. J., Zians, J. K. and Strathdee, S. A. (2005). Methamphetamine-using HIV-positive men who have sex with men: correlates of polydrug use. *J. Urban Health*, 82, I120–i126.

Pechansky, F., Woody, G., Inciardi, J. *et al.* (2006). HIV seroprevalence among drug users: an analysis of selected variables based on 10 years of data collection in Porto Alegre, Brazil. *Drug Alcohol Depend.*, 82, S109–13.

Petry, N. M. (2000). A comprehensive guide to the application of contingency. *Drug Alcohol Depend.*, 58, 9–25.

Petry, N. M. and Martin, B. (2002). Low-cost contingency management for treating cocaine- and opioid-abusing methadone patients. *J. Consult. Clin. Psychol.*, 70, 398–405.

Pilowsky, D. J., Hoover, D., Hadden, B. *et al.* (2007). Impact of social network charac-teristics on high-risk sexual behaviors among non-injection drug users. *Subst. Use Misuse*, 42, 1629–49.

Plankey, M. W., Ostrow, D. G., Stall, R. *et al.* (2007). The relationship between metham-phetamine and popper use and risk of HIV seroconversion in the Multicenter AIDS Cohort Study. *J. Acquir. Immune Defic. Syndr.*, 45, 85–92.

Prendergast, M. L., Podus, D., Chang, E. and Urada, D. (2002). The effectiveness of drug abuse treatment: a meta-analysis of comparison group studies. *Drug Alcohol Depend.*, 67, 53–72.

Prendergast, M. L., Podus, D., Finney, J. *et al.* (2006). Contingency management for treat-ment of substance use disorders: a meta-analysis. *Addiction*, 101, 1546–60.

Prinzleve, M., Haasen, C., Zurhold, H. *et al.* (2004). Cocaine use in Europe – a multi-cen-tre study: patterns of use in different groups. *Eur. Addiction Res.*, 10, 147–55.

Purcell, D. W., Parsons, J. T., Halkitis, P. N. *et al.* (2001). Substance use and sexual trans-mission risk behavior of HIV-positive men who have sex with men. *J. Subst. Abuse*, 13, 185–200.

Purcell, D. W., Moss, S., Remien, R. H. *et al.* (2005). Illicit substance use, sexual risk, and HIV-positive gay and bisexual men: differences by serostatus of casual partners. *J. Acquir. Immune Defic. Syndr.*, 19, S37–47.

Quimette, P. C., Finney, J. W. and Moos, R. H. (1997). Twelve-step and cognitive-behavioral treatment for substance abuse a comparison of treatment effectiveness. *J. Consult. Clin. Psychol.*, 65, 230–40.

Radcliffe, K. W., Ahmad, S., Gilleran, G. and Ross, J. D. C. (2001). Demographic and behavioural profile of adults infected with Chlamydia: a case–control study. *Sex. Transm. Infect.*, 77, 265–70.

Raj, A., Saitz, R., Cheng, D. M. *et al.* (2007). Associations between alcohol, heroin, and cocaine use and high risk sexual behaviors among detoxification patients. *Am. J. Drug Alcohol Abuse*, 33, 169–78.

Rakwar, J., Jackson, D., Maclean, I. *et al.* (1997). Antibody to *Haemophilus ducreyi* among trucking company workers in Kenya. *Sex. Transm. Dis.*, 24, 267–71.

Rawson, R. A., Huber, A., McCann, M. *et al.* (2002a). A comparison of contingency management and cognitive-behavioral approaches during methadone maintenance treatment for cocaine dependence. *Arch. Gen. Psychiatry*, 59, 817–24.

Rawson, R. A., Washton, A., Domier, C. P. and Reiber, C. (2002b). Drugs and sexual effects: role of drug type and gender. *J. Subst. Abuse Treatment*, 22, 103–8.

Rawson, R. A., Gonzales, R., Pearce, V. *et al.* (2008). Methamphetamine dependence and human immunodeficiency virus risk behavior. *J. Subst. Abuse Treatment*, e-pub ahead of print.

Rawstorne, P., Digiusto, E., Worth, H. and Zablotska, I. (2007). Associations between crystal methamphetamine use and potentially unsafe sexual activity among gay men in Australia. *Arch. Sex. Behav.*, 36, 646–54.

Reback, C. J., Larkins, S. and Shoptaw, S. (2003). Methamphetamine abuse as a barrier to HIV medication adherence among gay and bisexual men. *AIDS Care*, 15, 775–85.

Reback, C. J., Larkins, S. and Shoptaw, S. (2004). Changes in the meaning of sexual risk behaviors among gay and bisexual male methamphetamine abusers before and after drug treatment. *AIDS Behav.*, 8, 87–98.

Rhodes, S. D., Hergenrather, K. C., Yee, L. J. *et al.* (2007). Characteristics of a sample of men who have sex with men, recruited from gay bars and Internet chat rooms, who report methamphetamine use. *AIDS Patient Care STDs*, 21, 575–83.

Rhodes, T. (1996). Culture, drugs and unsafe sex: confusion about causation. *Addiction*, 91, 753–8.

Rhodes, T. and Stimson, G. V. (1994). What is the relationship between drug taking and sexual risk? Social-relations and social research. *Sociology Health Illness*, 16, 209–28.

Rhodes, T., Briggs, D., Kimber, J. *et al.* (2007). Crack-heroin speedball injection and its implications for vein care: qualitative study. *Addiction*, 102, 1782–90.

Robles, R. R., Marrero, C. A., Matos, T. D. *et al.* (1998). Factors associated with changes in sex behaviour among drug users in Puerto Rico. *AIDS Care*, 10, 329–38.

Ross, M. W., Mattison, A. M. and Franklin, D. R. (2003). Club drugs and sex on drugs are associated with different motivations for gay circuit party attendance in men. *Subst. Use Misuse*, 38, 1173–83.

Rusch, M., Lampinen, T. M., Schilder, A. and Hogg, R. S. (2004). Unprotected anal intercourse associated with recreational drug use among young men who have sex with men depends on partner type and intercourse role. *Sex. Transm. Dis.*, 31, 492–8.

SAMHSA (2006). *Results from the 2006 National Survey on Drug Use and Health: National Findings.* NSDUH Series. Rockville, MD: Substance Use and Mental Health Services Administration.

Sanchez, J., Gotuzzo, E. and Escamilla, J. (1996). Gender differences in sexual practices and sexually transmitted infections among adults in Lima, Peru. *Am. J. Public Health*, 86, 1098–107.

Sanchez, J., Finlayson, T., Drake, A. *et al.* (2006). Human immunodeficiency virus (HIV) risk, prevention, and testing behaviors – United States, National HIV Behavioral Surveillance System: Men who have Sex with Men, November 2003–April 2005. *Morbid. Mortal. Wkly Rep.*, 55, 1–16.

Scheidt, D. M. (1999). HIV risk behavior among alcoholic inpatients before and after treatment. *Addictive Behav.*, 24, 725–30.

Scheidt, D. M. and Windle, M. (1995). The alcoholics in treatment HIV risk (ATRISK) study: gender, ethnic and geographic group comparisons. *J. Stud. Alcohol*, 56, 300–8.

Schroeder, J. R., Epstein, D. H., Urnbricht, A. and Preston, K. L. (2006). Changes in HIV risk behaviors among patients receiving combined pharmacological and behavioral interventions for heroin and cocaine dependence. *Addictive Behav.*, 31, 868–79.

Semple, S. J., Patterson, T. L. and Grant, I. (2000). Psychosocial predictors of unprotected anal intercourse in a sample of HIV positive gay men who volunteer for a sexual risk reduction intervention. *AIDS Educ. Prev.*, 12, 416–30.

Semple, S. J., Patterson, T. L. and Grant, I. (2002). Motivations associated with methamphetamine use among HIV plus men who have sex with men. *J. Subst. Abuse Treatment*, 22, 149–56.

Semple, S. J., Patterson, T. L. and Grant, I. (2004). A comparison of injection and non-injection methamphetamine-using HIV positive men who have sex with men. *Drug Alcohol Depend.*, 76, 203–12.

Shafer, M., Hilton, J., Ekstrand, M. *et al.* (1993). Relationship between drug use and sexual behaviors and the occurence of sexually transmitted diseases among high-risk male youth. *Sex. Transm. Dis.*, 20, 307–13.

Shoptaw, S. and Reback, C. J. (2007). Methamphetamine use and infectious disease-related behaviors in men who have sex with men: implications for interventions. *Addiction*, 102, 130–5.

Shoptaw, S., Frosch, D., Rawson, R. A. and Ling, W. (1997). Cocaine abuse counseling as HIV prevention. *AIDS Educ. Prev.*, 9, 511–20.

Shoptaw, S., Reback, C. J., Frosch, D. L. and Rawson, R. A. (1998). Stimulant abuse treatment as HIV prevention. *J. Addictive Dis.*, 17, 19–32.

Shoptaw, S., Reback, C. J., Peck, J. A. *et al.* (2005). Behavioral treatment approaches for methamphetamine dependence and HIV-related sexual risk behaviors among urban gay and bisexual men. *Drug Alcohol Depend.*, 78, 125–34.

Shoptaw, S., Huber, A., Peck, J. *et al.* (2006). Randomized, placebo-controlled trial of sertraline and contingency management for the treatment of methamphetamine dependence. *Drug Alcohol Depend.*, 85, 12–15.

Shrier, L. A., Harris, S. K., Sternberg, M. and Beardslee, W. R. (2001). Associations of depression, self-esteem, and substance use with sexual risk among adolescents. *Prevent. Med.*, 33, 179–89.

Simbayi, L. C., Kalichman, S. C., Jooste, S. *et al.* (2004). Alcohol use and sexual risks for HIV infection among men and women receiving sexually transmitted infection clinic services in Cape Town, South Africa. *J. Stud. Alcohol*, 65, 34–442.

Sofuoglu, M. and Kosten, T. R. (2006). Emerging pharmacological strategies in the fight against cocaine addiction. *Expert Opin. Emerging Drugs*, 11, 91–8.

Spindler, H. H., Scheer, S., Chen, S. Y. *et al.* (2007). Wagra, methamphetamine, and HIV risk: results from a probability sample of MSM, San Francisco. *Sex. Transm. Dis.*, 34, 586–91.

Springer, A. E., Peters, R. J., Shegog, R. *et al.* (2007). Methamphetamine use and sexual risk behaviors in US High school students: Findings from a national risk behavior survey. *Prev. Sci.*, 8, 103–13.

Srisurapanont, M., Jarusuraisin, N. and Jittiwutikarn, J. (1999). Amphetamine withdrawal: II. A placebo-controlled, randomised, double-blind study of amineptine treatment. *Aust. NZ J. Psychiatry*, 33, 94–8.

Staines, G. L., Magura, S., Foote, J. *et al.* (2001). Polysubstance use among alcoholics. *J. Addictive Dis.*, 20, 53–69.

Stall, R. D., Paul, J. P., Barrett, D. C. *et al.* (1999). An outcome evaluation to measure changes in sexual risk-taking among gay men undergoing substance use disorder treatment. *J. Stud. Alcohol*, 60, 837–45.

Stall, R. D., Paul, J. P., Greenwood, G. *et al.* (2001). Alcohol use, drug use and alcohol-related problems among men who have sex with men: the Urban Men's Health Study. *Addiction*, 96, 1589–601.

St Lawrence, J. S., Crosby, R. A., Brasfield, T. A. and O'Bannon, R. E. III (2002). Reducing STD and HIV risk behavior of substance-dependent adolescents: a randomized controlled trial. *J. Consult. Clin. Psychol.*, 70, 1010–21.

Strathdee, S. A., Hogg, R. S., Martindale, S. L. *et al.* (1998). Determinants of sexual risk-taking among young HIV-negative gay and bisexual men. *J. Acquir. Immune Defic. Syndr.*, 19, 61–6.

Szwarcwald, C. L., Barbosa-Junior, A., Pascom, A. R. and De Souza-Junior, P. R. (2005). Knowledge, practices and behaviours related to HIV transmission among the Brazilian population in the 15–54 years age group, 2004. *AIDS*, 19, S51–8.

Talbot, E. A., Kenyon, T. A., Moeti, T. L. *et al.* (2002). HIV risk factors among patients with tuberculosis – Botswana 1999. *Intl J. STD AIDS*, 13, 311–17.

Tambour, S. and Quertemont, E. (2007). Preclinical and clinical pharmacology of alcohol dependence. *Fund. Clin. Pharmacol.*, 21, 9–28.

Teter, C. J. and Guthrie, S. K. (2001). A comprehensive review of MDMA and GHB: two common club drugs. *Pharmacotherapy*, 21, 1486–513.

Theall, K. P., Clark, R. A., Powell, A. *et al.* (2007). Alcohol consumption, art usage and high-risk sex among women infected with HIV. *AIDS Behav.*, 11, 205–15.

Thiede, H., Valleroy, L. A., Mackellar, D. A. *et al.* (2003). Regional patterns and correlates of substance use among young men who have sex with men in 7 US urban areas. *Am. J. Public Health*, 93, 1915–21.

Thomas, A. G., Brodine, S. K., Shaffer, R. *et al.* (2001). Chlamydial infection and unplanned pregnancy in women with ready access to health care. *Obstet. Gynecol.*, 98, 1117–23.

Tossmann, P., Boldt, S. and Tensil, M. D. (2001). The use of drugs within the techno party scene in European metropolitan cities. *Eur. Addiction Res.*, 7, 2–23.

UN (2007). *2007 World Drug Report*. Geneva: United Nations.

Van den Eeden, S. K., Habel, L. A. and Sherman, K. J. (1998). Risk factors for incident and recurrent condylomata acuminata among men. *Sex. Transm. Dis.*, 25, 278–84.

Vaudrey, J., Raymond, H. F., Chen, S. *et al.* (2007). Indicators of use of methamphetamine and other substances among men who have sex with men, San Francisco, 2003–2006. *Drug Alcohol Depend.*, 90, 97–100.

Vocci, F. and Ling, W. (2005). Medications development: successes and challenges. *Pharmacol. Therapeutics*, 108, 94–108.

Vuylsteke, B., Vandenbruaene, M., Vandenbulcke, P. *et al.* (1999). Chlamydia trachomatis prevalence and sexual behaviour among female adolescents in Belgium. *Sex. Transm. Infect.*, 75, 152–5.

Weber, A. E., Craib, K. J. P., Chan, K. *et al.* (2003). Determinants of HIV serconversion in an era of increasing HIV infection among young gay and bisexual men. *AIDS*, 17, 774–7.

Wechsberg, W. M., Lam, W. K. K., Zule, W. A. and Bobashev, G. (2004). Efficacy of a woman-focused intervention to reduce HIV risk and increase self-sufficiency among African-American crack abusers. *Am. J. Public Health*, 94, 1165–73.

Wechsberg, W. M., Luseno, W. K., Karg, R. S. *et al.* (2008). Alcohol, cannabis, and methamphetamine use and other risk behaviours among Black and Coloured South African women: a small randomized trial in the Western Cape. *Intl J. Drug Policy*, 19, 130–9.

Weiser, S. D., Leiter, K., Heisler, M. *et al.* (2006). A population-based study on alcohol and high-risk sexual behaviors in Botswana. *PloS Med.*, 3, 1940–8.

WHO (2004). *Global Status Report on Alcohol 2004*. Geneva: World Health Organization.

WHO (2007). *Alcohol and Injury in Emergency Departments*. Geneva: World Health Organization.

Windle, M. (1989). High-risk behaviors for AIDS among heterosexual alcoholics: a pilot study. *J. Stud. Alcohol*, 50, 503–7.

Wong, W., Chaw, J. K., Kent, C. K. and Klausner, J. D. (2005). Risk factors for early syphilis among gay and bisexual men seen in an STD clinic: San Francisco, 2002–2003. *Sex. Transm. Dis.*, 32, 458–63.

Woods, W. J., Lindan, C., Hudes, E. S. *et al.* (2000). HIV infection and risk behaviors in two cross-sectional surveys of heterosexuals in alcoholism treatment. *J. Stud. Alcohol*, 61, 262–6.

Woody, G., Donnell, D., Seage, G. R. *et al.* (1999). Non-injection substance use correlates with risky sex among men having sex with men: data from HIVNET. *Drug Alcohol Depend.*, 53, 197–205.

Woody, G., Gallop, R., Luborsky, L. *et al.* (2003). HIV risk reduction in the National Institute on Drug Abuse Cocaine Collaborative Treatment Study. *J Acquir. Immune Defic. Syndr.*, 33, 82–7.

Yadav, G., Saskin, Y., Ngugi, E. *et al.* (2005). Associations of sexual risk taking among Kenyan female sex workers after enrollment in an HIM prevention trial. *J. Acquir. Immune Defic. Syndr.*, 38, 329–34.

Zablotska, I. B., Gray, R. H., Serwadda, D. *et al.* (2006). Alcohol use before sex and HIV acquisition: a longitudinal study in Rakai, Uganda. *AIDS*, 20, 1191–6.

Zachariah, R., Spielmann, M. P., Harries, A. D. *et al.* (2003). Sexually transmitted infections and sexual behaviour among commercial sex workers in a rural district of Malawi. *Intl J. STD AIDS*, 14, 185–8.

Zenilman, J. M., Hook, E. W. III, Shepherd, M. *et al.* (1994). Alcohol and other substance use in STD clinic patients: relationships with STDs and prevalent HIV infection. *Sex. Transm. Dis.*, 21, 220–5.

Zuma, K., Gouws, E., Williams, B. and Lurie, M. (2003). Risk factors for HIV infection among women in Carletonville, South Africa: migration, clemography and sexually transmitted diseases. *Intl J. STD AIDS*, 14, 814–17.

Preventing HIV among sex workers

14

Bea Vuylsteke, Anjana Das, Gina Dallabetta and
Marie Laga

'Namaskar! Ami jaunokarmi.' [Greetings! I am a sex worker.]

This is how Kamla (name changed), a sex worker from a red light district in
Kolkata, India, greets visitors at the office of the sex workers' project, where she is
a peer educator. She has come a long way since she began sex work, a profession
which she did not admit to earlier as it was looked down upon by the general community and invited harassment from local goons and policemen. The empowerment
of sex workers has led to more effective HIV prevention in the community.

In every region of the world sex workers are a critical group in the HIV transmission dynamics, and effective interventions with sex workers are an important
component of comprehensive HIV control strategies. Sex workers are both at
high risk because of multiple sexual partners, and highly vulnerable because of
environmental and structural barriers that prevent them from accessing prevention services and having control over their activities. There is extensive global
experience in HIV prevention among sex workers, and an essential package of
proven effective interventions has been defined. Most interventions in sex workers, however, despite being effective, have had little impact on HIV transmission
dynamics in countries simply because they are implemented on such a small
scale that most sex workers who need prevention services do not have them
available. Addressing this "prevention gap" is the major challenge in HIV prevention in sex workers (Global HIV Prevention Working Group, 2007).

Sex workers are usually defined as men or women having sex in exchange for
money or its equivalent (Harcourt and Donovan, 2005). Sex work can be classified as "direct" (open, formal) or "indirect" (hidden, clandestine, informal).
Direct sex workers are typically men or women who do define themselves as
sex workers, and earn their living by selling sex. Indirect sex workers are men or

376

women for whom sex work is not the primary source of income. They do not self-identify as sex workers, and often work outside of known venues for sex work.

In many developing countries male purchase of commercial sex is a social norm, and married men purchase sexual services on a regular basis, with 20–40 percent of men acknowledging occasionally having sex with sex workers (Wilson *et al.*, 1991; Aklilu *et al.*, 2001; Morison *et al.*, 2001). Because of their high rate of partner change, sex workers have been considered an epidemiologic "core group" that accounts for a disproportionate amount of STI/HIV transmission in a community (Plummer *et al.*, 1991; Thomas and Tucker, 1996). As a result, interventions targeting sex workers have the potential to slow down the HIV epidemic in the general population. Some authors suggest that the relative importance of core groups to the spread and maintenance of HIV within the population will change over time, decreasing as HIV levels become higher in the general population (Wasserheit and Aral, 1996; Blanchard, 2002). However, the high levels of STIs and HIV among sex workers underline the importance of continuing interventions among sex workers and their clients. Even if the epidemic has spread beyond the core groups, prevention among people with high rates of partner change is likely to reduce transmission overall (Morris and Ferguson, 2006; Chen *et al.*, 2007). STI prevalence among miners in South Africa was found to decrease following an intervention among local sex workers (Steen *et al.*, 2000). High levels of condom use among sex workers and their clients in countries such as Benin, Thailand and Cambodia are likely to have slowed the epidemic (Rojanapithayakorn and Hanenberg, 1996; Morison *et al.*, 2001).

We will address the issues related to commercial sex in four sections in this chapter: (1) an overview of the epidemiology, HIV risk and vulnerability among sex workers; (2) the elements of effective interventions; (3) the intervention gap; and (4) some model programs illustrating these approaches.

Epidemiology, HIV risk and vulnerability among sex workers

High HIV prevalence levels have been found among sex workers all over the world where they face multiple risks, including multiple sexual partners, low levels of condom use and high STI prevalence. In addition, sex workers are often not in a position to control these risk factors, because of the environment and context in which they live.

Epidemiology of HIV among sex workers

Since the very beginning of the HIV epidemic, female sex workers have been among the populations most affected. As early as 1985, 62 percent of female sex workers in Nairobi, Kenya, were found to be HIV infected (Simonson *et al.*,

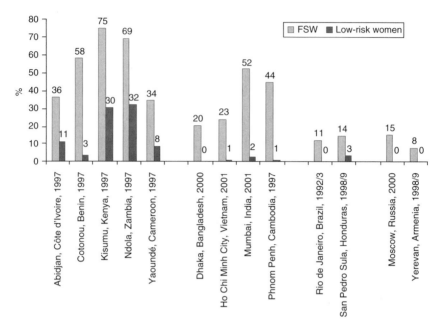

Figure 14.1 HIV prevalence in female sex workers and low-risk women in different cities.

1990), and in 1988 in Kinshasa, DRC, 35 percent of *femmes libres* tested positive for HIV (Nzila *et al.*, 1991). In the late 1980s and early 1990s, rates as high as 88 percent and 89 percent were found among sex workers in Butare, Rwanda, and in Abidjan, Côte d'Ivoire, respectively (Hunt, 1989; Ghys *et al.*, 2002).

In concentrated but also in generalized epidemics, HIV prevalence has been revealed as considerably higher among sex workers compared to the general population (see Figure 14.1). In Niger, 28 percent of female sex workers in the Tenere Dessert were found to be HIV-positive in 1997, while HIV seroprevalence among the Touareg population was reported at less than 1 percent (Gragnic *et al.*, 1998). In two cities in Papua New Guinea, 10 percent of the female sex workers were HIV-positive in 1998–1999 – 10 times the prevalence found among the general population (Mgone *et al.*, 2002). The four-cities study in sub-Saharan Africa found that the difference in prevalence remained, even with the high rates in the general population. The prevalence rates of HIV were 75 percent vs 30 percent in Kisumu (Kenya), 69 percent vs 23 percent in Ndola (Zambia), 55 percent vs 3 percent in Cotonou (Benin) and 34 percent vs 8 percent in Yaoundé (Cameroun) for female sex workers versus general population respectively (Morison *et al.*, 2001).

As expected, high HIV prevalence rates are accompanied with high levels of other STIs, as presented in Table 14.1. While not as numerous as female

Table 14.1 STI prevalence among female sex workers in selected countries

Country, place	Sex-worker population	Year	Syphilis %	NG[a] %	CT[b] %	HSV-2[c] %	HIV %	Ref.
Benin, Cotonou	Self-identified	1997–1998	4	12	4	91	55	Morison et al., 2001
Cameroun, Yaoundé	Self-identified	1997–1998	23	18	18	84	34	Morison et al., 2001
Kenya, Kisumu	Self-identified	1997–1998	11	13	8	94	75	Morison et al., 2001
Zambia, Ndola	Self-identified	1997–1998	42	15	9	87	69	Morison et al., 2001
South Africa, Johannesburg	Self-identified	1996–1997	25	23	8	–	46	Dunkle et al., 2005
Thailand, Chiang Rai	Brothel based	1991–1994	13	23	24	78	47	Limpakarnjanarat et al., 1999
China, Yunnan	Reeducation center	1999–2000	9	38	59	65	10	Chen et al., 2005
Indonesia, Kupang	Brothel based	1999	19	22	36	–[d]	–	Ford and Wirawan, 2005
India, Surat	Red light district	2000	23	17	8	–	43	Desai et al., 2003
Phillippines, Cebu	Registered and unregistered	1994–1995	–	23	22	–	–	Wi et al., 2006
Papua New Guinea	Self-identified	1998–1999	32	36	31	–	10	Mgone et al., 2002
Victoria, Australia	Brothel based	1999–2001	–	1	3	–	–	Morton, 2002
	Street	1999–2001	–	1	7	–	–	Morton, 2002
Russia, Moscow	Female detainees admitting SW	2001	41	29	28	–	3	Khromova, 2002
Bolivia, La Paz	Brothel based	1992–1995	15	26	17	–	–	Levine et al., 1998
Peru, Lima	Registered	1991–1992	2	8	14	82	1	Sanchez et al., 1998

Notes:
[a] Neisseria gonorrhoeae
[b] Chlamydia trachomatis
[c] HSV-2: Herpes simplex virus, type 2
[d] –: not tested

Table 14.2   HIV prevalence among male (MSW) and transgender sex workers (TSW)

Country, place	Population	Year	Number	HIV %	Ref.
Spain, 19 cities	MSW and TSW attending testing clinics	2000–2002	418	12	Belza, 2005
UK, London	MSW, dedicated clinic service	1994–1996 1997–1999 2000–2003	156 223 257	14 6 9	Sethi *et al.*, 2006
Italy, Rome	TSW, community recruited	1997–1998	40	20	Verster *et al.*, 2001
The Netherlands, Amsterdam	TSW	1993	25	24	Gras *et al.*, 1997
Thailand, Chiang Mai	MSW	2002	241	20	Kunawararak *et al.*, 1995
Indonesia, Jakarta	TSW, community recruited MSW, community recruited	2002	250	22 4	Pisani *et al.*, 2004
Côte d'Ivoire, Abidjan	MSW attending HIV testing services	2006	65	35	Personal communication (BV)

sex workers, some males and transgenders also sell sex, predominantly to men. Table 14.2 summarizes some HIV prevalence data among males and transgenders in different settings. Bisexual behavior is common among male sex workers, so they can act as a conduit for infections to pass between heterosexual, homosexual and bisexual populations (Morse *et al.*, 1991).

HIV risk among sex workers

The lifetime probability of a sex worker becoming infected by HIV is high due to multiple risk factors, including a high number and turnover of partners, low levels of condom use, high STI prevalence, unsafe practices such as douching and use of inappropriate lubricants, and injecting drug use.

The main focus of HIV prevention efforts among sex workers has been on increasing knowledge and access to condoms. In settings where no (or inadequate) prevention programs are implemented, condom use may be still very low. Only 15 percent and 13 percent respectively of the female sex workers in Gouangxi and Zhengzhou, China, for instance, reported consistent use of condoms with their clients (Wang *et al.*, 2005). Some 43 percent of transgender and 56 percent of male sex workers reported condom use with their most recent client in Indonesia (Pisani, 2004). Fortunately, these low levels of condom use are no longer the norm in many settings, at least for sex with paying clients. However, sex workers are usually less likely to use condoms with their non-paying partners than with their clients (Pickering *et al.*, 1993; Nguyen *et al.*, 2005). The level of intimacy and emotional ties with the sexual partner, regardless of assessed risk, seem an important factor for condom use in both studies.

STIs are another major factor affecting HIV transmission (Rottingen *et al.*, 2001). This relationship between STI and HIV transmission was first documented among sex workers in Kinshasa, showing that both ulcerative and non-ulcerative STI enhanced HIV transmission (Laga *et al.*, 1993). More recently, it became clear that Herpes Simplex Virus-2 (HSV-2), in particular, plays an important role in facilitating HIV transmission (Freeman *et al.*, 2007). The presence of HSV-2 and other genital infections among female bar/hotel workers in Moshe, Tanzania, led to an increased HIV incidence (Kapiga *et al.*, 2007). Because of their high rate of partner change, sex workers are highly exposed to STI, which in turn further increases their risk of and vulnerability to HIV. Incidence rates of STIs randing from 38 percent per year to 10 percent per month have been reported from Africa and Asia (Siraprapasiri *et al.*, 1991; Willerford *et al.*, 1993). Table 14.1 summarizes some of the STI and HIV prevalence rates in female sex workers in different countries. The prevalence of HIV may be still low in some settings, but, given the high prevalence of STIs and the fact that STIs facilitate the sexual transmission of HIV, coupled with high sexual partner change, the potential for the further spread of HIV is clearly present (van den Hoek *et al.*, 2001; Ford and Wirawan, 2005).

Unsafe practices such as douching, the use of vaginal drying agents and the use of inappropriate lubricants also play a role in increasing HIV risk among sex workers. Vaginal douching using substances such as soap, toothpaste or commercial antiseptics is common practice by many female sex workers, who often believe this may protect them against pregnancy, STIs or HIV (Fonck *et al.*, 2001; Reed *et al.*, 2001; Wang *et al.*, 2005). Vaginal drying agents are used in settings where there is a cultural preference for a dry rather than a lubricated vagina (Kun, 1998). However, there is evidence that frequent vaginal douching and vaginal drying agents may increase a woman's susceptibility to sexually transmitted agents (Mann *et al.*, 1988; Brown *et al.*, 1993; Zhang *et al.*, 1997; McClelland *et al.*, 2006). Mineral oil lubricants may affect condom integrity and cause condom rupture (Rojanapithayakorn and Goedken, 1995). Despite this, there is still a substantial proportion of sex workers using oil-based lubricants or no lubricants at all. In a study among male and female sex workers in the Netherlands, 9 percent used oil or Vaseline™ as a lubricant (de Graaf *et al.*, 1993). In a study in Jakarta, Indonesia,

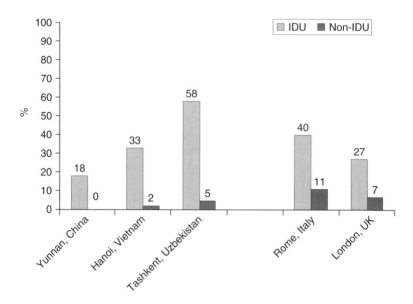

Figure 14.2 HIV prevalence in injecting drug using (IDU) and non-IDU sex workers in different settings.

only 13 percent of the transgender and 15 percent of the male sex workers had ever used water-based lubricant gel (Pisani *et al.*, 2004).

In some geographic areas, sharing unclean needles is a more important route of HIV transmission among sex workers than is unprotected sex. In Asia, Europe, North America and Latin America, there is a substantial overlap between injecting drug use (IDU) and commercial sex. Prevalence studies among sex workers in these regions have shown that HIV infection is much higher in IDU than in non-IDU sex workers, as shown in Figure 14.2. In addition to the HIV risk caused by sharing unclean equipment, one study showed that drug-using sex workers were about half as likely to use condoms as those who didn't use drugs (MAP, 2005).

Typology of sex work correlates with risk of HIV, and therefore is useful for program planners. When comparing direct versus indirect sex workers, registered versus clandestine sex workers, risk profiles differ considerably. A study in Vietnam showed that direct sex workers had more clients, reported less consistent condom use and more previous STIs, and had more IDU clients, and were thus more at risk for HIV infection than indirect sex workers (Minh *et al.*, 2004). In Senegal, unregistered sex workers had fewer clients, and a lower HIV prevalence but a higher syphilis and gonorhoeae prevalence as compared to registered sex workers (Laurent *et al.*, 2003). In Accra in 1997–1999, the prevalence of HIV among female sex workers was nearly 50 percent, varying from 26 percent among street-based sex workers to 74 percent among home-based sex workers working in red-light areas (Asamoah-Adu *et al.*, 2001). The prevalence of HIV among brothel-based compared to non-brothel-based female sex workers was 47 percent and 13 percent in a study in Thailand (Limpakarnjanarat *et al.*, 1999). Understanding those different types of sex work allows program managers to better target the interventions, giving priority to those at highest risk.

Vulnerability of sex workers

Different levels of economic need and working circumstances affect the sex workers' autonomy and ability to respond to health promotion messages (Harcourt and Donovan, 2005). Some of the sex workers in the informal or illegal settings operate in a covert or clandestine fashion, making them the hardest to reach with interventions, as they are scattered and usually deny involvement in sex work. Stigmatization, criminalization and violence, debt and financial need, and exploitation further contribute to the vulnerability of sex workers. Therefore, it is important to understand sex-work settings, the power structures surrounding it, and sex workers' needs when designing sex-worker interventions.

Sex workers are usually a stigmatized and marginalized group in society. They are often blamed for the breakdown of the traditional family, epidemics of STIs and HIV/AIDS, escalating crime, and the subversion of youth (Rekart, 2005). This results in the social isolation of sex workers, leading to their inability to access legal, social and health services for themselves and their family members. Sex workers often face a judgmental attitude in the service providers, which further decreases access to services. Male and transgender sex workers face a double stigma: that of being a sex worker, and that of being a man who has sex with men. In Spain, for instance, it is known that male sex workers use health services less than female sex workers, and that they show a higher tendency to hide their commercial sexual relations (Belza, 2005).

Sex work, or some aspect of it (such as soliciting), is illegal in many countries, resulting in the criminalization of sex workers (Vanwesenbeeck, 2001; UNAIDS, 2002). Even if prostitution is legal, sex workers can be treated as criminals. Criminalization leads to violence from clients, pimps and local goons, and police harassment. This may take the form of bribes to let off detained sex workers, and violence – including sexual violence and even gang rape at police stations. In Moscow, women are arrested under administrative codes for "petty hooliganism" or for not possessing the correct documents (Platt, 1998). This system is open to abuse by the police, who use the ambiguity of the legislation to enrich themselves financially through bribes, or by taking sexual services (Dehne *et al.*, 2000). Violence is found throughout the industry, and sometimes takes extreme forms. Sex workers around the world continue to be murdered, including about six each year in the United Kingdom (Goodyear and Cusick, 2007). Criminalization of sex work further leads to reduced access to services, poor self-esteem, drug use, loss of family and friends, and restrictions on travel, employment, housing and parenting (UNAIDS, 2002).

Risk-taking in sex workers has been statistically correlated with financial need (Van Wesenbeeck *et al.*, 1994). Young people sometimes enter sex work to support their families, but soon acquire personal debts for transportation, accommodation, clothes, cosmetics, condoms, food, medical care, fines, alcohol and injecting drugs. Sex workers in impoverished communities in developing countries are often forced to service large numbers of clients to maximize the profits of pimps and traffickers, or because they are unable to charge an appropriated fee for services (Harcourt and Donovan, 2005). In a study in Bali, Indonesia, the mean number of clients during the preceding week was 16 at low-price brothels, but only 3 in resort areas (Ford *et al.*, 1998). Desperate socio-economic conditions also drive the practice of engaging in unprotected intercourse for extra money. In a study in Kinshasa, DRC, more than one-fourth of the female sex workers (26 percent) engaged in unprotected intercourse for extra money (Ntumbanzondo *et al.*, 2006). These women were also significantly more likely to live or work in non-downtown (lower socio-economic) areas of Kinshasa. Brothels can hold sex workers in debt bondage, allowing them to keep just a small proportion of their earnings (WHO, 2001; UNAIDS, 2002). The *Shakti* program in Bangladesh reported that newcomers to the brothel, called *chukris*, were totally under the control of *sardarnis* – senior madams – and were not allowed to leave the brothel until their period of indenture was complete (Minh *et al.*, 2004). Financial needs may also force sex workers to move on their own to places where they are likely to get more clients. Their high mobility poses yet another challenge in providing them with sustained services.

Child prostitution, human trafficking for sex work, and the abuse of migrant sex workers are important examples of exploitation. Though women and girls in South Asia are trafficked for other purposes besides sexual exploitation, this remains the single largest category of exploitative trafficking crime throughout the world. The majority of victims are young girls and women from poor, illiterate families (UNDP, 2007). UNICEF has estimated that 1 million children enter the sex trade in Asia every year (UNICEF, 2000). Often, impoverished families of victims are tricked or coerced by traffickers, who pay them a fee up front and may promise to find their children respectable jobs as domestic or restaurant workers. A lot of young girls (and boys) are pushed into sex work at an age when they are most vulnerable to STIs and HIV owing to their biological immaturity. Recent studies revealed high HIV prevalence among repatriated sex-trafficked girls and women from Nepal (38 percent) and survivors of trafficking in Mumbai (23 percent) (Silverman, 2006, 2007). Sex tourism may increase the demand, and is often organized through tour operators (Sachs, 1994). Young girls are most in demand by clients, as they are deemed to be free from disease and, in some cultures, there is a misconception that STIs can be cured by having sex with a virgin (WHO, 2001; UNAIDS, 2002; Willis and Levy, 2002). Local traffickers and sex-tour operators are frequently involved with international crime syndicates for whom the sex trade is a major source of income. A human rights report from Thailand documents the role of police and other officials in the illicit sex trade, who go unpunished (Spaid, 1994). A report on girl trafficking in Nepal reveals that in spite of the government being a signatory to the UN convention on suppression of trafficking and local legislation to punish traffickers, widespread trafficking continues, with the disempowered victims discouraged from seeking justice by local politicians themselves involved in the business (Poudel and Carryer, 2000).

Migrant sex workers are at a particular disadvantage compared to local sex workers. Their access to services is restricted by their illegal migration status, language barriers, and lack of freedom when they are bonded (Cwikel *et al.*, 2006). Frequently, the victims are returned to their countries of origin when they are suspected to be HIV-infected. A report from Nepal says that about 60–70 percent of the sex workers returning from India are infected with HIV/STI (Poudel and Carryer, 2000).

Elements of effective interventions

There is ample evidence that targeted programs to reduce transmission of STIs/HIV infection among sex workers are feasible and effective (Laga *et al.*, 1994; Levine *et al.*, 1998; Steen *et al.*, 2000; Ghys *et al.*, 2001, 2002; Alary *et al.*, 2002; Wi *et al.*, 2006). Discussions in the previous parts of this chapter indicate that sex-worker interventions, to be effective, need to address both risk and vulnerability. Therefore, interventions should be implemented both at the individual sex-worker level, to reduce risk of infection, and at higher levels – policy, legal, programs – to reduce vulnerability and support and promote safer sex practices. However, there is no "one size fits all"; programs need to be tailored to the local situation and the sex workers' needs. Global experience suggests that a few elements are common to all effective sex-worker interventions (see Figure 14.3).

Individual sex-worker-level interventions

In order to reduce their individual risk for HIV infection, both prevention and care activities should be offered to sex workers. Prevention and care are most successful if delivered together; this is referred to as the prevention–care synergy (UNAIDS, 2002). Sex-worker interventions that combine risk reduction, condom promotion and improved access to STI treatment are effective in reducing HIV transmission in program participants (Laga *et al.*, 1994; Levine *et al.*, 1998; Steen *et al.*, 2000; Ghys *et al.*, 2001, 2002; Alary *et al.*, 2002; Wi *et al.*, 2006). Therefore, these elements should be part of an essential service package for sex workers. The package, however, may also include HIV counseling and

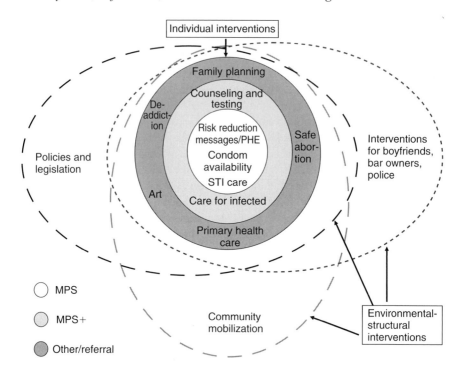

PHE: Peer health education
MPS: Minimum package of services
MPS+: Expanded minimum package of services

Figure 14.3 Elements of effective HIV prevention interventions for sex workers.

testing services, care and support for the HIV-infected, primary health-care services, and family planning and social services, according to local conditions and
needs. A combination of outreach by peer educators and clinic-based services is
an efficient way to deliver this package to the target population.

Outreach by peer education is a commonly used strategy to access high-risk,
hard-to-find populations. In a sex-work setting, peer educators are persons who
have worked as sex workers or are still actively involved in sex work (Rekart,
2005). They are knowledgeable about local conditions which influence work practices, their advice on safe sex practices in commercial sex settings is more acceptable, and it is easier for them to access sex workers – particularly the hard-to-reach
subgroups. As interventions mature and gain the confidence of the community,
these harder-to-reach subpopulations become more accessible. Peer education is
a powerful tool which can bring about individual behavior change and create a
supportive environment for practicing safe sex. Peer health educators may also

accompany sex workers for medical appointments and attend meetings with project staff to plan outreach and related activities, as is the case in Kyrgyzstan (UNAIDS, 2006a). Peer educators need to be trained and given supportive supervision on a regular basis (Brussa, 1998).

Sex-work related prevention messages and associated commodities (condoms and lubricants) are essential components of an effective service package. Prevention messages improve the basic knowledge, skills and practices related to STI/HIV transmission; and teach correct and consistent condom use, the use of lubricants and other safer sex practices, condom negotiation skills (especially with regular clients and boyfriends) and STI care-seeking behavior. Suitable materials adapted to the local sex-worker environment, such as flipcharts and flash cards, are used during interpersonal communication sessions or group education sessions. In London, UK, a narrative-based sex-worker-led, peer education resource tape containing extracts of sex workers talking candidly about their work have been produced (Rickard and Growney, 2001). Media such as theater, video and film also have been used. As an example, a very successful film in HIV education targeting female sex workers, *Amah Djah-foule*, has been produced in the Côte d'Ivoire and been distributed throughout the whole West African region. *Amah Djah-foule* is created in the tradition of a media genre called "entertainment-education" which combines a fictional story with educational content (Widmark, 2002, unpublished report).

Making condoms and lubricants available, and promoting their use, goes hand in hand with prevention messages. Good-quality male condoms should be made available at convenient locations at sex-work sites, as a free supply or socially marketed. There is evidence that water-based lubricants can reduce the incidence of condom breakage, and may be an acceptable and useful method for alleviating problems associated with the regular use of condoms (Voeller *et al.*, 1989; de Graaf *et al.*, 1993; Rojanapithayakorn and Goedken, 1995; Smith *et al.*, 1998). Female condoms, though expensive and not readily available at all places, are a female-controlled method, and as such are acceptable for sex workers, especially in situations where the client or partner refuses to use a male condom. A study in Thailand estimated a 17 percent reduction in the proportion of unprotected sexual acts when giving female sex workers the option of using the female condom if clients refused to use a male condom (Fontanet *et al.*, 1998). In the same study, the female condom proved to be more resistant to tearing than male condoms. Female condoms may also be used by male and transgender sex workers for anal sex (personal communication, BV).

STI management is the other major component of an essential service package for sex workers. Sex workers have high rates of STIs, many of them easily curable with antibiotics, and eliminating STIs reduces the efficiency of HIV transmission in the highest-risk commercial sex contacts – those where condoms are not used (Steen and Dallabetta, 2003). Diagnosis and treatment of symptomatic sex workers on syndromic grounds is probably the most commonly used clinical response (Steen and Dallabetta, 2003). However, infected sex workers, especially women, are frequently asymptomatic; therefore, it is important not to wait for symptoms. Regular screening should be done, involving monthly clinical check-ups and, where available and affordable, screening laboratory tests. Serological screening for syphilis and/or other STIs should be done in situations where these STI are prevalent (Steen *et al.*, 2006). Presumptive treatment of STIs in sex workers, a form of epidemiologic treatment, can be an effective short-term measure to rapidly reduce STI rates. Once prevalence rates have been brought down, however, other longer-term strategies are required (Steen and Dallabetta, 2003). Combinations of the above three approaches are possible that address both symptomatic and asymptomatic sex workers. In India, the India AIDS Initiative (Avahan, see below) supported clinics offer syndromic management, presumptive treatment, regular check-ups and syphilis screening (Steen *et al.*, 2006). Whatever strategy is used for STI diagnosis, health education and counseling services for promoting the 4Cs (consistent condom use, compliance with treatment, contact tracing, and counseling for special issues such as negotiating condom use and coping with incurable infections) should be provided along with STI diagnosis and treatment at the clinic site.

As sex workers are at high risk for HIV, a fair proportion of them are already infected, as is evident from the data given earlier in this chapter. HIV counseling and testing services are often an integral part of a sex-worker-service package, and are proposed at regular intervals, at least for HIV-negative sex workers or those who do not know their status. Knowing their HIV status may motivate sex workers to change or to keep safe sex behavior, and infected sex workers may access supportive counseling, treatment of opportunistic infections and antiretroviral therapy (ART). Providing ART to sex workers has the potential to have a huge impact on the AIDS epidemic, because HIV-infected sex workers have many partners and are frequently involved in high-risk social networks, so it may be possible to reduce sexual transmission risks to others in the community by significantly reducing their HIV load through effective ART. HIV-infected sex workers may be offered ART at the sex-worker clinic as part of the service package, or referred to a specialized treatment center. There are many challenges in providing ART to sex workers. Sex workers are typically mobile and have difficult or no family relationships, making adherence to ART more difficult (CARE International, 2001; McGreevey *et al.*, 2003). Moreover, we still understand relatively little about the way in which ART may influence the preventive behavior of male and female sex workers (see also Chapter 5).

Other health-care services may be part of the service package and contribute to a continuum of care for sex workers, including primary health care, family planning, and safe abortion and de-addiction services. Primary health-care services may increase the acceptability for sex workers of going for regular STI screening (Vuylsteke *et al.*, 2004). The low use of modern contraceptives and the high proportion of female sex workers having used abortion services in many settings, including in Italy, India and Cambodia, translates an unmet need for family planning services into part of a broader reproductive health services for sex workers (Verster *et al.*, 2001; Delvaux *et al.*, 2003; Pal *et al.*, 2003). Acceptability of services is probably best when offered as a comprehensive package in a sex-worker clinic; however, the main components of the essential package – peer prevention education, commodity provision and STI services – should not be overtaken by other services, and referral systems should be put in place to provide for services not available at the clinic.

Sex workers need accessible, acceptable and good-quality medical care, which may be provided either through exclusive sex-worker clinics or through integrated services at primary health-care centers. Whatever model is used, important characteristics of the service center include a suitable location, convenient opening times, affordable or free health care, confidentiality, and a non-judgmental attitude of service provides (Nyamuryekung'e *et al.*, 1997; Vuylsteke *et al.*, 2004). Promotion of clinic services in the community through peer education and strengthening of behavior change, and condom promotion messages in the clinic, further contribute to the prevention–care synergy.

Environmental–structural interventions

Programs targeting the individual risk of sex workers exclusively will have limited impact and sustainability. Environmental–structural interventions aimed at reducing the vulnerability of sex workers are needed to enable such workers to control their risk of infection (Sweat and Denison, 1995; Blankenship *et al.*, 2000; Kerrigan *et al.*, 2006; Sweat *et al.*, 2006). Three environmental–structural interventions have emerged in the context of sex work: community mobilization, interventions for the sex workers' environment, and government policy initiatives.

1. Community mobilization aims at promoting and sustaining safer sex and safe working conditions by increasing sex workers' control of their working environment. Self-organization also helps to overcome the problems of isolation and lack of self-esteem caused by marginalization and stigmatization. Studies have shown that sex workers who have the most control over their working conditions are the least vulnerable to violence, STIs and other health hazards, and have a more balanced view of the dangers faced in their lives (Whittaker and Hart, 1996). The range of services provided by the collectives is varied, including human rights, law reform, welfare services, facilitation for self-help groups in designing and implementing HIV/AIDS prevention programs, and collaboration with service providers to help ensure that sex-work interventions are appropriate. Some sex-worker organizations have evolved into powerful self-advocacy forces which challenge human rights violations and address sex workers' vulnerability. The role of female sex workers' collectives was evaluated in a project in Karnataka, India. A higher degree of collectivization

was associated with increased knowledge and higher reported condom use (Halli *et al.*, 2006). Collectivization had a positive impact in increasing knowledge and in empowering female sex workers to adopt safer sex practices, particularly with commercial clients. Another famous example of community mobilization is the Sonagachi Project in India. This project began in 1992 in a red-light district of Calcutta as a health promotion project to inform sex workers about AIDS, as well as to promote condom use and provide clinical STI services. The project has evolved into a multifaceted community effort to empower women not only to protect themselves from HIV but to also to fulfill their broader needs and aspirations (Population Council, 2002). From 1992 to 1998, rates of consistent condom use with clients jumped from 1 percent to 50 percent, while the prevalence of syphilis dropped from 25 percent to 11 percent, and HIV prevalence rates have remained below 10%.

2. A further enabling environment for reducing vulnerability of sex workers can be created through interventions for second and third parties in their environment, including boyfriends, clients, bar owners, pimps and the police. It has been shown in Cotonou, Benin, that clients of female sex workers are a reachable population, and that interventions targeting these men can also be successful in terms of increasing condom use and decreasing rates of STI (Lowndes *et al.*, 2000). In India, Avahan-supported sex-worker clinics also provide HIV prevention services to regular partners, while franchised private physicians situated around high-risk venues where sex is solicited provide STI services to men. Sex establishment owners and managers, such as pimps and madams, play a gatekeeper role, and hence interventions to promote safer sex need to be supported by the management (The Synergy Project, 2000). This can be accomplished by involving them through regular meetings and social events.

   Liaison between police and sex-work projects can have a number of benefits for sex workers. Advocacy and education activities with police may discourage police raids and violence against sex workers, and prevent project staff from being arrested or harassed. A peer-educator based intervention for police, aimed specifically at reducing the gang rape of sex workers, was launched in mid-1996 as part of a larger intervention with sex workers in Papua Guinea. Total condom use with casual and commercial sex partners rose from 49 percent to 70 percent, and the frequency of gang rape was halved, from 10 percent to 4.8 percent. Among sex workers, total condom use increased from 20 percent to 43 percent (Jenkins, 1997).

3. Legislative changes and the development of policy and frameworks can be part of promoting and creating an environment that reduces stigma and discrimination and supports vulnerability-reduction in sex workers. Legislation and policies can lead to a reduction in vulnerability by addressing access to prevention goods and services, trafficking and sex tourism, thus contributing to a safer sex industry. In a study in the Philippines, female bar workers from establishments where a condom-use policy existed were 2.6 times more likely

to consistently use condoms during sexual intercourse compared with workers from establishments that did not have such a policy in place (Morisky *et al.*, 2002). A famous example of a government policy initiative is the 100 percent condom-use program in Thailand. The key elements for the success of the program were a strong political commitment at both national and local levels; multisectoral cooperation among various sectors, including the involvement of the police and entertainment owners; adequate supply of good-quality condoms for the program; and integration of the program into existing infrastructure. The program has brought down rates of STIs as well as HIV prevalence (Hanenberg and Rojanapithayakorn, 1996; Rojanapithayakorn, 2003).

The WHO describes some of the key principles for HIV prevention, care and empowerment in diverse sex work settings in its Sex Work Toolkit, as illustrated in Table 14.3 (WHO, 2004).

Monitoring and evaluation of sex-worker interventions

Sex-worker interventions, as with any other interventions, require active monitoring that facilitates real-time decision-making to rapidly create an evidence base to guide programming activities, and evaluation to ensure that efforts are having impact. The main objectives of a sex-workers' intervention include increase in condom use and decrease of STI prevalence, which in turn decreases

Table 14.3   Key principles for HIV prevention, care, and empowerment of sex workers

Adopt non-judgmental attitude

Ensure that interventions do no harm

Respect sex workers' right to privacy, confidentiality, and anonymity

Respect sex workers' human rights and accord them basic dignity

Respect sex workers' views, knowledge, and life experiences

Include sex workers and, if appropriate, other community members in all stages of the development and implementation of interventions

Recognize that sex workers are usually highly motivated to improve their health and wellbeing, and that sex workers are part of the solution

Build capacities and leadership among sex workers to facilitate effective participation and community ownership

Recognize the role of clients and third parties in HIV transmission – i.e., targeting the whole sex-work setting, including clients and third parties, rather than only sex workers

Recognize and adapt to the diversity of sex work settings and of participating individuals

Source: WHO (2004).

HIV and STI transmission in all sex worker–client contacts. Therefore, coverage of the intervention, reported consistent condom use and STI/HIV prevalence should be measured at regular intervals in order to measure progress and to evaluate the outcome and impact of a sex-worker intervention.

It is essential to measure coverage, which is the proportion of sex workers contacted through outreach and their resulting use of clinic services. In order to quantify coverage as a percentage of sex workers reached, the numerator can be obtained from outreach reports and clinic registers (Steen *et al.*, 2006), but the denominator, which is the size of the target population, may be more difficult to estimate. The illicit nature of sex work means that people engaged in it frequently form a hidden population. It is often difficult to employ the usual social survey methodologies for obtaining accurate information on prevalence and geographic distribution. Therefore, alternative methods are needed, such as census, multiplier methods and capture–recapture. Most studies in the past have used mapping and census of sex workers to estimate their number (Vandepitte *et al.*, 2006). Census is an effective method if sex workers are well-defined and visible, which is not usually the case. The use of multiplier methods is an alternative strategy, which relies on having information from two sources that overlap in a known way. In Canada, for instance, a combination of HIV serodiagnostic information and data on HIV testing was used to estimate the population of injecting drug users and men who have sex with men (Archibald *et al.*, 2001). Capture–recapture is a more recent technique that was originally used to count and track animal populations. A detailed description of the method can be found elsewhere (Pisani, 2002). It was successfully used to estimate the number of sex workers in Diego-Suarez in Madagascar (Kruse *et al.*, 2003). Mapping and size-estimation should be done at regular intervals because of the fluid nature of sex work, and because, as programs become more trusted in the community, they are able to identify more sex workers. Geographic Information Systems may be an important tool in the future to monitor and analyze the coverage of sex-worker interventions and provide tools to examine spatial distributions and relationships involved in vulnerability to HIV (Welsh *et al.*, 2001; Ferguson and Morris, 2007).

In most surveys, self-reported condom use by sex workers is used as a measure for safe-sex behavior. This may not easily be obtained, especially in resource-poor situations, where low levels of education and high levels of inhibition surrounding sex and sexuality complicate any kind of formal recording. Since most sex workers are unable to recall condom use accurately for long periods, this information should be recorded for a relatively recent period – for example, the most recent sexual contact or the most recent working day for condom use with clients, and the last week for condom use with boyfriends (Ghys *et al.*, 2002). Methods to reduce reporting bias include the use of diaries, polling-box methods, self-administered questionnaires and audio computer-assisted interviewing (Tourangeau and Smith, 1996; Des Jarlais *et al.*, 1999; Ferguson *et al.*, 2006; Blanchard *et al.*, 2007). Reported condom use by clients provides another source for measuring safe-sex behavior in sex-worker-clients. Condom use reported by clients tends to be more reliable than self-reports by sex workers (social desirability bias). In Zimbabwe, Gambia and Benin, estimates of condom use by sex workers were 10–24 percent higher than clients' reports (Wilson *et al.*, 1989; Pickering *et al.*, 1993; Lowndes *et al.*, 2000, 2002; Alary *et al.*, 2002).

A primary challenge for surveillance of behavior characteristics and STI/HIV prevalence among sex-worker populations is obtaining "representative" samples, because no sampling frame exists for them. Facility-based surveys are rarely representative for the target population. However, if recruitment methods, questionnaires and STI assessment tools are standardized across surveys at different periods during the project, they may be a very useful way to follow up trends during interventions. For example, three serial cross-sectional studies were conducted among female sex workers in Cotonou, Benin, from 1993 to 1999. Increase in condom use (from 62.2 percent to 80.7 percent) and declining STI prevalence were measured among consecutive clinic attendees (Alary *et al.*, 2002). A similar method was used in La Paz, Bolivia, to measure a declining gonorrhea, syphilis and genital ulcer trend from 1992 through 1995 (Levine *et al.*, 1998).

Several methodologies have been developed to do probability-based sampling when sampling frames cannot be developed – time-location sampling, for instance, takes advantage of the fact that sex workers tend to gather or congregate at certain types of locations such as brothels, massage parlors and street corners in red-light districts (Magnani *et al.*, 2005). A series of female sex workers' population-based behavioral studies were carried out using time-location sampling, from 1991 to 1998, in Abidjan, Côte d'Ivoire (Ghys *et al.*, 2002). Sex-worker sites were selected by random cluster sampling proportional to the estimated number of female sex workers working at the site. Reported condom use with the most recent client increased from 63 percent in 1991 to 91 percent in 1997. Another relatively recent strategy is respondent-driven sampling. Respondent-driven sampling begins with a set number of non-randomly selected seeds (members of the target population). Seeds recruit their peers, who in turn recruit their peers in the study. This gives through successive waves of recruits which, it is argued, become increasingly more representative of the underlying population as the recruitment progresses (Simic *et al.*, 2006). Respondent-driven sampling was used to reach sex workers inaccessible by other sampling methods in various settings, including Vietnam, Papua New Guinea, Serbia, Montenegro and the Russian Federation (Johnston *et al.*, 2006; Simic *et al.*, 2006; Yeka *et al.*, 2006).

The intervention gap and the need to scale up sex-worker interventions

There is extensive global experience in HIV prevention among sex workers, and an essential package of proved effective interventions has been defined. Most interventions in sex workers, despite being effective, have had little impact on HIV transmission dynamics in countries simply because they are implemented on such a small scale that most sex workers who need prevention services do not have them available. Addressing this "prevention gap" is the major challenge in HIV prevention in sex workers. Small-scale successful interventions should now be translated into large-scale public health interventions at city, district or country level. Important issues for up-scaling arise, including maintaining the quality of interventions, sustainability, service delivery models, and key actors.

Maintaining the quality of interventions while up-scaling is one of the first challenges. Some argue that a slight drop-off in quality may be an inevitable stage of the process – that is, that in the interest of reaching a greater number of people, some sacrifice of quality is acceptable. The question then becomes whether the quality drops below an "acceptable" level, and indeed whether such a level can be specified. In this perspective, it is important to define an essential service package, to develop clear standards and tools, and to provide the necessary support to enable partners to reach and maintain the quality level in different settings.

Sustainability is yet another issue. A program must be built with sufficient financial, technical, social and political support so that it lasts over time. Both programmatic and organizational sustainability need to be considered. In order to obtain financial sustainability, an increased level of donor support is often necessary to increase the scale of operations. A study among 15 sex-worker programs in India showed that the cost per sex worker served decreases with an increasing number of sex workers served annually, but this has to be weighed against an associated modest trend of decrease in time spent with each sex worker in some programs (Dandona *et al.*, 2005). Another cost calculation in India resulted in an average cost varying with the scale of the project, and the authors conclude that estimates of resource requirement based on a constant average cost could underestimate or overestimate total costs (Guinness *et al.*, 2005). At a political level, sustainability depends on the degree to which the political climate is one that discourages or encourages HIV prevention among marginal groups such as sex workers. Forging strategic links between the government and implementing NGOs may contribute to the sustainability of the program.

Whether up-scaling of quality services for female sex workers is better obtained through specialized (vertical) clinics or through integrated services remains an important question. Sex workers have some specific needs regarding STI case management, preventive messages and condom promotion. Conventional health-care facilities may fail to meet these needs for several reasons, including a prejudicial attitude of health-care providers and ignorance about sex workers' problems. Specialized health services or "sex-worker only" clinics offer better opportunities for targeted educational sessions and health promotion, but they may not cover the whole target population and might not be very sustainable because they are too expensive (Vuylsteke *et al.*, 2004). There is therefore a need for alternative models, which integrate a comprehensive package of services for sex workers into the primary health-care system. In Côte d'Ivoire, a service package for sex workers is offered in primary health-care centers. Preliminary evaluation results indicate that the model is well accepted by the target population, and much less expensive than a dedicated sex-worker-only clinic (Bamba, 2007, unpublished report).

Another question concerns who should carry out up-scaling activities. Non-governmental organizations (NGOs) have traditionally provided services to marginalized communities, and are probably still best placed to help support outreach, peer education and condom promotion. However, as the medical services provided to sex workers widen, from STI to HIV care, NGOs will increasingly need to work in partnership with governments. In India, the Avahan project works through state-level NGO partners, which in turn contract with local NGOs to organize peer outreach, community mobilization and dedicated clinics for sex workers (Steen *et al.*, 2006). Linkages are made with government services to provide HIV testing, TB screening, HIV care and other services.

Capacity-building of implementing agencies is an essential part of obtaining both quality and sustainability of sex-worker programs. Although NGOs have traditionally played an important role in HIV prevention activities, few have the managerial and technical capacity necessary for effective implementation and expansion of sex-worker interventions (Steen *et al.*, 2006). Initial capacity-building involves both training technical staff at service centers and training community workers, including peer health educators. Other capacity-building includes increasing the organizational and managerial skills of implementing agencies, as well as their monitoring and evaluation capacities.

Model programs from around the world

In this section, sex-worker interventions from around the world are presented as an illustration of the principles in this chapter – namely, the implementation of the essential package of prevention, addressing structural and environmental constraints and increasing coverage to close the prevention gap.

Avahan, the India AIDS initiative

India has an estimated 2.5 million persons living with HIV, with six states accounting for 70 percent of the total cases (UNAIDS, 2006b). The heterogeneous epidemic is driven by sex work in four southern states and injecting drug use in two states in the north-east. The India AIDS Initiative (Avahan), funded by the Bill and Melinda Gates Foundation, is scaling up interventions with sex workers and other high-risk populations in India's six highest HIV-prevalence states. The intervention elements are consistent with India's Targeted Intervention strategy, and include condom promotion, STI treatment and behavior-change activities delivered through peer education, community centers, STI clinics (including primary care and basic HIV management) and linkages to government services. Avahan has a strong focus on social interventions for marginalized populations, including community mobilization efforts and advocacy at all levels to address structural and environmental barriers. After starting in 2004, by February 2007 there were 398 clinics and 545 community centers established in 571 towns or cities, and 7022 outreach workers/peer educators and 383 physicians were working within the program. Avahan provides HIV prevention services to about 290,000 female sex workers, high-risk men who have sex with men, and injection drug users, via more than 130 grass-roots NGOs. The program also provides prevention services to 6 million male clients in high-risk venues and across national highways in 120 high-risk locations in urban and peri-urban areas (personal communication, GD). Evaluation of the intervention is in accordance with different population needs, the density of the high-risk population, iteratively standardizing guidelines to define common standards for all NGOs, and systematic program and project management at multiple levels. Avahan, in conjunction with other partners in these states, has brought coverage of female sex workers to over 80 percent in the four high-prevalence states in the south, closing the prevention gap and demonstrating that prevention can be done at scale.

Project PAPO-HV, Côte d'Ivoire

Côte d'Ivoire is one of the countries in the West African region most affected by HIV/AIDS (UNAIDS, 2006b). In 1991, a prevention program directed to female sex workers, the Programme de Prévention et de Prise en charge des MST/SIDA chez les femmes libres et leurs Partenaires (PPP), was initiated by the National AIDS Control Program in three districts of the capital city, Abidjan. It was in this context that the Confidential Clinic, or the *Clinique de Confiance*, was created in November 1992 as a collaboration between the Centers for Disease Control and Prevention (CDC), the Côte d'Ivoire Ministry of Public Health, and the Institute of Tropical Medicine (ITM) in Antwerp, Belgium. It is an outpatient clinic that discreetly offers STI/HIV prevention and care to sex workers and their regular partners in Abidjan. From 1992 to 2001, condom use with clients increased from 20 percent to 85 percent, and HIV prevalence among clinic attendees decreased from 89 percent to 30 percent (Ghys *et al.*, 2002). The clinic has become an internationally known model and inspiration for other sex-worker interventions around the world (personal communication, ML).

In 2004, the clinic was also the model for broadening the scope of HIV/AIDS interventions in Côte d'Ivoire by way of the PAPO-HV Project (Assistance Project for Highly Vulnerable Populations). This project is financed by the President's Emergency Plan for AIDS Relief (PEPFAR) and implemented by Family Health International and its partner ITM. *Papo* in the local language is a word used to describe the protective covering or roof of a house. Seven prevention and care centers for sex workers in Abidjan, San Pedro, Gagnoa, Guiglo, Bouaké and Yamoussoukro are now fully operational. The PAPO-HV Project collaborates with local non-governmental organizations as actors for the implementation of the project. Services are free of charge, and include behavior change messages, condom promotion, STI treatment, HIV counseling and testing, and care and treatment for HIV-positive sex workers and their families. Prevention activities at the center and in the field are supported by local peer educators. As of October 2007, the PAPO-HV Project has provided prevention and care services to over 10,000 sex workers, 293 of who are now currently on ARV treatment (personal communication, BV).

TAMPEP, Europe

In Austria, Germany, Italy and the Netherlands, the Transnational AIDS/STD Prevention Among Migrant Prostitutes in Europe Project (TAMPEP) is working with 23 groups of female and transgender sex workers who have migrated from Africa, Eastern Europe, Latin America and Southeast Asia. TAMPEP uses cultural mediators and peer educators, and also offers prostitutes seminars, workshops and other field activities to empower them and create an environment that supports safer sex behavior. Because sex workers migrate, new peer educators are continuously trained. The most successful peer educators are leaders of their target group; exhibit some knowledge of health, educational talents, and excellent communication skills; and are highly ambitious and motivated. TAMPEP spends 2–3 months selecting, training and following up peer educators. Peer educators receive a small fee while undergoing training, and they participate in course design. The peer educators receive a certificate upon completion of the course. Cultural mediators conduct follow-up by supporting the peer educators as they assume their new role, providing additional information and materials, and facilitating contacts with public health personnel. Lessons learned during the program's first 5 years reveal that peer education programs should: (1) be part of a broader effort to improve conditions for migrant sex workers, (2) be conducted by autonomous community-based organizations, and (3) continuously adapt to change (Brussa, 1998).

COIN and CEPROSH, Dominican Republic

In 1995, two Dominican NGOs – Centro de Orientacion e Investigacion Integral (COIN), based in Santo Domingo, and Centro de Promocion y Solidaridad Humana (CEPROSH), based in Puerto Plata – began exploring the possibility of adapting elements of the Thai 100% condom program to the Dominican context, where peer education among sex workers began in the late 1980s. Both qualitative and quantitative research was conducted in Santo Domingo from 1996 to 1998 to help inform the process of adapting elements of the Thai model to the Dominican context. On the basis of the results of these studies, and through consultations with the local sex-worker organization, Moviemento de Mujeres Unidas (MODEMU), two environmental–structural intervention models were developed. Components of the intervention included:

1. Solidarity and collective commitment through quarterly workshops and monthly follow-up meetings with sex workers, establishment owners and managers, and other establishment employees.
2. Environmental cues through 100% condom posters, stickers, availability of condoms, disc-jockey messages, information booths, educational materials and interactive theater presentations.
3. Strengthening clinical services for sex workers through training of clinicians and inspectors.
4. Monitoring and encouraging adherence on a monthly basis.
5. Policy and regulation – a regional government policy requiring condom use between sex workers and clients was communicated to all participating sex establishment owners.

The latter component was implemented in Puerto Plata only because the political leadership necessary to implement a policy-level intervention was present in that city at the time of the study. After 1 year, significant increases in condom use with new clients (75 percent to 94 percent) were documented in Santo Domingo. In Puerto Plata, significant increases in condom use with regular partners (13 percent to 29 percent) and reductions in STI prevalence (29 percent to 16 percent) were documented, as were significant increases in sex workers' verbal rejections of unsafe sex (50 percent to 79 percent) and participating sex establishment's ability to achieve the goal of no STIs in routine monthly screenings of sex workers. These results show the importance of an integrated approach in mobilizing both communities and governments to confront HIV and STI-related vulnerability in the context of sex work (Kerrigan *et al.*, 2006).

Challenges for setting up HIV prevention programs for sex workers

There is no single, universal model for providing HIV prevention services to sex workers as illustrated in this chapter. The use of peer educators has been recognized as an effective strategy for reaching sex workers, particularly the hard-to-reach subgroups. An essential package for reducing risk in sex-worker interventions combines behavior-change messages, condom promotion and improved access to STI treatment. Water-based lubricants not only prevent vaginal irritation, but also decrease condom breakage. Regular STI screening should be performed, and presumptive treatment for STI should be applied in settings where STI prevalence is high. Community mobilization and involvement of sex workers in design and implementation of projects further contribute to controlling their working conditions and hence to reducing vulnerability. The package offered at a sex-worker clinic may be expanded with a range of other services to better meet sex worker's needs, including primary health care, HIV counseling and testing, ART, care and support for the HIV-infected, family planning, safe abortion, and social and legal services. Although the final package will depend on the setting, documentation and evaluation reports of pilot projects may be helpful to guide future sex-worker interventions.

Making this program work successfully on a large scale is one of the major challenges in HIV prevention for sex workers. More specifically, future challenges include increasing uptake and maintaining the quality of services, evaluation and adaptation of standards, and identification of the range of services to offer. An important step in ensuring quality while scaling up prevention programs for sex workers is the establishment of standards (clinical protocols, guidelines, standard operating procedures) as a basis for developing quality indicators. Evaluation of quality of sex-worker interventions should be performed, and networks of sex-worker clinics may be developed using a "quality label" based on an accreditation system in order to guarantee a continuum of quality care for sex workers who travel from one site to another.

HIV prevention programs targeting sex workers are one of the key interventions in reducing HIV spread in almost all settings. There is ample evidence now that targeted programs to reduce the transmission of HIV infection among sex workers are feasible, effective, and have led to successful risk reduction and decreased levels of infection. The challenge globally is to ensure that the new focus on HIV prevention needs includes high coverage of appropriate interventions for sex workers.

References

Aklilu, M., Messele, T., Tsegaye, A. *et al.* (2001). Factors associated with HIV-1 infection among sex workers of Addis Ababa, Ethiopia. *AIDS*, 15, 87–96.

Alary, M., Mukenge-Tshibaka, L., Bernier, F. *et al.* (2002). Decline in the prevalence of HIV and sexually transmitted diseases among female sex workers in Cotonou, Benin, 1993–1999. *AIDS*, 16, 463–70.

Archibald, C. P., Jayaraman, G. C., Major, C. *et al.* (2001). Estimating the size of hard-to-reach populations: a novel method using HIV testing data compared to other methods. *AIDS*, 15(Suppl. 3), S41–8.

Asamoah-Adu, C., Khonde, N., Avorkliah, M. *et al.* (2001). HIV infection among sex workers in Accra: need to target new recruits entering the trade. *J. Acquir. Immune Defic. Syndr.*, 28, 358–66.

Belza, M. J. (2005). Risk of HIV infection among male sex workers in Spain. *Sex. Transm. Infect.*, 81, 85–8.

Blanchard, J. F. (2002). Populations, pathogens, and epidemic phases: closing the gap between theory and practice in the prevention of sexually transmitted diseases. *Sex. Transm. Infect.*, 78(Suppl. 1), i183–8.

Blanchard, J. F., Halli, S., Ramesh, B. M. *et al.* (2007). Variability in the sexual structure in a rural Indian setting: implications for HIV prevention strategies. *Sex. Transm. Infect.*, 83(Suppl. 1), i30–6.

Blankenship, K. M., Bray, S. J. and Merson, M. H. (2000). Structural interventions in public health. *AIDS*, 14(Suppl. 1), S11–21.

Brown, R. C., Brown, J. E. and Ayowa, O. B. (1993). The use and physical effects of intra-vaginal substances in Zairian women. *Sex. Transm. Dis.*, 20, 96–9.

Brussa, L. (1998). Country watch: Europe. *Sex. Health Exch.*, 4, 5–7.

Buvé, A., Caraël, M., Hayes, R. J. *et al.* (2001). Multicentre study on factors determining differences in rate of spread of HIV in sub-Saharan Africa: methods and prevalence of HIV infection. *AIDS*, 15(Suppl. 4), S5–14.

CARE International (2001). A moving target: mobile populations need intense support to combat HIV/AIDS. Atlanta, GA: Position Paper.

Chen, L., Jha, P., Stirling, B. *et al.* (2007). Sexual risk factors for HIV infection in early and advanced HIV epidemics in sub-Saharan Africa: systematic overview of 68 epidemiological studies. *PLoS ONE*, 2, e1001.

Chen, X. S., Yin, Y. P., Liang, G. J. *et al.* (2005). Sexually transmitted infections among female sex workers in Yunnan, China. *AIDS Patient Care STDs*, 19, 853–60.

Cwikel, J. G., Lazer, T., Press, F. and Lazer, S. (2006). Sexually transmissible infections among illegal female sex workers in Israel. *Sex. Health*, 3, 301–3.

Dandona, L., Sisodia, P., Kumar, S. G. *et al.* (2005). HIV prevention programmes for female sex workers in Andhra Pradesh, India: outputs, cost and efficiency. *BMC Public Health*, 5, 98.

de Graaf, R., Vanwesenbeeck, I., van Zessen, G. *et al.* (1993). The effectiveness of condom use in heterosexual prostitution in The Netherlands. *AIDS*, 7, 265–9.

Dehne, K. L., Pokrovskiy, V., Kobyshcha, Y. and Schwartlander, B. (2000). Update on the epidemics of HIV and other sexually transmitted infections in the newly independent states of the former Soviet Union. *AIDS*, 14(Suppl. 3), S75–84.

Delvaux, T., Crabbe, F., Seng, S. and Laga, M. (2003). The need for family planning and safe abortion services among women sex workers seeking STI care in Cambodia. *Reprod. Health Matters*, 11, 88–95.

Desai, V. K., Kosambiya, J. K., Thakor, H. G. *et al.* (2003). Prevalence of sexually transmitted infections and performance of STI syndromes against aetiological diagnosis, in female sex workers of red light area in Surat, India. *Sex. Transm. Infect.*, 79, 111–15.

Des Jarlais, J., Paone, D., Milliken, J. *et al.* (1999). Audio-computer interviewing to measure risk behaviour for HIV among injecting drug users: a quasi-randomised trial. *Lancet*, 353, 1657–61.

Dunkle, K. L., Beksinska, M. E., Rees, V. H. *et al.* (2005). Risk factors for HIV infection among sex workers in Johannesburg, South Africa. *Intl J. STD AIDS*, 16, 256–61.

Ferguson, A. G. and Morris, C. N. (2007). Mapping transactional sex on the Northern Corridor highway in Kenya. *Health Place*, 13, 504–19.

Ferguson, A. G., Morris, C. N. and Kariuki, C. W. (2006). Using diaries to measure parameters of transactional sex: an example from the Trans-Africa highway in Kenya. *Culture Health Sex.*, 8, 175–85.

Fonck, K., Kaul, R., Keli, F. *et al.* (2001). Sexually transmitted infections and vaginal douching in a population of female sex workers in Nairobi, Kenya. *Sex. Transm. Infect.*, 77, 271–5.

Fontanet, A. L., Saba, J., Chandelying, V. *et al.* (1998). Protection against sexually transmitted diseases by granting sex workers in Thailand the choice of using the male or female condom: results from a randomized controlled trial. *AIDS*, 12, 1851–9.

Ford, K. and Wirawan, D. N. (2005). Condom use among brothel-based sex workers and clients in Bali, Indonesia. *Sex. Health*, 2, 89–96.

Ford, K., Wirawan, D. N. and Fajans, P. (1998). Factors related to condom use among four groups of female sex workers in Bali, Indonesia. *AIDS Educ. Prev.*, 10, 34–45.

Freeman, E. E., Orroth, K., White, R. *et al.* (2007). The proportion of new HIV infections attributable to HSV-2 increases over time: simulations of the changing role of sexually transmitted infections in sub-Saharan African HIV epidemics. *Sex. Transm. Infect.*, 83(Suppl. 1), i17–24.

Ghys, P. D., Diallo, M. O., Ettiegne-Traore, V. *et al.* (2001). Effect of interventions to control sexually transmitted disease on the incidence of HIV infection in female sex workers. *AIDS*, 15, 1421–31.

Ghys, P. D., Diallo, M. O., Ettiegne-Traore, V. *et al.* (2002). Increase in condom use and decline in HIV and sexually transmitted diseases among female sex workers in Abidjan, Côte d'Ivoire, 1991–1998. *AIDS*, 16, 251–8.

Global HIV Prevention Working Group (2007). Bringing HIV prevention to scale: an urgent global priority. Available at http://www.globalhivprevention.org/pdfs/PWG-HIV_prevention_report_FINAL.pdf.

Goodyear, M. D. and Cusick, L. (2007). Protection of sex workers. *Br. Med. J.*, 334, 52–3.

Gragnic, G., Julvez, J., Abari, A. and Alexandre, Y. (1998). HIV-1 and HIV-2 seropositivity among female sex workers in the Tenere Desert, Niger. *Trans. R. Soc. Trop. Med. Hyg.*, 92, 29.

Gras, M. J., van der Helm, T., Schenk, R. *et al.* (1997). HIV infection and risk behaviour among prostitutes in the Amsterdam streetwalkers' district; indications of raised prevalence of HIV among transvestites/transsexuals [in Dutch]. *Ned. Tijdschr. Geneeskd.*, 141, 1238–41.

Guinness, L., Kumaranayake, L., Rajaraman, B. *et al.* (2005). Does scale matter? The costs of HIV-prevention interventions for commercial sex workers in India. *Bull. WHO*, 83, 747–55.

Halli, S. S., Ramesh, B. M., O'Neil, J. *et al.* (2006). The role of collectives in STI and HIV/AIDS prevention among female sex workers in Karnataka, India. *AIDS Care*, 18, 739–49.

Hanenberg, R. and Rojanapithayakorn, W. (1996). Prevention as policy: how Thailand reduced STD and HIV transmission. *AIDScaptions*, 3, 24–7.

Harcourt, C. and Donovan, B. (2005). The many faces of sex work. *Sex. Transm. Infect.*, 81, 201–6.

Hunt, C. W. (1989). Migrant labor and sexually transmitted disease: AIDS in Africa. *J. Health Soc. Behav.*, 30, 353–73.

Jenkins, C. (1997). *Final Report to UNAIDS: Police and Sex Workers in Papua New Guinea.* Geneva: Joint United Nations Programme on HIV/AIDS.

Johnston, L. G., Sabin, K., Mai, T. H. and Pham, T. H. (2006). Assessment of respondent driven sampling for recruiting female sex workers in two Vietnamese cities: reaching the unseen sex worker. *J. Urban Health*, 83(Suppl.), i16–28.

Kapiga, S. H., Sam, N. E., Bang, H. *et al.* (2007). The role of herpes simplex virus type 2 and other genital infections in the acquisition of HIV-1 among high-risk women in northern Tanzania. *J. Infect. Dis.*, 195, 1260–9.

Kerrigan, D., Moreno, L., Rosario, S. *et al.* (2006). Environmental-structural interventions to reduce HIV/STI risk among female sex workers in the Dominican Republic. *Am. J. Public Health*, 96, 120–5.

Khromova, Y. Y., Safarova, E. A., Dubovskaya, L. K. *et al.* (2002). High rates of sexually transmitted diseases (STDs), HIV and risky behaviors among female detainees in Moscow, Russia. *XIVth International Conference on AIDS, Barcelona, Spain 6–10 July*, Abstract No. ThPeC7600.

Kruse, N., Behets, F. M., Vaovola, G. *et al.* (2003). Participatory mapping of sex trade and enumeration of sex workers using capture-recapture methodology in Diego-Suarez, Madagascar. *Sex. Transm. Dis.*, 30, 664–70.

Kun, K. E. (1998). Vaginal drying agents and HIV transmission. *Family Plan. Persp.*, 24, 93–4.

Kunawararak, P., Beyrer, C., Natpratan, C. *et al.* (1995). The epidemiology of HIV and syphilis among male commercial sex workers in northern Thailand. *AIDS*, 9, 517–21.

Laga, M., Manoka, A., Kivuvu, M. *et al.* (1993). Non-ulcerative sexually transmitted diseases as risk factors for HIV-1 transmission in women: results from a cohort study. *AIDS*, 7, 95–102.

Laga, M., Alary, M., Nzila, N. *et al.* (1994). Condom promotion, sexually transmitted diseases treatment, and declining incidence of HIV-1 infection in female Zairian sex workers. *Lancet*, 344, 246–8.

Laurent, C., Seck, K., Coumba, N. *et al.* (2003). Prevalence of HIV and other sexually transmitted infections, and risk behaviours in unregistered sex workers in Dakar, Senegal. *AIDS*, 17, 1811–16.

Levine, W. C., Revollo, R., Kaune, V. *et al.* (1998). Decline in sexually transmitted disease prevalence in female Bolivian sex workers: impact of an HIV prevention project. *AIDS*, 12, 1899–906.

Limpakarnjanarat, K., Mastro, T. D., Saisorn, S. *et al.* (1999). HIV-1 and other sexually transmitted infections in a cohort of female sex workers in Chiang Rai, Thailand. *Sex. Transm. Infect.*, 75, 30–5.

Lowndes, C. M., Alary, M., Gnintoungbe, C. A. *et al.* (2000). Management of sexually transmitted diseases and HIV prevention in men at high risk: targeting clients and non-paying sexual partners of female sex workers in Benin. *AIDS*, 14, 2523–34.

Lowndes, C. M., Alary, M., Meda, H. *et al.* (2002). Role of core and bridging groups in the transmission dynamics of HIV and STIs in Cotonou, Benin, West Africa. *Sex. Transm. Infect.*, 78(Suppl. 1), i69–77.

Magnani, R., Sabin, K., Saidel, T. and Heckathorn, D. (2005). Review of sampling hard-to-reach and hidden populations for HIV surveillance. *AIDS*, 19(Suppl. 2), S67–72 Pisani, E. (2002).

Mann, J. M., Nzilambi, N., Piot, P. *et al.* (1988). HIV infection and associated risk factors in female prostitutes in Kinshasa, Zaire. *AIDS*, 2, 249–54.

MAP (2005). *Sex Work and HIV/AIDS in Asia*. Map Report, 2005.

McClelland, R. S., Lavreys, L., Hassan, W. M. *et al.* (2006). Vaginal washing and increased risk of HIV-1 acquisition among African women: a 10-year prospective study. *AIDS*, 20, 269–73.

McGreevey, W., Alkenbrack, S. and Stover, J. (2003). Construction workplace interventions for prevention, care and support and treatment of HIV/AIDS. In: J. P. Moatti, B. Coriat, Y. Souteyrand *et al.* (eds), *Economics of AIDS and Access to HIV/AIDS Care in Developing Countries. Issues and Challenges*. Paris: ANRS, pp. 547–63.

Mgone, C. S., Passey, M. E., Anang, J. *et al.* (2002). Human immunodeficiency virus and other sexually transmitted infections among female sex workers in two major cities in Papua New Guinea. *Sex. Transm. Dis.*, 29, 265–70.

Minh, T. T., Nhan, D. T., West, G. R. *et al.* (2004). Sex workers in Vietnam: how many, how risky? *AIDS Educ. Prev.*, 16, 389–404.

Morisky, D. E., Pena, M., Tiglao, T. V. and Liu, K. Y. (2002). The impact of the work environment on condom use among female bar workers in the Philippines. *Health Educ. Behav.*, 29, 461–72.

Morison, L., Weiss, H. A., Buvé, A. *et al.* (2001). Commercial sex and the spread of HIV in four cities in sub-Saharan Africa. *AIDS*, 15(Suppl. 4), S61–9.

Morris, C. N. and Ferguson, A. G. (2006). Estimation of the sexual transmission of HIV in Kenya and Uganda on the trans-Africa highway: the continuing role for prevention in high risk groups. *Sex. Transm. Infect.*, 82, 368–71.

Morse, E. V., Simon, P. M., Osofsky, H. J. *et al.* (1991). The male street prostitute: a vector for transmission of HIV infection into the heterosexual world. *Soc. Sci. Med.*, 32, 535–9.

Morton, A. N., Tabrizi, S. N., Garland, S. M. *et al.* (2002). Will the legalisation of street sex work improve health? *Sex. Transm. Infect.*, 78, 309.

Nguyen, V. T., Nguyen, T. L., Nguyen, D. H. *et al.* (2005). Sexually transmitted infections in female sex workers in five border provinces of Vietnam. *Sex. Transm. Dis.*, 32, 550–6.

Ntumbanzondo, M., Dubrow, R., Niccolai, L. M. *et al.* (2006). Unprotected intercourse for extra money among commercial sex workers in Kinshasa, Democratic Republic of Congo. *AIDS'Care*, 18, 777–85.

Nyamuryekung'e, K., Laukamm-Josten, U., Vuylsteke, B. *et al.* (1997). STD services for women at truck stop in Tanzania: evaluation of acceptable approaches. *East Afr. Med. J.*, 74, 343–7.

Nzila, N., Laga, M., Thiam, M. A. *et al.* (1991). HIV and other sexually transmitted diseases among female prostitutes in Kinshasa. *AIDS*, 5, 715–21.

Pal, D., Raut, D. K. and Das, A. (2003). A study of HIV/STD infections amongst commercial sex workers in Kolkata (India). Part I: some socio-demographic features of commercial sex workers. *J. Commun. Dis.*, 35, 90–5.

Pickering, H., Quigley, M., Hayes, R. J. *et al.* (1993). Determinants of condom use in 24,000 prostitute/client contacts in The Gambia. *AIDS*, 7, 1093–8.

Pisani, E. (2002). Estimating the size of populations at risk for HIV. Issues and methods. UNAIDS/WHO Working Group on HIV/AIDS/STI Surveillance, UNAIDS/ 03.36E. Available at http://www.fhi.org/NR/rdonlyres/e66sj52tyha7m5dozchbbwp-txepfqr47vmvjxvyo2dy7trd2ne5giddtvkksddwrpwxatdgkprxwba/EstimatingSizePop.pdf.

Pisani, E., Girault, P., Gultom, M. *et al.* (2004). HIV, syphilis infection, and sexual practices among transgenders, male sex workers, and other men who have sex with men in Jakarta, Indonesia. *Sex. Transm. Infect.*, 80, 536–40.

Platt, L. (1998). Profile of Sex Workers in Moscow. Moscow: AIDS Infoshare.

Plummer, F. A., Nagelkerke, N. J., Moses, S. *et al.* (1991). The importance of core groups in the epidemiology and control of HIV-1 infection. *AIDS*, 5(Suppl. 1), S169–76.

Population Council (2002). The Sonagachi project: a global model for community development. *Horizons Report*, May, p.7.

Poudel, P. and Carryer, J. (2000). Girl-trafficking, HIV/AIDS, and the position of women in Nepal. *Gender Dev.*, 8, 74–9.

Reed, B. D., Ford, K. and Wirawan, D. N. (2001). The Bali STD/AIDS Study: association between vaginal hygiene practices and STDs among sex workers. *Sex. Transm. Infect.*, 77, 46–52.

Rekart, M. L. (2005). Sex-work harm reduction. *Lancet*, 366, 2123–34.

Rickard, W. and Growney, T. (2001). Occupational health and safety amongst sex workers: a pilot peer education resource. *Health Educ. Res.*, 16, 321–33.

Rojanapithayakorn, W. (2003). The 100 per cent condom use programme: a success story. *J. Health Management*, 5, 225–35.

Rojanapithayakorn, W. and Goedken, J. (1995). Lubrication use in condom promotion among commercial sex workers and their clients in Ratchaburi, Thailand. *J. Med. Assoc. Thai.*, 78, 350–4.

Rojanapithayakorn, W. and Hanenberg, R. (1996). The 100% condom program in Thailand. *AIDS*, 10, 1–7.

Rottingen, J. A., Cameron, D. W. and Garnett, G. P. (2001). A systematic review of the epidemiologic interactions between classic sexually transmitted diseases and HIV: how much really is known? *Sex. Transm. Dis.*, 28, 579–97.

Sachs, A. (1994). The last commodity: child prostitution in the developing world. *World Watch*, 7, 24–30.

Sanchez, J., Gotuzzo, E., Escamilla, J. *et al.* (1998). Sexually transmitted infections in female sex workers: reduced by condom use but not by a limited periodic examination program. *Sex. Transm. Dis.*, 25, 82–9.

Sethi, G., Holden, B. M., Gaffney, J. *et al.* (2006). HIV, sexually transmitted infections, and risk behaviours in male sex workers in London over a 10-year period. *Sex. Transm. Infect.*, 82, 359–63.

Silverman, J. G., Decker, M. R., Gupta, J. *et al.* (2006). HIV prevalence and predictors among rescued sex-trafficked women and girls in Mumbai, India. *J. Acquir. Immune Defic. Syndr.*, 43, 588–93.

Silverman, J. G., Decker, M. R., Gupta, J. *et al.* (2007). HIV prevalence and predictors of infection in sex-trafficked Nepalese girls and women. *J. Am. Med. Assoc.*, 298, 536–42.

Simic, M., Johnston, L. G., Platt, L. *et al.* (2006). Exploring barriers to "respondent driven sampling" in sex worker and drug-injecting sex worker populations in Eastern Europe. *J. Urban Health*, 83(Suppl.), i6–15.

Simonsen, J. N., Plummer, F. A., Ngugi, E. N. *et al.* (1990). HIV infection among lower socioeconomic strata prostitutes in Nairobi. *AIDS*, 4, 139–44.

Siraprapasiri, T., Thanprasertsuk, S., Rodklay, A. *et al.* (1991). Risk factors for HIV among prostitutes in Chiangmai, Thailand. *AIDS*, 5, 579–82.

Smith, A. M., Jolley, D., Hocking, J. *et al.* (1998). Does additional lubrication affect condom slippage and breakage? *Intl. J STD AIDS*, 9, 330–5.

Spaid, E. L. (1994). Thai police blamed in illicit sex trade. Human rights report. *Christian Science Monitor*, 31 January, 4. Available at http://www.popline.org/docs/1092/092891.html.

Steen, R. and Dallabetta, G. (2003). Sexually transmitted infection control with sex workers: regular screening and presumptive treatment augment efforts to reduce risk and vulnerability. *Reprod. Health Matters*, 11, 74–90.

Steen, R., Vuylsteke, B., DeCoito, T. *et al.* (2000). Evidence of declining STD prevalence in a South African mining community following a core-group intervention. *Sex. Transm. Dis.*, 27, 1–8.

Steen, R., Mogasale, V., Wi, T. *et al.* (2006). Pursuing scale and quality in STI interventions with sex workers: initial results from Avahan India AIDS Initiative. *Sex. Transm. Infect.*, 82, 381–5.

Sweat, M. D. and Denison, J. A. (1995). Reducing HIV incidence in developing countries with structural and environmental interventions. *AIDS*, 9(Suppl. A), S251–7.

Sweat, M., Kerrigan, D., Moreno, L. *et al.* (2006). Cost-effectiveness of environmental-structural communication interventions for HIV prevention in the female sex industry in the Dominican Republic. *J. Health Commun.*, 11(Suppl. 2), 123–42.

The Synergy Project & University of Washington Center for Health Education and Research (2000). Room for change: preventing HIV transmission in brothels. Transmission Settings, Part III – Brothels. Available at http://www.synergyaids.com/documents/Submodulebrothel.pdf).

Thomas, J. C. and Tucker, M. J. (1996). The development and use of the concept of a sexually transmitted disease core. *J. Infect. Dis.*, 174(Suppl. 2), S134–43.

Todd, C. S., Khakimov, M. M., Alibayeva, G. *et al.* (2006). Prevalence and correlates of human immunodeficiency virus infection among female sex workers in Tashkent, Uzbekistan. *Sex. Transm. Dis.*, 33, 496–501.

Tourangeau, R. and Smith, T. W. (1996). Asking sensitive questions: the impact of data collection mode, question format, and question context. *Public Opin. Q.*, 60, 275–304.

Tran, T. N., Detels, R., Long, H. T. *et al.* (2005). HIV infection and risk characteristics among female sex workers in Hanoi, Vietnam. *J. Acquir. Immune Defic. Syndr.*, 39, 581–6.

UNAIDS (2002). *Sex Work and HIV/AIDS*. Geneva: Joint United Nations Programme on HIV/AIDS.

UNAIDS (2006a). *HIV and Sexually Transmitted Infection Prevention Among Sex Workers in Eastern Europe and Central Asia*. Geneva: Joint United Nations Programme on HIV/AIDS.

UNAIDS (2006b). *Report on the Global AIDS Epidemic 2006*. Geneva: Joint United Nations Programme on HIV/AIDS, 2002.

UNDP (2007). Human Trafficking and HIV. Exploring Vulnerabilities and Responses in South Asia. Report 2007. New York, NY: UNDP.

UNICEF (2000). East Asia and Pacific. Children on the edge. Available at http://www.unicef.org/vietnam/resources_896.html.

van den Hoek, A., Yuliang, F., Dukers, N. H. *et al.* (2001). High prevalence of syphilis and other sexually transmitted diseases among sex workers in China: potential for fast spread of HIV. *AIDS*, 15, 753–9.

Vandepitte, J., Lyerla, R., Dallabetta, G. *et al.* (2006). Estimates of the number of female sex workers in different regions of the world. *Sex. Transm. Infect.*, 82(Suppl. 3), iii18–25.

Vanwesenbeeck, I. (2001). Another decade of social scientific work on sex work: a review of research 1990–2000. *Annu. Rev. Sex. Res.*, 12, 242–89.

Vanwesenbeeck, I., van Zessen, G., de Graaf, R. and Straver, C. J. (1994). Contextual and interactional factors influencing condom use in heterosexual prostitution contacts. *Patient Educ. Couns.*, 24, 307–22.

Verster, A., Davoli, M., Camposeragna, A. *et al.* (2001). Prevalence of HIV infection and risk behaviour among street prostitutes in Rome, 1997–1998. *AIDS Care*, 13, 367–72.

Voeller, B., Coulson, A. H., Bernstein, G. S. and Nakamura, R. M. (1989). Mineral oil lubricants cause rapid deterioration of latex condoms. *Contraception*, 39, 95–102.

Vuylsteke, B., Traore, M., Mah-Bi, G. *et al.* (2004). Quality of sexually transmitted infections services for female sex workers in Abidjan, Côte d'Ivoire. *Trop. Med. Intl Health*, 9, 638–43.

Wang, B., Li, X., Stanton, B. *et al.* (2005). Vaginal douching, condom use, and sexually transmitted infections among Chinese female sex workers. *Sex. Transm. Dis.*, 32, 696–702.

Wasserheit, J. N. and Aral, S. O. (1996). The dynamic topology of sexually transmitted disease epidemics: implications for prevention strategies. *J. Infect. Dis.*, 174(Suppl. 2), S201–13.

Welsh, M. J., Puello, E., Meade, M. *et al.* (2001). Evidence of diffusion from a targeted HIV/AIDS intervention in the Dominican Republic. *J. Biosoc. Sci.*, 33, 107–19.

WHO (2001). Sex Work in Asia. Geneva: World Health Organization.

WHO (2004). Sex Work Toolkit – targeted HIV/AIDS prevention and care in sex work settings. Available at http://who.arvkit.net/sw/en/contentdetail.jsp?ID=33&d=sw.00.03.

Wi, T., Ramos, E. R., Steen, R. *et al.* (2006). STI declines among sex workers and clients following outreach, one time presumptive treatment, and regular screening of sex workers in the Philippines. *Sex. Transm. Infect.*, 82, 386–91.

Willerford, D. M., Bwayo, J. J., Hensel, M. *et al.* (1993). Human immunodeficiency virus infection among high-risk seronegative prostitutes in Nairobi. *J. Infect. Dis.*, 167, 1414–17.

Willis, B. M. and Levy, B. S. (2002). Child prostitution: global health burden, research needs, and interventions. *Lancet*, 359, 1417–22.

Wilson, D., Chiroro, P., Lavelle, S. and Mutero, C. (1989). Sex worker, client sex behaviour and condom use in Harare, Zimbabwe. *AIDS Care*, 1, 269–80.

Wilson, D., Dubley, I., Msimanga, S. and Lavelle, L. (1991). Psychosocial predictors of reported HIV-preventive behaviour change among adults in Bulawayo, Zimbabwe. *Central Afr. J. Med.*, 37, 196–202.

Whittaker, D. and Hart, G. (1996). Research note: managing risks: the social organization of indoor sex work. *Sociology Health Illness*, 18, 399–414.

Yeka, W., Maibani-Michie, G., Prybylski, D. and Colby, D. (2006). Application of respondent driven sampling to collect baseline data on FSWs and MSM for HIV risk reduction interventions in two urban centres in Papua New Guinea. *J. Urban Health*, 83(Suppl.), i60–72.

Zhang, J., Thomas, A. G. and Leybovich, E. (1997). Vaginal douching and adverse health effects: a meta-analysis. *Am. J. Public Health*, 87, 1207–11.

Interventions with youth in high-prevalence areas

15

Quarraisha Abdool Karim, Anna Meyer-Weitz and Abigail Harrison

Adolescent development and behaviors vary within the time period that defines this stage of development, and differ by sex and setting. Diversity in programs and interventions is thus critical, as is targeting and designing interventions for specific time periods in adolescent transition to adulthood and autonomy. We need to understand better the predictive roles of increasing knowledge, self-efficacy skills, building self-esteem, creating hope in the future and access to services in reducing HIV incidence rates.

Despite the increasing vulnerability of young women for HIV infection, no interventions directed at sexually active young women and or their partner(s) have been developed to date. This represents a significant gap in our efforts to reduce HIV infection in sub-Saharan Africa and specifically in southern Africa.

More than two decades since the first reported cases of AIDS, preventing HIV infection continues to pose a major challenge globally. In southern Africa, where the HIV prevalence and incidence rates have reached unprecedented levels (UNAIDS, 2007a; Rehle *et al.*, 2007), sexual debut, especially for young women, poses a high risk for HIV acquisition (Pettifor *et al.*, 2005), with stable sexual relationships offering little protection. In these settings, traditional approaches to HIV risk-reduction are not adequate, and indeed can be misleading in terms of assessing individual risk.

Notwithstanding heterogeneity in distribution, burden of infection and/or modes of transmission of HIV infection, globally young people are an important focal point in terms of new HIV infections (UNAIDS, 2007a). In resource-constrained settings such as sub-Saharan Africa, they also have the most potential for altering epidemic trajectories that can reduce overall prevalence – as already observed in several countries, including Uganda, Kenya, Tanzania and Zimbabwe (Stoneburner and Low-Beer, 2004; Gregson *et al.*, 2006). This potential can be expanded to other countries by targeting the most at-risk or vulnerable young persons with the right intervention and/or combination of interventions.

The HIV risk in young women in sub-Saharan Africa is not surprising and reflects the significant gender disparities between men and women, with young women being at the lowest rung of social power but the highest risk of HIV (Abdool Karim, 2005). Strong moral and judgmental norms impact the types of interventions being developed and implemented in this target group, rather than evidence such as increasing rates of teenage pregnancies and HIV infection (Pettifor *et al.*, 2005; Department of Health, 2007). Notwithstanding need, there remains a paucity of evidence-based prevention interventions for reducing HIV incidence rates in resource-constrained settings, and specifically in young people at high risk in these settings.

There is growing recognition that, for HIV risk-reduction interventions to be effective, a good understanding of the epidemic for prioritizing the target population(s) in that setting is an essential starting point. Further, interventions that are customized to the needs of the target population are more likely to be effective. Also important is the need to monitor the impact of interventions with biological outcomes and the duration of sustainability of effect (UNAIDS, 2007b).

While biomedical interventions and advances in medical technology hold promise for reducing sexual acquisition of HIV, the majority of these approaches, such as microbicides, antiretroviral drugs for prevention, and vaccines, are still in early stages of clinical testing. Significantly, none of the clinical trials currently underway includes adolescents, despite the overwhelming epidemiological evidence of their HIV risk and their role as significant potential beneficiaries of such successful interventions. The very reasons cited for their exclusion – *viz.* cognitive ability to provide first-person consent for participation, feasibility of accruing and retaining adolescents, implications of physical stage of development or immune immaturity, adherence to product use and study procedures – are the reasons why they need to be included in product development at this early stage.

In contrast, most national AIDS programs in resource-constrained settings include a school based behavioral HIV risk-reduction intervention (UNAIDS, UNICEF & WHO, 2002). There is substantial variability in the target age group, content and desired outcomes, and most behavioral interventions have been developed and implemented with varying degrees of rigor (Paul-Ebhohimhen *et al.*, 2008). The assumptions that underlie these interventions mostly disregard the epidemiological data that unequivocally illustrate that all young people are not at equal risk of acquiring HIV infection, and therefore neither a single "cookie-cutter" approach to HIV risk-reduction nor a purely behavioral or biomedical intervention is likely to be effective.

Effectively addressing the determinants of young people's HIV risk could have a profound impact on adolescent HIV vulnerability and general health.

While there are limited data on protective factors (Meyer-Weitz and Steyn, 1999; UNAIDS, 2004) and few evidence-based interventions targeted at reducing HIV risk in young people in sub-Saharan Africa, there is a substantial body of knowledge that deepens understanding of HIV risk in young people. A review of these data is informative to guide efforts, including the development and testing of new interventions aimed at reducing HIV risk in young people, and young women in particular.

While there is growing recognition of the importance of young women in the epidemic in southern Africa, and the literature underscoring this has increased substantially in the past 2 years (Pettifor *et al.*, 2005; Singh *et al.*, 2006), less is known about how to intervene in this group. The intention of this chapter is to provide critical data and considerations for the design of interventions targeted at this group. It draws on best-practices principles that have evolved regarding how to intervene effectively in HIV prevention. An overview of the HIV epidemic in youth is provided, highlighting the importance of young women in this hyper-endemic setting. Data regarding the current knowledge about HIV risk in young women is presented, followed by a review of behavioral interventions targeted at young people in these settings that highlights the mismatch between young women's HIV risk and interventions being implemented. The chapter concludes with key considerations for the design of effective interventions targeted at this important population.

Epidemiology of HIV infection in youth

Youth are at the center of the HIV pandemic; 25 percent of HIV infections globally are in youth aged 15–24 years (UNAIDS, 2007a). Significantly, almost 50 percent of all *new* HIV infections in 2005 and 2006 occurred in this age group, translating to about 5000–6000 infections daily, or about 2.3 million HIV infections annually (UNAIDS, 2007a). Each year, new cohorts of young people are added to the large numbers of young men and women already at risk of acquiring HIV infection. Thus, there is an urgent need for interventions to minimize the risk of HIV infection in this age group.

The majority of HIV infections globally in young people are sexually acquired, and distributed unevenly across different regions. About two-thirds of all young people with HIV live in sub-Saharan Africa. In Asia, which is home to the next highest burden of HIV infection in young people, most of the infections occur in men with injecting drug use. In high-income countries, young people account for about one-third of new HIV infections, most of whom are young men who have sex with men.

While HIV epidemics in industrialized countries and parts of Asia and Eastern Europe are concentrated in certain subgroups of the population, the epidemic in sub-Saharan Africa is generalized. Southern Africa is at the epicenter of the

HIV pandemic, and has the highest number of people living with HIV/AIDS in the world (UNAIDS, 2007a). Despite efforts to increase access to antiretroviral treatment, AIDS is the leading cause of mortality among both women and men over 30 years of age (STATS SA, 2006). This increase in morbidity and mortality in women in their late twenties and thirties is shifting sexual coupling patterns to younger women, and further exacerbating HIV risk in women under 20 years of age.

In sub-Saharan Africa there are striking age and gender differences in the distribution of HIV infection in young people, reflecting intergenerational sexual coupling patterns (Abdool Karim and Abdool Karim, 2002). In South Africa, more than 20 percent of pregnant women aged 15–24 years attending prenatal clinics in this region are infected with HIV (Department of Health, 2007), and overall, about 76 percent of HIV-infected young people aged 15–24 are women. Young women in this age group are 1.3 to 12 times more likely to be infected than their male counterparts (Pettifor *et al.*, 2005; Rehle *et al.*, 2007; UNAIDS, 2007a).

Many social, economic, biological and political factors contribute to this greater vulnerability of young women in these settings (Strebel, 1996; Campbell and MacPhail, 2002). The immature genital tracts and cervix provide increased opportunity for the transmission of infection (Padian *et al.*, 1991), and other sexually transmitted infections also contribute to their high HIV burden (Wilkinson *et al.*, 2000; Coetzee and Johnson, 2005). Targeting sexually active young women is a high priority for risk-reduction interventions.

Young women and sexual risk

Adolescence refers to the developmental stage that spans the broad period from puberty into young adulthood, and is characterized by physical and emotional maturation. For the purposes of this chapter, we broadly consider young persons aged 10–24 as "adolescents". Additionally, recent global trends toward rising age at marriage and expanded educational and employment opportunities for young people have created an extended period of adolescence in many societies.

The advent of puberty and its associated cognitive, emotional and physiological development, coupled with the emergence of sexual expression and the desire to experiment, general risk-taking, high levels of mobility, instability and change, often lead to heightened vulnerability for young people. Because of their age and immature cognitive development, young people (in particular, young women) are frequently limited in their ability to negotiate sexual relationships and contraceptive use, and may also underestimate their own risk for HIV.

The background prevalence of HIV infection, age of sexual debut, types of relationships and sexual networking patterns, condom use, contraceptive-use patterns, pregnancy, and recreational drug or alcohol use are significant risk factors in this population. Knowledge, skills, self-efficacy, societal norms and values, gender, sexual aggression and coercion impact their ability to reduce their HIV risk. These risk factors and determinants of risk-reduction behaviors in the context of young people in southern and sub-Saharan Africa are elaborated below.

Factors influencing HIV risk

Historically, sexual initiation has been linked to marriage and subsequent child-bearing, and remains a marker of the transition from adolescence to adulthood (Bledsoe and Cohen, 1993). Early age at sexual debut places young women at particular risk for pregnancy and sexually transmitted infections including HIV/AIDS. Early sexual debut is a strong predictor of a life-course with more partners (White *et al.*, 2000; Harrison *et al.*, 2005), and is also more likely to be associated with HIV infection (Hallett *et al.*, 2006). In settings with high rates of HIV infection and generalized HIV epidemics, such as southern Africa, simply becoming sexually active places young people at substantial risk of HIV acquisition. Table 15.1 illustrates how, in one rural community in South Africa, HIV risk in young people under the age of 24 years has increased as the HIV epidemic has evolved and matured, and also shows how risk continues over the life-course of specific birth cohorts.

Throughout sub-Saharan Africa, the average age of sexual initiation falls between the ages of 15 and 17 for both young men and women – slightly older when compared to young men and women in industrialized countries (Eaton *et al.*, 2003; Simbayi *et al.*, 2004). Single-year interval age-disaggregated data from multiple sources demonstrate steadily increasing levels of sexual activity throughout the teen years (Eaton *et al.*, 2003), with younger age of sexual debut associated with greater HIV risk (MacPhail and Campbell, 2001; Brook *et al.*, 2006). A recently conducted national youth survey in South Africa (Figure 15.1) demonstrates that young women are most likely to become infected with HIV between the ages of

Table 15.1 Increase in HIV risk in young people under the age of 24 years in one rural community in South Africa, and risk over the life-course of specific birth cohorts

Age group	1992 (%)	1995 (%)	1998 (%)	2001 (%)
20–24	6.9	21.1	39.3	50.8
25–29	2.7	18.8	36.4	47.2
30–34	1.4	15.0	23.4	38.4
35–39	0.0	3.4	23.0	36.4

Source: Gouws and Abdool Karim (2005); reproduced with permission of Cambridge University Press.

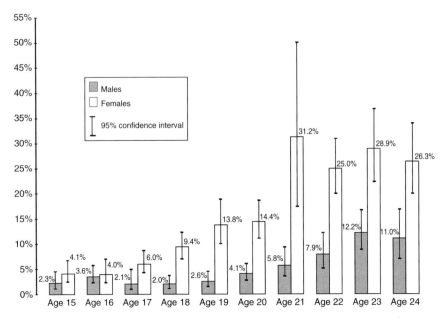

Figure 15.1 HIV prevalence by age and sex among 15- to 24-year-olds. In South Africa, one-third of women are infected with HIV by age 21, with infection most likely to occur between the ages of 18 and 21. Source: Pettifor *et al.* (2005); reproduced with permission of Lippincott Williams & Wilkins.

18–21; indeed by age 21 about a third of young women are already infected with HIV.

Delayed sexual debut

Abstinence, or refraining from sexual activity, is a common HIV/AIDS prevention strategy promoted for children and adolescents (Simbayi *et al.*, 2004). Data from qualitative studies indicate that adolescents prefer abstinence as a prevention strategy when family disapproval of sexual activity outside of marriage is high, but find it difficult to implement or sustain (Harrison *et al.*, 2001).

In practice, abstinence often entails either delaying sexual initiation for a limited period of time or practicing "secondary abstinence" – that is, a prolonged period without sexual activity amongst those who have previously been sexually active (Shisana and Simbayi, 2002; Cleland and Ali, 2006). Evidence for this approach remains sparse despite the widespread implementation of abstinence messages, particularly through life-skills education in schools and faith-based HIV risk-reduction interventions. Additionally, data suggest that in settings where the epidemic is generalized, postponement of sexual initiation simply delays the age of infection but does not in itself reduce rates of infection (Gouws, 2005; Pettifor *et al.*, 2005).

Partnership characteristics and patterns of sexual networking

Partnership characteristics, patterns of sexual networking and the age differences between partners are all embedded in gender-power dynamics that render adolescent women at highest HIV risk in sub-Saharan Africa. The practice of intergenerational sexual coupling between young women and older men, a common pattern of sexual networking in sub-Saharan Africa (Williams *et al.*, 2000; Wellings *et al.*, 2006), has been shown to increase women's risk of HIV infection substantially (Gregson *et al.*, 2002; MacPhail *et al.*, 2002). For example, in Zimbabwe, young women with sexual partners 5 or more years older than themselves were about seven times as likely to be HIV-infected as women with same-age partners (Gregson *et al.*, 2002). These age discrepancies contribute to HIV risk because the older partners might already be HIV-infected, or may engage in concurrent/ multiple partnership patterns that increase their own and partners' HIV risk. In South Africa such sexual partnering is an important factor, contributing to an HIV prevalence of 24.5 percent among young adult women (Pettifor *et al.*, 2005). Young people with multiple or concurrent partners also face increased HIV risk (Morris *et al.*, 1996), although gender and other factors strongly influence this association (Nnko *et al.*, 2004; Pettifor *et al.*, 2005). Concurrent relationships are more common in sub-Saharan Africa than one-time casual encounters, and the average duration of relationships is relatively long, resulting in tightly linked, overlapping sexual networks (Halperin and Epstein, 2004). There is a mismatch between self-report of concurrent partnerships and HIV prevalence. Young men report concurrent relationships about three times more frequently compared to young women (23 percent vs 8.8 percent), but HIV infection in young men remains low (Mnyika *et al.*, 1997; Meyer-Weitz *et al.*, 2003).

Most young men and women who are sexually active report a regular boyfriend or girlfriend but infrequent sexual intercourse (Pettifor *et al.*, 2004; Simbayi *et al.*, 2004). A recent multi-African country study found that half of teen women indicated that their last sexual activity with their current partner occurred longer than a month ago (Cleland and Ali, 2006). Low coital frequency among young people may relate to opportunity, mobility, distance and migration, and difficulties in spending time with partners, and makes understanding HIV acquisition based on mathematical models of transmission probability even more difficult to reconcile (Anderson *et al.*, 1986).

Condom use

Studies conducted in the early 2000s in several African countries demonstrated the steady increase of acceptability and use of condoms by young people (DHS,

2005; Cleland and Ali, 2006), particularly in settings where consistent messages are promoted and support for continued use is provided through access to free condoms (Goldstein *et al.*, 2005; Pettifor *et al.*, 2005).

Within a generalized HIV epidemic such as in southern Africa, very high levels of consistent condom use are required to reduce HIV incidence. A review of HIV risk studies conducted prior to 2000 in southern Africa (Eaton *et al.*, 2003) estimated that only about 20 percent of adolescents use condoms consistently. About 70 percent of youth report having ever used a condom, whereas about 50 percent report use of a condom at last coital encounter (Simbayi *et al.*, 2004; Pettifor *et al.*, 2005; Hoffman *et al.*, 2006; Maharaj and Cleland, 2006).

The partnership type strongly influences condom use, with condoms generally viewed as being less acceptable or desirable within long-term partnerships based on issues of love and trust (Worth, 1989; Rosengard *et al.*, 2001; Abdool Karim and Abdool Karim, 2002), but acceptable in casual relationships. Various obstacles to condom use include negative beliefs about and attitudes toward condoms, often grounded in traditional gender constructions (Meyer-Weitz *et al.*, 2003; Morojele *et al.*, 2006). Some studies have found that young people may also associate condom use with promiscuity and sexually transmitted infections, including HIV/AIDS (Buga *et al.*, 1996; Varga, 1997; Reddy *et al.*, 2000). Further, peer pressure or stigma about condom use inhibits actual use (Harrison *et al.*, 2001; MacPhail and Campbell, 2001; Boer and Mashamba, 2005).

Condom use for HIV prevention is also often linked to perceived risk for HIV/AIDS and other STI's. Contrary to expectation (Maharaj, 2006), condoms were less likely to be used by those young people in KwaZulu-Natal who considered themselves to be at medium to high risk than by those who believed that they were at low risk of HIV infection. Perceived risk for HIV/AIDS may be associated with inadequate knowledge about HIV/AIDS.

Recent studies have elucidated the determinants of condom use. Condom use at sexual debut and talking with one's first partner about condoms were the most significant predictors of condom use at last intercourse (Hendriksen *et al.*, 2007). Attitudes and perceived behavioral control were significantly related to the intention to use condoms (Jemmott *et al.*, 2007). In other studies, condom use at last sex has been associated with male sex, age at first sex, knowing a person infected with HIV/AIDS, and condom-use self-efficacy (Taylor *et al.*, 2007). Whilst acceptability of male condoms amongst young people is growing, self-efficacy for condom use remains a challenge. Condom self-efficacy is associated with a wide range of factors, including condom use at first sex, discussion with someone other than a parent about HIV, and knowing how to avoid HIV (Sayles *et al.*, 2006). Intentions, acceptance of sexuality, and gender are significant predictors of condom use (Bryan *et al.*, 2006), whereas gender power imbalance is negatively associated with condom use (Boer and Mashamba, 2007).

Contraception and pregnancy

There is growing evidence that HIV risk in young people has been superimposed on already poor sexual health outcomes in young people in sub-Saharan Africa. Notably high levels of unintended pregnancies among teenage and young adult women in the region underscore the need for young people to have dual protection against HIV and pregnancy through condom use (Kleinschmidt *et al.*, 2003; Pettifor *et al.*, 2005). However, common myths and concerns about modern contraceptive methods have led to delays in the adoption and use of these methods in some settings (Garenne *et al.*, 2000). Low levels of contraceptive use among young people who have recently become sexually active mean that opportunities for counselling regarding the importance of contraceptive use are missed, leaving young women vulnerable to both HIV and pregnancy and at risk of transmitting HIV to their children. Many family-planning providers still view condoms as a secondary fertility control option, and discourage its use (Abdool Karim *et al.*, 1992). The avoidance of male responsibility for contraception in general, including condom use, increases the likelihood of unprotected sex among youth (Hoosen and Collins, 2004). A recent study found that only 7 percent of contraceptive users reported use of dual methods, including the male condom (MacPhail *et al.*, 2007).

Alcohol and drug use

Not only is southern Africa home to a disproportionate burden of HIV infections; it also reports the highest rates of alcohol consumption globally. Alcohol and drug use among African youth have been linked to increased HIV prevalence, as young people under the influence of these substances are more likely to engage in unprotected sexual behavior (Flisher *et al.*, 1996; Simbayi *et al.*, 2004; Shisana *et al.*, 2004). The South African National Youth Health Risk Survey (Reddy *et al.*, 2003) reported that about half (49 percent) of all school learners had consumed at least one alcoholic drink in their lifetime, with 12.0 percent having had their first drink before the age of 13 years. In the month prior to the survey, 31.8 percent of learners had used alcohol on one or more days, while 23.0 percent had consumed five or more drinks within the space of a few hours on one or more days. Qualitative studies have indicated that youth perceive the use of substances as a major risk factor for engaging in unprotected sex, and that young women reported greater vulnerability to sexual aggression and exploitation while under the influence of alcohol or drugs (Morojele *et al.*, 2006). The use of substances as a risk factor for HIV infection needs to be specifically addressed in targeted interventions.

Reducing HIV risk

Knowledge is well-recognized as a prerequisite for behavior change. A 17-country survey between 1999 and 2003 (UNICEF, 2003) found that 80 percent of young women aged 15–24 years did not have sufficient knowledge of how to protect themselves and their partners from acquiring HIV. In contrast, in South Africa young people have adequate knowledge and awareness about HIV/AIDS in relation to modes of transmission and prevention options. Information about HIV risk does not include data on the high background prevalence of HIV infection in some populations and groups and how this impacts individual risk, but emphasizes risk from multiple-partner relationships. This results in poor internalization of information regarding personal risk. Peer influence and generalized mass-media messages contribute to misconceptions and uncertainty about personal risks. As important as information provided is what information is *not* provided. For example, there has been an increase in reporting of anal sex as a safer-sex practice, as messages regarding risk of transmission through peno-vaginal sex, but not anal sex, are covered – leading to many young people interpreting anal sex as being safe sex.

Gender

Unequal gender relations matter in terms of HIV risk; both the experience of sexual coercion and the practice of partnering with older men increase young women's risk for HIV infection (Gregson *et al.*, 2002; MacPhail *et al.*, 2002; Dunkle *et al.*, 2004).

Gender constructions are strongly influenced by socio-cultural norms that prescribe the appropriate gender roles for men and women and contribute to the pronounced gap between HIV awareness and practices, and the social processes that influence young women's disproportionate risk for HIV. Socio-cultural norms regarding sexual expression within the confines of accepted masculine and feminine behavior limit choices of protection from pregnancy and disease and increase the vulnerability of youth to HIV and other sexually transmitted infections (STIs), and also perpetuate longstanding gender inequalities (Hoosen and Collins, 2004; Reddy and Dunne, 2007).

Social and community norms are particularly important in African collectivist communities, where individual decisions about behavior are largely mediated by socio-cultural values and norms (Airhihenbuwa and Obregon, 2000). A number of studies have examined how gender, at a contextual level, determines the conduct of relationships at an interpersonal level with specific relevance to HIV risk for women (Ulin, 1992; Harrison *et al.*, 1997; Meyer-Weitz *et al.*, 1998). Conservative social norms remain dominant in many settings, making the acknowledgment of relationships and sexuality by young people difficult and thus limiting their ability to access prevention modalities or guidance.

In many societies, adolescence is a time when boys gain autonomy, mobility, opportunity and power, including in the sexual and reproductive realm, while girls' lives become more restricted (Mensch *et al.*, 1998). The position of young men in a world characterized by social disempowerment, failure and hopelessness brought about by poverty and limited opportunities strongly influences the development of the male gender identity, and notions about masculinity, sexuality and behaviors. Disempowerment, synonymous with a poor self-image and frustration, facilitates sexual aggression and related behaviors (Chant and Gutman, 2000). Socialization within a dominant male culture authorizes male sexual aggression (Kalichman *et al.*, 2005). Male peer-group support for sexual aggression as a way to control women who demonstrate "improper" agency sheds light on the subtleties and complexities of socialized male aggression, and provide challenges for intervention design.

Women's engagement in sex is influenced by their perceived gender-role expectations about love, sex, and compliance with male partners' desires (Ackerman and de Klerk, 2002; Kalichman *et al.*, 2005). Skills such as assertiveness and the ability to negotiate condom use are viewed as undesirable traits in women, and not compatible with a submissive gender role. Of note is that not all young women are passive and/or subservient. An increase in gender assertiveness has been observed in several recent studies. Of concern is that gender assertiveness in young women has been linked with an increase HIV risk because of their concurrent multiple sexual partners in order to meet a variety of consumerist needs, such as cash, cellphones, transport and clothes (Leclerc-Madlala, 2003). Less is known about men and how constructions of masculinity impact their health seeking behaviors, including those relating to sexual health, and knowledge of this is equally important when discussing adolescent women's risks.

While both men's and women's gender roles are changing, certain normative ideals regarding gender remain (Harrison *et al.*, 2006). Expectations of male control in decision-making and of holding power within relationships remain common. The most extreme manifestation of gender imbalance and male dominance is expressed in the form of gender-based violence. In South Africa, high levels of sexual coercion and violence against young women particularly has been well documented, both within interpersonal relations as well as random acts of violence against women (Jewkes and Abrahams, 2002; Jewkes *et al.*, 2002). Gender-based violence is associated with greater risk for HIV infection within Africa (Garcia-Moreno and Watts, 2000; Dunkle *et al.*, 2004). Sexual coercion within relationships, whether emotional, financial or physical, compounds women's inability to negotiate sexual activity or to protect themselves. Because young women's male partners are usually older, the power and maturity advantage that their male partners hold creates an environment conducive to sexual coercion. Thus, young women's power to negotiate safe sex is often inversely related to the age of their partner (Williams *et al.*, 2000). In fact, young women's multiple-level vulnerability effectively cedes control within relationships to the male partner.

417

Gender-power imbalances make informed decision-making and negotiation of safer sexual practices, and in particular condom use, difficult even when the risks of unprotected sex (i.e., HIV/AIDS and pregnancy) are understood.

Non-negotiation seems to be a common characteristic of adolescent relationships (Varga, 2003). The concepts of sexual "decision-making" and "negotiation" may therefore not always be present in youth and in particular in young women's cognitive and behavioral repertoires, thus presenting an important area for intervention.

Structural factors related to HIV risk

In southern Africa, young people's relationships occur in a context of social and family disruption and widespread poverty (Gilbert and Walker, 2002; Hunter, 2007). Poverty exacerbates the exchange of material goods or gifts within a relationship into transactional sex, whereby young women may depend on boyfriends for important needs, such as school fees, clothes, cash or mobile phones, or they may simply sell sex for survival (Luke and Kurz, 2002; Kuate-Defo, 2004).

A growing body of research emphasizes the multiple and complex ways in which adolescent sexuality and risk are shaped by societal or contextual factors. More recently, research has begun to examine community factors, encompassed in the concepts of community participation and social capital, and their impact on sexual risk.

Strong *social cohesion* at family and at neighborhood levels, as well as within schools, has been found to act as a protective factor for adolescent engagement in risk behaviors. Communities characterized by high levels of social cohesion seem to have a greater capacity to buffer against disruptive forces and exert social controls against threats to the well-being of the community (Sampson *et al.*, 1997). *Poverty* is often associated with the limited social capital that may increase the likelihood of inadequate support for HIV/AIDS preventive behaviors (Meyer-Weitz, 2005). There is association between risky sexual behaviors and low socio-economic contexts (MacPhail, 2001; Kalichman and Simbayi, 2003; Campbell *et al.*, 2005; Meyer-Weitz, 2005). Family poverty impacts on adolescent sexual risk behaviors through strained parent–child relationships (Brook *et al.*, 2006). Therefore, adolescents from these settings are more likely to associate with deviant peers and engage in risky behaviors.

Links between *education* and improved sexual health outcomes, such as a reduced incidence of teen pregnancy, have been observed globally. Few such analyses have been extended to HIV/AIDS. Available evidence suggests that school enrollment or attendance lowers sexual risk for youth (Pettifor *et al.*, 2008; Hargreaves *et al.*, 2008). Intuitively, this makes sense: those in school may be less likely to be involved in relationships, more positively engaged in other activities, and thus less likely to be sexually active. However, the causal mechanisms of this lowered risk are complex. Young women, for instance, may leave school because they are pregnant, thus making it appear as though those remaining in school are at lower risk. Comprehensive school programs aimed at developing a positive school climate and facilitating a shared identity and connectedness among pupils have been found to promote mental well-being, competence, social skills and school achievement, and others (Jane-Llopis *et al.*, 2005), that might influence the engagement in sexual risk behaviors.

Social bonding between school students, student–school connectedness, and a caring and supportive school climate are associated with a reduction in a range of risk behaviors in youth, such as aggression, delinquency, substance use and high-risk sexual behavior (Springer *et al.*, 2006). Understanding of the mediating role of school social cohesion in risk behaviors in African contexts is limited and would need further exploration. This is particularly pertinent in light of the greater socialization role schools need to play because of the erosion of existing family and support structures due to AIDS.

How young people *use their time* may strongly influence their sexual risk-taking behaviors. For instance, work and employment may occupy young people's time and reduce their ability to participate in sexual risk activities (Kaufman *et al.*, 2004). Also, some evidence suggests that young people who belong to community organizations may experience protective effects from doing so (Campbell *et al.*, 2002). These effects may be different for young men and women, with young women appearing to benefit most from membership in various organizations. For men, membership in sports organizations is most common, but the effect of participation on HIV risk is ambiguous, with some studies indicating that men who participate in sports are actually at increased risk (Campbell *et al.*, 2002).

Access to resources and services

Globally, *access to services* and accurate information about prevention and reproductive health for young people is inadequate. Throughout sub-Saharan Africa, this poor quality of care for young people encompasses limited access to services, poor reception and treatment from service providers, and non-availability of preventive methods that young people need – most notably, condoms (Wood and Jewkes, 2006). Limited condom and contraceptive accessibility in general, but also for youth in particular, has been reported to impact negatively on its use (Colvin, 1997; Gilmour *et al.*, 1999).

Lack of access for adolescents is not only limited to physical constraints, but also characterized by social inaccessibility. Service providers are often overtly hostile to serving young clients, particularly for their reproductive health needs (Wood and Jewkes, 2006). In fact, clinics are often conveniently located near schools, and thus could play an important role in meeting the prevention needs of young clients. However, personal motivation and access to care should be accompanied with the development of local resources that facilitate healthy choices. The establishment of separate youth-friendly services augurs well for improved youth accessibility to condoms and other reproductive health services. Nevertheless, access to counselling and testing services, condoms, treatment services for sexually transmitted infections, and harm-reduction strategies all remain limited for young people.

Socio-cultural norms and values

Too often, cultural practices that may be harmful to young people, such as virginity testing or initiation ceremonies, are promoted as having value in terms of safe sex or HIV prevention. An emphasis on virginity may be detrimental for young women, for instance, if they feel pressure to "prove their virginity" through sexual intercourse with a male partner, or pressured to practice anal sex to preserve their virginity (Harrison, 2008).

Some aspects of cultural practices related to sexuality, such as non-penetrative sex or the intergenerational instruction of young people in sexual matters by their elders, could have positive implications for HIV prevention if adapted to the needs of contemporary youth. This is of particular importance within the African context, where discussions about sexuality with young people are generally a taboo topic and occur only within a very specific context. Exploration and re-negotiation by young people are needed regarding the positive and negative aspects of cultural norms and practices, as many of them are strongly influenced by the societal values that promote them. Community concerns about sexuality education need to be considered, and ways in which these concerns can be accommodated should be explored through greater community involvement in youth programs. The development of indigenous institutions by communities to provide culturally sensitive sex counselling/education services for adolescent girls in Uganda have yielded success (Muyinda *et al.*, 2004), and should be explored within the southern African context. In addition, the combination of both traditional services and modern health and sexual education can provide a bridge between tradition and modern institutions (Muyinda, 2004). However, the incorporation of indigenous institutions in sexual counselling and education for youth requires careful planning and thoughtful processes to ensure that cultural practices and traditions that increase young people's vulnerability to HIV infection are not reinforced.

HIV prevention interventions in youth

As the severity of the HIV pandemic has heightened the vulnerability of both adolescents and the general population, intervention efforts have increased. Recent research highlights the need for comprehensive behavioral interventions that take into account the complex social context of adolescent sexuality and HIV risk. These interventions include an emphasis not only on individual-level outcomes, but also on modifying social norms to support behavioral change and addressing structural inequalities (e.g., gender) that contribute to high-risk sexual behavior (Wellings *et al.*, 2006).

Overview of behavioral interventions

We first turn our attention to a review of behavioral interventions. Interventions for young people are usually implemented in schools, through peer educators, community activities, mass-media campaigns and involvement of parents. They do not differentiate by sex or target those who are sexually active, and generally approach all young people as a homogenous group. In settings like South Africa, racial diversity, economic differences, quality of schooling and the age distribution in specific grades can vary substantially between schools depending on geographical location. Tables 15.2 and 15.3 provide an overview of recent behavioral interventions and their outcomes across sub-Saharan Africa over two time periods: the 1990s and the 2000s respectively.

Evaluation of these behavioral interventions has occurred with varying degrees of rigor, and their impact varies depending on coverage, target population and vehicle of delivery. To date, there are few behavioral interventions that have had a demonstrated impact on HIV incidence or other biological outcomes such as pregnancy and or other sexually transmitted infections. This reflects a general focus of behavioral interventions, regardless of setting, on assessing the short-term impact on proxy markers for HIV risk, such as knowledge, use of condoms at last coital encounter, abstinence, and partner reduction. In high-prevalence HIV settings such as South Africa, this represents a gap in youth interventions, given the need to focus on preventing HIV infection as a primary objective from the early teenage years onward. Notwithstanding this shortcoming, important trends and lessons on what works or does not work can be gleaned from a review of the literature that can inform design of much-needed new, effective and targeted interventions.

A distinction is made here between *programs*, which are broad-based efforts designed to reach a large population, and *interventions*, which are more narrowly focused efforts designed to be evaluated for efficacy. Below, we discuss programs and interventions that focus on individual, partnership (or dyadic) and structural levels of risk.

National HIV media-based prevention programs

Widespread, more generalized community and large-scale HIV/AIDS media interventions such as LoveLife and Soul City have been successful in creating

Table 15.2 Behavioral interventions to reduce HIV/AIDS and associated risk behaviors in adolescents and youth: sub-Saharan Africa, 1990–1999

Intervention	Location/ venue	Target group	Type of intervention	Content/focus	Design/outcomes	Impact/effectiveness
Dalrymple and du Toit (1993)	South Africa	Secondary-school students	School-based; Drama in AIDS Education, participatory drama delivered by teachers and youth	Knowledge, attitudes, behavior related to HIV/AIDS	Pre-post intervention assessment with 72 students	Increase in general knowledge of HIV/ AIDS and sexual risk
Kuhn et al. (1994)	South Africa	Secondary-school students	School-based; 2 weeks duration; during and after school; teacher-led, nurse-assisted	Raise HIV/AIDS awareness; provide condoms	1 intervention, 1 control school, pilot study, immediate follow-up	Improved attitudes toward PLWHAs; increased knowledge of condom use and condom intentions
Fawole et al. (1999)	Nigeria	Urban senior secondary students (n = 440)	Comprehensive health education over 6 weeks using lectures, film, role plays, stories, songs, essay contests	Methods of HIV prevention and demonstration of condom use, led by physician and 2 teachers	Controlled longitudinal design in 2 schools, with random allocation; 6 month follow-up	Improved attitudes to HIV/AIDS, increase in reported abstinence, no change in condom use, decrease in #partners
Fitzgerald et al. (1999); Stanton (1998)	Namibia	Secondary-school students, in out-of-school hours	My Future is My Choice, HIV risk reduction over 7 weeks (14 sessions), using narratives, games, facts, exercises and questions	Based on Protection Motivation theory, focused on reproductive biology, HIV/AIDS, substance use, relationship violence and communication skills	RCT, random assignment of students from 10 schools; 6 month follow-up	Improved attitudes toward abstinence and condom use; increased condom self-efficacy; increase in intentions to abstain and use condoms; increase in condom use at 12 mos.

Study	Country	Setting	Description	Content/Theory	Design	Results
Klepp et al. (1997)	Tanzania	Grade 6 and 7, primary school	School-based, conducted during school hours; 20 hours over 2–3 months	Based on Theory of reasoned action, social learning theory; HIV/AIDS, communication, transmission and prevention, delay in sexual debut, PLWHA	RCT; 6 intervention and 12 control schools, repeat cross-sectional surveys at 6 and 12 months	At one year follow-up: Increase in intentions to abstain; Increased general knowledge of HIV/AIDS and sexual risk; improved attitudes toward PLWHAs
Munodawafa et al. (1995)	Zimbabwe	5 rural secondary schools, Grades 9 and 10	School-based	Prevention of STD/AIDS/drug use; transmission and social issues, decision-making	Quasi-experimental design; 2 control and 3 intervention schools; immediate follow-up	Increase in general knowledge
MacLachlan et al. (1997)	Malawi	Secondary schools	AIDS education through active learning; school-based	Board game focused on sexual activity, abstinence	Matched control, no pre–post design	Increase in general knowledge
Rusakaniko et al. (1997)	Zimbabwe	Secondary schools	Focused on increasing knowledge of health issues through teachers and print material	Reproductive biology; STD/AIDS, responsible sexual behavior, pregnancy, contraception	Random sampling of students within selected schools	Measurement of HIV/AIDS knowledge; no impact
Shuey et al. (1999)	Uganda	Primary schools, ages 13–14	School-based, in-school activities, including School Health Club and peer led exercises with question box	Abstinence	Repeat cross-sectional design	Increase in general knowledge; impact on delay in sexual debut and number of partners
Visser (1996)	South Africa	Standards 6–9 (early secondary)	School-based, in classrooms, teacher implemented "kit" with various activities	Abstinence and condoms	Purposive sample, pre–post test with 187 students, no control	Increased knowledge (abstinence, condoms); attitudes toward PLWHAs; intentions to use condoms/abstain

Source: C. Baxter (personal communication).

Table 15.3 Behavioral interventions to reduce HIV/AIDS and associated risk behaviors in adolescents and youth: sub-Saharan Africa, 2000–2007

Intervention	Location/venue	Target group	Type of intervention	Content/focus	Outcomes/design	Impact/effectiveness
Agha (2002)	Zambia	Secondary-school students (boarding)	One 2-hour peer-led session using factual information, drama, skits	Safer sex, HIV prevention, abstinence and condom use	Quasi-experimental longitudinal panel with 1-week and 6-month post-intervention follow-up	Increase in abstinence knowledge and beliefs; reduction in number of partners
Cowan (2002)	Zimbabwe	30 communities with 82 secondary schools	Comprehensive adolescent reproductive health program	Regai Dzive Shiri, community intervention with components for in- and out-of-school youth, as well as adults	Community RCT, following extensive feasibility and pilot work, with HIV incidence as main outcome	Ongoing
Erulkar et al. (2004)	Kenya	Young people	Culturally consistent reproductive health program	Comprehensive reproductive health, with emphasis on safe-sex behavior change	36-month program with quasi-experimental design; project and control areas with pre–post assessments	Positive changes, differed by gender: *Girls*: increase in secondary absence, less likely to have had 3 or more partners. *Boys*: in intervention, more likely to use condoms. *Both* improved communication

Aaro et al. (2006), Ahmed et al. (2006)	South Africa/Tanzania	12- to 14-year-old school students	School-based sexuality education and teacher training	SATZ – school-based classroom curriculum focused on promotion of condom use and delay in sexual debut	Community RCT; intervention mapping based on modified theory of planned behavior	Ongoing
Wegner et al. (2007)	South Africa	Secondary-school age	School-based sexuality education plus leisure-time activities	Healthwise – focused on sexual risk and substance-use reduction through positive use of leisure time	Elements of life-skills training, time management, personal goal-setting	Ongoing
Mantell et al. (2006)	South Africa	Secondary-school students in Grades 8–10	14 sessions over 3-month period, delivered by teachers, peer educators, nurses	Mpondombili Project. HIV and pregnancy prevention, dual protection, gender equality, condoms, decision-making and refusal skills	Quasi-experimental design; 2 intervention, 2 comparison schools	Increased condom use and negotiation skills among girls; positive change in gender-role norms
Harvey et al. (2000)	South Africa	Grade 8 students	DramAide (drama in education) to increase AIDS awareness; drama workshops plus presentations	10-page booklet about HIV/AIDS prevention and transmission	RCT; 7 intervention and 7 control schools with random allocation	Attitudes toward PLWHAs; condom self-efficacy; increase in ever use of condoms

(Continued)

Table 15.3 (Continued)

Intervention	Location/venue	Target group	Type of intervention	Content/focus	Outcomes/design	Impact/effectiveness
James et al. (2005)	South Africa	Grade 11 secondary-school students	Knowledge, attitudes, communication and behavioral intentions related to HIV/AIDS	Photo novella, Laduma, based on health promotion and social learning theory, implemented in KwaZulu/Natal	RCT, 10 schools to control group, 9 schools in intervention group; assessments at intervention delivery and 6 weeks post-intervention	Improved attitudes among men toward PLWHAs; improved condom attitudes; increase in girls abstinence intentions; increase in condom intentions; no change in behavior
Jewkes et al. (2007)	South Africa	Late adolescents/ young adults	Stepping Stones, group-based learning using participatory learning theory	13×3-hour older peer-facilitated sessions: personal behavior, sex and love, sexual risk-reduction; gender-based violence and communication skills	Community-cluster RCT	15% reduction in HIV infections and 31% in HSV-2 in men after 2 years; fewer partners among men; reduction in % men reporting a casual partner or transactional sex
Karnell et al. (2006)	South Africa	Secondary school	HAPS – HIV and Alcohol	Implemented in KwaZulu/Natal,	9th-Grade students in 5 township schools, pilot study	At 18-month follow-up, significant

			Prevention Study. School-based, delivered by peer leaders and teachers, using audiotapes	with focus on alcohol prevention and sexual risk-reduction, based on US intervention		differences b/w intervention and control with regard to condom use intentions, alcohol use, and sex refusal
Kinsman *et al.* (2001)	Uganda	Late primary/early secondary	School-based, delivered by teachers in classrooms	Masaka HIV prevention, behavior change theory, focused on abstinence/delay	Part of larger RCT aimed at reducing HIV incidence; teacher training and implementation of comprehensive sex education	Improved attitudes and intentions toward abstinence; no impact on behavior
Okonofua *et al.* (2003)	Nigeria	3700+ students in intervention and control schools	Community participation, peer education, school health clubs, provider training	Focus on STD treatment-seeking behavior and STD prevalence	4 intervention, 8 control schools	Improved knowledge, condom use, partner awareness of STD, and treatment-seeking behavior; reduction in reported STI symptoms in past 6 months
Bhana *et al.* (2007)	South Africa	Young adolescents/pre-adolescents within families	Out-of-school approach to working with youth, intervention based on cartoon narrative for use with families	AmaQhawe Family Project	Family-based prevention in community participatory framework	Pilot results: improved parent/child communication, improved HIV/AIDS knowledge and attitudes, and decreased stigma

(Continued)

Table 15.3 (Continued)

Intervention	Location/venue	Target group	Type of intervention	Content/focus	Outcomes/design	Impact/effectiveness
Ross *et al.* (2007)	Tanzania	Primary school students, Years 5–7	Peer-assisted sexual health education in schools, youth-friendly services, and condom promotion in community	Mema kwa Vijana, comprehensive sexuality education, focused on condom use and delay in sexual debut	Community-cluster RCT	Improved attitudes and knowledge, as well as increased condom use, among intervention youth, but no impact on biological outcomes

awareness about HIV/AIDS in South Africa, and have been reported to bring about behavior change (Goldstein *et al.*, 2005; Pettifor *et al.*, 2005). While the literature points out the difficulties in measuring the efficacy of media interventions in changing behaviors, their use in creating public awareness is well documented. These interventions often have to compete with more appealing and subtle messages about substance use (alcohol, tobacco and other illicit drug use) and sex.

School-based interventions

School-based efforts remain a popular and common method of delivering behavioral intervention programs to young people, as schools represent an important setting for reaching great numbers of adolescents (though not all adolescents are enrolled in schools, and those who have left school for reasons including pregnancy and employment may represent a higher-risk group of young men and women). School-based interventions have also led to changes in knowledge, attitudes, and sometimes in sexual-risk behaviors. These programs have generally incorporated HIV prevention efforts into a broad-based "life skills" approach (James *et al.*, 2006). A number of school-based HIV prevention interventions have been conducted in sub-Saharan Africa (Table 15.2). There are several notable exceptions to this general approach to school-based interventions that encompass knowledge provision with short-term behavioral outcomes, including the Mpondombili and CAPs Interventions in South Africa and the Regai Dzive Shiri and MEMA kwa Vijana interventions in Zimbabwe and Tanzania respectively. The Mpondombili Intervention included youth peer educators, teachers and nurses in secondary schools in rural KwaZulu/Natal to deliver a 14-session curriculum aimed at promotion of dual protection, consistent condom use, and delay in sexual debut among Grade 8–10 pupils (Harrison, 2002; Mantell *et al.*, 2006). The HAPS project, used a combined sexual-risk and alcohol-awareness curriculum to address those interconnected risk factors among secondary school students in a pilot study that was an adaptation from a US-developed and -tested intervention (Karnell *et al.*, 2006). The Regai Dzive Shiri intervention in the rural Masvingo Province in Zimbabwe included pupils, parents, teachers and community leaders to determine acceptability of the intervention, and a questionnaire and urine sampling survey was undertaken among the pupils aged 12–18 years (*n* = 723, median age, 15 years) to determine the correlation between biological evidence of sexual experience and questionnaire responses (Cowan, 2002). The MEMA kwa Vijana ("Good things for young people") intervention had four components: community activities; teacher-led, peer-assisted sexual health education in Years 5–7 of primary school; training and supervision of health workers to provide "youth-friendly" sexual health services; and peer condom social marketing. Impacts on HIV incidence, herpes simplex virus 2 (HSV-2) and other sexual health outcomes were evaluated over approximately 3 years in 9645 adolescents recruited in late 1998 before entering Years 5, 6 or 7 of primary school. This inter-vention substantially improved knowledge, reported attitudes and some reported sexual behaviors, especially in boys, but had no consistent impact on biological outcomes within the 3-year trial period (Ross *et al.*, 2007).

Life-skills programs

Life-skills programs address basic knowledge about sexuality, HIV/AIDS and other STIs, as well as skills that include learning to manage peer pressure effec-tively, and improving communication and negotiation skills within relationships. The effectiveness of school-based prevention is also highly dependent on indi-viduals in the school to support and deliver the programs. In a recent pre-test/post-test control group design study (James *et al.*, 2006), the life-skills program in South Africa was only marginally successful in improving learners' knowledge about HIV/AIDS, and yielded no significant effect on safer-sex practices or on the psychosocial determinants of these sexual practices. It was also found that the life-skills program was seldom implemented as intended, and was being adapted based on school leadership commitment, teacher availability, skill and comfort level with the content; school attendance; and time and facilities available. In settings where the program was implemented with fidelity, significant positive changes in knowledge of HIV/AIDS and sexual risk were reported.

Peer education

Another popular school-based approach is peer education, due to recogni-tion of the influence of peers on adolescent decision-making and behavior. The effectiveness of peer education programs is dependent on the provision of ade-quate emotional and program support for peer counsellors as well as a supportive environment and community context (Campbell, 2003). These programs tend to be broad-based interventions similar in content to the life-skills programs. The goal of these interventions is to provide youth with HIV/AIDS knowledge and skills in terms of HIV transmission and prevention, and that the target group will translate the knowledge and skills gained into responsible and preventative behaviors.

It is noteworthy that while the school setting provides a captive audience for most interventions aimed at young people, teachers' personal HIV risk and values and norms towards sexuality impact the effectiveness of this approach. As teachers in sub-Saharan Africa increasingly leave school due to AIDS-related ill health or even death, it is possible that those remaining teachers will be forced to focus on only the essential educational skills, such as reading, writing and math-ematics, due to a shortage of qualified persons, as is being experienced in other sectors in society. Thus, an over-reliance on school-based approaches might be an increasingly unsustainable strategy. The use of voluntary community-based counsellors/educators and other community-level approaches may ultimately prove more viable in settings devastated by HIV/AIDS (Muyinda *et al.*, 2004).

Interventions involving parents

A strong adult protective shield for young people decreases their risk for HIV infection (Petersen *et al.*, 2005). While different parenting styles and practices have been linked to decreased risk, little information is available regarding the parenting styles and practices within an African context that would protect adolescents from HIV risk. Few HIV/AIDS interventions targeted at young people involve parents directly, and those that have included parents in school-based approaches have met with little success. A community-based intervention that specifically targets parents and their young pre-adolescent children in improving communication especially around HIV/AIDS, such as that implemented by (Bhana *et al.*, 2004), could provide some direction for interventions with older adolescents. Brook and colleagues (2006) found that stressors such as poverty increase difficulty in the parent–child relationship, and future interventions that incorporate parents should include components that will improve the quality of *parents'* lives as well as young people's lives, through improved knowledge and skills pertaining to the various challenges they may face – such as parent–child interactions, discipline, and stress management.

The challenged family constructions within the African context where female-headed households are often the norm (Preston-Whyte and Zondi, 1992), or where both parents are absent because they have died of AIDS or are migrant workers, will necessitate innovative use of supportive social relationships for youth.

Structural and community-level interventions

While community interventions within geographically-defined boundaries and focused on community mobilization or structural changes are more novel, innovative and creative, there is a paucity of data on effectiveness. For example, community interventions linked to youth-focused organizations, traditional kinship networks, or as part of community-wide events demonstrate gains in knowledge, skills and behavior change, and shifts in community norms.

There is a growing recognition that HIV prevention interventions must go beyond addressing sexual risk behaviors and their immediate determinants, and focus on broader societal and economic factors underlying HIV/AIDS risk. Several interventions have shown promise in reducing gender inequities and gender-based violence, and the HIV-associated risk behaviors that accompany it. "Stepping Stones" is a group-based intervention that uses participatory learning approaches, including critical reflection, role-playing and drama. It was first implemented in Uganda, and then adapted and implemented in South Africa

(Jewkes *et al.*, 2007). The program is delivered to groups over several weeks in a series of 13 3-hour sessions and 3 peer-group meetings, with facilitation from slightly older, trained peers. The sessions focus on personal behavior and what shapes it, sex and love, sexual risk reduction, HIV and STIs, gender-based violence, and communication skills. The program culminates in a community meeting in which a special request from the participants is made. The Stepping Stones intervention achieved a reduction of 15 percent in HIV infections and 31 percent in HSV-2 infections at the 2-year follow-up point. Although neither of these findings is statistically significant, they provide some evidence that the intervention had a beneficial impact on sexual behavior. In men, fewer partners were reported since the last interview at both the 12- and 24-month follow-up, and there was a significant reduction in the proportion of men reporting a casual partner or transactional sex during the follow-up period.

Success in reducing pregnancy among adolescent school-going girls via a structural intervention has also been reported from Kenya. Duflo and colleagues (2006) conducted an evaluation of three HIV prevention methods over a 3-year period among 70,000 students in Grades 6–8 in 328 Kenyan schools (Duflo *et al.*, 2006). The interventions included providing girls with school uniforms at no charge, conducting classroom debates and essay contests in which students wrote about why they should (or should not) be taught about condom use and HIV prevention, and a 40-minute course for 8th-Grade students that discussed HIV transmission statistics and their implications, along with a 10-minute video about the risks associated with relying on older men for money and gifts. Their findings showed that free uniforms had a direct, positive impact on school dropout and pregnancy rates, and that girls exposed to 1 year of information about the risks of partnering with older men were 65 percent less likely to become pregnant by an adult partner. The essay contests were found to increase condom use but not decrease sexual activity.

Another South African intervention, the IMAGE (Intervention with Microfinance for AIDS and Gender Equity) Project, while not specifically designed for young women, achieved a substantial reduction in intimate partner violence through a combined intervention of micro-finance for poor women and sexual risk-reduction and empowerment sessions designed to enhance partner communication. However, no impact on unprotected sex with non-spousal partners or on HIV incidence was observed. IMAGE was implemented across eight villages, four of which received the intervention at the outset of the program, with the other four receiving it 3 years later. In each village the intervention comprised a poverty-focused microfinance program, in which the poorest households were identified, women recruited and groups formed, and cycles of borrowing and repayment initiated. In addition, group ("center") meetings were held every second week with all the women, which provided opportunities to implement the "Sisters for Life" gender and HIV training program. This comprised two phases, structured training (including modules on gender, culture, sexual risk and women's

432

empowerment), and community mobilization, focused on group action plans (Pronyk *et al.*, 2006).

The approach adopted by the Population Council Adolescent Livelihoods program focuses on implementation of a curriculum aimed at building the social, economic and health capabilities of highly vulnerable youth (Hallman, 2006). This program follows an integrated approach, including efforts to reduce social isolation by bringing young people together in safe community spaces, enhancing financial "literacy", building social and economic bridges through informal savings groups as well as connections to role models and mentors, and increasing knowledge and skills related to sexuality and HIV prevention.

While none of these interventions have had an impact on HIV incidence, they appear to have successfully altered some of the underlying societal and structural determinants of HIV acquisition – namely gender inequalities and economic need.

Linking youth to HIV care and treatment

To date, there have been few interventions targeted at young people that specifically promote and/or expand access to HIV testing, care and treatment. Given the increase in public sector access to ARV treatment and care, and efforts to promote knowledge of status, prevention efforts that integrate HIV testing and care and treatment services are critical. The success of these initiatives hinges on addressing the prevailing HIV/AIDS stigma and discrimination that act as major barriers to knowledge of HIV status and health-seeking behaviors.

In terms of increasing access to and utilization of health services, training of health-service providers, improvements in clinic infrastructure combined with activities in the community, and provision of a defined package of services all impact health-service utilization patterns. Mass-media campaigns utilizing television and, to some extent, radio, posters and billboards are more effective for addressing specific issues, and are also effective in increasing knowledge, improving self-efficacy to use condoms, influencing social norms, increasing the amount of interpersonal communication and raising awareness of health services.

Out-of-school youth and youth in higher education

Effectiveness is enhanced when interventions work from young people's own perceived interests and needs – from employment opportunities to opportunities to talk about relationships and sexual health issues (projects such as TAP, and the Reposition Youth (TRY) Savings and Credit Scheme Project in Nairobi, Kenya; and the Young People's Sexual and Reproductive Health Project in Malawi). In higher-education institutions, innovative initiatives have aimed to diffuse HIV/AIDS across the university curriculum – for example, the integration of discussions into subjects such as law, agriculture, environment, etc. (University of

Botswana HIV/AIDS Project, Center for the Study of AIDS at Pretoria University in South Africa).

Lessons learned from interventions targeted at youth

Young people form a diverse group with varying needs within the transitional period to adulthood, and between each other depending on sexual experience, HIV risk based on geographical boundaries, and a range of individual and contextual factors that impact HIV risk and risk-reduction efforts.

Partnerships that enable the sharing of ideas and responsibilities are central to the success of interventions with young people. The importance of forming networks within a country and outside, as well as linking governments with civil society and forging networks between young people and the broader community, lay foundations for sustainability, as human and financial resource allocations are dependent on these linkages (Aggleton and Warwick, 2002).

Adults need support in working with youth. Not all adults have the skills and knowledge to educate young people about HIV/AIDS sexual and reproductive health issues (James *et al.*, 2004). The same is true for youth and community educators. Skills need to be developed in the use of experimental and experiential approaches to learning and development.

Awards and accreditation might be ways to prepare educators for their role in sexual and reproductive health (Aggleton and Warwick, 2002).

Research not only plays a role in social assessment during the initial stages of program planning, but may also play a role in the fostering of partnerships. Participatory evaluation research is another way in which partnerships and collective responsibility for interventions are sought (Meyer-Weitz and Sliep, 2005). Active involvement of young people in the articulation of their needs, and to inform program content, is essential. The social, political, financial and cultural contexts of young people will determine what they are able and willing to do (Aggleton and Warwick, 2002). Interactive and experiential learning opportunities can facilitate involvement, and enhance the internalization of new knowledge and skills. Interactive drama techniques (Sliep and Meyer-Weitz, 2003) have been used with great success in different contexts for peer and community education.

Risk behaviors cluster in individuals, and have common determinants influenced by the same risk and/or protective factors. Thus, by effectively addressing the determinants, a profound impact on adolescent health could be made. HIV/AIDS prevention efforts should be part of broader interventions directed at improving young people's sexual and reproductive health, and promote adolescent health and development.

Skills-based approaches directed to promoting young people's sexual and reproductive health should be followed (Aggleton and Warwick, 2002). Adolescent development and behavior change are influenced by a range of factors, including individual knowledge; attitudes; relationships with parents, caregivers and peers; schools; economic status; faith beliefs; and the entertainment media.

Behavioral outcomes, such as measures of condom use, injecting drug use, commercial sex and age at first sex, and number of sexual partners, are often used as indicators of program success. A human rights approach that is gender and culturally sensitive in which young people have a right to access to all the necessary information and skills that will enable them to protect themselves and their partners from HIV infection is a prerequisite (UNESCO, 2002) to enable young people to make informed choices about a range of sexual and reproductive health issues.

Community involvement and the use of indigenous institutions to provide culturally sensitive sex counselling/education services for adolescent girls in Uganda contributed to its success in HIV prevention (Muyinda *et al.*, 2004). Further, the combination of traditional services and modern health and sexual education provided a bridge between tradition and modernity.

The use of voluntary community-based counsellors/educators needs to be considered, and novel ways in which communities may reward them for their services could substantially improve the sustainability of prevention efforts in resource-constrained, hyper-endemic HIV settings.

Planning for sustainability requires dedicated political will and financial resources. While interventions might be commenced through funding from different agencies, local ownership and sustainability should be planned for from the initial onset of program planning and implementation (Aggleton and Warwick, 2002). In Kenya, the TAP project aims to become self-sustained through microfinance and strengthening of livelihood skills.

Conclusion

There is no "silver bullet", biomedical or behavioral, for decreasing HIV risk in youth. It is clear that no single intervention will be equally effective for all youth in all settings, and within a specific setting. Interventions that are based on good epidemiological data for prioritizing the target population, utilize behavioral and social theory appropriate for the target population and desired outcomes, consider contextual and structural realities of the target group, and capacity and infrastructure for sustainable delivery are more likely to be effective.

Adolescent development and behaviors vary within the time period that defines this stage of development, and differ by sex and setting. These behaviors are influenced by a range of factors, including individual self-esteem; skills; knowledge; beliefs; attitudes; relationships with parents, caregivers, peers and teachers; schools; economic status; faith beliefs; perceptions of rights; sense of future; and the media. Diversity in programs and interventions is thus critical, as is targeting and designing interventions for specific time periods in adolescent transition to adulthood and autonomy. Short-, medium- and long-term goals are critical for individuals being targeted, as well as for intervention design.

We need to understand better the predictive roles of increasing knowledge, self-efficacy skills, building self-esteem, creating hope in the future and access to services in reducing HIV incidence rates. HIV/AIDS prevention efforts need to form part of broader interventions directed at improving young people's overall health and well-being, and initiatives to address socio-economic and structural roots of risk. Additionally, program effectiveness is also enhanced when interventions incorporate young people's own perceived interests and needs – including employment opportunities, and opportunities to talk about relationships and sexual health issues. Partnerships that enable the sharing of ideas and responsibilities are central to success. The importance of forming networks within a country and beyond its borders, linking governments with civil society, as well as forging networks between young people and the broader community, lay foundations for sustainability, as human and financial resource allocations are dependent on these linkages.

Despite the increasing vulnerability of young women to HIV infection, no interventions directed at sexually active young women and or their partner(s) have been developed to date. This represents a significant gap in our efforts to reduce HIV infection in sub-Saharan Africa, and specifically in southern Africa. HIV prevention interventions for young, sexually active women that also involve their sexual networks would be instrumental in bringing about a negotiated change among men and women, and in particular change in the sexual dyad. Such an approach could also challenge traditional cultural views that reinforce existing gender inequalities, including women's expected submissiveness, their lack of power in sexual relationships pertaining to coercive sexual practices and negotiating safer sex, as well accepted masculine norms of multiple-partner relationships.

References

Abdool Karim, Q. (2005). Heterosexual transmission of HIV – the importance of a gendered perspective in HIV prevention. In: S. S. Abdool Karim and Q. Abdool Karim (eds), *HIV/AIDS in South Africa*. Cape Town: Cambridge University Press, pp. 243–461.

Abdool Karim, S. and Abdool Karim, Q. (2002). The evolving HIV epidemic in South Africa. *Intl. J. Epidemiol.*, 31, 37–40.

Abdool Karim, Q., Abdool Karim, S. S., Preston-Whyte, E. and Sankar, N. (1992). Reasons for lack of condom use among high school students. *S. Afr. Med. J.*, 82, 107–10.

Ackerman, L. and de Klerk, G. (2002). Social factors that make South African women vulnerable to HIV infection. *Health Care Women Intl*, 23, 163–72.

Aggleton, P. and Warwick, I. (2002). Education and HIV/AIDS prevention among young people. *AIDS Educ. Prev.*, 14, 263–7.

Agha, S. (2002). An evaluation of the effectiveness of a peer sexual health intervention among secondary-school students in Zambia. *AIDS Educ. Prev.*, 14, 269–81.

Airhihenbuwa, C. O. and Obregon, R. (2000). A critical assessment of theories/models used in health communication for HIV/AIDS. *J. Health Comm.*, 5(Suppl.), 5–15.

Anderson, R. M., Medley, G. F., May, R. M. and Johnson, A. M. (1986). A preliminary study of the transmission dynamics of the human immunodeficiency virus (HIV), the causative agent of AIDS. *IMA J. Math. Appl. Med. Biol.*, 3, 229–63.

Bhana, A., Petersen, I., Mason, A. *et al.* (2004). Children and youth at risk: adaptation and pilot study of the CHAMP (Amaqhawe) program in South Africa. *Afr. J. AIDS Res.*, 3, 33–41.

Bhana, A., Petersen, I., Bell, C. and McKay, M. (2007). Outcome and process evaluation of the CHAMP South Africa (Amaqhawe) family-based HIV prevention intervention. Paper presented at the *South African–United States Workshop on Behavioral HIV/STI Prevention and Mental Health Research*, Durban, 1–4 June.

Bledsoe, C. and Cohen, B. (1993). *Social Dynamics of Adolescent Fertility in Sub-Saharan Africa*. Washington, DC: National Academy Press.

Boer, H. and Mashamba, M. T. (2005). Psychosocial correlates of HIV protection motivation among black adolescents in Venda, South Africa. *AIDS Educ. Prev.*, 17, 590–602.

Boer, H. and Mashamba, M. T. (2007). Gender power imbalance and differential psychosocial correlates of intended condom use among male and female adolescents from Venda, South Africa. *Br. J. Health Psychol.*, 12, 51–63.

Brook, D. W., Morojele, N. K., Zhang, C. and Brook, J. (2006). South African adolescents: pathways to risky sexual behavior. *AIDS Educ. Prev.*, 18, 259–72.

Bryan, A., Kagee, A. and Broaddus, M. R. (2006). Condom use among South African adolescents: developing and testing theoretical models of intentions and behavior. *AIDS Behav.*, 10, 387–97.

Buga, G. A., Amoko, D. H. and Ncayiyana, D. J. (1996). Sexual behavior, contraceptive practice and reproductive health among school adolescents in rural Transkei. *Social Afr. Med. J.*, 86, 523–7.

Campbell, C. (2003). *Letting them Die: Why HIV/AIDS Intervention Programs Fail*. Oxford: James Curry Press.

Campbell, C. and MacPhail, C. (2002). Peer education, gender and the development of critical consciousness: participatory HIV prevention by South African youth. *Social Sci. Med.*, 55, 331–45.

Campbell, C., Williams, B. and Gilgen, D. (2002). Is social capital a useful conceptual tool for exploring community level influences on HIV infection? An exploratory case study from South Africa. *AIDS Care*, 14, 41–54.

Campbell, C., Foulis, C. A., Maimane, S. and Sibiya, Z. (2005). The impact of social environments on the effectiveness of youth HIV prevention: a South African case study. *AIDS Care*, 17, 471–8.

Chant, S. and Gutman, M. C. (2000). Mainstreaming men into gender and development: debates, reflections and experiences. Working Paper, Oxfam.

Cleland, J. and Ali, M. M. (2006). Sexual abstinence, contraception, and condom use by young African women: a secondary analysis of survey data. *Lancet*, 368, 1788–93.

Coetzee, D. and Johnson, L. (2005). Sexually transmitted infections. In: S. S. Abdool Karim and Q. Abdool Karim (eds), *HIV/AIDS in South Africa*. Cape Town: Cambridge University Press, pp. 193–202.

Colvin, M. (1997). *Sexually Transmitted Diseases. South African Health Review*. Durban: Health Systems Trust and Henry J. Kaiser Family Foundation.

Cowan, F. M. (2002). Adolescent reproductive health interventions. *Sex. Transm. Infect.*, 78, 315–18.

Dalrymple, L. and du-Toit, M. K. (1993). The evaluation of a drama approach to AIDS education. *Educational Psychol.*, 13, 147–54.

Department of Health (2007). *National HIV and Syphilis Antenatal Seroprevalence Survey in South Africa: 2007*. Pretoria: Department of Health: Epidemiology and Surveillance Cluster Health Information, Epidemiology, Evaluation and Research. Available at www.doh.gov.za (accessed December 2007).

DHS (2005). Demographic Health Survey website, www.measuredhs.com/.

Duflo, E., Dupas, P., Kremer, M. and Sinei, S. (2006). Education and HIV/AIDS prevention: evidence from a randomized evaluation in Western Kenya. World Bank Policy Research Working Paper 4024, October.

Dunkle, K. L., Jewkes, R. K., Brown, H. C. *et al.* (2004). Transactional sex among women in Soweto, South Africa: prevalence, risk factors and association with HIV infection. *Social Sci. Med.*, 59, 1581–92.

Eaton, L., Flisher, A. J. and Aaro, L. E. (2003). Unsafe sexual behavior in South African youth. *Social Sci. Med.*, 56, 149–65.

Erulkar, A. S., Ettyang, L., Onoka, C. *et al.* (2004). Behavioral impacts of a culturally consistent reproductive health program for young Kenyans. *Intl Family Plan. Persp.*, 30, 58–67.

Fawole, I. O., Asuzu, M. C., Oduntan, S. O. and Brieger, W. R. (1999). A school-based AIDS education program for secondary school students in Nigeria: a review of effectiveness. *Health Educ. Res.*, 14, 675–83.

Fitzgerald, A. M., Stanton, B. F., Terreri, N. *et al.* (1999). Use of Western-based HIV risk-reduction interventions targeting adolescents in an African setting. *J. Adolesc. Health*, 25, 52–61.

Flisher, A. J., Ziervogel, C. F., Chalton, D. O. *et al.* (1996). Risk-taking behavior of Cape Peninsula high-school students. Part X. Multivariate relationships among behaviors. *Social Afr. Med. J.*, 86, 1094–8.

Garcia-Moreno, C. and Watts, C. (2000). Violence against women: its importance for HIV/AIDS. *AIDS*, 14(Suppl. 3), S253–65.

Garenne, M., Tollman, S. and Kahn, K. (2000). Premarital fertility in rural South Africa: a challenge to existing population policy. *Studies Family Plan.*, 31, 47–54.

Gilbert, L. and Walker, L. (2002). Treading the path of least resistance: HIV/AIDS and social inequalities – a South African case study. *Social Sci. Med.*, 54, 1093–110.

Gilmour, E., Abdool Karim, S. S. and Fourie, H. J. (1999). Availability of condoms in Urban and Rural areas of Kwa-Zulu Natal, South Africa. *Sex. Transm. Dis.*, 27, 353–7.

Goldstein, S., Usdin, S., Scheepers, E. and Japhet, G. (2005). Communicating HIV and AIDS, what works? A report on the impact evaluation of Soul City's fourth series. *J. Health Comm.*, 10, 465–83.

Gouws, E. (2005). HIV incidence rates in South Africa. In: S. S. Abdool Karim and Q. Abdool Karim (eds), *HIV/AIDS in South Africa*. Cape Town: Cambridge University Press, pp. 67–78.

Gouws, E. and Abdool Karim, Q. (2005). HIV infection in South Africa: the evolving epidemic. In: S. S. Abdool Karim and Q. Abdool Karim (eds), *HIV/AIDS in South Africa*. Cape Town: Cambridge University Press.

Gregson, S., Nyamukapa, C. A., Garnett, G. P. *et al.* (2002). Sexual mixing patterns and sex-differentials in teenage exposure to HIV infection in rural Zimbabwe. *Lancet*, 359, 1896–903.

Gregson, S., Garnett, G. P., Nyamukapa, C. A. *et al.* (2006). HIV decline associated with behavior change in Eastern Zimbabwe. *Science*, 311, 664–6.

Hallett, T. B., Aberle-Grasse, J., Bello, G. *et al.* (2006). Declines in HIV prevalence can be associated with changing sexual behavior in Uganda, urban Kenya, Zimbabwe, and urban Haiti. *Sex. Transm. Infect.*, 82(Suppl. 1), i1–8.

Hallman, K. (2006). *Report on Adolescent Girls' Social Support and Livelihood Program Design Workshop, Nairobi, Kenya, 22–23 March*. San Francisco, CA: Population Council and University of California.

Halperin, D. T. and Epstein, H. (2004). Concurrent sexual partnerships help to explain Africa's high HIV prevalence: implications for prevention. *Lancet*, 364, 4–6.

Hargreaves, J. *et al.* (2008). The association between school attendance, HIV infection and sexual behaviour among young people in rural South Africa. *J. Epid. and Comm. Health.* 62(2), 113–9.

Harrison, A. (2002). The social dynamics of adolescent risk for HIV – using research findings to design a school-based intervention. *Agenda*, 53, 43–52.

Harrison, A., Wilkinson, D. and Lurie, M. (1997). From partner notification to partner treatment. *South Afr. Med. J.*, 87, 1055.

Harrison, A., Xaba, N. and Kunene, P. (2001). Understanding safe sex: gender narratives of HIV and pregnancy prevention by rural South African school-going youth. *Reprod. Health Matters*, 9, 63–71.

Harrison, A., Cleland, J., Gouws, E. and Frohlich, J. (2005). Early sexual debut among young men in rural South Africa: heightened vulnerability to sexual risk? *Sex. Transm. Infect.*, 81, 259–61.

Harrison, A., O'Sullivan, L. F., Hoffman, S. *et al.* (2006). Gender role and relationship norms among young adults in South Africa: measuring the context of masculinity and HIV risk. *J. Urban Health*, 83, 709–22.

Harrison, A. (2008). Hidden Love: Sexual Ideologies and Relationship Ideals among Rural South Africa. *J. Epid. and Comm. Health.* 62(2), 113–9.

Harvey, B., Stuart, J. and Swan, T. (2000). Evaluation of a drama-in-education program to increase AIDS awareness in South African high schools: a randomized community intervention trial. *Intl J. STD AIDS*, 11, 105–11.

Hendriksen, E. S., Pettifor, A., Lee, S. J. *et al.* (2007). Predictors of condom use among young adults in South Africa: the Reproductive Health and HIV Research Unit National Youth Survey. *Am. J. Public Health*, 97, 1241–8.

Hoffman, S., O'Sullivan, L. F., Harrison, A. *et al.* (2006). HIV risk behaviors and the context of sexual coercion in young adults' sexual interactions: results from a diary study in rural South Africa. *Sex. Transm. Dis.*, 33, 52–8.

Hoosen, S. and Collins, A. (2004). Sex, sexuality and sickness: discourses of gender and HIV/AIDS among KwaZulu-Natal women. *S. Afr. J. Psychology*, 34, 487–505.

Hunter, M. (2007). The changing political economy of sex in South Africa: the significance of unemployment and inequalities to the scale of the AIDS pandemic. *Social Sci. Med.*, 64, 689–700.

James, S., Reddy, S. P., Taylor, M. and Jinabhai, C. C. (2004). Young people, HIV/AIDS/STIs and sexuality in South Africa: the gap between awareness and behavior. *Acta Paed.*, 93, 264–9.

James, S., Reddy, P. S., Ruiter, R. A. *et al.* (2005). The effects of a systematically developed photo-novella on knowledge, attitudes, communication and behavioral intentions with respect to sexually transmitted infections among secondary school learners in South Africa. *Health Prom. Intl.*, 20, 157–65.

James, S., Reddy, P., Ruiter, R. A. *et al.* (2006). The impact of an HIV and AIDS life skills program on secondary school students in KwaZulu-Natal, South Africa. *AIDS Educ. Prev.*, 18, 281–94.

Jane-Llopis, E., Barry, M., Hosman, C. and Patel, V. (2005). Mental health promotion works: a review. *Prom. Educ.*(Suppl. 2), 9–25, 61, 67.

Jemmott, J. B., Heeren, G. A., Ngwane, Z. *et al.* (2007). Theory of planned behavior predictors of intention to use condoms among Xhosa adolescents in South Africa. *AIDS Care*, 19, 677–84.

Jewkes, R. and Abrahams, N. (2002). The epidemiology of rape and sexual coercion in South Africa: an overview. *Social Sci. Med.*, 55, 153–66.

Jewkes, R., Levin, J. and Penn-Kekana, L. (2002). Risk factors for domestic violence: findings from a South African cross-sectional study. *Social Sci. Med.*, 55, 1603–17.

Jewkes, R., Nduna, M., Levin, J. *et al.* (2007). *Evaluation of Stepping Stones: A Gender Transformative HIV Prevention Intervention.* Cape Town: Medical Research Council, Medical Research Council Policy Brief.

Kalichman, S. C. and Simbayi, L. C. (2003). HIV testing attitudes, AIDS stigma, and voluntary HIV counselling and testing in a black township in Cape Town, South Africa. *Sex. Transm. Infect.*, 79, 442–7.

Kalichman, S. C., Simbayi, L. C., Kaufman, M. *et al.* (2005). Gender attitudes, sexual violence, and HIV/AIDS risks among men and women in Cape Town, South Africa. *J. Sex Res.*, 42, 299–305.

Karnell, A. P., Cupp, P. K., Zimmerman, R. S. *et al.* (2006). Efficacy of an American alcohol and HIV prevention curriculum adapted for use in South Africa: results of a pilot study in five township schools. *AIDS Educ. Prev.*, 18, 295–310.

Kaufman, C. E., Clark, S., Manzini, N. and May, J. (2004). Communities, opportunities, and adolescents' sexual behavior in KwaZulu-Natal, South Africa. *Studies Family Plan.*, 35, 261–74.

Kinsman, J., Nakiyingi, J. K. A., Carpenter, L. *et al.* (2001). Evaluation of a comprehensive school-based AIDS education program in rural Masaka, Uganda. *Health Educ. Res. Theory Pract.*, 16, 85–100.

Kleinschmidt, I., Maggwa, B. N., Smit, J. *et al.* (2003). Dual protection in sexually active women. *S. Afr. Med. J.*, 93, 854–7.

Klepp, K. I., Ndeki, S. S., Leshabari, M. T. *et al.* (1997). AIDS education in Tanzania: promoting risk reduction among primary school children. *Am. J. Public Health*, 87, 1931–6.

Kuate-Defo, B. (2004). Young people's relationships with sugar daddies and sugar mummies: what do we know and what do we need to know? *Afr. J. Reprod. Health*, 8, 13–37.

Kuhn, L., Steinberg, M. and Mathews, C. (1994). Participation of the school community in AIDS education: an evaluation of a high school program in South Africa. *AIDS Care*, 6, 161–71.

Leclerc-Madlala, S. (2003). Transactional sex and the pursuit of modernity. *Social Dynamics*, 29, 1–21.

Luke, N. and Kurz, K. M. (2002). *Cross-generational and Transactional Sexual Relations in Sub-Saharan Africa: Prevalence of Behavior and Implications for Negotiating Safer Sexual Practices.* Washington, DC: International Center for Research on Women.

MacLachlan, M., Chimombo, M. and Mpemba, N. (1997). AIDS education for youth through active learning: a schoolbased approach from Malawi. *Intl J. Educ. Devel.*, 17, 41–50.

MacPhail, C. and Campbell, C. (2001). "I think condoms are good but, aai, I hate those things:" condom use among adolescents and young people in a Southern African township. *Social Sci. Med.*, 52, 1613–27.

MacPhail, C., Williams, B. G. and Campbell, C. (2002). Relative risk of HIV infection among young men and women in a South African township. *Intl J. STD AIDS*, 13, 331–42.

MacPhail, C., Pettifor, A. E., Pascoe, S. and Rees, H. V. (2007). Contraception use and pregnancy among 15–24 year old South African women: a nationally representative cross-sectional survey. BMC Med., 5, 31. Available at http://www.biomedcentral. com/1741-7015/5/31 (accessed March 2008).

Maharaj, P. (2006). Reasons for condom use among young people in KwaZulu-Natal: prevention of HIV, pregnancy or both? *Intl Family Plan. Persp.*, 32, 28–34.

Maharaj, P. and Cleland, J. (2006). Condoms become the norm in the sexual culture of college students in Durban, South Africa. *Reprod. Health Matters*, 14, 104–12.

Mantell, J. E., Harrison, A., Hoffman, S. *et al.* (2006). The Mpondombili Project: preventing HIV/AIDS and unintended pregnancy among rural South African school-going adolescents. *Reprod. Health Matters*, 14, 113–22.

Mensch, B. S., Bruce, J. and Greene, M. E. (1998). *The Uncharted Passage: Girls' Adolescence in the Developing World*. New York: Population Council.

Meyer-Weitz, A. (2005). Understanding fatalism in HIV/AIDS protection: the individual in dialogue with contextual factors. *Afr. J. AIDS Res.*, 4, 75–82.

Meyer-Weitz, A. and Sliep, Y. (2005). The evaluation of Narrative Theatre training: experiences of psychosocial workers in Burundi. *Intervention*, 3, 97–111.

Meyer-Weitz, A. and Steyn, M. (1999). *Situation Analysis: Existing Information, Education and Communication Strategies Regarding Adolescent Sexuality in the Piet Retief District, Mpumalanga*. Durban: Health Systems Trust.

Meyer-Weitz, A., Reddy, P., Weijts, W. *et al.* (1998). Socio-cultural contexts of sexually transmitted diseases in South Africa: implications for health education programs. *AIDS Care*, 10, S39–55.

Meyer-Weitz, A., Reddy, P., van den Borne, H. W. *et al.* (2003). Determinants of multi-partner behavior of male patients with sexually transmitted diseases in South Africa: implications for interventions. *Intl J. Men's Health*, 2, 149–62.

Mnyika, K. S., Klepp, K.-I., Kvale, G. and Ole, K. N. (1997). Determinants of high-risk sexual behavior and condom use among adults in the Arusha region, Tanzania. *Intl J. Sex. Transm. Dis. AIDS*, 8, 176–83.

Morojele, N. K., Brook, J. S. and Kachieng'a, M. A. (2006). Perceptions of sexual risk behaviors and substance abuse among adolescents in South Africa: a qualitative investigation. *AIDS Care*, 18, 215–19.

Morris, C. R., Araba-Owoyele, L., Spector, S. A. and Maldonado, Y. A. (1996). Disease patterns and survival after acquired immunodeficiency syndrome diagnosis in human immunodeficiency virus-infected children. *Ped. Infect. Dis. J.*, 15, 321–8.

Munodawafa, D., Marty, P. J. and Gwede, C. (1995). Effectiveness of health instruction provided by student nurses in rural secondary schools of Zimbabwe: a feasibility study. *Intl J. Nursing Studies*, 32, 27–38.

Muyinda, H., Nakuya, J., Whitworth, J. A. and Pool, R. (2004). Community sex education among adolescents in rural Uganda: utilizing indigenous institutions. *AIDS Care*, 16, 69–79.

Nnko, S., Boerma, J. T., Urassa, M. *et al.* (2004). Secretive females or swaggering males? An assessment of the quality of sexual partnership reporting in rural Tanzania. *Social Sci. Med.*, 59, 299–310.

Okonofua, F. E., Coplan, P., Collins, S. *et al.* (2003). Impact of an intervention to improve treatment-seeking behavior and prevent sexually transmitted diseases among Nigerian youths. *Intl J. Infect. Dis.*, 7, 61–73.

Padian, N. S., Shiboski, S. C. and Jewell, N. P. (1991). Female-to-male transmission of human immunodeficiency virus. *J. Am. Med. Assoc.*, 266, 1664–7.

Paul-Ebhohimhen, V. A., Poobalan, A. and van Teijlingen, E. R. (2008). A systematic review of school-based sexual health interventions to prevent STI/HIV in sub-Saharan Africa. *BMC Public Health*, 8, 4.

Petersen, I., Bhana, A. and McKay, M. (2005). Sexual violence and youth in South Africa: The need for community-based prevention interventions. *Child Abuse Neglect*, 29, 1233–48.

Pettifor, A. E., van der Straten, A., Dunbar, M. S. *et al.* (2004). Early age of first sex: a risk factor for HIV infection among women in Zimbabwe. *AIDS*, 18, 1435–42.

Pettifor, A. E., Rees, H. V., Kleinschimidt, I. *et al.* (2005). Young people's sexual health in South Africa: HIV prevalence and sexual behaviors from a nationally representative household survey. *AIDS*, 19, 1525–34.

Pettifor, A. *et al.* (2008). Keep them in school: the importance of education as a protective factor against HIV infection among young South African women. *Int. J. Epi.*

Preston-Whyte, E. and Zondi, M. (1992). Adolescent sexuality and its implications for teenage pregnancy and AIDS. *Continued Med. Educ.*, 9, 1389–94.

Pronyk, P. M., Hargreaves, J. R., Kim, J. C. *et al.* (2006). Effect of a structural intervention for the prevention of intimate-partner violence and HIV in rural South Africa: a cluster randomised trial. *Lancet*, 368, 1973–83.

Reddy, P., Meyer-Weitz, A., van den Borne, B. and Kok, G. (2000). Determinants of condom-use behavior among STD clinic attenders in South Africa. *Intl J. Sex. Transm. Dis. AIDS*, 11, 521–30.

Reddy, S. and Dunne, M. (2007). Risking it: young heterosexual femininities in South African context of HIV/AIDS. *Sexualities*, 10, 159–72.

Reddy, S. P., Panday, S., Swart, D. *et al.* (2003). *Umthenthe Uhlaba Usamila – The South African Youth Risk Behavior Survey 2002*. Cape Town: South African Medical Research Council.

Rehle, T., Shisana, O., Pillay, V. *et al.* (2007). National HIV incidence measures: new insights into the South African epidemic. *S. Afr. Med. J.*, 97, 194–9.

Rosengard, C., Adler, N. E., Gurvey, J. E. *et al.* (2001). Protective role of health values in adolescents' future intentions to use condoms. *J. Adolesc. Health*, 29, 200–7.

Ross, D. A., Changalucha, J., Obasi, A. I. *et al.* (2007). Biological and behavioral impact of an adolescent sexual health intervention in Tanzania: a community-randomized trial. *AIDS*, 21, 1943–55.

Rusakaniko, S., Mbizvo, M. T., Kasule, J. *et al.* (1997). Trends in reproductive health knowledge following a health education intervention among adolescents in Zimbabwe. *Central Afr. J. Med.*, 43, 1–6.

Sampson, R. J., Raudenbush, S. W. and Earls, F. (1997). Neighborhoods and violent crime: a multilevel study of collective efficacy. *Science*, 277, 918–24.

Sayles, J. N., Pettifor, A., Wong, M. D. *et al.* (2006). Factors associated with self-efficacy for condom use and sexual negotiation among South African youth. *J. Acquir. Immune Defic. Syndr.*, 43, 226–33.

Shisana, O. and Simbayi, L. (2002). *Nelson Mandela/HSRC Study of HIV/AIDS South African National HIV Prevalence, Behavioral Risks and Mass Media Household Survey 2002*. Pretoria: Human Sciences Research Council.

Shisana, O., Stoker, D., Simbayi, L. C. *et al.* (2004). South African national household survey of HIV/AIDS prevalence, behavioral risks and mass media impact – detailed methodology and response rate results. *S. Afr. Med. J.*, 94, 283–8.

Shuey, D. A., Babishangire, B. B., Omiat, S. and Bagarukayo, H. (1999). Increased sexual abstinence among in-school adolescents as a result of school health education in Soroti district, Uganda. *Health Educ. Res.*, 14, 411–19.

Simbayi, L. C., Chauveau, J. and Shisana, O. (2004). Behavioral responses of South African youth to the HIV/AIDS epidemic: a nationwide survey. *AIDS Care*, 16, 605–18.

Singh, J. A., Abdool Karim, S. S., Abdool Karim, Q. A. *et al.* (2006). Enrolling adolescents in research on HIV and other sensitive issues: lessons from South Africa. *PLoS Med.*, 3, e180.

Sliep, Y. and Meyer-Weitz, A. (2003). Strengthening social fabric through narrative theatre. *Intervention*, 1, 45–6.

Springer, A., Parcel, G., Baumler, E. and Ross, M. (2006). Supportive social relationships and adolescent health risk behavior among secondary school students in El Salvador. *Social Sci. Med.*, 62, 1628–40.

Stanton, B. F., Li, X., Kahihuata, J. *et al.* (1998). Increased protected sex and abstinence among Namibian youth following a HIV risk-reduction intervention: a randomized, longitudinal study. *AIDS*, 12, 2473–80.

STATS SA (2006). *Mortality and Causes of Death in South Africa, 2003 and 2004: Findings from Death Notification.* Pretoria: Statistics South Africa.

Stoneburner, R. L. and Low-Beer, D. (2004). Population-level HIV declines and behavioral risk avoidance in Uganda. *Science*, 304, 714–18.

Strebel, A. (1996). Prevention implications of AIDS discourses among South African women. *AIDS Educ. Prev.*, 8, 352–74.

Taylor, M., Dlamini, S. B., Nyawo, N. *et al.* (2007). Reasons for inconsistent condom use by rural South African high school students. *Acta Paed.*, 96, 287–91.

Ulin, P. R. (1992). African women and AIDS: negotiating behavioral change. *Social Sci. Med.*, 34, 63–73.

UNAIDS (2004). *Report on the Global AIDS Epidemic.* Geneva: Joint United Nations Program on HIV/AIDS.

UNAIDS (2007a). *2007 AIDS Epidemic Update.* Available at http://data.unaids.org/pub/EpiSlides/2007/2007_EpiUpdate_en.pdf (accessed 11-27-2007). Geneva: Joint United Nations Program on HIV/AIDS.

UNAIDS (2007b). *Practical Guidelines for Intensifying HIV Prevention.* Geneva: Joint United Nations Program on HIV/AIDS.

UNAIDS, UNICEF & WHO (2002). *Young People and HIV/AIDS: Opportunity in Crisis.* Geneva: UNAIDS UNICEF & WHO.

UNESCO (2002). *EFA Global Monitoring Report: Education for All. Is the World on Track?* Paris: UNESCO.

UNICEF (2003). *Briefing Notes on Gender and HIV/AIDS: UNICEF Guiding Policies and Strategies, Statistics, and Select Case Studies.* New York, NY: UNICEF.

Varga, C. A. (1997). Sexual decision-making and negotiation in the midst of AIDS: youth in KwaZulu-Natal South Africa. *Health Trans. Rev.*, 7, 45–67.

Varga, C. A. (2003). How gender roles influence sexual and reproductive health among South African adolescents. *Studies Family Plan.*, 34, 160–72.

Visser, M. (1996). Evaluation of the First AIDS Kit, the AIDS and lifestyle education program for teenagers. *S. Afr. J. Psychol.*, 26, 103–13.

Wellings, K., Collumbien, M., Slaymaker, E. *et al.* (2006). Sexual behavior in context: a global perspective. *Lancet*, 368, 1706–28.

White, R., Cleland, J. and Carael, M. (2000). Links between premarital sexual behavior and extramarital intercourse: a multi-site analysis. *AIDS Care*, 14, 2323–31.

Wilkinson, D., Abdool Karim, S. S., Williams, B. and Gouws, E. (2000). High HIV incidence and prevalence among young women in rural South Africa: developing a cohort for intervention trials. *J. Acquir. Immune Defic. Syndr.*, 23, 405–9.

Williams, B., Gouws, E., Colvin, M. *et al.* (2000). Patterns of infection: using age prevalence data to understand the epidemic of HIV in South Africa. *S. Afr. J. Sci.*, 96, 305–40.

Wood, K. and Jewkes, R. (2006). Blood blockages and scolding nurses: barriers to adolescent contraceptive use in South Africa. *Reprod. Health Matters*, 14, 109–18.

Worth, D. (1989). Sexual decision-making and AIDS: why condom promotion among vulnerable women is likely to fail. *Studies Family Plan.*, 20, 297–307.

Interventions with incarcerated persons

16

Ank Nijhawan, Nickolas Zaller, David Cohen and
Josiah D. Rich

Over the past three decades, the correctional system in the United States has had a
dramatic and unprecedented rise in census. Most of this increase is due to incarceration of two overlapping populations: those with the disease of addiction, and those
with other mental illnesses. Untreated, the natural history of these diseases leads to
behaviors that often result in interaction with the criminal justice system and, ultimately, incarceration. The rise in incarceration in the United States can be viewed
as a failure of society to otherwise address these two diseases. Incarceration is a
fundamentally flawed, extremely expensive and inefficient approach to these two
major societal problems. With this massive warehousing of people, we are embarking on a vast social experiment that will likely have dramatic and unanticipated
implications in the years to come, reaching far beyond the actual individuals being
incarcerated. One of these unanticipated consequences has been a disproportionate
number of imprisoned individuals with and at risk for HIV infection. The epidemic
of incarceration provides a tremendous opportunity to implement HIV prevention
measures in a population that is often marginalized by the US health-care system,
and poses a high risk of proliferating the HIV epidemic. This chapter reviews the
experience to date with interventions in the incarcerated setting that have relevance
to HIV prevention, and is predominantly focused upon the United States, which
has the highest rates of incarceration in the world.

Epidemiology: the epidemic of incarceration in the United States

At mid-year 2006, more than 7 million people lived under the jurisdiction of the
US correctional authorities, and more than 2.2 million inmates were incarcerated

444

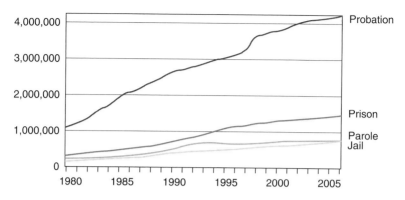

Figure 16.1 Adult correctional populations in the United States, 1980–2006. Source: US Bureau of Justice Statistics (2008).

in state and federal prisons (>1.5 million) and held in local jails (>750,000). As a proportion of the population, 753 of every 100,000 (or 1 of every 133) Americans were incarcerated (Sabol *et al.*, 2007). The 2006 statistics were the highest point of a precipitous 30-year rise in the rate of incarceration dating back to the "get tough on crime" legislation of the 1970s, which does not appear to be slowing (Figure 16.1). Altogether, these data represent both the greatest number and greatest proportion of individuals serving time in any country in the world (Beckwith and Poshkus, 2007).

The disproportionate rate of incarceration of minorities

The high incarceration rates in the United States have a disproportionate effect on minorities. According to the 2000 US census, at that time 12.3 percent of the population in the United States was African-American and 12.5 percent Hispanic/Latino (US Census Bureau, 2002). Of the more than 2 million men incarcerated at mid-year 2006, 40.9 percent were African-American and 20.9 percent were Hispanic/Latino (Sabol *et al.*, 2007). Similarly, of the 200,000 women incarcerated at mid-year 2005, 34 percent were African-American and 16 percent were Hispanic/Latino. In total, an estimated 4.8 percent of African-American men and 1.9 percent of Hispanic/Latino men were incarcerated, versus only 0.7 percent of Caucasian men. Most alarmingly, over 11 percent of African-American males between the ages of 25 and 34 were imprisoned at mid-year 2006. Similarly, African-American women were incarcerated in prison or jail at nearly four times the rate of Caucasian women, and Hispanic/Latina women were detained at nearly twice the rate of Caucasian women (Sabol *et al.*, 2007).

445

Substance use and dependence in incarcerated populations

As noted above, the epidemic of incarceration in the United States is inextricably linked to the national epidemic of substance dependence. An estimated 60–83 percent of all people in correctional facilities have used drugs at some point in their lives (Walter, 2001). According to the *2004 Survey of Inmates in State and Federal Correctional Facilities*, 53 percent of all state prisoners and 45 percent of all federal prisoners met the DSM-IV criteria for drug dependence, while 56 percent of state inmates and 50 percent of federal inmates reported drug use in the month prior to incarceration. Among state prisoners who were dependent on drugs or alcohol, 53 percent had had at least three prior sentences to probation or incarceration (Mumola and Karberg, 2006).

Participation rates in a range of in-house drug-dependence programs by drug-dependent inmates in state (39 percent) and federal (45 percent) prisoners reached new highs in 2004. These inmates mostly reported taking part in self-help groups, peer counseling and education programs. In 2005, it was estimated that one-third of male and one-half of female inmates needed residential treatment for substance dependence (Belenko and Peugh, 2005). However, only 15 percent in state and federal prisons received any form of substance-dependence treatment with a trained professional (Mumola and Karberg, 2006).

Overlapping risk factors: mental illness, homelessness, sexual abuse

Inmates in US correctional facilities often have a constellation of risk factors for HIV infection in addition to histories of substance dependence and abuse. Jails and prisons have become a storehouse for individuals with mental illness. Based on self-reported histories of mental health treatment or DSM-IV criteria for mental disorders, it was estimated that at mid-year 2005 more than half of all prison and jail inmates had a mental health problem. Approximately 50 percent of prison and jail inmates reported symptoms that met the criteria for mania, 25 percent reported symptoms of major depression, and 20 percent reported symptoms that met the criteria of a psychotic disorder. Only one in three state prisoners and one in six jail inmates with a mental health problem had received treatment since admission (James and Glaze, 2006).

In addition to mental illness, inmates frequently have histories of homelessness, physical and sexual abuse, and unstable family environments. These risk factors are often interrelated. Three-fourths of prison and local jail inmates who had a mental health problem also met the criteria for substance dependence (James and Glaze, 2006). Drug-dependent inmates (14 percent) were more than twice as likely as non-drug-dependent inmates (6 percent) to report being homeless during the year before admission to prison (Mumola and Karberg, 2006); 42 percent of drug-dependent inmates received public assistance growing up, 45 percent lived in single-parent homes, and 41 percent had a substance-using parent. In contrast, 31 percent of non-drug-dependent inmates received public assistance growing up, 39 percent grew up in single-parent homes, and 24 percent grew up with a substance-using parent (Mumola and Karberg, 2006). Drug-dependent state prisoners (23 percent) were also more likely to report a history of physical or sexual abuse (15 percent) than were non-drug-dependent inmates (Mumola and Karberg, 2006).

The epidemiology of HIV/AIDS in correctional settings

Over the past two decades, the burden of HIV/AIDS in the United States has become heavily concentrated in the correctional system. At the end of 2005, correctional authorities reported that a total of 20,888 state inmates (1.8 percent) and 1592 federal inmates (1.0 percent) were diagnosed with HIV infection or had confirmed AIDS (Maruschak, 2007). This represents a prevalence of HIV that is five times higher in US state and federal prisons than in the general population (Spaulding *et al.*, 2002). Similarly, the proportion of AIDS cases is three-and-a-half times higher in prisons than it is in the general population (Maruschak, 2004). Although fewer data are available for the US jail population, estimates place the prevalence of HIV/AIDS in US jails in the range of 2.1–2.5 percent (Maruschak, 2004). In Rhode Island, approximately one-third of all the individuals who tested positive statewide were identified at the Rhode Island Department of Corrections (Desai *et al.*, 2002). Overall, it is estimated that between 23 percent and 31 percent of people living with HIV/AIDS in the United States pass through the US correctional system each year (Hammett *et al.*, 2002). Several studies suggest that between one-third and one-half of these persons are not aware of their HIV status (Sabin *et al.*, 2001; Altice *et al.*, 2005).

People with known HIV infection in correctional settings are disproportionately female and of an ethnic minority. As a percentage of the total mid-year 2006 prison-custody population, female inmates (2.3 percent) had a higher prevalence of HIV than did male inmates (1.7 percent). Similarly, a survey of 900,000 state prison inmates in 2004 found that 2.0 percent of African-American and 1.8 percent of Hispanic/Latino inmates were HIV-positive, versus a 1.0 percent prevalence among white inmates. In total, African-Americans accounted for 53 percent and Hispanic/Latinos for 25 percent of known prison HIV/AIDS cases in 2004 (Maruschak, 2007; Sabol *et al.*, 2007).

At mid-year 2006, the correctional facilities in the northeast had the highest number of known HIV/AIDS cases as a percentage of their total incarcerated populations (3.9 percent), followed by the south (2.2 percent), the midwest (0.9 percent) and the west (0.7 percent). In total, 47 percent of known correctional HIV/AIDS cases were in the south, 29 percent were in the northeast, and the midwest and west accounted for approximately 10 percent of correctional HIV/AIDS cases each. Three states – New York, Florida and Texas – accounted for nearly half (49 percent) of the HIV/AIDS cases in state prisons.

Nationally, the prevalence of diagnosed HIV infection in state prisons decreased from 2.2 percent to 1.6 percent between 1997 and 2004. This decrease in HIV prevalence was uniform across race and gender, and was spurred largely by a reduction of new HIV infections among drug-offending inmates, whose HIV prevalence dropped from 2.9 percent in 1997 to 1.8 percent in 2004. Since 1999, the number of cases of HIV/AIDS has decreased moderately in the west and midwest, and has decreased by an impressive 35 percent in the northeast. However, the number of HIV/AIDS cases in a contingency of state prison systems in the deep south – Florida, Georgia, Louisiana, North Carolina and Mississippi – each rose between 27 and 57 percent during this time period (Maruschak, 2007).

Despite the reduction in HIV/AIDS-associated mortality in the past decade, HIV/AIDS remains a major cause of death in correctional facilities. It is estimated that, in 2005, 176 state inmates and 27 federal inmates died from AIDS-related causes. AIDS-related deaths accounted for 1 in 20 deaths reported in state prisons, and 7 percent of all deaths in federal prisons (Maruschak, 2007).

Hepatitis, STIs and tuberculosis

Hepatitis, tuberculosis and STIs other than HIV are also common in correctional settings. We found in Rhode Island that the prevalence rates of hepatitis B and hepatitis C among inmates were 20.2 percent and 23.2 percent, respectively (Macalino *et al.*, 2004). According to a national study, 29–43 percent of all those infected with hepatitis C virus and 40 percent of all those infected with tuberculosis passed through the US correctional system in 1997 (Hammett *et al.*, 2002).

Similarly, the overall prevalence in prisons and jails of syphilis (2.6–4.3 percent), gonorrhea (1.0 percent) and chlamydia (2.4 percent) were all substantially higher than in the US general population (Altice and Springer, 2005; CDC, 2007). Inmates with HIV are frequently co-infected with these diseases, and there is strong evidence that STIs increase the risk of HIV transmission (Sangani *et al.*, 2004). Controlling these interrelated epidemics is highly relevant to managing the HIV epidemic in correctional settings and in the community.

Intra-prison transmission of HIV

Early on in the HIV/AIDS epidemic, it was assumed by some that there would be high rates of transmission of HIV behind bars and that correctional facilities would serve as "incubators" for infection and fuel the epidemic through direct transmission. Studies to date have documented that transmission does exist, but high rates of transmission have not been found in the United States.

Due to the variety of high-risk behaviors that inmates engage in while incarcerated, such as male–male sex, drug use, and tattooing, the potential exists for HIV transmission between them (Krebs and Simmons, 2002; CDC, 2006a). Among former prisoners in Rhode Island, 31 percent of inmates who had a history of injection drug use prior to imprisonment had used illicit drugs in prison, and nearly half of these former prisoners had injected drugs while incarcerated (Clarke *et al.*, 2001). A 1982 study of sex within the US prison system, conducted by the Federal Bureau of Prisons, found that 30 percent of inmates in federal prisons engaged in homosexual activity while imprisoned (Nacci and Kane, 1982).

Despite the high-risk behavior that inmates engage in while imprisoned, numerous studies have shown intra-prison transmission of HIV to be relatively rare. One study estimated the rate of new HIV infections in inmates at 0.41 percent per year (Brewer *et al.*, 1988). A separate longitudinal study of 3837 inmates in a state prison system with 2.4 percent HIV prevalence showed only two seroconversions – a rate of new HIV infections of one conversion per 604 person-years (Horsburgh *et al.*, 1990). More recently, we found no new HIV infections among 446 inmates observed for 694 person-years in the Rhode Island correctional system despite a 1.8 percent HIV/AIDS prevalence (Macalino *et al.*, 2004).

Effects of incarceration on the community

The major public health danger posed by the high concentration of HIV-positive individuals in the US prison system may be the effect of their release into the community. In 1993, an estimated 7.2 million individuals were released from jails – a number which has almost certainly increased in the decade and a half since (Hammett *et al.*, 2002). One-fourth of the HIV-infected population is released from a prison or jail each year, representing approximately 250,000 HIV-positive persons re-entering the community after incarceration (Hammett *et al.*, 2002). When not adequately linked to primary care upon release, HIV-positive inmates receiving HAART while incarcerated are likely to see substantial increases in HIV viral load due to discontinuation of therapy post-release (Stephenson *et al.*, 2005). More alarmingly, one study showed that nearly 4 in 10 male inmates engage in risky sexual behavior within 6 months of release (MacGowan *et al.*, 2003). In another cohort study of HIV-positive inmates, 47 percent reported sexual activity shortly after release (mean 45 days). In the same study, 26 percent reported engaging in unprotected sex with their regular partners, and 33 percent of releases with regular partners reported having unprotected sex with HIV-seronegative partners (Stephenson *et al.*, 2006).

Prevention interventions

The United States Centers for Disease Control and Prevention (CDC) issued new HIV-testing guidelines in 2006 which recommend routine HIV testing as part of

449

clinical care in all health-care settings, including correctional health-care facilities (CDC, 2006b). Due to socio-economic factors such as mental illness, active substance abuse, unemployment and racial disparities in health care, inmates represent a population that often lies outside of the sphere of traditional health care. For many inmates, the correctional setting represents the first point of access to such a system (Glaser, 1998; Conklin *et al.*, 2000).

A variety of different correctional testing policies exist, including:

- mandatory testing;
- routine voluntary testing offered to all inmates;
- targeted voluntary testing based on risk assessment;
- diagnostic testing – offered in response to signs/symptoms of HIV infection;
- non-routine, voluntary testing at request of the inmate; and
- court-ordered testing.

The majority of correctional facilities offer HIV testing when requested by the inmate or when clinically recommended (Spaulding *et al.*, 2002). HIV testing is offered in all the federal and state prison systems within the United States. The Bureau of Justice Statistics (BJS) reported that, in 1999, 19 state prison systems employed mandatory testing at intake; 3 of these states also mandated testing upon release. The remaining 31 prison systems implemented various forms of non-voluntary HIV testing, with all but three offering either purely voluntary HIV testing or HIV testing in response to clinical indications (Maruschak, 2001).

In correctional settings that offer purely voluntary HIV testing, participation rates are generally low – in some cases, under 20 percent (Liddicoat *et al.*, 2006). Furthermore, in states that employ voluntary HIV-testing programs, prevalence estimates generated from voluntary testing have consistently been lower than those determined by blinded surveillance, suggesting that highest-risk individuals volunteer to be tested less readily (New York State AIDS Advisory Council, 1999). HIV-infected persons who do not undergo testing in voluntary programs in the correctional setting are likely to remain undiagnosed, given that many are marginalized from the health-care system in their communities. Thus, there is a large population of HIV-infected inmates who have undiagnosed or undisclosed HIV infection, who do not receive medical care, and will not otherwise be linked to medical and social services or receive prevention counseling upon release. While issues of stigma and obstacles to adequate health care may serve as barriers for inmates otherwise willing to be tested for HIV, these barriers will not simply disappear if fewer people are tested.

The case for routine opt-out testing

Many have championed routine opt-out testing as the preferred method of HIV testing, given its benefit to inmates, the correctional institution and the community

(Basu *et al.*, 2005). This method of testing establishes a norm in which HIV testing is routinely offered to all inmates and administered to all except those inmates who exercise their right to refuse. By removing the stigma associated with risk-based testing, routine opt-out methodology minimizes the psychological barriers to such testing (Simmons *et al.*, 2003). Furthermore, by eliminating the need for client-centered risk assessment and pre-test counseling, this approach has been shown to be a cost-effective method for testing a high volume of individuals (Phillips and Fernyak, 2000).

Routine, opt-out HIV testing has been shown to be feasible in jails as well as prisons, and has been associated with high rates of participation by inmates. In the Rhode Island correctional system, where routine opt-out testing has been administered at intake since 1991, 90 percent of inmates agree to be tested (Flanigan, 1995). Subsequently, approximately one-third of all persons in the entire state who have been diagnosed with HIV have been identified in the state correctional facility (Rich, Dickinson *et al.*, 1999). In a Massachusetts county prison, switching from non-routine voluntary testing to routine opt-out testing increased the rate of participation in testing from 18 percent to 73 percent. Additionally, of the inmates who had been tested in the previous year, nearly 80 percent had been tested in a correctional facility, demonstrating the public health significance of establishing HIV screening as standard practice in correctional settings for testing hard-to-reach populations (Liddicoat *et al.*, 2006).

Confidentiality

While every effort must be made to ensure confidentiality surrounding testing and counseling, the potential for abuses should not outweigh the potential for benefit to an inmate who tests positive for HIV infection and is therefore offered medical care while still incarcerated and linkage to medical care once released back into the community, and the opportunity to prevent the spread of HIV to others. Separating HIV-positive inmates from the rest of the incarcerated population, refusal of specific work assignments and other negative consequences that have been attributed to expansion of HIV testing in this setting should not serve as barriers to implementing routine testing policies. However, these practices are discriminatory, and it is the duty of all to exercise extreme vigilance to expose and prevent them.

Rapid HIV testing

With FDA approval of rapid HIV tests, new opportunities for correctional screening programs currently exist. Rapid HIV-testing technology can be implemented on site with minimal laboratory work, and results can be delivered in approximately 20 minutes in conjunction with result-specific post-test counseling and

451

risk-reduction interventions. This is particularly important in jail systems because high turnover rates are often associated with difficulties in implementing routine testing using standard HIV antibody testing, due to the 2-week time period necessary for results. Correctional testing programs that offer routine HIV testing to all persons entering jail ideally deliver test results prior to release, or in the community following release. It is now possible to deliver rapid test results immediately, in conjunction with post-test counseling, linkage to care, and risk-reduction interventions. Although, to date, rapid HIV testing has not been standardized in correctional settings, preliminary studies in the Rhode Island Department of Corrections have shown it to be feasible in a jail and acceptable to inmates (Beckwith *et al.*, 2007a). In a related survey of Rhode Island jail inmates, 88 percent said that they would prefer rapid HIV testing to standard HIV antibody testing in the jail setting (Beckwith *et al.*, 2007b). In fact, rapid testing may lead to increased testing overall, as a Maryland study found that 15 percent of female inmates and 22 percent of male inmates would not have been tested for HIV if the oral test had not been offered (Bauserman *et al.*, 2001).

Treatment and referral as secondary prevention

Health-care and prevention programs can be effectively administered in the incarcerated setting because clients are logistically easier to access, and are relieved from the financial burdens of medical care (Braithwaite and Arriola, 2003). In addition, diagnosing inmates while they are incarcerated affords the opportunity to develop a more comprehensive discharge plan that includes continuity of care post-release. Particularly effective are collaborations between community-based organizations and correctional facilities which involve service provision within the correctional setting and follow-up care post-release (Klein *et al.*, 2002). This model has been well demonstrated in Rhode Island through Project Bridge, which is based on collaboration between co-located medical and social-work staff. The primary goal of Project Bridge is to increase continuity of medical care for former inmates post-release through case management. In the first 3 years of the program, social workers were successfully able to follow 90 percent of enrolled offenders over 18 months. Of those in need of medical care, 75 percent received specialty medical care from community providers and 100 percent received HIV-related medical services. Of those requesting substance-dependence treatment, 67 percent consistently kept appointments at community substance-dependence treatment programs (Rich *et al.*, 2001). Project Bridge has demonstrated the feasibility of maintaining HIV-positive ex-offenders in medical care by providing ongoing case management services following release from incarceration.

Maintaining inmates in care during incarceration, and (often more challenging) after release, may well have a dramatic impact on reducing further spread of HIV both during incarceration and after release.

HIV prevention education for inmates

In response to the disproportionately high prevalence of HIV infection among inmates, correctional institutions need to focus more on providing comprehensive prevention services to this population. As Barry Zack writes in *Public Health Behind Bars*, "effective HIV prevention efforts targeted towards the incarcerated population will impact not only HIV incidence among inmates but will also provide them with the knowledge and skills necessary to prevent infection and/or transmission in the communities to which most inmates return" (Zack, 2007). Incarcerated settings offer an ideal opportunity to provide critical HIV-prevention education to high-risk individuals who are traditionally hard to reach, and who will in many cases be returning to communities that have been hardest hit by the HIV epidemic.

To date, most studies of HIV prevention interventions within the incarcerated setting have compared different interventions (el-Bassel *et al.*, 1995; St Lawrence *et al.*, 1997), employed a non-randomized study design (Grinstead *et al.*, 2001), or were performed without a control group (West and Martin, 2000). Current literature provides little evidence as to what constitutes a successful prison based HIV prevention intervention (Bryan, Robbins *et al.*, 2006). In a recent review of best-evidence-based interventions for HIV prevention by the CDC's HIV/AIDS Prevention Research Synthesis team, none of the 18 interventions reviewed between 2000 and 2004 were targeted towards incarcerated populations (Lyles *et al.*, 2007). In his review, Zack identified nine published studies on effective HIV prevention interventions, four of which found reductions in post-release HIV risk behavior and five of which found an impact on attitudes, self-efficacy and intentions (Zack, 2007).

A recent HIV prevention study by Bryan and colleagues examined the effects of racial and gender differences on HIV prevention behavior (Bryan *et al.*, 2006). In this study, the authors recruited 37 groups of inmates to participate in a multi-session HIV-prevention education intervention. The authors found that the intervention was associated with increases in HIV knowledge, condom attitudes, condom self-efficacy, condom intentions, self-efficacy for not sharing needles, peer education self-efficacy, peer education intentions, and peer education behavior (Bryan *et al.*, 2006). One important finding from this study was that among Hispanic/Latino participants there were lower gains in condom-use self-efficacy and in condom-use intentions, as well as lower gains in positive attitudes toward not sharing needles or tattooing equipment. As the authors point out, however, this may have been attributable to a lack of cultural appropriateness of the intervention, and/or that it was only given in English (Bryan *et al.* 2006). Ehrman focused on three effective HIV prevention program models which included community based organization partnerships and encompassed three primary domains: peer education (Centerforce in San Quintin State Prison, CA); discharge planning (Empowerment Through HIV/AIDS Information, Community and Services

(ETHICS) New York); and transitional case management (Transitional Services Unit, The Women's Prison Association, New York City). Collectively, these models demonstrate that interventions must be tailored to specific correctional institutions, that community-based organizations have an important role to play in prevention efforts, and must be creative, flexible and accommodating, and that prevention programs such as the one highlighted in the Ehrman's review are cost-effective (Ehrmann, 2002). Peer-led interventions are also being adapted in international correctional settings, such as South Africa (Sifunda *et al*., 2006).

Another prevention intervention that has shown promise is the Maryland Prevention Case Management (PCM) model, which combines individual counseling with group sessions. The PCM intervention focuses on several specific domains: personalizing HIV/AIDS risk and risk-reduction; transition to the community; condom and other preventative measures; and substance use (Bauserman *et al*., 2003). The Maryland intervention has been given to nearly 3000 inmates in more than 12 local jails throughout the state. The PCM model has been associated with increase in condom use, self-efficacy for injection drug and other substance use, and intentions to practice safer sex post-release (Bauserman *et al*., 2003). While the PCM participants may not have been completely representative of the entire prison and jail population of Maryland as a whole, the PCM model still demonstrated success in reaching inmates in need of HIV-prevention education.

There have been attempts in many correctional settings to integrate substance-dependence services with HIV prevention services. In an analysis of the National Treatment Improvement Evaluation Study, which included 1223 HIV-negative inmates in 9 correctional substance-dependence treatment programs, HIV prevention services within the nine respective correctional substance-dependence treatment programs were shown to be beneficial at reducing HIV-related risk behaviors among inmates expected to be released in the near future (Lubelczyk *et al*., 2002). Another novel intervention is the Jailbreak Health Project in Australia, which broadcasts a range of opinions, music and poetry from inmates to other inmates and the community. The project engages its listeners, inmates, former inmates and families of inmates in a variety of topics, including HIV and viral hepatitis prevention, and sexual health (Minc *et al*., 2007). Overall, it should be noted that any correctional prevention effort should focus on the following core components (Zack, 2007):

1. Type of intervention
2. Timing of the program
3. Content
4. The messenger.

Condoms in the incarcerated setting

Although the provision of condoms to inmates has been advocated to prevent the transmission of HIV and other sexually transmitted infections within correctional settings, this policy recommendation has been met with resistance in most correctional institutions (see also Chapter 18). One of the primary reasons for this is that sex (consensual or otherwise) in prisons and jails is generally prohibited, and the provision of condoms in such facilities would implicitly acknowledge that sex, including rape, does occur. This seeming contradiction, which has been coined the "catch 22 of condoms in US correctional facilities", should not override the public health considerations that condoms are an essential part of HIV/STI prevention, that there is a disproportionately high prevalence of HIV and STIs among correctional populations, that high-risk behavior for HIV/STI transmission is known to occur in correctional institutions, and that there exist several model programs which have been successful in distributing condoms to inmates (Tucker *et al.*, 2007).

In the United States, condoms are currently available in fewer than 1 percent of all correctional facilities (May and Williams, 2002). Locations where condoms are given to inmates include the Vermont and Mississippi state prison systems, and some New York City, Philadelphia, San Francisco and Washington DC jails (Braithwaite *et al.*, 1996; May and Williams, 2002). Washington DC has offered condoms to jail inmates since 1993, when the then Mayor Sharon Pratt Kelly's proposal to give condoms to inmates became policy (May and Williams, 2002). Globally, condoms are offered throughout prison systems in many countries including Canada, Australia, South Africa, Brazil and parts of Western Europe (May and Williams, 2002). A recent World Health Organization report on interventions to manage HIV prevention in prisons notes that the WHO recommended making condoms available to prisoners as early as 1993, and a 1991 WHO report found that 23 of 52 prison systems that were surveyed distributed condoms to prisoners (WHO, 2007a). The report issues five recommendations around the provision of condoms in the correctional setting (WHO, 2007a):

1. Condom distribution programs should be implemented and expanded to scale as soon as possible
2. Condoms should be made both readily and discreetly accessible to inmates
3. Water-based lubricant should be provided in addition to condoms
4. Education and information for inmates and correctional staff should be available prior to implementation of condom distribution programs
5. Women inmates should have access to both condoms and dental dams.

There is evidence that condom programs are feasible to implement and that inmates and correctional staff are supportive of such programs. The correctional system in New South Wales, Australia, began providing condoms to inmates in 1996. A 10-year analysis of this program revealed that many of the initial concerns of prison staff and inmates were not realized. For example, between 1996 and 2005 there was no increase in consensual male–male sex or in male sexual assaults (Yap *et al.*, 2007). And while there was use of condoms for non-sexual purposes (i.e., storage of contraband), there was not an associated increase in illegal drug activity (Yap *et al.*, 2007). In the United States, a survey in 2000–2001 in Washington DC demonstrated that 55 percent of jail inmates and 64 percent of correctional officers, respectively, agreed that it was a good idea to provide condoms to inmates (May and Williams, 2002). Additionally, 58 percent of inmates surveyed stated that they did not feel that providing condoms to inmates would result in an increase in sexual behavior (May and Williams, 2002). It is important to note that no correctional system which has allowed condom distribution to its inmates has reversed its decision to do so (WHO, 2007a).

Despite the WHO recommendation that all correctional systems in the world should allow for the distribution of condoms in their facilities, numerous barriers to implementing such measure still exist, including resistance from governments, prison staff, and inmates themselves. The most salient example of this resistance was in 1997, when prison guards went on a 3-day unofficial strike at St Catherine's District Prison and Kingston's General Penitentiary in Jamaica after Commissioner of Corrections John Prescod announced that condoms would be distributed. Sixteen prisoners were killed in the riots that ensued (*New York Times*, 1997). More recently, in a survey of Zambian prisoners, 68 percent reported being opposed to making condoms available to inmates (Simooya and Sanjobo, 2002). In Australia, despite the data provided by the New South Wales prison system, the majority of states still do not allow condom distribution (Yap *et al.*, 2007a). In the United States, numerous barriers continue to exist around the provision of condoms in incarcerated settings. Legislation was passed in California in 2006 that would have made condom distribution in correctional institutions in that state legal; however, Governor Arnold Schwarzenegger vetoed the legislation. On a more optimistic note, in 2007 Representative Barbara Lee (D-CA) reintroduced legislation in the US House of Representatives which would require federal correctional institutions to allow condom distribution and HIV/STI counseling and testing to all federal inmates (Dolinsky, 2007).

The silver lining to the "catch-22 of condoms in US correctional facilities" is that there is scant evidence for the effectiveness of condoms in these settings. Although the prevalence of infectious diseases is well documented in incarcerated populations, with a spectrum of sexual contact ranging from consenting adults to coercive sex to frank rape, it is not clear that condoms would be used in all of these situations. In addition, many inmates may not want to acknowledge MSM acts, or to be condemned for an activity that on some level they do not want to acknowledge even to themselves. Nevertheless, this is not a reason to continue to withhold condom availability.

Substance-dependence treatment in correctional facilities

In recognition of the significance of the epidemic substance use in the US correctional system, the Federal Bureau of Prisons issued recommendations for increased attention to addiction treatment in the Violent Crime Control and Law Enforcement Act of 1994, which requires drug treatment to be available to "all eligible prisoners prior to their release" (Walter, 2001). Overall, there are few data regarding the types of drug treatment programming that are effective within the correctional setting. While a recent survey found that 94 percent of federal prisons, 56 percent of state prisons and 33 percent of jails in the United States offer some sort of on-site substance-use treatment to inmates (SAMHSA, 2002), the definition of "treatment" is not uniform across correctional settings (Mears *et al.*, 2003). In the 1990s, a US panel of experts was assembled to produce the Center For Substance Abuse Treatment's (CSAT) *Planning for Alcohol or Drug Abuse Treatment for Adults in the Criminal Justice System*, which outlined a group of core principles around drug treatment in the correctional setting. These principles are as follows (Vigdal and Stadler, 1992):

1. Treatment should not represent a substitute for punishment or sanctions
2. Treatment should be universally available as needed for persons with drug treatment needs
3. Alcohol- or drug-abuse treatment services should be tailored to the needs of the specific offender, based on a thorough assessment at jail or prison intake
4. Offender supervision should continue once an individual enters treatment
5. Offenders should remain accountable to the sentencing judge or probation/ parole authorities (Vigdal and Stadler, 1992).

The most common types of substance-use related programs in correctional facilities are Alcoholics Anonymous (AA), Narcotics Anonymous (NA), and education and awareness programs (Mears *et al.*, 2003). Substance-use education and awareness programs are established in 74 percent of prisons in the US (Taxman *et al.*, 2007). Services for inmates with substance-use histories available within correctional facilities also encompass several other modalities of varying involvement. Among these are detoxification; drug testing; individual and group counseling; outpatient, drug free; milieu therapy (intensive counseling and separate living conditions); family therapy; in-patient short-term; residential; pharmacological therapy; and transitional services (Mears *et al.*, 2003). Education and low-intensity group counseling still account for 42 percent of all substance-use treatment services in correctional facilities, with more intensive services offered much less frequently (Taxman *et al.*, 2007). Often inmates are faced with an inadequate choice with respect to substance-use services: either too limited and short-term, such as self-help and education programs, or too intensive and expensive, such as residential programs or TCs (Belenko and Peugh, 2005). This may explain why only 24 percent of inmates who are in need of treatment reported having received drug treatment since their admittance into a state prison, among those surveyed in a 1997 Survey of Inmates in State Correctional Facilities (Belenko and Peugh, 2005).

Therapeutic communities (TCs) in prisons and jails

With the creation of the Residential Substance Abuse Treatment (RSAT) initiative in 1994, modified Therapeutic Communities (TC) within prisons were promoted as a novel form of substance-dependence therapy in correctional settings. In Therapeutic Communities, inmates are housed separately from the general population, and participate in several months of intensive rehabilitation, self-help and peer groups, drug-abuse education classes and professional counseling. Therapeutic Communities are the most common substance-use program that is offered for more than 90 days (Taxman *et al.*, 2007). There is some evidence, both nationally and internationally, to suggest that Therapeutic Communities are associated with reductions in post-release drug use and recidivism (Lipton, 1995; Pearson and Lipton, 1999). However, the effectiveness of these programs has recently been questioned. A report by the California Office of the Inspector General reported that US$1 billion had been spent on treatment programs for incarcerated persons, without evidence of effectiveness (Cate, 2007). It should also be noted that there is little evidence showing that Therapeutic Communities are significantly better than other types of residential treatment (Smith *et al.*, 2006). In addition, one of the primary challenges with providing more intensive treatment services to substance-dependent inmates is that these services may counteract the retributive and incapacitating nature of correctional facilities (Taxman *et al.*, 2007).

Opiate replacement therapy (ORT) in prisons and jails

Despite the documented evidence of methadone's beneficial effects, including reduced heroin use (Dolan *et al.*, 2003a), crime (Marsch, 1998), recidivism and HIV risk behaviors (Longshore *et al.*, 1993; Metzger *et al.*, 1993), the linkage of heroin-addicted inmates into a methadone maintenance treatment (MMT) program at the time of community re-entry has not been widely implemented. Although only a few opiate replacement therapy programs exist in prisons or jails around the world, the potential benefits of implementing such programs have also been well documented. Project KEEP, on Rikers Island, New York, successfully initiated methadone treatment within the prison setting starting in 1987. Project KEEP annually performed approximately 4000 admissions to methadone treatment, and referred inmates directly to outpatient methadone treatment on release, with 80 percent of participants reporting to their clinic (Fallon, 2001). However, long-term retention in treatment proved problematic, especially for those people without an immediate source of funds (Tomasino *et al.*, 2001). For a period of time, Project KEEP was the only methadone maintenance program for incarcerated individuals in the United States. Recently, there have been several attempts in the United States to incorporate ORT in the correctional setting. A small-scale trial of levo-alpha-acetylmethadol (LAAM) conducted in Baltimore, MD, among male inmates with pre-incarceration heroin dependence demonstrated that ORT in corrections facilitated access to ORT upon release (Kinlock *et al.*, 2002). A trial of MMT prior to release from incarceration in Baltimore also found that initiating MMT within the incarcerated setting was associated with access to MMT upon release (Kinlock *et al.*, 2007).

Despite the successes of providing ORT to inmates in Baltimore and Riker's Island, the vast majority of correctional facilities in the US do not offer MMT. We conducted a survey of methadone practices in state and federal prison systems, and found that less than half of prison facilities reported provision of any opiate replacement therapy for detoxification. Maintenance treatment was only provided for pregnant women. Additionally, discharge planning for opiate-addicted inmates often fails to refer them to ORT (Rich *et al.*, 2005). There is some support for new ORT initiatives, however. For example, an innovative program at the Maryland Department of Corrections plans to start ORT for heroin-addicted inmates just prior to release, with community follow-up. In February of 2002, Bernalillo County, NM, announced the opening of the nation's first public-health office inside a county jail that will initiate a MMT program as part of its patient services. In March of 2002, the New Mexico Medical Society became the first statewide medical society to endorse prison and jail-based opioid replacement treatment (SAMHSA, 2002). Kinlock and colleagues have demonstrated that initiation of methadone maintenance prior to release is beneficial to inmates with heroin addiction (Kinlock *et al.*, 2008).

Internationally, correctional facilities in Canada, Australia, many parts of Western Europe, parts of Eastern Europe, Iran, and Indonesia offer ORT to inmates (WHO, 2007b). In addition, other countries are in the process of or are planning on implementing ORT programs within correctional settings (Moller, 2005). However, many prison administrators have not been receptive to providing ORT to inmates due to philosophical opposition to ORT and fears about potential diversion, violence or other security concerns (Magura *et al.*, 1992). Resistance by correctional institutions persists despite the fact that, overall in the international community, ORT programs in the incarcerated setting have been shown to be effective in reducing drug use, recidivism, mortality, and rates of HIV and HCV infection, and facilitating continuity of treatment (WHO, 2007b).

Needle exchange in correctional facilities

The institution of needle-exchange programs within correctional facilities has tremendous public health potential, given the high prevalence of injection drug use that is known to persist behind bars. The first documented syringe-exchange program in a correctional facility began in Switzerland in 1994, at the Hindelbank Women's Prison. Results from this early experiment showed a 0 percent incidence of new HIV and viral hepatitis infections, decreased needle-sharing, no increase in drug consumption, and no use of needles as weapons (UNAIDS, 1997). In 1995, two German prisons implemented a syringe pilot project that demonstrated that such a program was feasible, that it had no associated increase in drug consumption, led to no incidents of the use of needles as weapons, and was associated with improvements in overall health and health knowledge in participating inmates (Jacob and Stover, 2000). Similarly, other studies of needle-exchange programs in Switzerland, Spain and Germany have found no associated increases in drug consumption, have caused no new cases of HIV or viral hepatitis infections, have led to decreases in needle-sharing and have resulted in no reported cases of needles being used as weapons (Dolan *et al.*, 2003b; Okie, 2007). Dolan and colleagues' review focused on 19 correctional syringe-exchange programs throughout the world. Evaluation of all 19 programs indicated that they are feasible to implement (Dolan *et al.*, 2003b).

The successes of the pilot needle-exchange programs in correctional facilities influenced UNAIDS in 1997 to recommend syringe exchange and making full-strength bleach available to sterilize syringes and needles in all correctional facilities (UNAIDS, 1997). However, despite the data on correctional syringe-exchange programs and the successes of such programs throughout the world, numerous obstacles remain to implementing syringe-exchange programs in incarcerated settings. In the US, similar to the issue of condom distribution, the illegality of drug use while incarcerated has stymied the conversation regarding allowing clean needles and other forms of harm reduction for injection drug users in correctional institutions. As with the provision of condoms, needle exchange in US correctional facilities will likely have to overcome considerable resistance before it is implemented on a large scale.

Needle exchange and tattooing

Similar to injection drug use behind bars, tattooing in prison is common, and carries a high risk of transmission of blood-borne pathogens. Tattooing equipment is often not sterilized after use, and is frequently shared between inmates. A cross-sectional survey in the Australian correctional system revealed that, of inmates with tattoos, 41 percent had received them in prison, 27 percent of whom reported sharing a tattoo needle and 42 percent of whom reported sharing ink (Hellard *et al.*, 2007). A study of injection drug users in New Mexico found that those who had received tattoos in prison had much higher rates of hepatitis B and C, with odds ratios of 2.3 and 3.4, respectively (Samuel *et al.*, 2001). In addition, case reports of HIV and hepatitis C transmission related to tattoos received in prison have been documented (Doll, 1988; Tsang *et al.*, 2001).

Despite the risk of transmission of hepatitis and HIV, providing prisoners with safe tattooing equipment has been met with political opposition. A safer tattooing initiative in six Canadian prisons was stopped early by the Public Safety Minister, who reported that it "failed to conclusively determine that the health and safety of staff members, inmates and the general public would be protected by maintaining this program" (Kondro, 2007). Critics of providing needles in prison for tattooing have raised similar safety concerns to those of critics of correctional syringe-exchange programs; in particular, that needles implicitly condone injection drug use, or could be used by inmates as weapons.

STD (STI) testing and treatment in the incarcerated setting

The epidemics of HIV and other sexually transmitted infections are closely intertwined due to their similar risk-behaviors, and there are compelling reasons to screen for and treat STIs as an HIV prevention intervention in correctional settings. The presence of an STI itself increases the risk of acquiring HIV (Sexton *et al.*, 2005). In addition, both ulcerative and non-ulcerative sexually transmitted infections are associated with an increased risk of HIV transmission (Cohen, 2004), and there is some evidence from population-based control studies to suggest that treatment of STIs decreases the risk of HIV transmission (Sangani *et al.*, 2004). There is also evidence that the coordination of testing for both HIV and STIs may result in higher rates of testing overall (MacGowan *et al.*, 2006). Because both chlamydia and gonoccoccal infections can be asymptomatic, may not be detected without routine screening, and are highly prevalent, correctional facilities provide an ideal location to carry out such screening.

In a CDC/HRSA-funded pilot program in five county jails in 2000 aimed at enhancing HIV/STI screening as part of the multisite Corrections Demonstration Project, 78 percent of inmates who tested positive for chlamydia, gonorrhea or syphilis received treatment for their infection (Arriola *et al.*, 2001). While this program demonstrated the feasibility of STI screening and treatment in correctional settings, STI screening in correctional facilities is still not standard practice due largely to lack of funding for STI control. However, specific populations show markedly higher prevalence of STIs, and subsequently more recent research has called for optimizing STI screening in correctional settings by targeting the highest-risk populations (Barry *et al.*, 2007). A 7-year study in 14 US juvenile detention centers, comprised of 33,615 female inmates, found the prevalence of chlamydia and gonorrhea to be 15.6 and 5.1 percent, respectively (Kahn *et al.*, 2005). As a point of comparison, the prevalence of chlamydia and gonorrhea among young adults in the general population is 4.2 and 0.4 percent, respectively (Miller *et al.*, 2004). Additional studies have shown similarly disproportionate prevalence of chlamydia and gonorrhea among female inmates under the age of 30 and male inmates who identify as having engaged in MSM sexual activity (Chen *et al.*, 2003; Barry *et al.*, 2007).

461

Urine screening for chlamydia and gonorrhea

New urine tests for chlamydia and gonorrhea have made it substantially easier to screen for these infections in correctional settings. A pilot study that encompassed jails in Chicago, Baltimore and Birmingham (Alabama) showed the feasibility and acceptability of such testing by female inmates. Of the female inmates who were approached, 87–92 percent agreed to be tested, and 61–85 percent of female inmates who tested positive for an STI subsequently received treatment in jail or in the community upon release (Mertz *et al.*, 2002). Just as the development of rapid HIV-testing technology has opened the door to wider testing practices in incarcerated settings, so too has urine testing led to more comprehensive STI screening among inmates. However, the major obstacles regarding funding, personnel and infrastructure still remain.

HIV/AIDS in international prisons

Outside of the United States, China (1.56 million) and the Russian Federation (1.2 million) have the largest prison populations in the world (Alexandrova, 2003; ICPS, 2008). With an incarceration rate of 628 per 100,000 citizens, the Russian Federation rivals the United States regarding the proportion of its citizenry incarcerated. Similar incarceration rates can be found in Georgia, Kazakhstan and other former Soviet Union (FSU) nations (ICPS, 2008). Due in large part to lack of national HIV/AIDS education programs, as well as generally slow or non-existent governmental responses to their respective burgeoning HIV/AIDS epidemics, both China and FSU countries are anticipated to have – and are likely currently undergoing – explosive epidemics of HIV/AIDS. However, relatively little is known about how HIV/AIDS is being managed in correctional facilities in China and FSU nations, due to the limited information disclosed by their respective governments.

China did not publicly acknowledge the existence of an HIV/AIDS epidemic until 2001. In 2007, the Chinese government estimated that 650,000 people in the country had been diagnosed with HIV infection, and that another 500,000 were infected but unaware of their status (Gill and Okie, 2007). The annual budget for HIV/AIDS prevention has grown from US$12.5 million in 2001 to US$185 million in 2007 (*People's Daily*, 2005). Approximately half of China's HIV-positive population contracted HIV through injection drug use. In 2005, the Chinese government authorized the creation of methadone maintenance programs, as well as needle-exchange programs. By 2006, there were 320 methadone maintenance programs throughout the country (Gill and Okie, 2007).

Chinese governmental policy currently appears to be one of isolation and containment for prisoners with HIV/AIDS. The prevalence of HIV in the prison populations in the Yunan province is estimated to be 3 percent (Dolan *et al.*,

2007). Given that commercial sex work and injection drug use are both illegal and widespread in China, and that both are associated with HIV infection, it is likely that many HIV-infected individuals are incarcerated throughout the country (Tucker and Ren, 2008). In 2004, the Chinese Ministry of Health began conducting nationwide HIV testing of its prison population (*China Daily*, 2004). In Beijing, prisoners who test positive for HIV/AIDS are all sent to the Jinzhong prison, and receive free medical treatment both during and after their incarceration (Goldkorn, 2005). Similarly, the government of the Guangdong province has announced plans to build two special jails to hold inmates with HIV/AIDS, in an attempt to halt the spread of the virus (BBC, 2005). Unfortunately, prison personnel in many cases have the same misconceptions about HIV/AIDS as exists in the civilian population, and there have been reports of HIV/AIDS prisoners being treated inhumanely in these facilities (Xinhau News Agency, 2007). Furthermore, establishing separate HIV/AIDS prisons creates tremendous potential for abuse. For instance, authorities in the central province of Henan are reportedly preparing a prison facility outside of Shangqui for civilian AIDS patients. During a session of China's parliament in March of 2006, the Shangqui municipal Communist Party Secretary purportedly ordered all county officials to submit lists of local HIV/AIDS patients who had been to Beijing to voice complaints about their treatment by local officials, in order to target them for incarceration (Yuan, 2006). Incarceration of civilians with HIV/AIDS is not without precedent, and in fact was even instituted in the United States in the early days of the epidemic (Rohter, 1993). Although reports of human rights violations in China are difficult to verify at this time, the existence of these rumors highlights the importance of increased openness by the Chinese government, of increased national HIV/AIDS education, of further investigation by international human rights organizations.

Similar to China, top Russian Federation officials did not acknowledge the existence of a full-blown HIV/AIDS epidemic until 2005 (Bransten, 2005). Political commitment to tackling HIV/AIDS within Russia is growing, however. In 2006, over US$100 million was allocated to tackling the epidemic (Moran and Jordaan, 2007). Although the official figures put the number of infections in Russia in 2006 at around 375,000, actual figures are probably closer to 1.5 million (Moran and Jordaan, 2007). The HIV/AIDS epidemic has been observed by to be "predominantly … among urban, young, male injecting drug users and their sexual partners" and, subsequently, that 80 percent of the epidemic is in people under 30 (UNDP, 2004). The vastness of the IDU epidemic in combination with Russia's harsh policy of imprisonment for drug use means that correctional facilities will, and likely already do, bear a significant burden of the epidemic (HIV/AIDS Policy Law Review, 2004). HIV prevalence in Russian prisons is estimated in different regions to be between 0.8 and 4.8%. HIV prevalence in IDU prisoners in St Petersburg is 46 percent, and even higher prevalence of HIV is found in prisons in other FSU countries: Estonia (88–90 percent), Ukraine (up to 26 percent; 83 percent of cases are in IDUs), and Romania (13 percent), to cite several sources (Dolan *et al.*, 2007). It is estimated that, already, 15–20 percent of all people living with HIV/AIDS in Russia are detained in prisons or other correctional facilities (Alexandrova, 2003).

In one qualitative study of injection drug users in Moscow, Volgograd and Barnaul in western Siberia – 77 percent of whom reported spending time incarcerated for offenses related to their injection drug use – drugs were perceived to be readily available in correctional institutions. In this same group, sterile injection equipment was reported to be scarce in prisons, and was frequently shared among large groups. Punishment for possession of sterile injection equipment was reported to place further constraints on harm-reduction practices. The IDUs in the study also reported having the perception of safety due to the belief that other prisoners around them were HIV-negative. They assumed that all prisoners are tested upon entry, and that those found to be HIV-positive would be segregated. Although limited in scope, this study suggests that HIV prevention education, harm reduction and intervention is non-existent in some Russian prisons (Sarang *et al.*, 2006). More alarmingly, other studies have suggested that the chances of survival of young men from disadvantaged backgrounds in Russia may actually be improved by being incarcerated (Bobrik *et al.*, 2005). The responsibility will fall heavily on FSU prison systems to manage the treatment of HIV/AIDS cases in each FSU country, and to institute a full range of HIV prevention programs.

As is the case in the United States, correctional institutions in China and FSU nations have the potential to exacerbate spread of HIV/AIDS if the problem is mismanaged or neglected. It should also be stressed that China and FSU present only a piece of the global picture. Other countries have high HIV prevalence in prisons – Brazil (12–17 percent), Jamaica (26 percent), Argentina (7 percent), India (1.8 percent, 9.5 percent in women), Pakistan (2.7 percent), Vietnam (28.4 percent) and Malaysia (6 percent), to name several (Dolan *et al.*, 2007). Taking a more optimistic outlook, however, a tremendous opportunity exists to institute internationally the HIV/AIDS education and prevention interventions that have been established in the United States over the past 30 years.

Future directions

The epidemic of incarceration in the United States shows no signs of lessening in the near future. If a change to the *status quo* is to be enacted, more attention must be paid by the public and by policy-makers to the detrimental effects of incarceration on our society as a whole, and a stronger emphasis must be placed on developing more constructive strategies for treating the underlying epidemics of poverty, addiction and mental illness that for many individuals are the root cause of incarceration in the first place. In the meantime, the staggering volume of imprisoned individuals in the United States that are predominantly persons of color, from impoverished, urban neighborhoods, who have high rates of addiction, mental illness, HIV and other sexually and blood-borne diseases, presents a public health opportunity to medically serve and to curb the HIV epidemic in an otherwise hard-to-reach population. Making the most of this opportunity is among the most important areas of public health on which to focus our attentions.

Acknowledgments

This work was made possible, in part, through the support of grant number P30-AI-42853 from the National Institutes of Health, Center for AIDS Research (NIH/CFAR); grants number 5T32DA13911, RO1 DA 018641 and P30 DA013868 from the National Institute on Drug Abuse, National Institutes of Health (NIDA/NIH); and grant number 6H79TI14562 from the Substance Abuse and Mental Health Services Administration, Center for Substance Abuse Treatment (SAMHSA/CSAT).

References

Alexandrova, A. (2003). Russia: new criminal process code promises a more tolerant incarceration policy. *Can. HIV AIDS Policy Law Rev.*, 8(1), 54.

Altice, F. L. and Springer, S. A. (2005). Management of HIV/AIDS in correctional settings. In: K. H. Mayer and H. F. Pizer (eds), *The AIDS Pandemic, Impact on Science and Society*. San Diego, CA: Elsevier, pp. 449–87.

Altice, F. L., Marinovich, A., Khoshnood, K. *et al.* (2005). Correlates of HIV infection among incarcerated women, implications for improving detection of HIV infection. *J. Urban Health*, 82, 312–26.

Arriola, K. R., Braithwaite, R. L., Kennedy, S. *et al.* (2001). A collaborative effort to enhance HIV/STI screening in five county jails. *Public Health Rep.*, 116, 520–9.

Barry, P. M., Kent, C. K., Scott, K. C. *et al.* (2007). Optimising sexually transmitted infection screening in correctional facilities, San Francisco, 2003–2005. *Sex. Transm. Infect.*, 83, 416–18.

Basu, S., Smith-Rohrberg, D., Hanck, S. *et al.* (2005). HIV testing in correctional institutions, evaluating existing strategies, setting new standards. *AIDS Public Policy J.*, 20, 3–24.

Bauserman, R. L., Ward, M. A., Eldred, L. *et al.* (2001). Increasing voluntary HIV testing by offering oral tests in incarcerated populations. *Am. J. Public Health*, 91, 1226–9.

Bauserman, R. L., Richardson, E., Ward, M. *et al.* (2003). HIV prevention with jail and prison inmates, Maryland's Prevention Case Management program. *AIDS Educ. Prev.*, 15, 465–80.

BBC (2005). China plans jails for HIV inmates. Available at http://news.bbc.co.uk/2/hi/asia-pacific/4434446.stm (accessed 02-27-2008).

Beckwith, C. and Poshkus, M. (2007). Perspective, HIV behind bars, meeting the need for HIV testing, education, and access to care. In: *Infectious Diseases in Corrections Report*. A. S. De Groot. Providence: A. S. De Groot, IDCR, 9(17).

Beckwith, C. G., Atunah-Jay, S., Cohen, J. *et al.* (2007a). Feasibility and acceptability of rapid HIV testing in jail. *AIDS Patient Care STDs*, 21, 41–7.

Beckwith, C. G., Cohen, J., Shannon, C. *et al.* (2007b). HIV testing experiences among male and female inmates in Rhode Island. *AIDS Read.*, 17, 459–64.

Belenko, S. and Peugh, J. (2005). Estimating drug treatment needs among state prison inmates. *Drug Alcohol Depend.*, 77, 269–81.

Bobrik, A., Danishevski, K., Eroshina, K. *et al.* (2005). Prison health in Russia, the larger picture. *J. Public Health Policy*, 26, 30–59.

Braithwaite, R. L. and Arriola, K. R. (2003). Male prisoners and HIV prevention, a call for action ignored. *Am. J. Public Health*, 93, 759–63.

Braithwaite, R. L., Hammett, T. M. and Mayberry, R. M. (1996). *Prison and AIDS, A Public Health Challenge.* San Francisco, CA: Jossey–Bass.

Bransten, J. (2005). Russia, government shows signs of acknowledging country's AIDS epidemic. Available at http://www.rferl.org/featuresarticle/2005/03/fbc22c99-d547-4859-9f15-b22516631bfc.html (accessed 03-11-2008).

Brewer, T. F., Vlahov, D., Taylor, E. *et al.* (1988). Transmission of HIV-1 within a statewide prison system. *AIDS*, 2, 363–7.

Bryan, A., Robbins, R. N., Ruiz, M. S. *et al.* (2006). Effectiveness of an HIV prevention intervention in prison among African Americans, Hispanics, and Caucasians. *Health Educ. Behav.*, 33, 154–77.

Canadian HIV/AIDS Policy Law Review (2004). Russian Federation, Battle not over in drug-law changes. *Can. HIV AIDS Policy Law Rev.*, 9, 34.

Cate, M. L. (2007). The state's substance abuse programs do not reduce recidism, yet cost the state $140 million per year. Available at http://www.oig.ca.gov/press-rlse/pdf/prlse_022107.pdf (accessed 01-03-2008).

CDC (2006a). HIV transmission among male inmates in a state prison system – Georgia, 1992–2005. *Morbid. Mortal. Wkly Rep.*, 55, 421–6.

CDC (2006b). Revised recommendations for HIV testing of adults, adolescents, and pregnant women in health-care settings. *Morbid. Mortal. Wkly Rep.*, 55(RR–14), 1–17.

CDC (2007). *Sexually Transmitted Disease Surveillance, 2006.* Atlanta, GA: US Department of Health and Human Services, Centers for Disease Control and Prevention.

Chen, J. L., Bovee, M. C. and Kerndt, P. R. (2003). Sexually transmitted diseases surveillance among incarcerated men who have sex with men – an opportunity for HIV prevention. *AIDS Educ. Prev.*, 15(Suppl. A), 117–26.

China Daily (2004). China to start HIV test on prison population. Available at http://www.chinadaily.com.cn/english/doc/2004-11/25/content_394887.htm (accessed 02-27-2008).

Clarke, J. G., Stein, M. D., Hanna, L. *et al.* (2001). Active and former injection drug users report of HIV risk behaviors during periods of incarceration. *Subst. Abuse*, 22, 209–16.

Cohen, M. S. (2004). HIV and sexually transmitted diseases, lethal synergy. *Top. HIV Med.*, 12, 104–7.

Conklin, T. J., Lincoln, T. and Tuthill, R. W. (2000). Self-reported health and prior health behaviors of newly admitted correctional inmates. *Am. J. Public Health*, 90, 1939–41.

Desai, A. A., Latta, E. T., Spaulding, A. *et al.* (2002). The importance of routine HIV testing in the incarcerated population, the Rhode Island experience. *AIDS Educ. Prev.*, 14(Suppl. B), 45–52.

Dolan, K. A., Shearer, J., MacDonald, M. *et al.* (2003a). A randomised controlled trial of methadone maintenance treatment versus wait list control in an Australian prison system. *Drug Alcohol Depend.*, 72, 59–65.

Dolan, K., Rutter, S. and Wodak, A. D. (2003b). Prison-based syringe exchange programmes, a review of international research and development. *Addiction*, 98, 153–8.

Dolan, K., Kite, B., Black, E. *et al.* (2007). HIV in prison in low-income and middle-income countries. *Lancet Infect. Dis.*, 7, 32–41.

Dolinsky, A. (2007). US proposed federal legislation to allow condom distribution and HIV testing in prison. *HIV AIDS Policy Law Rev.*, 12, 30–1.

Doll, D. C. (1988). Tattooing in prison and HIV infection. *Lancet*, 1, 66–7.

Ehrmann, T. (2002). Community-based organizations and HIV prevention for incarcerated populations, three HIV prevention program models. *AIDS Educ. Prev.*, 14(Suppl. B), 75–84.

el-Bassel, N., Ivanoff, A., Schilling, R. F. *et al.* (1995). Preventing HIV/AIDS in drug-abusing incarcerated women through skills building and social support enhancement, preliminary outcomes. *Social Work Res.*, 19, 131–41.

Fallon, B. M. (2001). The Key Extended Entry Program (KEEP), from the community side of the bridge. *Mt Sinai J. Med.*, 68, 21–7.

Flanigan, T. P. (1995). HIV testing in prison. *Lancet*, 345, 390.

Gill, B. and Okie, S. (2007). China and HIV – a window of opportunity. *N. Engl. J. Med.*, 356, 1801–5.

Glaser, J. B. (1998). Sexually transmitted diseases in the incarcerated. An underexploited public health opportunity. *Sex. Transm. Dis.*, 25, 308–9.

Goldkorn, J. (2005). Prisoners in Beijing to get HIV testing and free treatment. Available at http://www.danwei.org/breaking_news/prisoners_in_beijing_to_get_hi.php (accessed 02-28-2008).

Grinstead, O., Zack, B. and Faigeles, B. (2001). Reducing postrelease risk behavior among HIV seropositive prison inmates, the health promotion program. *AIDS Educ. Prev.*, 13, 109–19.

Hammett, T. M., Harmon, M. P. and Rhodes, W. (2002). The burden of infectious disease among inmates of and releases from US correctional facilities, 1997. *Am. J. Public Health*, 92, 1789–94.

Hellard, M. E., Aitken, C. K. and Hocking, J. S. (2007). Tattooing in prisons – not such a pretty picture. *Am. J. Infect. Control*, 35, 477–80.

Horsburgh, C. R., Jr., Jarvis, J. Q., McArther, T. *et al.* (1990). Seroconversion to human immunodeficiency virus in prison inmates. *Am. J. Public Health*, 80, 209–10.

ICPS (2008). *World Prison Brief*. International Center for Prisoner Studies. Available at http://www.kcl.ac.uk/depsta/law/research/icps/worldbrief/.

James, D. J. and Glaze, L. E. (2006). *Mental Health Problems of Prison and Jail Inmates*. U.S. Department of Justice Document NCJ 213600.

Kahn, R. H., Mosure, D. J., Blank, S. *et al.* (2005). *Chlamydia trachomatis* and *Neisseria gonorrhoeae* prevalence and coinfection in adolescents entering selected US juvenile detention centers, 1997–2002. *Sex. Transm. Dis.*, 32, 255–9.

Kinlock, T. W., Battjes, R. J. and Schwartz, R. P. (2002). A novel opioid maintenance program for prisoners, preliminary findings. *J. Subst. Abuse Treat.*, 22, 141–7.

Kinlock, T. W., Gordon, M. S., Schwartz, R. P. *et al.* (2007). A randomized clinical trial of methadone maintenance for prisoners, results at 1-month post-release. *Drug Alcohol Depend.*, 91, 220–7.

Kinlock, T. W., Gordon, M. S., Schwartz, R. P. *et al.* (2008). A study of methadone maintenance for male prisoners, 3-month post-release outcomes. *Criminal Justice Behav.*, 35, 34–47.

Klein, S. J., O'Connell, D. A., Devore, B. S. *et al.* (2002). Building an HIV continuum for inmates, New York State's criminal justice initiative. *AIDS Educ. Prev.*, 14(Suppl B.), 114–23.

Kondro, W. (2007). Prison tattoo program wasn't given enough time. *Can. Med. Assoc. J.*, 176, 307–8.

Krebs, C. P. and Simmons, M. (2002). Intraprison HIV transmission: an assessment of whether it occurs, how it occurs, and who is at risk. *AIDS Educ. Prev.*, 14(Suppl. B), 53–64.

Liddicoat, R. V., Zheng, H., Internicola, J. *et al.* (2006). Implementing a routine, voluntary HIV testing program in a Massachusetts county prison. *J. Urban Health*, 83, 1127–31.

Lipton, D. S. (1995). *The Effectiveness of Treatment for Drug Abusers Under Criminal Justice Supervision.* Washington, DC: The National Institute of Justice.

Longshore, D., Hsieh, S., Danila, B. *et al.* (1993). Methadone maintenance and needle/syringe sharing. *Intl. J. Addict.*, 28, 983–96.

Lubelczyk, R. A., Friedmann, P. D., Lemon, S. C. *et al.* (2002). HIV prevention services in correctional drug treatment programs, do they change risk behaviors? *AIDS Educ. Prev.*, 14, 117–25.

Lyles, C. M., Kay, L. S., Crepaz, N. *et al.* (2007). Best-evidence interventions: findings from a systematic review of HIV behavioral interventions for US populations at high risk, 2000–2004. *Am. J. Public Health*, 97, 133–43.

Macalino, G. E., Vlahov, D., Sanford-Colby, S. *et al.* (2004). Prevalence and incidence of HIV, hepatitis B virus, and hepatitis C virus infections among males in Rhode Island prisons. *Am. J. Public Health*, 94, 1218–23.

MacGowan, R. J., Margolis, A., Gaiter, J. *et al.* (2003). Predictors of risky sex of young men after release from prison. *Intl J. STD AIDS*, 14, 519–23.

MacGowan, R. J., Eldridge, G., Sosman, J. M. *et al.* (2006). HIV counseling and testing of young men in prison. *J. Correct. Healthcare*, 12, 203–13.

Magura, S., Rosenblum, A. and Joseph, H. (1992). Evaluation of in-jail methadone maintenance, preliminary results. *NIDA Res. Monogr.*, 118, 192–210.

Marsch, L. A. (1998). The efficacy of methadone maintenance interventions in reducing illicit opiate use: HIV risk behavior and criminality, a meta-analysis. *Addiction*, 93, 515–32.

Maruschak, L. M. (2001). *HIV in Prisons and Jails, 1999.* US Department of Justice Document NCJ 187456.

Maruschak, L. M. (2004). *HIV in Prisons and Jails, 2002.* US Department of Justice Document NCJ 205333.

Maruschak, L. M. (2007). *HIV in Prisons, 2005.* US Department of Justice Document NCJ 218915.

May, J. P. and Williams, E. L. Jr. (2002). Acceptability of condom availability in a US jail. *AIDS Educ. Prev.*, 14(Suppl. B), 85–91.

Mears, D. P., Winterfield, L., Hunsaker, J. *et al.* (2003). *Drug Treatment in the Criminal Justice System: The Current State of Knowledge.* Washington, DC: The Urban Institute.

Mertz, K. J., Schwebke, J. R., Gaydos, C. A. *et al.* (2002). Screening women in jails for chlamydial and gonococcal infection using urine tests: feasibility, acceptability, prevalence, and treatment rates. *Sex. Transm. Dis.*, 29, 271–6.

Metzger, D. S., Woody, G. E., McLellan, A. T. *et al.* (1993). Human immunodeficiency virus seroconversion among intravenous drug users in- and out-of-treatment: an 18-month prospective follow-up. *J. Acquir. Immune Defic. Syndr.*, 6, 1049–56.

Miller, W. C., Ford, C. A., Morris, M. *et al.* (2004). Prevalence of chlamydial and gonococcal infections among young adults in the United States. *J. Am. Med. Assoc.*, 291, 2229–36.

Minc, A., Butler, T. and Gahan, G. (2007). The Jailbreak Health Project – incorporating a unique radio programme for prisoners. *Intl J. Drug Policy*, 18, 444–6.

Moller, L. (2005). Substitution therapy in prisons: a review of international experience. *Presentation at HIV/AIDS in Prisons in Ukraine – From Evidence to Action, Prevention and Care, Treatment, and Support*, 1–2, November, Kiev.

Moran, D. and Jordaan, J. A. (2007). HIV/AIDS in Russia, determinants of regional prevalence. *Intl J. Health Geogr.*, 6, 22.

Mumola, C. J. and Karberg, J. C. (2006). *Drug Use and Dependence, State and Federal Prisons, 2004.* US Department of Justice Document NCJ 213530.

Nacci, P. and Kane, T. (1982). *Sex and Sexual Aggression in Federal Prisons*. Washington, DC: Federal Bureau of Prisons.

New York State AIDS Advisory Council (1999). Report on HIV/AIDS services in New York state correctional facilities. Available at www.health.state.ny.us/diseases/aids/workgroups/aac/docs/servicescorrectional.pdf (accessed 12-12-2007).

New York Times (1997). Jamaica tries to quiet 2 prisons after riots. *New York Times*, 25 August 25. Available at http://query.nytimes.com/gst/fullpage.html?sec=health&res=9903E3D71E3EF936A1575BC0A961958260&fta=y (accessed 12-14-2008).

Okie, S. (2007). Sex, drugs, prisons, and HIV. *N. Engl. J. Med.*, 356, 105–8.

Pearson, F. S. and Lipton, D. S. (1999). A meta-analytical review of the effectiveness of corrections-based treatment for drug abuse. *Prison J.*, 79, 384–410.

People Daily (2005). Spending on HIV-AIDS prevention set to double. Available at http://english.peopledaily.com.cn/200512/28/eng20051228_231288.html (accessed 03-10-2008).

Phillips, K. A. and Fernyak, S. (2000). The cost-effectiveness of expanded HIV counselling and testing in primary care settings, a first look. *AIDS*, 14, 2159–69.

Rich, J. D., Dickinson, B. P., Macalino, G. *et al.* (1999). Prevalence and incidence of HIV among incarcerated and reincarcerated women in Rhode Island. *J. Acquir. Immune Defic. Syndr.*, 22, 161–6.

Rich, J. D., Holmes, L., Salas, C. *et al.* (2001). Successful linkage of medical care and community services for HIV-positive offenders being released from prison. *J. Urban Health*, 78, 279–89.

Rich, J. D., Boutwell, A. E., Shield, D. C. *et al.* (2005). Attitudes and practices regarding the use of methadone in US state and federal prisons. *J. Urban Health*, 82, 411–19.

Rohter, L. (1993). Out of HIV prison: the misery is lightened for a handful of Haitians. *New York Times*, available at http://query.nytimes.com/gst/fullpage.html?res=9F0CE3DF133CF933A15755C0A965958260 (accessed 03-06-20080).

Sabin, K. M., Frey, R. L. Jr., Horsley, R. *et al.* (2001). Characteristics and trends of newly identified HIV infections among incarcerated populations: CDC HIV voluntary counseling, testing, and referral system, 1992–1998. *J. Urban Health*, 78, 241–55.

Sabol, W. J., Minton, T. D. and Harrison, P. M. (2007). *Prison and Jail Inmates at Midyear, 2006*. US Department of Justice Document NCJ 217675.

SAMHSA (2002). DASIS Report: substance abuse services and staffing in adult correctional facilities. Available at http://www.oas.samhsa.gov/2k2/justice/justice.htm (accessed 12-05-2007).

Samuel, M. C., Doherty, P. M., Bulterys, M. *et al.* (2001). Association between heroin use, needile sharing and tattoos received in prison with hepatitis B and C positivity among street-recruited injecting drug users in New Mexico, USA. *Epidemiol. Infect.*, 127, 475–84.

Sangani, P., Rutherford, G. and Wilkinson, D. (2004). Population-based interventions for reducing sexually transmitted infections, including HIV infection. *Cochrane Database Syst. Rev.*, 2, CD001220.

Sarang, A., Rhodes, T., Platt, L. *et al.* (2006). Drug injecting and syringe use in the HIV risk environment of Russian penitentiary institutions: qualitative study. *Addiction*, 101, 1787–96.

Sexton, J., Garnett, G. and Rottingen, J. A. (2005). Meta-analysis and metaregression in interpreting study variability in the impact of sexually transmitted diseases on susceptibility to HIV infection. *Sex. Transm. Dis.*, 32, 351–7.

Sifunda, S., Reddy, P. S., Braithwaite, R. *et al.* (2006). Access point analysis on the state of health care services in South African prisons: a qualitative exploration of correctional health care workers' and inmates' perspectives in Kwazulu-Natal and Mpumalanga. *Social Sci. Med.*, 63, 2301–9.

Simmons, E., Lally, M. A. and Flanigan, T. P. (2003). Routine, not risk-based, human immunodeficiency virus testing is the way to go. *J. Infect. Dis.*, 187, 1024.

Simooya, O. and Sanjobo, N. (2002). Infections and risk factors in entrants to Irish prisons. Study in Zambia showed that robust response is needed in prisons. *Br. Med. J.*, 324, 850.

Smith, L. A., Gates, S. and Foxcroft, D. (2006). Therapeutic communities for substance related disorder. *Cochrane Database Syst. Rev.*, 1, CD005338.

Spaulding, A., Stephenson, B., Macalino, G. *et al.* (2002). Human immunodeficiency virus in correctional facilities, a review. *Clin. Infect. Dis.*, 35, 305–12.

St Lawrence, J., Eldridge, G. D., Shelby, M. C. *et al.* (1997). HIV risk reduction for incarcerated women, a comparison of brief interventions based on two theoretical models. *J. Consult. Clin. Psychol.*, 65, 504–9.

Stephenson, B. L., Wohl, D. A., Golin, C. E. *et al.* (2005). Effect of release from prison and re-incarceration on the viral loads of HIV-infected individuals. *Public Health Rep.*, 120, 84–8.

Stephenson, B. L., Wohl, D. A., McKaig, R. *et al.* (2006). Sexual behaviours of HIV-seropositive men and women following release from prison. *Intl J. STD AIDS*, 17, 103–8.

Taxman, F. S., Perdoni, M. L. and Harrison, L. D. (2007). Drug treatment services for adult offenders: the state of the state. *J. Subst. Abuse Treat*, 32, 239–54.

Tomasino, V., Swanson, A. J., Nolan, J. *et al.* (2001). The Key Extended Entry Program (KEEP): a methadone treatment program for opiate-dependent inmates. *Mt Sinai J. Med.*, 68, 14–20.

Tsang, T. H., Horowitz, E. and Vugia, D. J. (2001). Transmission of hepatitis C through tattooing in a United States prison. *Am. J. Gastroenterol.*, 96, 1304–5.

Tucker, J. D. and Ren, X. (2008). Sex worker incarceration in the People's Republic of China. *Sex. Transm. Infect.*, 84, 34–5, discussion 36.

Tucker, J. D., Chang, S. W. and Tulsky, J. P. (2007). The Catch-22 of condoms in US correctional facilities. *BMC Public Health*, 7, 296.

UNAIDS (1997). *Prison and AIDS, Technical Update*, UNAIDS Best Practice Collection. Geneva: UNAIDS.

UNDP (2004). Reversing the Epidemic, Facts and Policy Options. Bratislava: United Nations Development Programme, p. 12,

US Bureau of Justice Statistics (2008). Adult correctional populations, 1980–2006. Available at http://www.ojp.usdoj.gov/bjs/glance/corr2.htm (accessed 04-26-2008).

US Census Bureau (2002). *US Census, 2000*. Washington, DC: US Department of Commerce, Economics and Statistics Administration.

Vigdal, G. L. and Stadler, D. W. (1992). Comprehensive system development in corrections for drug-abusing offenders, the Wisconsin Department of Corrections. *NIDA Res. Monogr.*, 118, 126–41.

Walter, J. P. (2001). *Drug Policy Clearinghouse Fact Sheet, Drug Treatment in the Criminal Justice System*. Washington, DC: Office of National Drug Control Policy.

West, A. D. and Martin, R. (2000). Perceived risk of AIDS among prisoners following educational intervention. *J. Offender Rehab.*, 32, 75–104.

WHO (2007). Interventions to address HIV in prison: drug dependence treatment. Evidence for Action Technical Papers, available at http://www.who.int/hiv/idu/oms_ea_drug_treatment_df.pdf (12-12-2008).

Xinhau News Agency (2007). Jailed with HIV – a struggle against despair. Available at http://www.china.org.cn/english/China/216554.htm (accessed 02-12-2008).

Yap, L., Butler, T., Richters, J. *et al.* (2007). Do condoms cause rape and mayhem? The long-term effects of condoms in New South Wales' prisons. *Sex. Transm. Infect.*, 83, 219–22.

Yuan, F. (2006). China's Henan sets up prison for AIDS patients. Available at http://www.rfa.org/english/news/2005/04/06/china_AIDS/ (accessed 02-27-2008).

Zack, B. (2007). HIV prevention: behavioral interventions in correctional settings. In: R. Greifinger (ed.), *Public Health Behind Bars: From Prisons to Communities*. New York, NY: Springer Science + Business Media, pp. 156–73.

The Center for Prisoner Health and Human Rights may be reached at www.prisonerhealth.org

Preventing mother-to-child transmission of HIV

17

James A. McIntyre and Glenda E. Gray

The first descriptions of AIDS in children and of possible transmission from mother-to-child were published in the early 1980s (Oleske *et al.*, 1983; Cowan *et al.*, 1984). Remarkable progress has been made in the years since then in the prevention of mother-to-child transmission (PMTCT), leading to extremely low rates of infection in well-resourced countries – one of the major advances in HIV prevention.

UNAIDS estimates that around 420,000 children (350,000–540,000) were newly infected with HIV in 2007, most through mother-to-child transmission, and 90 percent in sub-Saharan Africa (UNAIDS, 2007). In well-resourced settings, the standard of care for HIV-positive pregnant women has become the provision of combination antiretroviral therapy in pregnancy, ongoing for those women who require it, and the use of combination antiretroviral prophylaxis throughout the pregnancy for those with higher CD4+ counts, together with the use of replacement feeding. This has resulted in a drop in mother-to-child transmission rates to below 2 percent (Newell and Thorne, 2004; Fowler *et al.*, 2007). In most low-resource settings this combination therapy has not been available, and estimates of transmission are much higher, with an estimated average rate of 26 percent in the 33 most affected countries – more than 10-fold higher than the rates in better-resourced settings (World Health Organization, 2007). To contextualize these estimates, this is equivalent to approximately 1 child infected every second day in the United States, 1 every day in Europe, 2 a week in Asia, and over 1000 a day in Africa (McIntyre and Lallement, 2008).

472

Progress in preventing mother-to-child transmission of HIV

Three phases of progress in PMTCT can be identified. In the first decade, most research focused on the identification of the risk factors for transmission, including transmission through breastfeeding, the timing of transmission and the relationship to risk factors such as prolonged rupture of membranes and maternal health (Mofenson and McIntyre, 2000; Fowler *et al.*, 2007). This research led to the identification of three major interventions which have been responsible for the reduction in risk: the use of antiretroviral therapy, the avoidance of breastfeeding, and the use of elective Caesarean section.

The second phase encompassed the significant advances in preventing transmission through the use of antiretrovirals. The introduction of the first antiretroviral drug, zidovudine (AZT, ZDV), and the subsequent use of this in pregnant women led to the landmark Pediatric AIDS Clinical Trial Group Protocol 076 (PACTG076) Study, published in 1994. In this double-blind, randomized, placebo-controlled trial, a regimen of oral zidovudine (ZDV) given prenatally, intrapartum and postpartum reduced transmission from mother to child by 67.5 percent, from 22.6 percent in the placebo group to 7.6 percent in the ZDV group (Connor *et al.*, 1994). In the years that followed, trials of shorter-course treatments in low-resource settings, with zidovudine alone (Shaffer *et al.*, 1999; Wiktor *et al.*, 1999) or zidovudine and lamivudine (Petra Study Team, 2002), demonstrated that these regimens were effective in reducing transmission to around 10–15 percent (Leroy *et al.*, 2005). In 1999, the HIVNET 012 trial, using a very simple regimen of a single dose of nevirapine for mothers at the onset of labor and a single dose to the infant within 72 hours of birth, demonstrated a reduction of 42 percent compared to a week-long course of zidovudine stated in labor (Guay *et al.*, 1999; Jackson *et al.*, 2003). This simple and inexpensive regimen was soon recommended by the international agencies, and it has enabled the expansion of PMTCT services in low-resourced settings, reaching several million mothers and infants by 2007 (WHO, 2007).

The third phase of development of PMTCT strategies has been a combination of a push towards universal access to PMTCT services, although this is still far from accomplished in many low-resource, high-prevalence settings, coupled with a renewed focus on reducing transmission through breastfeeding. These remain the major challenges to achieving similar transmission reductions in low-resource settings to those seen in the US and Europe.

Factors affecting mother-to-child transmission

Transmission of HIV from mother to child can occur antepartum, during labor and delivery, or postpartum through breast-milk transmission. In the absence

473

of treatment or prophylaxis, most transmission is thought to occur during the intrapartum period. The proportions of transmission at different stages are changed by the use of prophylactic interventions, with combination antiretroviral therapy almost eliminating antepartum and intrapartum transmission, and short-course antiretrovirals having less effect on antepartum transmission, although achieving a major reduction in intrapartum transmission (Newell, 1998; Moodley *et al.*, 2003; Jourdain *et al.*, 2007). Where antepartum and intrapartum transmission have been reduced by antiretroviral prophylaxis, breastfeeding becomes the major source of transmission in low-resource settings (Kourtis *et al.*, 2006).

Factors which increase the risk of mother-to-child transmission are maternal stage of disease, high viral load, low CD4+ count, use of antiretroviral therapy, mode of delivery, duration of rupture of membranes, increased genital secretion of HIV, the practice of breastfeeding, and other factors such as prematurity (Magder *et al.*, 2005). High viral load is the factor which most consistently correlates with transmission, as evidenced by the success of antiretroviral treatment in pregnancy in reducing transmission (European Collaborative Study, 1992; Martinez *et al.*, 2006; Jourdain *et al.*, 2007; Warszawski *et al.*, 2008). Prior to the use of antiretroviral therapy, several maternal and obstetrical factors were identified which facilitated transmission – including advanced maternal HIV infection with a clinical diagnosis of AIDS, low CD4+ T-lymphocyte count, sexually transmitted infections, significantly higher transmission rates in the first-born twin compared to the second-born twin, increasing duration of ruptured membranes during labor, preterm birth, maternal drug use, vitamin A deficiency, and female gender of the infant – but the relative importance of these appears to be less with the use of antiretroviral prophylaxis (Newell, 2003; McIntyre, 2008). Caesarean section before labor was shown to reduce the risk of transmission compared to vaginal delivery in the randomized European Mode of Delivery trial, in which transmission in women randomized to deliver vaginally was 10.5 percent, compared to 1.8 percent in women who were randomized to deliver by elective Caesarean section (European Mode of Delivery Collaboration, 1999).

Principles of prevention of MTCT: a comprehensive approach

The optimal prevention of mother-to-child transmission of HIV involves much more than interventions in pregnancy, and should be seen as part of broader HIV prevention efforts. The World Health Organization has recommended a four-part comprehensive approach to preventing mother-to-child transmission (WHO, 2002). These components include:

1. Primary prevention of HIV infection
2. Prevention of unintended pregnancies among HIV-infected women
3. Prevention of HIV transmission from HIV-infected mothers to their infants

4. Care, treatment and support for HIV-infected mothers, their children and families.

While the latter two have been become widely accepted as integral parts of PMTCT services, more focus is needed on the issues of primary prevention in young adults and in the provision of reproductive health and contraceptive services for HIV-positive women (Cates, 2006; Morrison and Cates, 2007). Additionally, to fully implement an integrated strategy of providing antiretroviral therapy (ART) for women with low CD4+ counts or clinical symptoms of AIDS and a prophylactic regimen for other HIV-positive women, the antenatal services and health workers need to be able to undertake clinical assessment for HIV signs and symptoms and obtain a baseline CD4+ cell count. This level of care is not widely available in maternity settings in many low-resource settings, and needs to be strengthened, along with the infrastructure for antiretroviral drug supplies and access to early diagnostic HIV testing for exposed children.

The importance of HIV testing in pregnancy

The use of antiretroviral strategies in pregnancy and modifications of infant feeding to reduce the risk of transmission rely on the identification of HIV-positive women. First, antenatal care services must be available and accessible at an early stage of pregnancy to identify HIV-positive mothers and to provide antiretroviral regimens. These services need to provide HIV testing and counseling, which must be acceptable to pregnant women to ensure uptake. High levels of acceptance of testing in pregnancy have been demonstrated in many high HIV-prevalence, low-resource areas in Africa and Asia (Manzi *et al.*, 2005; McIntyre, 2005), but in many others the low uptake of testing is one of the major reasons for the failure of PMTCT programs, as HIV testing is the entry point to care (Bajunirwe and Muzoora, 2005; Shetty *et al.*, 2005; Urassa *et al.*, 2005). The reasons for this are multifactorial, including fear and stigmatization, lack of service infrastructure, staff attitudes and stigma, service delivery failures, staff attitudes and cost. This situation is not limited to low-resource settings. In the US, the percentage of pregnant women tested for HIV was 41 percent in 1995 and only 60 percent by 1998 (Lansky *et al.*, 2001). This low uptake resulted in recommendations for routine testing in pregnancy, or an "opt-out" universal approach where all pregnant women are informed that HIV testing will be performed in the routine prenatal tests unless the mother declines this testing (Chou *et al.*, 2005). This has formed the basis of a number of current US recommendations, and has been shown to increase uptake to as much as 98 percent (Jamieson *et al.*, 2007). A similar trend has been seen in some African countries. In Botswana, the initial uptake of testing in pregnancy was low, estimated at only 21 percent of antenatal clinic attendees in 2001 (Rakgoasi, 2005). The proportion of HIV-infected pregnant women in public sector PMTCT service sites who knew their HIV status increased from 48 percent to 78 percent in an 8-month period following the introduction of routine testing in 2004, and has continued to rise to above 90 percent (CDC, 2004; Creek *et al.*, 2007a). In Lilongwe, Malawi, HIV-testing uptake increased from 45 percent to 73 percent when rapid, same-day testing was instituted, and to 99 percent after opt-out testing was instituted (Moses *et al.*, 2008). Similar results have been reported in Cameroon (Welty *et al.*, 2005) and other areas, and in both developed and developing countries PMTCT programs are increasingly moving to a routine "opt-out" testing approach (Mulder-Folkerts *et al.*, 2004; van't Hoog *et al.*, 2005; Perez *et al.*, 2006). The routine offer of HIV testing in pregnancy could make a significant difference to the success of PMTCT programs, although there may be other remaining challenges in these programs, including difficulties in delivering results, and the fear of recognition by hospital staff from the community (Giuliano *et al.*, 2005; Stringer *et al.*, 2005). With improved access to HIV testing, many more women will enter pregnancy aware of their HIV infection. This has been shown to be as high as 60–80 percent of HIV-positive women in the US and UK (Thorne and Newell, 2005; US Public Health Service, 2005), and also applies in some African countries where testing services have been in place for several years; this could facilitate the use of more complex PMTCT regimens earlier in pregnancy.

Preventing mother-to-child transmission in high-resource settings: PMTCT advances

The major PMTCT research findings were rapidly translated into practice in the United States and in Europe. By 1985, recommendations from the CDC advised against breastfeeding, and most HIV-positive women have used replacement feeding since then (CDC, 1985; Fowler *et al.*, 2007). Following the results of the PACTG076 study the use of zidovudine was swiftly incorporated into practice guidelines, and as the use of combination antiretroviral treatment was introduced this was also adopted for pregnant women. Within a decade this resulted in an 80 percent reduction in the number of infants infected annually in the US (CDC, 2006), the number of infected infants peaking in 1992 at close to 2000 and now standing at less than 200 per year, with transmission rates below 2 percent (McKenna and Hu, 2007). The rate of transmission in the US Women and Infants Transmission Study dropped from 18.1 percent in 1990–1992 to 1.6 percent in 1999–2000, with an increase in the proportion of antepartum infections from 27 percent to 80 percent, reflecting the success of ARV interventions in reducing intrapartum infection (Magder *et al.*, 2005).

A similar effect was seen in Europe, where the proportion of HIV-infected pregnant women receiving antenatal antiretroviral therapy increased from 5 percent to 92 percent between 1997 and 2003, reducing the transmission rate in 2003 to 0.99 percent (95% CI 0.32–2.30) (European Collaborative Study, 2005). In

Western Europe, overall transmission rates have dropped to below 2 percent, with most HIV-infected pregnant women receiving triple antiretroviral therapy, and two-thirds delivering by Caesarean section (European Collaborative Study, 2006). The epidemic in Western Europe has moved from being predominantly among intravenous drug users to more heterosexually-acquired transmission, in part reflecting the increasing proportion of immigrants in the total numbers of HIV-positive pregnant women. In Eastern Europe, PMTCT service coverage is less advanced and available drug regimens are more restricted, with more use of single-dose nevirapine, and transmission rates are slightly higher at around 4 percent.

The use of triple-combination antiretroviral therapy for PMTCT does not appear to have an adverse effect on mothers' future prognosis. An observational cohort study in the United States showed that pregnancy was associated with a lower risk of disease progression in women followed from 1997 to 2004, possibly due to a beneficial effect of the short-term antiretroviral therapy (Tai *et al.*, 2007). There are few data available on whether it would be better to continue antiretroviral treatment in women postpartum, irrespective of CD4+ cell counts, and this is an area for future research.

Remaining PMTCT challenges in high-resource settings

While pediatric HIV infection has not been eradicated in these situations, it has been controlled to very low numbers. Several challenges remain – including the need to recognize and treat those women living in vulnerable circumstances who do not currently access prenatal care and HIV testing, or do so too late to benefit from antiretroviral interventions. Moving to a routine offer of HIV testing for all pregnant women will go some way to achieving this, as will the provision of testing in labor, if needed (Jamieson *et al.*, 2007). In Europe many HIV-infected pregnant women now come from marginalized immigrant populations (Floridia *et al.*, 2006), and similar issues complicate the delivery of PMTCT care. Late identification and treatment of HIV-positive pregnant women in these communities reduces the chances of averting transmission. In a French review of infections occurring in children despite their mothers receiving antiretrovirals, the duration of antiretroviral therapy and the resulting drop in viral load was the major determining factor, emphasizing the importance of timely identification of HIV-positive women (Warszawski *et al.*, 2008).

Another reason for inadequate care and vulnerability in well-resourced settings is drug use in HIV-positive women. These women may be less likely to attend seek prenatal care (Minkoff *et al.*, 1990), and be reluctant to discuss their drug use because of fear of legal consequences or stigma (British HIV Association, 2005). Adherence to antiretroviral regimens is another challenging factor in all HIV-positive women. Where women are on pre-existing antiretroviral treatment, adherence may drop during pregnancy; several studies have also shown poorer adherence in the postpartum period, where there are many other physical and emotional demands on the women, and the motivation to be adherent in order to protect the infant may be less. There may be better adherence when drugs are being used only for prophylaxis, although the need to start treatment early in pregnancy may make pre-treatment and adherence counseling more difficult, especially where the HIV diagnosis has been made for the first time during pregnancy.

Preventing MTCT in low-resource settings: advances

Following the successful results of the PACTG076 trial, a number of studies were initiated to compare shorter, less expensive antiretroviral regimens for the prevention of mother-to-child transmission, which would be more feasible in low-resource settings. These trials (Table 17.1) demonstrated that the use of short-course zidovudine, either alone (Shaffer *et al.*, 1999; Wiktor *et al.*, 1999) or in combination with lamivudine (Petra Study Team, 2002), could reduce transmission to around 10–15 percent (Leroy *et al.*, 2005). A subsequent study in Thailand, PHPT-1, showed that zidovudine prophylaxis was most effective when started early in the third trimester, but that the efficacy of shorter courses could be improved by giving 6 weeks of treatment to the infant, rather than the standard 1 week (Lallemant *et al.*, 2000).

The major impact on PMTCT services in low-resource settings has come from the use of nevirapine. The HIVNET 012 study first showed that a single dose of nevirapine in the mother during labor and a single dose to the infant within the first days of life could reduce transmission by 40 percent for up to 18 months (Guay *et al.*, 1999; Jackson *et al.*, 2003). The low cost, relative ease of administration and safety of this regimen has made it the central focus of most PMTCT services in low-resource settings, reaching close to two million women and infants since 2000. In 2004, the results of the PHPT-2 study in Thailand demonstrated transmission rates of 2 percent in a non-breastfeeding population with the addition of the single dose nevirapine regimen to a base regimen of zidovudine given from 28 weeks' gestation (Lallemant *et al.*, 2004). This combined use of zidovudine and peripartum nevirapine has become the first-line recommendation in the 2006 WHO guidelines for PMTCT (WHO, 2006a). In the Côte d'Ivoire, the addition of the single-dose maternal and infant nevirapine to a zidovudine regimen from 36 weeks' gestation gave transmission rates of 6.5 percent at 6 weeks (95% CI 3.9–9.1), and adding these to a zidovudine and lamivudine regimen reduced transmission to 4.7 percent (95% CI 2.4–7.0) (Dabis *et al.*, 2005).

Consolidating research into practice guidelines

These shorter, more straightforward regimens have been the cornerstone of PMTCT programs in low-resource settings. When they were initially researched

Table 17.1 Selected key PMTCT antiretroviral regimen trials

Trial (citation)	Site/setting	Number	Regimen				Outcome	Policy implications
			Antepartum	Intrapartum	Postpartum	Infant		
PACTG 076 (Connor et al., 1994)	USA and France Non-breastfeeding	477	From 14 to 34 weeks: Arm 1: ZDV 100 mg 5 ×/d from Arm 2: Placebo	Arm 1: Intravenous ZDV infusion Arm 2: Placebo	No ARV	ZDV 2 mg/kg q.i.d. × 6 weeks	68% efficacy in reducing transmission: 8.3% vs 25.5% at 18 months	First demonstration of ARV efficacy in reducing transmission – rapidly adopted into practice in well-resourced settings
Thailand CDC (Shaffer et al., 1999)	Bangkok, Thailand Non-breastfeeding	392	From 36 weeks: Arm 1: ZDV 300 mg b.i.d. from 36 wk Arm 2: Placebo	Arm 1: ZDV 300 mg 3-hourly Arm 2: Placebo	No ARV	No ARV	50% efficacy in reducing transmission: 9.4% vs 18.9% at 6 months	First demonstration of short-course ZDV efficacy in non-breastfeeding mothers
Côte d'Ivoire CDC (Wiktor et al., 1999)	Abidjan, Côte d'Ivoire Breastfeeding	280	From 36 weeks: Arm 1: ZDV 300 mg b.i.d. Arm 2: Placebo (stopped Feb 1998)	Arm 1: ZDV 300 mg 3-hourly Arm 2: Placebo	No ARV	No ARV	37% efficacy in reducing transmission: 16.5% vs 26.1% at 3 months	First demonstration of short-course ZDV efficacy in breastfeeding mothers

(continued)

Table 17.1 (Continued)

Trial (citation)	Site/setting	Number	Regimen Antepartum	Regimen Intrapartum	Regimen Postpartum	Regimen Infant	Outcome	Policy implications
PETRA study (Petra Study Team, 2002)	South Africa, Uganda, Thailand Breastfeeding	1797	From 36 weeks: Arm 1: ZDV 300 mg b.i.d. + 3TC 150 mg b.i.d. Arm 2: Placebo Arm 3: Placebo Arm 4: Placebo (stopped Feb 1998)	Arm 1: ZDV 300 mg 3-hourly + 3TC 150 mg 12-hourly Arm 2: ZDV 300 mg 3-hourly + 3TC 150 mg 12-hourly Arm 3: ZDV 300 mg 3-hourly + 3TC 150 mg 12-hourly Arm 4: Placebo	Arm 1: ZDV 300 mg b.i.d. + 3TC 150 mg b.i.d. × 7 days Arm 2: ZDV 300 mg b.i.d. + 3TC 150 mg b.i.d. × 7 days Arm 3: Placebo Arm 4: Placebo	Arm 1: ZDV 4 mg/kg b.i.d. + 3TC 2 mg/kg b.i.d. × 7 days Arm 2: ZDV 4 mg/kg b.i.d. + 3TC 2 mg/kg b.i.d. × 7 days Arm 3: Placebo Arm 4: Placebo	At 6 weeks, 63% efficacy of Arm 1, 42% efficacy for Arm 2; no significant difference by 18 months: At 6 weeks Arm 1, 5.7%; Arm 2, 8.9%; Arm 3, 14.2%; Arm 4, 15.3% and at 18 months: Arm 1, 14.9%; Arm 2, 18.1%; Arm 3, 20.0%; Arm 4, 22.2%	Demonstration of efficacy of short-course combination regimens, but reductions no longer significant at 18 months due to breastfeeding transmissions
HIVNET 012 (Guay et al., 1999) (Jackson et al., 2003)	Kampala, Uganda Breastfeeding	626	Arm 1: No ARV Arm 2: No ARV Arm 3: No ARV	Arm 1: NVP 200 mg × 1 Arm 2: ZDV 300 mg 3 hrly Arm 3: Placebo (Stopped Feb 1998)	No ARV	Arm 1: NVP 2 mg/kg × 1 at 48–72 h Arm 2: ZDV 4 mg/kg b.i.d. × 7 days Arm 3	Transmission 11.8% in NVP arm compared to 20.0% in ZDV arm at 6–8 weeks, 15.7% in NVP	Demonstration of efficacy of single maternal intrapartum NVP and infant NP dose in reducing x

						vs 25.8% in ZDV arm at 18 months	transmission. Rapidly adopted into expanding PMTCT programs in low-resource settings.
					Placebo (stopped Feb 1998)		
PHPT-2 (Lallemant et al., 2004)	Thailand Non-Breastfeeding	1844	From 28 weeks: Arm 1: ZDV 300 mg b.i.d Arm 2: ZDV 300 mg b.i.d. Arm 3: ZDV 300 mg b.i.d.	Arm 1: ZDV 300 mg 3 hrly Arm 2: ZDV 300 mg 3 hrly + NVP 200 mg × 1 Arm 3: ZDV 300 mg 3 hrly + NVP 200 mg × 1	No ARV	Arm 1: ZDV 2 mg/kg q.i.d. × 7 days Arm 2: ZDV 2 mg/kg q.i.d. × 7 days Arm 3: ZDV 2 mg/kg q.i.d. × 7 days + NVP 6 mg × 1 at 48–72 hours	Arm 1 discontinued in June 2002 due to significantly higher transmission (6.3% vs 2.1% in Arm 2 and 1.1% in Arm 3). Final analysis Arm 3: transmission rates of 2.8% in Arm 2 vs 2.0% Arm 3	Demonstration of very low transmission rates, in non-breastfeeding mothers, with antenatal ZDV combined with intrapartum and infant NVP. First-line recommendation in 2006 WHO guidelines

and introduced, the prospect of access to affordable antiretroviral treatment in these settings seemed remote, if not impossible. Access to ongoing treatment is particularly important for pregnant women with low CD4+ counts or symptomatic AIDS, who are more likely to transmit to their infants (Thorne and Newell, 2004; Coovadia *et al.*, 2007), whose infants are more likely to die (Taha *et al.*, 1996; Newell *et al.*, 2004) and who are themselves more likely to develop resistant virus following nevirapine use (Martinson *et al.*, 2004). There has been a dramatic increase in the availability of treatment and the numbers of people receiving antiretrovirals since 2002, with the World Health Organization estimating that more than two million people in low- and middle-income countries were receiving antiretroviral therapy by December 2006 – close to 30 percent of the estimated 7 million in need of treatment (WHO, 2007).

This increased access to care has changed the recommendations for optimal PMTCT services in low-resource settings towards a framework in which the PMTCT intervention is part of a continuum of care of mother and child, and where ongoing antiretroviral therapy is provided to those in need. This is reflected in the WHO PMTCT guidelines from 2006 onwards, which set out an approach starting with access to ongoing antiretroviral treatment for those who need it and a short-course treatment regimen for those who do not yet require ongoing treatment (WHO, 2006a). The 2006 WHO guidelines recommend that women with CD4+ counts less than 200/mm^3, or Stage 3 or 4 clinical disease, start and continue antiretroviral treatment, and also recommend initiation of ongoing antiretroviral treatment in pregnant women with CD4+ counts between 200 and 350/mm^3, where this is feasible. Initial reports from Thailand and from several parts of Africa have confirmed the safety and feasibility of antiretroviral treatment for pregnant women in these low-resource settings (Phanuphak *et al.*, 2005; Thomas *et al.*, 2005; Tonwe-Gold *et al.*, 2005). It is likely that the provision of triple combination antiretroviral prophylaxis will increase to women as PMTCT prophylaxis, irrespective of CD4+ count, in more low-resource settings in the future.

The possible impact of nevirapine resistance

One of the major concerns about the use of nevirapine in the short-course PMTCT regimens is the selection of non-nucleoside reverse transcriptase inhibitor (NNRTI) resistant virus (McIntyre, 2006). The drug has a long half-life, detectable up to 20 days following the single dose given in labor (Cressey *et al.*, 2005; Muro *et al.*, 2005), and a low barrier for resistance, with only one point mutation required in the viral codon. This selection of resistant virus has been demonstrated in mothers a number of studies, ranging from 15 percent to 75 percent, using standard population sequencing techniques, following single-dose nevirapine alone (Eshleman *et al.*, 2003, 2004a, 2004b, 2005; Martinson *et al.*, 2004; Chaix *et al.*, 2007), and around 30 percent where nevirapine is added to a zidovudine regimen (Jourdain *et al.*, 2004; Chaix *et al.*, 2005). It may also happen in around 15 percent of women following nevirapine-containing triple antiretroviral prophylaxis, stopped after delivery (Cunningham *et al.*, 2002; Lyons *et al.*, 2005). Selection of resistant virus is more likely with subtype C virus, low CD4+ and high viral loads (Flys *et al.*, 2006), and the K103N mutation has been the most commonly seen mutation in mothers. The use of more sensitive techniques to detect resistant viral mutations, such as allele-specific real-time PCR and similar assays, shows that minority resistant viral populations can be detected in up to 80 percent of exposed women (Flys *et al.*, 2005; Johnson *et al.*, 2005; Palmer *et al.*, 2006). High rates of resistant virus are also seen in nevirapine-exposed, HIV-infected infants, ranging from 20 percent to 87 percent (Eshleman *et al.*, 2005; Kurle *et al.*, 2007). In a South African study of 53 HIV-infected infants exposed to the HIVNET 012 regimen, 24 (45.3 percent) had detectable resistance at their first visit, with the most frequent mutations being Y181C (75 percent), K103N (25 percent) and Y188C (12 percent) (Martinson *et al.*, 2007).

Strategies to reduce the risk of the selection of resistance by providing other antiretroviral cover during at least part of the prolonged half-life of nevirapine have been shown to be successful. The addition of 4 or 7 days of zidovudine and lamivudine following intrapartum nevirapine reduces the prevalence of NNRTI-resistant virus from 60 percent to around 10 percent (McIntyre *et al.*, 2005). In the DITRAME-plus study where women received zidovudine and lamivudine from 32 weeks' gestation and for 3 days postpartum, with intrapartum nevirapine, a rate of NNRTI resistance of 1.14 percent (95% CI 0.03–6.17) was seen, although lamivudine-resistant mutations were found in 8.33 percent (95%CI 3.66–15.7) (Chaix *et al.*, 2005). A study in Zambia utilized a single dose of tenofovir and emtricitabine, given with the nevirapine dose, in women with CD4+ counts above 200/mm^3 who had also received an antenatal zidovudine regimen, and noted a 53 percent reduction in detectable resistance at 6 weeks' postpartum (Chi *et al.*, 2007a).

The date available from observational studies suggest that there is relatively little impact of this selection of resistance on the success of nevirapine-containing antiretroviral treatment regimens if treatment is started more than 6 months after pregnancy (Zijenah *et al.*, 2005; Chi *et al.*, 2007b; Lockman *et al.*, 2007; WHO, 2007). While these results are reassuring, the long-term impact on treatment outcomes is not yet fully known, and programs should be encouraged to identify women in need of treatment early in pregnancy and provide this, rather than giving short-course nevirapine-containing regimens in these cases.

Infant feeding: mother-to-child transmission through breastfeeding

Breast-milk transmission is an important cause of HIV infection in children, accounting for up to half of the global infections (John-Stewart, 2007), or more than 200,000 infections in children annually. Reductions in the risk of MTCT which can be achieved by antiretroviral strategies may be completely negated by the effect of breastfeeding, but this risk needs to be balanced against the risk of morbidity and mortality caused by replacement feeding in low-resource settings. The safe prevention of transmission through breastfeeding and the improvement of HIV-free survival in exposed children remains one of the most difficult issues in PMTCT.

The exact mechanism of transmission in breast milk is not fully understood. Infection may take place through the entry of cell-free virions or of cell-associated HIV. Both these forms of HIV have been detected in colostrum and breast milk, although breast-milk viral load is generally lower than that in plasma (Rousseau *et al.*, 2003; Lehman and Farquhar, 2007; Walter *et al.*, 2008). Cell-associated virus is suggested to be more highly correlated with the risk of early transmission, and both cell-associated and cell-free with later transmission (Koulinska *et al.*, 2006). This may have implications for the effectiveness of antiretroviral treatment strategies to reduce breast-milk transmission, since antiretrovirals have been shown to suppress cell-free, but not cell-associated, virus (Shapiro *et al.*, 2005). Cell-free virus could penetrate the mucosal lining of the gastro-intestinal tract of infants by infecting inter-epithelial dendritic cells and be sampled by M cells of the Peyer's patches, or it could enter the submucosa directly by the kinds of mechanisms which allow intact proteins to traverse the immature mucosal barrier. Alternatively virus could penetrate through damaged mucosal foci. An *in vitro* study showed that HIV-infected immunocytes stimulated enterocytes to engulf HIV particles present on their surfaces (Bomsel, 1997), and it appears there may be both cell-associated and cell-free mechanisms for this transmission. These mucosal factors may explain the protective effect of exclusive breastfeeding over mixed breastfeeding, which has now been documented in several studies (Coutsoudis *et al.*, 2001; Iliff *et al.*, 2005; Kuhn *et al.*, 2007).

Risk factors for breast-milk transmission

Breastfeeding approximately doubles the risk of MTCT, and in populations where breastfeeding continues into the second year this gives an additional absolute risk of 15–20 percent (Fowler and Newell, 2002; Newell, 2006; Kourtis *et al.*, 2007). The risk of infection through breast-milk continues through the duration of breastfeeding. In the Breastfeeding and HIV International Transmission Study (BHITS) meta-analysis, 42 percent of HIV-infected children for whom the time of infection was known had been infected via breastfeeding after they had reached 4 weeks of age (Breastfeeding and HIV International Transmission Study Group, 2004), and in a large study in Zimbabwe, the ZVITAMBO Study, 68 percent of HIV transmission in infants who had tested negative at 6 weeks occurred after 6 months, through continued breastfeeding (Iliff *et al.*, 2005). A Malawi study suggested that 85 percent of transmission could have been avoided if women had weaned at 6 months (Taha *et al.*, 2007). The BHITS analysis also showed a relatively constant risk of infection of around 0.9 percent per month after the first month, although previous studies had suggested a higher risk in the first 6 months to a year (Miotti *et al.*, 1999). The impact of the duration and pattern of breastfeeding as a major determinant of postnatal transmission was shown in an analysis that pooled data from two studies conducted in West and South Africa (Becquet *et al.*, 2008). In infants who were exposed to more than 6 months of breastfeeding the risk of transmission doubled (95% CI 1.2–3.7), with a transmission rate of 8.7 percent (95% CI 6.8–11 percent) as compared to a transmission rate of 3.9 percent (95% CI 2.3–6.5) in infants with less than 6 months' exposure to breastfeeding. Exposure to solids in the first 2 months of life resulted in a 2.9-fold (95% CI 1.1–8.0) increased risk of transmission ($P = 0.04$). Shortening the period of breastfeeding or postponing the introduction of solids may be useful strategies in settings where breast-milk is essential for infant survival.

The risk of transmission through breast-milk thus depends on the age of complete cessation of breastfeeding, as does the risk of the potential complications of stopping breastfeeding (John-Stewart, 2007). Other risk factors for breast-milk transmission include the stage of maternal disease, with an increased risk in mothers with advanced disease, low CD4+ counts or high viral loads (John *et al.*, 2001; Coovadia *et al.*, 2007); breast-health, including a possible increased risk with sub-clinical mastitis (Rollins *et al.*, 2004; Kantarci *et al.*, 2007; Kourtis *et al.*, 2007); and the modality of feeding – exclusive breastfeeding or mixed feeding (Coovadia *et al.*, 2007).

Interventions to reduce breast-milk transmission

One of the major factors contributing to the very low rates of vertical transmission in well-resourced settings has been the avoidance of breastfeeding (Sansom *et al.*, 2007). This approach is much more difficult in less-resourced countries, where the more significant risks of replacement feeding have to be balanced against the risk of HIV transmission (Wilfert and Fowler, 2007). The World Health Organization recommends avoidance of breastfeeding where replacement feeding is "Acceptable, Feasible, Affordable, Sustainable and Safe" – now referred to as the AFASS criteria (World Health Organization, 2001). Replacement feeding has become the norm for HIV-positive women in America and Europe, in Brazil, Thailand and some other parts of Asia, and has been shown to be achievable and safe for women in some settings in Africa (Coetzee *et al.*, 2005; Leroy *et al.*, 2007; Njom Nlend *et al.*, 2007; Plipat *et al.*, 2007). For many other women living in poorly-resourced, high HIV-prevalence settings it is not currently feasible, because replacement feeding carries a high risk of morbidity to the baby. Where this is the case, exclusive breastfeeding carries a lower risk of HIV transmission than "mixed feeding", and is the recommendation (Coutsoudis *et al.*, 1999; Holmes and Savage, 2007). However, mixed feeding is the norm in most settings, and exclusive breastfeeding, where no other food or water is given to the child, is uncommon without major support interventions. If the promotion and support of exclusive breastfeeding is to be a successful intervention to reduce transmission, it will require staff and support strategies which do not currently exist in most places. Although the WHO guidelines initially recommended exclusive breast-feeding to 3–6 months of age, with rapid weaning at this time followed by the use of alternative foods (WHO, 2001), the accumulation of evidence of the potential risks of early weaning has prompted a change in these guidelines to recommend exclusive breastfeeding for the first 6 months (WHO, 2006b), and to make this the first recommendation in low-resource settings unless the AFASS requirements are met. The additional risk of late postnatal transmission of around 1 percent per month has to be balanced against the risk of severe morbidity or mortality from replacement feeding in unsafe circumstances, which may exceed the HIV risk (Coovadia *et al.*, 2007; Creek *et al.*, 2007b). Increasing evidence is accumulating that early weaning and the use of replacement feeding carries risks (Thior *et al.*, 2005; Sinkala *et al.*, 2007), just as it does earlier if the AFASS criteria are not met, and that continued breastfeeding, with additional complementary feeds, may be safer in these situations in terms of overall survival.

Unfortunately, the AFASS criteria, whether assessed immediately post-delivery or at the time of weaning, relate mainly to the socio-economic circumstances of mothers and are largely not within their power to change. Partly because of this, the field effectiveness of using the criteria to guide infant-feeding choices has not been optimal (Bland *et al.*, 2007; Jackson *et al.*, 2007; Chopra and Rollins, 2008). Health workers may be reluctant to advise strongly in one or other direction for infant feeding, or may do so incorrectly, based on their own level of knowledge, stigma or bias. A South African study demonstrated that three factors were associated with successful replacement feeding, even in very poor environments, which could be used to guide infant-feeding decisions. These were the availability of electricity or gas as fuel, the availability of piped water, and disclosure of maternal HIV status (Doherty *et al.*, 2007).

One of the most elusive goals in PMTCT has been a solution which preserves the benefits of breastfeeding without risk of transmission. One possible alternative approach could be the provision of antiretroviral prophylaxis – either given to breastfed infants, or by providing antiretroviral therapy to breastfeeding mothers (Gaillard *et al.*, 2004; Kourtis *et al.*, 2007). Post-exposure prophylaxis given to infants has been shown to reduce transmission risk (Taha *et al.*, 2004; Gray *et al.*, 2005). The provision of 6 to 14 weeks of nevirapine to breastfed infants has shown reduced transmission and mortality by 6 weeks of life, although this was less marked later. Studies conducted in Ethiopia, India and Uganda using 6 weeks' treatment with nevirapine confirmed that neonatal prophylaxis administered during breastfeeding reduced HIV transmission.

In the 6-week extended nevirapine (SWEN) trial, mothers received nevirapine in labor and infants nevirapine 2 mg/kg at birth, and infants were then randomized to receive NVP 5 mg daily from day 8 of life to day 48, or multivitamins. At 6 weeks of age, infants receiving SWEN had a 50 percent lower risk of sero-conversion as compared to the control arm (2.5 percent vs 5.27 percent, $P = 0.009$). At 6 months of age this was no longer significant, with a 20 percent reduction (6.9 percent vs 9 percent, $P = 0.16$). When mortality was combined with HIV infection, SWEN reduced the risk of HIV infection or death at 6 months of age (8.1 percent vs 11.6 percent, $P = 0.028$) (Sastry and The Six Week Extended Dose Nevirapine (SWEN) Study Team, 2008). In an open label, controlled three-arm study conducted in Malawi, more than 3000 infants who were uninfected at birth were randomized to receive either single-dose nevirapine (sd-NVP) and 1 week of AZT (control); sd-NVP, 1 week of AZT and nevirapine for 14 weeks (Ext NVP); or nevirapine and AZT for 14 weeks (ExtNVP/ZDV) (Taha *et al.*, 2008). The neonatal prophylaxis reduced breast-milk transmission in the first year of life, with transmission rates of 5.1 percent, 1.7 percent and 1.6 percent at 6 weeks in the SdNVP, Ext NVP and EXTNVP/ZDV arms respectively. At 6 months, the rates had risen to 10.1 percent, 4.0 percent and 5.2 percent in these arms. However, this protection was no longer evident at 2 years of age, with rates of 14.5 percent, 11.2 percent and 12.3 percent, showing that ongoing breastfeeding without prophylaxis appears to reverse this effect. Further research is examining the role of longer courses of nevirapine prophylaxis to infants.

Maternal antiretroviral treatment has been shown to reduce the viral load in breast milk (Shapiro *et al.*, 2005), and several observational trials have shown promise in reducing transmission during breastfeeding (Arendt *et al.*, 2007; Kilewo *et al.*, 2007). The Kisumu Breastfeeding Study, conducted in Kenya, assessed the efficacy of AZT/3TC and nevirapine or nelfinavir initiated antenatally from 34 weeks' gestation and given for 6 months post-partum to prevent breast-milk transmission of HIV (Thomas *et al.*, 2008). Infants received single-dose nevirapine. Overall, 29/502 (5.9 percent) of children were infected by 1 year of age, with 24/29 (83 percent) of infants being infected by 6 months of age while mothers were still on ART. An additional five children were infected after 6 months via breastfeeding after maternal ART had been discontinued. Although maternal ART during the breastfeeding period reduced transmission substantially, HIV-infected infants exposed to antiretroviral therapy were at risk of developing resistance, and this appeared to increase over time from 33 percent at 2 weeks to 67 percent at 6 months.

If further randomized trials confirm the success of this approach in reducing transmission, and the safety for mothers of such interrupted treatment, it could be a powerful option where access to antiretrovirals is available. The feasibility may be limited by the affordability and infrastructure demands to deliver antiretroviral treatment to large numbers of women and by the social effects of disclosure and support for women on treatment, and this evidence will be required before wide-scale introduction of such a policy. Effective ways to make either exclusive replacement feeding or exclusive breastfeeding safer are essential to further reduce mother-to-child transmission of HIV.

From research to implementation

The major research advances in PMTCT and the success in controlling transmission in high-resource settings have not been matched in low-resource, high HIV-prevalence settings, where it is estimated that PMTCT interventions reach less than 10 percent of HIV-infected women, even with the simplest single-dose nevirapine interventions. Estimates of the proportion of HIV-infected pregnant women receiving antiretroviral prophylaxis in 2005 ranged from less than 1 percent to 54 percent in sub-Saharan Africa, with an average regional coverage of 11 percent (8–15 percent). Coverage was 75 percent (38–95 percent) in Eastern Europe and Central Asia, 24 percent (13–46 percent) in Latin America and the Caribbean, 5 percent (3–10 percent) in East, South and South-East Asia, and less than 1 percent in North Africa and the Middle East (WHO, 2007). True success in reducing mother-to-child transmission cannot be achieved without a major effort to increase this coverage, and to increase the availability of antiretroviral treatment for those in need. Such an effort should include the use of routine HIV testing in maternity settings, and better integration of PMTCT and treatment services, more efficient use of health workers and health infrastructure, and improved systems of drug supply (McIntyre and Lallemant, 2008).

Future research directions include the need to focus on vaccines for the prevention of mother-to-child transmission (Cunningham and McFarland, 2008), on the urgent need to reduce transmission through breastfeeding, and on operational research into better ways to deliver existing effective interventions. The optimal reduction in mother-to-child transmission will be achieved when the PMTCT interventions in pregnancy are integrated into a comprehensive HIV prevention and care framework, including circumcision services and the eventual introduction of effective vaccines or microbicides. Much progress has been made in the field of PMTCT, reflecting remarkable research and health-service efforts, but significant challenges remain to reach the majority of HIV-infected pregnant women worldwide who still have no access to these advances, and to further reduce the risk of transmission.

References

Arendt, V., Ndimubanzi, P., Vyankandondera, J. *et al.* (2007). AMATA Study: effectiveness of antiretroviral therapy in breastfeeding mothers to prevent post-natal vertical transmission in Rwanda. Cross-track Session. In: *4th IAS Conference on HIV Pathogenesis, Treatment and Prevention, Sydney, 22–25 July 2007*, Abstract no. TUAX102.

Bajunirwe, F. and Muzoora, M. (2005). Barriers to the implementation of programs for the prevention of mother-to-child transmission of HIV: a cross-sectional survey in rural and urban Uganda. *AIDS Res. Ther.*, 2, 10.

Becquet, R., Bland, R., Leroy, V. *et al.* (2008). Duration and pattern of breastfeeding and postnatal transmission of HIV: pooled analysis of individual data from a west and South African cohort study. In: *15th Conference on Retroviruses and Opportunistic Infections, 3–6 February, Boston, MA*, Abstract No. 46.

Bland, R. M., Rollins, N. C., Coovadia, H. M. *et al.* (2007). Infant feeding counselling for HIV-infected and uninfected women: appropriateness of choice and practice. *Bull. World Health Org.*, 85, 289–96.

Bomsel, M. (1997). Transcytosis of infectious human immunodeficiency virus across a tight human epithelial cell line barrier. *Nature Med.*, 3, 42–7.

Breastfeeding and HIV International Transmission Study Group (2004). Late postnatal transmission of HIV-1 in breast-fed children: an individual patient data meta-analysis. *J. Infect. Dis.*, 189, 2154–66.

British HIV Association (2005). *Guidelines for the Management of HIV Infection in Pregnant Women and the Prevention of Mother-to-child Transmission of HIV*. London: The British HIV Association.

Cates, W. Jr (2006). Contraception and prevention of HIV infection. *J. Am. Med. Assoc.*, 296, 2802, author reply 2802.

CDC (1985). Recommendations for assisting in the prevention of perinatal transmission of human T-lymphotropic virus type III/lymphadenopathy-associated virus and acquired immunodeficiency syndrome. *Morbid. Mortal Wkly Rep.*, 34, 721–32.

CDC (2004). Introduction of routine HIV testing in prenatal care – Botswana. *Morbid. Mortal. Wkly Rep.*, 53, 1083–6.

CDC (2006). Achievements in public health. Reduction in perinatal transmission of HIV infection – United States, 1985–2005. *Morbid. Mortal. Wkly Rep.*, 55, 592–7.

Chaix, M. L., Dabis, F., Ekouevi, D. *et al.* (2005). Addition of 3 days of ZDV + 3TC postpartum to a short course of ZDV + 3TC and single-dose NVP provides low rate of NVP resistance mutations and high efficacy in preventing peri-partum HIV-1 transmission: ANRS DITRAME Plus, Abidjan, Côte d'Ivoire. In: *12th Conference on Retroviruses and Opportunistic Infections, Boston, 22–25 February*, Abstract No. 72LB.

Chaix, M. L., Ekouevi, D. K., Peytavin, G. *et al.* (2007). Impact of nevirapine (NVP) plasma concentration on selection of resistant virus in mothers who received single-dose NVP to prevent perinatal human immunodeficiency virus type 1 transmission and persistence of resistant virus in their infected children. *Antimicrob. Agents Chemother.*, 51, 896–901.

Chi, B. H., Sinkala, M., Mbewe, F. *et al.* (2007a). Single-dose tenofovir and emtricitabine for reduction of viral resistance to non-nucleoside reverse transcriptase inhibitor drugs in women given intrapartum nevirapine for perinatal HIV prevention: an open-label randomised trial. *Lancet*, 370, 1698–705.

Chi, B. H., Sinkala, M., Stringer, E. M. *et al.* (2007b). Early clinical and immune response to NNRTI-based antiretroviral therapy among women with prior exposure to single-dose nevirapine. *AIDS*, 21, 957–64.

Chopra, M. and Rollins, N. (2008). Infant feeding in the time of HIV: assessment of infant feeding policy and programmes in four African countries scaling up prevention of mother to child transmission programmes. *Arch. Dis. Child.*, 93, 288–91.

Chou, R., Smits, A. K., Huffman, L. H. *et al.* (2005). Prenatal screening for HIV: a review of the evidence for the US Preventive Services Task Force. *Ann. Intern. Med.*, 143, 38–54.

Coetzee, D., Hilderbrand, K., Boulle, A. *et al.* (2005). Effectiveness of the first district-wide program for the prevention of mother-to-child transmission of HIV in South Africa. *Bull. WHO*, 83, 489–94.

Coffie, P., Ekouevi, D., Chaix, M. L. *et al.* (2007). Short-course zidovudine and lamivudine or single-dose nevirapine-containing PMTCT compromises 12-month response to HAART in African Women, Abidjan, Côte d'Ivoire (2003–2006). In: *14th Conference on Retroviruses and Opportunistic Infections, Los Angeles, 25–28 February*, Abstract No. 93LB.

Connor, E. M., Sperling, R. S., Gelber, R. *et al.* (1994). Reduction of maternal–infant transmission of human immunodeficiency virus type 1 with zidovudine treatment. Pediatric AIDS Clinical Trials Group Protocol 076 Study Group. *N. Engl. J. Med.*, 331, 1173–80.

Coovadia, H. M., Rollins, N. C., Bland, R. M. *et al.* (2007). Mother-to-child transmission of HIV-1 infection during exclusive breastfeeding in the first 6 months of life: an intervention cohort study. *Lancet*, 369, 1107–16.

Coutsoudis, A., Pillay, K., Spooner, E. *et al.* (1999). Influence of infant-feeding patterns on early mother-to-child transmission of HIV-1 in Durban, South Africa: a prospective cohort study. South African Vitamin A Study Group. *Lancet*, 354, 471–6.

Coutsoudis, A., Pillay, K., Kuhn, L. *et al.* (2001). Method of feeding and transmission of HIV-1 from mothers to children by 15 months of age: prospective cohort study from Durban, South Africa. *AIDS*, 15, 379–87.

Cowan, M. J., Hellmann, D., Chudwin, D. *et al.* (1984). Maternal transmission of acquired immune deficiency syndrome. *Pediatrics*, 73, 382–6.

Creek, T. L., Ntumy, R., Seipone, K. *et al.* (2007a). Successful introduction of routine opt-out HIV testing in antenatal care in Botswana. *J. Acquir. Immune. Defic. Syndr.*, 45, 102–7.

Creek, T., Arvelo, W., Kim, A. *et al.* (2007b). Role of infant feeding and HIV in a severe outbreak of diarrhea and malnutrition among young children, Botswana, 2006. In: *14th Conference on Retroviruses and Opportunistic Infections, Los Angeles, 25–28 February*, Abstract No. 770.

Cressey, T. R., Jourdain, G., Lallemant, M. J. *et al.* (2005). Persistence of nevirapine exposure during the postpartum period after intrapartum single-dose nevirapine in addition to zidovudine prophylaxis for the prevention of mother-to-child transmission of HIV-1. *J. Acquir. Immune Defic. Syndr.*, 38, 283–8.

Cunningham, C. and McFarland, E. (2008). Vaccines for prevention of mother to child transmission of HIV. *Curr. Opin. HIV AIDS*, 3, in press.

Cunningham, C. K., Chaix, M. L., Rekacewicz, C. *et al.* (2002). Development of resistance mutations in women receiving standard antiretroviral therapy who received intrapartum nevirapine to prevent perinatal human immunodeficiency virus type 1 transmission: a substudy of pediatric AIDS clinical trials group protocol 316. *J. Infect. Dis.*, 186, 181–8.

Dabis, F., Bequet, L., Ekouevi, D. K. *et al.* (2005). Field efficacy of zidovudine, lamivudine and single-dose nevirapine to prevent peripartum HIV transmission. *AIDS*, 19, 309–18.

Doherty, T., Chopra, M., Jackson, D. *et al.* (2007). Effectiveness of the WHO/UNICEF guidelines on infant feeding for HIV-positive women: results from a prospective cohort study in South Africa. *AIDS*, 21, 1791–7.

Eshleman, S. H., Jones, D., Guay, L. *et al.* (2003). HIV-1 variants with diverse nevirapine resistance mutations emerge rapidly after single dose nevirapine: HIVNET 012.

XIIth International HIV Drug Resistance Workshop: Basic Principles and Clinical Implications. July 2003, Cabo San Lucas, Mexico. (Abstract)

Eshleman, S. H., Guay, L. A., Mwatha, A. *et al.* (2004a). Characterization of nevirapine resistance mutations in women with subtype A vs. D HIV-1 6–8 weeks after single-dose nevirapine (HIVNET 012). *J. Acquir. Immune Defic. Syndr.*, 35, 126–30.

Eshleman, S. H., Guay, L. A., Mwatha, A. *et al.* (2004b). Comparison of nevirapine (NVP) resistance in Ugandan women 7 days vs 6–8 weeks after single-dose nvp prophylaxis: HIVNET 012. *AIDS Res. Hum. Retroviruses*, 20, 595–9.

Eshleman, S. H., Hoover, D. R., Chen, S. *et al.* (2005). Resistance after single-dose nevirapine prophylaxis emerges in a high proportion of Malawian newborns. *AIDS*, 19, 2167–9.

European Collaborative Study (1992). Risk factors for mother-to-child transmission of HIV-1. European Collaborative Study. *Lancet*, 339, 1007–12.

European Collaborative Study (2005). Mother-to-child transmission of HIV infection in the era of highly active antiretroviral therapy. *Clin. Infect. Dis.*, 40, 458–65.

European Collaborative Study (2006). The mother-to-child HIV transmission epidemic in Europe: evolving in the East and established in the West. *AIDS*, 20, 1419–27.

European Mode of Delivery Collaboration (1999). Elective caesarean-section versus vaginal delivery in prevention of vertical HIV-1 transmission: a randomised clinical trial. The European Mode of Delivery Collaboration. *Lancet*, 353, 1035–9.

Floridia, M., Ravizza, M., Tamburrini, E. *et al.* (2006). Diagnosis of HIV infection in pregnancy: data from a national cohort of pregnant women with HIV in Italy. *Epidemiol. Infect.*, 134, 1120–7.

Flys, T., Nissley, D. V., Claasen, C. W. *et al.* (2005). Sensitive drug-resistance assays reveal long-term persistence of HIV-1 variants with the K103N nevirapine (NVP) resistance mutation in some women and infants after the administration of single-dose NVP: HIVNET 012. *J. Infect. Dis.*, 192, 24–9.

Flys, T. S., Chen, S., Jones, D. C. *et al.* (2006). Quantitative analysis of HIV-1 variants with the K103N resistance mutation after single-dose nevirapine in women with HIV-1 subtypes A, C, and D. *J. Acquir. Immune Defic. Syndr.*, 42, 610–13.

Fowler, M. G. and Newell, M. L. (2002). Breast-feeding and HIV-1 transmission in resource-limited settings. *J. Acquir. Immune Defic. Syndr.*, 30, 230–9.

Fowler, M. G., Lampe, M. A. and Jamieson, D. J. (2007). Reducing the risk of mother-to-child human immunodeficiency virus transmission: past successes, current progress and challenges, and future directions. *Am. J. Obstet. Gynecol.*, 197, S3–9.

Gaillard, P., Fowler, M. G., Dabis, F. *et al.* (2004). Use of antiretroviral drugs to prevent HIV-1 transmission through breast-feeding: from animal studies to randomized clinical trials. *J. Acquir. Immune Defic. Syndr.*, 35, 178–87.

Giuliano, M., Magoni, M., Bassani, L. *et al.* (2005). A theme issue by, for, and about Africa: results from Ugandan program preventing maternal transmission of HIV. *Br. Med. J.*, 331, 778.

Gray, G. E., Urban, M., Chersich, M. F. *et al.* (2005). A randomized trial of two postexposure prophylaxis regimens to reduce mother-to-child HIV-1 transmission in infants of untreated mothers. *AIDS*, 19, 1289–97.

Guay, L., Musoke, P., Fleming, T. *et al.* (1999). Intrapartum and neonatal single-dose nevirapine compared with zidovudine for prevention of mother-to-child transmission of HIV-1 in Kampala, Uganda: HIVNET 012 randomised trial. *Lancet*, 354, 795–802.

Holmes, W. R. and Savage, F. (2007). Exclusive breastfeeding and HIV. *Lancet*, 369, 1065–6.

Iliff, P. J., Piwoz, E. G., Tavengwa, N. V. *et al.* (2005). Early exclusive breastfeeding reduces the risk of postnatal HIV-1 transmission and increases HIV-free survival. *AIDS*, 19, 699–708.

Jackson, D. J., Chopra, M., Doherty, T. M. *et al.* (2007). Operational effectiveness and 36 week HIV-free survival in the South African program to prevent mother-to-child transmission of HIV-1. *AIDS*, 21, 509–16.

Jackson, J. B., Musoke, P., Fleming, T. *et al.* (2003). Intrapartum and neonatal single-dose nevirapine compared with zidovudine for prevention of mother-to-child transmission of HIV-1 in Kampala, Uganda: 18-month follow-up of the HIVNET 012 randomised trial. *Lancet*, 362, 859–68.

Jamieson, D. J., Clark, J., Kourtis, A. P. *et al.* (2007). Recommendations for human immunodeficiency virus screening, prophylaxis, and treatment for pregnant women in the United States. *Am. J. Obstet. Gynecol.*, 197, S26–32.

John, G. C., Nduati, R. W., Mbori-Ngacha, D. A. *et al.* (2001). Correlates of mother-to-child human immunodeficiency virus type 1 (HIV-1) transmission: association with maternal plasma HIV-1 RNA load, genital HIV-1 DNA shedding, and breast infections. *J. Infect. Dis.*, 183, 206–12.

John-Stewart, G. C. (2007). Breast-feeding and HIV-1 transmission: how risky for how long? *J. Infect. Dis.*, 196, 1–3.

Johnson, J. A., Li, J. F., Morris, L. *et al.* (2005). Emergence of drug-resistant HIV-1 after intrapartum administration of single-dose nevirapine is substantially underestimated. *J. Infect. Dis.*, 192, 16–23.

Jourdain, G., Ngo-Giang-Huong, N., Le Coeur, S. *et al.* (2004). Intrapartum exposure to nevirapine and subsequent maternal responses to nevirapine-based antiretroviral therapy. *N. Engl. J. Med.*, 351, 229–40.

Jourdain, G., Mary, J. Y., Coeur, S. L. *et al.* (2007). Risk factors for in utero or intrapartum mother-to-child transmission of human immunodeficiency virus type 1 in Thailand. *J. Infect. Dis.*, 196, 1629–36.

Kantarci, S., Koulinska, I. N., Aboud, S. *et al.* (2007). Subclinical mastitis, cell-associated HIV-1 shedding in breast milk, and breast-feeding transmission of HIV-1. *J. Acquir. Immune Defic. Syndr.*, 46, 651–4.

Kilewo, C., Karlsson, K., Ngarina, M. *et al.* (2007). Prevention of mother-to-child transmission of HIV-1 through breastfeeding by treating mothers prophylactically with triple antiretroviral therapy in Dar es Salaam, Tanzania – the MITRA PLUS Study. In: *4th IAS Conference on HIV Pathogenesis, Treatment and Prevention. Sydney, 22–25 July*, Abstract No. TUAX101.

Koulinska, I. N., Villamor, E., Chaplin, B. *et al.* (2006). Transmission of cell-free and cell-associated HIV-1 through breast-feeding. *J. Acquir. Immune Defic. Syndr.*, 41, 93–9.

Kourtis, A. P., Lee, F. K., Abrams, E. J. *et al.* (2006). Mother-to-child transmission of HIV-1: timing and implications for prevention. *Lancet Infect. Dis.*, 6, 726–32.

Kourtis, A. P., Jamieson, D. J., de Vincenzi, I. *et al.* (2007). Prevention of human immunodeficiency virus-1 transmission to the infant through breastfeeding: new developments. *Am. J. Obstet. Gynecol.*, 197, S113–122.

Kuhn, L., Sinkala, M., Kankasa, C. *et al.* (2007). High uptake of exclusive breastfeeding and reduced early post-natal HIV transmission. *PLoS ONE*, 2, e1363.

Kurle, S. N., Gangakhedkar, R. R., Sen, S. *et al.* (2007). Emergence of NNRTI drug resistance mutations after single-dose nevirapine exposure in HIV type 1 subtype C-infected infants in India. *AIDS Res. Hum. Retroviruses*, 23, 682–5.

Lallemant, M., Jourdain, G., Le Coeur, S. *et al.* (2000). A trial of shortened zidovudine regimens to prevent mother-to-child transmission of human immunodeficiency virus type 1. Perinatal HIV Prevention Trial (Thailand) Investigators. *N. Engl. J. Med.*, 343, 982–91.

Lallemant, M., Jourdain, G., Le Coeur, S. *et al.* (2004). Single-dose perinatal nevirapine plus standard zidovudine to prevent mother-to-child transmission of HIV-1 in Thailand. *N. Engl. J. Med.*, 351, 217–28.

Lansky, A., Jones, J. L., Frey, R. L. and Lindegren, M. L. (2001). Trends in HIV testing among pregnant women: United States, 1994–1999. *Am. J. Public Health*, 91, 1291–3.

Lehman, D. A. and Farquhar, C. (2007). Biological mechanisms of vertical human immunodeficiency virus (HIV-1) transmission. *Rev. Med. Virol.*, 17, 381–403.

Leroy, V., Sakarovitch, C., Cortina-Borja, M. *et al.* (2005). Is there a difference in the efficacy of peripartum antiretroviral regimens in reducing mother-to-child transmission of HIV in Africa? *AIDS*, 19, 1865–75.

Leroy, V., Sakarovitch, C., Viho, I. *et al.* (2007). Acceptability of formula-feeding to prevent HIV postnatal transmission, Abidjan, Cote d'Ivoire: ANRS 1201/1202 Ditrame Plus Study. *J. Acquir. Immune Defic. Syndr.*, 44, 77–86.

Lockman, S., Shapiro, R. L., Smeaton, L. M. *et al.* (2007). Response to antiretroviral therapy after a single, peripartum dose of nevirapine. *N. Engl. J. Med.*, 356, 135–47.

Lyons, F. E., Coughlan, S., Byrne, C. M. *et al.* (2005). Emergence of antiretroviral resistance in HIV-positive women receiving combination antiretroviral therapy in pregnancy. *AIDS*, 19, 63–7.

Magder, L. S., Mofenson, L., Paul, M. E. *et al.* (2005). Risk factors for in utero and intrapartum transmission of HIV. *J. Acquir. Immune Defic. Syndr.*, 38, 87–95.

Manzi, M., Zachariah, R., Teck, R. *et al.* (2005). High acceptability of voluntary counselling and HIV-testing but unacceptable loss to follow up in a prevention of mother-to-child HIV transmission program in rural Malawi: scaling-up requires a different way of acting. *Trop. Med. Intl Health*, 10, 1242–50.

Martinez, A. M., Hora, V. P., Santos, A. L. *et al.* (2006). Determinants of HIV-1 mother-to-child transmission in Southern Brazil. *An. Acad. Bras. Cienc.*, 78, 113–21.

Martinson, N., Morris, L., Gray, G. *et al.* (2004). HIV resistance and transmission following single-dose nevirapine in a PMTCT Cohort. In: *11th Conference on Retroviruses and Opportunistic Infections, San Francisco, 8–11 February*, Abstract No. 38.

Martinson, N. A., Morris, L., Gray, G. *et al.* (2007). Selection and persistence of viral resistance in HIV-infected children after exposure to single-dose nevirapine. *J. Acquir. Immune Defic. Syndr.*, 44, 148–53.

McIntyre, J. (2005). Preventing mother-to-child transmission of HIV: successes and challenges. *Br. J. Obstet. Gynaecol.*, 112, 1196–203.

McIntyre, J. A. (2006). Controversies in the use of nevirapine for prevention of mother-to-child transmission of HIV. *Expert Opin. Pharmacother.*, 7, 677–85.

McIntyre, J. (2008). Managing pregnant patients. In: R. Dolin, H. Masur and M. Saag (eds), *AIDS Therapy*, 3rd edn. Philadelphia, PA: Churchill Livingstone Elsevier, pp. 595–635.

McIntyre, J. A. and Lallemant, M. (2008). The prevention of mother-to-child transmission of HIV: are we translating scientific success into programmatic failure? *Curr. Opin. HIV AIDS*, 3, in press.

McIntyre, J. A., Martinson, N., Gray, G. E. *et al.* (2005). Addition of short course Combivir to single dose Viramune for the prevention of mother to child transmission of HIV-1 can significantly decrease the subsequent development of maternal and paediatric NNRTI-resistant virus. In: *3rd IAS Conference on HIV Pathogenesis and Treatment, Rio de Janeiro, 24–27 July*, Abstract No. TuFoO2O42005.

McKenna, M. T. and Hu, X. (2007). Recent trends in the incidence and morbidity that are associated with perinatal human immunodeficiency virus infection in the United States. *Am. J. Obstet. Gynecol.*, 197, S10–16.

Minkoff, H. L., McCalla, S., Delke, I. *et al.* (1990). The relationship of cocaine use to syphilis and human immunodeficiency virus infections among inner city parturient women. *Am. J. Obstet. Gynecol.*, 163, 521–6.

Miotti, P. G., Taha, T. E., Kumwenda, N. I. *et al.* (1999). HIV transmission through breast-feeding: a study in Malawi. *J. Am. Med. Assoc.*, 282, 744–9.

Mofenson, L. M. and McIntyre, J. A. (2000). Advances and research directions in the prevention of mother-to-child HIV-1 transmission. *Lancet*, 355, 2237–44.

Moodley, D., Moodley, J., Coovadia, H. *et al.* (2003). A multicenter randomized controlled trial of nevirapine versus a combination of zidovudine and lamivudine to reduce intrapartum and early postpartum mother-to-child transmission of human immunodeficiency virus type 1. *J. Infect. Dis.*, 187, 725–35.

Morrison, C. S. and Cates, W. Jr (2007). Preventing unintended pregnancy and HIV transmission: dual protection or dual dilemma? *Sex. Transm. Dis.*, 34, 873–5.

Moses, A., Zimba, C., Kamanga, E. *et al.* (2008). Prevention of mother-to-child transmission: program changes and the effect on uptake of the HIVNET 012 regimen in Malawi. *AIDS*, 22, 83–7.

Mulder-Folkerts, D. K., van den Hoek, J. A., van der Bij, A. K. *et al.* (2004). [Less refusal to participate in HIV screening among pregnant women in the Amsterdam region since the introduction of standard HIV screening using the opting-out method]. *Ned. Tijdschr. Geneeskd*, 148, 2035–7.

Muro, E., Droste, J. A., Hofstede, H. T. *et al.* (2005). Nevirapine plasma concentrations are still detectable after more than 2 weeks in the majority of women receiving single-dose nevirapine: implications for intervention studies. *J. Acquir. Immune Defic. Syndr.*, 39, 419–21.

Newell, M. L. (1998). Mechanisms and timing of mother-to-child transmission of HIV-1. *AIDS*, 12, 831–7.

Newell, M. L. (2003). Antenatal and perinatal strategies to prevent mother-to-child transmission of HIV infection. *Trans. R. Soc. Trop. Med. Hyg.*, 97, 22–4.

Newell, M. L. (2006). Current issues in the prevention of mother-to-child transmission of HIV-1 infection. *Trans. R. Soc. Trop. Med. Hyg.*, 100, 1–5.

Newell, M. L. and Thorne, C. (2004). Antiretroviral therapy and mother-to-child transmission of HIV-1. *Expert Rev. Anti. Infect. Ther.*, 2, 717–32.

Newell, M. L., Coovadia, H., Cortina-Borja, M. *et al.* (2004). Mortality of infected and uninfected infants born to HIV-infected mothers in Africa: a pooled analysis. *Lancet*, 364, 1236–43.

Njom Nlend, A., Penda, I., Same Ekobo, C. *et al.* (2007). Is exclusive artificial feeding feasible at 6 months post partum in Cameroon urban areas for HIV-exposed infants? *J. Trop. Pediatr.*, 53, 438–9.

Oleske, J., Minnefor, A., Cooper, R. Jr *et al.* (1983). Immune deficiency syndrome in children. *J. Am. Med. Assoc.*, 249, 2345–9.

Palmer, S., Boltz, V., Martinson, N. *et al.* (2006). Persistence of nevirapine-resistant HIV-1 in women after single-dose nevirapine therapy for prevention of maternal-to-fetal HIV-1 transmission. *Proc. Natl Acad. Sci. USA*, 103, 7094–9.

Perez, F., Zvandaziva, C., Engelsmann, B. and Dabis, F. (2006). Acceptability of routine HIV testing ("opt-out") in antenatal services in two rural districts of Zimbabwe. *J. Acquir. Immune Defic. Syndr.*, 41, 514–20.

Petra Study Team (2002). Efficacy of three short-course regimens of zidovudine and lamivudine in preventing early and late transmission of HIV-1 from mother to child in Tanzania, South Africa, and Uganda (Petra Study): a randomised, double-blind, placebo-controlled trial. *Lancet*, 359, 1178–86.

Phanuphak, N., Apornpong, T., Intarasuk, S. *et al.* (2005). Toxicities from nevirapine in HIV-infected nales and females, including pregnant females with various CD4 cell counts. In: *12th Conference on Retroviruses and Opprtunistic Infections, Boston, 22–25 February*, Abstract No. 21.

Plipat, T., Naiwatanakul, T., Rattanasuporn, N. *et al.* (2007). Reduction in mother-to-child transmission of HIV in Thailand, 2001–2003: results from population-based surveillance in six provinces. *AIDS*, 21, 145–51.

Rakgoasi, S. D. (2005). HIV counselling and testing of pregnant women attending antenatal clinics in Botswana, 2001. *J. Health Pop. Nutr.*, 23, 58–65.

Rollins, N., Meda, N., Becquet, R. *et al.* (2004). Preventing postnatal transmission of HIV-1 through breast-feeding: modifying infant feeding practices. *J. Acquir. Immune Defic. Syndr.*, 35, 188–95.

Rousseau, C. M., Nduati, R. W., Richardson, B. A. *et al.* (2003). Longitudinal analysis of human immunodeficiency virus type 1 RNA in breast milk and of its relationship to infant infection and maternal disease. *J. Infect. Dis.*, 187, 741–7.

Sansom, S. L., Harris, N. S., Sadek, R. *et al.* (2007). Toward elimination of perinatal human immunodeficiency virus transmission in the United States: effectiveness of funded prevention programs, 1999–2001. *Am. J. Obstet. Gynecol.*, 197, S90–95.

Sastry, J. and The Six Week Extended Dose Nevirapine (SWEN) Study Team (2008). Extended-dose nevirapine to 6 weeks of age for infants in Ethiopia, India, and Uganda: a randomized trial for prevention of HIV transmission through breastfeeding. In: *15th Conference on Retroviruses and Opportunistic Infections, 3–6 February, Boston*, Abstract No. 43.

Shaffer, N., Chuachoowong, R., Mock, P. A. *et al.* (1999). Short-course zidovudine for perinatal HIV-1 transmission in Bangkok, Thailand: a randomised controlled trial. Bangkok Collaborative Perinatal HIV Transmission Study Group. *Lancet*, 353, 773–80.

Shapiro, R. L., Ndung'u, T., Lockman, S. *et al.* (2005). Highly active antiretroviral therapy started during pregnancy or postpartum suppresses HIV-1 RNA, but not DNA, in breast milk. *J. Infect. Dis.*, 192, 713–19.

Shetty, A. K., Mhazo, M., Moyo, S. *et al.* (2005). The feasibility of voluntary counselling and HIV testing for pregnant women using community volunteers in Zimbabwe. *Intl J. STD AIDS*, 16, 755–9.

Sinkala, M., Kuhn, L., Kankasa, C. *et al.* (2007). No benefit of early cessation of breast-feeding at 4 months on HIV-free survival of infants born to HIV-infected mothers in Zambia: The Zambia Exclusive Breastfeeding Study (ZEBS). In: *14th Conference on Retroviruses and Opportunistic Infections, Los Angeles, 25–28 February*, Abstract No. 74.

Stringer, J. S., Sinkala, M., Maclean, C. C. *et al.* (2005). Effectiveness of a city-wide program to prevent mother-to-child HIV transmission in Lusaka, Zambia. *AIDS*, 19, 1309–15.

Taha, T. E., Miotti, P., Liomba, G. *et al.* (1996). HIV, maternal death and child survival in Africa. *AIDS*, 10, 111–12.

Taha, T. E., Kumwenda, N. I., Hoover, D. R. *et al.* (2004). Nevirapine and zidovudine at birth to reduce perinatal transmission of HIV in an African setting: a randomized controlled trial. *J. Am. Med. Assoc.*, 292, 202–9.

Taha, T. E., Hoover, D. R., Kumwenda, N. I. *et al.* (2007). Late postnatal transmission of HIV-1 and associated factors. *J. Infect. Dis.*, 196, 10–14.

Taha, T. E., Thigpen, M., Kumwenda, N. *et al.* (2008). Extended infant post-exposure prophylaxis with antiretroviral drugs significantly reduces postnatal HIV transmission: The PEPI-Malawi Study (2008). In: *15th Conference on Retroviruses and Opportunistic Infections, February 3–6, Boston*, Abstract No. 42LB.

Tai, J. H., Udoji, M. A., Barkanic, G. *et al.* (2007). Pregnancy and HIV disease progression during the era of highly active antiretroviral therapy. *J. Infect. Dis.*, 196, 1044–52.

Thior, I., Lockman, S., Smeaton, L. *et al.* (2005). Breast-feeding with 6 months of infant zidovudine prophylaxis vs formula-feeding for reducing postnatal HIV transmission and infant mortality: a randomised trial in Southern Africa. In: *12th Conference on Retroviruses and Opportunistic Infections, Boston, 22–25 February*, Abstract No. 75LB.

Thomas, T., Amornkul, P., Mwidau, J. *et al.* (2005). Preliminary findings: incidence of serious adverse events attributed to nevirapine among women enrolled in an ongoing trial using HAART to prevent mother-to-child HIV transmission. In: *12th Conference on Retroviruses and Opportunistic Infections, 22–25 February, Boston*, Abstract No. 809.

Thomas, T., Masaba, R., Ndivo, R. *et al.* (2008). Prevention of mother-to-child transmission of HIV-1 among breastfeeding mothers using HAART: The Kisumu Breastfeeding Study, Kisumu, Kenya, 2003–2007. In: *15th Conference on Retroviruses and Opportunistic Infections, 3–6 February, Boston*, Abstract No. 45aLB.

Thorne, C. and Newell, M. L. (2004). Prevention of mother-to-child transmission of HIV infection. *Curr. Opin. Infect. Dis.*, 17, 247–52.

Thorne, C. and Newell, M. L. (2005). Managing mother-to-child transmission of HIV. *Women's Health*, 1, 1–15.

Tonwe-Gold, B., Ekouevi, D., Rouet, F. *et al.* (2005). Highly active antiretroviral therapy for the prevention of perinatal HIV transmission in Africa: mother-to-child HIV transmission plus, Abidjan, Côte d'Ivoire, 2003–2004. In: *12th Conference on Retroviruses and Opportunistic Infections, Boston, 22–25 February*, Abstract No. 785.

UNAIDS (2007). *AIDS Epidemic Update: December 2007*. Geneva: Joint United Nations Program on HIV/AIDS (UNAIDS) and World Health Organization (WHO).

Urassa, P., Gosling, R., Pool, R. and Reyburn, H. (2005). Attitudes to voluntary counselling and testing prior to the offer of nevirapine to prevent vertical transmission of HIV in northern Tanzania. *AIDS Care*, 17, 842–52.

US Public Health Service (2005). *Public Health Service Task Force Recommendations for Use of Antiretroviral Drugs in Pregnant HIV-1-Infected Women for Maternal Health and Interventions to Reduce Perinatal HIV-1 Transmission in the United States*, November 17.

van't Hoog, A. H., Mbori-Ngacha, D. A., Marum, L. H. *et al.* (2005). Preventing mother-to-child transmission of HIV in Western Kenya: operational issues. *J. Acquir. Immune Defic. Syndr.*, 40, 344–9.

Walter, J., Kuhn, L. and Aldrovandi, G. (2008). Advances in basic science understanding of mother-to-child HIV-1 transmission. *Curr. Opin. HIV AIDS*, 3, in press.

Warszawski, J., Tubiana, R., Le Chenadec, J. *et al.* (2008). Mother-to-child HIV transmission despite antiretroviral therapy in the ANRS French Perinatal Cohort. *AIDS*, 22, 289–99.

Welty, T. K., Bulterys, M., Welty, E. R. *et al.* (2005). Integrating prevention of mother-to-child HIV transmission into routine antenatal care: the key to program expansion in Cameroon. *J. Acquir. Immune Defic. Syndr.*, 40, 486–93.

Wiktor, S. Z., Ekpini, E., Karon, J. M. *et al.* (1999). Short-course oral zidovudine for prevention of mother-to-child transmission of HIV-1 in Abidjan, Cote d'Ivoire: a randomised trial. *Lancet*, 353, 781–5.

Wilfert, C. M. and Fowler, M. G. (2007). Balancing maternal and infant benefits and the consequences of breast-feeding in the developing world during the era of HIV infection. *J. Infect. Dis.*, 195, 165–7.

WHO (2001). New data on the prevention of mother-to-child transmission of HIV and their policy implications: conclusions and recommendations. WHO Technical Consultation on behalf of the UNFPA/UNICEF/WHO/UNAIDS inter-Agency Task Team on Mother-to-Child Transmission of HIV, October 11–13, 2000. WHO/RHR/01.28ed. Geneva: World Health Organization.

WHO (2002). Strategic approaches to the prevention of HIV infection in infants. Report of a WHO Meeting, Morges, Switzerland, 20–22 March. Geneva: World Health Organization.

WHO (2006a). Antiretroviral drugs for treating pregnant women and preventing HIV infection in infants in resource-limited settings: towards universal access. Recommendations for a Public Health Approach. Geneva: World Health Organization.

WHO (2006b). WHO HIV and Infant Feeding Technical Consultation. Held on behalf of the Inter-agency Task Team (IATT) on Prevention of HIV Infections in Pregnant Women, Mothers and their Infants Geneva, October 25–27, 2006. Consensus Statement.

WHO (2007). Towards Universal Access: Scaling Up Priority HIV/AIDS Interventions in the Health Sector: Progress Report, April 2007. Geneva: World Health Organization.

Zijenah, L., Kadzirange, G., Rusakaniko, S. *et al.* (2005). Community-based generic antiretroviral therapy following single-dose nevirapine or short-course AZT in Zimbabwe. In: *12th Conference on Retroviruses and Opportunistic Infections, Boston, 22–25 February*, Abstract No. 632.

Structural and technical issues in HIV prevention

PART

III

Harm reduction, human rights and public health

18

Chris Beyrer, Susan G. Sherman and Stefan Baral

We are now well into the third decade of the HIV/AIDS pandemic. Despite vigorous efforts on the part of the international community, and an unprecedented level of research across an enormous variety of fields, we have seen HIV/AIDS become the most severe infectious disease epidemic of modern times. AIDS began as almost uniformly fatal, and the early decades of the 1980s and 1990s will be remembered in the cities of the Western world and the villages of the global South as a time of unremitting loss. That has changed, due to the extraordinary success of the research effort to develop effective antiretroviral agents. The new era in treatment began in 1996: from one ineffective agent, to equally useless two-drug regimens, we now have more than 5 *classes* of drugs, more than 20 agents, and single-dose formulations have made AIDS care a once-a-day manageable therapy. However, the HIV epidemic, by which we mean the spread of new infections among those at risk, continues. In many settings, HIV spread is either accelerating or remains stable at unacceptably high levels. Despite a relatively long list of prevention tools with good evidence for efficacy in HIV prevention, this domain has been marked by many failures and only a few successes – by a tragic and ongoing history of inability to implement the tools we have to control HIV spread. How can this be?

HIV is spread through sexual and reproductive behaviors and by blood and blood-products exposures, including sharing of injection equipment among drug users. The history of other sexually transmitted diseases suggests that most societies have been challenged in dealing pragmatically with diseases spread by sexual behavior – syphilis, for example, is curable with about 5 cents' worth of penicillin, and yet in 2007 there is recrudescent syphilis in the US, the UK, Russia and much of the Former Soviet Union. In much of Africa, syphilis has never been under good control. And this is primarily because it has proven difficult for politicians, and for society more generally, to deal with sexual behavior, including the kinds of sexual behavior most likely to cause disease spread, such as adolescent sexual activity, commercial sex, multiple partnerships, and sex between men. The same holds true for injecting drug users – a group highly stigmatized virtually everywhere, and well before the HIV epidemic began. In the case of HIV this has proven especially problematic, since HIV seemed initially so concentrated in urban gay and bisexual men, among men who practiced anal intercourse, among sex workers and their clients, and among those injecting drugs like heroin and cocaine. These groups were all stigmatized before HIV/AIDS, as were the behaviors in which they engaged. The conflation of AIDS with deviant, sinful and abnormal behavior was early and consistent – and herein lay the difficulty for prevention. Could public health reach out and respond to individuals and communities at risk through behavior the public found at best distasteful, and at worst criminal? What support would there be for the "rights" of deviant individuals to continue their risk-taking, but with the risks reduced through public expenditures? These attitudes were captured rather succinctly by Senator Jesse Helms, Republican, of South Carolina, when he attempted to stop congressional funding for rectal microbicides – a research agenda he said was committed "to making sodomy safer". If anal sex is how one defines "sodomy", the Senator had it right – that was precisely the point of the research effort. And those working on HIV prevention research were right to be concerned about this kind of rhetoric and thinking. Pragmatic approaches to HIV prevention for many at risk would face these kinds of challenges repeatedly. Too often, HIV prevention research, programs and funding have been the losers in these contentious struggles.

It is not simply that governments have been reluctant to fund prevention programs with evidence for efficacy in "making sodomy safer". We have seen all too many examples where programs have been implemented with little or no evidence for efficacy, but with ideological, moral or religious appeal. So the US included a stipulation known as the "Prostitution Pledge" – that all recipients of US PEPFAR dollars must have policies opposing prostitution – though the efficacy of this approach had not been studied. In contrast, Russia has long opposed any form of substitution therapy for opiate addiction, one of the cornerstones for HIV prevention for this population (Masenior and Beyrer, 2007). This was not based on any moral squeamishness or religious opposition to methadone, but to the Soviet-era ideology that substitution therapy itself was harmful. This ideological position, now so outmoded in the era of AIDS, has proven more difficult than changing the mind of Senator Helms, who late in his career became an enthusiastic supporter of AIDS funding for Africa.

Nowhere has the difficulty in implementing programs been more marked than in HIV prevention for men who have sex with men in developing countries. In the West, gay and bisexual men generated many of the early and effective prevention responses for themselves – knowing full well that they had little political support and everything to lose by waiting for public health authorities to act in their interests. Organizations were born such as Gay Men's Health Crisis, The Terence Higgins Trust in the UK, San Francisco's Project Inform and many others, and these led the prevention effort. These were largely organizations which built on the strength of organized and informed gay communities with experience in advocating for civil liberties. These were empowered people fighting for their lives. In contrast, 83 UN Member States still criminalize sex between consenting adults of the same gender, more than half of all African states do so, and 10 have the death penalty for homosexual relations between consenting adults of the same gender (ILGA, 2007) – including Pakistan, Saudi Arabia, Iran, Nigeria and Sudan. Yet UNAIDS has stated that vulnerability to HIV infection is dramatically increased where sex between men is criminalized (UNAIDS, 2006). USAID has concurred, stating that criminalization and homophobia limit MSM access to HIV prevention, information, commodities, treatment and care. And faced with these legal and social sanctions MSM are excluded, or exclude themselves, from sexual health and welfare (UNAIDS, 2006). Fewer than 1 in 10 MSM worldwide have any access to prevention services – and this in the *third* decade of HIV/AIDS, not the first (UNAIDS, 2007).

As with men who have sex with men, so with drug users. Pragmatic and evidence-based approaches to HIV spread among drug users fall under the broad category of *harm reduction*. This has proven equally challenging to implement in many settings. Harm reduction is a set of policies and programs that address the issue of drug use with the aim of reducing its associated harms. Its focus is on reducing the risks associated with drug use, while recognizing the many users are likely to continue using. This is in stark contrast to "zero tolerance" policies or law-enforcement approaches as the primary response to drug use (Lenton and Single, 1998). The principals of harm reduction date back to Britain's Rolleston Report, published in 1926, which adopted an approach to opiate dependence that included the possibility of medically maintaining the addict. This principle has guided the British system of opiate addiction for over 50 years (Strang *et al.*, 1997).

Needle exchange or methadone maintenance are two commonly cited harm-reduction interventions, and both are pragmatic and evidence-based (Hilton *et al.*, 2001; Wodak and Cooney, 2006). The spread of HIV among injection drug users drove the spread of needle exchanges and, more broadly, the concept of harm reduction throughout Europe, Australia and North America (Stimson, 2007). In Europe, Australia and Canada, harm reduction has successfully been adopted as the guiding philosophy for state and federal HIV prevention efforts targeting IDUs and, in some cases, other high-risk populations such as sex workers and prison inmates. Throughout much of the world, harm reduction is synonymous with HIV prevention interventions that focus on structural and contextual factors beyond the individual as targets of change. Such macrofactors include laws regarding drug paraphernalia, condom distribution in

Thai brothels, or prison conditions (Rojanapithayakorn and Hanenberg, 1996; Burris *et al.*, 2004; WHO, 2005).

Review of harm reduction interventions among vulnerable populations

Intravenous drug users (IDUs) remain one of the most vulnerable groups for HIV infection. There are an estimated 13 million IDUs worldwide, of which greater than three-fourths reside in low- and middle-income countries (LMIC). Intravenous drug use accounted for an estimated 10 percent of global HIV infections in 2005, but for one-third of HIV infections outside sub-Saharan Africa. The years since 2005 have been associated with an increase in illicit opium production, principally in Afghanistan, and thus there is little likelihood of decreases in injection drug use due to supply-side changes in the near future. Due to the twin epidemics of injection drug use and HIV, there is speculation that during the next decade the global epicenter of the HIV pandemic may shift from sub-Saharan Africa to Asia. Within Asia and Eastern Europe, HIV spread among IDUs has been episodic and explosive, but, with few exceptions, these epidemics continue to grow in magnitude and are at best stable at high-transmission frequencies. A recent study of IDU in Togliatti City in Russia showed that more than one in two was HIV-positive, the vast majority of which were unaware of their status. Just a few years earlier, the prevalence of HIV among IDUs was essentially zero, giving credence to how fast HIV can spread via parenteral transmission in the absence of harm-reduction strategies (Rhodes *et al.*, 2002). Similar high incidence rates have been shown in multiple other Asian and Eastern European settings, including Thailand, China, Indonesia, Vietnam, Malaysia, Ukraine and Belarus. While incidence rates are not always available, indicators that HIV epidemics among IDUs are even more widespread include studies showing HIV prevalence of greater than 20 percent among a sample of IDUs in low- and middle-income countries, including Libya, Serbia and Montenegro, Nepal, Tajikistan, Kazakhstan, Kenya, Tanzania, Ghana and Nigeria, as well as in many high-income countries (Aceijas *et al.*, 2004).

Though there is no cure, the medical therapy for HIV works; however, in many regions of the world, including the Former Soviet Union, only a small minority of those who inject drugs are able to access this treatment (Burrows, 2006). Thus, while it is important to increase antiretroviral coverage of IDUs already infected with HIV, there should also be an equally great focus on the prevention of HIV infection (Baral *et al.*, 2007a). Prevention strategies for HIV infection can be divided into two main groups: harm-reduction strategies and a protective vaccine against HIV infection. And while it is important to include IDUs as study participants for HIV vaccine trials, harm-reduction strategies are currently the only proven way of preventing HIV infection (Beyrer *et al.*, 2007). The immediate goal of harm-reduction strategies is not to reduce the use of drugs; instead it is to reduce the associated risk of HIV infection with drug use. Ultimately, the decrease in harm is mediated by providing clean needles and syringes to drug users, and/or by helping drug users transition from injecting drugs to the use of opiate analogues that are ingested orally. However, there is still much conjecture over what are the most effective ways to implement these harm-reduction strategies.

Harm-reduction interventions among IDUs

The first needle and syringe exchange program (NSP) was started in response to a 1984 hepatitis b outbreak among IDUs in Amsterdam (Buning, 1991). After the success of this initial program, NSPs were initiated in many high-income countries (HIC) settings, and by the 1990s several reviews underwritten by the United States government had concluded that these programs were effective and safe in reducing HIV spread among IDUs. The introduction of NSP was delayed in many low- and middle-income country settings, likely because of government fears that support for harm reduction would be interpreted as causing emasculation of the policies of the United Nations Office of Drugs and Crime. However, where these programs *were* introduced, they were often successful in limiting high-risk injecting behaviors of program participants. A study prospectively following clients of a NSP from 1991 to 1994 showed a significant increase in HIV-related knowledge, a decrease in needle-sharing and no new HIV infections in the last 2 years of the survey among 141 IDUs (Peak *et al.*, 1995). Alex Wodak and Annie Cooney recently published the most comprehensive international review to date of the evidence surrounding NSPs, concluding that these programs are causally related to a reduction of HIV infection among IDUs. These programs are also extremely cost-effective, ethical, and legal according to the UNODC (Department of Health and Ageing, 2002; Kleinig, 2006; Wodak and Cooney, 2006). Though NSPs mediate limited interaction of IDUs with outreach workers, safe drug consumption rooms allow the possibility of delivering comprehensive health-care services to drug users.

There is evidence that injecting in public environments, such as shooting galleries and crack houses, is associated with a higher risk of blood-borne virus (BBV) acquisition (Fuller *et al.*, 2003). This increased risk has found to be mediated by increased odds of sharing injection devices with other injectors, or using otherwise previously used injection equipment. Supervised injection facilities (SIF), one type of safe drug consumption rooms, is an intervention aimed at mitigating the increased risk of BBV acquisition associated with public or semi-public injection. These are legally sanctioned facilities where IDUs have access to the necessary tools to minimize the harm of injecting previously obtained drugs, including clean needles, disinfecting equipment, and health-care staff (Fry, 2003). There are 65 SIF in 27 cities in 8 countries across Europe, Australia and Canada, and they have been intensely evaluated. Overwhelmingly, the evidence has shown these programs to have beneficial outcomes, including decreased sharing of injecting equipment, decreased frequency of public injection, reductions in fatal overdose, and increased usage of opiate substitution therapy (Wood *et al.*, 2003, 2005, 2006; Kerr *et al.*, 2005, 2006; Rhodes *et al.*, 2006). Depending on the level of geographic concentration of drug users in a city, safe drug consumption rooms are most effectively designed as either store-front or mobile operations, similar to NSPs. The major obstacle to these is not a lack of evidence of technical efficacy, but rather a lack of political will in consenting to the autonomous use of its citizens of drugs determined to be illicit. And it has been suggested that instead of more studies evaluating derived public benefits and drug user outcomes, greater attention be paid to the ethical and political dimensions of these initiatives (Fry *et al.*, 2006). Notably, in October 2007, San Francisco city health officials started taking the steps needed to open up a SIF based on the evidence of secondary public health benefit (Associated Press, 2007).

Comprehensive harm-reduction programs targeting IDUs include more than just those focused on the dissemination of clean needles. Interventions for IDUs also target HIV/AIDS education, self-management skills such as learning to cope with drug cravings, and sex-related risk reduction including condom distribution (Burrows, 2006). Educational and therapeutic sessions regarding these topics can either be one-on-one or, more commonly, in groups facilitated by a trained leader. A recent meta-analysis of 37 RCTs evaluating 49 behavioral HIV risk-reduction interventions among 10,190 IDUs found that, compared to controls, intervention participants statistically significantly decreased both IDU and non-IDU, increased drug-treatment entry, increased condom use, and decreased trading sex for drugs (Copenhaver *et al.*, 2006). Most of the interventions in this meta-analysis consisted of, on average, eight sessions of 87 minutes duration (range 33–300 minutes) over the course of 29 days (range 1–113 days). Notably, the beneficial outcomes in terms of condom usage were subject to decay over time, while those related to injection drug use were not.

In very few cities are harm-reduction programs scaled to meet the needs of the total population of IDUs in the proposed catchment area for those services. For example, InSite, the safe injection facility in Vancouver, has only 12 injection booths, with a capacity of 850 injections per 18-hour day of operation, to serve the city's estimated 12,000 injectors. Thus, while there is convincing evidence that these programs provide benefit to the people who use them, it is more difficult to show benefit on a city level. That said, a few studies have demonstrated ecological benefit to NSPs. Hurley and colleagues reviewed 81 cities worldwide with HIV seroprevalence data from more than 1 year and NSP implementation details, and then calculated the rate of change of HIV seroprevalence by regression analysis. With this analysis, they found that seroprevalence increased by 5.9 percent per year in the 52 cities without NSP, and decreased by 5.8 percent per year in the 29 cities with NEPs – giving an average annual reduction in seroprevalence of 11 percent in cities with NEPs (95% CI −17.6 to −3.9, *P* = 0.004) (Hurley *et al.*, 1997). Such results lend credence to the idea that scaling up comprehensive harm-reduction programs may result in better control of global HIV epidemics among IDUs, and in some cases even begin reversing these epidemics.

The HIV epidemic among sex workers

Sex work is a grim and dangerous occupation. A number of individual and struc-tural factors have been associated with entry to sex work. The individual-level determinants include low self-esteem, childhood neglect and abuse, and at times mental health issues (Dunlap *et al.*, 2003; Goodyear and Cusick, 2007). Structural factors such as poverty, political instability, gender inequality, and a lack of edu-cation have also been demonstrated to predict initiation of sex work (Balos and Fellows, 1999; Beyrer, 2001; Cusick, 2006b). Sex workers face increased expo-sure and vulnerability to infectious diseases, and abuse from pimps, police and clients. Furthermore, there are links between trafficking and debt bondage and violence and adverse health outcomes. Finally, in unregulated sex-work set-tings, there is the potential for the regular abuse of adolescents and children (International Harm Reduction Development Program, 2007). Some have posited benefits associated with sex work, including positive body image, job satisfaction and autonomy (Manopaiboon *et al.*, 2003; Cusick, 2006b).

Sex workers have consistently been shown to have high rates of both curable and incurable STI. Curable STI are generally bacterial in nature, and include gon-orrhea, chlamydia and syphilis. Incurable STI are generally caused by viruses such as human papilloma virus, herpes simplex virus (HSV) and HIV. Those STI including herpes and syphilis which can cause genital ulcerative disease can increase the per coital act transmission rate of HIV by as much as 300 times. A recent review of STIs among sex workers indicated that prevalence of active ulcers is 10 percent or more, and over 30 percent have serological evidence of active syphilis infection (Steen and Dallabetta, 2003). Furthermore, over one-third of sex workers have either active gonorrhea or chlamydia infections, and more than 60 percent have chronic HSV infection. In the context of such high rates of STI and low rates of condom use, sex workers continue to be one of the most vul-nerable groups for HIV infection. Indeed, studies have consistently shown that, in the context of emerging epidemics, sex work can drive HIV epidemics. A pro-spective study completed at the Clinique de Confiance in Abidjan, Cote d'Ivoire, demonstrated an HIV prevalence among 356 SW in 1992 of 89 percent (Ghys *et al.*, 2002). In this study, HIV infection was associated with duration of sex work (OR 1.16, 95% CI 1.08–1.25) and number of clients (OR 1.20; 95% CI 1.09–1.32). Similar prevalence rates among sex workers have been seen in numer-ous studies across Africa, including 50.3 percent HIV prevalence among 145 sex workers in KwaZulu-Natal, and 58.2 percent (95% CI 53.4–62.9) HIV prevalence among 426 sex workers in Burkina Faso (Lankoande *et al.*, 1998; Ramjee *et al.*, 1998). Though there is some regional variance in the burden of HIV among SW, it is, almost without exception, disproportionately high compared to the general population.

Harm-reduction strategies among sex workers

Early efforts to reduce the prevalence of sex work focused on punitive laws targeting sex workers, and were only successful in making this vulnerable population more difficult to locate and treat. Only with the advent of antibiotics which had the potential to cure the majority of early bacterial STIs did these epidemics reverse. With the emergence of incurable STIs among sex workers, these same strategies have been employed; however, in the absence of curative therapy they have met with little success (Steen and Dallabetta, 2003; Goodyear and Cusick, 2007). Evidence-based harm-reduction therapies focused on increasing condom use during transaction sex, improving diagnosis and treatment of bacterial STIs, and reducing demand for sex work using risk-reduction messages targeting men (Celentano *et al.*, 1998, 2000; Ghys *et al.*, 2001).

Interventions targeting the increased usage of condoms were initially designed as cognitive-behavioral therapy meant to change an individual's attitudes, knowledge and behaviors. These early strategies were not adjusted for higher-level factors, such as poverty and a dearth in condom negotiation skills and power, that contextualize and partially explain individual behaviors (Morisky *et al.*, 2006). Successful condom campaigns have integrated structural factors, including working conditions and employer's attitudes towards condoms, along with individually focused interventions. The Thai 100% Condom Program was a national government policy mandating condom usage in all brothel-based commercial sex acts from 1991 on. This program was successful in that it focused on increasing condom use among all Thai brothel-based sex workers by targeting the brothel management as well as the sex workers themselves (Rojanapithayakorn and Hanenberg, 1996; Hanenberg and Rojanapithayakorn, 1998). In a study of 288 sex workers in the Dominican Republic, having structural support for condom use was a significant predictor of condom usage (OR 2.16; CI 1.18–3.97). Structural support can take many forms, and in this study involved the level of access to condoms in the establishment, a clear policy that condoms be used at all times on dates with clients by the brothel management, and continual reminders from the management about condom use (Kerrigan *et al.*, 2003). Numerous other studies from around the globe have consistently report that comprehensive programs targeting multiple levels of determinants of condom usage among sex workers have been successful in increasing condom usage and should be adopted (Ghys *et al.*, 2001, 2002; Alary *et al.*, 2002).

The HIV epidemic among MSM

The initial presentation of HIV in many regions of the world was among populations of men who have sex with men (MSM) (Caceres *et al.*, 2005). Male-to-male
sexual contact remains an important route of HIV-1 spread more than a quarter of
a century later. It is important to note that MSM is a phrase describing a particular behavior, rather than actually being the defining characteristic of a population.
Thus, MSM includes all of the following subpopulations: self- and non-self-
identified homosexuals, bisexuals, male sex workers, transgenders, and other
country-specific populations of MSM. Each of these populations has its own risk
status and HIV epidemic dynamics. Practically, it will be a long time before the
differential risk status of these various populations can be accurately described,
given the prevailing social climate and limited dedicated funding. Recent data
indicate that these epidemics of HIV are not limited to the high-income countries
in which they were initially described. A limited set of recent reports has demonstrated high HIV prevalence rates among MSM from a number of low- and
middle-income country (LMIC) settings (Bautista *et al.*, 2004; van Griensven
et al., 2005; Wade *et al.*, 2005; EuroHIV, 2006; World Bank, 2006). And among
certain low- or middle-income countries, such as Thailand, Cambodia and
Senegal, characterized by relatively low and declining HIV prevalence rates
among heterosexual populations, studies have suggested that there are concentrated HIV epidemics among MSM, suggesting a "dislinked" epidemic pattern
(Girault *et al.*, 2004; Beyrer *et al.*, 2005, 2007; Wade *et al.*, 2005; *Morbidity and
Mortality Weekly Report*, 2006a). Finally, a recent systematic review demonstrated
that while indeed there is a dearth of data (especially from Africa and the former
Soviet Union), where studied, MSM have had consistently high prevalence rates
of HIV (Baral *et al.*, 2007b). However, most data available evaluating determinants of risk for HIV among MSM are derived from high-income countries, and
thus this evidence will form the focus of discussion here. Among high-income
countries, available evidence suggests that structural risks are important in defining HIV risk for any one individual MSM. Individual-level acquisition risks have
focused on the highest probability exposure – unprotected anal intercourse (UAI) –
and specifically on correlates of receptive anal intercourse (Koblin *et al.*, 2006).
Use of party drugs such as methamphetamines and alkyl nitrates (poppers) has
been associated with heightened sexual exposure among MSM in several settings
(Beyrer *et al.*, 2005; Koblin *et al.*, 2006). Also, as with men who only report sex
with women, HIV transmission in MSM is associated with genitourinary disease,
being uncircumcised, high frequency of male partners, and high lifetime number
of male partners (Buchbinder *et al.*, 2005). Finally, being a black or minority ethnic (BME) man who has sex with men in developed country settings is associated
with higher risk of HIV in comparison to white MSM (Harawa *et al.*, 2004). A
critical review of the evidence examining the racial differential seen in the HIV

epidemic among MSM suggested that the increased HIV prevalence seen among black MSM is most likely due to a decreased proportion having been tested and knowing their HIV status, and higher rates of STIs facilitating HIV transmission. A similar trend among minorities, though of lesser magnitude, has been observed in other high-income countries, including the UK and Canada. Having particularly high-risk individuals, such as male sex workers (MSW) and transgenders, in a sexual network will put all the members of that network at increased risk of infection (Beyrer *et al.*, 2005), and a high prevalence of sexually transmitted infections will increase the probability of HIV transmission within the network. At the community level, access to preventive services, voluntary counseling and testing, and ARV will contribute to the level of risk of any particular community of MSM. Finally, an advancing HIV-epidemic stage is responsible for adding risk to all of these lower-order determinants of HIV infection.

Harm-reduction interventions among MSM

The latest evidence tells us that while UAI is less prevalent than in the earlier phases of the HIV pandemic, it remains a common activity, especially among HIV-positive MSM. Of 20 cross-sectional studies of sexual risk behavior among MSM, 14 studies demonstrated that UAI is significantly more common among HIV-positive men compared to HIV-negative men or those never tested (van Kesteren *et al.*, 2007). These results are not limited to any one region, with consistent prevalence of sexual risk behaviors observed in Europe, North America and Australia. In a study of nearly 5000 MSM in Amsterdam, being HIV-negative or never tested were both strongly associated with decreased UAI at last sexual encounter (OR 0.24, 95% CI 0.14–0.42; OR 0.18, 95% CI 0.11–0.32, respectively. Notably, even after correction for known serosorting, HIV-positive men were more likely to have UAI. In this study, other statistically significant determinants of UAI at last sexual encounter were being less educated, not being of Dutch origin, and being younger (Hospers *et al.*, 2005). Similar results were seen among over 10,000 MSM from San Francisco, and subgroup analysis demonstrated that among HIV-positive MSM, higher odds of UAI were predicted by being white (OR 1.8, 95% CI 1.3–2.5), and being older than 30 (OR 1.9, 95% CI 1.3–2.7). Being white was also associated with higher odds of UAI among HIV-negatives, though notably the prevalence of UAI was increasing both among HIV-positive and HIV-negative MSM (Chen *et al.*, 2003). Combined anti-retroviral therapy (CART) has been tremendously successful in managing the clinical manifestations of HIV infection, and in places with consistent access to these medications this trend of increasing UAI and hence decreased risk avoidance has been seen. Since 1996, the Brazilian public health-care infrastructure has worked to ensure universal access to medications needed to treat HIV, resulting in huge improvements in the quality and length of life of those affected. With this decreased

HIV-related morbidity and mortality there has been increased optimism among high-risk groups in Brazil, including MSM. A recent cross-sectional study from Sao Paulo demonstrated that the most optimistic study participants were nearly two times more likely to have had UAI than the least optimistic MSM (da Silva *et al.*, 2005).

The fact that UAI is common among MSM, and indeed becoming more common, suggests that there is a desperate need for the intensification of specific prevention strategies aimed at increasing condom use among MSM. While there is still some uncertainty as to which interventions are most effective at increasing condom use, their failure to curb the HIV epidemic is secondary to limited access of MSM to these programs. A recent systematic review and meta-analysis including 16,224 men in 38 experimental and observational studies demonstrated that, compared to controls with no interventions, study groups reduced UAI by 27 percent (95% CI 15–37%) (Herbst *et al.*, 2005; Johnson *et al.*, 2005). Moreover, in an additional 16 studies where MSM were given targeted prevention strategies, UAI decreased by 17 percent (95% CI 5–27%) more than in MSM who received standard HIV prevention measures. Prevention strategies tend to work better when targeting community-level risks rather that individual-level ones, and functioned equally well independent of the proportion of minorities included. Globally, between 5 and 10 percent of MSM have access to programs such as these, and even then the majority of these are in high-income countries (UNAIDS, 2005, 2006). However, where studies of interventions targeting MSM have been completed in low- or middle-income country settings, they have consistently demonstrated both need and efficacy (Amirkhanian *et al.*, 2003, 2005; Choi *et al.*, 2004; Operario *et al.*, 2005). Similar to MSM in high-income countries, data from low- to middle-income countries suggest that interventions should target network- or community-level factors to potentiate increased condom usage (Choi *et al.*, 2006). Though prevention strategies targeting MSM have been shown to be effective across country income levels, the benefit of these interventions has been subject to decay over time, indicating that such interventions should be ongoing to preserve increased condom usage.

The HIV epidemic among incarcerated populations

Incarcerated populations remain amongst the most vulnerable for HIV infection worldwide because the burden of infectious disease in entrants to the penal system is significantly higher than in the general population, and in prison there is a heightened risk environment (Small *et al.*, 2005; see also Chapter 16). In concert with a growing number of incarcerated people – approximately 2.4 percent of the entire US population is either currently incarcerated or awaiting trial – this is a group in need of increased attention in order to comprehensively battle the HIV pandemic (Polonsky *et al.*, 1994; Fox *et al.*, 2005). Numerous studies have established that the burden of HIV in prison is disproportionately high in both high- and lower-income settings. In the US, estimates range from a 4- to 10-fold increase in the prevalence of HIV among currently incarcerated populations compared to the general population (Vlahov *et al.*, 1991; *Morbidity and Mortality Weekly Report*, 2006b; Okie, 2007). A recent study among 1877 participants (1547 men, 330 women) across 13 remand facilities in Ontario, Canada, demonstrated an HIV prevalence of 2.0% (95% CI 1.3–2.8) (Calzavara *et al.*, 2007). A similar study completed among 1607 inmates (1357 men, 250 women) in Quebec demonstrated an overall HIV prevalence of 3.4% (95% CI 2.5–4.3) (Poulin *et al.*, 2007). Both of these estimates are approximately 10 times higher than the UNAIDS HIV prevalence estimate of 0.3% (95% CI 0.2–0.5) among reproductive-age adults in Canada (UNAIDS, 2006). The majority of information related to HIV prevalence in prisons has been garnered from studies and surveillance in high-income settings; however, a systematic review of HIV prevalence in prison in 75 low- and middle-income countries has recently been published (Dolan *et al.*, 2007). HIV prevalence was found to be greater than 10 percent in 20 countries, often related to IDU, while certain settings have been able to demonstrate significant HIV incidence in prison (Horsburgh *et al.*, 1990; Simbulan *et al.*, 2001; Ruiz *et al.*, 2002; Alizadeh *et al.*, 2005; Drobniewski *et al.*, 2005; Kushel *et al.*, 2005; Dolan *et al.*, 2007). The conclusions of this review are that HIV prevalence is high in many settings, and there is an urgent need for systematic HIV surveillance and introduction of evidence-based prevention strategies.

Harm-reduction strategies among incarcerated populations

Whether in high- or lower-income settings, political realities have made evaluating harm-reduction strategies in prison difficult. As such, the proposed benefits of these harm-reduction strategies are based on theory, extrapolated from evidence obtained in non-incarcerated populations, and occasionally from observational studies in politically progressive contexts. The incidence of HIV is likely mostly explained by injection drug use, male sexual contact and, occasionally, tattooing. Thus, mitigating the risk of these activities with proven harm-reduction strategies would likely result in a lowered risk environment behind bars. The least politically charged harm-reduction intervention is education, and even this has shown efficacy in the prison context (Dolan *et al.*, 2004). To minimize risk of disease transmission with male sexual contact, the first step has to be increasing the availability of condoms to inmates (Betteridge, 2004). With greater condom availability, further research can be completed to assess the most effective method of increasing condom usage among male inmates. The first NSP inside a prison was established in 1992 in Switzerland; within 1 year the program had distributed 5335 syringes to 110 inmates and essentially resulted in the elimination of needle-sharing (Nelles and Harding, 1995; Dolan *et al.*, 2003). While NSPs have been legal in Canada for many years, Correctional Services Canada (CSC) has pro-hibited inmates from possessing syringes because of the concern that these will be used as weapons against staff or other inmates (Small *et al.*, 2005). However, to date there has not been one reported assault with a syringe in a correctional facility with needle exchange (Hellard and Aitken, 2004). Today there are needle-exchange programs in Switzerland, Germany, Spain and Moldova, using a variety of models of needle distribution, including by prison staff, dispensers, or even having the inmates themselves in charge of the distribution (Nelles *et al.*, 1998). In many settings, such as Canada, where NSP have not been allowed in prison, another option is the provision of bleach for the sterilization of syringes, which has also shown some benefit in prevention transmission of HIV (Dolan *et al.*, 1998; Small *et al.*, 2005). Notably, it is much more difficult to achieve a sterile syringe using only bleach, and thus this can create a false sense of security among inmates sharing injection paraphernalia (WHO, 2005). The ultimate risk-reduction mechanism in terms of IDU is to potentiate the termination of inmate injecting by offering opiate substitution therapy to those interested; this, when operationalized, was found to decrease both re-incarceration and mortality (Dolan *et al.*, 2005).

Interplay between harm reduction, human rights and public health

How do pragmatic public health interventions, including harm reduction, interact with human rights approaches? First, it is clear that access to a minimal standard of health care must mean appropriate and relevant health care for individual and communities. If a life-threatening illness is spreading in a community, and the means to prevent it are available and well understood, it is a fundamental abro-gation of the rights to health, and to life itself, to be denied access to preven-tion services. In the context of HIV, rights-based approaches and harm-reduction strategies are powerfully synergistic. This is most especially true in the realm of stigmatized and vulnerable minorities. International human rights standards and conventions, including the 1976 International Covenant on Economic, Social and Cultural Rights, and its General Comment 14 on health rights, make clear that the prevention, treatment and control of epidemic diseases is the responsibility of states and the right of all persons. This includes, of course, persons whose risks for an epidemic disease are, like engaging in male-to-male sex, selling sex and injecting drugs, considered socially unacceptable. Universal rights are universal, and sexual or substance-using minorities are not excluded *a priori*.

Harm reduction and human rights are two parallel but distinct conceptual frameworks that share core values. While human rights goals tend to be aspira-tional, harm reduction has been focused on pragmatic service delivery, and on meeting people in need "where they are", and not where society would like them to be. Nevertheless the crux of harm reduction is the attempt to provide services to those in need, and in this sense harm reduction can be seen as an attempt to fulfill a human right – that of access to care – and to protect that most basic of all human rights, that of life itself. Both are rooted in the most fundamental princi-pal, human dignity, which cannot be divested from anyone by social opprobrium, labels of deviance, or criminalization of consensual adult sexual activity:

> Whereas recognition of the inherent dignity and of the equal and inalienable rights of all members of the human family is the foundation of freedom, justice and peace in the world ... Now, therefore, the General Assembly proclaims this Universal Declaration of Human Rights as a common standard of achievement for all peoples and all nations, to the end that every individual and every organ of society, keeping this Declaration constantly in mind, shall strive by teaching and education to pro-mote respect for these rights and freedoms and by progressive measures ...

> Universal Declaration of Human Rights (UDHR), adopted by the UN General Assembly, 10 December 1948

On the heels of World War II, the first Universal Declaration of Human Rights (UNHR) was adopted by the UN General Assembly without one dissenting vote. The human right to health is recognized in numerous international documents that have been ratified over the past 60 years by the United Nations General Assembly (Organization of American States, 1988; UN General Assembly, 1948, 1966, 1976, 1979, 1989). Article 25.1 of the UDHR affirms that "everyone has the right to a standard of living adequate for the health of himself and of his family, includ-ing food, clothing, housing and medical care and necessary social services." The International Covenant on Economic, Social and Cultural Rights (ICESCR), passed 3 January 1976, provides the most comprehensive article on the right to health in international human rights law. General Comment 12, amended in 2000, provides a definition of the right to health as "the right of everyone to the enjoy-ment of the highest attainable standard of health", and requires states to take all necessary steps for "the prevention, treatment, and control of epidemic, endemic, occupational and other diseases". Article 12.2 details various examples of a state's obligations to ensure individuals' right to health, including the right to:

> The prevention, treatment and control of epidemic, endemic, occupational and other diseases requires the establishment of prevention and education programmes for behaviour-related health concerns such as sexually transmitted diseases, in particu-lar HIV/AIDS, and those adversely affecting sexual and reproductive health, and the promotion of social determinants of good health. . . . . The control of diseases refers to States' individual and joint efforts to, *inter alia*, make available relevant technologies, using and improving epidemiological surveillance and data collection on a disaggregated basis, the implementation or enhancement of immunization pro-grammes and other strategies of infectious disease control.

> UN General Assembly, 1976

514

As clearly articled in Article 12 of ICESCR, States are responsible for preventing disease in terms of creating a public health and legal environment conducive to disease prevention. HIV/AIDS has underscored the link between the promotion of the public's health and the rights of humans. The state's responsibility for disease prevention is particularly challenged in regard to its role in prevention among some of the populations most infected and affected by HIV/AIDS – namely commercial sex workers and IDUs. The very illegal nature of the behaviors that place these populations in disproportionate risk for HIV/AIDS challenges the state's ability to fulfill its role of disease prevention.

By the very definition of health rights provided in these UN doctrines, human rights and *public* health are inextricably linked on several levels. First, there is an innate tension between policies that are geared to protect the public's health, and individuals' rights that might be restricted in order to promote the public's health. In such an instance, the protection of the public's health must outweigh concerns for individual rights. Human rights can provide the tools for assessing the just application and efficacy of health-related policies meant to promote health across a population. A human rights analysis is useful and often necessary in understanding the distribution of, access to and need for health services in a given population. The HIV epidemic has been a very sobering example of the interconnectedness of the respect for human rights and the spread of disease (Malinowska-Sempruch and Gallagher, 2004). The 2001 UN General Assembly Special Session on HIV/AIDS (UNGASS) identified human rights violations as a major factor in the spread of HIV/AIDS (UNGASS, 2001). Such violations occur on structural levels through policies, procedures and laws. One such example would be limiting the access to clean and sterile syringes to IDUs, through such policies as drug paraphernalia laws that criminalize the possession, sale or distribution of disposable syringes (Burris *et al.*, 2004).

Human rights violations occur when subgroups of a population, most often the most socially and economically marginalized, are systematically denied health care or, by design, are unable or not willing to access specific health services because of the outcomes. Such a form of denial of access occurs in the situation in which HIV testing is available but the results are not confidential. A human rights perspective is vital in assessing whether public health policy is being enacted as intended, or is not designed in a way that recognizes the special needs of populations in need.

A two-decade example of the tension between human rights, public health and criminalization statutes is that of syringe distribution in the United States. As mentioned, there are numerous states that criminalize the possession, distribution or sale of disposable syringes (Burris *et al.*, 2004). Syringe possession or, more broadly, "drug paraphernalia laws", have been found to be associated with increased syringe-related HIV risk behaviors. In a nutshell, the more restrictive laws that exist regarding syringes, the more difficult it is to obtain clean and sterile syringes, and the less likely it is that individuals will inject safely. Specifically, elevated HIV risk can be the result of the expense of syringes on the black market, or the fear of carrying equipment – both of which result in minimizing access to clean and sterile syringes. A survey of 42 NSPs in 35 cities in 18 US states revealed that the street price of syringes depended on individual state's laws governing the possession of syringes by people who use drugs. Prices were lowest when there was no law regarding syringe possession, significantly higher when there was an unenforced law and highest when there was an enforced law against their possession by people who inject drugs (Rich *et al.*, 2000). In a study of regional comparisons of syringe-sharing, Calsyn and colleagues (1991) found that less syringe-sharing was reported in areas where the purchase and possession of needles are illegal. It can be argued that, in the context of syringe possession laws, the legality of drug use is more valued that the health and human rights of injection drug users (Calsyn *et al.*, 1991). Many cities have rectified this legal and public health conflict by either rejecting paraphernalia laws or declaring a public health state of emergency which suspends such laws.

Discussion

The tension between human rights, public health and harm reduction is not limited to the US, nor is it limited to service delivery targeting IDUs. Indeed, the major barriers to harm-reduction strategy implementation in sex workers are rooted in policy rather than a lack of evidence of efficacy. Ecologic studies have demonstrated poorer health outcomes for sex workers where their work is criminalized compared to where this is a regulated industry. In Nevada, USA, decriminalization resulted in a reduction of associated harm; similar outcomes were observed in the Netherlands (Cusick, 2006a; Goodyear and Cusick, 2007). As was seen with the 100% Condom Campaign in Thailand, when a government engaged and began regulating the brothels, there were numerous secondary benefits (Hanenberg and Rojanapithayakorn, 1996, 1998). However, governments in lower-income settings that want to engage this business have now been placed at a further disadvantage because of the US Anti-Prostitution Pledge (APP). The APP states that if an organization or country is to receive AIDS-related funds from the President's Emergency Plan for AIDS relief (PEPFAR), or from the US government as a whole, that grantee must have a policy specifically opposing prostitution and sex trafficking, and also provide a certification of compliance to not provide services to sex workers (Masenior and Beyrer, 2007). As many governments in lower-income settings rely on US funding, this pledge has likely resulted in the termination of numerous programs providing services to sex workers, with Brazil a notable exception. A documented example is the Lotus Project in Cambodia, which was managed by Médecins Sans Frontières and provided services and treatment for sex workers, funded by USAID. In the context of heightened awareness of programs working with prostitutes, as well as diminished funding from USAID secondary to the APP, the Lotus Project eventually terminated (Masenior and Beyrer, 2007). Unfortunately, it is difficult to enumerate the likely increase in incidence of blood-borne virus infection, including HIV, secondary to this requirement.

Pragmatists consider practical consequences or real effects to be vital components of both meaning and truth, and would thus support the introduction of evidence-based harm-reduction strategies. That said, not all harm-reduction strategies are practical, as these programs tend to be limited because of structural realities such as program coverage, funding, and political will. The themes of policy-based limitations of harm-reduction program implementation established with the examples of IDU and sex workers also hold true for the other vulnerable populations examined in this chapter – MSM and incarcerated populations. In 2007, 83 member states of the United Nations still criminalize consensual same-sex acts among adults, severely limiting the service provision to MSM and the broader lesbian, gay, bisexual and transgender (LGBT) populations (ILGA, 2007). PEPFAR has even sponsored abstinence-only programs in Uganda that have also advocated the criminalization of service delivery for LGBT persons in that country, furthering limiting community groups aiming to provide evidence-based harm-reduction services (HRW, 2007). Most countries, including every state in the US, have laws prohibiting sex between adult residents of correctional institutions. However, the evidence suggests that up to 30 percent of inmates are sexually active while incarcerated, resulting in outbreaks of bacterial STIs and, as described earlier, occasional documentation of HIV incidence (Weinbaum *et al.*, 2003). We described earlier the policies in place limiting harm-reduction strategies targeting IDU while incarcerated; similar themed policies also limit harm reduction targeting male sexual contact behind bars. Consequently, in the US, only two state prison systems (Vermont and Mississippi) and five city correctional systems (New York City, Philadelphia, San Francisco, Washington DC and Los Angeles) simply provide condoms to inmates. Notably, not a single juvenile correctional system in the country is known to provide condoms (Weinbaum *et al.*, 2003). Wide adoption of evidence-based harm reduction for vulnerable populations would, in principle, both serve the public good, in terms of disease burden and health-care costs, and fulfill the human rights owed to each citizen. In short, the best scenarios for public health and human rights are when they are mutually supportive – a critical test met by evidence-based harm-reduction strategies.

References

Aceijas, C., Stimson, G. V., Hickman, M. and Rhodes, T. (2004). Global overview of injecting drug use and HIV infection among injecting drug users. *AIDS*, 18, 2295–303.
Alary, M., Mukenge-Tshibaka, L., Bernier, F. *et al.* (2002). Decline in the prevalence of HIV and sexually transmitted diseases among female sex workers in Cotonou, Benin, 1993–1999. *AIDS*, 16, 463–70.

Alizadeh, A. H., Alavian, S. M., Jafari, K. and Yazdi, N. (2005). Prevalence of hepatitis C virus infection and its related risk factors in drug abuser prisoners in Hamedan–Iran. *World J. Gastroenterol.*, 11, 4085–9.

Amirkhanian, Y. A., Kelly, J. A., Kabakchieva, E. *et al.* (2003). Evaluation of a social network HIV prevention intervention program for young men who have sex with men in Russia and Bulgaria. *AIDS Educ. Prev.*, 15, 205–20.

Amirkhanian, Y. A., Kelly, J. A., Kabakchieva, E. *et al.* (2005). A randomized social network HIV prevention trial with young men who have sex with men in Russia and Bulgaria. *AIDS*, 19, 1897–905.

Associated Press (2007). San Francisco Considers Injection Room. Associated Press, 19 October.

Balos, B. and Fellows, M. L. (1999). A matter of prostitution: becoming respectable. *NY Law Rev.*, 74, 1220–303.

Baral, S., Sherman, S. G., Millson, P. and Beyrer, C. (2007a). Vaccine immunogenicity in injecting drug users, a systematic review. *Lancet Infect. Dis.*, 7, 667–74.

Baral, S., Sifakis, F., Cleghorn, F. and Beyrer, C. (2007b). Elevated risk for HIV infection among men who have sex with men in low- and middle-income countries 2000–2006: a systematic review. *PLoS Med.*, 4, e339.

Bautista, C. T., Sanchez, J. L., Montano, S. M. *et al.* (2004). Seroprevalence of and risk factors for HIV-1 infection among South American men who have sex with men. *Sex. Transm. Infect.*, 80, 498–504.

Betteridge, G. (2004). Bangkok 2004. Prisoners' health and human rights in the HIV/AIDS epidemic. *HIV AIDS Policy Law Rev.*, 9, 96–9.

Beyrer, C. (2001). Shan women and girls and the sex industry in Southeast Asia; political causes and human rights implications. *Social Sci. Med.*, 53, 543–50.

Beyrer, C. (2007). HIV epidemiology update and transmission factors: risks and risk contexts – 16th International AIDS Conference epidemiology plenary. *Clin. Infect. Dis.*, 44, 981–7.

Beyrer, C., Sripaipan, T., Tovanabutra, S. *et al.* (2005). High HIV, hepatitis C and sexual risks among drug-using men who have sex with men in northern Thailand. *AIDS*, 19, 1535–40.

Beyrer, C., Baral, S., Shaboltas, A. *et al.* (2007). The feasibility of HIV vaccine efficacy trials among Russian injection drug users. *Vaccine*, 25, 7014–16.

Buchbinder, S. P., Vittinghoff, E., Heagerty, P. J. *et al.* (2005). Sexual risk, nitrite inhalant use, and lack of circumcision associated with HIV seroconversion in men who have sex with men in the United States. *J. Acquir. Immune Defic. Syndr.*, 39, 82–9.

Buning, E. C. (1991). Effects of Amsterdam needle and syringe exchange. *Intl J. Addict*, 26, 1303–11.

Burris, S., Blankenship, K. M., Donoghoe, M. *et al.* (2004). Addressing the "risk environment" for injection drug users, the mysterious case of the missing cop. *Milbank Q.*, 82, 125–56.

Burrows, D. (2006). Advocacy and coverage of needle exchange programs: results of a comparative study of harm reduction programs in Brazil, Bangladesh, Belarus, Ukraine, Russian Federation, and China. *Cad. Saude Publica*, 22, 871–9.

Caceres, C., Konda, K. and Pecheny, M. (2005). *Review of the Epidemiology of Male Same-Sex Behavior in Low and Middle-Income Countries.* Geneva: UNAIDS.

Calsyn, D. A., Saxon, A. J., Freeman, G. and Whittaker, S. (1991). Needle-use practices among intravenous drug users in an area where needle purchase is legal. *AIDS*, 5, 187–93.

Calzavara, L., Ramuscak, N., Burchell, A. N. *et al.* (2007). Prevalence of HIV and hepatitis C virus infections among inmates of Ontario remand facilities. *Can. Med. Assoc. J.*, 177, 257–61.

Celentano, D. D., Nelson, K. E., Lyles, C. M. *et al.* (1998). Decreasing incidence of HIV and sexually transmitted diseases in young Thai men: evidence for success of the HIV/AIDS control and prevention program. *AIDS*, 12, F29–36.

Celentano, D. D., Bond, K. C., Lyles, C. M. *et al.* (2000). Preventive intervention to reduce sexually transmitted infections: a field trial in the Royal Thai Army. *Arch. Intern. Med.*, 160, 535–40.

Chen, S. Y., Gibson, S., Weide, D. and McFarland, W. (2003). Unprotected anal intercourse between potentially HIV-serodiscordant men who have sex with men, San Francisco. *J. Acquir. Immune Defic. Syndr.*, 33, 166–70.

Choi, K., Pan, Q., Ning, Z. and Gregorich, S. (2006). Social and sexual network characteristics are associated with HIV risk among men who have sex with men (MSM) in Shanghai, China. In: *XVIth International AIDS Conference, Toronto, 13–18 August*, Abstract no. TUPE0470.

Choi, K. H., McFarland, W. and Kihara, M. (2004). HIV prevention for Asian and Pacific Islander men who have sex with men, identifying needs for the Asia Pacific region. *AIDS Educ. Prev.*, 16, v–vii.

Copenhaver, M. M., Johnson, B. T., Lee, I. C. *et al.* (2006). Behavioral HIV risk reduction among people who inject drugs, meta-analytic evidence of efficacy. *J. Subst. Abuse Treat.*, 31, 163–71.

Cusick, L. (2006a). Widening the harm reduction agenda: from drug use to sex work. *Intl J. Drug Policy*, 17, 3–11.

Cusick, L. (2006b). Sex workers to pay the price, a street sex worker responds to new government strategy. *Br. Med. J.*, 332, 362.

da Silva, C. G., Goncalves, D. A., Pacca, J. C. *et al.* (2005). Optimistic perception of HIV/AIDS, unprotected sex and implications for prevention among men who have sex with men, Sao Paulo, Brazil. *AIDS*, 19(Suppl. 4), S31–36.

Department of Health and Ageing (2002). *Return on Investment in Needle and Syringe Programs in Australia*. Canberra: Commonwealth Department of Health and Ageing.

Dolan, K. A., Wodak, A. D. and Hall, W. D. (1998). A bleach program for inmates in NSW: an HIV prevention strategy. *Aust. NZ J. Public Health*, 22, 838–40.

Dolan, K., Rutter, S. and Wodak, A. D. (2003). Prison-based syringe exchange programmes, a review of international research and development. *Addiction*, 98, 153–8.

Dolan, K. A., Bijl, M. and White, B. (2004). HIV education in a Siberian prison colony for drug dependent males. *Intl J. Equity Health*, 3, 7.

Dolan, K. A., Shearer, J., White, B. *et al.* (2005). Four-year follow-up of imprisoned male heroin users and methadone treatment, mortality, re-incarceration and hepatitis C infection. *Addiction*, 100, 820–8.

Dolan, K., Kite, B., Black, E. *et al.* (2007). HIV in prison in low-income and middle-income countries. *Lancet Infect. Dis.*, 7, 32–41.

Drobniewski, F. A., Balabanova, Y. M., Ruddy, M. C. *et al.* (2005). Tuberculosis, HIV seroprevalence and intravenous drug abuse in prisoners. *Eur. Respir. J.*, 26, 298–304.

Dunlap, E., Golub, A. and Johnson, B. D. (2003). Girls' sexual development in the inner city, from compelled childhood sexual contact to sex-for-things exchanges. *J. Child Sex. Abuse*, 12, 73–96.

EuroHIV (2006). *HIV/AIDS Surveillance in Europe: Mid-year Report 2005*. Saint-Maurice: Institut de Veille Sanitaire, pp. 21–35.

Fox, R. K., Currie, S. L., Evans, J. *et al.* (2005). Hepatitis C virus infection among prisoners in the California state correctional system. *Clin. Infect. Dis.*, 41, 177–86.

Fry, C. L. (2003). Safer injecting facilities in Vancouver, considering issues beyond potential use. *Can. Med. Assoc. J.*, 169, 777–8.

Fry, C. L., Cvetkovski, S. and Cameron, J. (2006). The place of supervised injecting facilities within harm reduction, evidence, ethics and policy. *Addiction*, 101, 465–7.

Fuller, C. M., Vlahov, D., Latkin, C. A. *et al.* (2003). Social circumstances of initiation of injection drug use and early shooting gallery attendance, implications for HIV intervention among adolescent and young adult injection drug users. *J. Acquir. Immune Defic. Syndr.*, 32, 86–93.

Ghys, P. D., Diallo, M. O., Ettiegne-Traore, V. *et al.* (2001). Effect of interventions to control sexually transmitted disease on the incidence of HIV infection in female sex workers. *AIDS*, 15, 1421–31.

Ghys, P. D., Diallo, M. O., Ettiegne-Traore, V. *et al.* (2002). Increase in condom use and decline in HIV and sexually transmitted diseases among female sex workers in Abidjan, Cote d'Ivoire, 1991–1998. *AIDS*, 16, 251–8.

Girault, P., Saidel, T., Song, N. *et al.* (2004). HIV, STIs, and sexual behaviors among men who have sex with men in Phnom Penh, Cambodia. *AIDS Educ. Prev.*, 16, 31–44.

Goodyear, M. D. and Cusick, L. (2007). Protection of sex workers. *Br. Med. J.*, 334, 52–3.

Hanenberg, R. and Rojanapithayakorn, W. (1996). Prevention as policy: how Thailand reduced STD and HIV transmission. *AIDScaptions*, 3, 24–7.

Hanenberg, R. and Rojanapithayakorn, W. (1998). Changes in prostitution and the AIDS epidemic in Thailand. *AIDS Care*, 10, 69–79.

Harawa, N. T., Greenland, S., Bingham, T. A. *et al.* (2004). Associations of race/ethnicity with HIV prevalence and HIV-related behaviors among young men who have sex with men in 7 urban centers in the United States. *J. Acquir. Immune Defic. Syndr.*, 35, 526–36.

Hellard, M. E. and Aitken, C. K. (2004). HIV in prison, what are the risks and what can be done? *Sex. Health*, 1, 107–13.

Herbst, J. H., Sherba, R. T., Crepaz, N. *et al.* (2005). A meta-analytic review of HIV behavioral interventions for reducing sexual risk behavior of men who have sex with men. *J. Acquir. Immune Defic. Syndr.*, 39, 228–41.

Hilton, B. A., Thompson, R., Moore-Dempsey, L. and Janzen, R. G. (2001). Harm reduction theories and strategies for control of human immunodeficiency virus, a review of the literature. *J. Adv. Nursing*, 33, 357–70.

Horsburgh, C. R. Jr, Jarvis, J. Q., McArther, T. *et al.* (1990). Seroconversion to human immunodeficiency virus in prison inmates. *Am. J. Public Health*, 80, 209–10.

Hospers, H. J., Kok, G., Harterink, P. and de Zwart, O. (2005). A new meeting place, chatting on the Internet, e-dating and sexual risk behavior among Dutch men who have sex with men. *AIDS*, 19, 1097–101.

HRW (2007). Letter to US Global AIDS Coordinator about "abstinence-only" funding and homophobia in Uganda. New York, NY: Human Rights Watch.

Hurley, S. F., Jolley, D. J. and Kaldor, J. M. (1997). Effectiveness of needle-exchange programmes for prevention of HIV infection. *Lancet*, 349, 1797–800.

ILGA (2007). State-sponsored homophobia: a world survey of laws prohibiting same sex activity between consenting adults. D. Ottosson (ed), Brussels: ILGA.

International Harm Reduction Development Program (2007). *Women, Harm Reduction, and HIV*. New York, NY: OSI.

Johnson, W. D., Holtgrave, D. R., McClellan, W. M. *et al.* (2005). HIV intervention research for men who have sex with men, a 7-year update. *AIDS Educ. Prev.*, 17, 568–89.

Kerr, T., Tyndall, M., Li, K. *et al.* (2005). Safer injection facility use and syringe sharing in injection drug users. *Lancet*, 366, 316–18.

Kerr, T., Stoltz, J. A., Tyndall, M. *et al.* (2006). Impact of a medically supervised safer injection facility on community drug use patterns, a before and after study. *Br. Med. J.*, 332, 220–2.

Kerrigan, D., Ellen, J. M., Moreno, L. *et al.* (2003). Environmental–structural factors significantly associated with consistent condom use among female sex workers in the Dominican Republic. *AIDS*, 17, 415–23.

Kleinig, J. (2006). Thinking ethically about needle and syringe programs. *Subst. Use Misuse*, 41, 815–25.

Koblin, B. A., Husnik, M. J., Colfax, G. *et al.* (2006). Risk factors for HIV infection among men who have sex with men. *AIDS*, 20, 731–9.

Kushel, M. B., Hahn, J. A., Evans, J. L. *et al.* (2005). Revolving doors, imprisonment among the homeless and marginally housed population. *Am. J. Public Health*, 95, 1747–52.

Lankoande, S., Meda, N., Sangare, L. *et al.* (1998). Prevalence and risk of HIV infection among female sex workers in Burkina Faso. *Intl J. STD AIDS*, 9, 146–50.

Lenton, S. and Single, E. (1998). The definition of harm reduction. *Drug Alcohol Rev.*, 17, 213–19.

Malinowska-Sempruch, K. and Gallagher, S. (2004). *War on Drugs, HIV/AIDS and Human Rights*, 1st edn. New York, NY: IDEA.

Manopaiboon, C., Bunnell, R. E., Kilmarx, P. H. *et al.* (2003). Leaving sex work, barriers, facilitating factors and consequences for female sex workers in northern Thailand. *AIDS Care*, 15, 39–52.

Masenior, N. F. and Beyrer, C. (2007). The US anti-prostitution pledge, First Amendment challenges and public health priorities. *PLoS Med.*, 4, e207.

Morbidity and Mortality Weekly Report (2006a). HIV prevalence among populations of men who have sex with men – Thailand, 2003 and 2005. *Morbid. Mortal. Wkly Rep.*, 55, 844–8 (http://www.cdc.gov/MMWR/preview/mmwrhtml/mm5531a2.htm).

Morbidity and Mortality Weekly Report (2006b). HIV transmission among male inmates in a state prison system – Georgia, 1992–2005. *Morbid. Mortal. Wkly Rep.*, 55, 421–6 (http://www.cdc.gov/MMWR/preview/mmwrhtml/mm5515a1.htm).

Morisky, D. E., Stein, J. A., Chiao, C. *et al.* (2006). Impact of a social influence intervention on condom use and sexually transmitted infections among establishment-based female sex workers in the Philippines, a multilevel analysis. *Health Psychol.*, 25, 595–603.

Nelles, J. and Harding, T. (1995). Preventing HIV transmission in prison, a tale of medical disobedience and Swiss pragmatism. *Lancet*, 346, 1507–8.

Nelles, J., Fuhrer, A., Hirsbrunner, H. and Harding, T. (1998). Provision of syringes, the cutting edge of harm reduction in prison? *Br. Med. J.*, 317, 270–3.

Okie, S. (2007). Sex, drugs, prisons, and HIV. *N. Engl. J. Med.*, 356, 105–8.

Operario, D., Nemoto, T., Ng, T. *et al.* (2005). Conducting HIV interventions for Asian Pacific Islander men who have sex with men: challenges and compromises in community collaborative research. *AIDS Educ. Prev.*, 17, 334–6.

Organization of American States (1988). Additional Protocol to the American Convention on Human Rights in the Area of Economic, Social and Cultural Rights ("Protocol of San Salvador"). *OAS Treaty Series*, art. 10 edn.

Peak, A., Rana, S., Maharjan, S. H. *et al.* (1995). Declining risk for HIV among injecting drug users in Kathmandu, Nepal, the impact of a harm-reduction programme. *AIDS*, 9, 1067–70.

Polonsky, S., Kerr, S., Harris, B. *et al.* (1994). HIV prevention in prisons and jails, obstacles and opportunities. *Public Health Rep.*, 109, 615–25.

Poulin, C., Alary, M., Lambert, G. *et al.* (2007). Prevalence of HIV and hepatitis C virus infections among inmates of Quebec provincial prisons. *Can. Med. Assoc. J.*, 177, 252–6.

Ramjee, G., Karim, S. S. and Sturm, A. W. (1998). Sexually transmitted infections among sex workers in KwaZulu-Natal, South Africa. *Sex. Transm. Dis.*, 25, 346–9.

Rhodes, T., Lowndes, C., Judd, A. *et al.* (2002). Explosive spread and high prevalence of HIV infection among injecting drug users in Togliatti City, Russia. *AIDS*, 16, F25–31.

Rhodes, T., Kimber, J., Small, W. *et al.* (2006). Public injecting and the need for "safer environment interventions" in the reduction of drug-related harm. *Addiction*, 101, 1384–93.

Rich, J. D., Foisie, C. K., Towe, C. W. *et al.* (2000). High street prices of syringes correlate with strict syringe possession laws. *Am. J. Drug Alcohol Abuse*, 26, 481–7.

Rojanapithayakorn, W. and Hanenberg, R. (1996). The 100% condom program in Thailand. *AIDS*, 10, 1–7.

Ruiz, J. D., Molitor, F. and Plagenhoef, J. A. (2002). Trends in hepatitis C and HIV infection among inmates entering prisons in California, 1994 versus 1999. *AIDS*, 16, 2236–8.

Simbulan, N. P., Aguilar, A. S., Flanigan, T. and Cu-Uvin, S. (2001). High-risk behaviors and the prevalence of sexually transmitted diseases among women prisoners at the women state penitentiary in Metro Manila. *Social Sci. Med.*, 52, 599–608.

Small, W., Kain, S., Laliberte, N. *et al.* (2005). Incarceration, addiction and harm reduction, inmates experience injecting drugs in prison. *Subst. Use Misuse*, 40, 831–43.

Steen, R. and Dallabetta, G. (2003). Sexually transmitted infection control with sex workers, regular screening and presumptive treatment augment efforts to reduce risk and vulnerability. *Reprod. Health Matters*, 11, 74–90.

Stimson, G. V. (2007). "Harm reduction–coming of age", a local movement with global impact. *Intl J. Drug Policy*, 18, 67–9.

Strang, J., Griffiths, P. and Gossop, M. (1997). Heroin in the United Kingdom: different forms, different origins, and the relationship to different routes of administration. *Drug Alcohol Rev.*, 16, 329–37.

UN General Assembly (1948). Universal Declaration of Human Rights. Resolution 217 A (III), A/910 art. 25(1) edn.

UN General Assembly (1966). International Convention on the Elimination of All Forms of Racial Discrimination. Resolution 2106 (XX), annex, 20 UN GAOR Suppl. (No. 14) art 47, UN Doc. A/6014 (1966a) art. 5(e) edn.

UN General Assembly (1976). International Covenant on Economic, Social and Cultural Rights (ICESCR). Resolution 2200A (XXI), 21 UN GAOR Suppl. (No. 16) art 49, UN Doc. A/6316 (1966b) art. 12(1) edn.

UN General Assembly (1979). International Convention on the Elimination of All Forms of Discrimination against Women. Resolution 34/180, 34 UN GAOR Suppl. (No. 46) art. 193, UN Doc. A/34/46 (1979) arts. 11(f). and 12 edn.

UN General Assembly (1989). Convention on the Rights of the Child. Resolution 44/125, annex, 44 UN GAOR Suppl. (No. 49) art 167, UN Doc. A/44/49 (1989) art. 24 edn.

UNAIDS (2005). *Update on the Global HIV/AIDS Pandemic*. Geneva: UNAIDS.

UNAIDS (2006). *2006 Report on the Global AIDS Epidemic*. Geneva: UNAIDS.

UNAIDS (2007). *Practical Guidelines for Intensifying HIV Prevention*. Geneva: UNAIDS.

UNGASS (2001). *Global Crisis – Global Action*. United Nations General Assembly Twenty-sixth Special Session Doc: A/s-26/L.2, adopted 27 June, New York.

van Griensven, F., Thanprasertsuk, S., Jommaroeng, R. *et al.* (2005). Evidence of a previously undocumented epidemic of HIV infection among men who have sex with men in Bangkok, Thailand. *AIDS*, 19, 521–6.

van Kesteren, N. M., Hospers, H. J. and Kok, G. (2007). Sexual risk behavior among HIV-positive men who have sex with men, a literature review. *Patient Educ. Couns.*, 65, 5–20.

Vlahov, D., Brewer, T. F., Castro, K. G. *et al.* (1991). Prevalence of antibody to HIV-1 among entrants to US correctional facilities. *J. Am. Med. Assoc.*, 265, 1129–32.

Wade, A. S., Kane, C. T., Diallo, P. A. *et al.* (2005). HIV infection and sexually transmitted infections among men who have sex with men in Senegal. *AIDS*, 19, 2133–40.

Weinbaum, C., Lyerla, R. and Margolis, H. S. (2003). Prevention and control of infections with hepatitis viruses in correctional settings. *Morbid. Mortal. Wkly Rep. Recomm. Rep.*, 52, 1–36.

WHO (2005). Status Paper on Prisons, Drugs and Harm Reduction. London: WHO.

Wodak, A. and Cooney, A. (2006). Do needle syringe programs reduce HIV infection among injecting drug users: a comprehensive review of the international evidence. *Subst. Use Misuse*, 41, 777–813.

Wood, E., Kerr, T., Spittal, P. M. *et al.* (2003). An external evaluation of a peer-run "unsanctioned" syringe exchange program. *J. Urban Health*, 80, 455–64.

Wood, E., Kerr, T., Stoltz, J. *et al.* (2005). Prevalence and correlates of hepatitis C infection among users of North America's first medically supervised safer injection facility. *Public Health*, 119, 1111–15.

Wood, E., Tyndall, M. W., Zhang, R. *et al.* (2006). Attendance at supervised injecting facilities and use of detoxification services. *N. Engl. J. Med.*, 354, 2512–14.

World Bank (2006). Socioeconomic Impact of HIV/AIDS in Ukraine. Washington, DC: World Bank.

HIV testing and counseling

19

Julie A. Denison, Donna L. Higgins and
Michael D. Sweat

The first licensed test for HIV became available in 1985, 4 years after the initial *Morbidity and Mortality Weekly Report* on the epidemic that described cases of *Pneumocystis carina* pneumonia in Los Angeles (CDC, 1981; US Food and Drug Administration, 2008). Treatment for HIV was not available at the time, and fear and stigma abounded. In this context, the HIV test emerged as much more than a purely diagnostic tool. The test process, for example, involved counseling on risk reduction and behavior change. Expansion of HIV testing, and thus growing numbers of people who were aware that they were infected with HIV, also supported political mobilization and advocacy among those infected and affected by the virus. In the two decades since then, the rationale for HIV testing and counseling programs has expanded in many ways, and now also includes gaining access to care and treatment. Overall, the delivery of HIV testing and counseling has never been static and has continually evolved in response to changing social and political environments and technological advances. In this chapter, we provide a brief overview of how HIV testing and counseling evolved and the different purposes it serves. We then explore the two decades of evidence regarding the behavioral effects of HIV testing and counseling, followed by an introduction to the different delivery models for providing this service. We end with a discussion of emerging issues and challenges around HIV testing and counseling in an era of rapid global scale-up and expansion of HIV/AIDS treatment.

History of HIV testing and counseling

It is critical to understand the history of HIV testing in order to examine objectively the assumptions behind this service, and to understand how HIV testing and counseling might continue to evolve and respond to different epidemic and social contexts. Since its inception, the HIV test has been viewed differently from most other medical diagnostic tools. The initial public health reason for promoting HIV testing in the United States was to protect the blood supply; however, two concerns existed. First, it was believed that members of population groups who were experiencing the initial burden of HIV, such as men who have sex with men (MSM), would turn to blood banks as the only place where they might learn their HIV status. The test, however, could not detect HIV infection during the window period before the body developed HIV antibodies. Thus, infected blood could remain in the blood supply, making it a less than ideal location for testing individuals at risk for infection. Secondly, fear abounded regarding confidentiality. In an environment characterized by stigma, a lack of HIV treatment to suppress the virus, and fear that the health system would disclose an HIV-positive status to insurance companies, government agencies and employers, the debate about how to provide testing was intense. As described in the book *And the Band Played On*, organizations like the gay advocacy group Lambda Legal Defense Fund questioned how the US government, without provisions for confidentiality, could "release a test that could have such devastating impact on so many American lives", while CDC staff asked "how could these people threaten to halt a test that could clearly save lives?" (Shilts, 1987). In response to these public health and political concerns the US government supported the development of testing sites separate from blood banks, where individuals could test specifically to learn their HIV status.

In this context, The US Centers for Disease Control and Prevention (CDC) and the World Health Organization (WHO) provided initial guidance documents on HIV testing that emphasized HIV prevention counseling, and confidentiality (CDC, 1986, 1987; WHO, 1990). The WHO document helped to solidify the role of counseling by stating that counseling should be "an integral part of all HIV testing" programs, and defined counseling as an ongoing dialogue and relationship between client or patient and counselor (WHO, 1990). A CDC technical update reinforced the concept that counseling associated with HIV testing should be an interactive rather than didactic process (CDC, 1993). That same year, the WHO issued a statement on testing and counseling that used the term "voluntary" for the first time, and emphasized that testing without informed consent is "ineffective and unethical" (WHO, 1993). Shortly thereafter, the CDC published the first standards and guidelines on HIV counseling, testing and referral (CDC, 1994). From these initial guidance documents and experiences emerged a model of testing and counseling termed "voluntary counseling and testing" (VCT), which included pre- and post-test counseling with a focus on client-centered counseling, and the prevention of HIV transmission.

It was this VCT model of HIV testing and counseling that was promoted heavily as a prevention tool in the mid-1990s, both within the US and in developing countries hardest hit by the HIV epidemic (UNAIDS, 1997, 2000a). New testing technologies, particularly rapid tests, were available by 1994, which made the provision of VCT in resource constrained settings more feasible. Treatment options for HIV-related diseases were also expanding, and zidovudine (AZT) was found to prevent mother-to-child transmission (PMTCT) of HIV. In this light, HIV testing slowly transitioned from a procedure that primarily assuaged fears of infection, and prevention of further transmission, to a diagnostic gateway to treatment. In recognition of the importance of HIV testing and counseling as an entry point for treatment, the UN General Assembly Special Session (UNGASS) on HIV/AIDS in 2001 included in the Declaration of Commitment on HIV/AIDS that all countries by 2005 should have a wide range of prevention activities, including expand access to voluntary and confidential counseling and testing (United Nations, 2001).

There is substantial consensus in the public health community that too few people living with HIV – only an estimated 8–25 percent – know their HIV status (WHO, UNAIDS & UNICEF, 2007). There is also consensus that testing is an important gateway for accessing care and treatment. However, there has been considerable, often heated debate on how specifically to provide HIV testing, and what outcomes can be attributed to HIV testing and counseling services. In 2001, the CDC updated their 1994 standards to recognize that providers need flexibility in implementing the guidelines, "given their particular client base, setting HIV prevalence level, and available resources" (CDC, 2001). Essentially, this policy change questioned whether the individual-focused VCT model of HIV testing – developed in the West when treatment was not available and stigma ran high – was appropriate in a context more than a decade later in which treatment is accessible and prevalence is much higher. In particular, the debate focused on the role of counseling in HIV testing. Was counseling associated with an HIV test a critical element for behavior change to occur, or was simply knowing one's status enough? In 2004, UNAIDS and the WHO issued a policy statement on HIV testing and counseling that presented the mechanisms and conditions for what they termed "provider-initiated routine testing" (UNAIDS & WHO, 2004). In 2006, the CDC recommended routine testing in all medical settings in the US, and in 2007 the WHO released the latest guidance on HIV testing and counseling that focuses on provider-initiated testing (Branson *et al.*, 2006; WHO & UNAIDS, 2007).

This brief historical overview of HIV testing and counseling illustrates the evolving role the HIV test has had as a prevention and care tool. Overall, the reasons for taking a test and for promoting knowledge of one's HIV status are varied, and include gaining knowledge for behavior change and future decisions (e.g., getting married, having children), testing as a human rights issue, government planning, political mobilization, and access to treatment. The next section will examine the existing evidence on the effectiveness of HIV testing and counseling in terms of prevention and care.

HIV testing and counseling and behavior change

Since the 2001 UNGASS commitment to expand access to VCT (United Nations, 2001), the estimated number of people using HIV testing and counseling services in more than 70 surveyed countries increased from 4 million persons in 2001 to 16.5 million in 2005 (UNAIDS, 2006). This increase reflects a commitment taken at the international, national, community and individual levels. For example, in Kenya more than 500 VCT sites, both stand-alone and integrated into public health facilities, were established in 5 years (Marum *et al.*, 2006). In Malawi, the Ministry of Health holds annual HIV Testing and Counseling (HTC) weeks to increase the number of people aware of their HIV status (PEPFAR, 2007). During the 2007 HTC week, approximately 1367 sites, including mobile services, offered HTC. As a result, 186,631 Malawians learned their HIV status, 71 percent of whom were learning this for the first time. In Nigeria, The Global HIV/AIDS Initiative Nigeria (GHAIN) project has supported more than 79 counseling and testing sites, and has counseled and tested over a million people (FHI, 2008). Such advances are occurring at a slower pace for pregnant women. Data from 70 resource-limited countries found that only 10 percent of pregnant women had received an HIV test (WHO, UNAIDS & UNICEF, 2007). Corresponding with the increased uptake of HIV testing and counseling has been a rapid increase in the number of HIV-infected patients starting antiretroviral therapy (ART), from 240,000 in 2001 to 1.3 million in 2005 (UNAIDS, 2006). While the numbers of people undergoing HIV testing and counseling and accessing ART have greatly increased, three important questions remain:

1. What strategies can be implemented to reach the estimated 80 percent of people who are infected with HIV and remain unaware of their HIV status?
2. How can people tested for HIV be effectively linked to appropriate prevention, care and treatment services?
3. To what degree does HIV testing and counseling impact HIV risk behaviors and prevention?

Changing sexual risk behaviors

While the number of people learning their HIV status and accessing HIV prevention, care, and treatment is critical, it remains important to examine the effectiveness of HIV counseling and testing in reducing HIV risk behaviors. By combining personalized counseling with knowledge of one's HIV status, HIV testing and counseling is believed to motivate people to change their sexual behaviors to prevent the transmission of the virus. More than two decades of research exists regarding the impact of the service on risk behaviors. An early review of HIV testing and counseling data examined findings from 50 manuscripts and conference abstracts that were published between 1986 and 1990 (Higgins *et al.*, 1991). The authors presented the available data by four subgroups: men who have sex with men (MSM), injection drug users (IDUs), pregnant women, and other heterosexuals. While data from MSM and IDUs showed clear reductions in sexual risk behaviors, the authors questioned whether such changes were due to secular changes or, for IDUs, due to drug treatment. As such, the specific effects of HIV counseling and testing on sexual risk behaviors of MSM and IDUs were inconclusive. The paucity of evidence then available also suggested that HIV counseling and testing had no impact on pregnancy decisions among infected women, including pregnancy rates and termination. The most consistent finding was from four longitudinal studies among discordant couples that showed increased condom use and safe sexual practices among couples with at least one infected member.

Wolitski *et al.* (1997) later conducted a second review of HIV counseling and testing, examining data from 35 manuscripts and abstracts published between 1990 and 1996. This review used the same inclusion criteria as the Higgins review and presented both sexual and health-seeking behaviors by four subpopulations: MSM, IDUs, women and heterosexual couples, and mixed populations. This review also found that the effects of HIV counseling and testing on risk behaviors varied by research population and design, with the most consistent sexual behavior change evidence from studies conducted among discordant couples. Authors of both these initial reviews discussed how studies of HIV counseling and testing were rarely designed specifically to assess the behavioral impact of learning one's HIV status.

The first meta-analysis of VCT efficacy data was published in 1999 (Weinhardt *et al.*, 1999). Meta-analyses combine data from different studies to quantitatively determine the impact of the intervention on the outcome of interest. Results from this meta-analysis supported the provision of HIV counseling and testing intervention as an effective secondary behavior-change strategy, but only among those infected with HIV. Receiving HIV counseling and testing did not have an impact on the sexual behaviors of HIV-negative recipients compared to untested participants. Further analyses also found that greater reduction in unprotected sex occurred among older versus younger recipients, and among people who sought HIV counseling versus those who were offered the service. In addition, significantly more change occurred among populations in high- compared to low-prevalence settings. Such variation in the effectiveness of HIV counseling and testing among different populations and settings was an important finding as the provision of the service expanded to new locations and epidemiologic settings. Overall, the main message from these studies was that the greatest risk-reduction appears to occur among people infected with HIV, either individuals or discordant couples.

While these reviews provided valuable information on HIV counseling and testing, few studies from developing countries were either available or matched the reviews' inclusion criteria. As a result, the majority of studies included in these reviews were conducted in developed countries in North America and Europe. Thus, a key question remained – is HIV testing and counseling an effective HIV prevention strategy in developing countries, where the majority of people with HIV live? The debate around this question has intensified over the years, with two distinctive perspectives illustrating opposite ends of the spectrum. On one side is an argument that previous meta-analyses and rigorous studies show very limited behavior change due to HIV counseling and testing, and that change predominantly occurs among those living with HIV. Interestingly, a more recent community-based trial conducted in Uganda found no difference between participants who learned their HIV test result and those who did not in terms of sexual risk behaviors or HIV incidence. This Ugandan study (Matovu *et al.*, 2005), together with Weinhardt's meta-analysis article (Weinhardt *et al.*, 1999), are often cited as evidence supporting the view that HIV testing and counseling is not effective in reducing risk behavior. At the other end of the spectrum is an argument that HIV testing and counseling does lead to behavior change and does not intensify social harms, as shown by a randomized controlled trial that examined the efficacy of VCT in three developing countries (Kenya, Tanzania, and Trinidad). This trial, known as the VCT Efficacy Trial, found that participants randomly assigned to the VCT arm had a greater proportional reduction in unprotected sex with non-primary partners than participants assigned to a health information arm (men, 35 percent vs 13 percent; women, 39 percent vs 17 percent; VCT vs health information) (VCT Efficacy Study Group, 2000). The intensity of this debate continues in an environment of rapid scale-up of HIV-testing services internationally.

To address this issue, investigators at the World Health Organization, the Medical University of South Carolina and the Johns Hopkins Bloomberg School of Public Health (JHSPH) conducted an updated meta-analysis examining VCT data from developing countries (Denison *et al.*, 2008a). This meta-analysis focused only on VCT services that provided both pre- and post-test counseling, were conducted in developing countries, and had either pre-/post- or comparison arm designs. Seven studies met the inclusion criteria, some of which were analyzed in earlier meta-analyses. Three studies were from Africa, three from Asia, and one had study sites in both Africa and the Caribbean. Pooled data from these seven studies found that VCT recipients were significantly less likely to engage in unprotected sex when compared to behaviors before receiving VCT, or as compared to participants who had not received VCT (OR 1.69; 95%CI 1.25–2.31). Similar to earlier meta-analyses and syntheses (Higgins *et al.*, 1991; Wolitski *et al.*, 1997; Weinhardt *et al.*, 1999), it was found that the significant effect on unprotected sex from VCT was driven primarily by studies conducted among HIV-infected persons or discordant couples. Data from three studies provided inconclusive evidence that VCT has no significant effect on the number of sex partners (OR 1.22; 95%CI 0.89–1.67).

Overall, the pooled data in this meta-analysis provide objective evidence that VCT has a moderate and significant effect on increasing protected sex. It is also reassuring that none of the included studies found a significant increase in risk-taking behaviors among VCT participants. These data contribute valuable (although far from definitive) evidence regarding the effects and potential harms attributable to VCT. Interestingly, the implementation of VCT across the included studies in this meta-analysis varied greatly. For example, recruitment strategies included researcher-, provider- and client-initiated methods in clinic, community and home settings; the counseling provided included couple, individual and group approaches. Such differences may affect study outcomes – as found in the Weinhardt meta-analysis, in which more behavior change occurred among participants who actively sought HIV counseling and testing compared to those who were approached by researchers (Weinhardt *et al.*, 1999).

Several key issues can be gleaned from these review articles and meta-analyses. First, the greatest reductions in risk following HIV counseling and testing appear to occur among people living with HIV (PLWH). Second, the effect of VCT seems to vary widely within and across studies. Third, the strength of these reviews is limited by the fact that included studies were often not designed to examine effectiveness of the service. Fourth, the long-term effects of HIV counseling and testing remain unknown. Finally, the greatest amount of data available on the efficacy of HIV testing and counseling is from studies among adults in mainly urban areas. Data on VCT among other groups, such as adolescents (Denison *et al.*, 2008b) and drug-using populations, as well as in more diverse settings, including peri-urban and rural areas, are needed.

Cost-effectiveness

Various studies have estimated the cost-effectiveness of HIV voluntary counseling and testing (VCT), with results varying by setting and population group. In general, HIV VCT has typically been found to be cost-effective relative to other prevention interventions, but primarily when targeted to populations with high prevalence of HIV. For example, one study in the United States in 1994 on the provision of VCT to hospital inpatients found that the cost per HIV infection averted was US$753 million when prevalence was 1 percent, but dropped to US$8353 when HIV prevalence approached 10 percent (Lurie *et al.*, 1994). Another US study showed VCT to be cost-saving when provided to prison inmates, and resulted in improved linkages to HIV-related health care (Varghese *et al.*, 2001). In addition to cost analyses of the traditional VCT model of HIV testing, Holtgrave (2007) recently compared the cost-effectiveness of targeted HIV counseling and testing to that of opt-out HIV testing. In this regard, the opt-out strategy was based on providing HIV testing without the requirement for risk assessment and counseling in all health-care encounters in the US for persons aged 13 to 64 years. Assuming 1 percent prevalence of HIV, he estimated that, for the same cost, targeted HIV testing would detect 3.3 times more HIV infections and prevent almost 4 times as many HIV infections, as compared to an opt-out testing program. Moreover, under the targeted HIV counseling and testing program the cost per HIV infection averted was $59,383 as compared to $237,149 for the opt-out approach. Again, this analysis was conducted for the United States, and assuming a relatively low HIV prevalence of 1 percent. Additional analysis would be needed to examine the cost-effectiveness of opt-out testing programs in higher HIV-prevalence settings.

In developing countries with high HIV prevalence several cost analyses of VCT have also been conducted, all finding HIV VCT to be a cost-effective intervention. One cost analysis based on a randomized trial of the efficacy of VCT conducted in Kenya and Tanzania in 2000 found the cost per HIV infection averted to be US$249 in Kenya and US$346 in Tanzania, with an associated cost per disability life-year saved of $12.77 and $17.78 respectively. The study also revealed that recurrent costs, such as labor and rent, account for 73 percent of the costs of providing VCT (Sweat *et al.*, 2000). Another cost-effectiveness analysis of VCT in Tanzania found the cost per HIV infection ranged from US$92 to $170, with an associated cost per disability life-year saved ranging from $4.72 to $8.72 (Thielman *et al.*, 2006). Greater cost-effectiveness was achieved as the program matured over time, and fees were waived, resulting in greater numbers of clients and thus more efficient use of resources. In Kenya, Forsythe and colleagues conducted a careful cost analysis of integration of HIV VCT into rural health clinic services (Forsythe *et al.*, 2002). They found that the addition of VCT added US$6800 in costs per year per clinic, translating to $16 per client. They also estimated that the cost could be reduced by half if existing clinic staff provided the counseling. Clients in this study reported that they would be willing to pay approximately US$2 for VCT, which offered the potential for even further cost reductions. With the expansion of AIDS care and treatment in developing countries there has also been a concomitant expansion of VCT to provide pathways to care for HIV-infecting persons, and donors have evermore viewed VCT as both a cost-effective preventative intervention as well as a means to diagnose HIV infection and route those infected to emerging treatment services.

HIV testing and counseling models

HIV testing and counseling, particularly the VCT model that emerged from the US, initially focused on HIV prevention with a very individualistic approach. As the testing technology improved and the need and demand increased, however, the provision of HIV testing has been adapted to fit different contexts and cultures, with a shifting emphasis on primarily identifying people living with HIV and less emphasis on counseling on HIV prevention and living with the virus.

Below is a more detailed discussion of the various models being implemented and debated today. Cutting across these models, however, is consensus that all HIV testing should be confidential, and accompanied by informed consent and post-test counseling (WHO & UNAIDS, 2007).

Voluntary counseling and testing (VCT)

The VCT model of HIV testing is client-initiated, and consists of pre- and post-test counseling. CDC/UNAIDS guidelines recommend that during pre-test counseling the counselor and client discuss the test process, assess the client's risk behaviors, discuss coping strategies related to receipt of test results, review prevention options, and reaffirm the decision to test for HIV (UNAIDS, 2000a; CDC, 2001). In the post-test counseling session clients receive their HIV status, discuss risk-reduction strategies and disclosure of test results, and receive appropriate referrals for care and support. While pre- and post-test counseling is the hallmark of VCT, the implementation of the VCT model has varied greatly. These variations include the development of stand-alone VCT facilities that only provide HIV counseling and testing (e.g., VCT Efficacy Study, 2000), integrating VCT into formal health-care services, particularly ANC, STI and FP clinics (Farquhar, 2004; FHI, 2004a, 2007a), providing VCT through community-based organizations and services, and providing VCT in people's homes (Matovu *et al.*, 2005; Bateganya *et al.*, 2007). Mobile VCT units have also evolved in an effort to bring VCT to people in communities who may not seek the service at formal health-care settings (Morin *et al.*, 2006). It is also important to note that the recommended counseling modality associated with VCT was originally strongly influenced by a client-centered psychotherapeutic approach pioneered by Rogers in 1951, and which later became dominant theme of modern western psychological counseling (Rogers, 1951, 1959, 1964). This client-centered orientation recommended for HIV counseling and testing asserted that the client's needs and concerns should primarily drive the counseling session, that the client possessed the capacity to best decide on actions to be taken, that direct recommendations for action by the counselor should be avoided, and that the role of the counselor was to facilitate self-awareness from within the client. This strategy also puts heavy emphasis on the counseling as an engine for risk reduction and coping, and requires moderate time and a skilled counselor to be effective.

In practical terms, however, how VCT is provided, especially with regard to the counseling, varies considerably, and includes individual one-on-one counseling, couple counseling with both members of a partnership receiving pre and/ or post-test counseling together, and group pre-test counseling. The counselors themselves also encompass a wide range of expertise and experiences, ranging from medical staff such as doctors or nurses to trained members of the community and/or positive-living groups. These variations represent the evolving nature of VCT as it expanded from resource-rich to resource-limited settings. In addition, the terminology used to describe this model of testing and counseling has also changed with time, and includes, for example, VCT, VCCT (voluntary and confidential counseling and testing) and client-initiated counseling and testing. This shift in terminology reflects the changes in thinking around the original VCT model and the importance of different components of HIV testing – mainly, is it voluntary, who initiates it, how is confidentiality maintained, and when/how is counseling provided?

Provider Initiated Testing and Counseling (PITC)

While VCT was developed initially to help people learn their HIV status and as an HIV-prevention tool to promote behavior change, PITC emerged in an effort to rapidly identify HIV-infected patients in the clinic setting in order to engage them in HIV care and treatment. As such, counseling plays a less pronounced role. In PITC, the health-care provider offers HIV testing to patients who attend formal health care (WHO and UNAIDS, 2007). Pre-test information, but not counseling, is mandatory, and the WHO stresses that all clients should provide informed consent to undergo the test. Two different approaches – opt-in and opt-out – exist for informing patients about the test. The opt-in approach is when a provider specifically offers patients an HIV test and the patient explicitly agrees. The opt-out approach is when the test is routinely performed unless the patient explicitly refuses or declines to take the test (UNAIDS Reference Group on HIV and Human Rights, 2007). The WHO/UNAIDS guideline recommends opt-out, testing except for vulnerable populations who may not be able to refuse a test (WHO & UNAIDS, 2007). PITC provides post-test counseling when the patient receives his or her test result. For HIV-negative persons, post-test counseling consists of an explanation of the test result, basic advice on HIV prevention methods, and the provision of male and female condoms and their use (WHO & UNAIDS, 2007). Counseling for HIV-positive persons focuses on helping the person to cope with the result, facilitate access to treatment, care and prevention services, and facilitate disclosure (WHO & UNAIDS, 2007). In practice, who provides the pre-test information and the post-test counseling may vary in clinic settings (e.g., doctors, clinicians, lay people).

The WHO and UNAIDS provide further guidance on which health facilities should provide PITC according to the typology of the HIV epidemic and the context where the facility exists. For example, in countries with low-level or concentrated HIV epidemics, health-care providers should recommend HIV testing to only two groups – patients who exhibit symptoms of HIV, and children who are known to have been perinatally exposed. In addition, it is recommended that countries should consider (based on context) initiating PITC in health-care settings that serve high-risk populations, and including sexually transmitted infection clinics, antenatal clinics and TB services. It is also recommended that where HIV is firmly established in the general population, PITC should be implemented in all health facilities that have adequate resources and are supported by an enabling environment, including access to basic HIV prevention, care and treatment. Priority health settings for PITC in generalized epidemics include medical inpatient and outpatient facilities, as well as services for antenatal clinics, sexually transmitted infection clinics, most-at-risk population, children under 10, reproductive health and family planning settings, and for adolescents and surgical services.

HIV self-testing

Self-testing for HIV encompasses two very distinct approaches. The first is based upon an FDA-approved home sample collection test kit (Firth, 2007). Approved by the FDA in 1996, patients use the test kits to collect a sample of blood that they send to a laboratory for testing. They then receive their test result, and counseling, by telephone. This approach requires reliable mail and phone systems, and has not yet been widely implemented in developing countries.

The second model is home self-testing, and mirrors home pregnancy tests. Patients collect the sample, run the test and then interpret the results in the privacy of their own home. The US FDA has not yet approved home self-test kits (http://www.fda.gov/CbER/infosheets/hiv-home2.htm). Concerns regarding the use of HIV self-tests include mechanisms for ensuring the provision of counseling and linking infected individuals with care, and the potential for forced or coerced testing of individuals (Spielberg *et al.*, 2003; Kachroo, 2006; Firth, 2007).

Mandatory screening of blood product and organs

The most widely endorsed example of mandatory HIV testing is screening donated blood. Since the development of an HIV test in 1985, the United States has screened all donated blood, reducing the risk of HIV transmission from blood and blood products to 1:420,000 by 1995 (IOM, 1995). This improvement in blood safety is attributable in part to advances in testing technologies that have shortened the window period between infection and test detection (Lackritz *et al.*, 1995). Persons donating blood are, in many countries, also screened for behavioral risk for HIV, and made aware that their blood will be tested.

Mandatory and compulsory testing of individuals

Mandatory testing, for the purposes of determining eligibility for visas, work permits, immigration, insurance and military service, as well as mandatory testing of newborns, prisoners and other specific groups, is common globally. This type of mandatory testing often has punitive implications for persons identified as HIV-positive. For example, pre-marital screening in some areas is used as a tool to deny PLWHA from marrying (Uneke *et al.*, 2007). Overall, without the provision of counseling and access to care and treatment, mandatory testing is seen as an ineffective approach for changing HIV risk behaviors (UNAIDS & WHO, 2004), and there are concerns that it discourages people from seeking testing. HIV testing linked to individuals without a process of informed consent and counseling is also deemed unethical (UNAIDS, 1997), and human rights violations due to mandatory or compulsory testing have been documented (Human Rights Watch, 2006). Concern has also been raised about the implementation of an opt-out provider-initiated approach, as described above. The UNAIDS Reference Group on HIV and Human Rights issued a statement and recommendations on scaling up HIV testing and counseling which states that, in practice, PITC may result in a greater number of people being tested without their informed and voluntary consent. In settings where there is a power imbalance between test provider and client, the voluntary nature of HIV testing may be compromised, as the client may feel compelled to consent to the provider's offer. (UNAIDS Reference Group on HIV and Human Rights, 2007).

This blurring of the line between mandatory testing and opt-out routine testing, which should be offered in the context of informed consent, confidentiality and post-test counseling, is one that warrants ongoing vigilance as the provision of PITC continues to expand.

Expanding HIV testing and counseling to reach specific populations

Context matters, and it is not a surprise, given the epidemiology of HIV transmission and the variations in social and cultural settings, that one model of HIV testing and counseling will not reach all of the people infected with or exposed to HIV. In order to expand access, adaptations to of the above VCT and PITC models have been made in order better to reach specific groups. Below is a brief review of some of the approaches and issues in expanding HIV testing and counseling to specific populations. It remains important when examining the role of HIV testing for specific groups that the test is viewed as part of the HIV prevention and care continuum, and not in isolation of access to care, antiretroviral (ARV) drugs and psycho-social support.

Prevention of mother-to-child transmission (PMTCT)

Infants are at risk of contracting HIV from their HIV-positive mothers during pregnancy, birth and breast-feeding. It is estimated that without any interventions, between 20 percent and 45 percent of infants may become infected (De Cock *et al.*, 2000). Mother-to-child transmission of HIV in resource-rich settings like the United States has decreased substantially with the introduction of PMTCT that includes HIV screening of pregnant women and access to antiretroviral therapies. In developing countries, however, PMTCT efforts have not been very successful, with only an estimated 10 percent of pregnant women learning their HIV status, and an estimated 1400 children who become infected with HIV daily (WHO & UNICEF, 2007). In response, clinics that serve pregnant women and women of reproductive health, including ANC and family planning (FP) clinics, have started providing either VCT or PITC services (see Table 19.1).

Even in these women-focused settings, uptake of VCT and opt-in PITC often remains low (WHO & UNICEF, 2007). Reasons for low uptake include, first and foremost, that large proportions of women in resource-constrained settings do not attend ANC during their pregnancy. Those that do attend ANC clinics often encounter overwhelmed health-care systems and staff unable to provide them with an HIV test and counseling. Studies have also found that women often decline to test for HIV for fear of partner opposition, or potential stigma and discrimination (Medley *et al.*, 2004; Homsy *et al.*, 2007; Kominami *et al.*, 2007; Okonkwo *et al.*, 2007). The HIV test, however, is critical for identifying women in need of ART to prevent transmission to their child. Job aids have been developed (CDC, WHO, UNICEF & USAIDS, 2005) and research undertaken (e.g., comparing HIV testing uptake among women offered opt-in vs opt-out PITC, Chandisarewa *et al.*, 2007) to strengthen HIV testing as a key component of PMTCT strategies worldwide.

Couple- and family-based HIV testing and counseling

HIV risk behaviors always involve at least one other person. As such, the need to broaden the focus of HIV testing from individuals to couples and families has received growing emphasis. Several studies and review syntheses have found that couples counseled and tested together experience greater reduction in risk behaviors than people who are tested and counseled individually (Higgins *et al.*, 1991; Kamenga *et al.*, 1991; Weinhardt *et al.*, 1999; VCT Efficacy Study Group, 2000; Grinstead *et al.*, 2001; Allen *et al.*, 1992a, 2003). The reported change in risk behavior in these studies is both dramatic and significant. For example, among 963 discordant couples in Lusaka, Zambia, self-reported condom use had increased from 3 percent to 80 percent after 1 year (Allen *et al.*, 2003). Corroborating this self-reported information, there is evidence that HIV counseling and testing for HIV-discordant couples is associated with reduced seroconversion rates of negative partners (e.g., Allen *et al.*, 1992a, 1992b).

Clearly, couples' testing and counseling strategies have enormous benefit. However, special training is needed for providers to facilitate discussion and

Table 19.1   Number and percentage of ANC facilities providing PMTCT services

Region	Total number of health facilities providing ANC, 2006	No. of facilities providing ANC that provide HIV testing and counseling, 2006	Percentage of ANC facilities providing HIV testing and counseling	No. of facilities providing ANC which also provide minimum package of PMTCT services, 2006	Percentage of ANC facilities that provide the minimum package of PMTCT services*
East and Southern Africa	27,328	10,484	38%	10,185	37%
West and Central Africa	19,249	2,325	12%	2,420	13%
East Asia and the Pacific	15,489	3,394	22%	2,178	14%
C. and S. America and the Caribbean	100,092	80,351	80%	4,916	5%
Central and Eastern Europe and the Caucuses	24,789	13,504	54%	10,344	42%
South Asia	46,609	4,708	10%	2,462	5%
Middle East and North Africa	4,707	15	<1%	9	<1%
Total	238,263	114,781	48%	32,514	14%

Source: WHO, UNAIDS & UNICEF (2008).

disclosure among couples. The CDC has developed such training (http://www.cdc.gov/nchstp/od/gap/CHCTintervention).

Some faith-based communities in Ghana and Nigeria have instituted pre-marital testing as a requirement for marriage. As it is sometimes a mandatory condition for marriage, concerns have been expressed that pre-marital HIV screening and subsequent disclosure of HIV status could result in individuals or couples experiencing stigma and discrimination. HIV researchers and practitioners instead call for the provision of VCT with an emphasis on consent, confidentiality and counseling, and guidance for the religious leaders and health-care workers based in faith-based organizations on maintaining confidentiality of those individuals and couples who test HIV-positive (Luginaah *et al.*, 2005; Uneke *et al.*, 2007).

Some HIV programs are taking a family-centered approach to HIV by offering testing, treatment and other services to all members of a single household in one location. One such program in Uganda provides home-based VCT to household members of people who are starting treatment for HIV. They found that VCT was "well accepted and resulted in the detection of a large number of previously undiagnosed HIV infections and HIV-discordant relationships" (Were *et al.*, 2006). This approach holds great promise in helping people to learn their HIV status; however, concerns have also been expressed over the ability of family members to give consent for HIV testing and counseling without being coerced by other family members, and the ability for people to guard the results of their test results if they are not yet ready to disclose this information.

Adolescent/youth testing

Data from several countries reveal that large proportions of adolescents want to test for HIV, including 70 percent of Zambia's 15- to 19-year-olds (Baryarama *et al.*, 2000; DAPP, 2001; Boswell and Baggaley, 2002; Measures DHS, 2003). Less than 7 percent of Zambian youth though, know their HIV status – a trend also seen in other countries. Despite the fact that young people between the ages of 15 and 24 years account for half of new HIV infections worldwide (WHO, 2005a), little is known about the social context and motivating factors behind youth seeking HIV testing and counseling. As presented in an article by Denison *et al.* (2008b), US-based research has shown that peers and families often influence adolescent health behaviors, including condom and alcohol use (Perrino *et al.*, 2000; Tinsley *et al.*, 2004). Peers tend to spend large amounts of time together, and are seen as sources of information and pressure for adolescents to engage in risk behaviors. Family factors, such as family communication, positive parent–child relationships and parental monitoring, are also related to adolescents' risk and protective behaviors (Perrino *et al.*, 2000; Tinsley *et al.*, 2004). Only a few studies have explored the social context of HIV testing among youth, though, with most research tending to focus on individual cognitive measures (e.g., VCT/HIV knowledge; risk perceptions) (FHI, 2006, 2007b) and service delivery issues (e.g., client satisfaction; cost) (Boswell and Baggaley, 2002; Finger, 2002; McCauley, 2004a, 2004b).

A study in East Africa, however, did find that youth informed families and peers of their plans to test for HIV (Horizons, 2001). A program in Tanzania also found that more youth tested after efforts to involve parents and community members in the VCT strategy (Likwelile, 2004). Another study in Zambia explored how adolescents involve their families, friends and sex partners when making decisions about seeking HIV voluntary counseling and testing (VCT) and disclosing their HIV-status (Denison *et al.*, 2008b). This study presented findings from qualitative in-depth interviews among HIV-tested adolescents, and found that: (1) almost half of the youth turned to family members for advice or approval prior to seeking VCT; (2) a disapproving reaction from family members or friends often discouraged youth from attending VCT until they found someone supportive; (3) informants often attended VCT alone or with a friend, but rarely with a family member; and (4) disclosure was common to family and friends, infrequent to sex partners, and not linked to accessing care and support services. Qualitative research in South Africa also found that adolescents would disclose their HIV status to supportive family members (MacPhail *et al.*, 2008). These research findings reinforce efforts to expand HIV prevention programs beyond individual-level approaches to address relational, community and environmental factors (Sweat and Denison, 1995; Tawil *et al.*, 1995; DiClemente and Wingood, 2000; Rotheram-Borus *et al.*, 2004), and to specifically include families in HIV prevention research (Brown *et al.*, 2000; Pequegnat and Szapocznik, 2000; Pequegnat *et al.*, 2001; Rotheram-Borus, *et al.*, 2005). These research findings also support the provision of confidential HIV testing and counseling services for adolescents that do not require parental permission. Several informants from the Zambia study chose not to disclose to their families for fear of stigma, and very few participants visited an HIV-testing site with a family member (Denison *et al.*, 2008b). Overall, negative consequences from disclosure have been rare in research studies among adults, although they do occur, and include violence and abandonment (De Zoysa *et al.*, 1995; Grinstead *et al.*, 2001; Maman *et al.*, 2003). The outcomes for disclosure among adolescents, who may be at a distinctly different developmental phase than adults, require further investigation. As such, family involvement should not be required for accessing VCT or related services, but rather seen as an opportunity, when appropriate, to support youth before and after VCT.

In summary, research findings among adolescents support the need to go beyond the individual focus of HIV testing and counseling to investigate the mechanisms by which the social context influence adolescent testing behaviors. Determining mechanisms and innovative approaches to connect adolescents back into their families and communities, and into care and treatment services, are imperative, given the high rates of HIV infection and low rates of HIV testing among adolescents.

Other settings and populations

The section above presents just a few of the issues involved in reaching specific populations with HIV testing and counseling. Other groups not discussed in detail include patients attending STI and TB clinics, the provision of HIV testing and counseling in the workplace (FHI, 2004b), as well as reaching IDUs, people in rural settings and people in conflict settings. This list is not exhaustive, and is not meant to exclude or prioritize the focus of HIV testing. What this list does do, however, is illustrate the need to apply WHO/UNAIDS guiding principles for the provision of HIV testing and counseling that respect internationally recognized human rights. Within such a framework, VCT and PITC models of HIV testing can be adapted to the context where the service is being provided in order to best serve that community (see "Enabling environment" section below).

Continuing challenges and emerging issues

With more than 20 years of experience with HIV testing and counseling, many lessons have been learned and many challenges still confront us. The following section highlights some of the issues that require attention when developing and implementing HIV testing and counseling, as well as emerging issues.

Enabling environment

Testing technology and the outpouring of international support for HIV testing and counseling have in many ways outpaced the environment where testing is becoming available. An enabling environment should reduce the existence of compulsory testing and mitigate negative consequences of learning one's HIV status. The rapid expansion of HIV testing, however, has resulted in the use of rapid tests in settings where laws, policies and procedures to protect and support people infected and affected by HIV do not exist. The WHO and UNAIDS PITC guidance document outlines the basic social, policy, legal and service elements needed to support a successful and ethical PITC program (WHO & UNAIDS, 2007). Many of these elements are applicable to VCT as well as PITC, and include a process for obtaining informed consent, maintaining confidentiality, and ensuring appropriate linkages to prevention, care and treatment services. Studies have found that violence, abandonment and other negative consequences related to HIV infection and testing and counseling occur infrequently, and the majority of testing clients report positive life events associated with learning and disclosing their HIV status (Grinstead *et al.*, 2001; Maman, *et al.*, 2003). Ensuring these elements, however, can minimize the potential for social harm (WHO & UNAIDS, 2007). The WHO has also published recommendations on how to address violence against women in the context of expanding HIV testing and counseling programs (WHO, 2006). Taking the context into consideration, and implementing HIV testing and counseling with the three Cs (informed consent, confidentiality and post-test counseling) should reduce negative consequences including stigma, violence and increased sexual activity.

Quality assurance/quality control

Quality assurance and quality control are also critical components of HIV testing and counseling. Generally speaking, quality assurance is a set of activities that is carried out to establish standards and to monitor and improve performance (Brown *et al.*, 2000). Quality assurance measures ensure that the basic components of VCT and PITC are provided in a standard measurable fashion within and across programs. Aspects of quality assurance include technical competence, access, effectiveness of service delivery norms and guidelines, interpersonal relations, efficiency, and continuity of services and care (Brown *et al.*, 2000). Importantly, for HIV testing and counseling, quality assurance efforts ensure that the testing and counseling procedure is explained in a clear manner so that informed consent can be obtained, confidentiality assured, and quality counseling provided.

Quality control measures are also essential to ensure the reliability and accuracy of diagnostic/laboratory testing for HIV. This is essential, because testing requires the correct implementation of a series of processes and procedures. The WHO, CDC and PEPFAR recommend that a quality system that addresses all aspects of testing be instituted to keep errors to a minimum. Very specific regimens are recommended to develop, implement and maintain a quality system for HIV testing. Global recommendations and training have been developed by the WHO and CDC. Regional and country adaptations are also available (WHO, 2005b).

Monitoring and evaluation (M&E)

Another important aspect of HIV testing and counseling is the establishment of an effective and simple monitoring and evaluation system that supports the provision of quality services (UNAIDS, 2000b; UNAIDS & World Bank, 2002; WHO & UNAIDS, 2007). A strong monitoring system routinely collects essential information that is utilized by program management and staff to improve and support service provision (see also Chapter 21). Monitoring routine inputs, processes and outputs, including but not limited to service delivery components (e.g. number of clients tested, number counseled by type of counseling per day/week/month etc), cost issues, and laboratory functioning and results, can help to identify areas of strength and weakness and establish progress towards specific service-delivery goals, including scaling up. Staff burnout and client satisfaction and waiting time are also critical to assess through regular evaluation activities (e.g. staff meetings, client exit and staff interviews, and "secret shoppers"). Periodic and more rigorous evaluations of HIV testing and counseling programs are important to assess the outcomes and impact of such programs. For example, evaluations can test and inform strategies for improving access to HIV testing and counseling programs, increasing access to and uptake of referrals to HIV-related prevention, treatment, care and support services, and measuring social impact (e.g., on rates of disclosure, on stigma and discrimination, and on adverse outcomes). Chapter 21 of this book covers program monitoring and evaluation in detail; however, the authors wish to stress the importance of monitoring and evaluation systems as a necessary component for ensuring the quality of all aspects of HIV testing and counseling programs, including the laboratory. Additionally, it is crucial that scientists develop and carry out operational research studies to help improve the delivery of HIV testing and counseling programs, and the subsequent linkages to other programs.

Acute infections and prevention for positives

Acute infection refers to the first few weeks after HIV transmission, when a person experiences high HIV viral load but still tests negative for the HIV antibody (West, 2007). Following the acute infection stage is a period of several months, characterized as a recent infection, when high but declining viral loads exist. During recent infections, however, tests are able to detect HIV antibodies (West *et al.*, 2007). A combination of biology (high viral loads) and behaviors (HIV risk behaviors) makes acute and recent infections a time period when HIV can be efficiently transmitted. A study in Uganda showed high rates of HIV transmission during early stages of the infection (Wawer, 2005). Identifying people in these early stages of HIV infection is possible due to viral antigen and viral RNA tests that can detect the presence of HIV within 9 to 14 days following infection (West *et al.*, 2007). Concerns over the use of testing for acute or recent HIV infection have been raised and include the feasibility of case detection, and partner referral (Cassell and Surdo, 2007). Despite such concerns, research efforts have started to yield results – such as one study in Malawi that developed a risk-score algorithm for detecting acute HIV infection in sub-Saharan Africa (Powers *et al.*, 2007).

Overall, as history has repeatedly shown, utilizing these HIV tests will have direct implications for HIV care and prevention. Early detection provides an opportunity to engage the person living with HIV in health care. Early detection also provides an opportunity for preventing new infections through HIV prevention efforts targeted at a specific time when a PLWHA can most efficiently transmit, and thereby most efficiently prevent, HIV transmission to an uninfected partner. Traditionally, HIV prevention programs have focused on supporting

HIV-negative persons to remain free of infection. The importance of HIV prevention for positives, however, has slowly become more prominent, reflecting how tightly woven care and prevention truly are, and how both must be addressed in order to slow the spread of HIV. How to work with PLWHA to ensure linkages to prevention activities, sustained engagement in care and adherence to antiretroviral therapy needs to be addressed, as new testing technologies emerge. Testing for acute and recent infection also has implications for research efforts by allowing HIV research studies to assess the timing of infection, and HIV incidence as an intervention outcome measure.

Advancing HIV testing and counseling

Many challenges have been overcome and accomplishments achieved since the introduction of the first HIV test in 1985. Refinements and innovation in testing technology, including the development of rapid tests, have made it feasible to reach communities previously deemed impossible. The implications and potential outcomes of learning one's test result have also expanded with the introduction and scale-up of ART. Overall, science has repeatedly shown us the importance of HIV testing and counseling in furthering HIV prevention efforts, and in engaging people living with HIV in care and treatment.

Despite these successes, many policy and environmental challenges lie ahead. Reinforcing the seriousness of these challenges are the often repeated facts that the majority of those living with HIV remain unaware of their infection status, and the number of new infections each year greatly outpaces any advances made in increasing access to ART.

In this chapter we have provided a brief history of HIV testing and counseling so that we may better understand the different models of HIV testing that exist, as well as current debates surrounding these approaches. We have also reviewed some of the continuing challenges and emerging issues related to HIV testing and counseling, including quality assessment and monitoring and evaluation. Some key lessons gleaned from more than two decades of HIV testing experience include the need to: (1) ensure that HIV testing and counseling occurs in an enabling environment where human rights are respected; (2) create a stronger evidence base on the efficacy of different models of HIV testing and counseling and to determine the differential behavioral impact of counseling versus testing in changing behavior; and (3) conduct and improve monitoring, evaluation and quality assurance of HIV testing and counseling services to assure the provision of quality services. Each of these issues requires careful consideration of who the designated beneficiaries of HIV testing and counseling are, how best to reach them, and how to link them to appropriate prevention and care services. As efforts to expand HIV testing and counseling services continue, we can expect and hope for changes in the context and technology that surround knowing one's

HIV status. With focused intentions and international cooperation, such changes can strengthen the provision of HIV testing and counseling as a prevention and care strategy for individuals, families and communities worldwide.

References

Allen, S., Tice, J., Van de Perre, P. *et al.* (1992a). Effect of serotesting with counseling on condom use and seroconversion among HIV discordant couples in Africa. *Br. Med. J.*, 304, 1605–9.

Allen, S., Serufilira, A., Bogaerts, J. *et al.* (1992b). Confidential HIV testing and condom promotion in Africa: Impact on HIV and gonorrhea rates. *J. Am. Med. Assoc.*, 268, 3338–43.

Allen, S., Meinzen-Derr, J., Kautzman, M. *et al.* (2003). Sexual behavior of HIV discordant couples after HIV counseling and testing. *AIDS*, 17, 733–40.

Baryarama, F., Bunnell, R., Namwebya, J. H. *et al.* (2000). Declining HIV prevalence but continued transmission among youth seeking HIV testing in Uganda. *13th International AIDS Conference, 9–14 July, Durban*, Abstract No. MoPeC2419.

Bateganya, M. H., Abdulwadud, O. A. and Kiene, S. M. (2007). Home-based HIV voluntary counseling and testing in developing countries (review). *Cochrane Database Syst. Rev.*, 17, CD006493.

Boswell, D. and Baggaley, R. (2002). Voluntary Counseling and Testing (VCT) and Young People Family. Arlington, VA: Health International.

Branson, B. M., Handsfield, H. H., Lampe, M. A. *et al.* (2006). Revised recommendations for HIV testing of adults, adolescents, and pregnant women in health-care settings. *Morbid. Mortal. Wkly Rep.*, 55, 1–17.

Brown, L. D., Franco, L. M., Rafeh, N. and Hatzell, T. (2000). Quality Assurance of Health Care in Developing Countries. USAID. Available at http://www.qaproject.org/pubs/PDFs/DEVCONT.pdf (accessed 03-25-2008).

Brown, L. K., Lourie, K. J. and Pao, M. (2000). Children and adolescents living with HIV and AIDS: a review. *J. Child Psychol. Psychiatry*, 41, 81–96.

Cassell, M. M. and Surdo, A. (2007). Testing the limits of case finding for HIV prevention. *Lancet*, 7, 491–5.

CDC (1981). Kaposi's sarcoma and Pneumocystis pneumonia among homosexual men – New York City and California. *Morbid. Mortal. Wkly Rep.*, 30, 305–8.

CDC (1986). Current trends additional recommendations to reduce sexual and drug abuse-related transmission of Human T-Lymphotripic Virus type III/Lymphadenopathy-Associated Virus. *Morbid. Mortal. Wkly Rep.*, 35, 152–5.

CDC (1987). Public Health Service guidelines for counseling and antibody testing to prevent HIV infection and AIDS. *Morbid. Mortal. Wkly Rep.*, 36, 509–15.

CDC (1993). Technical guidance on HIV counseling. *Morbid. Mortal. Wkly Rep.*, 42, 11–17.

CDC (1994). *HIV Counseling, Testing and Referral Standards and Guidelines.* Atlanta, GA: US Department of Health and Human Services, Public Health Service.

CDC (2001). Revised guidelines for HIV counseling, testing and referral and revised recommendations for HIV screening of pregnant women. *Morbid. Mortal. Wkly Rep.*, 50, 1–58.

CDC, WHO, UNICEF & USAIDS (2005). Testing and counseling for prevention of mother to child transmission of HIV (TC for PMTCT): support tools. Available at http://www.womenchildrenhiv.org/wchiv?page=vc-10-00 (accessed 03-29-2008).

Chandisarewa, W., Stranix-Chibanda, L., Chirapa, E. *et al.* (2007). Routine offer of ante-natal HIV testing ("opt-out" approach) to prevention mother to child transmission of HIV in urban Zimbabwe. *Bull. WHO*, 85, 843–50.

DAPP (2001). Hope Humana Annual Report 2001. Ndola: DAPP in Zambia.

Denison, J. A., O'Reilly, K. R., Schmid, G. P. *et al.* (2008a). HIV voluntary counseling and testing and behavioral risk reduction in developing countries: a meta-analysis, 1990–2005. *AIDS Behav.*, 12, 363–73.

Denison, J. A., McCauley, A. P., Dunnett-Dagg, W. A. *et al.* (2008b). The HIV testing experiences of adolescents in Ndola, Zambia: do families and friends matter. *AIDS Care*, 20, 101–5.

De Zoysa, I., Phillips, K. A., Kamenga, M. C. *et al.* (1995). Role of HIV counseling and testing in changing risk behavior in developing countries. *AIDS*, 9(Suppl. A), S95–101.

DiClemente, R. J. and Wingood, G. M. (2000). Expanding the scope of HIV prevention for adolescents: beyond individual-level interventions. *J. Adolesc. Health*, 26, 377–8.

Farquhar, C., Kiarie, J. N., Richardson, B. A. *et al.* (2004). Antenatal couple counseling increases uptake of interventions to prevent HIV-1 transmission. *J. Acquir. Immune Defic. Syndr.*, 37, 1620–6.

FHI (2004a). Integrating family planning into VCT services: the feasibility of integration is demonstrated in Africa and the Caribbean. *Network*, 23, 12. Available at http://www.fhi.org/en/RH/Pubs/Network/v23_3/index.htm (accessed 03-25-2008).

FHI (2004b) place of work HIV/AID programs: an action guide for managers. Available at http://www.fhi.org/en/HIVAIDS/pub/guide/Workplace_HIV_program_guide.htm (accessed 03-28-2008).

FHI (2006). Voluntary HIV counseling and testing services for youth and linkages with other reproductive health services in Tanzania. Youth Research Working Paper No. 5, YouthNet. Research Triangle Park, NC: Family Health International.

FHI (2007a). Family Health Research 1(1), FHI & USAID, North Carolina. Available at http://www.fhi.org/en/RH/Pubs/fhr/v1_1/index.htm (accessed 03-30-2008).

FHI (2007b). Voluntary HIV counseling and testing services for youth and linkages with other reproductive health services in Haiti. Youth Research Working Paper No. 6, YouthNet. Research Triangle Park, NC: Family Health International.

FHI (2008). GHAIN ProfileFamily Arlington, VA: Health International. Available at http://www.fhi.org/en/HIVAIDS/country/Nigeria/res_ghainprofile (accessed 04-05-2008).

Finger, W. (2002). HIV: Voluntary counseling and testing, YouthLens No 3. YouthNet Family Research Triangle Park, NC: Family Health International. YouthLens No 3. YouthNet.

Firth, L. (2007). HIV self-testing: a time to revise current policy. *Lancet*, 369, 243–5.

Forsythe, S., Arthur, G., Ngatia, G. *et al.* (2002). Assessing the cost and willingness to pay for voluntary HIV counselling and testing in Kenya. *Health Policy Plan.*, 17, 187–95.

Grinstead, O. A., Gregorich, S. E., Choi, K. H. and Coates, T. (2001). Positive and neg-ative life events after counselling and testing: the Voluntary HIV-1 Counselling and Testing Efficacy Study. *AIDS*, 15, 1045–52.

Higgins, D. L., Galavotti, C., O'Reilly, K. R. *et al.* (1991). Evidence for the effects of HIV antibody counseling and testing on risk behaviors. *J. Am. Med. Assoc.*, 266, 2419–29.

Holtgrave, D. (2007). Costs and consequence of the US Centers for Disease Control and Prevention's Recommendations for opt-out HIV testing. *PLoS Med.*, 4(6).

Homsy, J., King, R., Malamba, S. S. *et al.* (2007). The need for partner consent is a main reason for opting out of routine HIV testing for prevention of mother-to-child trans-mission in a rural Ugandan Hospital. *J. Acquir. Immune Defic. Syndr.*, 44, 366–9.

Horizons (2001). HIV Voluntary Counseling and Testing among Youth: Results from an Exploratory Study in Nairobi, Kenya, and Kampala and Masaka, Uganda. Horizons Final Report. Washington, DC: Population Council.

Human Rights Watch (HWR) (2006). AIDS Conference: Drive for HIV testing must respect rights. Available at http://hrw.org/english/docs/2006/08/09/canada13944.htm (accessed 03-29-2008).

IOM (1995). HIV and the Blood Supply: An Analysis of Crisis Decisionmaking. L. B. Leveton, H. C. Sox Jr. and M. A. Stoto (eds), Washington, DC: Institute of Medicine. National Academy Press.

Kachroo, S. (2006). Promoting self-testing for HIV in developing countries: potential benefits and pitfalls. *Bull. WHO*, 84, 999–1000.

Kamenga, M., Ryder, R. W., Jingu, M. *et al.* (1991). Evidence of marked sexual behavior change associated with low HIV-1 seroconversionin 149 married couples with discordant HIV-1 serostatus: experience at an HIV counseling center in Zaire. *AIDS*, 5, 61–7.

Kominami, M., Kawata, K., Ali, M. *et al.* (2007). Factors determining prenatal HIV testing for prevention of mother to child transmission in Dar Es Salaam, Tanzania. *Pediatr. Intl.*, 49, 286.

Lackritz, E. M., Satten, G. A., Aberle-Grasse, J. *et al.* (1995). Estimated risk of transmission of the human immunodeficiency virus by screened blood in the United States. *N. Engl. J. Med.*, 333, 1721–5.

Likwelile, O. (2004). Challenges in Voluntary Counseling and Testing for Youth: The Case of Karago Refugee Camp, Kibondo, Tanzania. Paper presented at the *Youth and Health: Generation on the Edge, 31st Annual Conference*. Washington, DC: Global Health Council.

Luginaah, I. N., Yiridoe, E. K. and Taabazuing, M. M. (2005). From mandatory to voluntary testing: balancing human rights, religious and cultural values, and HIV/AIDS prevention in Ghana. *Social Sci. Med.*, 61, 1689–700.

Lurie, P., Avins, A. L., Philips, K. A. *et al.* (1994). The cost-effectiveness of voluntary counseling and testing of hospital inpatients for HIV infection. *J. Am. Med. Assoc.*, 272, 1832–8.

MacPhail, C. L., Pettifor, A., Coates, T. and Rees, H. (2008). "You must do the test to know your status": attitudes to HIV voluntary counseling and testing for adolescents among South African youth and parents. *Health Educ. Behav.*, 35, 87–104.

Maman, S., Mbwambo, J. K., Hogan, N. M. *et al.* (2003). High rates and positive outcomes of HIV-serostatus disclosure to sexual partners: reasons for cautious optimism from a voluntary counseling and testing clinic in Dar es Salaam, Tanzania. *AIDS Behav.*, 7, 373–82.

Marum, E., Taegtmeyer, M. and Chebet, K. (2006). Scale-up of voluntary HIV counseling and testing in Kenya. *J. Am. Med. Assoc.*, 296, 859–62.

Matovu, J. K., Gray, R. H., Makumbi, F. *et al.* (2005). Voluntary HIV counseling and testing acceptance, sexual risk behavior and HIV incidence in Rakai, Uganda. *AIDS*, 19, 503–11.

McCauley, A. (2004a). Equitable Access to HIV Counseling and Testing for Youth in Developing Countries: A Review of Current Practice. Horizons Report. Washington, DC: Population Council.

McCauley, A. (2004b). Attracting Youth to Voluntary Counseling and Testing Services in Uganda: Horizons Research Summary. Horizons Report. Washington, DC: Population Council.

Measures DHS (2003). Zambia Demographic and Health Survey, 2001–2002: Lusaka, Zambia Calverton, MD: Central Statistical Office, Central Board of Health.

Medley, A., Garcia-Moreno, C., McGill, S. and Maman, S. (2004). Rates, barriers and outcome of HIV serostatus disclosure among women in developing countries: implications for prevention of mother to child transmission programmes. *Bull. WHO*, 82, 299–307.

Morin, S. F., Khumalo-Sakutukwa, G., Charlebois, E. D. *et al.* (2006). Removing barriers to knowing HIV status: same-day mobile HIV testing in Zimbabwe. *J. AIDS*, 41, 218–24.

Okonkwo, K. C., Reich, K., Alabi, A. I. *et al.* (2007). An evaluation of awareness: attitudes and beliefs of pregnant Nigerian women toward voluntary counseling and testing for HIV. *AIDS Patient Care STDs*, 21, 252–60.

PEPFAR (2007). Malawi Ministry of Health holds second annual HIV counseling and testing week (September 2007). Available at http://www.pepfar.gov/press/92651.htm (accessed 03-27-2008).

Pequegnat, W. and Szapocznik, J. (eds). *Working with Families in the Era of HIV/AIDS.* Thousand Oaks, CA: Sage Publications.

Pequegnat, W., Bauman, L., Bray, J. *et al.* (2001). Measurement of the role of families in prevention and adaptation to HIV/AIDS. *AIDS Behav.*, 5, 1–19.

Perrino, T., Gonzalez-Soldevilla, A., Pantin, H. and Szapocznik, J. (2000). The role of families in adolescent HIV prevention: a review. *Clin. Child Fam. Psychol. Rev.*, 3, 81–96.

Powers, K. A., Miller, W. C., Pilcher, C. D. *et al.* (2007). Improved detection of acute HIV-1 infection in sub-Saharan Africa: development of a risk score algorithm. *AIDS*, 21, 2237–42.

Rogers, C. R. (1951). Client-centered Therapy; its Current Practice, Implications, and Theory. Boston, MA: Houghton-Mifflin.

Rogers, C. R. (1959). The theory of therapy, personality, and interpersonal relationships as developed in the client-centered framework. In: S. Koch (ed.), *Psychology: The Study of Science,* Vol. 3, *Formulations of the Person and the Social Contexts.* New York, NY: McGraw-Hill, pp. 184–256.

Rogers, C. R. (1964). Towards a modern approach to values: the valuing process in the mature person. *J. Abnormal Social Psychol.*, 68, 160–7.

Rotheram-Borus, M. J., Flannery, D., Lester, P. and Rice, E. (2004). Prevention for HIV-positive families. *J. Acquir. Immune Defic. Syndr.*, 37(Suppl. 2), S133–4.

Rotheram-Borus, M. J., Flannery, D., Rice, E. and Lester, P. (2005). Families living with HIV. *AIDS Care*, 17, 978–87.

Shilts, R. (1987). And the Band Played On: Politics, People, and the AIDS Epidemic New York, NY: St. Martin's Press, p. 521.

Spielberg, F., Levine, R. O. and Weaver, M. (2003). Home self-testing for HIV: directions for action research in developing countries. Available at http://www.synergyaids.com/SynergyPublications/AdvancesThroughHomeSelfTest3Oct03.pdf (accessed 03-30-2008).

Sweat, M. D. and Denison, J. A. (1995). Reducing HIV incidence in developing countries with structural and environmental interventions. *AIDS*, 9(Suppl. A), S251–7.

Sweat, M., Gregorich, S., Sangiwa, G. *et al.* (2000). Cost-effectiveness of voluntary HIV-1 counseling and testing in reducing sexual transmission of HIV-1 in Kenya and Tanzania. *Lancet*, 356, 113–21.

Tawil, O., Verster, A. and O'Reilly, K. R. (1995). Enabling approaches for HIV/AIDS prevention: can we modify the environment and minimize the risk? *AIDS*, 9, 1299–306.

Thielman, N. M., Chu, H. Y., Ostermann, J. *et al.* (2006). Cost-effectiveness of free HIV voluntary counseling and testing through a community-based AIDS service organization in Northern Tanzania. *Am. J. Public Health*, 96, 114–19.

Tinsley, B. J., Lees, N. B. and Sumartojo, E. (2004). Child and adolescent HIV risk: familial and cultural perspectives. *J. Fam. Psychol.*, 18, 208–24.

UNAIDS (1997). UNAIDS Policy on HIV Testing and Counselling Geneva: Joint United Nations Programme on HIV/AIDS.

UNAIDS (2000a). Voluntary counselling and testing (VCT). *UNAIDS Best Practice Collection, Technical Update*. 1–12. Geneva: Joint United Nations Programme on HIV/AIDS.

UNAIDS (2000b). Geneva: *Tools for Evaluating HIV Voluntary Counseling and Testing* Joint United Nations Programme on HIV/AIDS. Available at http://www.cpc. unc.edu/measure/publications/unaids-00.17e/tools/unaidsvct.pdf (accessed 04-05-2008).

UNAIDS (2006). 2006 Report on the Global AIDS Epidemic. Geneva: Joint United Nations Programme on HIV/AIDS.

UNAIDS Reference Group on HIV and Human Rights (2007). Statement and recommendations on scaling up HIV testing and counselling. Available at http://www.unaids.org/ en/PolicyAndPractice/HumanRights/20070601_reference_group_HIV_human_rights. asp (accessed 03-30-2008).

Uneke, C. J., Alo, M. and Ogbu, O. (2007). Mandatory pre-marital HIV testing in Nigeria: the public health and social implications. *AIDS Care*, 19, 116–21.

United Nations (UN) (2001). Resolution adopted by the General Assembly: S-26/2 declaration of commitment on HIV/AIDS. Twenty-sixth special session, A/RES/S-26-2.

Varghese, B. and Peterman, T. A. (2001). Cost-effectiveness of HIV counseling and testing in US prisons. *J Urban Health*, 78, 304–12.

VCT Efficacy Study Group (2000). Efficacy of voluntary HIV-1 counselling and testing in individuals and couples in Kenya, Tanzania, and Trinidad: a randomised trial. *Lancet*, 356, 103–12.

Wawar, M. J., Gray, R. H., Sewankambo, N. K. *et al.* (2005). Rates of HIV-1 transmission per coital act, by stage of HIV-1 infection, in Rakai, Uganda. *J. Infect. Dis.*, 191, 1403–9.

Weinhardt, L. S., Carey, M. P., Johnson, B. T. and Bickham, N. L. (1999). Effects of HIV counseling and testing on sexual risk behavior: a meta-analytic review of published research, 1985–1997. *Am. J. Public Health*, 89, 1397–405.

Were, W. A., Mermin, J. H., Wamai, N. *et al.* (2006). Undiagnosed HIV infection and couple HIV discordance among household members of HIV-infected people receiving antiretroviral therapy in Uganda. *J. Acquir. Immune Defic. Syndr.*, 43, 91–5.

West, G. R., Corneli, A. L., Best, K. *et al.* (2007). Focusing HIV prevention on those most likely to transmit the virus. *AIDS Educ. Prev.*, 19, 275–88.

WHO (1990). Guidelines for Counselling about HIV infection and Disease. Geneva: World Health Organization.

WHO (1993). Statement from the Consultation on Testing and Counselling for HIV Infection: Geneva, 16–18 November 1992. Geneva: World Health Organization.

WHO (2005a). Child and adolescent health and development. Available at http://who.int/ child-adolescent-health/hiv.htm (accessed 02-28-2007).

WHO (2005b). *Guidelines for Assuring the Accuracy and Reliability of HIV Rapid Testing: Applying a Quality System Approach*. Geneva: World Health Organization.

WHO (2006). *Addressing Violence against Women in HIV Testing and Counselling: A Meeting Report*. Geneva: World Health Organization.

WHO & UNAIDS (2007). *Guidance on Provider-initiated HIV Testing and Counseling in Health Facilities*. Geneva: WHO & UNAIDS.

WHO & UNICEF (2007). Guidance on global scale-up of the prevention of mother to child transmission of HIV: towards universal access for women, infants and young children and eliminating HIV and AIDS among children. WHO & UNICEF with the

Interagency Task Team (IATT) on Prevention of HIV Infection in Pregnant Women, Mothers and their Children. Available at http://www.who.int/hiv/mtct/PMTCT_enWEBNov26.pdf (accessed 03-29-2008).

WHO, UNAIDS & UNICIF (2007). *Towards Universal Access: Scaling Up Priority HIV/ AIDS Interventions in the Health Sector: Progress Report, April 2007*. Geneva: World Health Organization.

WHO, UNAIDS & UNICIF (2008). *Towards Universal Access: Scaling Up Priority HIV/ AIDS Interventions in the Health Sector: Progress Report, April 2008*. Geneva: World Health Organization.

Wolitski, R. J., MacGowan, R. J., Higgins, D. L. and Jorgensen, C. M. (1997). The effects of HIV counseling and testing on risk-related practices and help-seeking behavior. *AIDS Educ. Prev.*, 9(3 Suppl.), 52–67.

Structural interventions in societal contexts

20

Suniti Solomon and Kartik K. Venkatesh

HIV interventions that fundamentally alter the social context within which risk-taking behaviors occur are necessary for long-term, sustainable HIV prevention. There is certainly a need for structural interventions that address the needs of historically marginalized populations, which are often at great risk for HIV infection. Many current social-cognitive and cognitive-behavioral models that address HIV-related stigma may not be feasible in resource-limited settings, or operate using a highly individualized framework, which may be alien in more collective societies, such as in many parts of Africa, Asia and Latin America (Parker and Aggleton, 2003). Though targeted HIV prevention programs are underway across the developing world, most of these efforts have historically been targeted at behavior-change communication and condom distribution. A potential shortcoming of these modalities is that they can fail to address environmental factors that can organically lead to community-based social changes. Additionally, these targeted behavioral strategies can be inadequate because individuals invariably return to larger social settings where risk-taking practices are a normative influence.

Venue-based approaches: wine shops to red-light districts

Due to the sensitivity of openly discussing sex in public and the social stigma associated with HIV in many traditional societies, understanding the extent to which sexuality and sexual health are discussed in community venues can inform the design of HIV prevention interventions. Community-based health promotion relies on the communication of messages that help inform and empower individuals to take control of their own health. Communication networks consist of individuals linked to each other by the information they exchange. Utilizing peers can be an effective strategy to access a community to bring about changes in social norms and behavior. Community mobilization, advocacy and social change can serve as a localized intervention strategy to transform the ways in which individuals and communities respond to HIV and AIDS (Parker and Aggleton, 2003). In the context of HIV prevention programs, the communication of interest is sexual health, which can take the form of talking about safe sex, promoting behavior change and encouraging testing for HIV.

When attempting to achieve societal-level change, a venue-based approach to addressing communities is an efficacious model when the audience of interest can be clearly identified and reached. However, a major difficulty with this prevention approach is identifying an appropriate venue. In Thailand, the government established an innovative 100 percent condom campaign in 1991 to enforce universal condom use in all commercial sex establishments. As part of this national program, sex workers were given condoms and health officials would routinely visit these establishments to check compliance with the campaign. By 1994, it was estimated that 90 percent of Thai sex workers used condoms for protection, and the number of men presenting to government clinics for STD treatment dropped by 90 percent between 1989 and 1995 (Hanenberg and Rojanapithayakorn, 1996).

Sex is a taboo subject for open, candid discussion within many societies. Negotiating safe sex can be hampered by multiple risk-taking behaviors, such as drug use and alcohol. An emerging area of research interest is the synergistic relationship between alcohol use and HIV transmission. Studies from diverse regional settings have documented the connection between high-risk sexual behavior and alcohol use (Madhivanan *et al.*, 2005; Coldiron *et al.*, 2007). In South Africa, risk-reduction counseling about alcohol use among clients of STD clinics led to increased condom use and reductions in unprotected intercourse, suggesting the value of long-term structural interventions targeting alcohol use in sexual contexts (Kalichman *et al.*, 2007a). More recently in India there have been increasing calls for prevention efforts to focus on the risk-taking behaviors of heterosexual men, who are a very heterogeneous and mobile population. A study in Chennai identified wine shops as appropriate HIV prevention venues for heterosexual men due to the reported high rates of hazardous alcohol use in India, preliminary research demonstrating that wine-shop clients engaged in significantly higher sexual risk behaviors compared to the community sample, and the association between alcohol and transmission of HIV/STDs and sex with female sex workers (Sivaram *et al.*, 2007). Wine shops, which are commercially licensed establishments, provide a unique venue to target an important yet difficult to access risk-group of men, particularly male clients of sex workers. Additionally, such venues are ideally suited for prevention, due to the convenience and feasibility of the wine-shop premises for housing an intervention. The fact that participants confirmed the role of friends in obtaining information about sex and as personal confidants confirmed the value of the community popular opinion leader (CPOL) model in this population (Sivaram *et al.*, 2004). These wine shops are primarily patronized by men, who come with their friends to drink and socialize. Women rarely drink there; however, high-risk women frequently meet men outside the wine shops. Formative research allowed for understanding the prevention needs and high-risk practices of patrons of wine shops that could then be integrated into a theory-driven culturally relevant intervention.

Despite the increasing realization for the need of wider social interventions in many parts of the developing world, most HIV prevention models continue to be driven towards individual behavior change, rather than also including community-level factors in HIV transmission as well as power relations within society (Parker and Aggleton, 2003; Castro and Farmer, 2005). The Sonagachi Project, based in Calcutta, India, targeted female sex workers within a geographically circumscribed red-light district (Jana *et al.*, 2004). This model employed a multi-level, multifaceted approach towards HIV prevention, incorporating community (having a high status advocate, addressing environmental barriers and resources), group (changing social relationships) and individual factors (improving skills and competencies related to HIV prevention and treatment). Interestingly, the model that developed over time was not initially theory driven, but rather grew out of what began as an STD clinic targeting female sex workers as a program to improve their occupational safety and occupational health. It became clear over time that the marital and familial aspirations of these women were the same as in mainstream society. This program went beyond creating an environment in which these women had easy access to condoms, STD treatment and HIV testing, but also addressed larger structural issues related to the women's marginalization, including illiteracy, inadequacy of loan programs and no formal unionization. The question of stigma associated with the community of sex workers was initially addressed by the credibility of the professional staff, who lent status to the issue of the health of sex workers. The program became sustainable at a group level by building relationships between those in the target population, including between sex workers and stakeholders and between the initial change agents (professional staff) and the sex workers. Over time, organizing as an occupational employment group became an empowering experience for the women and reframed many of their HIV-related economic problems from being due to their own perceived social deficiencies to their being disenfranchised workers.

Social-network based approaches: prevention targeted at negotiating safe sex

Many prevention programs target individuals; however, in many traditional societies where the overwhelming majority of adults are married and HIV transmission occurs primarily through heterosexual intercourse, couple-based counseling can also be an important preventive strategy. Though the risk of HIV transmission may decrease in steady partnerships after early sexual contacts, high-plasma RNA viremia and STDs may place the HIV-uninfected partner at continued high risk for infection. In traditionally pronatal societies, couples may face a diverse set of deterrents to HIV prevention – primarily, the pressure to bear children. Couple-based HIV prevention models can improve open communication between genders, which can reduce the practice of blaming women for infections, infertility, and other sexual and family issues, and encourage both partners to be involved with making decisions (Solomon *et al.*, 2003). Through the facilitation of a trained counselor, the counseling session can promote open discussions between husband and wife about difficult and stigmatized sexual topics, including discussing personal sexual behavior and connecting personal risk to HIV. Due to the centrality of family within many cultures, emphasizing the potential impact of HIV on the family can be an effective means for men and women to understand their vulnerabilities to HIV, and reinforce the responsibility of the couple to openly discuss their sexual practices to preserve the family. A family-based approach can avoid blaming the husband or the wife, and instead reinforces the collective economic and social consequences of HIV infection for parents and children.

HIV-discordant couples are an important population to target for HIV prevention; however partner notification can be hampered due to one's denial of HIV status and lack of disclosure due to HIV-related stigma. A study in Uganda among serodiscordant couples identified various misconceptions about discordance, such as the concept of a hidden infection not detectable, belief in protection by God, and belief in immunity (Bunnell *et al.*, 2005). These explanations for discordance can reinforce denial of HIV risk for the negative partner, and hence potentially increase the chance of transmission. In another study in Uganda, home-based voluntary counseling and testing of household members of individuals initiating ART was socially accepted, and led to the detection of a large number of previously undiagnosed HIV infections and serodiscordant relationships (Were *et al.*, 2006). Further studies are needed to understand how couples deal with HIV infection, in order to optimize the potential of greater voluntary HIV counseling and testing of couples.

Since sex and sexuality are often considered topics too sensitive for public discussion and policy implementation in traditional societies, understanding the extent to which sex and sexuality are currently discussed by couples and broader social networks can inform the design of HIV prevention interventions. Understanding how, when and with whom communication about sex occurs, and the information that is exchanged, can allow for the development of effective, culturally acceptable preventive measures. A study in the slums of Chennai identified four key social networks, which consisted of married men, married women, unmarried men and unmarried women. Within each of these groups, key opinion leaders who were local members of the community served as credible sources regarding information about HIV/AIDS prevention, STDs and sex, for their peers

(Sivaram *et al.*, 2005); however the content of the communication between these groups differed significantly. Unmarried men discussed a range of topics regarding sex and sexual health, and practiced behaviors that placed themselves at increased risk for HIV, which made this group a particularly efficacious group on which to target prevention interventions.

In order to effectively utilize opinion leaders within social groups, interventions can design methods to train these leaders in HIV prevention messages that reflect scientific facts and dispel myths about transmission. These messages can then be delivered within appropriate social contexts accounting for the cultural nuances in which discussions about sex and sexuality occur. Additionally, inter-network diffusion of ideas may be possible through interventions that facilitate discussions and questioning between opinion leaders of different networks.

Daughters, wives, and mothers: the impact of subordinating women in the home and community

Cascade of dilemmas faced by a young Indian woman:

Her parents arrange for her marriage $\rightarrow$ She faithfully performs her duties in her in-laws home $\rightarrow$ Her husband does not want to have a child $\rightarrow$ She silently obeys everyone $\rightarrow$ Her longing for a child is unknown $\rightarrow$ Her parents and in-laws insist on seeing their grandchild $\rightarrow$ Society frowns upon her and calls her "barren" $\rightarrow$ She turns to her husband in despair $\rightarrow$ She is pregnant and all are happy $\rightarrow$ She tests HIV-positive

The social construct of gender, a product of larger religious, colonial and nationalist discourses, can make women highly vulnerable to HIV infection within traditionally patriarchal societies. Gender-based imbalances and cultural restrictions with the outcome of curtailing women's independence can also have the outcome of offering sexual freedom to men. Though not encouraged, it is socially permissible for men in many societies to have pre-marital and extra-marital sexual encounters. Simultaneously, the intense familial pressure for women to maintain their virginity until marriage can be great. In patriarchal social systems, marriage often serves as the institution that gives a woman permission to participate in sexual activity and gain a culturally sanctioned identity as a married woman within her husband's household.

In many parts of the developing world, public health officials throughout the early years of the HIV epidemic held that only female sex workers, homosexuals and intravenous drug users (IVUs) could contract HIV, largely ignoring the risk of HIV infection to married women. In the ensuing decades, the prevalence of HIV among sex workers has stabilized due to targeted interventions, empowerment strategies and increased condom use. However, this simplistic and short-sighted mindset of who was at risk for HIV further stigmatized already socially marginalized populations, and failed to account for the fact that women could be placed at increased risk for HIV infection due to the risk-taking behaviors of their partners.

Marriage, monogamy, motherhood and widowhood

Married women with lifelong single partners have increasingly become a larger proportion of the HIV-infected population. In India, these married monogamous women are primarily put at risk for HIV infection from their husbands, from whom it is most likely they acquired the infection (Gangakhedkar *et al.*, 1997; Newmann *et al.*, 2000). Women are often detected with HIV during their first pregnancy, and the majority within the first 2 years of marriage. Many women may not perceive themselves to be at risk for HIV due to a strong belief in marriage as an institution and their husband as the maintainer of familial well-being. The married monogamous housewife, the traditional symbol of Indian womanhood, has become the new face of the AIDS epidemic in India.

Within a paternalistic social order, different values are often placed on the births of male and female children. In India, parents celebrate the birth of a son as the future family heir, and begin to worry about how they will pay the dowry with the birth of a daughter. As a woman matures, she is expected to maintain a culture of silence when confronted with questions of her own sexuality; worse, if she openly discusses her sexuality, she can be considered promiscuous. Hence, the ability of a woman to access information about her sexual health can be hampered by fear of damaging her reputation.

Due to the tremendous social and familial pressure to marry, young women often agree to early "timely" marriages to avoid the stigma of being an unmarried grown woman. Men may be aware that the risk behavior they engage in places them at increased risk for HIV infection and even of their own HIV status, but may also fear the implications of remaining single and unmarried. A common form of documented violence against women in Africa and South Asia is that when an HIV-infected husband dies, the family of the diseased husband may try to grab the material belongings and children from the surviving wife (Mendenhall *et al.*, 2007). Though often illegal by law, this form of property grabbing can cause women to remarry with their husband's brother, and thus transmit the virus. When the husband dies the widow is subject to the will of whoever is the head of the household, because power and authority often reside with the male members of the household (Luginaah *et al.*, 2005). Further studies are needed to assess various coping mechanisms facing windows directly or indirectly affected by the HIV/AIDS epidemic.

Prevention programs must examine the impact of placing responsibility and blame for sexual activity, marriage and reproduction solely on the shoulders of women, while relieving men of the responsibility of sexual risk behavior and violence (Solomon *et al.*, 2003). Given the high risk of HIV transmission to women from their spouses in many parts of Africa and Asia, strategies that inform uneducated women who work at home about STDs and HIV may decrease rates of HIV transmission. The increasing feminization of the AIDS epidemic across the developing world necessitates increased prevention studies involving microbicides and other female-controlled HIV prevention strategies. However, consistent use of a microbicide would still require that women have sexual power, including the ability to influence the timing and occurrence of sex, and to communicate about microbicide use with their husbands or use it clandestinely.

Addressing cultural vulnerability to HIV infection requires that prevention efforts confront entrenched gender norms, which is neither simple nor quick but rather requires a long-term, multifaceted community-based approach. Men and women may hesitate to adopt behavior changes that may reduce their risk of HIV infection due to real or perceived threats to their culturally sanctioned roles and relationships. While Indian men have claimed that they could have prevented their infection had they been aware of the infection and transmission risk factors, married and widowed women have felt that they had no way to avoid infection and felt dependent on their doctors and family members to provide them with access to information (Tarakeshwar *et al.*, 2006). For men, admitting ignorance about sexual matters can be an affront to their masculinity. Conceptions of masculinity that promote promiscuity and even sexual violence can cause men to experiment with unsafe sexual practices. A lack of information about sex and sexual health can lead men to believe various myths associated with HIV transmission and prevention behaviors.

Socially-driven HIV interventions that alter the status of women

Prevention efforts aimed at decreasing the risk of HIV infection to women must examine the role of women within a wider set of social interactions, involving their spouses, families and communities. Women need sexuality education long before they get married, and hence the need for school-based sexuality education is necessary in the formative years of both boys and girls. HIV prevention strategies aimed at women can be difficult in societies where a large proportion of women may be illiterate, and social norms dictate that women cannot make decisions for themselves. A poor understanding of reproductive health may further impair the ability of women to make decisions for themselves. Within the context of delivering clinical care, gaining the informed consent of a client can be a major barrier to HIV testing. Tradition may dictate that such questions as deciding on medical testing be decided by another family member, such as the husband or in-laws.

While the biological vulnerability of women to HIV infection has been examined at depth, the socio-cultural and structural factors that place women at increased risk for HIV infection have often been overlooked. HIV prevention messages have consistently focused on two key messages: practice mutual monogamy, and use condoms. These messages assume that individuals can modify their behaviors. Community gender norms can sanction domestic violence against women, and can interfere with the ability of women independently to adopt HIV-preventive behaviors (Go *et al.*, 2003). Given the choice between the immediate threat of domestic violence from their spouses and the hypothetical threat of HIV infection, many women may resign themselves to the sexual demands and indiscretions of their spouses, even though this places them at increased risk for HIV. Studies from South Africa have documented that women are placed at great risk for STIs and HIV infection due to sexual assault at the hands of abusive men (Dunkle *et al.*, 2004; Kalichman *et al.*, 2007b). Men have been unlikely to change sexually risky behavior and sexually violent actions against women when the prevailing social norms foster these very attitudes. Interventions aimed at reducing the risk of HIV transmission in these situations must address hostile attitudes towards women as well as the risk of sexual assault of women and the sexual risk behavior of men. To further understand the risk of HIV transmission posed to women, studies are needed that examine wider societal constructions of masculinity, male dominance within relationships, and the dynamics of partner violence.

Educating married women about the risk of HIV infection from their husbands will require providing greater educational and financial opportunities that can empower women to discuss openly their husbands' risk-taking behaviors in order to protect themselves and their families from HIV infection. An interventional model aimed at improving the economic status of women has been to provide microcredit to women. The goal has been to provide these women with the potential to expand their own autonomy and hence be able to independently question repressive gender norms. South African women provided with loans via a microfinance program showed decreased rates of intimate partner violence, which enabled women to challenge the acceptability of violence, receive better treatment from their partners, leave abusive relationships, and raise public awareness about intimate partner violence (Kim *et al.*, 2007). Another intervention in South Africa assessed the impact of a microfinance-based strategy to prevent HIV infection and intimate-partner violence. This intervention confronted structural issues, including household economic well-being, social capital, and empowerment, demonstrating a significant decrease in intimate-partner violence perpetrated against women over time (Pronyk *et al.*, 2006). The findings of these studies suggest that increasing women's economic participation can have the impact of reducing gender inequity and violence. Programs that attempt to change the structural context in which risk-taking behavior takes place have the potential to change risk environments for the acquisition of HIV.

Breastfeeding and culture

The majority of mother-to-child HIV infection occurs in developing countries that have been unable to implement interventions already in place in the industrialized world. Perenatal transmission accounts for over 90 percent of infections among HIV-infected children. In developing countries, popular literature has for decades waxed eloquent about "breast is best". The portrayal of the purity of breast-milk contrasts sharply with the reality that it can be infectious. For many women, not to breastfeed raises major cultural questions that women need to answer not only for themselves, but also for the whole family, and even the community. When an HIV-infected mother decides to not breastfeed her child, she potentially jeopardizes the confidentiality of her HIV status. Though increasingly as part of antenatal care, HIV-infected mothers do receive counseling about the risks posed by breastfeeding, the decision is ultimately left to the mothers whether to use formula milk.

Reduction of HIV during lactation is one of the most pressing public health dilemmas confronting researchers, policy-makers and HIV-infected women. A great challenge in many resource-limited settings is the lack of an adequate maternal health infrastructure, illustrated by unacceptably high rates of maternal mortality. Major barriers to utilizing formula feed over breast-milk can be the availability of safe water, as well as unhygienic circumstances that can be associated with increased morbidity and mortality from infectious diseases and malnutrition (De Cock *et al.*, 2000). Beyond the economic implications, many women may be unaware of mother-to-child transmission (MTCT), and report that their partner may express negative attitudes if they were not to breastfeed, such as thinking that the mother is harming the baby, that she is not a good mother, that she has HIV, and that she has been unfaithful (Rogers *et al.*, 2006). Ongoing counseling of women and their family members by health-care staff can allay the concerns about formula-feeding as an unnatural practice. Though women may view the use of nevirapine as a useful and effective means to fulfill their social responsibility of bearing a child free from HIV infection, greater interventions must be developed around counseling husbands and family members. These HIV prevention efforts must be fully integrated within the fabric of current maternal, child-care and family welfare programs that are already in place.

The time period immediately after the birth of the child can serve as a crucial point to discuss and re-counsel the mother about infant feeding practices and ultimately provide support to the mother's infant-feeding decision. Women who test HIV-negative can be counseled about ways to decrease the future risk of infection. For women who test HIV-positive, a policy of offering short-term peripartum antiretroviral therapy needs to be implemented. As part of the counseling session, HIV-infected women must be offered clear instructions on the alternatives to breastfeeding, including the hygienic preparation of infant food. In resource-limited settings, it has been shown that the risk of HIV transmission/death of babies given formula-feed is linked to not having electricity and running water, and disclosing HIV status (Doherty *et al.*, 2007). Hence counseling women about different feeding options for their infants should include appropriate consideration of individual and environmental factors in order to select the most appropriate infant-feeding choice.

As the majority of women may deliver their babies at home and not access a health-care center for follow-up infant care, innovative home-based programs need to be developed that potentially utilize family-planning clinic workers and NGO outreach staff. A number of trials have been underway over the past few years in various sites across Africa to assess different strategies to decrease HIV transmission during the breastfeeding period, including prophylactic antiretroviral therapy, passive and active immune-based approaches, exclusive breastfeeding followed by early weaning, and micronutrient supplementation (Fowler and Newell, 2002). Part of developing these interventions is understanding local breastfeeding practices, such as the common practice of mixed feeding, where breast-milk may be combined with other liquids. Further research is needed to test interventions that can decrease HIV transmission during the period of lactation for those HIV-infected women who decide to breastfeed.

Socially-driven public health responses

Doctors refuse to deliver baby of HIV positive woman – husband forced to deliver the child

THE HINDU (South Indian Daily English Newspaper), 30 JUNE 2007
Meerut, Uttar Pradesh: In a shocking incident, a man was forced to deliver the baby of his HIV positive wife in a government hospital here after doctors allegedly refused to help because of the woman's health condition. Sunita alleged that her painter husband, Raees Abbas, was given instructions by the doctors at the Meerut Medical College and Hospital on how to deliver the baby as they watched. "My husband cut the umbilical cord," she said. The incident which sparked an outrage among the medical fraternity is being investigated by the Uttar Pradesh government which said that according to a preliminary probe the Head of the Gynaecology Department and the doctor on call were guilty of negligence. It is however yet to be conclusively established that treatment was denied because the patient was HIV positive. The incident came to light when Mr Abbas complained to the hospital. Medical Education Minister Lalji Verma said action would be taken if any doctor was found guilty of negligence. Mr Abbas said he and his wife came to the hospital after the woman went into labour. The woman, who gave birth to a baby boy, alleged that the doctors refused to take her to the delivery room and instead her husband had to deliver their baby. "We faced a lot of problems. The doctors were not ready to help. My husband delivered the baby and cut the umbilical cord." The Head of the Gynaecology Department, Abhilasha Gupta, said she refused to "blindly believe" what is being said against the doctors. Mr Abbas said that after the delivery, no doctor asked about his wife's health.

Stigma in the medical establishment

Health-care workers play a central role in the prevention of HIV. Prevention programs that are closely linked to a larger public health system are more likely to be sustainable and effective in the long term. Despite various theoretical models of how stigma operates in the social science literature, our current understanding of how stigma may be impacting the quality of health-care delivered to HIV-infected individuals remains very limited. Currently in most resource-limited settings there is an alarming lack of adequately trained health-care personnel to offer HIV-related health services. For instance, in many large Indian medical institutions, pregnant women and surgical patients routinely undergo HIV antibody testing. Patients found to be HIV-positive can be refused medical care, in which cases the test results are rarely discussed with patients via a follow-up risk assessment. Few hospitals deliver care to HIV-infected individuals, which is likely due to the reluctance on the part of health-care workers owing to personal judgments and prejudice, an inaccurate perception of occupational risk, and the idea that HIV-negative patients will refuse to share health-care facilities with HIV-infected patients. HIV-infected patients who experience stigma within the health-care setting may be more likely to forgo continued clinical care as a result of these early negative interactions and may be less likely to disclose their HIV status to health-care providers, hampering their clinical care (Sayles *et al.*, 2007).

Interventions that aim to increase accurate and up-to-date information of HIV among physicians and health-care workers need to address the stigma associated with HIV within the medical establishment. Continuing medical education (CME) is a relatively new concept outside the developed world, and is often not mandatory in the few places where it exists. Additionally, the current information that physicians have about the treatment of HIV is often limited to the dated and potentially biased views of pharmaceutical representatives. Increases in public sector HIV and STI care require an increased commitment to adequately train health-care personnel in delivering these services, and greater sensitization to the needs of marginalized communities at highest risk for HIV.

Challenges to sex education and communication

In many conservative societies, cultural and religious inhibitions may hamper the wider implementation of sex education interventions. School-based programs have been shown to be an effective strategy in reducing risky sexual behavior in developing countries (Kirby *et al.*, 2006). In the case of Uganda, some researchers have argued that a sustained ABC (Abstinence, Be faithful, and Condom use)-based approach has been responsible for the decreased incidence of HIV in recent years. It is likely that ABC-related behavior change was a product of extensive social mobilization and strong political leadership, both of which integrated gender-related interventions to advance women's status as part of the ABC strategy (Murphy *et al.*, 2006). The social and political efforts of Ugandan President Museveni included messages of promoting sexual behavior change and equity between men and women, affirmative action policies that allowed women to participate in politics, enforcement of laws against sex with minors, sexuality education for youth with gender equity messages, and increased economic credit schemes for women. The success of Uganda's HIV prevention approach suggests the need for programs to address pervasive social and gender inequities that lead to the subordination of women and encourage men to take sexual risks to demonstrate their masculinity (Murphy *et al.*, 2006). The Ugandan approach of behavior change was locally tailored to prevalent attitudes and beliefs about sexual behaviors with an overall grassroots approach, without a single policy line being pushed too strongly (Parkhurst and Lush, 2004).

In many developing countries, ABC approaches have been far too narrowly conceived and do not address underlying social inequities that create fertile ground for the continued spread of HIV. In India, there has been an increasing political will for greater prevention education within schools. A National Parliamentarian Forum has been constituted to address HIV prevention programs through school-based adolescent education programs and a national campaign to raise awareness about STDs and treatment. However, India's traditional prevention campaign has been based on the ABC model (Solomon *et al.*, 2004). Often these governmental campaigns may not be successful due to often moralistic and judgmental overtones. In societies where women have limited control over sexual relationships, they may not be able to insist on abstinence. The ABC message of "being faithful" as a preventive measure fails to acknowledge the reality that, throughout many parts of Africa and Asia, married monogamous women are one of the highest-risk groups, and it can dangerously lead many women to believe they are not at risk for HIV because they are remaining faithful to their spouses. An effective health education program, particularly at the formative age of schoolchildren, can help increase awareness in the community with the aim of decreasing the stigmatization associated with HIV in society. Students need to be given non-contradictory messages and teachers need to be trained to provide adequate and objective counseling to youth who may be sexually active. Ideally, these programs would be interactive, allowing for an open candid discussion of sex and sexuality, which has usually not been politically palatable in traditionally conservative societies.

Antenatal testing and fertility pressures

In traditional societies, social and cultural norms make the childbearing role of a woman paramount, even if it potentially places the woman and her newborn child at an increased health risk of HIV infection. In pronatal societies, HIV testing has been shown to be socially acceptable within the context of antenatal care where the welfare of the unborn child is of primary importance (Shankar *et al.*, 2006). Bearing children becomes a social responsibility for women to begin soon after marriage in order to prove their fertility and gain respect within their families and communities. Women are often forced to balance preventing HIV transmission to their baby with having to face the social repercussions of disclosure to their families if they test HIV-positive. Women who test HIV-positive through antenatal testing may fear a breach of confidentiality of their HIV status by hospital staff to their families, self-stigmatization in a general ward or when giving formula-feed to their babies, and discrimination after delivery when their newborns may be isolated within pediatric wards (Mawar *et al.*, 2007). The potential for discrimination within the health-care system and the fear of disclosure of their HIV status to their families may dissuade many women from seeking HIV antenatal testing services.

Antenatal care facilities need to be equipped with voluntary counseling and testing (VCT) services so that women can know their HIV serostatus and those women who do test HIV-positive can be provided with a timely intervention. Recent interventional studies in different African settings have suggested that introducing an opt-out strategy for antenatal testing may have a far-reaching public health impact, be acceptable to pregnant mothers, and even lead to greater uptake than current VCT without detectable undesirable consequences (Perez *et al.*, 2006; Creek *et al.*, 2007). Risk-reduction counseling (RRC) has been shown to be an effective means of limiting the sexual spread of HIV among high-risk heterosexuals (Solomon *et al.*, 2006) and, with the increasing availability of VCT services in resource-limited settings, it is possible that RRC can be coupled with VCT as a culturally appropriate intervention program. Despite the fact that an opt-out approach may normalize and integrate the process of HIV testing into normal medical care, associated social factors remain with both VCT and an opt-out approach, including stigma and post-test counseling, and staffing must be considered. A cause for concern with antenatal testing is that women may be tested and find out they are HIV infected before their husbands, which can lead to women being blamed for bringing HIV into the family. Additionally, access to prenatal care varies drastically across the developing world, and access to even basic health care remains a problem for many women. Women may ignore their own health due to other competing demands that require familial resources. Even if prenatal care is available, it raises the question of whether these services are culturally sensitive to the needs and priorities of pregnant women.

In resource-limited settings, the implementation of prevention strategies has involved establishing and strengthening the maternal and child-care infrastructure. These programs require considerable resources in order to provide adequate prenatal and interpartum care, prenatal HIV counseling and testing, and antiretroviral drugs. Additionally, laboratory-capacity building is essential in order to provide rapid testing so that women can find out their result on the same day blood is drawn. Interventions that include HIV testing should be combined more broadly with other simple protocols, such as vaginal cleansing during labor and multivitamin and multinutrient supplementation. In resource-limited settings most women accessing antenatal testing have iron deficiency anemia due to malnutrition, and hence providing early and correctable treatment for anemia as part of antenatal care can be initiated with zidovudine (Sinha *et al.*, 2007). Women may not want to reveal their diagnosis to family members, and hence may need options to adhere to treatment without revealing their diagnosis. An option may be to combine antiretrovirals with vitamin supplementation in order to preserve patient confidentiality.

In order to increase the utilization of antenatal testing by women, ongoing sensitization through women-centered group education, including pre- and post-test counseling, can be a useful form of social support. The choice of an intervention should be specific to the setting and the time of first presentation for antenatal care. Creating a confidential health-care atmosphere can begin with simple protocols, such as providing a unique identification number to each woman. If a woman presents late in her pregnancy, a regimen providing medication is likely the most effective strategy. Well-trained and effective counselors can be an important link to wives and their husbands consenting to antenatal testing. Once the HIV test result is known, follow-up counseling can provide increased knowledge about the implications of the test results, such as how best to continue the pregnancy and ways to prevent women from being blamed for HIV infection. Couple-friendly antenatal testing may be an option to involve the spouses of women. Since many women deliver babies outside of health-care facilities in resource-limited settings, interventions may need to be accessible at home.

Targeting the individual in HIV prevention: condoms and antiretrovirals

Interventions at a larger societal scale, such as pre-exposure antiretroviral prophylaxis, early diagnosis and treatment of STDs, and male circumcision, may lead to population-level reductions in HIV risk, but these efforts cannot be successful without individual-level behavioral change. A mainstay of individualized HIV risk-reduction worldwide has been the promotion of consistent condom use. Despite the relative availability of condoms in many developing world settings, along with other barrier and non-barrier contraceptives, the community uptake of these contraceptives has often been limited (Serwadda *et al.*, 1995; Roth *et al.*, 2001). Reasons for low condom use may include stigma regarding sexual activity, and misinformation about HIV and STD transmission. Repeated studies have documented the inability for Indian married women to negotiate condom use with their male partners, which underscores the point that safer sex practices have not been the norm for Indian women due to a lack of perceived risk and a lack of being able to negotiate sex. However, due to targeted sustained HIV interventions for the last 15 years among high-risk groups, 80 percent consistent condom use has been observed among female sex workers in Western and Southern India.

Efforts in traditional societies need to be geared towards removing the social barriers associated with condom use. Major reasons Indians have reported not using condoms is due to the social stigma and the lack of privacy in stores (Roth *et al.*, 2001). Interventions addressing the practicalities of purchasing condoms also involve confronting broader social concerns, such as empowerment, gender equity and condom negotiation skills, as well as individual and community capacity-building. The increased and consistent use of condoms may thus require efforts that aim to ease the social costs of and constraints on safe behavior. In social contexts where marital communication about condoms, fidelity and marital sex are initiated and controlled by husbands, prevention interventions need to adequately target married men. Husbands may only use condoms when they perceive themselves to be at risk for HIV infection (Roth *et al.*, 2001). Wife-initiated discussions of using condoms can raise suspicions of her possible infidelity and disrupt normative gender roles in which women are to respond to their husband's sexual urges, and female-initiated discussions about sex are inappropriate. Using barrier methods or non-penetrative sex as safer modes of intercourse for couples can be seen as an affront to motherhood where a woman's value often resides in her number of children (Solomon *et al.*, 2003).

Though condom use has remained low in India, many women undergo sterilization during their peak reproductive years after completing their families. Since 1970 there has been an intriguing gender reversal among the ratio of men to women seeking sterilization, from 2:1 in 1970–1971 to 1:31 in 1994–1995; however during that same time period, sterilization rates increased from 8 percent to 30.2 percent of all contraceptive methods used (Gopalan and Shiva, 2000). Government programs encourage couples to be sterilized after completing their families, but in the majority of cases it is the woman who comes forward for sterilization. In couples where either the man or the women is sterilized, the motivation to use condoms may be low, which has implications for HIV prevention efforts that target married couples in societies with high rates of sterilization. Prevention efforts must portray condoms as a means of having safe sex to avoid HIV transmission, rather than solely as a means to avoid pregnancy.

Antiretroviral agents have increasingly become readily available in the developing world. Based on data collected in Africa and Thailand, it is also well known that highly active antiretroviral therapy (HAART) reduces the viral load to undetected levels and hence prevents the transmission of the virus. In resource-limited settings, HAART is generally initiated based on World Health Organization (WHO) treatment guidelines: CD4 cell count <200 cells/μl, or the presence of an AIDS-defining illness. The Health Prevention Trials Network (HPTN) 052 of the National Institutes of Health is a preventive research project evaluating antiretroviral therapy plus HIV primary care vs HIV primary care alone to prevent the sexual transmission of HIV in serodiscordant couples (National Institutes of Health, 2007). This interventional study that is currently underway will add to our understanding of how antiretroviral therapy can make HIV-infected individuals less infectious.

Methods of assessing drug adherence used in Western cultures may require adaptation for non-Western social settings. Qualitative research methods may be effective in identifying culturally specific determinants of adherence in diverse settings, and consequently lay the groundwork for greater adherence counseling and intervention strategies. A qualitative study in South India had the aim of accessing culturally specific barriers and facilitators to adherence in order to maximize the benefit of ART and reduce HIV transmission. In sero-concordant couples where only one partner has ready access to antiretroviral therapy, they shared their drugs with the other spouse. Beyond the direct cost of drugs, a major barrier to adherence was privacy, as patients were not able to purchase antiretrovirals at local pharmacies near their homes due to the fear of stigma (Safren *et al.*, 2005). The stigma associated with HIV can make disclosure of one's serostatus to family members difficult, which can lead to problems with taking pills in public, getting familial support for HIV care, and adherence support (Kumarasamy *et al.*, 2005). From this study, enlisting assistance from family members to provide adherence support can be an important aspect of an effective drug adherence strategy.

The continued process of increasing access to antiretrovirals requires a careful assessment of cultural issues, such as the patient–physician relationship, the potential stigma regarding disclosure of HIV status at work and home, and family structure. Assessing patient adherence to antiretroviral treatment can be a difficult task in resource-limited settings. In order to achieve optimal adherence, various social interventions can be used, including directly observed treatment, family counseling, and intensive patient education. It is possible that inadequate adherence could be addressed through using family members as directly administered antiretroviral therapy (DAART) interventionists, due to the cost of adherence interventions and the historic success of directly observed therapy (DOT) for tuberculosis management. A study conducted in rural Uganda demonstrated that a comprehensive home-based model of AIDS care resulted in excellent retention to care and adherence to antiretroviral therapy in a population with otherwise limited access to health-care services and transportation (Weidle *et al.*, 2006). In addition to providing free antiretrovirals with home delivery, this comprehensive program provided patients with medicine companions, counseling, personal adherence plans, and weekly visits by field officers. In geographic settings where transportation constraints lead patients not to adhere to antiretroviral treatment, home-based delivery, even on a monthly or quarterly basis, may be able to overcome this major barrier. Additionally, the home-based intervention in Uganda led to positive social outcomes, such as increasing emotional support from family and the community, and strengthening of relationships with spouses (Apondi *et al.*, 2007). This approach could provide a venue for couples counseling to allow for disclosure and to minimize domestic violence. However, a home-mediated DAART program would still have to overcome barriers related to stigma and disclosure. If a DAART outreach worker were employed, this individual may not have to be readily identifiable in order to minimize the risk of disclosure of the patient's HIV infection.

Socially relevant HIV prevention

As the HIV epidemic has spread outside of key risk-group populations into the wider population across the developing world, there has been an increased need for prevention efforts to address wider structural and social correlates that are associated with the continued spread of HIV among vulnerable populations. In light of the increasing feminization of the HIV epidemic globally, there is certainly great need for interventions that examine the cultural and historic vulnerability of women to HIV infection within often patriarchal societies with rampant gender inequities. For instance, in India it is now widely understood that the greatest risk of HIV infection for a woman is being married, suggesting a strong risk posed by the risk-taking behaviors of her spouse. As HIV has increasingly become a disease associated with poverty and social inequity, it is necessary for prevention interventions to delve beyond conventional limited interventions, and begin to address wider social and economic risk factors for HIV infection.

The stigma associated with HIV can cause infected men and women to live a life of fear. Disclosure of one's HIV status is a double-edged sword, as revealing one's HIV status can open up the potential for greater social support, but can also lead to increased stigmatization, discrimination, and the loss of personal relationships. The inability to disclose one's HIV status can lead to problems with taking pills in public and attaining support from family members for HIV care and adherence. Individuals may internalize stigma, leading them not to disclose their status and then potentially closing themselves off to possible sources of support.

In traditionally conservative societies where most marriages are arranged and men and women do not reveal their HIV status due to stigma, a strong partner-notification system could prevent acquisition of HIV from infected spouses. Disclosure can be particularly complicated for women who generally seek medical care only after receiving permission from family members. Women rarely attend a clinic without being accompanied by a family member. In India, owing to a strong family and social support network that can include neighbors and friends, there can be some degree of "shared confidentiality" in which breach of confidentiality, to a certain degree, among these networks is expected (Chandra *et al.*, 2003). Though women may not perceive themselves to be at risk for HIV and be willing to be tested, many women may be concerned about confidentiality and disclosing their HIV serostatus due to negative reactions from their husbands, parents and the community (Rogers *et al.*, 2006). Within social systems in which women are often blamed for bringing HIV to the family, men may find it easier to disclose their positive serostatus to family members.

With the expansion of HIV-testing services comes the need to strengthen testing and counseling protocols with referral systems for clients testing HIV-positive to established treatment and support facilities. The lack of adequate referral of a client who tests HIV-positive can lead some patients to seek care at inappropriate medical establishments, experience depression, have suicidal ideation, and even engage in risky behavior with intent to transmit the infection. Psycho-social counseling is crucial for infected individuals and their caregivers in order for them to cope positively and constructively with HIV. Past prevention efforts have clearly demonstrated that fear-inducing messages about AIDS that emphasize the fatality of contracting HIV and the acquisition of HIV due to immoral behavior are counterproductive, and may only increase the stigmatization experienced by infected individuals, doing little to promote the adoption of safer behaviors. Greater integration of HIV-infected individuals into their families and communities can change common misconceptions associated with the illness.

The example of the AIDS epidemic in the developing world and the history of prevention programs highlight the reality that AIDS is often less a medical problem than a problem of culture, development and economics. Short-term solutions, such as treatment of sexually transmitted diseases, condom promotion, blood safety, and drug de-addiction programs, though possibly bringing about quick reductions in HIV incidence, will need to be effectively complemented by more long-term preventive strategies that aim at fundamental structural change. However, bringing about structural changes within societies requires different and innovative approaches based on varying cultural, social and economic realities. Patterns of sexual risk-taking behavior often lie at the fault lines where traditional social norms and modernity collide. Within a South Indian setting, men and women may be separated from each other during certain inauspicious times of the year. Newly married women are taken to their mother's home, and the men are left alone without the extended family support structure of the past. YRG CARE has documented that some of these men have acquired HIV after seeking extramarital relationships in the absence of their wives. In the rapidly changing developing world, questions of social vulnerability and risk behavior are dynamic processes reflecting the intersections of tradition and modernity.

HIV prevention strategies will need to account for how local cultural practice and historical changes run into socio-economic modernization. Simply stated, prevention interventions that work in South Asia may not be appropriate in Africa or Latin America. Additionally, bringing about structural change of deeply entrenched social norms may require broad coalitions between health-care systems, NGOs, civil society and national governments. Though social interventions may not provide immediate results, they offer the potential for long-term, sustained reductions in HIV transmission by fundamentally altering the relationships between individuals.

References

Apondi, R., Bunnell, R., Awor, A. *et al.* (2007). Home-based antiretroviral care is associated with positive social outcomes in a prospective cohort in Uganda. *J. Acquir. Immune Defic. Syndr.*, 44, 71–6.

Bunnell, R., Nassozi, J., Marum, E. *et al.* (2005). Living with discordance: knowledge, challenges, and prevention strategies of HIV-discordant couples in Uganda. *AIDS Care*, 17, 999–1012.

Castro, A. and Farmer, P. (2005). Understanding and addressing AIDS-related stigma: from anthropological theory to clinical practice in Haiti. *Am. J. Public Health*, 95, 53–9.

Chandra, P. S., Deepthivarma, S. and Manjula, V. (2003). Disclosure of HIV infection in South India: patterns, reasons and reactions. *AIDS Care*, 15, 207–15.

Coldiron, M., Stephenson, R., Chomba, E. *et al.* (2007). The relationship between alcohol consumption and unprotected sex among known HIV-discordant couples in Rwanda and Zambia. *AIDS Behav.*, (e-pub ahead of publication).

Creek, T., Ntumy, R., Seipone, K. *et al.* (2007). Successful introduction of routine opt-out HIV testing in antenatal care in Botswan. *J. Acquir. Immune Defic. Syndr.*, 45, 102–7.

De Cock, K., Fowler, M. G., Mercier, E. *et al.* (2000). Prevention of mother-to-child HIV transmission in resource-poor countries: translating research into policy and practice. *J. Am. Med. Assoc.*, 283, 1175–82.

Doherty, T., Chopra, M., Jackson, D. *et al.* (2007). Effectiveness of the WHO/UNICEF guidelines on infant feeding for HIV-positive women: results from a prospective cohort study in South Africa. *AIDS*, 21, 1791–7.

Dunkle, K., Jewkes, R. K., Brown, H. C. *et al.* (2004). Gender-based violence, relationship power, and risk of HIV infection in women attending antenatal clinics in South Africa. *Lancet*, 363, 1415–21.

Fowler, M. and Newell, M. L. (2002). Breast-feeding and HIV-1 transmission in resource-limited settings. *J. Acquir. Immune Defic. Syndr.*, 30, 230–9.

Gangakhedkar, R., Bentley, M. E., Divekar, A. D. *et al.* (1997). Spread of HIV infection in married monogamous women in India. *J. Am. Med. Assoc.*, 278, 2090–2.

Go, V. F., Sethulakshmi, C. J., Bentley, M. E. *et al.* (2003). When HIV-prevention messages and gender norms clash: the impact of domestic violence on women's HIV risk in slums of Chennai, India. *AIDS Behav.*, 7, 263–72.

Gopalan, S. and Shiva, M. (2000). *National Profile on Women, Health and Development: Country Profile – India*. New Delhi: Voluntary Health Association of India and World Health Organization.

Hanenberg, R. and Rojanapithayakorn, W. (1996). Prevention as policy: how Thailand reduced STD and HIV transmission. *AIDScaptions*, 3, 24–7.

Jana, S., Basu, I., Rotheram-Borus, M. J. and Newman, P. A. (2004). The Sonagachi Project: a sustainable community intervention program. *AIDS Educ. Prev.*, 16, 405–14.

Kalichman, S., Simbayi, L. C., Vermaak, R. *et al.* (2007a). HIV/AIDS risk reduction counseling for alcohol using sexually transmitted infections clinic patients in Cape Town, South Africa. *J. Acquir. Immune Defic. Syndr.*, 44, 594–600.

Kalichman, S., Simbayi, L. C., Cain, D. *et al.* (2007b). Sexual assault, sexual risks and gender attitudes in a community sample of South African men. *AIDS Care*, 19, 20–7.

Kim, J., Watts, C. H., Hargreaves, J. R. *et al.* (2007). Understanding the impact of a microfinance-based intervention on women's empowerment and the reduction of intimate partner violence in South Africa. *Am. J. Public Health*, 97, 1794–802.

Kirby, D., Obasi, A. and Laris, B. A. (2006). The effectiveness of sex education and HIV education interventions in schools in developing countries. *WHO Technical Report Series*, 938, 103–50.

Kumarasamy, N., Safran, S. A., Raminani, S. R. *et al.* (2005). Barriers and facilitators to antiretroviral medication adherence among patients with HIV in Chennai, India: a qualitative study. *AIDS Patient Care STDs*, 19, 526–37.

Luginaah, I., Elkins, D., Maticka-Tyndale, E. *et al.* (2005). Challenges of a pandemic: HIV/AIDS-related problems affecting Kenyan widows. *Social Sci. Med.*, 60, 1219–28.

Madhivanan, P., Hernandez, A., Gogate, A. *et al.* (2005). Alcohol use by men is a risk factor for the acquisition of sexually transmitted infections and human immunodeficiency virus from female sex workers in Mumbai, India. *Sex. Transm. Dis.*, 32, 685–90.

Mawar, N., Joshi, P. L., Sahay, S. *et al.* (2007). Concerns and experiences of women participating in a short-term AZT intervention feasibility study for prevention of HIV transmission from mother-to-child. *Culture Health Sex.*, 9, 199–207.

Mendenhall, E., Muzizi, L., Stephenson, R. *et al.* (2007). Property grabbing and will writing in Lusaka, Zambia: an examination of wills of HIV-infected cohabiting couples. *AIDS Care*, 19, 369–74.

Murphy, E., Greene, M. E., Mihailovic, A. and Olupot-Olupot, P. (2006). Was the "ABC" approach (abstinence, being faithful, using condoms) responsible for Uganda's decline in HIV? *PLoS Med.*, 3, e379.

NIH (2007). *Health Prevention Trials Network*. Bethesda, MD: National Institutes for Health.

Newmann, S., Sarin, P., Kumarasamy, N. *et al.* (2000). Marriage, monogamy and HIV: a profile of HIV-infected women in south India. *Intl J. STD AIDS*, 11, 250–3.

Parker, R. and Aggleton, P. (2003). HIV and AIDS-related stigma and discrimination: a conceptual framework and implications for action. *Social Sci. Med.*, 57, 13–24.

Parkhurst, J. and Lush, L. (2004). The political environment of HIV: lessons from a comparison of Uganda and South Africa. *Social Sci. Med.*, 59, 1913–24.

Perez, F., Zvandaziva, C., Engelsmann, B. and Dabis, F. (2006). Acceptability of routine HIV testing ("opt-out") in antenatal services in two rural districts of Zimbabwe. *J. Acquir. Immune Defic. Syndr.*, 41, 514–20.

Pronyk, P., Hargreaves, J. R., Kim, J. C. *et al.* (2006). Effect of a structural intervention for the prevention of intimate-partner violence and HIV in rural South Africa: a cluster randomised trial. *Lancet*, 368, 1973–83.

Rogers, A., Meundi, A., Amma, A. *et al.* (2006). HIV-related knowledge, attitudes, perceived benefits, and risks of HIV testing among pregnant women in rural southern India. *AIDS Patient Care STDs*, 20, 803–11.

Roth, J., Krishnan, S. P. and Bunch, E. (2001). Barriers to condom use: results from a study in Mumbai (Bombay), India. *AIDS Educ. Prev.*, 13, 65–77.

Safren, S., Kumarasamy, N., James, R. *et al.* (2005). ART adherence, demographic variables and CD4 outcome among HIV-positive patients on antiretroviral therapy in Chennai, India. *AIDS Care*, 17, 853–62.

Sayles, J., Ryan, G. W., Silver, J. S. *et al.* (2007). Experiences of social stigma and implications for healthcare among a diverse population of HIV-positive adults. *J. Urban Health*, 84, 14–28.

Serwadda, D., Gray, R. H., Wawer, M. J. *et al.* (1995). The social dynamics of HIV transmission as reflected through discordant couples in rural Uganda. *AIDS*, 9, 745–50.

Shankar, A., Pisal, H., Patil, O. *et al.* (2006). Women's acceptability and husband's support of rapid HIV testing of pregnant women in India. *AIDS Care*, 15, 871–4.

Sinha, G., Choi, T. J., Nayak, U. *et al.* (2007). Clinically significant anemia in HIV-infected pregnant women in India is not a major barrier to zidovudine use for prevention of maternal-to-child transmission. *J. Acquir. Immune Defic. Syndr.*, 45, 210–17.

Sivaram, S., Srikrishnan, A. K., Latkin, C. A. *et al.* (2004). Development of an opinion leader-led HIV prevention intervention among alcohol users in Chennai, India. *AIDS Educ. Prev.*, 16, 137–49.

Sivaram, S., Johnson, S. L., Bentley, M. E. *et al.* (2005). Sexual health promotion in Chennai, India: key role of communication among social networks. *Health Prom. Intl*, 20, 327–33.

Sivaram, S., Johnson, S., Bentley, M. E. *et al.* (2007). Exploring "wine shops" as a venue for HIV prevention interventions in Urban India. *J. Urban Health*, 84, 563–76.

Solomon, S., Beck, J., Chaguturu, S. K. *et al.* (2003). Stopping HIV before it begins: issues faced by women in India. *Nat. Immunol.*, 4, 719–21.

Solomon, S., Chakraborty, A., D'Souza, R. and Yepthomi, R. (2004). A review of the HIV epidemic in India. *AIDS Educ. Prev.*, 63, 155–69.

Solomon, S., Solomon, S., Masse, B. R. *et al.* (2006). Risk reduction is associated with decreased HIV transmission-associated behaviors in high-risk Indian heterosexuals. *Journal of Acquired Immune Deficiency Syndrome*, 42, 478–83.

Tarakeshwar, N., Krishnan, A. K., Johnson, S. *et al.* (2006). Living with HIV infection: perceptions of patients with access to care at a non-governmental organization in Chennai, India. *Culture Health Sex.*, 8, 407–21.

Weidle, P., Wamai, N., Solberg, P. *et al.* (2006). Adherence to antiretroviral therapy in a home-based AIDS care programme in rural Uganda. *Lancet*, 368, 1587–94.

Were, W., Mermin, J. H., Wamai, N. *et al.* (2006). Undiagnosed HIV infection and couple HIV discordance among household members of HIV-infected people receiving antiretroviral therapy in Uganda. *J. Acquir. Immune Defic. Syndr.*, 43, 91–5.

Evaluating HIV/AIDS programs in the US and developing countries

21

Jane T. Bertrand, David R. Holtgrave and Amy Gregowski

Over the past 25 years tens of billions of dollars have been invested in programs to prevent the transmission of HIV, to treat people living with HIV and AIDS (PLWHA), and to care for those affected by the epidemic, such as orphans and vulnerable children. Have these programs made any difference? What elements or strategies have proven more successful than others? What has been the return on investment of these programs? How could we improve existing programs to make them more effective?

Without evaluation, we are at a loss to answer any of these questions. This chapter provides an overview of the issues related to evaluating HIV/AIDS programs. It compares the approaches and tools used to evaluate programs in the United States and developing countries. Further, it outlines the major challenges to evaluating programs.

Issues in defining the evaluation design

Types of evaluations

Three primary types of evaluation are formative, process and summative (Rossi *et al.*, 2005). Formative evaluation relates to assessment conducted prior to the intervention that is used in guiding the design of the evaluation in terms of content, audiences, messages, logistics and related factors. Process evaluation describes several different types of activities that track the implementation

571

of a given intervention. Process evaluation answers one or more of the following questions: Was this program implemented according to plan? Did it reach the intended recipients? What was the reaction of those receiving the services of the program (client satisfaction)? By contrast, summative evaluation focuses on results: did the expected change occur? Was this change attributable to the program intervention (often called *impact assessment*)? What was the public health return on investment (economic evaluation)? And how does the program compare to other medical or public health programs in terms of cost per QALY (or DALY) saved? This chapter focuses heavily on summative results that seek to answer the questions: Did the desired change occur? Did the intervention or program cause the change to occur? And what was the return on investment to achieve this change?

Use of quantitative and qualitative methods for measuring effectiveness

Both quantitative and qualitative methods are in widespread use among evaluators to answer questions about HIV/AIDS programs. Qualitative methods are particularly useful in formative and process evaluation to examine the question "why?" For example, focus groups and in-depth interviews reveal the existing beliefs, social norms, perceptions, apprehensions and other psychosocial factors that bear directly on the desired behavior change in question. Similarly, qualitative methods can be effective in process evaluation to obtain client satisfaction and reactions to a given intervention. Qualitative methods also have a role in summative evaluation, primarily to explain why a given program did or did not achieve the expected results.

Quantitative methods (surveys, surveillance, service statistics and others) have a role to play in all three types of evaluation: formative (e.g., to establish existing levels of knowledge, identify barriers, and obtain a baseline against which to measure change); process (e.g., to quantify the volume of outputs that a program has produced as one indicator of performance, such as number of trainings conducted or number of pamphlets distributed); and especially summative research, because it yields a measure of the extent of change. Also, quantitative methods lend themselves to the use of statistical techniques to tease out causal inferences.

Internal versus external evaluation

Program administrators often debate whether to carry out an evaluation from within their organization or bring in an external evaluator. The advantage of internal evaluation (using one's own staff to conduct the evaluation) is their far greater familiarity with the program, their sense of ownership in the results, and the greater likelihood that they will feed the results back into the next program cycle. However, internal evaluation is often considered to be less objective, and staff may not have the same high-level analytic skills that the organization can obtain from an outside consultant.

Conversely, external evaluation is perceived to be more objective. The external evaluators are often knowledgeable about similar programs elsewhere, and generally have strong quantitative and analytic skills required for sophisticated statistical analysis. Yet they may have difficulty in grasping the nuances of the intervention, and they are no longer present when it comes time to apply the results to modifying the intervention or program.

Two factors often influence the decision regarding "internal versus external": availability of funding for evaluation, and required level of methodological/analytic sophistication. For example, monitoring service utilization requires an understanding of the process, attention to detail, and diligence in ensuring completeness of data collection, but it is not methodologically complex. By contrast, experimental and quasi-experimental designs used to evaluate interventions at multiple levels (see "Evaluating interventions at multiple levels", below) require advanced statistical training. In general, monitoring trends does not require advanced training, whereas teasing out causal inference does.

Stakeholders

Stakeholders have a vested interest in the results from program evaluation. Stakeholders may include the donor agency, the managerial and front-line staff of the implementing organization, the beneficiaries from the program, and other community members, among others. The interests of different stakeholders often diverge, creating a challenge for the evaluator trying to address the needs of these different groups.

It is essential for the evaluator to work closely with these diverse stakeholders from the design stage through the presentation of final results. (Some stakeholders will want to be more involved than others, but most will appreciate the opportunity to be included.) Such involvement from the start increases the chances that these diverse groups will accept and "own" the final results. Working with stakeholder groups can be particularly important in evaluating programs among marginalized groups that may have distrust for persons in positions of authority, especially where the stakeholders' work or life circumstance is highly stigmatized if not illegal.

Objectives and indicators

The first question to address in designing an evaluation is: what is the objective of the intervention? The program objective is a determining factor in the choice of indicators to be used, the population to be involved in the evaluation, and

573

the study design. It is useful to develop a conceptual framework (often called a logic model) that illustrates how a given intervention is expected to achieve the stated objective of the program, for both evaluative and programmatic purposes (e.g., developing consensus among all stakeholders regarding what the program is expected to do). For example, if the objective of a given program is to increase the utilization of VCT (Voluntary Counseling and Testing) services, then the conceptual framework should reflect the pathway of changes one would expect leading up to the actual act of getting tested (e.g., knowledge of the benefits, knowledge of the source of service, attitudes regarding getting the service, access, cost, self-efficacy, overcoming stigma). These concepts on the conceptual framework influence the selection of indicators used in tracking results related to the effectiveness of the intervention.

Types of summative evaluation

This chapter focuses on three types of summative evaluation that are widely used to track progress and measure impact. To understand how these categories fit into the grand scheme of evaluation, it is useful to review a framework developed by Rugg *et al.* (2004) (see Figure 21.1). The three categories described in this chapter correspond to the top two levels on the Rugg diagram, as follows:

1. Monitoring and evaluating national programs (Rugg's terminology):
   - monitoring utilization (e.g., VCT, PMTCT)
   - evaluating interventions at multiple levels
2. Determining collective effectiveness (Rugg's terminology):
   - monitoring outcomes at the national level

For each of these three categories of evaluation, we discuss the purpose and importance of this type of M&E; key indicators; common approaches to data collection and study designs, and illustrative examples from both the United States and developing countries.

Monitoring service utilization (e.g., VCT, PMTCT)

Service statistics are often the most basic measure of "results", because they capture what is occurring in the service delivery sites related to HIV prevention. Moreover, most health delivery services collect these data as part of routine management, and thus the data are relatively inexpensive to obtain (as compared to population-based surveys).

These data are useful to establish trends over time (e.g., increases in the uptake of VCT). They also serve to compare performance across districts or other units of analysis.

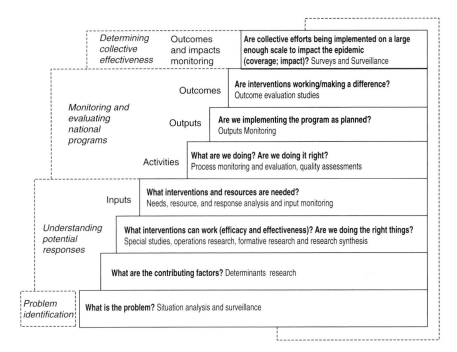

Figure 21.1 A public health questions approach to HIV/AIDS M&E. Source: Rugg *et al.* (2004), reproduced with permission.

Indicators

The most appropriate indicators depend on the nature and objectives of the service. For example, if the service is VCT testing, then the key indicator is the number of VCT tests carried out. An important variation on this is the percentage of VCT conducted in which the person returns to receive the test results.

In the case of programs designed to prevent the transmission of HIV from mother to child, leading indicators include:

- percentage of pregnant women tested for HIV;
- percentage of seropositive women that deliver an HIV-negative baby.

Indicators of service utilization have two major limitations. First, they are "numerator data" – that is, we may know that country X conducted 100,000 VCT tests, but this statistic has relatively little meaning as an absolute number if we don't have some sense of the "denominator" (to obtain a rough estimate of the percentage of the population that has been tested). In countries with good census data, it is often possible and useful to convert these numerator data to percentages based on estimates of the population size among the relevant group.

Second, the proliferation of indicators across agencies, countries and programs is mind-boggling. To date, some 18 HIV/AIDS indicator manuals are in circulation; they contain over 400 indicators, many of which relate to service utilization.

Data collection/study design

Data collection takes the form of collecting information at each service delivery point and aggregating it over time (e.g., monthly, quarterly, annual reports) or up the levels of the system (e.g., county/state/national or district/province/national). In most cases the data are presented in descriptive form to illustrate trends over time, across geographical units or by type of service. However, service statistics collected at regular intervals (e.g., by month, quarter or year) lend themselves to time-series analysis. If an identifiable intervention has taken place at a given point in time, these data may allow for some causal inferences. Figure 21.2 presents the data on condom sales in Ghana before and after the start of the communication campaign ("Stop AIDS Love Life"). Although we can not rule out the effects of "history" – some other event unrelated to the campaign that may have influenced the outcome – the findings present fairly plausible evidence that the campaign influenced behavior.

Examples from the US context

In the United States, the Centers for Disease Control and Prevention (CDC) have developed two systems for the purposes of collecting service utilization or output data. One system is devoted to the tracking of the delivery of HIV counseling

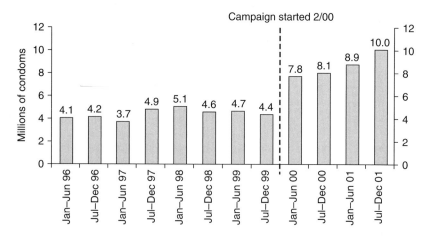

Figure 21.2 Time-series analysis: condom sales in Ghana.

and testing services in CDC-funded sites. This system utilizes client-level forms, which can be completed easily by the service provider and then scanned in via optical readers to allow for data analysis. These forms are colloquially referred to as the "CT bubble forms" because they involve filling in small circles to indicate client characteristics. The bubble forms provide information that can be aggregated at the site, state and national levels (CDC, 2006a). Indeed, the CDC issues an annual report which summarizes the number of persons receiving HIV counseling and testing via their funded programs; further, this report summarizes (among other factors) HIV serostatus results, client testing history, client risk behaviors, as well as client socio-demographics. Moreover, the form assesses whether an individual client returns for his or her result. While this information is basic, it is extremely useful. It provides local and national information that can guide service delivery choices related to service targeting, need for scale-up, and quality improvement.

A second system, knows as PEMS (Program Evaluation and Monitoring System) is now being fully implemented by the CDC. PEMS is designed to monitor key indicators of importance to the CDC (and the nation). These indicators include a variety of aspects of the delivery of specific types of HIV prevention services, as well as information about community planning, capacity building, reporting compliance, and surveillance activities. According to the CDC's website, PEMS was rolled out in segments between 2004 and 2006 (CDC, 2006b). PEMS collects and reports individual clients' service receipt experiences, as well as organizational characteristics. Further, PEMS allows evaluators to compare actual services delivered to the service delivery goals of local organizations/communities. The process information addresses basic questions of accountability (i.e., who got what type of service, when, where and how). The matching of service delivery information to planning targets allows for mid-course corrections to improve the effectiveness of the resources being utilized at a local level (and in the aggregate, nationally).

Examples from the developing country context

In the developing world, one widely known set of service utilization measures is the PEPFAR (the President's Emergency Plan for AIDS Relief) indicators, which are required of countries receiving US government funding for HIV/AIDS. The policy of "Three One's" (one policy, one program and one M&E system) has led to efforts to "harmonize" indicators – that is, develop a common set of indicators that different donors are willing to accept.

PEPFAR is the major source of funding for HIV/AIDS programming in developing countries (The White House, 2007). As a requirement for receiving PEPFAR funding, countries receiving at least US$ 1 million must report at least annually on a series of 46 indicators of service utilization (also termed *output-level indicators*); the 15 focus countries must report semi-annually. Most of these are program-level indicators that measure some aspect of service utilization, falling into one of four categories: number of organizations that provided TA (technical assistance), number of service outlets, number of individuals served, and number of people trained. The guidelines for reporting on these indicators give explicit instructions to avoid double-counting (e.g., individuals and sites). Also, the guidelines differentiate "downstream funding" (direct support of activities that can be associated with counts of uniquely identified individuals receiving treatment, care or support at a unique service delivery point that receives USG funding) from "upstream support", which refers to funding for capacity building and service strengthening that occurs apart from or at higher levels than specific service delivery points.

Evaluating interventions at multiple levels

Comprehensive interventions designed to curb the AIDS epidemic require multiple levels of interventions, consistent with the social ecological framework. A range of evaluation tools exists to track progress and measure change at these different levels. Ultimately, change must occur at the individual level to halt the spread of HIV/AIDS. While the interventions designed to bring about that change operate at different levels, the indicators used to measure this change are often the same – that is, the behavioral practices most directly linked to preventing the transmission of HIV (e.g., abstain from sex, reduce the number or avoid concurrency of partners, use condoms, avoid contaminated needles).

A framework for levels of causation of HIV incidence described by Sweat and Denison (1995) is useful in outlining the multiple levels at which interventions can take place. This framework includes four levels: superstructural, structural, environmental and individual (Sweat and Denison, 1995). Superstructural interventions happen at the international or national level, resulting in changes to overarching issues such as poverty that may then reduce the incidence of HIV infections indirectly. Empirical research supports the link between poverty and high rates of HIV. Yet relatively few HIV prevention programs target poverty as the means of reducing HIV, but rather focus on the proximate determinants (and thus we don't discuss evaluation of such interventions in this chapter). Structural interventions refer to changes at a macro-level – such as legislation or policies – that deter harmful practices somewhat independently of the volition of the individual (e.g., policies requiring and enforcing 100 percent condom use among brothel owners in brothels). Environmental or institutional interventions aim to change the behavior of groups by creating an environment in which people make decisions more conducive to healthy behaviors (e.g., making condoms more available, improving care for STIs, reducing stigma). These types of interventions aim to influence behavior of an individual by capitalizing on existing contextual dynamics. And finally, individual-level interventions target the behavior of persons within a given society or population, within the environment in which their behavior occurs.

Evaluations of program effectiveness attempt to determine a causal relationship between the intervention and the outcome, be it among a population or subgroup in a defined geographical area or among a group of clients at the program level.

Indicators used to track progress

In the context of research studies or clinic-based evaluations, it has been possible to use HIV incidence or prevalence as the outcome measure. Indeed, the public health specialists increasingly call for biological markers as outcomes in HIV program evaluation. However, the vast majority of evaluations of field-based programs to date – across the different levels of intervention – have used behavioral outcomes. The indicators relate directly to the program objectives and may include psycho-social factors ("initial outcomes"), although most program managers and donors are far more interested in changes in actual behavior ("intermediate outcomes") – for example:

1. Initial (psycho-social) outcomes:
   - knowledge of where to get tested for HIV
   - confidence (self-efficacy) to be able to negotiate condom use
   - correct knowledge about HIV transmission
2. Intermediate (behavioral) outcomes:
   - percentage of young people aged 15–19 that have yet to initiate sexual relations
   - percentage of women aged 15–49 that have had only one sexual partner in the past 12 months
   - percentage of men aged 15–19 that used a condom at last sex with a casual partner
   - percentage of injection drug users reporting the use of sterile injecting equipment the last time they injected

For certain concepts (e.g., knowledge of HIV/AIDS, stigma), evaluators often use composite indices made up of several individual items. For example, one composite indicator for stigma developed in connection with the Demographic and Health Survey (DHS) includes four items measuring the respondents' attitudes of acceptance toward people with HIV in different settings (Measure DHS, 2007). It is useful to adopt indices that have been developed and tested by other credible sources (e.g., the DHS) as a means of increasing comparability and ensuring reliable instruments.

To illustrate the point that the same indicator may be applicable in evaluating interventions at different levels of the social ecological framework, we take the example of condom use. For example, the following three interventions operating at different levels could all use the same outcome indicator, such as condom use at last sex:

1. 100 percent condom brothels (structural)
2. Access to AIDS prevention at worksites (environmental)
3. Clinic-based HIV VCT (individual).

Data collection/study design

To test for a causal relationship between the intervention and the outcome, evaluators use experimental designs, quasi-experimental designs and/or advanced analytic techniques to tease out causal inferences (Fisher and Foreit, 2002). The gold standard for establishing causality in clinical medicine is the true experimental design (often referred to as a randomized controlled trial, or RCT). This design lends itself to situations where it is possible to randomize individuals, communities, schools, clinics, or similar units of analysis to treatment groups, which is often *not* the case with "full coverage programs" intended to reach all members of the population.

In a chapter on program evaluation, it is useful to differentiate between two types of RCTs. One involves applying an experimental design to an actual program that is operating in a given population to determine its effectiveness in achieving its objective (e.g., to increase the number of HIV tests performed in a given clinic, to improve client satisfaction through special training of personnel). The evaluation takes place in the context of an ongoing program under normal conditions. The second involves pilot-testing a new approach which, if shown to be effective, may be introduced into the program or replicated more widely. We would consider this second type of RCT to be research (or operations research), rather than program evaluation *per se*. The three community-based trials on male circumcision in South Africa, Uganda and Kenya would fall in this category, as would much of NIH-funded research on interventions.

With regard to data collection, surveys can be administered in several different ways, depending on what is most appropriate for the setting and the people participating. ACASI (computer-assisted interviewing) is a useful method when asking sensitive questions, but may not be appropriate among people unfamiliar with computers.

Face-to-face interviewing allows the interviewer to assure that the participant understands the questions but may introduce social-desirability bias, while self-administered surveys may reduce such bias but introduce issues in survey comprehension.

Examples of interventions at the different levels

Sweat (2006) coordinated an ambitious project, involving numerous colleagues, to review the literature and conduct some meta-analyses on a variety of different

HIV prevention approaches in developing countries (e.g., mass media, VCT, treatment programs, abstinence, psychological support, needle-exchange programs, social marketing). This body of work provides the most comprehensive analysis of program evaluation results to date.

In this section, we highlight several studies that illustrate the different levels of evaluation consistent with an ecological approach.

Structural-level program

An intervention aimed at reducing HIV and STI risks among female commercial sex workers in the Dominican Republic compared an environmental-level intervention to a combined environmental- and structural-level intervention. Both arms of the intervention promoted consistent condom use among sex workers and their partners: the environmental intervention focused on community solidarity by working with sex workers and establishment owners, while the combined intervention also included government policies ensuring and regulating condom use in the establishments. The results of a cross-sectional survey showed that the combined environmental and structural intervention was a predictor for consistent condom use among sex workers (Kerrigan *et al.*, 2003).

Environmental-level programs

Several HIV prevention projects in the US have focused on working with particular subgroups that are at high risk for infection, including injection drug users and women involved in commercial sex work. One such intervention carried out in Baltimore city operated at both the environmental and individual levels. The project's environmental component targeted the economic situation of drug-using women who were also involved in sex work, while the individual component focused directly on reducing risk behavior. This intervention, called the Jewel project, combined sessions on risk-reduction and the production and sales of jewelry made by program participants. The women participating were surveyed at baseline and at 3 months' post-intervention, and reported significant reductions in median number of sex partners and in receiving money or drugs for sex (Sherman *et al.*, 2006).

Another environmental-level intervention, conducted in the late 1980s, used key opinion leaders in gay communities in three cities to reduce HIV risk behavior by modifying community norms. This study, based on the diffusion of innovation theory, randomly allocated one community (Biloxi, Mississippi) to the intervention condition with opinion leaders, and two communities to the control condition. Bar tenders at 4 clubs identified 22 opinion leaders, who then identified friends to be trained by the study team to be behavior change "endorsers". A total of 39 opinion leaders were trained, and promoted risk-reduction practices among their peers. Significant decreases were found in a 2-month period in the intervention city in the proportion of men who engaged in unprotected anal intercourse and in the number of men with more than one sex partner, while a significant increase in condoms use was observed. No such changes were found in the comparison communities (Kelly *et al.*, 1991).

Individual-level program

The evaluation of a national roll-out of a primary school HIV prevention program in Kenya used a pre- and post-intervention study design (a type of quasi-experimental design) to measure self-efficacy, sexual behavior and condom use among upper primary-school students receiving the intervention in one province. The effects of this intervention were evaluated and understood using a mix of quantitative and qualitative methods of the program in the Nyanza Province of Kenya. Quantitative surveys and focus-group discussions were conducted at baseline and after the program had been delivered in 40 matched schools. Significant program effects were found among sexually experienced girls, with an increased likelihood of reporting being able to say "no" to sex. Boys who were virgins at the start of the program experienced high condom self-efficacy associated with high exposure to the program (Maticka-Tyndale *et al.*, 2007).

Tracking outcomes at the national level (based on individual behavior or biomarkers)

In this case, the "intervention" is the national program designed to curb the spread of HIV. This includes the sum total of activities conducted by all players – government, NGOs, faith-based and other organizations – that in any way contribute to decreasing the transmission of HIV. Although many implementing agencies might not identify themselves as part of the "national program", in fact for evaluation purposes their activities potentially contribute to the outcome measure of interest (decreasing HIV prevalence or incidence). This corresponds to the "determining collective effectiveness" on the Rugg *et al.* (2004) framework in Figure 21.1.

Many policy-makers, health officials and donors are most interested in answering basic questions: where do we stand on key indicators, such as HIV prevalence by age group or by ethnic group? Is the trend improving or worsening over time?

Data on trends (e.g., HIV prevalence) at the national level inform policy decisions regarding the allocation of resources (for example, the selection of the 15 countries that received the first round of PEPFAR funding). Trend data also provide a measure of success (or failure) to curb the spread of the epidemic at the national level and/or by geographical, ethnic or age group. As such, trend data on outcomes form a vital part of evaluation of HIV programs. However, because multiple factors contribute to these rates (outcomes), rarely is it possible to attribute change in a given indicator to a specific intervention or program, based simply on the trend data.

Indicators used to track progress

Ideally, evaluators would like to measure the incidence of HIV transmission. However, given the challenges of doing so reliably on a representative sample of the population at the national level (for logistical, ethical and financial reasons), many accept proxy indicators.

Two widely used indicators are:

1. Prevalence of HIV among adults 15–49 years old
2. Prevalence of HIV among pregnant women aged 15–24 in antenatal clinics

Other useful indicators include percentage of adults aged 15–49 who have been tested for HIV (or by subgroup: age, special subgroup, such as pregnant women), and percentage of HIV-positive pregnant women receiving a complete course of antiretroviral prophylaxis to reduce the risk of mother-to-child transmission

With regard to HIV incidence, specific research projects have been able to collect incidence data at the population level by establishing a surveillance system, such as in Mwanza (Tanzania), Masaka and Rakai (both in Uganda) (Orroth *et al.*, 2005). Whereas these data-collection efforts are representative of a given population, they fall short of constituting a national sample.

The CDC has invested substantial effort in developing tests that can measure incidence in cross-sectional population samples; specifically, they have applied the BED assay in a number of countries. However, this method has not yet produced the desired level of accuracy, because of the need to determine the "window period" in which it is detecting recent infections, and misclassification of some late-stage infections and people on ART as recent infections (UNAIDS, 2005). Methods to adjust for these problems are being developed, but have not yet been standardized. Work continues on the development of the BED as a means of measuring incidence at the population level, because of the great importance attached to being able to measure and track this indicator (J. Stover, personal communication 3 March 2008).

Data collection/study design

Until recent years, the primary source of estimating HIV prevalence in the developing world was data on the percentage of pregnant women presenting at antenatal services who tested positive for HIV. However, in recent years a number of countries (30 in sub-Saharan Africa, Asia and the Caribbean since 2001) have conducted population-based surveys with HIV prevalence measurement among a representative sample of adult men and women. These surveys, adjusted for non-response and other biases, provide improved data to estimate national prevalence. However, they do not yield reliable estimates of HIV prevalence among populations at higher risk of HIV infection in countries with concentrated epidemics or trends (UNAIDS & WHO, 2007).

To approximate a measure of incidence, researchers have used sentinel site data (antenatal clinics) and focused specifically on pregnant women aged 15–24. This population is largely HIV-negative before the age of 15, and thus prevalence among women aged 15–24 tends to capture new infections. Comparative studies have shown that extrapolations from sentinel site surveillance tend to overestimate HIV prevalence at the national level when compared with data from population-based surveys. For example, data from surveys gave only 80 percent of the HIV prevalence yielded by antenatal clinic statistics in both urban and rural areas of selected African countries (UNAIDS, 2007a).

In the US, the CDC has supported the development of testing strategies (the BED assay, cited above) that can be used to detect whether an HIV infection is recent ("acute") or has occurred some more distant time in the past. This very innovative approach utilizes HIV tests with two different levels of sensitivity; if the more sensitive test is positive but the less sensitive test is negative, then the infection is more likely to be recent. Generally speaking, this is the type of approach now being used by the CDC to better estimate HIV incidence in the US at the national level; previously, the best available estimate of HIV incidence in the US was based on a literature-based, synthetic approach.

Examples from the US context

In the United States, policy-makers have followed national trends in HIV prevalence/incidence with great interest, in order to understand the epidemiology of HIV transmission and to track progress in curbing the spread of the epidemic. In addition, two projects by Holtgrave and colleagues have sought to demonstrate a causal linkage between HIV prevention efforts and HIV incidence in the United States, as well as to estimate the associated costs. In the first project, Holtgrave (2002) estimated what the HIV incidence curve in the US would have been, had HIV prevention programs not been in place since the beginning of the 1980s. This approach is based on economic analysis, and involves constructing a "counterfactual". The counterfactual is then compared to what has been observed. The difference between the observed and counterfactual HIV incidence curves in the US allows for an estimation of the number of HIV infections prevented by the prevention programs delivered. The best available estimates indicate that in the US, between 1985 (the time when new infections peaked in the US) and 2000, HIV prevention efforts in the nation averted between 200,000 and 1.5 million HIV infections. Even though the cost of prevention programs during this period was just over $10 billion (including federal, state and private sources), the medical care costs averted by the prevention programs were higher than the service delivery costs (i.e., the programs were cost-saving to society).

In a second project, Holtgrave and Kates (2006) used lag-regression analysis to determine the temporal, correlational relationship between annual HIV prevention investments in the US (represented as the CDC's HIV prevention budget adjusted for inflation) and annual HIV incidence. Trend lines showed a "mirror image" between the CDC's HIV prevention budget and HIV incidence over the course of the epidemic in the US. However, the question remained: is it money that drives incidence, or does incidence drive investment levels? The researchers sought to answer this two-pronged question. They found that from the beginning of the epidemic until the mid-1980s, incidence predicted investment in the following 1 or 2 years; it was as if society were responding to the HIV emergency with increasing levels of investment. But beginning in the mid-1980s onward, investment began to predict incidence 1 or 2 years hence. This provides correlational evidence that, since the mid-1980s, the nation "gets what it pays for" in terms of HIV-incidence reduction.

Examples from the developing country context

The most comprehensive trend data for monitoring global progress are the UNGASS indicators, which are published once every 2 years to report on the epidemic and response. The system includes 25 national-level indicators (in four categories: national commitment and action, national programs, knowledge and behavior, and impact) as well as four global indicators that track resource allocations. A total of 137 countries submitted indicator data for the 2006 report. These indicators inform programmatic decisions, guide reviews of the National Strategic Plan, inform resource mobilization efforts (e.g., Global Fund proposal development), serve as an advocacy tool, and report progress toward the Declaration of Commitment at the global level (UNAIDS, 2007b).

Most developing countries have prevalence data, based either on a representative sample of the adult population (male and female) or on a sample of pregnant women attending prenatal services. Where such data are available on a yearly basis over an extended period of time, it is possible to observe trends in prevalence (or other indicators) at the national level, as shown in Figure 21.3 for Uganda. However, these trends do not in themselves indicate the reasons or causes for the changes over time. Indeed, multiple authors have written extensively on the question of what caused the decrease in HIV prevalence in Uganda (Kilian *et al.*, 1999; Hogle *et al.*, 2002; Singh *et al.*, 2003; Stoneburner and Low-Beer, 2004). Hallert *et al.* (2006) used mathematical modeling to demonstrate that the decline in prevalence in Uganda, urban Kenya, Zimbabwe and urban Haiti could not have occurred in the absence of safer sexual practices (i.e., consistent condom use, median age at first sex, age-specific use of condoms, age-specific mean number of sexual partners in the last year, and age of spouse

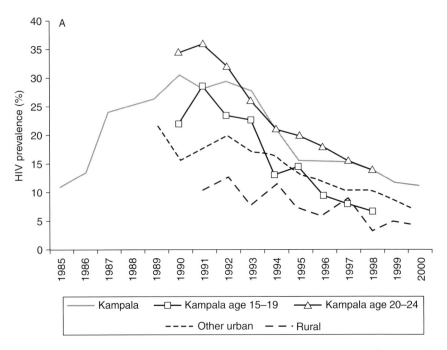

Figure 21.3 HIV prevalence in Uganda: 1985–2000. Source: Stoneburner and Low-Beer (2004), reproduced with permission.

or last partner). This type of research is extremely useful in understanding the dynamics of the epidemic. However, the findings from this study do not show the extent to which each specific practice changed, or that the change in that practice linked to a particular intervention.

Tracking trends in prevalence and other key indicators is very useful in determining the direction of change and in cross-national comparisons, but it does not reveal anything definitive about the influence of programs on these trends. However, for policy-makers and program managers less interested in the details of casual inference and threats to validity, these trend data often fulfill their needs for "evaluation".

Major challenges in evaluating HIV/AIDS programs

The challenges to evaluating HIV/AIDS programs include issues common to the evaluation of social programs in general, as well as some specific to HIV/AIDS programs.

Difficulty of attribution

To determine causation (i.e., to establish that a specific program or intervention caused a measurable change in behavior or health-status outcome), the often preferable study design is a randomized control trial (RCT). This type of study design is possible when the researcher can control who does and does not get the intervention, either at the individual level or at the community/school/clinic level. Yet many HIV/AIDS programs constitute "full coverage programs", in which all members of a given population or subgroup are potentially exposed to an intervention (e.g., a mass media program on HIV/AIDS prevention). Quasi-experimental designs and multivariate analytic techniques allow researchers to draw some causal inferences in such situations, but it is difficult to isolate the effects of a specific set of interventions (e.g., all activities that constitute the national program) on the key outcome indicators. Hence, there is some tension between the ideal study design (if the intervention delivery were perfectly controllable) and the ability to answer urgent, important policy and programmatic questions. Indeed, this tension between internal validity threat control and external relevance is at the forefront of HIV evaluation challenges.

One group that is grappling with the issues of measuring impact in HIV prevention programs at the national level is the Technical Evaluation Reference Group (TERG) of the Global Fund. Specifically, the group will collect and analyze primary data in eight countries, as well as analyzing secondary data in 12 countries to assess the reduction in the burden of HIV, tuberculosis and malaria associated with the collective scale-up of prevention and treatment activities by all partners. This analysis will use data from surveys, surveillance, research studies, service statistics, vital statistics, financing and administrative records; it will include synthesis and modeling (Social & Scientific Systems, Inc., 2006). This remains an extremely ambitious task, and the results are not yet forthcoming.

The CDC and NIH have also assembled working groups to address the issues of causal attribution for studies in field settings, and they share a common theme. Both are attempting to uphold the highest standard of scientific rigor, but recognize that at times a randomized controlled trial is impossible (e.g., the intervention cannot be randomized), prohibitively expensive or ethically problematic. Hence, rigorous designs are needed to help protect against threats to internal validity while at the same time offering excellent external validity. The NIH and CDC seek in these papers to offer highly rigorous alternatives to the RCT (Des Jarlais *et al.*, 2004; West *et al.*, 2008).

Dearth of evaluation using cost data

Although most intervention programs track cost data as part of program management, often such information is not available in a form that lends itself to cost analysis. Alternatively, the data exist but researchers do not take advantage to apply any form of cost analysis to a given evaluation. One possible reason has been the difficulty of simply measuring program effectiveness; researchers have not taken the additional step of determining cost-effectiveness or return on investment (ROI). However, this is beginning to change. Hornberger *et al.* (2007) carefully reviewed the literature on cost–utility analyses in HIV/AIDS. They found 106 papers that met basic standards of scientific rigor. Of these, 26 were dated 1997 or earlier; the other 80 were published in 1998 or later. Hence, the field has seen a relative explosion of growth in this arena in the past decade, although the majority of these papers relate to developed countries. What remains is for further uptake of well-accepted standards of economic evaluation practice into HIV/AIDS studies (Bertrand and Hutchinson, 2006; Holtgrave, 2006). HIV/AIDS does pose some special challenges for economic evaluation. For instance, Holtgrave has noted that many HIV/AIDS programs are focused toward racial/ethnic minority communities, and some economic evaluation techniques, if not carefully applied, could introduce methodological discrimination (e.g., using a life table for African-American populations in the US would seem to indicate that preventing an infection in an African-American rather than a white person would save less years of life and thereby be less cost-effective). This type of methodological discrimination is, of course, to be avoided (Holtgrave, 2004).

Lack of a single behavioral indicator that links to HIV incidence

Evaluators would ideally like to measure the objective of HIV/AIDS prevention programs: (reducing) incidence. As described earlier, measuring *incidence* is very difficult for financial, logistical and ethical reasons, and thus many use *prevalence* as a proxy.

In addition, evaluators would like to identify an indicator (or cluster of indicators) that satisfactorily captures behaviors that link to the long-term outcomes (e.g., incidence or prevalence). For example, in the field of family planning, contraceptive prevalence is highly correlated with fertility rates (except in countries with a high level of abortion). As such, this indicator of a behavioral practice represents an excellent measure for evaluating program effectiveness. By contrast, in the case of HIV/AIDS, no single behavioral indicator comprehensively captures or predicts with certainty HIV prevalence or incidence. As a result, evaluators use multiple indicators to capture different aspects of program effectiveness. The data collection is far more onerous and the results far less satisfying than if a single behavioral indicator linked strongly to HIV incidence (or prevalence). This problem does not have any quick "methodological fix", but rather reflects the nature of the epidemic.

Conclusion

Throughout the world, and even in some relatively low-prevalence countries like the US, we see the occurrence of HIV infection every minute. Even in the era of HAART, multiple people die hourly of AIDS in far too many nations. Sadly, this occurs even though we have outstanding tools already at our disposal to prevent and treat this deadly disease. What we need to know is how better to target and scale-up our evidence-based tools; strong program evaluation can help us to do just that. It is not overstating the case to assert that rigorous program evaluation can literally save lives.

References

Bertrand, J. (2006). Special Issue on cost-effectiveness analysis. *J. Health Commun.*, 11(Suppl. 2), 1–173.

CDC (2006a). HIV counseling and testing at CDC-supported sites-United States, 1999–2004. Available at http://www.cdc.gov/hiv/topics/testing/reports.htm.

CDC (2006b). Evaluation. Available at http://www.cdc.gov/hiv/topics/evaluation/.

Des Jarlais, D. C., Lyles, C. and Crepaz, N. and the TREND Group (2004). Improving the reporting quality of nonrandomized evaluations of behavioral and public health interventions: the TREND statement. *Am. J. Public Health*, 94, 361–6.

Fisher, A. and Foreit, J. (2002). *Handbook for Operations Research for HIV/AIDS Programs*. New York, NY: Population Council.

Hallett, T. B., Aberle-Grasse, J., Bello, G. *et al.* (2006). Declines in HIV prevalence can be associated with changing sexual behaviour in Uganda, Urban Kenya, Zimbabwe, and Urban Haiti. *Sex. Transm. Infect.*, 82(Suppl. 1), i1–8.

Hogle, J. A., Green, E., Nantulya, V. *et al.* (2002). *What Happened in Uganda? Declining HIV Prevalence, Behavior Change, and the National Response*. Washington, DC: TvT Associates, Synergy Project.

Holtgrave, D. R. (2002). Estimating the effectiveness and efficiency of HIV prevention efforts in the US using scenario and cost-effectiveness analysis. *AIDS*, 16, 2347–9.

Holtgrave, D. R. (2004). HIV prevention, cost–utility analysis, and race/ethnicity: Methodological considerations and recommendations. *Med. Decis. Making*, 24, 181–91.

Holtgrave, D. R. and Curran, J. W. (2006). *Handbook of Economic Evaluation of HIV Prevention Programs*. New York, NY: Springer.

Holtgrave, D. R. and Kates, J. (2006). HIV incidence and CDC's HIV prevention budget: an exploratory correlational analysis. *Am. J. Prev. Med.*, 32, 63–7.

Hornberger, J., Holodniy, M., Robertus, K. *et al.* (2007). A systematic review of cost–utility analyses in HIV/AIDS: implications for public policy. *Med. Decis. Making*, 27, 789–821.

Kelly, J. A., St Lawrence, J. S., Diaz, Y. E. *et al.* (1991). HIV risk behavior reduction following intervention with key opinion leaders of population: an experimental analysis. *Am. J. Prev. Med.*, 81, 168–71.

Kerrigan, D., Ellen, J. M., Moreno, L. *et al.* (2003). Environmental–structural factors significantly associated with consistent condom use among female sex workers in the Dominican Republic. *AIDS*, 17, 415–23.

Kilian, A. H., Gregson, S., Ndyanabangi, B. *et al.* (1999). Reductions in risk behaviour provide the most consistent explanation for declining HIV-1 prevalence in Uganda. *AIDS*, 13, 391–8.

Maticka-Tyndale, E., Wildish, J. and Gichuru, M. (2007). Quasi-experimental evaluation of a national primary school HIV intervention in Kenya. *Eval. Program. Plan.*, 30, 172–86.

Measure DHS (2007). HIV/AIDS Survey Indicators Database. Available at http://www.measuredhs.com/hivdata/ind_detl.cfm?ind_id=6&prog_area_id=3 (11-27-2008).

Orroth, K. K., Korenromp, E. L., White, R. G. *et al.* (2005). Higher risk behaviour and rates of sexually transmitted diseases in Mwanza compared to Uganda may help explain HIV prevention trial outcomes. *AIDS*, 17, 2653–9.

Rossi, P. H., Lipsey, M. H. and Freeman, H. E. (2005). *Evaluation: A Systematic Approach, 7th edn.* Thousand Oaks, CA: Sage.

Rugg, D., Carael, M., Boerma, J. T. and Novak, J. (2004). Global advances in monitoring and evaluation of HIV/AIDS: from AIDS case reporting to program improvement. New directions for evaluation. *Wiley Periodicals*, 103, 33–48.

Sherman, S. G., German, D., Cheng, Y. *et al.* (2006). The evaluation of the JEWEL project: an innovative economic enhancement and HIV prevention intervention study targeting drug using women involved in prostitution. *AIDS Care*, 18, 1–11.

Singh, S., Darroch, J. E. and Bankole, A. (2004). A, B and C in Uganda: the roles of abstinence, monogamy and condom use in HIV decline. *Reprod. Health Matters*, 12, 129–31.

Social & Scientific Systems, Inc. (2006). Technical background document on the scale and scope of the five-year evaluation of the global fund to fight AIDS, tuberculosis and malaria. Available at http://www.theglobalfund.org/en/files/terg/14_tech_report.pdf.

Stoneburner, R. L. and Low-Beer, D. (2004). Population-level declines and behavioral risk avoidance in Uganda. *Science*, 304, 714–18.

Sweat, M. (organizer) (2006). The importance of evidence: Johns Hopkins–WHO systematic reviews of behavioral interventions for HIV/AIDS in developing countries. XVIth International AIDS Conference, Toronto, Canada: organized satellite session.

Sweat, M. D. and Denison, J. A. (1995). Reducing HIV incidence in developing countries with structural and environmental interventions. *AIDS*, 9(Suppl. A), S251–7.

The White House: President's Emergency Plan for AIDS Relief (2007). *Indicators Reference Guide. FY 2007 Reporting/FY 2008 Planning.* Washington, DC.

UNAIDS (2005). Available at http://data.unaids.org/pub/EPISlides/2006/statement_bed_policy_13dec05_en.pdf.

UNAIDS (2007a). Comparing adult antenatal-clinic based HIV prevalence with prevalence from national population based surveys in sub-Saharan Africa. UNAIDS presentation. Available at http://data.unaids.org/pub/Presentation/2007/survey_anc_2007_en.pdf (accessed 11-17-2007).

UNAIDS (2007b). Monitoring and evaluation for accountability and global progress tracking. Available at http://search.unaids.org/Results.aspx?q = ungass+indicators& o=html&d=en&l=en&s=false&x=10&y=3 (accessed 03-06-2008).

UNAIDS & WHO (2007). 2007 AIDS Epidemic Update. Geneva: UNAIDS.

West, S. G., Duan, N., Pequegnat, W. *et al.* (2008). Alternatives to the randomized trial. *Am. J. Public Health*, in press.

Adapting successful research studies in the public health arena: going from efficacy trials to effective public health interventions

22

Kevin A. Fenton, Richard J. Wolitski,
Cynthia M. Lyles and Sevgi O. Aral

We should be winning in HIV prevention. There are effective means to prevent every mode of transmission; political commitment on HIV has never been stronger; and financing for HIV programs in low-and-middle income countries increased six-fold between 2001 and 2006. . . . If comprehensive HIV prevention were brought to scale, half of the infections projected to occur by 2015 could be averted.

(Global HIV Prevention Working Group; June 2007)

Since the recognition of human immunodeficiency virus (HIV) as the cause of acquired immune deficiency syndrome (AIDS), considerable resources – both human and economic – have been allocated to the development and evaluation of efficacious interventions to halt the transmission of the virus and the progression of the disease. For an even longer period, researchers have worked to develop and evaluate efficacious interventions to prevent the transmission of other sexually transmitted pathogens and to treat and cure the diseases caused by them. The interventions studied have included behavioral interventions, medical treatments,

591

vaccines, vaginal microbicides and male circumcision. A recent review of trials of interventions to prevent sexual transmission of all sexually transmitted infections (STI), including HIV, identified 83 trials of individual-, group- or community-level interventions and concluded that although many interventions have been found to be effective against STI, including HIV, too few have been replicated, widely implemented or carefully evaluated for effectiveness in other settings (Manhart and Holmes, 2005).

In order to stop the spread of an infection in populations, the right intervention must be delivered to the right people at the right scale, the delivery of interventions must be sustained, and the adherence of individuals must be ensured. The gap between the development and evaluation of an efficacious intervention and the implementation of the correct mix of interventions at the right scale in populations, so as to achieve population-level impact, is considerable.

Even in the case of individual-level biomedical interventions, such as drug therapy, efficacy under clinical trial conditions may differ importantly from effectiveness in real-life, usual clinical conditions. A number of factors may contribute to differences between efficacy under ideal trial conditions and effectiveness under real-life conditions. Many people may be screened before a few who fit the inclusion criteria are chosen to be included in a study, yet the results of the study are often applied to the very people who were excluded (Marley, 2000). The population studied in drug trials tends to be young, male and white; individuals tend to suffer from a single condition and use a single treatment. Most patients do not fit the description of study populations. They often have multiple illnesses, take multiple medications, and may be younger or older or of a different gender or ethnicity than trial subjects. A treatment is effective if it works in real life in non-ideal circumstances. In real life, medications will be used in doses and frequencies which were never studied, and in patient groups which were never assessed in the trials. Drugs may be used in combination with other medications that have not been tested for interactions. As Marley (2000) comments, "Effectiveness cannot be measured in controlled trials, because the act of inclusion into a study is a distortion of usual practice". It is important to note that drugs are licensed for use based on the results of controlled trials, but withdrawn from use because of observational data that would not be acceptable for licensing purposes. The gap between efficacy and effectiveness of interventions may be even wider for social and behavioral, group- or community-level interventions.

Evaluating efficacy, effectiveness, and impact of biological interventions on HIV and STI incidence

The strongest evidence for the efficacy of an intervention is provided by randomized controlled trials (RCTs) with appropriate controls and randomization algorithms, adequate statistical power, and appropriate statistical analysis

(Kramer, 2003). Randomized trials have been generally accepted as the gold reference standard not only for biomedical interventions but also for behavioral and social interventions (Oakley, 2000). Some experts, while accepting RCTs as the gold standard in the evaluation of medical technologies (such as medications and vaccines), suggest that randomized trials may not be the most appropriate evaluation approach for public health interventions (Barreto, 2005). Randomized trials evaluating intervention effects on HIV incidence are difficult to conduct; nearly all such trials have failed to show protection from HIV infection, and a few biomedical interventions have shown increases in risk of HIV acquisition (Gray and Wawer, 2007). In some of these trials, the intervention itself was not efficacious or was potentially harmful (Pitisuttithum *et al.*, 2006; Check, 2007). Other trials assessed the efficacy of intercourse-related interventions such as condoms, diaphragms or microbicides, and compliance with intercourse-related interventions is often poor (Gray and Wawer, 2007).

Compliance is also poor with other user-dependent interventions. A trial of daily tenofovir pre-exposure prophylaxis (PREP) reported product use for only 69–74 percent of study days (Peterson *et al.*, 2007). More recently, a trial of a diaphragm and lubricant gel, with condoms, failed to show efficacy; HIV incidence was not lower in the intervention group compared to the control (condoms only) group (Padian *et al.*, 2007). In this study, compliance with self-reported use of the diaphragm and gel was around 73 percent. In addition, self-reported condom use at last intercourse was substantially higher in the control (85 percent) compared to the intervention (54 percent) group. Poor compliance (or adherence) compromises statistical power of an experimental design to show efficacy. It is noted that assessment of a user-dependent method of prevention is driven by both efficacy and actual use of the product (Padian *et al.*, 2007), and by the effects of using the product on other protective or risk behaviors (Gray and Wawer, 2007). Often the use of one prevention method results in discontinued use of other prevention methods, referred to as risk compensation. Condom migration (reduction in condom use following microbicides introduction) is a well-known example of this phenomenon.

The importance of adherence to intervention in the evaluation of biomedical intervention efficacy has become increasingly clear in recent years as additional efficacy trials of HIV prevention have failed to decrease HIV acquisition. These findings suggest that one-time, provider-delivered interventions that are temporally removed from sexual arousal and intercourse may be more successful in decreasing HIV acquisition. Male circumcision is one example of such interventions, and has been shown to be efficacious in three randomized trials (Auvert *et al.*, 2005; Bailey *et al.*, 2007; Gray *et al.*, 2007).

Adherence is an important issue in the evaluation of efficacy of interventions, but it is not the only one. Recruitment of study subjects, retention of study subjects through the course of the trial, adequate coverage for follow-up measurement, and avoidance of contamination between intervention and control conditions are all important parameters which, if not adequately implemented, may threaten successful measurement of efficacy.

Cluster-randomized trials, where pairs of groups or sometimes whole communities are randomly assigned to intervention and control conditions, introduce additional concerns in the measurement of intervention efficacy. Population movements which threaten the "closed population" status of clusters, population movement between intervention and control communities, and contamination across intervention and control communities emerge as important considerations in this context (Hayes *et al.*, 2000).

Evaluation of intervention effectiveness in real-life conditions often involves a different set of important parameters. Effectiveness of an efficacious intervention in real-life settings is influenced by fidelity to the original intervention, adaptation of the intervention to the particular population or context as needed and without sacrificing fidelity, achievement of adequate scale-up, maintenance of the intervention over time, and uptake of the intervention by the target population (Rietmeijer and Gandleman, 2007). A focus on effectiveness in real-life situations necessarily shifts attention from internal validity as the predominant concern (which is the case in efficacy trials) to external validity. Recently, public health literature has reflected an increasing interest in effectiveness of interventions. One example of such interest is the RE-AIM model (Glasgow *et al.*, 2003). The RE-AIM model includes Reach, Effectiveness, Adoption, Implementation and Maintenance as its key dimensions; it is a multidimensional conceptual model that intends to focus attention on important steps and key elements in the successful translation of research to practice. Evaluation of effectiveness involves monitoring the key parameters of program evaluation, including inputs, activities, outputs and outcomes.

Evaluation of the impact of interventions at the population level is more complicated. "Impact" is not clearly defined in the literature, and the definition apparently varies across authors. Most experts in the field agree that impact refers to the overall effect of an intervention at the population level, including direct effects on persons exposed to the intervention and indirect effects on others in the community whose health outcomes are affected by the health outcomes of persons exposed to the intervention (Aral, 2007). Thus, following wide-scale implantation, population-level impact of an intervention depends on coverage; the magnitude of indirect effects which may vary by the characteristics of the population and the transmission dynamics of the infection. In the STI/HIV area, prevention impact has been defined as a function of the efficacy of the intervention in a specific subpopulation, the contribution of the specific subpopulation to the health outcome of the population, and the achieved effective coverage of the intervention in the specific subpopulation (St Louis and Holmes, 1999).

Evaluation of the population-level impact of an intervention is complicated, since at any point in time multiple interventions are implemented to curb the spread of HIV or other STI. It is difficult to attribute any changes in HIV incidence or prevalence to any one intervention. In the presence of multiple interventions, it is important to consider synergistic or antagonistic interactions between interventions, potential duplication across interventions, potential saturation, incremental impact of each additional intervention, and the potential for diminishing marginal returns (Aral *et al.*, 2007). Recently, broader scope approaches to public health evaluation and health impact assessment have been proposed (Cole and Fielding, 2007). Health impact assessment (HIA) is a combination of methods to examine formally the potential health effects of a proposed policy, program or project. Over the past decade, it has received attention internationally as a cross-sectoral approach to promoting health (Cole and Fielding, 2007).

Evaluating efficacy, effectiveness, and impact of behavioral interventions on HIV and STI risk behavior

Because of the many challenges described above, the majority of trials of HIV behavioral interventions have evaluated the efficacy of these interventions on self-reported behavioral outcomes. Laboratory- or clinically-based assessments of HIV/STI outcomes have been coupled with self-reported data in a number of studies. However, this remains the exception rather than the norm due to the large numbers of participants that are needed to ensure adequate statistical power.

The efficacy of behavioral interventions to prevent HIV/STI risk behavior, and in some cases STI incidence, has been demonstrated in multiple qualitative and quantitative reviews of rigorously evaluated HIV behavioral interventions (e.g., Kalichman *et al.*, 1996; Manhart and Holmes, 2005; Albarracin and Durantini, 2006; Noar, 2008). These reviews have shown that behavioral interventions reduce risk behaviors in a wide range of populations, including drug users (Semaan *et al.*, 2002), heterosexual adults (Neumann *et al.*, 2002), sexually active adolescents (Mullen *et al.*, 2002), men who have sex with men (Johnson *et al.*, 2002; Herbst *et al.*, 2005, 2007a), African-Americans (Darbes *et al.*, 2003), Hispanics (Herbst *et al.*, 2007b), STD clinic patients (Crepaz *et al.*, 2007) and persons living with HIV (Crepaz *et al.*, 2006; Johnson *et al.*, 2006). Although data are limited for some populations, a number of reviews and meta-analyses have also demonstrated the ability of behavioral interventions to reduce STI acquisition (Neumann *et al.*, 2002; Manhart and Holmes, 2005; Crepaz *et al.*, 2006; Noar and Zimmerman, 2008).

Based on the strength of this evidence, the Centers for Disease Control and Prevention (CDC) and other organizations have recommended the implementation of efficacious interventions in prevention programs (Lyles *et al.*, 2007). In addition, the CDC has provided training and technical assistance to accelerate the diffusion of these interventions into the practices of providers working in healthcare and community-based settings (Collins *et al.*, 2006), and has supported increased process and outcome monitoring of prevention programs to better assess the effectiveness of these programs as implemented in the real world.

The CDC's experience provides insight into the challenges associated in moving efficacious behavioral intervention into program practice and the numerous issues that may affect the effectiveness and impact of these interventions. We focus our attention on these issues and challenges in the remainder of this chapter.

Promoting the successful implementation of science-based HIV prevention: the CDC's approach

The core mission of the CDC is to protect the health and safety of people in the United States and throughout the world, provide reliable health information, and improve health through strong partnerships. As part of this mission, the CDC has the lead role in efforts to prevent HIV infection and AIDS in the United States. Improving the effectiveness of HIV prevention programs is a key part of these efforts. In 2007, the CDC provided approximately $640 million to support a range of HIV prevention programs conducted by health departments, community-based organizations and other entities. The vast majority of this funding is provided to state and local health departments and community-based organizations (CBOs). These agencies use CDC funds to support a range of HIV prevention efforts, including HIV counseling and testing, interventions for persons living with HIV, and interventions for persons who are at risk of contracting HIV.

The allocation of resources provided to health departments by the CDC is guided by plans developed by local community planning groups (CPGs). Each plan is designed to meet the needs of populations most affected by HIV, and prioritizes specific HIV prevention strategies to meet local needs. The plan is responsive to and anticipates the course of the HIV epidemic in the local community, assesses and prioritizes local HIV prevention needs, and identifies science-based interventions to meet those needs (Valdiserri *et al.*, 1997; CDC, 1999).

Early in the HIV epidemic, the urgent need to provide HIV prevention information and programs led CPGs to rely on "homegrown" interventions that typically had not been rigorously evaluated but were based on theories of behavior change, formative research findings, or the developers' experience with the target population. Since this time, the number of rigorously evaluated HIV prevention interventions has grown considerably, but this information is often not readily accessible to CPGs and organizations wishing to adopt science-based HIV prevention programs. In the first decade of the epidemic, this information was primarily disseminated through presentations at scientific meetings and peer-reviewed journal articles. Reports at scientific conferences and in scholarly journals make information about new prevention technologies available to the research community, but do little to make interventions accessible to CPGs, health departments and community-based organizations that are responsible for planning and implementing prevention programs (Goldstein *et al.*, 1998; Kelly *et al.*, 2000a, 2000b).

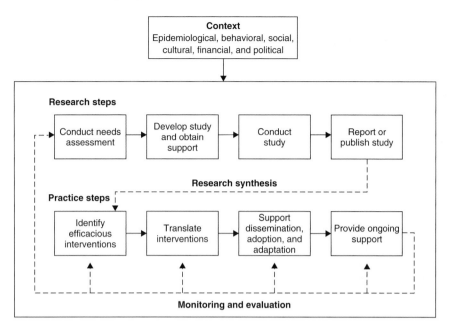

Figure 22.1 Research-to-practice framework for intervention research and technology transfer.

To address this problem, the CDC has developed a range of activities that are designed to improve the ability of local CPGs, health departments and community-based organizations to identify, implement and maintain effective HIV prevention programs. These activities are grounded in a conceptual framework that describes the movement of interventions from research to practice (Figure 22.1). This framework represents a minor modification of a model originally developed by Sogolow and colleagues (2000). Four research-to-practice processes, all of which are influenced by the context in which they occur, are described by the model:

1. Research steps that begin with collecting data that inform the development of an intervention trial and end with the dissemination of results from the trial
2. Research synthesis
3. Program practice steps that begin with the identification of HIV prevention interventions and end with ongoing activities that support their implementation
4. Feedback and evaluation data from implementation of intervention activities that inform future research and prevention efforts.

The broader context in which research and prevention activities are conducted profoundly affects the development, transfer and implementation of interventions that seek to prevent the further spread of HIV. According to Sogolow and colleagues (1996: 23), this context includes "the biological, behavioral, social, cultural, and political events that influence all aspects of the framework". The context can include priorities established by funding agencies that determine the types of research and programs which receive funding, influences that affect the types of prevention messages or strategies that are employed, changes in the profile of persons most affected by the HIV epidemic, advances in HIV treatment and prevention, and a myriad of other factors. As the context changes and evolves, so too must researchers and prevention providers in order to ensure the existence of their funded activities, to respond to the concerns of various stakeholders, and to better meet the needs of the communities they serve.

Research activities are typically undertaken by university faculty in response to a need for the development of an intervention to meet unmet needs identified through formal or informal needs assessments. The complexity of these assessments can vary widely. Ideally, they involve reviews of the research literature, examination of local data and programs, and input from key stakeholders such as community members and providers who serve the community (Higgins *et al.*, 1996). The results of the needs assessment provide a foundation for the development of an intervention and the design of an intervention trial, and often form the evidence base the researcher needs to obtain funding for the full trial. Once the financial and other resources needed for the study are obtained, the research is implemented and ultimately leads to research findings that are reported in peer-reviewed journals, presented at professional conferences, or disseminated in other ways.

The large number of intervention trials for various populations requires a systematic process for the distillation of lessons learned from the evaluation of HIV interventions. Well-established procedures for conducting systematic reviews of diverse research literatures provide the tools that are needed to synthesize this information. The rigorous synthesis of this information provides a much needed empirical basis for the developing future interventions and understanding the effects of existing interventions. The use of meta-analysis allows for the quantitative synthesis of intervention trials to determine their overall efficacy, assess their efficacy within specific subgroups that are at increased risk of HIV infection, and identify intervention strategies and characteristics that are most strongly associated with reductions in HIV risk.

In this framework, research synthesis provides a crucial link between research and practice by identifying interventions that have been evaluated, and by providing an empirical basis for comparing the relative efficacy of interventions and the quality of the evaluation methods used. Identification is not as easy as just finding publications that describe an intervention as having positive outcomes. Differences in intervention approaches, the methods used to evaluate various interventions, the outcomes used to determine efficacy, and approaches to data analysis require further critical examination of evaluation results in order to identify and characterize those interventions with the strongest evidence of efficacy.

Once identified, the next step in the process is the translation of intervention materials and procedures into materials and learning activities that give prevention providers the tools and skills they need to successfully implement an intervention. These materials are often substantially different than those used by researchers and their staff in the original evaluation of an intervention, and need to be easily understood and used by persons that do not have specialized knowledge of the theory and behavior-change principles underlying the intervention. Depending on the complexity of the intervention, in-person training may be needed to give prevention providers the knowledge and skills needed to conduct the intervention. The successful transfer of efficacious interventions into program practice often requires ongoing technical assistance in order to ensure the success of the intervention and its maintenance over time. Various models exist for providing initial training and ongoing support, including structured in-person training, follow-up technical assistance at the provider's location, on-line or telephone-based training and technical assistance, and peer-to-peer networking (Kelly *et al.*, 2000a, 2000b; Collins *et al.*, 2006; Hamdallah *et al.*, 2006; Peterson and Randall, 2006; Shea *et al.*, 2006; Rietmeijer and Gandelman, 2007; Taveras *et al.*, 2007).

Evaluation of intervention implementation and technology transfer activities and the use of the information from evaluation efforts are essential components of the research-to-practice framework. Monitoring of program implementation, soliciting participant feedback, and assessing the ability of prevention program to affect key outcomes (such as behavioral intention, attitudes, and behavior) are important not only to improve the delivery of prevention services in a given community, but also to inform future research and practice steps. This information can stimulate the development of new research hypotheses, identify unmet needs, or document problems with the structure or content of evidence-based interventions that can be addressed in future research. Feedback from evaluation activities also informs program practice by providing information about interventions that are feasible in real-world settings – the types of information, training, and support that prevention providers need to successfully implement and maintain science-based programs. Without this feedback, efforts to develop and transfer effective HIV interventions will likely fail because they are not responsive to the needs and experiences of prevention providers and the communities they serve.

Research-to-practice activities in the CDC's Division of HIV/AIDS Prevention

At the CDC's National Center for HIV/AIDS, Viral Hepatitis, STD, and TB Prevention, the movement of HIV interventions from research trials to prevention programs across the United States is carried out through programs in the Division of HIV/AIDS Prevention that seek to better integrate science and practice. These programs focus on the identification of efficacious interventions, development of provider-friendly materials needed to implement interventions, training and dissemination of efficacious interventions, technical assistance and capacity-building, and program monitoring and evaluation. In the following sections, we provide an overview of these programs and discuss their role in the process of moving research into practice.

Prevention Research Synthesis Project

In 1996, the CDC established the HIV/AIDS Prevention Research Synthesis (PRS) Project to provide an ongoing mechanism for quantitatively assessing the efficacy of HIV prevention interventions, describing characteristics of interventions and studies associated with efficacy, finding interventions with strong evidence of efficacy, and identifying gaps in intervention research (Sogolow *et al.*, 2002). PRS conducts comprehensive computerized and hand searches of published and in press reports that describe the results of HIV intervention trials (Lyles *et al.*, 2007). These searches have yielded more than 500 reports of evaluation studies that assessed the ability of behavioral interventions to reduce HIV or STI risk or disease incidence (Sogolow *et al.*, 2002; Lyles *et al.*, 2007).

In order to quantitatively assess the combined effects of these interventions and identify those with strong evidence of efficacy, PRS staff review and code these studies using standardized procedures. The codes summarize key information about each report, including characteristics of the intervention, the population studied, the research design, how well the study was implemented, and the results of the study. The codes provide the data for meta-analyses that subject the coded information from many different studies to quantitative analyses that summarize the effects of HIV/STI interventions and identity intervention and study characteristics that are most strongly associated with significant positive outcomes.

In order to identify interventions with the strongest evidence of efficacy, PRS uses criteria that were developed based on reviews of the research literature and consultation with national experts (Sogolow *et al.*, 2002; Flores and Crepaz, 2004; Lyles *et al.*, 2007). At present, PRS uses these criteria to determine whether interventions provide best or promising evidence of efficacy. Criteria used to establish best-evidence of efficacy include the strength of positive outcomes, quality of the study design, quality of the study implementation and analysis (Lyles *et al.*, 2006, 2007). The promising-evidence criteria assess the same domains, but the threshold for satisfying some criteria is relaxed (CDC, 2007).

As of August 2007, PRS had identified 49 individual and small-group interventions that met either best- or promising-evidence criteria (CDC, 2007). The interventions identified by PRS are based on a number of behavior change-theories, and use a variety of behavior-change strategies (such as risk assessment, motivational enhancement, skills training) that are delivered in different ways (e.g., individual counseling sessions, small groups, educational videos). Most of the interventions focus on subpopulations that have been disproportionately affected by HIV, and include interventions for heterosexual adults, at-risk youth, men who have sex with men, and drug-using men and women, as well as interventions for persons living with HIV. A current list and description of these interventions can be found at www.cdc.gov/hiv/topics/research/prs.

Replicating Effective Programs Project

Materials needed to support the implementation of interventions identified by PRS are developed by the Replicating Effective Programs (REP) Project, which began in 1996. REP develops plain-language, provider-friendly intervention packages that are designed to provide the resources prevention providers need to plan for, implement and maintain successful interventions in their communities (Neumann and Sogolow, 2000; Eke *et al.*, 2006). The intervention packages include:

- information that helps agencies decide if a given intervention meets their needs and is consistent with their organizational goals and capacity;
- an intervention manual;
- supporting materials (e.g., video tapes, posters, handouts) that are used when conducting the intervention; and
- tools and guidance that assist agencies with evaluating and maintaining the intervention.

REP packages are developed in partnership between CDC staff and the original researchers, or organizations they are collaborating with (Kegeles *et al.*, 2000; Neumann and Sogolow, 2000; Eke *et al.*, 2006). The development of these packages is typically guided by input from advisory boards that are comprised of prevention providers, experienced trainers, members of the population served by the intervention, and other stakeholders. These materials are then pilot-tested by prevention providers who are naïve to the intervention, to provide feedback on the REP package's usability, clarity, and completeness. This feedback is used to refine the materials and to develop the final package, which is disseminated in the next research-to-practice step by the CDC's Diffusion of Effective Interventions Project.

As of April 2008, 11 interventions had been packaged by REP and 10 additional interventions were at various phases of the translation or pilot-testing process. The interventions packaged by REP address the needs of a wide range of populations at risk for HIV, including at-risk women, men who have sex with men, injection drug users, incarcerated men, at-risk youth, STD clinic patients and persons living with HIV. Additional information about these interventions and the REP packages can be found at: www.cdc.gov/hiv/topics/prev_prog/rep.

Disseminating effective behavioral interventions

Merely making the information and tools needed to implement an intervention is not sufficient to successfully diffuse an intervention and ensure that it is delivered as intended. The Diffusion of Effective Behavioral Interventions (DEBI) Project works to disseminate interventions with best- and promising-evidence of efficacy into prevention practice through national training, technical assistance and capacity-building (Collins *et al.*, 2006). The DEBI Project, which was initiated in 2002, is based on a bidirectional approach to the research-to-practice process that draws on the experiences of researchers, trainers, and prevention providers.

The DEBI Project uses multiple strategies to move interventions into practice. This process starts with developing a diffusion strategy for each intervention, and a marketing strategy. The marketing strategy includes an assessment of intervention needs perceived by prevention providers, and the capacity of potential adopters for a given intervention. DEBI interventions are marketed to prevention providers through information and resources available on the Internet (see www. effectiveinterventions.org), satellite- and web-based broadcasts, and presentations at national meetings. In addition, adoption of DEBI interventions is encouraged, and some cases required, through various CDC recommendations and funding announcements.

Training and technical assistance for 12 DEBI interventions was available as of August 2007. Training for each intervention is available nationwide from a variety of trainers and technical assistance providers that are experienced at working with health department and CBO staff, and understand the challenges associated with implementing successful prevention programs. At present, DEBI training sessions are in-person multiday events that use adult learning principles to give participants a clear understanding of the theory and behavior-change principles underlying the intervention, familiarize participants with the structure and content of the intervention, teach them how to use the intervention manual and materials, and provide them with opportunities to develop and refine the skills needed to conduct the intervention. The structure, content and duration of the DEBI training sessions vary considerably, depending on the nature and complexity of the intervention and the level of skill needed to conduct the intervention. As of 30 June 2007, more than 8600 persons representing 3719 agencies had attended one of the 470 DEBI training sessions that have been conducted.

Capacity building and technical assistance

The successful adoption, implementation and maintenance of HIV prevention interventions are significantly improved when local organizations have the experience, skills and resources needed to implement these interventions, and when ongoing support and technical assistance are available (Miller, 2001; Gandelman *et al.*, 2006; Collins *et al.*, 2007). Organizations with greater capacity (including well-trained staff with necessary skill sets, prior experience in implementing similar interventions, an established track record working with the target population, stable financial resources and a supportive management structure) are likely to be more successful implementing evidence-based prevention programs than are organizations with less capacity. Even when sufficient organizational capacity is present, training and technical assistance play important roles in the successful adoption, implementation and maintenance of evidence-based HIV prevention interventions. For example, Kelly and colleagues (2000a, 2000b) found that ongoing technical assistance resulted in more frequent adoption and use of evidence-based interventions than did only providing intervention manuals, or providing manuals plus formal training without ongoing technical assistance.

In addition to the manuals developed by REP and the training provided by DEBI, the CDC supports the implementation and maintenance of evidence-based HIV prevention interventions through efforts to strengthen organizational capacity and train prevention providers in core areas and skills, and by providing ongoing technical assistance. These activities are conducted by national, regional and non-governmental organizations, contractors and private sector organizations, and focus on four priority areas (Taveras *et al.*, 2007):

1. Improving the capacity of community-based organizations to develop and sustain organizational infrastructures that support the delivery of effective HIV prevention services and interventions
2. Improving the capacity of community-based organizations and health departments to adapt, implement and evaluate effective HIV prevention interventions
3. Improving the capacity of racial and ethnic minority communities and organizations to implement models that will increase access to, and utilization of, HIV prevention interventions and services
4. Improving the capacity of community planning groups and health departments to include HIV-infected and HIV-affect racial and ethnic minority participants in the community planning process, and to increase parity, inclusion, and representation on community planning groups.

The CDC's HIV capacity-building activities are guided by conceptual models, prior research, input from prevention and capacity-building providers, and evaluation results (Collins *et al.*, 2007; Nu'Man *et al.*, 2007). The approach taken by the CDC recognizes the strengths that both capacity-building providers and the providers of local HIV prevention services bring to HIV prevention efforts, which rejects the "expert-to-non-expert" approach that characterize some technology transfer efforts (Nu'Man *et al.*, 2007). This is accomplished by the identification and prioritization of capacity-building needs in collaboration with local prevention providers in order to analyze and categorize these needs, develop strategies to build capacity, apply these strategies, and monitor progress and reassess needs over time.

Feedback from monitoring and evaluation

The monitoring and evaluation of capacity-building activities and HIV prevention programs provide important sources of feedback that are critical for the continuous improvement of HIV prevention efforts. As shown in Figure 22.1, the feedback from monitoring and evaluation activities should lead to the development of new research questions, and evaluation data should be used to inform the delivery of HIV prevention programs. Initial evaluation results based on responses from 52 organizations receiving capacity-building services supported by the CDC indicate that these services were viewed favorably and were perceived to have had a positive influence on the planning, monitoring and evaluation of local HIV prevention activities (Nu'Man *et al.*, 2007). Evaluation of the CDC's capacity-building activities has also identified challenges to achieving its goals, including an inability to access training in a timely manner; a need for more in-depth training on implementing, adapting, and monitoring evidence-based interventions; a need to increase the number of skilled capacity-building consultants; and a need for culturally relevant materials in Spanish (Ayala *et al.*, 2007; Sheth *et al.*, 2007). This feedback has provided important information that the CDC is using in its continuous quality improvement efforts in order to further refine its ability to strengthen the capacity of local agencies.

Key information that helps improve agency capacity and efforts to build this capacity also comes from the monitoring and evaluation of HIV prevention programs and services. Such information allows local providers and funding agencies to identify gaps between intended and actual program delivery, and to assess their ability to reach the target audience, the perceptions of intervention participants, and the effects of HIV prevention programs. A detailed discussion of the CDC's efforts to monitor and evaluate HIV prevention programs is beyond the scope of this chapter, but interested readers can find this information elsewhere (see Thomas *et al.*, 2006).

Effectiveness and adaptation

Two fundamental issues confront those who seek to use evidence-based interventions to improve HIV prevention program practice and prevent the further spread of HIV in vulnerable communities. These are the ability of interventions tested in rigorous trials to yield similar results when implemented in real-world settings (i.e., replication and effectiveness), and the ability of these interventions to be adapted to meet the needs of local communities and prevention providers (Rietmeijer and Gandelman, 2007). In research trials, intervention implementation is carefully monitored (usually by staff who developed the intervention and are intimately familiar with the theory and goals behind each intervention component), implemented by highly trained staff, and delivered to members of a well-defined study sample who usually receive monetary and non-monetary incentives for participating in intervention and/or research activities. It cannot be assumed that the efficacy demonstrated in such trials will always translate into effectiveness when interventions are used in real-world settings by local prevention providers.

Further complicating this issue is the fact that interventions are often adapted by local prevention providers, which may also affect effectiveness. Providers adapt interventions to meet local needs for a number of reasons (Miller, 2001; Bell *et al.*, 2007), including the fact that interventions do not exist for some groups that are disproportionately affected by HIV. Other reasons for adaptation include: a need to make the intervention work in the settings that the agency works in; a need to adjust to the realities of the staffing and resources that are available; and a need to adapt existing interventions to address scientific advances, emerging behavioral trends, or changes in the environmental, social, and cultural contexts of the communities being served.

Adapting an intervention to improve its relevance for a specific population and setting may serve to maintain or enhance its effectiveness. However, the limits of adaptation and the best way to go about adapting interventions are not currently known. Adaptation may threaten the ability of an intervention to achieve reductions in HIV risk if core elements responsible for an intervention's efficacy are eliminated or substantively changed. The complexity of adapting HIV interventions is reflected in the numerous challenges reported by those who have adapted interventions or studied the adaptation process (Miller, 2001; McKleroy *et al.*, 2006; Bell *et al.*, 2007; Rietmeijer and Gandelman, 2007). In response to these challenges, researchers have begun to document the adaptation process and are assessing the effects of adapted interventions (Miller, 2001; Harshbarger *et al.*, 2006; Hitt *et al.*, 2006; McKleroy *et al.*, 2006; Rebchook *et al.*, 2006; Bell *et al.*, 2007).

Evaluation research has shown that results achieved in efficacy trials can be replicated and that it is possible for adapted interventions to result in positive outcomes (e.g., Kelly *et al.*, 1992, 1997; Kegeles *et al.*, 1999; Sikkema *et al.*, 2000; Jones *et al.*, 2005; Somerville *et al.*, 2006; Villarruel *et al.*, 2006). However, evaluation research has also shown that this does not always happen. Some studies that have sought to replicate adapted or unaltered interventions have not achieved positive results (e.g., Collins *et al.*, 1999; Elford *et al.*, 2001; Flowers *et al.*, 2002; Bell *et al.*, 2007). An illustration of these issues is provided in Box 22.1, which discusses the experiences of various groups using and adapting the Popular Opinion Leader intervention.

Emerging issues and challenges

There are many challenges faced when translating research into practice. Some of the challenges related to technical, organizational and capacity issues have been discussed elsewhere (Neumann and Sogolow, 2000). In the final section of this

---

Box 22.1 Case study: Popular Opinion Leader intervention

The Popular Opinion Leader (POL) intervention was originally developed by Jeff Kelly and colleagues for use with gay and bisexual men who frequented gay bars in small cities in the southern United States (Kelly, 2004; Kelly *et al.*, 1991, 1992, 1997). In the intervention, popular opinion leaders in a well-defined target population are identified, recruited, and trained to have conversations with their peers that encourage safer sex norms, identify the advantages of safer sex, and communicate practical ways to initiate and maintain safer behaviors. After initial training, the popular opinion leaders meet weekly to improve their knowledge and ability to successfully engage their peers in risk reduction conversations. Cues in the environment, such as posters and buttons worn by popular opinion leaders, are also key elements of the intervention that are used to trigger the risk reduction conversations.

The first POL study, which was conducted in 1989, found that men in the intervention city reported significant reductions in unprotected anal intercourse and multiple sex partners as well as a significant increase in condom use compared to men in two comparison cities that did not receive the intervention (Kelly *et al.*, 1991). These findings were later replicated by the original researchers when the intervention was introduced in the comparison cities (Kelly *et al.*, 1992) and in a larger study of gay men in small cities that expanded this research beyond the southern region of the United States (Kelly *et al.*, 1997). Since that time, the intervention has been widely disseminated by CDC as part of its DEBI program and has been successfully adapted for other populations in the United States, including low-income women (Sikkema *et al.*, 2000), young Latino migrant men (Somerville *et al.*, 2006), African-American men who have sex with men (Jones *et al.*, 2005), and men in New York City bars frequented by male sex workers (Miller *et al.*, 1998).

---

chapter, we discuss those challenges related to selecting the most cost-effective EBIs for translation, and the effectiveness and applicability of the packaged EBIs once they move into practice.

Research synthesis

Research synthesis is the critical link between research and practice, and in HIV prevention it rigorously summarizes HIV behavioral prevention research findings and transforms them into evidence-based recommendations that can be used by national, state and local agencies for HIV prevention planning and decision-making

(Lyles *et al.*, 2006). Although progress has been made with translating research into practice, many previously identified EBIs have already been (or are in the process of being) disseminated to prevention providers across the nation. Key research questions remain, including intervention effectiveness, cost-effectiveness, and the effects of adaptation. Looking forward, there will be opportunities to strengthen collaboration between the national, state and local prevention agencies to evaluate EBIs as they are implemented in practice, and also to synthesize the findings from these evaluation studies to build on the current evidence and ultimately maximize the intervention's effectiveness (Lyles *et al.*, 2006).

Economic issues

Despite the ability to provide critical information for decision-making, many efficacy reviews, including the PRS review, have generally not considered economic issues in its evidence-based recommendations (Lyles *et al.*, 2006). Cost-effectiveness analyses can, for example, compare the costs, benefits or utilities of different interventions, or assess the net benefits of an intervention at the societal level. Information on these economic parameters helps to provide data for understanding the return on investment and, consequently, assist decision-makers in optimizing resource allocation (Holtgrave and Anderson, 2004; Pignone *et al.*, 2005; Lyles *et al.*, 2006). Critical appraisal of the economics of HIV behavioral interventions is underused and will be implemented along with future PRS research synthesis activities (Lyles *et al.*, 2006).

Effectiveness in the community

As stated at the beginning of this chapter, effectiveness is the extent to which an intervention works in the real world, as opposed to efficacy, which is the extent to which an intervention works under the optimal conditions of a highly controlled research environment (Flay, 1986). Consequently, the efficacy of an intervention does not always guarantee its effectiveness in practice (Flay, 1986; Glasgow *et al.*, 2003). The PRS efficacy reviews are based on the intervention's effects within relatively rigorous and controlled research environments; however, the effectiveness of most of these EBIs is unknown (Lyles *et al.*, 2006). Experimental or quasi-experimental designs are among of the best methods for evaluating these interventions, because other factors (such as implementation, fidelity, availability and acceptance) can be monitored but will naturally vary given the real-world context (Flay, 1986; Kelly *et al.*, 2000a, 2000b; Glasgow *et al.*, 2003; Lyles *et al.*, 2006). Any evaluation component should ideally include high-quality performance and outcome measures, and a future broad-based evaluation system could provide a systematic way to collect outcome monitoring or evaluation data for those organizations with sufficient funding and capacity to conduct effectiveness studies (Lyles *et al.*, 2006).

Adapting interventions

As EBIs are translated into practice, they will be directly applicable and ready for immediate implementation by prevention programs (Lyles *et al.*, 2006). However, several populations at greatest need of effective prevention tools (e.g., MSM of color, substance-using MSM, transgendered persons) are not the target or focus of the previously identified EBIs (CDC, 1999, revised 2001; Kay *et al.*, 2003; Lyles *et al.*, 2006) or in the recently identified best-evidence behavioral interventions (Lyles *et al.*, 2007). As a result, there remain many communities for which these intervention packages may not be directly applicable, since many social, cultural or contextual factors within communities may not be not appropriately addressed by the intervention. In these situations, a more efficient, effective and cost-effective way to proceed involves adapting the EBI to meet the community-specific unmet prevention needs (Lyles *et al.*, 2006). Guidelines on how to adapt an intervention have been developed, but have been tested in only a limited number of applied research settings (McKleroy *et al.*, 2006; Wingood and DiClemente, 2008). Operational research on the adaptation process and the effectiveness of adapted interventions in real-world settings should help provide greater insights into the best way to adapt interventions while preserving their effects (Lyles *et al.*, 2006).

Geography and existing community structures

Geographical contexts may influence the relevance and applicability of interventions as well as the cultural acceptability (Bell *et al.*, 2007). Applying effective programs in new places may necessitate different adaptations for timing, number and length of program sessions. They need to fit in to ongoing organizational structures and schedules. For example, the lives of adolescents in the United States may vary greatly depending on the type of area in which they live (Bell *et al.*, 2007). The greater dispersal of rural and suburban youth compared to urban youth will influence and complicate the methods and opportunities for delivering school-based prevention programs, as well as interventions outside of the school setting. The greater transportation needs of rural and suburban students will impact upon their ability to access interventions, and may increase their dependence on others (e.g., parents, other adults) for participation in interventions when the venues are far from home. Another example involves the different considerations and challenges that might surround the implementation of intervention for homeless or drug-using populations, depending upon the geographic location or site of implementation. Concerns regarding infrastructural integrity, accessibility and security will be paramount for both staff and participants. In addition, the chaotic or inconsistent use of services by the intervention participants may influence how the intervention is implemented, when it is offered, how and whether participants are pre-screened and selected, and whether and what type of incentives are used.

Another element of geographic context is the need to obtain local support for successful program implementation. Numerous examples abound of community resistance to the establishment of local programs for drug users or the homeless ("not in my backyard"); local parental concerns about the content of interventions targeting youth; or community mistrust of local or federal government-funded interventions targeting sub-sectors of that group (e.g., urban poor). However, attention to the geographic context is also needed even when local stakeholders have provided theoretical support for a known intervention (Bell *et al.*, 2007). Sometimes, despite best efforts being made to adapt interventions to be more culturally appropriate and sensitive, differences in the receptivity of communities may be the result of all communities not being equally receptive to a particular intervention, of the openness of communities to receive the advice of outside experts; or of changing community priorities (Bell *et al.*, 2007). In these circumstances, greater attention to community cultures and infrastructures and the involvement of community leaders will be required (see following section).

Community cultures and program content

Understanding local community cultures and ensuring that these are reflected within the adaptation of an intervention is crucial. Using inappropriate contextual references or communication examples that clash with local cultural norms is counterproductive, and should be avoided wherever possible. Community and youth councils and advisory boards can be helpful in identifying areas where changes in the intervention may be necessary. They may also assist the researcher in obtaining and assuring local support for the program (Bell *et al.*, 2007). Ultimately, investigators may be able to make changes to programs or intervention in order to improve relevance and acceptability with community cultures. These may involve changing or adjusting communication styles; using locally relevant examples or contextually appropriate narratives; introducing participants to culturally relevant attitudes; ensuring that key staff are bilingual; and adapting programs to address emergent needs or specific problems within the new target population (Bell *et al.*, 2007).

Promising evidence behavioral interventions

HIV prevention organizations may face difficulties in trying to implement EBIs in situations that differ fundamentally from those originally studied in the research supporting the EBIs (Lyles *et al.*, 2006). In these instances, the best-evidence behavioral intervention packages may simply not be applicable, and adapting these interventions will not meet the needs of the community. Therefore, alternatives based on weaker scientific evidence – for example, EBIs that fall short of meeting the criteria for best evidence for scientific demonstration of efficacy but that appropriately address a community's needs and show some potential that the intervention could reduce HIV-related risk behaviors – may need to be considered. Lyles *et al.* (2007) identified a number of interventions that potentially fall within this category of behavioral interventions with potential efficacy, and have defined a second tier of less rigorously evaluated interventions (promising-evidence interventions) with evidence of efficacy. These promising interventions and other interventions that have been less fully evaluated may in fact have the ability significantly to affect HIV risk behavior. In some cases such interventions may even be more efficacious than existing EBIs, but their efficacy has not yet been demonstrated in a rigorous trial (Lyles *et al.*, 2006). There is a need for guidance to assist communities to identify interventions that do not currently meet the highest standards for evidence-based interventions and to match these interventions with high-risk gap populations that could benefit from them. Care should be taken, however, to underscore that the evidence of the effectiveness of these promising-evidence interventions is weaker, and that a commitment to ongoing monitoring and process and outcome evaluation of their effectiveness is required. In the future, partnering with local and federal research agencies to fund and support operational research or more robust efficacy and effectiveness studies will be required to accelerate the generation of evidence available to support their implementation.

Conclusions

The tool-kit of evidence-based HIV prevention strategies is expanding, and alongside this is a growing acknowledgment that the adaptation of evidence-based interventions (EBIs) is increasingly critical. As we look to the future, more robust guidance will be required to facilitate and expedite the availability of appropriate and effective interventions for all populations by providing implementers with guidance regarding adapting EBIs for the unique circumstances of the agency and target population. Adaptations will need to be done as expediently and cost-effectively as possible, while maintaining the scientific integrity of the interventions. McKelroy *et al.* (2006) have proposed such an adaptation guidance, which presents a model to increase consistency in terms of how the adaptation of EBIs is discussed and operationalized. The hope is that this will allow for greater consistency of approach; generalization and synthesis across disciplines; and facilitate interchange between researchers and implementers.

In addition to standardizing approaches for adaptation, consideration should be given to investigating and identifying core elements of interventions, and to determining what elements are most likely responsible for addressing risk factors, behavioral determinants and risk behaviors during the empirical research. This is particularly urgent as the pace and diversity of adaptations increase, and so too the attendant pressures to ensure that multiple components of an intervention are made culturally and contextually relevant. Having an improved understanding of the core elements and internal logic of the intervention throughout the development of the intervention would strengthen and expedite future dissemination and adaptation efforts. It would also ensure greater transparency in the structure and delivery of interventions, thereby improving the understanding of implementers (and prevention partners) as to why an intervention is effective and what needs to be preserved in its local evolution.

In this chapter we have highlighted ways in which multiple elements and considerations must be brought together for a prevention program to be replicated and to produce behavior change in a new environment. Although many of the structural and cultural challenges may be addressed by implementing theory-driven amendments to interventions, local constraints and realities will also influence the nature and form of these adaptations. Care will need to be taken in moving forward as our understanding develops regarding the impact of changes, amendments and additions on the effectiveness of interventions. For example, Mayer *et al.* (1986) argue that "additions" to a program do not necessarily eviscerate its effectiveness, but that "adaptations" or "alterations" may. In addition to concerns regarding program effectiveness, other issues that need to be considered include the impact of moving effective programs across geographic regions and social contexts. Research studies are now demonstrating the important curricular and programmatic elements needed to facilitate such changes, and it will be essential that these are incorporated into guidance for program implementers. In all of these circumstances, a commitment to ongoing monitoring and evaluation of adapted interventions will be required.

In summary, adequately addressing these issues will require increased collaboration between researchers and practitioners who are responsible for implementing HIV prevention interventions. These collaborations are critical to improving our understanding of the use and effectiveness of HIV interventions as they are delivered in community-based settings. The potential rewards of such collaborations are considerable, and hold much promise for improving our ability to reduce the further spread of HIV in the United States and beyond.

Experience with the POL intervention also illustrates the challenges associated with replicating evidence-based interventions. Two studies conducted in the United Kingdom found that the POL intervention did not lead to significant reduction in risk among men attending gay gyms in London (Elford *et al.*, 2001) or men attending gay bars in Glasgow (Flowers *et al.*, 2002). The failure of POL in these settings raised questions regarding the definition of core elements of the POL intervention that are responsible for its efficacy, the need to achieve key programmatic objectives in the faithful replication of an intervention, and the ability of evidence-based interventions to remain feasible and efficacious in other settings and times (Elford *et al.*, 2004; Kelly, 2004). In the case of POL, the inability of the UK researchers to replicate findings from the US may have been due to inadequate implementation of the intervention (i.e., too few popular opinion leaders were recruited, too few conversations took place), implementation in larger social settings that did not facilitate intimate conversations about safer sex and diffusion of safer sex norms, and, possibly, differences in the receptivity of popular opinion leaders and community members to HIV-related issues as a result of prior exposure to HIV information and the reduced impact of the epidemic due to the availability of HIV treatment (Elford *et al.*, 2004; Kelly, 2004).

These studies show the importance of clearly identifying the core elements of HIV prevention interventions and implementing these elements with fidelity. Even if this is done, however, it is not certain that interventions will achieve the same effects as they did in the original intervention trial or subsequent evaluations. These findings highlight the importance of ongoing program monitoring to assess the delivery of HIV prevention interventions and their fidelity to the original intervention protocol. They also show the need for outcome monitoring to ensure that interventions are achieving their intended outcomes. Evaluation data from CDC-funded projects will provide additional information about the effectiveness of POL in diverse communities in the United States in coming years. The ability of this intervention to be adapted to the needs of very diverse communities across the globe is currently being tested in a five-country trial, which will provide new insights into the potential flexibility of the POL intervention and its ability to be generalized to new cultures, risk populations and times (NIMH Collaborative HIV/STD Prevention Trial Group, 2007).

Acknowledgment

The authors thank Patricia Jackson for her outstanding support in the preparation of this article.

Disclaimer

The findings and conclusions in this chapter are those of the authors, and do not necessarily represent the views of the Centers for Disease Control and Prevention.

References

Albarracin, D. and Durantini, M. R. (2006). Empirical and theoretical conclusions of an analysis of outcomes HIV-prevention interventions. *Curr. Dir. Psychological Sci.*, 15, 73–8.

Aral, S. O. (2007). Sexually transmitted disease prevention: from efficacy-to-effectiveness; from effectiveness-to-impact. Presented at the *17th Meeting of the International Society for STD Research (ISSTDR) and the 10th International Union against STIs (IUSTI), 29 July 2007, Seattle, Washington.*

Aral, S. O., Lipshutz, J. A. and Douglas, J. M. Jr (2007). Introduction. In: S. O. Aral, J. M. Douglas Jr and J. A. Lipshutz (eds), *Behavioral Interventions for Prevention and Control of Sexually Transmitted Diseases.* New York, NY: Springer Science+ Business Media, pp. ix–xviii.

Auvert, B., Taljaard, D., Lagarde, E. *et al.* (2005). Randomized controlled intervention trial of male circumcision for reduction of HIV infection risk: The ANRS 1265 Trial. *PLoS Med.*, 2, e298.

Ayala, G., Chión, M., Díaz, R. M. *et al.* (2007). Acción Mutua (Shared Action): a multi-pronged approach to delivering capacity-building assistance to agencies serving Latino communities in the United States. *J. Public Health Manage. Pract.*, January(Suppl.), S33–9.

Bailey, R. C., Moses, S., Parker, C. B. *et al.* (2007). Male circumcision for HIV prevention in young men in Kisumu, Kenya: a randomized controlled trial. *Lancet*, 369, 643–56.

Barreto, M. L. (2005). Efficacy, effectiveness, and the evaluation of public health interventions. *J. Epidemiol. Community Health*, 59, 345–6.

Bell, S. G., Newcomer, S. F., Bachrach, C. *et al.* (2007). Challenges in replicating interventions. *J. Adolesc. Health*, 40, 514–20.

CDC (1999). *Guidance: HIV Prevention Community Planning for HIV Prevention Cooperative Agreement Recipients.* Atlanta, GA: Centers for Disease Control and Prevention.

CDC (2007). HIV/AIDS Prevention Research Project. Available at: http://www.cdc.gov/hiv/topics/research/prs/ (accessed 08-11-2007).

Check, E. (2007). Scientists rethink approach to HIV gels. *Nature*, 446, 12.

Cole, B. L. and Fielding, J. E. (2007). Health impact assessment: a tool to help policy makers understand health beyond health care. *Annu. Rev. Public Health*, 28, 393–412.

Collins, C., Kohler, C., DiClemente, R. and Wang, W. Q. (1999). Evaluation of the exposure effects of a theory-based street outreach HIV intervention on African-American drug users. *Eval. Prog. Plan.*, 22, 279–93.

Collins, C., Harshbarger, C., Sawyer, R. and Hamdallah, M. (2006). The Diffusion of Effective Behavioral Interventions Project: development, implementation, and lessons learned. *AIDS Educ. Prev.*, 18(Suppl. A), 5–20.

Collins, C., Phields, M. E. and Duncan, T. for the Science Application Team (2007). An agency capacity model to facilitate implementation of evidence-based behavioral interventions by community-based organizations. *J. Public Health Manage. Pract.*, January(Suppl.), S16–23.

Crepaz, N., Lyles, C. M., Wolitski, R. J. *et al.* for the HIV/AIDS Prevention Research Synthesis (PRS) Team (2006). Do prevention interventions reduce HIV risk behaviours among people living with HIV? A meta-analytic review of controlled trials. *AIDS*, 20, 143–57.

Crepaz, N., Horn, A. K., Rama, S. M. *et al.* for the HIV/AIDS Prevention Research Synthesis Team (2007). The efficacy of behavioral interventions in reducing HIV risk sex behaviors and incident sexually transmitted diseases in black and Hispanic sexually transmitted disease clinic patients in the United States: a meta-analytic review. *Sex. Transm. Dis.*, 34, 319–32.

Darbes, L. A., Crepaz, N., Lyles, C. M. *et al.* (2003). Meta-analysis of HIV prevention interventions in African-American heterosexuals in the US. Paper presented at the *National HIV Prevention Conference, Atlanta, 27 July.*

Eke, A. N., Neumann, M. S., Wilkes, A. L. and Jones, P. L. (2006). Preparing effective behavioral interventions to be used by prevention providers: the role of researchers during HIV prevention trials. *AIDS Educ. Prev.*, 18(Suppl. A), 44–58.

Elford, J., Bolding, G. and Sherr, L. (2001). Peer education has no significant impact on HIV-risk behaviours among gay men in London. *AIDS*, 15, 535–8.

Elford, J., Bolding, G. and Sherr, L. (2004). Popular opinion leader in London: a response to Kelly. *AIDS Care*, 16, 151–8.

Flay, B. R. (1986). Efficacy and effectiveness trials (and other phases of research) in the development of health promotion programs. *Prev. Med.*, 15, 451–74.

Flores, S. A. and Crepaz, N. for the HIV Prevention Research Synthesis Team (2004). Quality of study methods in individual- and group-level HIV intervention research: critical reporting elements. *AIDS Educ. Prev.*, 16, 341–52.

Flowers, P., Hart, G., Williamson, L. *et al.* (2002). Does bar-based, peer-led sexual health promotion have a community-level effect among gay men in Scotland? *Intl J. STD AIDS*, 13, 103–8.

Gandelman, A. A., Desantis, L. M. and Rietmeijer, C. A. (2006). Assessing community needs and agency capacity–an integral part of implementing effective evidence based interventions. *AIDS Educ. Prev.*, 18(Suppl. A), 32–43.

Glasgow, R. E., Lichtenstein, E. and Marcus, A. C. (2003). Why don't we see more translation of health promotion research to practice? Rethinking the efficacy-to-effectiveness transition. *Am. J. Public Health*, 93, 1261–7.

Global HIV Prevention Working Group (2007). Bringing HIV prevention to scale: an urgent global priority. Available at www.GlobalHIVPrevention.org (accessed 03-30-2008).

Goldstein, E., Wrubel, J., Faigeles, B. and DeCarol, P. (1998). Sources of information for HIV prevention program managers: a national survey. *AIDS Educ. Prev.*, 10, 63–74.

Gray, R. H. and Wawer, M. J. (2007). Randomised trials of HIV prevention. *Lancet*, 370, 200–1.

Gray, R. H., Kigozi, G., Serwadda, D. *et al.* (2007). Male circumcision for HIV prevention in men in Rakai, Uganda: a randomized trial. *Lancet*, 368, 657–66.

Hamdallah, M., Vargo, S. and Herrera, J. (2006). The VOICES/VOCES success story: effective strategies for training, technical assistance and community-based organization implementation. *AIDS Educ. Prev.*, 18(Suppl. A), 171–83.

Hayes, R. J., Alexander, N. D. E., Bennet, S. and Cousins, S. N. (2000). Design and analysis issues in cluster-randomized trials of interventions against infectious diseases. *Stat. Methods Med. Res.*, 9, 95.

Herbst, J. H., Sherba, R. T., Crepaz, N. *et al.* (2005). A meta-analytic review of HIV behavioral interventions for reducing sexual risk behavior of men who have sex with men. *J. Acquir. Immune Defic. Syndr.*, 39, 228–41.

Herbst, J. H., Fielding, J. E., Abraido-Lanza, A. *et al.* (2007a). Recommendations for use of behavioral interventions to reduce the risk of sexual transmission of HIV among men who have sex with men. *Am. J. Prev. Med.*, 32(Suppl. S), S36–7.

Herbst, J. H., Kay, L. S., Passin, W. F. *et al.* for the HIV/AIDS Prevention Research Synthesis Team (2007b). A systematic review and meta-analysis of behavioral interventions to reduce HIV risk behaviors of Hispanics in the United States and Puerto Rico. *AIDS Behav.*, 11, 25–47.

Higgins, D. L., O'Reilly, K., Tashima, N. *et al.* (1996). Using formative research to lay the foundation for community level HIV prevention efforts: an example from the AIDS Community Demonstration Projects. *Public Health Rep.*, 111(Suppl. 1), 28–35.

Hitt, J. C., Robbins, A. S., Galbraith, J. S. *et al.* (2006). Adaptation and implementation of an evidence-based prevention counseling intervention in Texas. *AIDS Educ. Prev.*, 18(Suppl. A), 108–18.

Holtgrave, D. R. and Anderson, T. (2004). Utilizing HIV transmission rates to assist in prioritizing HIV prevention services. *Intl J. STD AIDS*, 15, 789–92.

Johnson, B. T., Carey, M. P., Chaudoir, S. R. and Reid, A. E. (2006). Sexual risk reduction for persons living with HIV: research synthesis of randomized controlled trials, 1993 to 2004. *J. Acquir. Immune Defic. Syndr.*, 41, 642–50.

Johnson, W. D., Hedges, L. V., Ramirez, G. *et al.* (2002). HIV prevention research for men who have sex with men: a systematic review and meta-analysis. *J. Acquir. Immune Defic. Syndr.*, 30(Suppl. 1), S118–29.

Jones, K., Gray, P., Wang, T. *et al.* (2005). Evaluation of a community-level peer-based HIV prevention intervention adapted for young black men who have sex with men (MSM). Paper presented at the *XVIth International AIDS Conference, Toronto, 13–18 August*.

Kalichman, S. C., Carey, M. P. and Johnson, B. T. (1996). Prevention of sexually transmitted HIV infection: A meta-analytic review of behavioral outcome literature. *Ann. Behav. Med.*, 18, 6–15.

Kegeles, S. M., Hays, R. B., Pollack, L. M. and Coates, T. J. (1999). Mobilizing young gay and bisexual men for HIV prevention: a two-community study. *AIDS*, 13, 1753–62.

Kegeles, S. M., Rebchook, G. M., Hays, R. B. *et al.* (2000). From science to application: The development of an intervention package. *AIDS Educ. Prev.*, 12(Suppl. A), 62–74.

Kelly, J. A. (2004). Popular opinion leaders and HIV prevention peer education: resolving discrepant findings, and implications for the development of effective community programmes. *AIDS Care*, 16, 139–50.

Kelly, J. A., St Lawrence, J. S., Diaz, Y. E. *et al.* (1991). HIV risk behavior reduction following intervention with key opinion leaders of population: an experimental analysis. *Am. J. Public Health*, 81, 168–71.

Kelly, J. A., St Lawrence, J. S., Stevenson, L. Y. *et al.* (1992). Community AIDS/HIV risk reduction: the effects of endorsements by popular people in three cities. *Am. J. Public Health*, 82, 1483–9.

Kelly, J. A., Murphy, D. A., Sikkema, K. J. *et al.* (1997). Randomised, controlled, community-level HIV-prevention intervention for sexual-risk behaviour among homosexual men in US cities. *Lancet*, 350, 1500–5.

Kelly, J. A., Heckman, T. G., Stevenson, L. Y. *et al.* (2000). Transfer of research-based HIV prevention interventions to community service providers: Fidelity and adaptation. *AIDS Educ. Prev.*, 12(Suppl. A), 87–98.

Kelly, J. A., Somlai, A. M., DiFranceisco, W. J. *et al.* (2000). Bridging the gap between science and practice: Transferring effective research-based HIV prevention interventions to community AIDS service providers. *Am. J. Public Health*, 90, 1082–8.

Kramer, M. S. (2003). Randomized trials and public health interventions: time to end the scientific double standard. *Clin. Perinatol.*, 330, 351–61.

Lyles, C. M., Crepaz, N., Herbst, J. H. and Kay, L. S. *et al.* for the HIV/AIDS Prevention Research Synthesis Team (2006). Evidence-based HIV behavioral prevention from the perspective of the CDC's Prevention Research Synthesis Team. *AIDS Educ. Prev.*, 18(Suppl. A), 21–31.

Lyles, C. M., Kay, L. S., Crepaz, N. *et al.* for the HIV/AIDS Prevention Research Synthesis Team (2007). Best-evidence interventions: findings from a systematic review of HIV behavioral interventions for US populations at high risk, 2000–2004. *Am. J. Public Health*, 97, 133–43.

Manhart, L. E. and Holmes, K. K. (2005). Randomized controlled trials of individual-level, population-level, and multilevel interventions for preventing sexually transmitted infections: what has worked? *J. Infect. Dis.*, 191(Suppl. 1), S7–24.

Marley, J. (2000). Efficacy, effectiveness, efficiency. *Austr. Prescriber*, 23, 114–15.

Mayer, J. P., Blakely, C. H. and Davidson, W. S. II (1986). Social program innovation and dissemination: a study of organizational change processes. *Pol. Stud. Rev.*, 6, 273–86.

McKleroy, V. S., Galbraith, J. S., Cummings, B. *et al.* (2006). Adapting evidence-based behavioral interventions for new settings and target populations. *AIDS Educ. Prev,* 18(Suppl. A), 59–73.

Miller, R. L. (2001). Innovation in HIV prevention: organizational and intervention characteristics affecting program adoption. *Am. J. Comm. Psychol.,* 29, 621–47.

Miller, R. L., Klotz, D. and Eckholdt, H. M. (1998). HIV prevention with male prostitutes and patrons of hustler bars: replication of an HIV preventive intervention. *Am. J. Community Psychol.,* 26, 97–131.

Mullen, P. D., Ramírez, G., Strouse, D. *et al.* (2002). Meta-analysis of the effects of behavioral HIV prevention interventions on the sexual risk behavior of sexually experienced adolescents in controlled studies in the United States. *J. Acquir. Immune Defic. Syndr.,* 30(Suppl. 1), S94–105.

Neumann, M. S. and Sogolow, E. D. (2000). Replicating effective programs: HIV/AIDS prevention technology transfer. *AIDS Educ. Prev.,* 12(Suppl. A), 35–48.

Neumann, M. S., Johnson, W. D., Semaan, S. *et al.* (2002). Review and meta-analysis of HIV prevention intervention research for heterosexual adult populations in the United States. *J. Acquir. Immune Defic. Syndr.,* 30(Suppl. 1), S106–17.

NIMH Collaborative HIV/STD Prevention Trial Group (2007). Methodological overview of a five-country community-level HIV/sexually transmitted disease prevention trial. *AIDS,* 21(Suppl. 2), S3–18.

Noar, S. (2008). Behavioral interventions to reduce HIV-related sexual risk behavior: review and synthesis of meta-analytic evidence. *AIDS Behav.,* 12, 335–53.

Noar, S. and Zimmerman, R. S. (2008). Health behavior theory and cumulative knowledge regarding health behaviors: are we moving in the right direction? *Health Educ, Res.,* in press.

Nu'Man, J., King, W., Bhalakia, A. and Criss, S. (2007). A framework for building organizational capacity integrating planning, monitoring, and evaluation. *J. Public Health Management Pract.,* January(Suppl.), S24–32.

Oakley, A. (2000). A historical perspective on the use of randomized trials in social science settings. *Crime Delinq.,* 46, 315–29.

Padian, N. S., van der Straten, A., Ramjee, G. *et al.* (2007). Diaphragm and lubricant gel for prevention of HIV acquisition in southern African women; a randomized controlled trial. *Lancet,* 370, 251–61.

Peterson, A. S. and Randall, L. M. (2006). Utilizing multilevel partnerships to build the capacity of community-based organizations to implement effective HIV prevention interventions in Michigan. *AIDS Educ. Prev.,* 18(Suppl. A), 83–95.

Peterson, L., Taylor, D., Roddy, R. *et al.* (2007). Tenofovir disoproxil fumerate for prevention of HIV infection in women: a phase 2, double-blind, randomized, placebo-controlled trial. *PLoS Clin. Trials,* e27.

Pignone, M., Saha, S., Hoerger, T. *et al.* (2005). Challenges in systematic reviews of economic analyses. *Ann. Intern. Med.,* 21, 1073–9.

Pitisuttithum, P., Gilbert, P., Gurwith, M. *et al.* (2006). Randomized, double-blind, placebo-controlled efficacy trial of a bivalent recombinant glycoprotein 120 HIV-1 vaccine among injection drug users in Bangkok, Thailand. *J. Infect. Dis.,* 194, 1661–71.

Rebchook, G. M., Kegeles, S. M., Huebner, D. *et al.* (2006). Translating research into practice: the dissemination and initial implementation of an evidence-based HIV prevention program. *AIDS Educ. Prev.,* 18(Suppl. A), 119–36.

Rietmeijer, C. A. and Gandelman, A. A. (2007). From best practices to better practice: adopting model behavioral interventions in the real world of STD/HIV prevention. In: S. O. Aral and J. M. Douglas Jr (eds), *Behavioral Interventions for Prevention and Control of Sexually Transmitted Diseases.* New York, NY: Springer Science + Business Media, LLC, pp. 500–14.

Semaan, S., Des Jarlais, D. C., Sogolow, E. *et al.* (2002). A meta-analysis of the effect of HIV prevention interventions on the sex behaviors of drug users in the United States. *J. Acquir. Immune Defic. Syndr.*, 30(Suppl. 1), S73–93.

Shea, M. A., Callis, B. P., Cassidy-Stewart, H. *et al.* (2006). Diffusion of effective HIV prevention interventions–lessons from Maryland and Massachusetts. *AIDS Educ. Prev.*, 18(Suppl. A), 96–107.

Sheth, L., Operario, D., Latham, N. and Sheoran, B. (2007). National-level capacity-building assistance model to enhance HIV prevention for Asian and Pacific Islander communities. *J. Public Health Management*, January(Suppl.), S40–48.

Sikkema, K. J., Kelly, J. A., Winett, R. A. *et al.* (2000). Outcomes of a randomized community-level HIV prevention intervention for women living in 18 low-income housing developments. *Am. J. Public Health*, 90, 57–63.

Sogolow, E., Kay, L. S., Doll, L. S. *et al.* (2000). Strengthening HIV prevention: application of a research-to-practice framework. *AIDS Educ. Prev.*, 12(Suppl. A), 21–32.

Sogolow, E., Peersman, G., Semaan, S. *et al.* (2002). The HIV/AIDS Prevention Research Synthesis Project: scope, methods, and study classification results. *J. Acquir. Immune Defic. Syndr.*, 30(Suppl. 1), S15–29.

Somerville, G. G., Diaz, S., Davis, S. *et al.* (2006). Adapting the popular opinion leader intervention for Latino young migrant men who have sex with men. *AIDS Educ. Prev.*, 18(Suppl. A), 137–48.

St Louis, M. E. and Holmes, K. K. (1999). Conceptual framework or STD/HIV prevention and control. In: K. K. Holmes, P. F. Sparling, P.-A. Mardh *et al.* (eds), *Sexually Transmitted Diseases*, 3rd edn. New York, NY: McGraw Hill, pp. 1239–53.

Taveras, S., Duncan, T., Gentry, D. *et al.* (2007). The evolution of the CDC HIV Prevention Capacity-building Assistance Initiative. *J. Public Health Management Pract.*, January(Suppl.), S8–15.

Thomas, C. W., Smith, B. D. and DeAgüero, L. W. (2006). The Program Evaluation and Monitoring System: a key source of data for monitoring evidence-based HIV prevention program processes and outcomes. *AIDS Educ. Prev.*, 18(Suppl. A), 74–80.

Valdiserri, R. O., Robinson, C., Lin, L. S. *et al.* (1997). Determining allocations for HIV-prevention interventions: assessing a change in federal funding policy. *AIDS Public Policy J.*, 12, 138–48.

Villaruel, A. M., Jemoott, J. B. III and Jemmott, L. S. (2006). A randomized controlled trial testing an HIV prevention intervention for Latino youth. *Arch. Ped. Adolesc. Med.*, 160, 772–7.

Wingood, G. M. and DiClemente, R. J. (2008). The ADAPT_ITT model: s novel method of adapting evidence-based HIV interventions. *J. Acquir. Immune Defic. Syndr.*, 47(Suppl. 1), S40–46.

Index

An asterisk (*) before a page spread indicates mentions on each page covered, rather than unbroken discussion

Aaron Diamond AIDS Research Center (ADARC), 74
Abacavir, 110, 112
    in non-occupational post-exposure prophylaxis, 122
ABC (Abstinence, Be Faithful, Condom use), 15, 20–1, 560–1
    couples, 241, 242
    CVCT as part of, 242
    program, 85
Abstinence, 412
    secondary, 412–13
    see also ABC
Acamprosate, for alcohol use disorders, 342, 348
ACASI (computer-assisted interviewing), 580
Acquired immune deficiency syndrome, see AIDS entries
ACT UP, 172
Acute HIV infection (AHI), 108–9, 131–2, 194, 542
    antiretroviral therapy (ART), 130–1
Acyclovir, 21, 100
Adaptation, effectiveness and, 604–5
Add Health, 351
Adeno-associated virus (AAV) vectors, 65
Adenovirus 5 (Ad5) virus vectors, 62–4, 79–80
Adherence, 183
    to intervention, 592, 593–4
    predictors of non-adherence, 183
Adjuvants, 66–7

Adolescence, 417
    definition, 410
    HIV testing, 538–9
    see also Young women; Youth
Aerobic exercise training, 176
AFASS criteria, 485, 486
Afghanistan, opium production, 504
Africa
    alcohol consumption, 345
    antenatal testing, 562
    breastfeeding period trials, 559
    cohabiting couples, 241
    death penalty for homosexual relations, 503
    HAART, 564
    HIV-infected pregnant women receiving antiretroviral prophylaxis, 482, 488
    male circumcision, 146–7, 149–52, 157
    mother-to-child infection rates, 472
    replacement for breastfeeding, 485
    sex workers, 11
    syphilis control, 501–2
    testing in pregnancy, 475
    VCT data analysis studies, 529
    violence against women, 555
    widowhood and remarriage, 555
    women with HIV, 4, 15
    youth testing, 539
    see also Central Africa; East Africa; North Africa; Southern Africa; Sub-Saharan Africa; West Africa; and individual countries

African-Americans, 213, 216, 231
  women and crack cocaine, 355
Age, injection drug users, 314, 317
Agence Nationale de Recherches sur le
    SIDA (ANRS), 73, 74
  ANRS VAC 18 trial, 73
AHI, *see* Acute HIV infection
AIDS (acquired immune deficiency
    syndrome)
  as problem of culture, development and
    economics, 567
  stigma, 173
  see also HIV; HIV infection
AIDS Epidemic Updates, 11
AIDS prevention research and payoffs,
    171–195
  1983-1985 period, 171–72
  1986-1991 period, 172–7
  1992-1997 period, 177–84
  1998-2004 period, 184–92
  2005-2008 period, 192–5
  conclusions, 195
AIDS Research Centers, 173–4
  funding, 173
AIDS risk reduction model (ARRM), 213
AIDS Vaccine Advocacy Coalition
    (AVAC), 74
Alcohol, 344–8
  binge drinking, 344, 346
  and HIV transmission, 551
  per capita consumption, 344
  see also Alcohol use
Alcohol myopia, 344
Alcohol use
  cessation of, 346–7
  disorders, 340
  HIV risk, 344–6
  studies, 345–6
  and transmission risk, 248–9
  treatment, 346–8
  youth, 415
Alcoholics Anonymous (AA), 343, 457
Alcoholism, relapse prevention, 343
Alkyl nitrates, 509
Americans with Disabilities Act, 172
Americas

epidemics, 16–17
  men who have sex with men, 11
  replacement for breastfeeding, 485
  see also Central America; Latin
    America; North America; South
    America; *and individual countries*
Amphetamine, 340, 349
Amprenavir, 112
Anal intercourse (AI), unprotected (UAI),
    90, 284
  men who have sex with men (MSM),
    509, 510–11
  see also Unprotected anal or vaginal
    intercourse
ANC clinics, *see* Antenatal care clinics
Anemia, 563
Angola, people living with HIV/AIDS, 14
Animal studies
  adenovirus 5 (Ad5) vaccine, 62–3, 64
  ART for post-exposure prophylaxis,
    115–17, 119
  attenuated virus tests, 59, 62
  cellular response tests, 60
  infection studies, 87, 88
  microbicides and rectal safety, 90
  pre-exposure prophylaxis, 123–5, 132
  sexual transmission, 40–1
  VSV vaccine, 65
ANRS VAC 18 trial, 73
Antenatal care (ANC) clinics
  antenatal testing, 561–3
  couple-friendly testing, 563
  providing PMTCT services, 536, 537
  sampling, 13
Anti-Prostitution Pledge (APP), 502, 516
Antibody-dependent cellular cytotoxicity
    (ADCC), 44
Antiretroviral drugs, 22–3
  in the genital tract, 111, 112
Antiretroviral prophylaxis, 486, 488
  breastfeeding and, 486–8
Antiretroviral therapy (ART), 35, 53, 58,
    107–45, 564–5
  for acute HIV infection, 130–1
  conclusions, 133–4
  in the developing world, 98

ecological prevention benefits, 114
effect on discordant couples, 113
effect on infectiousness, 113–15
future strategies, 131–3
highly active (HAART), *see* Highly active antiretroviral therapy
for HIV-infected individuals, 130
mother-to-child transmission, 115
newer agents, 132–3
patient adherence to, 565
pharmacology, 109–12
for PMTCT, 473, 474, 475, 476, 477, 486
for post-exposure prophylaxis (PEP), 115–18
in prevention of HIV transmission, 112–15
as public health prevention, 126–8
resistance to, 131
sex workers, 388
and sexual behaviors, 128–31
studies, 113–14
triple-combination, for PMTCT, 477
Antisocial personality disorder (ASPD), injection drug users, 318
Anxiety, 233–4
Argentina
HIV epidemic, 17
HIV prevalence in prisons, 464
Armed Forces Research Institute of Medical Sciences (AFRIMS), 73, 74
Armenia, female sex workers/low risk women, HIV prevalence, 378
ART, *see* Antiretroviral therapy
Asia
commercial sex, injection drug use, 382
HIV pandemic, 504
HIV prevalence among IDUs, 310
HIV prevalence surveys, 583–4
HIV in young people, 409
injection drug use, 311, 382
male circumcision, 146–7, 158
methadone maintenance programs, 328
mother-to-child infection rates, 472
replacement for breastfeeding, 485
sex workers, 11

sub-pandemics, 15–16
testing in pregnancy, 475
VCT data analysis studies, 529
see also Central Asia; East Asia; South Asia; Southeast Asia; *and individual countries*
Assisted Reproduction Clinics, 37
Atazanavir, 110, 112, 134
in non-occupational post-exposure prophylaxis, 122
Australia
condom availability in prisons, 455, 456
drug users' organizations, 324
female sex workers/low risk women, HIV prevalence, 379
harm reduction, 503
HIV prevalence among IDUs, 310, 311
Jailbreak Health Project, 454
male circumcision, 158
men who have sex with men, 11
methamphetamine use, 353
needle exchange and tattooing in prison, 460
opiate replacement therapy (ORT) for prison inmates, 459
persons living with HIV/AIDS, 17
pharmacies, 326–7, 328
preventive vaccine trials, 69, 73
public health responses, 16
sexual risk behaviors, 510
supervised injection facilities (SIF), 505–6
unprotected anal intercourse in men who have sex with men, 128–9
Austria, TAMPEP Project, 397
Avahan (India AIDS initiative), 388, 390, 395–6
Aventis Pasteur, 73
AZT, *see* Zidovudine
Azurocidin, 43

Balanitis, 152
Bangladesh
brothels, 384
female sex workers/low risk women, HIV prevalence, 378

Bangui definition, 13
Barebacking, 315
BED assay, 583, 584
Behavior change, 169–70, 180–1
    advocacy, 24
    testing and counseling in HIV, 527
Behavioral interventions, 185
    evaluation trials, 595–6
Behavioral interventions at individual
        level, 203–39
    AIDS risk reduction model (ARRM),
        213
    behavioral learning model, 206, 228,
        229–30
    biomedical model, 205–6
    cognitive perspective, 207
    communication perspective, 207
    conclusions, 232–5
    efficacy trials, 213–32
    health belief model (HBM), 207, 207–8,
        223, 232
    health decision model (HDM), 212–13
    information–motivation–behavioral
        (IMB) skills model, 211
    protection motivation model, 207, 210
    relapse prevention (RP) model, 212, 227
    self-regulation theory, 211
    social cognitive theory (SCT), *see*
        Social cognitive theory
    summary, 232–5
    theoretical models, 205–13
    theory of planned behavior (TPB),
        209–10
    theory of reasoned action (TRA), 207,
        209, 232
    transtheoretical model (TTM), 211–12
Behavioral learning model, 206
    injection drug users, 228
    women, 229–30
Behavioral skills training
    injection drug users, 228
    women, 229–30
Behaviors
    antecedent, 171
    see also Sexual risk behaviors
Belarus

HIV incidence among IDUs, 504
HIV prevalence among IDUs, 311
Belgium, vaccine research organizations,
    74
Belmont Report, 77–8, 99
Benin
    clients of female sex workers, 390
    condom use, 393
    female sex workers/low risk women,
        HIV prevalence, 378, 379
    microbicide study, 99–100
BHITS (Breastfeeding and HIV
        International Transmission Study),
        484, 485
Bill and Melinda Gates Foundation, 76,
        395
Binge drinking, 344, 346
Biojector, 66
Biological interventions, evaluating, 592–6
Biomedical HIV prevention, 100, 192–5
Biomedical model of behavioral
        intervention, 205–6
Blacks, 216, 231
Bloodborne virus (BBV) acquisition, 505,
        517
Bolivia
    condom use, 393
    female sex workers/low risk women,
        HIV prevalence, 379
Botswana
    adult HIV prevalence, 15
    condom use, 1
    oral pre-exposure prophylaxis trials, 125
    risk behaviors, 15
    sexual debut, 1
    tenofovir trial, 125
    testing in pregnancy, 475
Brazil
    condom availability in prisons, 455
    decreased HIV-related morbidity,
        510–11
    female sex workers/low risk women,
        HIV prevalence, 378
    HIV incidence, 1
    HIV prevalence in prisons, 464
    public health responses, 16–17

replacement for breastfeeding, 485
syringe exchange programs, 324
Breastfeeding
antiretroviral prophylaxis and, 486–8
avoidance of, 473, 485
and culture, 557–9
interventions to reduce MTCT, 485–8
Kisumu Breastfeeding Study, 487
maternal antiretroviral treatment, 487
mixed feeding, 486
mother-to-child transmission (MTCT)
through, 473, 483–8
risk factors, 484–5
Breastfeeding and HIV International
Transmission Study (BHITS), 484,
485
BufferGel, 89, 93, 102
Bulgaria, methadone maintenance
programs, 328
Buprenorphine
for opiate disorders, 342
as treatment for opioid dependence in
IDUs, 329–30
Burkina Faso
HIV prevalence among sex workers, 508
male circumcision, 147
Burma, new infections, 3

C31G (Savvy), 88
Caesarean section, 473, 474, 477
California Partner Study, 285
Cambodia
ABC strategy, 20
declining epidemic, 16
female sex workers/low risk women,
HIV prevalence, 378
HIV among men who have sex with
men, 509
HIV incidence decline, 1
Lotus Project, 516–17
microbicide study, 99
new infections, 3
pre-exposure prophylaxis trials aborted,
125
Cameroon
condom use, 1

female sex workers/low risk women,
HIV prevalence, 378, 379
male circumcision, 147
microbicide study, 99
pre-exposure prophylaxis trials aborted,
125
pre-exposure tenofovir trials, 126
risk behaviors, 15
sexual debut, 1
testing in pregnancy, 476
Canada
ART resistance, 131
ART study, 114
condom availability in prisons, 455
Correctional Services Canada (CSC),
513
harm reduction, 503
HIV prevalence, 512
HIV prevalence among IDUs, 311
HIV prevalence in prisons, 512
MSM STI prevalence, 510
needle exchange and tattooing in
prisons, 460–1
new infections, 4
opiate replacement therapy (ORT) for
prison inmates, 459
pharmacies, 327
preventive vaccine trials, 69
supervised injection facilities (SIF),
505–6
syringes in prisons, 513
Canary-pox vectors, 64–5, 71
Cancer, penile, 155
Cannabis, 340, 358
CAPRISA 004 study, 102
CAPRISA tenofovir gel trial, 132
Caribbean
ANC facilities providing PMTCT
services, 537
HIV prevalence surveys, 583–4
HIV-infected pregnant women receiving
antiretroviral prophylaxis, 488
new infections, 4
preventive vaccine trials, 69
stable epidemic, 16
VCT data analysis study, 529

Carraguard, 88, 89, 101–2, 187
CAS (Correctional Services Canada), 513
Cathelicidins, 43
Cathepsin G, 43
CDC, *see* Centers for Disease Control;
    Consensus Development Conference
Cellulose sulfate, 88, 99–100, 103, 187
Center for AIDS Interventions Research
    (CAIR), 180
Center for AIDS Prevention Studies
    (CAPS), 173
Center for Biopsychosocial Study of
    AIDS, 173
Center for HIV Identification, Prevention,
    and Treatment Services (CHIPTS),
    181
Center for Interdisciplinary Research on
    AIDS (CIRA), 183
Center for NeuroAIDS Preclinical Studies,
    173
Center for Substance Abuse Treatment
    (CSAT), 457
Centers for Disease Control (CDC), 68,
    74, 244, 449–50, 576–7, 595–6
  ART guidelines, 118
  capacity building, 602–3
  core mission, 596
  Diffusion of Effective Behavioral
    Interventions (DEBI) Project, 178,
    191, 602, 603, 606
  Diffusion of Effective Interventions
    Project, 601
  disseminating effective behavioral
    interventions, 602
  effectiveness and adaptation, 604–5
  feedback from monitoring and
    evaluation, 604
  guidelines, 118, 173, 525, 526
  implementation approach, 596–605
  Prevention Research Synthesis (PRS)
    Project, 600–1
  Replicating Effective Programs (REP)
    Project, 601, 603
  research-to-practice activities, 599–605
  technical assistance, 602–3
  use of funds, 596

  website, 601
Central Africa
  ANC facilities providing PMTCT
    services, 537
  see also Africa; *and individual countries*
Central African Republic
  condom use, 1
  risk behaviors, 15
  sexual debut, 1
Central America
  ANC facilities providing PMTCT
    services, 537
  HIV epidemics, 16
  new infections, 4
  see also Americas; *and individual
    countries*
Central Asia
  commercial sex workers, 317
  HIV-infected pregnant women receiving
    antiretroviral prophylaxis, 488
  see also Asia; *and individual countries*
Central Europe
  ANC facilities providing PMTCT
    services, 537
  see also Europe; *and individual
    countries*
Centro de Orientación e Investigación
    Integral (COIN), Dominican
    Republic, 397–8
Centro de Promoción y Solidaridad
    Humana (CEPROSH), Dominican
    Republic, 397–8
CEPROSH (Centro de Promoción y
    Solidaridad Humana), Dominican
    Republic, 397–8
Cervical ectopy, 39
Cervical transformation zone, 39
Cervicovaginal secretions, 36–7
Chad
  condom use, 1
  risk behaviors, 15
  sexual debut, 1
Chat room partners, 190–1
Child prostitution, 384
Childhood sexual abuse, 290
Children, desire for, 248–9

China
    commercial sex workers, 318
    drug trafficking and abuse, 330
    female sex workers/low risk women,
      HIV prevalence, 379
    HIV incidence among IDUs, 504
    HIV prevalence among IDUs, 312
    HIV/AIDS in prisons, 462–3, 464
    IDU/non-IDU sex workers, HIV
      prevalence, 382
    injecting drug users, 11, 16
    injection drug use, 311
    new infections, 3
    sex work, 16
Chlamydia, 507
    male circumcision and, 155
    urine screening for, 462
    in US correctional settings, 448, 461
Chlamydia trachomatis, 38, 232
    prevalence among female sex workers,
      379
Cialis, 356
Circumcision, 85, 100–1
    see also Male circumcision
Clinton Foundation, 2
Cluster-randomized trials, 594
Cocaine, 188, 340, 348, 353–6
    crack cocaine, 317, 329, 354, 355
    HIV prevention, 355–6
    HIV risk, 354–5
    MSM and, 354–5
    pharmacotherapy research, 356
Cochrane review, 147
Cognitive behavioral therapy (CBT), 224,
    342–3, 352
    culturally sensitive, 352
    MSM, 225, 226
    relapse prevention, 343
Cognitive perspective of behavioral
    intervention, 207
Cohabitation, *see* Couples
COIN (Centro de Orientación e
    Investigación Integral), Dominican
    Republic, 397–8
Collaborative HIV/STD Prevention Trial,
    194–5

Combined antiretroviral therapy (CART),
    510
Commercial sex workers (CSWs),
    injection drug users (IDUs), 317–18
Communication perspective of behavioral
    intervention, 207
Community Advisory Boards (CABs), 173
Community cultures, intervention program
    content, 609
Community planning groups (CPGs),
    596–7
Community popular opinion leader
    (CPOL), 552
Compartment, 33
Compliance, tenofovir study, 593
Comprehensive counseling, 293
Computer-assisted interviewing, ACASI,
    580
Condom use, 563–4
    couples voluntary counseling and
      testing (CVCT), 241–2
    self-efficacy, 414
    sex workers, 377, 387, 392–3
    young people, 413–14
    see also ABC
Condoms
    availability to incarcerated persons in
      the US, 454–6
    female, 181, 387
Confidentiality, shared, 566
Conformational masking, 55
Consensus Development Conference
    (CDC), 178
Contingency management (CM), 343, 352
Contraception, young women, 415
Correctional Services Canada (CSC), 513
Corrections Demonstration Project, 461
Cost-effectiveness of intervention
    programs, 185–6
Côte d'Ivoire
    condom use, 1
    female sex workers/low risk women,
      HIV prevalence, 378
    HIV prevalence among sex workers, 507
    male (MSW)/transgender (TSW) sex
      workers, HIV prevalence, 380

Côte d'Ivoire (*contd*)
  PAPO-HV Project, 396–7
  PMTCT antiretroviral trials, 478, 479
  risk behaviors, 15
  sex workers' behavioral studies, 393
  sexual debut, 1
Council for International Organizations
    of Medical Sciences (CIOMS)
    Guidelines, 78
Counseling, 173
  comprehensive, 293
  couple-based, 552–3
  safer sex, 189
  sexual risk behavior among PLWHA,
    291–4, 296–8
  in vaccine trials, 78–9
  see also Couples voluntary counseling
    and testing; Testing and counseling
    in HIV; Voluntary counseling and
    testing
Couple-based counseling, 552–3
Couple-based HIV testing and counseling,
    536–8
Couples
  concordant HIV-negative, 242, 254
  concordant HIV-positive, 242, 254
  definition, 249
  HIV-discordant, 242, 254–5, 553
  prevention through behavior change,
    240–1
Couples VCT centers, 244
Couples voluntary counseling and testing
    (CVCT), 240–66
  administration methods, 250
  alcohol use, 248–9
  centers, 244
  child care onsite, 250
  client interviews, 255–6
  clinics, *see* CVCT clinics; Rwanda
    Zambia HIV Research Group
  conclusion, 259–60
  condom use, 241–2
  controversies, 249–51
  counselor assessments and feedback,
    256
  counselor mentoring system, 257

desire for children, 248–9
development of, 243–4
evaluation of services, 255–7
evolution of, 257–9
financial barriers, 247–8
importance of, 242–3
Influential Network Agents (INAs), 246,
    247
Influential Network Leaders (INLs),
    246, 247
integration with other health-care
    services, 258
logistical barriers, 247–8, 250
vs male circumcision, 257
mobile units, 258
monitoring of services, 255–7
obstacles to, 244–9
promotion strategies, 246–7
psychological factors, 248
in Rwanda, 243–4, 246–8, 250, 258–9
structural barriers to, 245
topics for consideration, 256–7
in Zambia, 244, 246–8, 250, 258–9
CpG (unmethylated cytosine-guanine
    dinucleotides), 67
CPGs (community planning groups),
    596–7
Crack cocaine, 354, 355
  smoking, 317, 329
Crystal meth, 187–8
Crystal Meth Anonymous, 352
"CT bubble forms", 577
CVCT, *see* Couples voluntary counseling
    and testing
CVCT clinics
  activities and timetable, 251
  group HIV education and discussion,
    252
  post-test counseling (results session),
    253–4
  pre-test counseling and risk assessment,
    252–3
  risk reduction counseling, 254–5
Cyanovirin, 101
Cyanovirin-N, 88
Cytokine receptor antagonists, 44

Cytokines, proinflammatory, 44, 45
Czech Republic, syringe exchange
      programs, 324

Dale and Betty Bumpers Vaccine Research
      Center (VRC), 63, 75
Dapirivine (TMC-120), viral resistance, 95
DEBI (Diffusion of Effective Behavioral
      Interventions) Project, 178, 191, 602,
      603, 606
Declaration of Commitment on HIV/
      AIDS, 526, 585
Declaration of Geneva, 99
Declaration of Helsinki, 77, 99
Defensins, 43, 45
Delavirdine, 112
Democratic Republic of the Congo (DRC),
      sex workers in, 384
Demographic and Health Surveys (DHS),
      13, 147, 579
Depression, 233–4
      among gay men, 273, 274
      and methamphetamine, 352
Developed nations, HIV increasing in, 2
Developing countries
      evaluating HIV/AIDS programs, 571–90
      monitoring service utilization, 577–8
      peer educators, 324
      prevalence/incidence of HIV, 583–4,
            585–6
      syringe exchange programs, 324
      voluntary counseling and testing (VCT)
            in, 529, 531
      see also individual countries
Didanosine, 110, 112
      in non-occupational post-exposure
            prophylaxis, 122
Diffusion of Effective Behavioral
      Interventions (DEBI) Project, 178,
      191, 602, 603, 606
Diffusion of Effective Interventions
      Project, 601
Dipping, 270
Directly administered antiretroviral therapy
      (DAART), 565
Directly observed therapy (DOT), 94

Disclosure/non-disclosure of HIV status,
      285–6
Discrimination, 433
      AIDS-related, 172–3
Disulfiram, for alcohol disorders, 342,
      347
DITRAME-plus study, 483
Division of Acquired Immunodeficiency
      Syndrome (DAIDS), 73, 74
DNA vaccines, 65–6
Dominican Republic
      Centro de Orientación e Investigación
            Integral (COIN), 397–8
      Centro de Promoción y Solidaridad
            Humana (CEPROSH), 397–8
      Movimento de Mujeres Unidas
            (MODEMU), 398
      reducing risk in female sex workers, 581
      sex workers' condom use, 508
      stable epidemic, 16
Drug Addiction Treatment Act, 329
Drug use
      in HIV-positive women, 477
      youth, 415
      see also Injection drug use/users;
            Non-injection substance use/users;
            Recreational drug use; *and individual
            drugs*

East Africa
      ANC facilities providing PMTCT
            services, 537
      male circumcision, 158, 160
      see also Africa; Sub-Saharan Africa;
            and individual countries
East Asia
      ANC facilities providing PMTCT
            services, 537
      see also Asia; *and individual countries*
Eastern Europe
      ANC facilities providing PMTCT
            services, 537
      commercial sex workers, 317
      condom availability in prisons, 455
      HIV epidemics, 16, 504
      HIV incidence, 3–4

Eastern Europe (*contd*)
HIV-infected pregnant women receiving
antiretroviral prophylaxis, 488
injection drug use, 11, 311
methadone maintenance programs,
328
young people, 409
see also Europe; *and individual
countries*
Ecstasy (methylenedioxy
methamphetamine: MDMA), 188,
340, 359
Ecuador, oral pre-exposure prophylaxis
trials, 125
EDGE intervention, 188
Education, 418–19
entertainment-education, 387
incarcerated populations, 453–4, 457,
512
in Indian schools, 561
sex workers, 386–7
sexual, 420, 560–1
see also Psychoeducation; Schools
Efavirenz, 110, 112
in non-occupational post-exposure
prophylaxis, 122–3
Effectiveness
and adaptation, 604–5
in the community, 607–8
definition, 607
vs efficacy, 592, 607
evaluating, 592–6
quantitative/qualitative methods, 572
Effectiveness testing, in research, 204
Efficacy
definition, 607
vs effectiveness, 592, 607
evaluating, 592–6
reviews, economic issues, 607
Efficacy trials
behavioral interventions, 213–32
replication of results, 605
Electroporation, 66
Employment, AIDS-related discrimination,
172–3
Employment policies, 176
Emtricitabine, 110, 112, 134

in non-occupational post-exposure
prophylaxis, 122, 123
in pre-exposure clinical trials, 125
Enfuvirtide, 112
Entertainment-education, 387
Ephedrine, 349
Erectile dysfunction (ED), 188
medications, 356
Estonia, HIV/AIDS in prisons, 463
Ethical considerations in HIV preventative
vaccine research, 76, 78
Ethiopia, nevirapine treatment studies, 486
Ethnic/racial minorities, behavioral
intervention research, 223, 231–2
Ethnographic assessment models, 175
Europe
antenatal antiretroviral therapy, 476–7
drug users' organizations, 324
harm reduction, 503
HIV incidence, 3–4
HIV prevalence among IDUs, 310
immigrant populations, 477
injecting drug use and commercial sex,
382
mother-to-child infection rates, 472
replacement for breastfeeding, 485
sexual risk behaviors, 510
supervised injection facilities (SIF),
505–6
Transnational AIDS/STD Prevention
Among Migrant Prostitutes in Europe
Project (TAMPEP), 397
see also Central Europe; Eastern
Europe; Western Europe; *and
individual countries*
European Agency for the Evaluation of
Medicinal Products (EMEA), 96
European Mode of Delivery Trial, 474
European Project on Non-occupational
Post-Exposure Prophylaxis for HIV
(EURO-NONOPEP), 118
European Union, 74
Evaluation of HIV/AIDS programs
at multiple levels, 578–86
behavioral outcomes, 579
causation determination, 587
challenges in, 586–8

conclusion, 589
cost data, 587–8
data collection, 580, 583–6
economic evaluation, 572, 588
environmental-level programs, 581–2
evaluation types, 571–2
formative evaluation, 571, 572
impact assessment, 572
indicators, 573–4
individual-level programs, 582
internal vs external, 572–3
national level, 582–6
objectives, 573–4
process evaluation, 571, 572
progress indicators, 579–80
psycho-social outcomes, 579
stakeholders, 573
structural-level programs, 581
study design, 580, 583–6
summative evaluation, 571, 572, 574–86
trend data, 582–3
in US and developing countries, 571–90
Evidence-based HIV prevention strategies,
   603, 610
Evidence-based interventions (EBIs), 606,
   607–8, 609–10
Evidence-based research, Onken's model,
   203–4
EXPLORE study, 178, 189, 226

Faithfulness, *see* ABC
Family Health International, 396–7
Family planning (FP) clinics, 536
Family-based HIV testing and counseling,
   536–7
Feedback, from monitoring and evaluation,
   604
Female condoms, 181, 387
Female genital tract
   acquired immune response, 43
   anatomy, 34–5
   antiretroviral drug ins, 111, 112
   secretions, 44
Fertility pressures, 561–3
Financial literacy, 433
Food and Drug Administration (FDA), 96
   clinical intervention trial standards, 203

Foreskin, 152–4, 155
   as infection site, 38–9, 42
   see also Male circumcision
Former Soviet Union (FSU)
   HIV/AIDS in prisons, 462, 463–4
   injection drug use, 311
   intravenous drug users, 504
   new epidemics, 3
   syphilis in, 501
   see also individual countries
Formula-feeding, 558
Fos-amprenavir, 110
Fossa navicularis, 39
4Cs, 388
France
   PMTCT antiretroviral trials, 479
   preventive vaccine trials, 73
   vaccine research organizations, 74

Gambia, condom use, 393
Gamma-hydroxybutrate (GMA), 340, 359
Gay men, 267–80
   challenges in HIV prevention work, 270,
      276–7, 278–9
   community viral load, 274–5
   creating self-knowledge, 272–3
   demographic diversity, 273
   depression, 273, 274
   efficacy of HIV prevention, 269
   efficacy into effectiveness, 275
   informational needs, 270
   oral sex, 270
   partner violence, 274
   prevention strategies, 276–9
   psychosocial health problems,
      273–4
   responsibility for HIV prevention,
      271–2
   risk reduction strategies, 270–1
   safer sex trial, 225–6
   substance abuse, 273, 274
Gay Men's Health Crisis, 503
GCP (Good Clinical Practice) guidelines,
   76
GEM (gender relations, economic
   empowerment, migratory behavior) of
   male partners, 85

Gender
  injection drug users, 314–15
  roles and identity, 416–18
Gender-based violence, 417, 431
Gender equality, 23
Genetic inserts, 67–8
  design approaches, 67–8
Genital antibodies, 44
Genital secretions, 108–9
  HIV in, 32–7, 44
Genital tract
  antiretroviral drugs in, 111, 112
  see also Female genital tract; Male
    genital tract
Genital warts, 155
Genitourinary dysplasia (GUD), 151,
  154–5, 156–7
Geographic Information Systems, 392
Germany
  needle exchange in correctional
    facilities, 460
  needle-exchange programs, 513
  TAMPEP Project, 397
Ghana
  condom sales, 576
  female sex workers HIV prevalence,
    382–3
  HIV prevalence among IDUs, 504
  male circumcision, 147
  pre-exposure tenofovir trials, 126
  pre-marital testing, 538
GHB (gamma-hydroxybutrate), 340, 359
Global Aids Fund, 98
Global Campaign for Microbicides, 187
Global Fund, 585, 587
Global HIV/AIDS Initiative Nigeria
  (GHAIN) project, Nigeria, 527
Global HIV/AIDS Vaccine Enterprise, 72,
  74
Global influenza, genetic diversity, 68
Glycoprotein gp120, 68
Glycosolation, 55
Gonorrhea, 155, 507
  urine screening for, 462
  in US correctional settings, 448, 461
Good Clinical Practice (GCP) guidelines/
  standards, 76

in clinical research, 96, 99
Granulocyte-macrophage stimulating
  factor (GM-CSF), 67
GUD (genitourinary dysplasia), 151,
  154–5, 156–7
Guidelines
  from Centers for Disease Control
    (CDC), 170, 173
  occupational/non-occupational post-
    exposure prophylaxis, 118

HAART, *see* Highly active antiretroviral
  therapy
Haemophilus ducreyi, 39
  male circumcision and, 155
HAI (health impact assessment), 595
Haiti
  HIV epidemic, 16
  HIV prevalence decline, 585
HAPS Intervention, 429
Harm reduction, 503–4
  goal of, 504–5
  human rights and, 513–16
  IDUs, 503, 505–7
  incarcerated populations, 512–13, 517
  interventions, 504–7
  men who have sex with men (MSM),
    510–11
  sex workers, 508
  see also Methadone maintenance;
    Needle exchange
Health belief model (HBM), 207, 207–8
  ethnic/racial minorities, 223, 232
Health-care providers, *see* Pharmacies
Health decision model (HDM) of behavior
  change, 212–13
Health impact assessment (HAI), 595
Health Prevention Trials Network (HPTN),
  052, 564
Healthy Living Project (HLP), 178, 190,
  294
Healthy Relationships study, 289
Hepatitis, in US correctional settings,
  448
Hepatitis B vaccine, 62
Heroin, 340
Herpes simplex virus (HSV), 39, 507

Herpes simplex virus type 2 (HSV-2), 381
  prevalence among female sex workers, 379
Heterosexuals
  behavioral intervention research, 215, 222–3
  reviews of HIV testing and counseling data, 527–8
HHV-6, 176
Highly active antiretroviral therapy (HAART), 35, 177, 182–3, 188, 564
  effect on HIV transmission, 113
  for gay men, 274, 277
  relationship with sexual risk behavior, 286–8
Highly active retroviral prevention (HARP), 24
HIV (human immunodeficiency virus)
  factors affecting genital levels of, 35–7
  in genital secretions, 32–7, 44
  interventions concept, test of, 172–7
  mutation rate, 67
  seroconversion, 351
  types, 67–8
  viral diversity, 67–8
HIV Center for Clinical and Behavioral Studies, 173
HIV Cost Services Utilization Study (HCSUS), 283
HIV incidence, 14
  behavioral indicators, 588
  framework for causation levels, 578–9
  and HIV prevention budget, 585
  measuring, 588
  observed vs counterfactional, 584
  see also under individual countries and regions
HIV infection
  acute, 108–9, 130–2, 194, 542
  factors associated with decreased risk of, 31–2
  factors associated with increased risk of, 31
  population-based surveys, 14
  prevention, *see* HIV prevention
  risk groups, 14, 172, 176
  surveillance methodology, 13–14

transmission, *see* HIV transmission
  trends in, 11–13
  see also AIDS
HIV Neurobehavioral Research Center (HNRC), 173
HIV prevalence, 14, 583–4
  see also under individual countries and regions
HIV prevention
  biomedical, 100, 192–5
  components, 17
  conceptual framework, 17–19
  conclusions, 24
  education, *see* Education
  the future, 8, 23–4
  primary/secondary/tertiary, 17
HIV Prevention Trials Network (HPTN), 115
HIV transmission, 107–8
  ART in prevention of, 112–15
  concentration response association, 107–8
  modes of, 55–6
  probability, 4
  sexual transmission, *see* Sexual transmission of HIV
HIV Vaccine Trials Network (HVTN), 73, 75
  HVTN 204 trial, 73
HIVNET 012 trial, 473, 478
HIV-SIV (SHIV) virus, 60, 65
HIV/STD operational research, 170
HIV/STD prevention research, 170
Homelessness, incarcerated persons, 446–7
Homophobia, 315
Homosexual men, 114, 128
Honduras
  behavior change, 17
  female sex workers/low risk women, HIV prevalence, 378
Hong Kong, methadone maintenance programs, 328
HPV, *see* Human papilloma virus
HSV, *see* Herpes simplex virus
Human immunodeficiency virus, *see* HIV *entries*
Human papilloma virus (HPV), 38, 39
  male circumcision and, 155

Human rights
  harm reduction and, 513–16
  public health and, 515
  violations, as factor in spread of HIV/
    AIDS, 515
Hungary
  methadone maintenance programs, 328
  syringe exchange programs, 324
HVTN 204 trial, 73
Hypersexuality, 350

IAVI A002 trial, 73
IDUs (injection drug users), *see* Injection
  drug use/users
IMAGE (Intervention with Microfinance
  for AIDS and Gender Equity) Project,
  432–3
IMB (information–motivation–behavioral)
  skills model, 211
Immune system
  cellular, 54, 55
  humoral branch, 54, 55
  viral control, 55–6
Immunology, 54–6
Impact
  evaluation of, 594–5
  health impact assessment(HAI), 595
  prevention impact, 594
Incarcerated populations
  condom provision, 517
  education, 453–4, 457, 512
  harm-reduction strategies among,
    512–13
  HIV among, 511–13
Incarceration in the US, 444–62
  condoms, 454–6
  confidentiality, 451
  drug-dependence programs, 446
  effects of, on the community, 449
  the future, 464–5
  hepatitis, 448
  HIV prevention education, 453–4
  HIV prevention interventions, 449–62
  HIV risk factors, 446–7
  HIV testing policies, 450–2
  HIV/AIDS epidemiology, 447–8

HIV/AIDS-associated mortality, 448
  homelessness and, 446–7
  interventions with, 444–71
  intra-prison HIV transmission, 448–9
  mental illness and, 446
  of minorities, 445, 447
  needle exchange, 459–60
  opiate replacement therapy (ORT),
    458–9
  rapid HIV testing, 451–2
  routine opt-out HIV testing, 450–1
  sexual abuse and, 446–7
  STD (STI) testing and treatment, 461
  STIs, 448
  substance dependence treatment, 456–7
  substance use dependence and, 446
  substance use education and awareness
    programs, 457
  treatment and referral as secondary
    prevention, 452
  tuberculosis, 448
  urine screening for chlamydia and
    gonorrhea, 462
  women inmates, 447
  see also Prisons
Incidence, *see* HIV incidence
India
  AIDS initiative (Avahan), 395–6
  Avahan (India AIDS initiative), 388,
    390, 395–6
  condom use among female sex workers,
    563
  denial of treatment, 559
  drug adherence study, 565
  drug users' organizations, 324
  epidemics, 16
  female sex workers/low risk women,
    HIV prevalence, 378, 379
  female sex-workers' collectives, 389–90
  heterosexual men, 551
  HIV prevalence among IDUs, 311
  HIV prevalence in prisons, 464
  HIV risk for married women, 566
  male circumcision, 154
  methadone maintenance programs, 328
  microbicide study, 99–100

nevirapine treatment studies, 486
new infections, 3
peer educators, 324
people living with HIV/AIDS, 14
prevention education, 561
sex-trafficking, 384
shared confidentiality, 566
social networks, 553–4
Sonagachi Project, 390, 552
sterilization campaigns, 161
sterilization of women, 564
syringe exchange programs, 324
Indinavir, 111, 112
Indonesia
    female sex workers/low risk women,
        HIV prevalence, 379
    HIV incidence among IDUs, 504
    HIV prevalence among IDUs, 312
    injecting drug users, 16
    male (MSW)/transgender (TSW) sex
        workers, HIV prevalence, 380
    men who have sex with men, 16
    new infections, 3
    opiate replacement therapy (ORT) for
        prison inmates, 459
    sex workers' clients, 384
Infections, stopping the spread of, 592
Infectiousness, antiretroviral therapy
    (ART) effect on, 113–15
Influential Network Agents (INAs), CVCT
    and, 246, 247
Influential Network Leaders (INLs),
    CVCT and, 246, 247
Influenza, genetic diversity, 68
Information–motivation–behavioral (IMB)
    skills model, 211
Informed consent
    testing and counseling in HIV, 525, 532,
        533, 535, 540, 541
    of trial volunteers, 77, 99, 100
Injection drug use/users (IDUs), 305–39
    age and, 314, 316, 317
    anti-drug legislation, 330
    with antisocial personality disorder, 318
    behavioral intervention research, 216,
        220, 227–8

commercial sex workers (CSWs),
    317–18
demographic characteristics, 313–16
discrimination, 318
drug use behaviors, 316–17
environment/neighbourhood and, 320,
    321
epidemiology of HIV infection, 307–12
factors contributing to, 306
the future, 330–1
gender and, 314–15
harm reduction interventions, 503,
    505–7
HIV prevalence/incidence globally,
    310–12
HIV prevalence/incidence in US,
    307–10
intervention strategy factors, 312–21
interventions, 321–30
miscellaneous factors, 320
MSM (men who have sex with men),
    315–16
newly initiated, 314, 317
opioid substitution pharmacotherapy,
    328–9
partner violence, 315
peer education programs, 321
peer outreach, 319, 321–2, 322, 323,
    324
political context, 305–7
prevention strategy factors, 312–21
psychosocial factors, 318
race/ethnicity, 308–10, 313
racial segregation and, 306–7
reviews of HIV testing and counseling
    data, 527–8
sex workers, 382
sexual behaviors, 317–18
sexual orientation and, 315–16
smoking crack cocaine, 317
social context, 305–7
social environmental factors, 318–19
social network interventions, 322–4
social networks and, 319–20
socioeconomic status and, 319
stigma, 316, 318

Injection drug use/users (IDUs) (*contd*)
substance abuse treatment, 328–30
syringe access, 324–8
syringe exchange programs (SEPs), 306,
308, 324–6
WSW (women who have sex with
women), 316
zero tolerance policy, 330
Injection facilities, supervised (SIF),
505–6
InSite, 506
Institute of Tropical Medicine (ITM), 396
Interferons type, 1, 43, 44
Intergenerational sexual coupling, 413
International AIDS Vaccine Initiative
(IAVI), 59, 73, 75
International Covenant on Economic,
Social and Cultural Rights (ICESCR),
513, 514–15
International research, 192–5
Internet-based research, 190–1
Intervention with Microfinance for AIDS
and Gender Equity (IMAGE) Project,
432–3
Interventions
adapting, 605, 608
behavioral, 595–6
biological, 592–5
evaluation of impact, 594–5
geographical contexts, 608–9
implementation of, 604–5
vs programs, 421
promising-evidence behavioral
interventions, 609–10
studies, 591-2
Interviewing
computer-assisted (ACASI), 580
see also Motivational interviewing
Iran
death penalty for homosexual relations,
503
opiate replacement therapy (ORT) for
prison inmates, 459
Israel, male circumcision, 158
Italy
IDU/non-IDU sex workers, HIV
prevalence, 382

male (MSW)/transgender (TSW) sex
workers, HIV prevalence, 380
TAMPEP Project, 397
ITM (Institute of Tropical Medicine), 396

Jailbreak Health Project, 454
Jails, *see* Incarcerated populations;
Incarceration in the US; Prisons
Jamaica
condom availability in prisons, 456
HIV prevalence in prisons, 464
Jewel Project, 581

Kazakhstan
HIV prevalence among IDUs, 504
injection drug use, 311
Kenya, 408
behavioral interventions in adolescents
and youth, 424
female sex workers/low risk women,
HIV prevalence, 378, 379
HIV prevalence among IDUs, 504
HIV prevalence/incidence decline, 1,
15, 585
injection drug use, 312
Kisumu Breastfeeding Study, 487
male circumcision, 21, *148–52,
*154–9, 161
people living with HIV/AIDS, 14
primary school prevention program,
582
Reposition Youth (TRY) Savings and
Credit Scheme Project, 433
TAP project, 433, 435
VCT cost-effectiveness, 531
VCT Efficacy Trial, 529
voluntary counseling and testing, 527
Ketamine, 340, 359
Kisumu Breastfeeding Study, 487
Knowledge
attitudes and behaviors (KABs), 171–2
of HIV protection, 416
KwaZulu-Natal
alcohol use, 347
HIV prevalence among sex workers,
508
Mpondombili Intervention, 429

Lactobacilli, 45
  vaginal, 43
Lactoferrin, 43, 44, 45
Lamivudine, 110, 111, 112, 134, 473,
  478
  in non-occupational post-exposure
    prophylaxis, 122, 123
Lamivudine triphosphate, 112
Langerhans cells, 38, 39, 42–3, 44–5,
  153
Latin America
  HIV prevalence among IDUs, 310
  HIV-infected pregnant women receiving
    antiretroviral prophylaxis, 488
  injecting drug use and commercial sex,
    382
  injection drug use, 312
  methadone maintenance programs, 328
  see also Americas; *and individual*
    countries
Latino/Hispanic culture, 213, 215, 216,
  231
Latvia, methadone maintenance programs,
  328
Lesotho
  adult HIV prevalence, 15
  male circumcision, 147
Levitra, 356
Levo-alpha-acetylmethadol (LAAM) trial,
  459
Libya, HIV prevalence among IDUs, 504
Life-skills programs, school-based
  interventions, 430
Lithuania, methadone maintenance
  programs, 328
Liverpool School of Tropical Medicine,
  244
Lopinavir, 110, 112, 134
  in non-occupational post-exposure
    prophylaxis, 122
Lotus Project, Cambodia, 516–17
LSD (lysergic acid diethylamide), 359
Lubricants
  mineral oil, 381–2
  water-based, 382, 387, 398–9
Lysergic acid diethylamide (LSD), 359
Lysozyme, 43, 44, 45

Macaques, *see* Animal studies
Macedonia, methadone maintenance
  programs, 328
Major Histocompatibility Complex (MHC)
  antigens, 44
Malawi
  acute HIV infection, 108, 542
  behavioral interventions in adolescents
    and youth, 423
  breastfeeding study, 484–5
  HIV prevalence/incidence decline, 1, 15
  HIV Testing and Counseling (HTC)
    weeks, 527
  male circumcision, 147
  pre-exposure prophylaxis trials aborted,
    125
  testing in pregnancy, 476
  Young People's Sexual and Reproductive
    Health Project, 433
Malaysia
  HIV incidence among IDUs, 504
  HIV prevalence in prisons, 464
  injection drug use, 311
Male circumcision, 21, 39, 85, 100–1,
  146–66, 258
  acceptability of, 158–9
  age at, 148, 161
  behavioral disinhibition, 159, 160, 161
  biological evidence for protective
    effects, 152–4
  as biomedical prevention strategy, 193
  complication rates, 157–8
  vs CVCT, 257
  effects on sexual satisfaction and
    function, 158
  effects on STI acquisition, 154–7
  efficacy, 593
  and frequency of HIV exposure, 154
  heterosexual HIV in men, 146–8
  HIV risk, 146–7
  and HIV/STI infection in women,
    156–7
  men who have sex with men, 148–9
  modelling effects of/cost effectiveness,
    159–60
  neonatal, 155, 158
  penile cancer, 155

Male circumcision (*contd*)
  prevalence, 158
  as protection for women, 194
  reasons for, 148
  risk compensation, 159
  safety of, 157–8
  scale-up of programs, 160–2
  trials, 149–52
  see also under individual countries
Male genital tract
  acquired immune response, 43
  anatomy, 32–3
  antiretroviral drug ins, 111
  secretions, 44
Maraviroc, 111, 132–3, 134
  in non-occupational post-exposure
    prophylaxis, 122
Marijuana, 358–9
Mathematical modeling, vs evidence-based
    research, 128
Médecins Sans Frontières, 516
Media-based prevention programs, 421,
    429
Medication event monitoring systems
    (MEMS), 94
MEMA kwa Vijana intervention, 429,
    429–30
Men who have sex with men (MSM)
  behavioral intervention research, 213,
    214, 217–19, 224–7
  black or minority ethnic (BME) men,
    509–10
  circumcision, 148–9
  harm reduction interventions, 510–11
  HIV among, 509–11
  injection drug users, 315–16
  reviews of HIV testing and counseling
    data, 527–8
  unprotected anal intercourse (UAI), 509,
    510–11
Menses, 108
Mental illness, incarcerated persons,
    446
Merck, 73
  closure of studies, 85, 103
Meta-analysis, intervention trials, 598
Methadone

beneficial effects, 458
  for opiate disorders, 342
Methadone maintenance programs, 314,
    328–9, 503
  countries established in, 328
Methadone maintenance treatment
    (MMT), in prisons, 458, 459
Methamphetamine, 187, 188, 340, 349–50,
    356, 509
  HIV prevention, 352–3
  HIV risk, 350–2
  and MSM, 226, 350, 351, 351–2
  pharmacotherapy research, 353
Methylenedioxymethamphetamine
    (MDMA) (Ecstasy), 340, 359
Mexico, methamphetamine production,
    349–50
Mice, *see* Animal studies
Microbicide development, 88–91, 103
  challenges to, 93–6
  the future, 103
  the "pipeline", 91–3, 101
  see also Microbicide trials
Microbicide trials
  adherence, 93–4
  CAPRISA 004 study, 102
  clinical trial retention, 93
  informed consent, 99, 100
  ongoing, 89, 101
  pre-exposure prophylaxis (prep) studies,
    African, 98
  pregnancy and, 94–5
  premature termination, 99, 103
  rectal safety studies, 90–1
  reimbursement of volunteers in
    developing world, 98
  "safari"/"parachute" research, 98
  socio-cultural perspectives in the
    developing world, 96–100
  VOICE (Vaginal and Oral Interventions
    to Control the Epidemic) study, 102
  see also Microbicide development
Microbicide Trials Network (MTN), 102
Microbicides, 32, 45, 85–106
  access to, 102
  biological rationale, 86–8
  combination, 89, 103

definition, 85
development, *see* Microbicide
  development
inducing viral resistance, 95–6
mechanism of action, 91–3
rapid evolution of research, 101
rectal safety, 87, 90–1
reverse transcriptase (RT) inhibitors,
  89, 101
topical, as pre-exposure prophylaxis,
  126
trials, *see* Microbicide trials
Middle East
  ANC facilities providing PMTCT
    services, 537
  HIV prevalence among IDUs, 310
  injection drug use, 312
  neonatal circumcision, 158
  pregnant women receiving antiretroviral
    prophylaxis, 488
Midwest AIDS Biobehavioral Research
  Center, 171
Mineral oil lubricants, 381–2
MIV-150 (RT inhibitor), 102
MMT (methadone maintenance treatment),
  in prisons, 458, 459
MODEMU (Movimento de Mujeres
  Unidas), 398
Modified vaccinia Ankara (MVA) virus
  vectors, 64, 65
Moldova, needle-exchange programs, 513
Monitoring and evaluation, feedback from,
  604
Monitoring service utilization, 574–8
  data collection, 576
  examples from developing countries,
    577–8
  examples from US context, 576–7
  indicators, 575–6
  study design, 576
Monkey meat, 2–3
Montenegro
  HIV prevalence among IDUs, 504
  respondent-driven sampling of sex
    workers, 393
Mother-to-child transmission (MTCT),
  177

ART preventing, 115, 133
factors affecting, 473–4
future research, 488
HIV testing in pregnancy, 475–6
practice guidelines, 478, 482
prevention of, *see* Prevention of
  mother-to-child transmission
through breastfeeding, 473, 483–8
transmission rates, 472
twins, 474
Motivational enhancement therapy (MET),
  see Motivational interviewing
Motivational interviewing (MI), 207, 212,
  343
  injection drug users, 228
  MSM, 224–5, 226
  women, 229
Movimento de Mujeres Unidas
  (MODEMU), 398
Mozambique
  adult HIV prevalence, 15
  people living with HIV/AIDS, 14
Mpondombili Intervention, 429
MSM, *see* Men who have sex with men
MTCT, *see* Mother-to-child transmission
Mucosal epithelial layers, 38
Myanmar, declining epidemic, 16

N-9 (nonoxynol-9), 86, 88, 90, 103
Naltrexone, for alcohol use disorders, 342,
  347–8
Namibia
  adult HIV prevalence, 15
  behavioral interventions in adolescents
    and youth, 422
  condom use, 1
  risk behaviors, 15
  sexual debut, 1
Narcotics Anonymous (NA), 343, 457
National AIDS Behavioral Survey
  (NABS), 174
National Health and Social Life Survey
  (NORC), 174–5
National HIV Behavioral Surveillance
  (NHBS) System, 360
National Institute of Allergy and Infectious
  Diseases (NIAID), 75

National Institute on Drug Abuse (NIDA), 183, 331
National Institute of Mental Health (NIMH), 244
  AIDS Research Centers, 173–4
  Collaborative HIV/STD Prevention Trial, 194–5
  Healthy Living Project (HLP), 178, 190
National Institutes of Health (NIH), 68, 75, 170, 178, 587
National Survey on Drug Use and Health (NSDUH), 344
National Treatment Improvement Evaluation Study, 454
Needle exchange, 503
  in correctional facilities, 459–60
  programs, 172, 178, 185
  and tattooing, 460–1
Needle-sharing, 382
Needle and syringe exchange programs (NSPs), 505, 512, 513, 516
Nef gene, deletions in, 55
Neisseria gonorrhoeae, prevalence among female sex workers, 379
Nelfinavir, 112, 487
Nepal
  HIV prevalence among IDUs, 504
  methadone maintenance programs, 328
  migrant sex workers infected, 385
  peer educators, 324
  sex-trafficking, 384, 385
  syringe exchange programs, 324
Netherlands
  decriminalization of sex workers, 516
  male (MSW)/transgender (TSW) sex workers, studies, 380
  MSM study, 510
  needle and syringe exchange programs, 505
  pharmacies, 326–7
  preventive vaccine trials, 73
  TAMPEP Project, 397
Neurocognitive skills, 176
Nevirapine, 110, 112, 177, 258, 477, 478
  for breastfed infants, 486, 487
  in prevention of vertical transmission, 133

resistance, 482–3
  6-week extended nevirapine (SWEN) trial, 487
New Mexico Medical Society, 459
New York State Expanded Syringe Access Demonstration Program (ESAP), 327
New York strain of vaccinia (NYVAC) virus vectors, 65
New Zealand
  gay men, 225–6
  male circumcision, 158
  men who have sex with men, 11
  public health responses, 16
Nigeria
  behavioral interventions in adolescents and youth, 422, 427
  death penalty for homosexual relations, 503
  Global HIV/AIDS Initiative Nigeria (GHAIN) project, 527
  HIV prevalence among IDUs, 504
  male circumcision, 157
  microbicide study, 99–100
  people living with HIV/AIDS, 14
  pre-exposure prophylaxis trials aborted, 125
  pre-exposure tenofovir trials, 126
  pre-marital testing, 538
Nitrite use, 348, 357–8, 358
NNRTI (non-nucleoside reverse transcriptase inhibitor) resistant virus, 482–3
Non-adherence, predictors of, 183
Non-injection substance use/users, 340–75
  conclusions, 360
  contingency management (CM), 343, 352
  HIV risk studies, 341
  intervention, pharmacotherapy/ behavioral, 342–4
  motivational enhancement therapy (motivational interviewing), 343
  non-alcohol substance use, 348–59
  personality factors, 341
  twelve-step programs for addiction problems, 343, 352
Non-nonoxynol, 9, 187

Non-nucleoside reverse transcriptase
inhibitors (NNRTIs), 109, 110, 112
NNRTI resistant virus, 482–3
Non-occupational post-exposure
prophylaxis, 118–23, 182
antiretroviral selection for, 119–23
clinical studies, 118–23
ease of therapy, 123
European Project on Non-occupational
Post-Exposure Prophylaxis for HIV
(EURO-NONOPEP), 118
impact of ART on sexual behavior,
128–9
implementation guidelines, 118
toxicity rates, 122
Nonoxynol-9 (N-9), 86, 88, 90, 103
North Africa
ANC facilities providing PMTCT
services, 537
HIV-infected pregnant women receiving
antiretroviral prophylaxis, 482
injection drug use, 312
neonatal circumcision, 158
North America
harm reduction, 503
injecting drug use and commercial sex,
382
male circumcision, 158
men who have sex with men, 17
persons living with HIV/AIDS, 17
preventive vaccine trials, 69, 73
sexual risk behaviors, 510
see also Americas; *and individual
countries*
Nucleic acid amplification testing (NAAT),
22, 194
Nucleoside/nucleotide reverse transcriptase
inhibitors (NRTIs), 110, 112
phosphorylation of analogue RTIs,
111–12
Nuremberg Code, 77, 99

Oceania, epidemics, 16–17
Office of Drugs and Crime (United
Nations), 505
Onken's model of evidence-based research,
203–4

Operant conditioning, 206
Opiate replacement/substitution therapy
(ORT), in prisons, 458–9, 513
Opiates, 340
Opioid substitution pharmacotherapy,
injection drug users, 328–9
Opium production, 504
OPTIONS project, 293, 296
Oral sex
gay men, 270
HIV transmission via, 31
Orphans, 186–7

Pacific, ANC facilities providing PMTCT
services, 537
PACTG076 (Pediatric AIDS Clinical Trial
Group Protocol 076) study, 473, 476,
478
Pakistan
death penalty for homosexual relations,
503
HIV prevalence in prisons, 464
injecting drug users, 16
men who have sex with men, 16
new infections, 3
PAPO-HV Project, Côte d'Ivoire, 396–7
Papua New Guinea
female sex workers/low risk women,
HIV prevalence, 379
gang rape of sex workers, 390
respondent-driven sampling of sex
workers, 393
Parents, interventions involving, 431
Partnership for Health (PfH), 292–3,
296
Pathogen-associated molecular patterns
(PAMPs), 43
Pattern-recognition receptors (PRRs), 43
PAVE 100 study, 66, 71–2
Pediatric AIDS Clinical Trial Group
Protocol 076 (PACTG076) study, 473,
476, 478
Peer networkers, 323
PEMS (Program Evaluation and
Monitoring System), 577
Penile cancer, 155
Penile urethra, 39

PEPFAR, *see* President's Emergency Plan for AIDS Relief

Peptide vaccines, 62

Persons living with HIV/AIDS (PLWHA)
  disclosure of HIV status, 285–6
  reducing sexual risk behavior, 281–304
  reducing transmission risk, 288–95
  sex-partner characteristics, 283–6
  transmission risk behaviors, 283–6
  see also Sexual risk behaviors among PLWH

Peru
  acyclovir adherence, 21
  female sex workers/low risk women, HIV prevalence, 379
  male circumcision, 149
  oral pre-exposure prophylaxis trials, 125
  substance use, 349

Phambili study, 70–1, 73

Pharmacies, syringe access, 326–8

Pharmacotherapy trial standards, 203

Philippines
  condom use, 390–1
  female sex workers/low risk women, HIV prevalence, 379
  syringe exchange programs, 324

Phosphodiesterase type 5 (PDE-5) inhibitors, 188

Phospholipase A2, 43

PLWHA, *see* Persons living with HIV/AIDS

PMTCT, *see* Prevention of mother-to-child transmission

Pneumocystis carinii pneumonia, 524

POL (Popular Opinion Leader) intervention, 606

Poland
  methadone maintenance programs, 328
  syringe exchange programs, 324

Polyactide-coglycolide (PLG), 67

Polydrug use, 356–7

Polymeric microspheres, 67

Poppers, 357, 359, 509

Popular Opinion Leader (POL) intervention, 606

Population Council Adolescent Livelihoods program, 433

Positive STEPs demonstration project, 296

Post-exposure prophylaxis (PEP), 182, 194
  antiretroviral therapy (ART) for, 115–18
  ART in animal models, 115–17, 119
  ART following occupational exposure, 117–18
  see also Non-occupational post-exposure prophylaxis

Poverty alleviation, 23

Pox viruses, 64–5

Praneem, 91

Pre-exposure prophylaxis (PrEP), 123–6, 132, 194
  animal models, 123–5, 132
  antiretroviral, 102
  ART and sexual behavior, 129–30
  clinical trials, ongoing and proposed, 124, 125–6
  studies, 98, 99
  topical microbicides, 126

Pregnancy
  as challenge to microbicide trials, 94–5
  HIV testing in, 475–6
  impact of ART on, 133
  learning HIV status, 536
  review of HIV testing and counseling data, 527–8
  young women, 415, 432

PrEP, *see* Pre-exposure prophylaxis

President's Emergency Plan for AIDS Relief (PEPFAR), 2
  ART in Africa, 98
  developing countries, 577–8
  indicators, 577
  PAPO-HV Project, 396
  Prostitution Pledge, 502, 516
  and Uganda, 517
  and VCT, 244, 311

Prevalence, *see* HIV prevalence; *and other individual infections*

Prevention
  impact, 594
  media-based programs, 421, 429
  prevention programs in the clinical setting, 295–7
  tailoring strategies, 4–5

see also AIDS prevention research; HIV prevention
Prevention case management, 293, 454
  PCM model, 454
Prevention of mother-to-child transmission (PMTCT), 472–97
  antenatal care (ANC) clinics providing PMTCT services, 536, 537
  antiretroviral regimen trials, 479–81
  in high-resource settings, 476–8
  in low-resource settings, 478–82
  monitoring service utilization, 574–8
  principles of, 474–5
  progress phases, 473
  strategy implementation, 488
  testing and counseling, 535–6
  zidovudine (AZT), 526
Prevention Research Synthesis (PRS) project, 178, 600–1, 607
Prisons
  drug-dependence programs, 446
  HIV infection in, 448
  international, HIV/AIDS in, 462–4
  opiate replacement therapy (ORT), 458–9
  therapeutic communities (TCs) in, 458
  see also Incarcerated populations; Incarceration in the US
PRO 2000, 89, 93, 102
Progesterone therapy, 41
Program Evaluation and Monitoring System (PEMS), 577
Program implementation, local support for, 609
Programs, vs interventions, 421
Project Bridge, 452
Project Inform, 503
Project KEEP, 458
Project LIGHT (Living in Good Health Together), 185–6
Project Respect, 184–5, 189
Promising-evidence behavioral interventions, 609–10
Prostitution Pledge, 502, 516
Protease inhibitors (PIs), 110
  in genital tract, 111
  therapy, 287

Protection motivation model, 207, 210
Provider-initiated routine testing, 526
Provider Initiated Testing and Counseling (PITC), 533–4
PRS (Prevention Research Synthesis) Project, 178, 600–1, 607
Pseudo-ephedrine, 349
Psychoeducation
  injection drug users, 228
  women, 229–30
Psychosocial variables, 233–4
Public health, human rights and, 515
Public health responses, socially-driven, 559–63

Race/ethnicity, injection drug users, 308–10, 313
Racial minorities, behavioral intervention research, 223, 231–2
Racial segregation, injection drug use, 306–7
Randomized controlled trials (RCTs), 580, 587, 592–3
  multisite, multipopulation, 177–84
RE-AIM model, 594
Recreational drug use, 187
Rectal intercourse, infection risk, 31
Rectal safety, microbicides and, 87, 90–1
Rectal secretions, 36
Rectum, anatomy, 36
Regai Dzive Shiri intervention, 429
Relapse prevention (RP), 343
  men who have sex with men, 227
  model, 212
Replicating Effective Programs (REP) Project, 601, 603
Replicating Effective Programs Plus (REP+), 178, 179
Reposition Youth (TRY) Savings and Credit Scheme Project, 433
Research
  effectiveness testing, 204
  emerging issues and challenges, 605–12
  empirical, 175
  evidence-based, Onken's model, 203–4
  international, 192–5

Research (*contd*)
  intervention research, 175
  translating into practice, 591–617
Research synthesis, 598–9, 606–7
Research trials
  cluster-randomized trials, 594
  recruitment and retention of study
    subjects, 593–4
  replication of results of efficacy trials,
    605
  university faculties undertaking, 598
  see also Randomized controlled trials;
    Vaccine trials
Reservoir, 33
Residential Substance Abuse Treatment
  (RSAT) initiative, 458
Reverse transcriptase (RT) inhibitors, 89,
  101
Risk behaviors, 130, 171, 185
  increasing, 188
  reduction, 189
Risk of HIV infection
  lack of perceived risk, 245
  reducing, 416–20
  risk compensation, 593
  risk factors, 20–3, 171–72
  risk groups, 14, 172, 176
  risk settings, 171
  risk-reduction counseling (RRC), 562
  sexual mixing rates, 20
  structural factors relating to young
    people, 418–19
  see also Sexual risk behaviors
Ritonavir, 110, 112
Rolleston Report, 503
Romania, HIV/AIDS in prisons, 463
Russia
  female sex workers/low risk women,
    HIV prevalence, 378, 379
  intravenous drug user study, 504
  substitution therapy for opiate addiction,
    502
  syphilis in, 501
  syringe exchange programs, 324
  see also Former Soviet Union
Russian Federation
  commercial sex workers, 317

growing HIV epidemic, 16
HIV prevalence among IDUs, 310
HIV/AIDS in prisons, 462, 463–4
respondent-driven sampling of sex
  workers, 393
RV 144 trial, 73
Rwanda
  condom use, 1
  couples voluntary counseling and
    testing (CVCT), 243–4, 246–8, 250,
    258–9
  HIV in female sex workers, 378
  risk behaviors, 15
  sexual debut, 1
  study of discordant couples, 113
Rwanda Zambia HIV Research Group
  (RZHRG), 244, 251, 256–7
  model of mutual disclosure, 249
  see also CVCT clinics

Safer-sex counseling, 189
Saquinavir, 110, 112, 177
Saudi Arabia, death penalty for
  homosexual relations, 503
Savvy (C31G), 88
Schools
  influence of, 418–19
  involving parents, 431
  life-skills programs, 430
  peer education, 430–1
  prevention education in India, 561
  prevention programs, 582
  school-based interventions, 422–8,
    429–33
  structural and community-level
    interventions, 431–3
Science-based HIV prevention, 596–605
Self-efficacy, 208, 208–9
Self-regulation theory, 211
Semen, 32, 36–7
  alkalinity, 43
Senegal
  ABC strategy, 20
  condom use, 1
  HIV among men who have sex with
    men, 509
  risk behaviors, 15

sexual debut, 1
unregistered sex workers, 382
Sentinel surveillance system, 13, 14
Serbia
  HIV prevalence among IDUs, 504
  respondent-driven sampling of sex
    workers, 393
Sex education and communication
  challenges to, 560–1
  school-based programs, 560–1
Sex-partner characteristics, 283–6
Sex tourism, 384–5
Sex-trafficking, 384
Sex work
  human trafficking for, 384
  legality, 383
Sex workers, 376–406
  antiretroviral therapy (ART) for, 388
  census, 392
  challenges for setting up HIV prevention
    programs, 398–9
  community mobilization, 389–90
  condom use, 377, 387, 392–3, 563
  definition, 376
  direct/indirect, 376–7, 382–3
  effect of criminalization of work, 516
  effective interventions, 385–93
  environmental–structural interventions,
    389–91
  epidemiology of HIV, 377–80
  estimating number of, 392
  exploitation of, 384
  financial need, 384
  government policy, 390–1
  harm reduction strategies, 508
  high levels of STIs and HIV, 377
  HIV among, 507–8
  intervention coverage measurement
    methods, 392
  intervention gap, 393–5
  intervention monitoring and evaluation,
    391–3
  intervention sustainability, 394, 395
  interventions at individual level, 385–9
  key principles for HIV prevention, care
    and empowerment, 391
  male (MSW), 380

migrant, 384, 385
model intervention programs, 395–8
peer education, 386–7
prevention gap, 394
prevention messages, 387
prevention–care synergy, 385, 389
respondent-driven sampling, 393
risk of HIV, 380–3
scaling up of intervention services,
  394–5
screening, 388
STI management, 387–8
time-location sampling, 393
transgender (TSW), 380
unsafe practices, 381–2
vulnerability, 383–5
Sexual abuse
  in childhood, 290
  history of, in incarcerated persons,
    446–7
Sexual education, 420, 560–1
Sexual risk behaviors
  ART and, 128–31
  changing, HIV testing and counseling,
    527–30, 540, 541
  injection drug users, 317–18
  PLWHA, *see* Sexual risk behaviors
    among PLWHA
  post-exposure prophylaxis, 128–9
  pre-exposure prophylaxis, 129–30
Sexual risk behaviors among PLWHA
  conclusions, 298–9
  counseling, 291–4, 296–8
  effectiveness of reduction programs,
    294–5
  reducing, 281–304
  reduction approaches within HIV
    primary-care setting, 291–4
  relationship with HAART, 286–8
  repeated assessments, 291
  youth, 290
Sexual transmission of HIV, 113, 127
  animal studies, 40–1
  biology of, 31–52
  establishing infection, 37–40
  explant tissue models, 41–3
  the future, 44–5

Sexual transmission of HIV (*contd*)
  host immune defense, 43–4
  in vitro studies, 41–3
  mechanisms, 44–5
  natural regulation factors, 45
  summary, 44–5
  transmission rates, 31
Sexually transmitted diseases, reproductive
    rate model, 170
Sexually transmitted infections (STIs)
  among sex workers, 507–8
  curable, 507
  effects of male circumcision on, 154–7
  increasing, associated risk factors, 31
  incurable, 507
  treatment services, 21
  in US correctional settings, 448
Shared confidentiality, 566
SHIV (HIV–SIV) virus, 60, 65
Sildenafil, for erectile dysfunction, 188
Simian immunodeficiency virus (SIV)
    infection, 116–17
SIV/HIV chimeric viruses (SHIV), 116–17
6-week extended nevirapine (SWEN) trial,
    487
Slovak Republic, methadone maintenance
    programs, 328
Slovenia, methadone maintenance
    programs, 328
SLP1, 44, 45
Social action model, MSM, 225
Social cognitive theory (SCT), 207, 208–9
  behavioral approaches, 323
  ethnic/racial minorities, 232
  injection drug users, 228
  men who have sex with men, 224–5,
    226
  women, 229, 230
Social learning theory
  injection drug users, 228
  MSM, 224–5, 225, 226
  women, 229
Socially relevant HIV prevention, 566–7
Societal contexts
  family-based approach, 553
  interventions in, 550–70
  public health responses, 559–63

social network-based approaches, 552–4
venue-based approaches, 550–2
Socioeconomic status (SES), injection
    drug users, 319
Sonagachi Project, 390, 552
South Africa
  acyclovir adherence, 21
  adolescents' HIV status, 539
  adult HIV prevalence, 15
  alcohol counseling, 551
  behavioral interventions in adolescents
    and youth, 422, 423, 425–6, 427
  breastfeeding studies, 485
  condom availability in prisons, 455
  female sex workers/low risk women,
    HIV prevalence, 379
  HAPS Intervention, 429
  male circumcision, 21, 149, 151, 152,
    155, 157–8, 159
  media-based prevention programs, 421,
    429
  microbicide study, 99–100
  microfinance program strategy, 557
  Mpondombili Intervention, 429
  National Youth Health Risk Survey, 415
  people living with HIV/AIDS, 15
  PMTCT antiretroviral trials, 480
  post-exposure prophylaxis, 129
  preventive vaccine trials, 70–1, 73
  sexual assault victims, 129
  vaccine research organizations, 75
  violence against women, 557
  women and substance use, 349
  young people's increase in HIV risk, 411
South African Aids Vaccine Initiative
    (SAAVI), 73, 75
South America
  ANC facilities providing PMTCT
    services, 537
  HIV epidemics, 16–17
  new infections, 4
  preventive vaccine trials, 69, 73
  see also Americas; *and individual
    countries*
South Asia
  ANC facilities providing PMTCT
    services, 537

violence against women, 555
widowhood and remarriage, 555
see also Asia; *and individual countries*
Southeast Asia
injection drug use, 11
see also Asia; *and individual countries*
Southern Africa
ANC facilities providing PMTCT
services, 537
epicenter of HIV pandemic, 409–10
HIV in young people, 409–10
male circumcision, 148, 158, 160
see also individual countries
Soviet Union, *see* Former Soviet Union
Spain
male (MSW)/transgender (TSW) sex
workers, HIV prevalence, 380
needle exchange in correctional
facilities, 460
needle-exchange programs, 513
Spermicides, 86
Stavudine, 112
STEP study, 56, 59, 60, 69–71, 73
"Stepping Stones", 431–2
Sterilization campaigns, 161
"Sterilizing immunity", 54
Stigma, 433, 565, 566
IDUs, 316, 318
in the medical establishment, 560
Stop AIDS Love Life, 576
Sub-pandemics, 14–17
Sub-Saharan Africa
acute HIV infection study, 542
ART strategy, 132
children with HIV/AIDS, 15
concurrent relationships, 413
HIV-infected pregnant women receiving
antiretroviral prophylaxis, 488
HIV pandemic, 504
HIV prevalence, 97, 583–4
HIV in young people, 408, 409–10
infection rates, 5
intergenerational sexual coupling,
413
interventions with youth, 407–43
male circumcision, 21, 146–7, 149–52,
157, 159, 159–60

mother-to-child transmission of HIV,
472
regional differences, 15
sexual initiation age, 411
sub-pandemics, 11, 14–15
women, 85
see also individual countries
Substance abuse
gay men, 273, 274
treatment, 328–30
Substance use
education and awareness programs for
incarcerated persons in the US, 457
non-alcohol, 348–57
Subunit vaccines, 62
Sudan, death penalty for homosexual
relations, 503
SUMIT study, 289
Summative evaluation, 571, 572
monitoring service utilization, 574–8
types of, 574–86
Supervised injection facilities (SIF), 505–6
Swaziland, adult HIV prevalence, 15
SWEN (6-week extended nevirapine) trial,
487
Switzerland
HIV prevalence among IDUs, 311
needle exchange in correctional
facilities, 459–60
needle-exchange programs, 513
needle and syringe exchange program in
prison, 512
vaccine research organizations, 75
Syndemic, 274
Syphilis, 190, 501–2, 507
male circumcision and, 155
MSM and methamphetamine use, 351
prevalence among female sex workers,
379, 390
in US correctional settings, 448
Syringe access, 324–8
New York State Expanded Syringe
Access Demonstration Program
(ESAP), 327
through pharmacies and health-care
providers, 326–8
Syringe distribution in the US, 515–16

Syringe exchange programs (SEPs), 306,
  308, 324–6
  laws and regulations, 326
Syringes, bleach for sterilization, 513

Tadalafil, for erectile dysfunction, 188
Tailored gay-specific group,
  methamphetamine use, 352
Tajikistan, HIV prevalence among IDUs,
  504
TAMPEP (Transnational AIDS/STD
  Prevention Among Migrant
  Prostitutes in Europe Project), 397
Tanzania, 408
  behavioral interventions in adolescents
    and youth, 423, 425, 428
  HIV prevalence among IDUs, 504
  male circumcision, 148
  MEMA kwa Vijana intervention, 429,
    429–30
  surveillance system for incidence data,
    583
  treatment of STIs, 21
  VCT cost-effectiveness, 531
  VCT Efficacy Trial, 529
  youth testing, 539
TAP project, 433, 435
Targeted Genetics, 73
Task Force on Community Prevention
  Services, 269
Tattooing, in prisons, 460–1
Technical Evaluation Reference Group
  (TERG), 587
Technology transfer, 184–92
Tenofovir, 88, 110, 111, 112, 117, 134
  CAPRISA tenofovir gel trial, 132
  in non-occupational post-exposure
    prophylaxis, 122, 123
  in pre-exposure clinical trials, 125–6
  as pre-exposure prophylaxis, 126
  pre-exposure prophylaxis (PREP) trials,
    compliance study, 593
  viral resistance, 95
Tenofovir diphosphate, 111–12
Terence Higgins Trust, 503
TERG (Technical Evaluation Reference
  Group), 587

Testing
  antenatal, 561–3
  unlinked anonymous (UAT), 13
  see also Couples voluntary counseling
    and testing; Voluntary counseling and
    testing
Testing and counseling in HIV, 524–49
  adolescent/youth testing, 538–9
  and behavior change, 527
  changing sexual risk behaviors, 527–30
  client-initiated, 533
  couple and family-based, 536–8
  in developing countries, 529
  guidelines, 173
  history of, 524–6
  informed consent, 525, 532, 533, 535,
    540, 541
  knowledge of HIV status, 526, 527
  models, *see* Testing and counseling
    models in HIV
  pre-marital screening, 535, 538
  prevention of mother-to-child
    transmission (PMTCT), 535–6
  provider-initiated routine testing, 526
  reaching specific populations, 535–40
  reviews of data, 527–9
  see also Voluntary counseling and
    testing
Testing and counseling models in HIV,
    531–5
  acute infections, 542
  advancing, 543–4
  challenges in, 540–3
  emerging issues, 540–3
  enabling environment, 540–1
  HIV prevention for positives, 542–3
  HIV self-testing, 534
  mandatory screening of blood products
    and organs, 534
  mandatory testing of individuals, 534–5
  miscellaneous, 540
  monitoring and evaluation (M&E),
    541–2
  Provider Initiated Testing and
    Counseling (PITC), 533–4
  quality assurance and control, 541
  recent infections, 542

see also Voluntary counseling and
   testing
Testis, 33–4
Thai AIDS Vaccine Evaluation Group, 73
Thai MoH, 73
Thai Prime-Boost Trial, 71
Thailand
   ABC strategy, 20
   concentration response study, 107–8
   condom campaign, 551
   condom distribution in brothels, 503
   condom program, 508, 516
   condom use, 387, 391
   declining epidemic, 16
   direct vs indirect sex workers, 383
   female sex workers/low risk women,
      HIV prevalence, 379
   HAART, 564
   HIV among men who have sex with
      men, 509
   HIV incidence, 1
   HIV incidence among IDUs, 504
   HIV prevalence among IDUs, 311
   male (MSW)/transgender (TSW) sex
      workers, HIV prevalence, 380
   methadone maintenance programs, 328
   new infections, 3
   oral pre-exposure prophylaxis trials, 125
   PMTCT antiretroviral trials, 479, 480,
      481, 482
   preventive vaccine trials, 69, 73
   replacement for breastfeeding, 485
   sex-trafficking, 385
   syringe exchange programs, 324
   vaccine research organizations, 74
   zidovudine prophylaxis study, 478
Theory of planned behavior (TPB),
   209–10
Theory of reasoned action (TRA), 207,
   209
   ethnic/racial minorities, 232
TLC (Together Learning Choices)
   program, 290
TMC-120 (dapirivine), viral resistance, 95
Together Learning Choices (TLC)
   program, 290
Toll-like receptors (TLRs), 67

Trafficking, for sex work, 384
Transcytosis, 42
Transnational AIDS/STD Prevention
   Among Migrant Prostitutes in Europe
   Project (TAMPEP), 397
Transtheoretical model (TTM), 211–12
Treponema pallidum, 39
Trials, *see* Research trials
Trinidad, VCT Efficacy Trial, 529
Tuberculosis
   directly-observed therapy, 565
   in US correctional settings, 448
Turkey, circumcision program
   complications rate, 157

UAI, *see* Unprotected anal intercourse
UAT, *see* Unlinked anonymous testing
UAVI, *see* Unprotected anal or vaginal
   intercourse
Uganda, 408
   ABC strategy, 15, 20, 560–1
   alcohol use, 345
   behavioral interventions in adolescents
      and youth, 423, 427
   early stages of HIV infection, 542
   HIV prevalence decline, 585–6
   HIV testing and counseling, 529
   home-based AIDS care, 565
   home-based voluntary counseling and
      testing, 538, 553
   male circumcision, 21, *147–56, *158–61
   microbicide study, 99–100
   nevirapine treatment studies, 486
   patients initiating ART, 113
   penile cancer, 155
   PMTCT antiretroviral trials, 480
   preventive vaccine trials, 73
   risk behaviors, 15
   serodiscordant couples, 553
   study of serodiscordant couples, 107
   surveillance system for incidence data,
      583
Ukraine
   HIV incidence among IDUs, 504
   HIV prevalence among IDUs, 310–11
   HIV/AIDS in prisons, 463
   increasing trends, 16

United Kingdom (UK)
  IDUs, HIV prevalence, 310, 311
  male (MSW)/transgender (TSW) sex
    workers, HIV prevalence, 380
  MSM STI prevalence, 510
  pharmacies, 326–7, 328
  POL intervention studies, 611–12
  sex workers, HIV prevalence, 382
  syphilis in, 501
United Nations
  Office of Drugs and Crime, 505
  UN General Assembly Special Session
    (UNGASS) on HIV/AIDS, 526, 585
  UN Population Fund (UNFPA), CVCT
    evaluation kits, 255
  UNAIDS CVCT evaluation kits, 255
  UNAIDS guidelines on HIV testing,
    525, 540
United States (US)
  acyclovir adherence, 21
  adult correctional populations in, 445
  alcohol/drug use, 345
  ART use by homosexual men, 114
  cocaine use, 353–4
  condom provision to incarcerated
    populations, 517
  correctional system, *see* Incarceration in
    the US; Prisons
  decriminalization of sex workers, 516
  evaluating HIV/AIDS programs, 571–90
  HIV increasing in, 2
  HIV prevalence among incarcerated
    populations, 512
  HIV prevalence/incidence among IDUs,
    307–10, 311
  incarceration, *see* Incarceration in the
    US; Prisons
  male circumcision, 146–7
  marijuana use, 358
  methamphetamine production, 349
  monitoring service utilization, 577–8
  mother-to-child infection rates, 472
  needle and syringe exchange programs,
    516
  new infections, 4
  oral pre-exposure prophylaxis trials, 125

  pharmacies, 327–8
  PMTCT antiretroviral trials, 479
  post-exposure prophylaxis, 129
  pre-exposure prophylaxis study, 194
  prevalence/incidence trends, 584–5
  preventive vaccine trials, 69
  syphilis in, 501
  syringe exchange programs, 324–5
  vaccine research organizations, 74–5
  VCT cost-effectiveness, 530–1
Universal Declaration of Human Rights
  (UDHR), 514
Unlinked anonymous testing (UAT), 13
Unmethylated cytosine-guanine
  dinucleotides (CpG), 67
Unprotected anal intercourse (UAI), 90,
  284
  men who have sex with men (MSM),
    509, 510–11
Unprotected anal or vaginal intercourse
  (UAVI), 283, 284, 285
  prevalence, 290, 292–3
Urethritis, 33, 36
  non-gonococcal, 155
Urine screening for chlamydia and
  gonorrhea, 462
US Military HIV Research Program
  (USMHRP), 75
Uzbekistan, IDU/non-IDU sex workers,
  HIV prevalence, 382

Vaccine development
  approaches, 56–7, 59–61
  challenges for, 56–9
  clinical trial design, 58–9
  conclusions, 80
  efficacy trials, *see* Vaccine trials
  to elicit cellular response, 60–1
  to elicit neutralizing antibodies, 59–60
  evaluations, 59–61
  hurdles to finding effective vaccines,
    53–4
  key outcomes for success, 56–7
  multiclade issues, 76–7
  success measures, 57–8
  see also Vaccine trials

Vaccine trial volunteers
  care of infected volunteers, 79
  counseling, 78–9
  informed consent, 77
  social harms and benefits, 78
Vaccine trials, 53–4, 68–79
  advanced design, 58–9
  community involvement, 78
  confidentiality, 78
  costs, 72–6
  Data and Safety Monitoring Boards
    (DSMBs), 77
  delivery methods, 66–7
  early-phase trials, 72
  efficacy trials, 71–2
  ethical considerations, 76
  Ethical considerations in HIV
    preventative vaccine research, 78
  Ethics Committees, 77
  funding, 72–6
  the future, 79–80
  heterologous prime-boost strategies, 66
  Institutional Review Boards (IRBs), 77
  ongoing, 71–2, 73
  organizations involved in, 72–5
  oversight of, 77–8
  past trials, 69–71
  PAVE 100 study, 66, 71
  Phambili study, 70–1, 73
  review, 76–7
  STEP study, 56, 59, 60, 69–71, 73
  test-of-concept trials, 59
  Thai Prime-Boost Trial, 71
  upcoming, 71–2
  VAX trials, 69
  VaxGen trials, 60, 62
  volunteers, *see* Vaccine trial volunteers
  see also Vaccine development
Vaccines, 32, 53–84
  action mechanisms, 54
  adjuvants, 66–7
  clade-specific, 76–7
  development, *see* Vaccine development
  DNA vaccines, 65–6
  effectiveness measurements, 57–8
  genetic inserts, 67–8

hepatitis B, 62
  ideal, 57
  live-attenuated/killed viruses, 59, 61–2
  peptide, 62
  research, *see* Vaccine trials
  subunit, 62
  T-cell-based, 56
  trials, *see* Vaccine trials
  see also individual vaccines
Vaccinia, 65
Vagina, lactobacilli, 43
Vaginal douching, 381
Vaginal drying agents, 381
Vaginal secretions, acidity, 43
Vaginosis, bacterial, 43
Vardenafil, for erectile dysfunction, 188
VAX trials, 69
VaxGen trials, 60, 62, 73
VCCT (voluntary and confidential
  counseling and testing), 533
VCT, *see* Voluntary counseling and testing
VE (vaccine effectiveness) measurement,
  57–8
Vesicular stomatitis virus (VSV) vectors,
  65
Viagra, 356
Vietnam
  direct vs indirect sex workers, 382
  female sex workers/low risk women,
    HIV prevalence, 378
  HIV incidence among IDUs, 504
  HIV prevalence among IDUs, 311, 312
  HIV prevalence in prisons, 464
  IDU/non-IDU sex workers, HIV
    prevalence, 382
  injection drug use, 16, 311
  men who have sex with men, 16
  methadone maintenance programs, 328
  new infections, 3
  respondent-driven sampling of sex
    workers, 393
  syringe exchange programs, 324
Violence
  against women, WHO
    recommendations, 540–1
  gender-based, 417, 431

Violent Crime Control and Law Enforcement Act, 457
Viral vectors, 62–5
Virginity, young women, 420
VOICE (Vaginal and Oral Interventions to Control the Epidemic) study, 102
Voluntary and confidential counseling and testing (VCCT), 533
Voluntary counseling and testing (VCT), 181, 525–6, 531–3
  in antenatal care facilities, 562
  cost-effectiveness of, 530–1
  couples vs individual, 22
  in developing countries, 529, 531
  effect on increasing protected sex, 530
  efficacy data, 528–9
  enabling environment, 540
  home-based, 553
  individually-focused model, 526
  key issues arising, 530
  mobile units, 532
  monitoring service utilization, 574–8
  quality assurance and control, 541
  role of, 22
  VCT Efficacy Trial, 529
  see also Couples voluntary counseling and testing; Testing and counseling in HIV
Volunteers for research trials, microbicide trials, 98
  see also Vaccine trial volunteers
VRC, 73

Walter Reed Army Institute of Research (WRAIR), 73, 75
War conditions, affecting risk behaviors, 320
Warts, genital, 155
Water-based lubricants, 382, 387, 398–9
West Africa
  ANC facilities providing PMTCT services, 537
  breastfeeding studies, 485
  neonatal circumcision, 158
  pre-exposure tenofovir, 129
  see also individual countries
Western Europe

condom availability in prisons, 455
HIV epidemics, 16–17
HIV incidence, 3–4
  men who have sex with men, 11
  opiate replacement therapy (ORT) for prison inmates, 459
  persons living with HIV/AIDS, 17
  see also Europe; *and individual countries*
WiLLOW study, 289
Wine shops, 551–2
Women, 228–30
  behavioral intervention research, 220–3
  ethnic/racial minorities, 223
  HIV infection in Africa, 4, 15
  increased percentage of new HIV cases, 192
  infection and male circumcision, 156–7
  lack of power in sexual relationships, 408, 410, 413, 417, 436
  marriage, 555
  motivational interviewing, 229
  partner violence, 315
  presence of children, 314
  prevention strategies aimed at, 556–7
  subordination of, 554–9
  violence against, WHO recommendations, 540–1
  vulnerability of, 556–7
  see also Breastfeeding; Pregnancy; Young women
Women and Infants Transmission Study, 476
Women who have sex with women (WSW), injection drug users, 316
Women's Interagency HIV Study, 287
World Health Organization (WHO), 75
  Bangui definition, 13
  guidelines on HIV testing, 525, 540
  preventing mother-to-child transmission, guidelines, 474–5, 482, 485, 486
WSW (women who have sex with women), injection drug users, 316

Young People's Sexual and Reproductive Health Project, 433
Young women

contraception, 415
factors influencing HIV risk, 411–12
gender assertiveness, 417
partnership characteristics, 413
pregnancy, 415, 432
reducing HIV risk, 416–20
sexual networking, 413
sexual risk, 410–15
virginity, 420
see also Women
Youth
access to resources and services, 419–20
alcohol use, 415
behavioral interventions, 421, 422–8
conclusion, 435–6
condom use, 413–14
drug use, 415
epidemiology of HIV infection,
409–10
health behavior of HIV-positive, 290
in higher education, 433–4
HIV care and treatment, 433
HIV prevention interventions, 420–34
HIV testing, 538–9
interventions in high-prevalence areas,
407–43
lessons learned from interventions,
434–5
out-of-school, 433–4
school-based interventions, 422–8,
429–33
socio-cultural norms and values, 420
see also Adolescence; Young women
Youth Risk Behavior Survey, 351

Zambia
acyclovir adherence, 21
adolescent VCT, 538, 539
adult HIV prevalence, 15
behavioral interventions in adolescents
and youth, 424
concentration response study, 107–8
condom distribution in prisons, 456
condom use, 1
couples voluntary counseling and
testing (CVCT), 244, 246–8, 250,
258–9
female sex workers/low risk women,
HIV prevalence, 378, 379
male circumcision, 156
NNRTI study, 483
preventive vaccine trials, 73
risk behaviors, 15
self-reported condom use, 536
sexual debut, 1
study of discordant couples, 113
testing and counseling, 536
see also Rwanda Zambia HIV Research
Group
Zero grazing, 20
Zidovudine (AZT), 109, 110, 111, 112,
117, 172
monotherapy, 113
in non-occupational post-exposure
prophylaxis, 122, 123, 134
in occupational post-exposure
prophylaxis, 182
pregnant women, 473, 476, 478
in prevention of mother-to-child
transmission, 133, 177, 526
Zidovudine triphosphate, 112
Zimbabwe, 408
acyclovir adherence, 21
age of partners, 423
behavioral interventions in adolescents
and youth, 423, 424
condom use, 393
HIV prevalence/incidence, 1, 15, 585
people living with HIV/AIDS, 14
Regai Dzive Shiri intervention, 429
ZVITAMBO study, 484–5
ZVITAMBO study, 484–5